Handbook of Formal Argumentation
Volume 2

Handbook of Formal Argumentation
Volume 2

Edited by
Dov Gabbay
Massimiliano Giacomin
Guillermo R. Simari
Matthias Thimm

© Individual authors and College Publications 2018. All rights reserved.

ISBN 978-1-84890-336-4

College Publications
Scientific Director: Dov Gabbay
Managing Director: Jane Spurr

http://www.collegepublications.co.uk

Cover produced by Laraine Welch

All rights reserved. No part of this publication may be reproduced, stored in a retrieval system or transmitted in any form, or by any means, electronic, mechanical, photocopying, recording or otherwise without prior permission, in writing, from the publisher.

Foreword

This volume is the second step in the initiative aimed at creating the Handbook of Formal Argumentation, continuing the community effort to produce a series of volumes containing survey articles and personal views of recognized researchers in the field. In consequence, this effort's central goal is to help students and researchers interested in contributing to Formal Argumentation to access both state-of-the-art and future research perspectives in the field, thus stimulating the work in the area by addressing progress in existing research lines, describing open problems, and presenting emerging topics.

During the preparation of this second volume, the authors met in a workshop to discuss "Current Trends in Formal Argumentation", in Bertinoro, Italy, at the Bertinoro international Center for informatics (BiCi), November 4-6, 2019. During the workshop, the attendants presented drafts of the contents of the future chapters giving rise to many suggestions and improvements in the coordination among the topics to be covered, completing the process of envisioning the new volume in the series.

The chapters have been reviewed, and the final versions presented here have considered the suggestions made by the specialists. It is important to note that the reviewers' feedback provided an essential contribution to the final versions, and the process culminated with the manuscripts submitted by the authors in the winter of 2020/21 and offered here.

The work contained in this volume of the Handbook can be broadly thought of as part of the general area of formal argumentation, being related to extensions to abstract argumentation, dynamics and dialogues, and meta investigations. In what follows, we will provide summaries of the chapters that are part of this volume.

In *Higher-Order Interactions (Bipolar or not) in Abstract Argumentation: a State of the Art*, C. Cayrol, A. Cohen, M-C. Lagasquie-Schiex, start recalling the essential elements of abstract argumentation, then introducing higher-order attacks and summarizing five existing approaches for these attacks. They continue with a brief introduction to traditional bipolar argumentation frameworks and their three variants related to the three possible types of support. Using the different frameworks presented, the authors introduce extended frameworks using higher-order interactions and analyze some contributions in structured argumentation. The computational issues and applications are also described and analyzed, and a comparative synthesis of all the presented approaches is included.

The chapter *Joint Attacks and Accrual in Argumentation Frameworks*, by A. Bikakis, A. Cohen, W. Dvořák, G. Flouris, and S. Parsons, considers the case where multiple arguments jointly attack another, introduced as "joint attacks". This possibility of analyzing joint attacks represents an extension of abstract argumentation with added expressive power. Various works considering joint attacks are analyzed from various perspectives, which include abstract and structured frameworks. Also, guidelines for future research considering current research on the subject are presented.

Preference in Abstract Argumentation by S. Kaci, L. van der Torre, S. Vesic, and S. Villata examines the role of comparative preference in abstract argumentation. First, four known reductions that provide semantics to preference-based argumentation frameworks are surveyed, and ten principles for these semantics are discussed. Alternative semantics that are not based on reductions, principles in the context of symmetric attack, the relation to structured argumentation, and the dynamics of preference and argumentation are presented. Some of these principles are novel for the subject. An examination of the abstract semantics based on these four reductions and ten principles is included. Later, in the second part of the chapter, an analysis of the various research challenges regarding preference in abstract argumentation is given.

In *Collective Acceptability in Abstract Argumentation*, D. Baumeister, D. Neugebauer, J. Rothe explore and survey the various approaches to collective acceptability in multi-agent argumentation, which is related to the problem of collective decision-making in the field of computational social choice that collects contributions from social choice theory, theoretical computer science, and artificial intelligence. Also, the chapter describes practical methods for structural aggregation of argumentation frameworks and presents their properties.

Value-based Argumentation, by K. Atkinson and T. Bench-Capon, presents an extension of abstract argumentation known as Value-based

Argumentation Framework (VAF), its motivations, a formal description, and its properties. The notion of Audience-Specific VAF that incorporates to the framework the focus in an audience is presented. Also, an argumentation scheme and its associated critical questions and some of the applications of value-based argumentation that have been implemented are included.

In *Weighted Argumentation*, S. Bistarelli and F. Santini introduce Weighted Argumentation Frameworks (WAFs), summarizing different critical points of their formalization, developing weight-related concepts such as relaxation of attacks, new semantics based on weighted acceptability and relaxation, and real-world applications related to information coming from social networks and reviewing platforms.

The chapter *Probabilistic Argumentation: A Survey* by A. Hunter, S. Polberg, N. Potyka, T. Rienstra, and M. Thimm is devoted to present an overview of the recent advances in including probabilities in argumentation. Probability is one of the fundamental ways for introducing a representation of uncertainty in the argumentation process, offering a device for assessing the level of this uncertainty. The constellations and the epistemic approaches are two of the fundamental ways of introducing probabilities in abstract argumentation. Also, attaching a probability distribution over labellings gives a form of probabilistic argumentation that overlaps with the two mentioned. The chapter includes a discussion of other possible alternatives for including probabilities in argumentation.

Enforcement in Formal Argumentation by R. Baumann, S. Doutre, J-G. Mailly, and J. P. Wallner, offers an overview of the notion of enforcement in abstract argumentation. The authors center their presentation on extension enforcement, its general characterization, and how it can be algorithmically achieved. The premise assumed is that the various changes applied to the structure of the argumentation framework, and to the semantics associated, should be minimal. The complexity of enforcement, and associated algorithms, and a discussion on the feasibility of this approach are presented.

In *Argumentation-Based Dialogue*, E. Black, N. Maudet, and S. Parsons discuss dialogue as a fundamental part of argumentation, since dialogue introduces a dialectical basis for ascertaining the set of acceptable arguments. In argumentation-based dialogues, participants exchange arguments, and the argumentation mechanisms are employed to discover what is acceptable when the exchange ends. The required elements to conduct argumentation-based dialogues are discussed, and open issues are considered.

In *Strategic Argumentation*, G. Governatori, M. Maher, F. Olivieri, study games where players have perfect information of the moves players make; however, the information on the possible moves (arguments) that other players have available is incomplete. The authors look at games using logically structured arguments and games using abstract arguments, showing that playing these games can be computationally hard. Also, they consider how corruption can affect the argumentation games, and examine forms of countering it.

On the Incremental Computation of Semantics in Dynamic Argumentation by G. Alfano, F. Parisi, S. Greco, G. I. Simari, and G. R. Simari, examines the efficiency of recomputing extensions of abstract argumentation frameworks and warranted literals from defeasible knowledge bases in dynamic environments. An incremental algorithmic solution is presented, making use of an initial extension of a framework and an update with the aim of identifying a subset of the framework enough to compute an extension after the update. The incremental technique for the computation of extensions of abstract argumentation frameworks is considered, exploring how transferred concepts can be employed in the computation of warranted literals in Defeasible Logic Programming.

In *Logic-Based Approaches to Formal Argumentation*, by O. Arieli, A. Borg, J. Heyninck, and C. Strasser, the logical foundations of Dung-style argumentation frameworks are presented. Two perspectives on logic-based methods in the context of argumentation theory are offered. First, a survey of logic-based instantiations of argumentation frameworks is introduced, along with their properties and relations, and then logical methods for studying argumentation dynamics are reviewed. The work is focused to Tarskian logics, based on propositional languages and the associated constructive semantics or syntactic rule-based systems.

In the chapter *Empirical Cognitive Studies about Formal Argumentation* by F. Cerutti, M. Cramer, M. Guillaume, E. Hadoux, A. Hunter, and S. Polberg carries out a description, comparison, and discussion of empirical cognitive studies that have been conducted to test the relationship between human behaviour and the formal models of abstract and structured argumentation, acknowledging their different methodological approaches. Moreover, the relevance and potential benefits of considering these studies for formal argumentation are also discussed, reviewing open questions worth further research in this area.

To conclude this foreword, we would like to thank the authors of this volume for their contributions, the reviewers, and the colleagues for their valuable help in providing comments, suggestions, critiques, and encouragement during the development. The works included have accomplished our two essential objectives by first providing material for the researcher coming to the area of argumentation and facilitating their acquisition of the elements to have a good view of the work at the forefront of research, which represented our second goal. In closing, we would like to give special thanks to College Publications and, with great emphasis, we are thankful to Jane Spurr for her steadfast and invaluable continued support.

Dov M. Gabbay
Massimiliano Giacomin
Guillermo R. Simari
Matthias Thimm

CONTENTS

ARTICLES

Extensions to Abstract Argumentation

1 Higher-Order Interactions (Bipolar or not) in Abstract Argumentation: A State of the Art 15
Claudette Cayrol, Andrea Cohen, Marie-Christine Lagasquie-Schiex

2 Joint Attacks and Accrual in Argumentation Frameworks 131
Antonis Bikakis, Andrea Cohen, Wolfgang Dvořák, Giorgos Flouris, Simon Parsons

3 Preference in Abstract Argumentation 211
Souhila Kaci, Leendert van der Torre, Srdjan Vesic, Serena Villata

4 Collective Acceptability in Abstract Argumentation 249
Dorothea Baumeister, Daniel Neugebauer, Jörg Rothe

5 Value-based Argumentation 299
Katie Atkinson, Trevor Bench-Capon

6 Weighted Argumentation 355
Stefano Bistarelli, Francesco Santini

7 Probabilistic Argumentation: A Survey 397
Anthony Hunter, Sylwia Polberg, Nico Potyka, Tjitze Rienstra, Matthias Thimm

II Dynamics and Dialogues

8 Enforcement in Formal Argumentation 44
  Ringo Baumann, Sylvie Doutre, Jean-Guy Mailly, Johannes P. Wallner

9 Argumentation-based Dialogue 51
  Elizabeth Black, Nicolas Maudet, Simon Parsons

10 Strategic Argumentation 57
  Guido Governatori, Michael J. Maher, Francesco Olivieri

11 On the Incremental Computation of Semantics in Dynamic Argumentation 66
  Gianvincenzo Alfano, Sergio Greco, Francesco Parisi, Gerardo I. Simari, Guillermo R. Simari

III Meta Investigations

12 Logic-Based Approaches to Formal Argumentation 71
  Ofer Arieli, AnneMarie Borg, Jesse Heyninck, Christian Straßer

13 Empirical Cognitive Studies About Formal Argumentation 85
  Federico Cerutti, Marcos Cramer, Mathieu Guillaume, Emmanuel Hadoux, Anthony Hunter, Sylwia Polberg

Part I

Extensions to Abstract Argumentation

CHAPTER 1

HIGHER-ORDER INTERACTIONS (BIPOLAR OR NOT) IN ABSTRACT ARGUMENTATION: A STATE OF THE ART

CLAUDETTE CAYROL
IRIT, Université Toulouse 3
Toulouse, France

ANDREA COHEN
Institute for Computer Science and Engineering, CONICET-UNS
Dept. of Computer Science and Engineering, Universidad Nacional del Sur
Bahía Blanca, Argentina
ac@cs.uns.edu.ar

MARIE-CHRISTINE LAGASQUIE-SCHIEX
IRIT, Université Toulouse 3
Toulouse, France
lagasq@irit.fr

Abstract

In Dung's seminal work, an argumentation framework was defined by a set of abstract arguments and a binary (and also abstract) relation between these arguments, called attack relation and expressing conflicts between arguments. Due to its simplicity and the power of its abstraction, this representation has been intensively used by the community for over 25 years. Another advantage of this approach is the ease with which we can extend the framework, weighting arguments or attacks, using priorities or pre-orderings on the sets of arguments, considering that these interactions are no longer binary ones over the set of arguments (*e.g.* collective attacks), adding new kinds of interactions (*e.g.* supports), and proposing that the targets of these interactions can also be interactions themselves (*i.e.* higher-order interactions).

These last two points are the core of this chapter, in which we present a survey of the proposed approaches existing around the notion of higher-order interactions (attacks and supports) in an abstract argumentation framework.

1 Introduction

Argumentation has become an essential paradigm for knowledge representation and, especially, for reasoning from contradictory information [Amgoud and Cayrol, 2002; Dung, 1995] and for formalizing the exchange of arguments between agents in, *e.g.* , negotiation [Amgoud et al., 2000] (see [Rahwan and Simari, 2009] for a general overview on the role of argumentation in AI). Formal abstract frameworks have greatly eased the modelling and study of argumentation. For instance, a Dung's argumentation framework (AF) [Dung, 1995] consists of a collection of arguments interacting with each other through an attack relation, enabling to determine "acceptable" sets of arguments called *extensions*.

A natural generalization of Dung's argumentation frameworks consists in allowing higher-order attacks (also called recursive attacks in the relevant literature) that target other attacks. Here is an example from the legal domain, borrowed from [Arisaka and Satoh, 2017].

Example 1 ([Arisaka and Satoh, 2017]). *The lawyer says that the defendant did not have intention to kill the victim (Argument b). The prosecutor says that the defendant threw a sharp knife towards the victim (Argument a). So, there is an attack from a to b, denoted by α. And the intention to kill should be inferred. Then, the lawyer says that the defendant was in a habit of throwing the knife at his wife's foot once drunk. This latter argument (Argument c) is better considered as attacking the attack from a to b, rather than argument a itself (so there is now another attack from c to α, denoted by β). Now the prosecutor's argumentation seems no longer sufficient for proving the intention to kill. This example is represented as a recursive framework in Figure 1.*

1 - Higher-Order Interactions in Abstract Argumentation

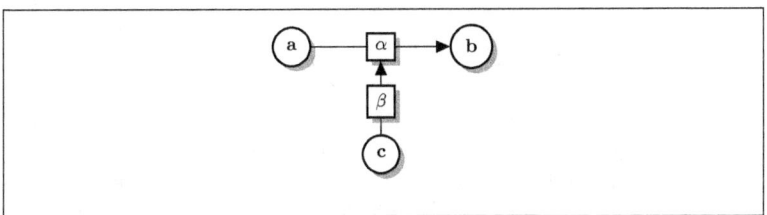

Figure 1: An acyclic recursive framework: arguments are vertices in circles, attacks are directed edges labelled by their name in squares

The idea of encompassing attacks to attacks in abstract argumentation frameworks was first considered in [Barringer *et al.*, 2005] in the context of an extended framework handling argument strengths and their propagation. Then, a semantics for *recursive frameworks* was introduced in [Modgil, 2009b], motivated by the fact that attacks to attacks come from preferences between conflicting arguments. More recently, recursive frameworks have been studied in [Baroni *et al.*, 2011] under the name of *AFRA* (Argumentation Framework with Recursive Attacks), extending Modgil's work by considering higher-order attacks and not only second-order attacks (interactions can be either attacks between arguments or attacks from an argument to another attack at any level). Then, the *AFRA* has been extended in order to handle recursive support interactions together with recursive attacks [Cohen *et al.*, 2015; Cohen *et al.*, 2016]. Another variant of *AFRA*, called *RAF* has been proposed in [Cayrol *et al.*, 2017] and extended in turn to take into account for support interactions. Similar works have proposed to handle recursive frameworks through the definition of a Meta-Argumentation Framework. The idea goes back to [Boella *et al.*, 2008a; Boella *et al.*, 2009b; Gabbay, 2009a; Gabbay, 2009b].

A common point of all these approaches for taking into account higher-order attacks, and then higher-order supports, is the fact that they somehow change the role that attacks play in Dung's frameworks. Moreover, in addition to accounting for the acceptance status of arguments in the framework, some of these works go further by also extending the traditional notion of extension from Dung's AFs to also account for the acceptance of sets of interactions (either attacks or supports). In this chapter many different approaches are presented, trying to highlight their key points and establishing comparisons between them. In order to do this presentation, some choices have been made.

The first one is to present each approach using the main definitions and results given in its seminal paper. Sometimes it occurs that, for a same line of research, many other variants are produced (for completing something that was missing, for adapting it to a specific context, for correcting some undesired behaviours, etc). In such cases, the presentation of each variant is not detailed. Indeed, in this survey, we want to give the most synthetic point of view of each approach (to the extent we can) in order to compare them.

The second choice is the presentation frame we follow for each approach with higher-order interactions: first the definition of the framework (the basic components), second its semantics (extension-based then labelling-based, when provided) and finally some other elements that may exist, such as translation mechanisms; moreover, some comparison points with the approaches presented previously in the chapter will also be given. Of course, this presentation frame will be adjusted since the degree of attention received from the scientific community varies depending on the approach (for instance, labelling-based semantics do not exist for some approaches whereas for others there is no translation mechanism, and so on).

The third choice is the organization of the chapter itself.

- In Section 2, we first recall the cornerstone of the abstract argumentation, the *first-order abstract argumentation framework* defined by Dung.

- Section 3 describes the main contributions on abstract argumentation frameworks using *higher-order attacks*. In this section the reader can find the EAF proposed by Modgil, the $HLAF$ discussed by Gabbay, the $AFRA$ defined by Baroni *et al*, the inductive approach introduced by Hanh *et al* and the RAF presented by Cayrol *et al*. Section 3 ends with a succinct subsection summarizing all the comparison points between the five approaches that are presented throughout this section.

- Section 4 contains a succinct presentation of *bipolar first-order argumentation frameworks* with three variants: the general support (Cayrol *et al*), the necessary support (Nouioua *et al*) and the evidential support (Oren *et al*). A short subsection is included at the end of this section, linking the first-order argumentation frameworks presented there with other approaches: structured argumentation systems that take supports into account; also, works using support relations for performing legal reasoning, for mining arguments and relations from debates, and for identifying arguments and their relations in an empirical study.

- Then, the works taking into account *higher-order attacks and supports* are presented in Section 5. In this section, another kind of support (the deductive one) is discussed since it is directly introduced as a component in an higher-order framework by Boella *et al*; then, for the necessary support, two approaches are presented: the *ASAF* and the *RAFN* respectively defined by Cohen *et al* and Cayrol *et al*; and finally, we study the *REBAF* defined by Cayrol *et al* for the evidential support. As in Section 3, Section 5 ends with a succinct subsection that summarizes all the comparison points between the four approaches that can be found throughout this section.

- Section 6 is dedicated to some computational issues and applications.

- A comparative synthesis is presented in Section 7 covering all the presented approaches.

- Finally, we conclude in Section 8.

Figure 2 shows how the reader can explore the presentation of each type of approach among the sections of the chapter.

2 Dung's approach: a first-order abstract argumentation framework

In this section, we will introduce the abstract argumentation framework proposed in [Dung, 1995], the corner stone of most of the developments in abstract argumentation for the past 25 years.

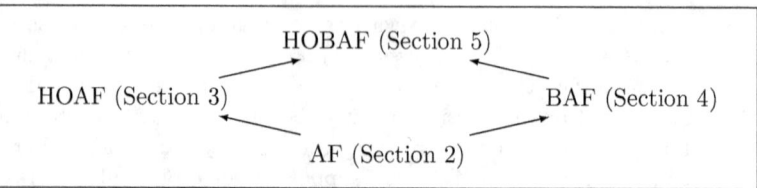

Figure 2: Roadmap for this chapter (HO: Higher-Order, AF: argumentation framework, BAF: bipolar AF)

As defined in [Dung, 1995], an (abstract) argumentation framework is characterized by a set of abstract entities called arguments and a conflict relation among them.

Definition 1 (Def. 2 in [Dung, 1995]). *An argumentation framework (AF) is a pair $\langle Ar, att \rangle$, where Ar is a set of arguments and $att \subseteq Ar \times Ar$.*

For any two arguments $a, b \in Ar$, the meaning of $(a, b) \in att$ is that a attacks b or, equivalently, that a is an attacker of b. Also, an AF can be graphically represented through a directed graph, where the nodes depict the arguments and the edges correspond to the attack relation.

Dung then moves forward to formally characterizing the outcome of an AF, expressed in terms of sets of accepted arguments or *extensions*. As different outcomes may be obtained under different criteria, referred to as *semantics*, Dung started by proposing some basic semantic notions.

Definition 2 (Defs. 5 and 6 in [Dung, 1995]). *Let $\langle Ar, att \rangle$ be an AF and $S \subseteq Ar$:*

- *S is conflict-free iff there are no arguments $a, b \in S$ such that $(a, b) \in att$.*

- *An argument $a \in Ar$ is acceptable w.r.t. S iff for each argument $b \in Ar$ such that $(b, a) \in att$, there exists an argument $c \in S$ such that $(c, b) \in att$.*

- *S is admissible iff it is conflict-free and each argument in S is acceptable w.r.t. S.*

To illustrate these notions, let us consider the following example.

Example 2. *The AF $\langle \{a,b,c,d,e,f\}, \{(a,b),(b,a),(c,a),(e,d),(d,e),(e,f)\}\rangle$ can be represented by the graph illustrated below:*

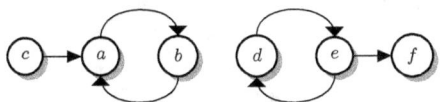

Some examples of conflict-free sets of this AF are $\varnothing$, $\{a\}$, $\{b\}$, $\{c\}$, $\{d\}$, $\{e\}$, $\{f\}$, $\{b,c\}$, $\{d,f\}$ and $\{a,e\}$.

Regarding the notion of acceptability, for instance, argument c is acceptable w.r.t. any set of arguments since it is unattacked. Also, arguments d and f are acceptable w.r.t. the set $\{d\}$. Then, since the set $\{d,f\}$ is conflict-free, it is also an admissible set of AF. In contrast, the set $\{a\}$ is not admissible because, even though it is conflict-free and it defends a against the attack from b, it does not defend a against the attack from c.

As part of the definition of acceptability semantics for AF, [Dung, 1995] introduced the characteristic function, $F_{AF} : 2^{Ar} \mapsto 2^{Ar}$, where $F_{AF}(S) = \{a \mid a \text{ is acceptable w.r.t. } S\}$. Then, the complete, preferred, grounded and stable semantics for AFs are defined using the notion of extension as follows.

Definition 3 (Defs. 7, 13, 20 and 23 in [Dung, 1995]). *Given $AF = \langle Ar, att \rangle$ and $S \subseteq Ar$:*

- *S is a preferred extension of AF iff it is a maximal (w.r.t. set inclusion) admissible set of AF.*

- *S is a stable extension of AF iff it is conflict-free and $\forall a \in Ar \backslash S$, $\exists b \in S$ such that $(b,a) \in att$.*

- *S is the grounded extension of AF iff it is the least fixed point of F_{AF}.*

- *S is a complete extension of AF iff it is an admissible set and $\forall a \in Ar$ such that a is acceptable w.r.t. S, $a \in S$.*

A series of results surrounding the basic semantic notions and the characterization of different semantics are formalized in [Dung, 1995], some of which establish a relationship between sets of extensions obtained under different semantics. On the one hand, Dung's *Fundamental Lemma* shows that given any two arguments a and a' which are acceptable w.r.t. an admissible set S, the set $S' = S \cup \{a\}$ is also admissible, and a' is acceptable w.r.t. S'. Then, it is also shown that the characteristic function of an AF is monotonic (w.r.t. set inclusion). Then, amongst the results over the different semantics, we can highlight the following:

- Each preferred extension is also a complete extension, but not vice-versa.

- Every stable extension is also a preferred extension but not vice-versa.

- A complete extension is a fixed point of the characteristic function of AF.

- The grounded extension is the least (w.r.t. set inclusion) complete extension.

- Every argumentation framework possesses a grounded extension and at least one preferred extension. This is not the case for stable extensions.

The different semantics proposed in [Dung, 1995], as well as some of their relationships, are illustrated below.

Example 2 (cont'd) *The grounded extension of AF is $\{b, c\}$, whereas the preferred (also, stable) extensions are $\{b, c, d, f\}$ and $\{b, c, e\}$.*

Finally, it is worth mentioning that many additional semantics for AFs have been proposed in the literature, as well as alternative characterizations in terms of *labellings* (see [Baroni et al., 2018a] for an overview). However, in this chapter we will focus on the complete, preferred, stable and grounded semantics (referred to as the *Dung semantics* or the *classical semantics*) since they are the ones covered by the approaches to higher-order interactions considered in this chapter.

3 The premises for higher-order interactions: higher- order argumentation frameworks

To our best knowledge, the first work in which the idea of higher-order interactions appears has been presented is [Barringer et al., 2005]. In that article, generalized argumentation networks are presented considering the following points: nodes are arguments, arrows are interactions with two possible cases (attacks or supports), each element of these networks (nodes and arrows) are valued, and the interactions are used in order to propagate these values. Note that the notion of support used in that work is not clearly defined and seems not to correspond to any of the types of support presented in Section 2. In this context, higher-order interactions (from an argument to an interaction)[1] are introduced only in order to influence the value of the target interaction; such a propagation process is described through some examples. Nevertheless, no semantics (extension-based or labelling-based) is formally defined.

Following this seminal work, many different approaches have been developed with, at least at the beginning, a focus on higher-order attacks (so without taking into account the support relation) and a strong link to the notion of "valuation" (in the most general sense, so values or preferences). This is for instance the case of the *Extended Argumentation Framework* (*EAF*) that is proposed in [Modgil, 2007; Modgil, 2009b].

3.1 The Extended Argumentation Framework (*EAF*)

The aim of this approach is to explicitly represent the impact of the preferences between arguments in the argumentation framework by the introduction of attacks that target other attacks. These "second-order attacks" are then used in the definition of the defeat relation (the attack relation refined by preferences between arguments), that is in turn used in the computation of semantics. The formal definition of an *EAF* issued from [Modgil, 2009b] is the following:

Definition 4 (Def. 4 in [Modgil, 2009b]). *An* Extended Argumentation Framework *(EAF) is a tuple* $\langle Ar, att, att2 \rangle$ *such that:*

---

[1] Note that the possibility of having an attack as a source of an attack is also evoked in [Barringer et al., 2005] but not really used.

1. Ar is a set of arguments,

2. $att \subseteq Ar \times Ar$ is a set of "simple attacks" (i.e. binary attacks between arguments),

3. $att2 \subseteq Ar \times att$ is a set of attacks targeting simple attacks,

4. if $(a, (b, c))$ and $(a', (c, b)) \in att2$ then (a, a') and $(a', a) \in att$.

As in Dung's framework, an EAF can be represented using a directed graph in which nodes correspond to arguments, and edges to attacks (solid arrows for simple attacks – elements of att – and double-pointed arrows for attacks to attacks – elements of $att2$ –).

This definition can be illustrated using an example also issued from [Modgil, 2009b]:

Example 3 (Introduction example in [Modgil, 2009b]). *Consider two people exchanging arguments about the weather forecast:*

Argument a: *Today will be dry in London since the BBC forecast sunshine.*

Argument b: *Today will be wet in London since CNN forecast rain.*

Argument c: *But the BBC are more trustworthy than CNN.*

Argument c': *However, statistically CNN are more accurate forecasters than the BBC.*

Argument e: *And basing a comparison on statistics is more rigorous and rational than basing a comparison on your instincts about their relative trustworthiness.*

Here, c and c' do not attack a nor b. They attack the attacks between a and b: c by saying that a is preferred to b, and c' by saying that b is preferred to a. Moreover, the same behaviour occurs with e, which attacks the attack from c to c' (by saying that c' is preferred to c). This example can be represented by the following EAF:

1 - Higher-Order Interactions in Abstract Argumentation

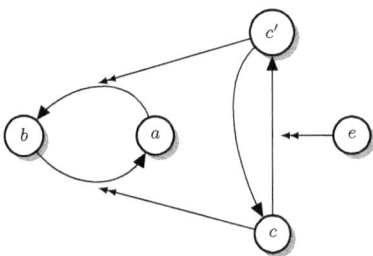

Then, using the *att2* relation, the notion of conflict-freeness can be refined and the notion of defeat, related to a given set of arguments, can be introduced:

Definition 5 (Defs. 5 and 6 in [Modgil, 2009b]). *Let $\langle Ar, att, att2 \rangle$ be an EAF and $S \subseteq Ar$.*

S is conflict-free *iff $\forall a, b \in S$, if $(a, b) \in att$, then $(b, a) \notin att$ and $\exists c \in S$ such that $(c, (a, b)) \in att2$.*

The argument a defeats *the argument b w.r.t. S (denoted by $a \rightarrow^S b$) iff $(a, b) \in att$ and there exists no argument $c \in S$ such that $(c, (a, b)) \in att2$.*

Note that each unattacked attack originates a defeat w.r.t. any set. Another interesting point is the fact that an argument and its attacker can belong to the same conflict-free set if the attack between them is not a symmetrical one and is attacked by an element of the set. Moreover, even if the notion of defeat is not directly used in the definition of conflict-free sets, both notions are related: a conflict-free set cannot contain elements involved in a defeat.

In order to refine the concept of acceptability, an additional notion is introduced in [Modgil, 2009b]: the reinstatement set (informally, the set of defeats that is able to reinstate a given defeat using a given set of arguments).

Definition 6 (Def. 7 in [Modgil, 2009b]). *Let $\langle Ar, att, att2 \rangle$ be an EAF and $S \subseteq Ar$.*

Consider the set of defeats $R^S = \{a_1 \rightarrow^S b_1, \ldots, a_n \rightarrow^S b_n\}$. R^S is a reinstatement set *for the defeat $c \rightarrow^S d$ iff :*

1. $c \to^S d \in R^S$,

2. for $i = 1 \ldots n$, $a_i \in S$,

3. $\forall a_i \to^S b_i \in R^S$, $\forall b'$ such that $(b', (a_i, b_i)) \in att2$, $\exists a' \to^S b' \in R^S$.

Then, semantics for EAF are defined in much the same way as for Dung's framework but using the defeat relation in place of the attack relation and also the reinstatement set for defining the notion of acceptability.

Definition 7 (Defs. 8 and 9 in [Modgil, 2009b]). *Let $\langle Ar, att, att2 \rangle$ be an EAF and $S \subseteq Ar$.*

An argument $a \in Ar$ is acceptable w.r.t. S iff $\forall b$ such that $b \to^S a$, $\exists c \in S$ such that $c \to^S b$ and there is a reinstatement set for $c \to^S b$.

Let S be a conflict-free set of arguments, then:

- *S is an admissible extension iff every argument in S is acceptable w.r.t. S.*

- *S is a preferred extension iff S is a $\subseteq$-maximal admissible extension.*

- *S is a complete extension iff each argument which is acceptable w.r.t. S is in S.*

- *S is a stable extension iff $\forall b \notin S$, $\exists a \in S$ such that $a \to^S b$.*

Example 3 (cont'd) *With this example, we can illustrate the previous definitions. Considering the notion of defeat, there are 4 possible defeats (one for each element of att):*

- *$b \to^S a$ with any S that does not contain c (i.e. if S contains c, then the attack from b to a is not a defeat w.r.t. that set).*

- *$a \to^S b$ with any S that does not contain c'.*

- *$c \to^S c'$ with any S that does not contain e.*

- *$c' \to^S c$ with any S, since the attack (c', c) is never attacked.*

Note that e is never defeated, since it is never attacked. Note also that the two-length cycle between c and c' has been "broken" by the attack issued from e: the attack from c' to c is always a defeat, whereas the attack from c to c' is a defeat only w.r.t. sets that do not contain e. The same thing occurs for the two-length cycle between a and b, since these attacks become defeats w.r.t. sets with different constraints.

Concerning the notion of conflict-freeness, some examples follow. The set $\{a,b\}$ (resp. $\{c,c'\}$) is not conflict-free since there is a symmetrical attack between these arguments; whereas the set $\{e,c,a\}$ is conflict-free since there is no attack between these arguments.

The notion of acceptability w.r.t. a given set can be illustrated using the set $S = \{e, c', b\}$: e (resp. c') is acceptable w.r.t. S since it is unattacked (resp. since there is no defeat targeting c' w.r.t. S which contains e); c (resp. a) is not acceptable w.r.t. S since c' cannot be attacked by a defeat w.r.t. S (resp. since b cannot be attacked by a defeat w.r.t. S).

And finally, following Definition 7, one can conclude that the set $S = \{e, c', b\}$ is an admissible, preferred, complete and stable extension of the EAF.

The particular case of an EAF with an empty $att2$ relation easily shows that EAFs are a conservative generalization of AFs. Indeed, if $att2 = \varnothing$, then the defeat and attack relations coincide and the reinstatement set can be reduced to a singleton (the attack used for defending the acceptability of the argument against a given attack).

Of course, when the $att2$ relation is not empty, and even if EAFs can inherit some properties from AFs (for instance, the fact that preferred extensions are also complete but not vice-versa, see [Polberg, 2016]), they also have some specifics in terms of semantics: the characteristic function of EAF is not, in general, monotonic and so the definition of the grounded extension differs.

Definition 8 (Defs. 10-11 in [Modgil, 2009b]). *Let $EAF = \langle Ar, att, att2\rangle$, $S \subseteq Ar$, and 2^{ArC} denote the set of all conflict-free subsets of Ar. The characteristic function F_{EAF} of EAF is defined as follows:*

$$F_{EAF} : 2^{ArC} \mapsto 2^{Ar}$$

$$F_{EAF}(S) = \{a \mid a \text{ is acceptable w.r.t. } S\}$$

For any EAF $\langle Ar, att, att2\rangle$ the following sequence of subsets of Ar can be defined:

- $F^0 = \emptyset$
- $F^{i+1} = F(F^i)$

Then, the grounded extension of an EAF can be defined in terms of the sequence in the preceding definition as long as the EAF is finitary:

Definition 9 (Defs. 11-12 in [Modgil, 2009b]). Let $EAF = \langle Ar, att, att2\rangle$. EAF is said to be finitary iff $\forall a \in Ar$ the set $\{b \mid (b, a) \in att\}$ is finite, and $\forall (a, b) \in att$ the set $\{c \mid (c, (a, b)) \in att2\}$ is finite.

If EAF is finitary and $F^0 = \emptyset$, $F^{i+1} = F(F^i)$, then $\bigcup_{i=0}^{\infty}(F^i)$ is the grounded extension of EAF.

Example 3 (cont'd) The EAF corresponding to the weather example is clearly finitary. Then, $F^1 = \{e\}$, since e is the only unattacked argument. Now, as shown previously, $c' \to^S c$ w.r.t. any set of arguments, in particular, $F^1 = \{e\}$. In contrast, c does not defeat c' w.r.t. $F^1 = \{e\}$, and c' has no other attackers. As a result, c' is acceptable w.r.t. $\{e\}$ and $F^2 = \{e, c'\}$. Then, since a does not defeat b w.r.t. $F^2 = \{e, c'\}$, and b has no other attackers, it holds that $F^3 = \{e, c', b\}$ is the grounded extension of EAF.

Note that some of the previous definitions were slightly improved since the publication of [Modgil, 2009b] in order to take into account some new constraints or to correct some undesired behaviours (see for instance the definition of conflict-freeness given in Def. 13 of [Modgil and Prakken, 2010], where the authors establish a link between structured argumentation systems and EAFs).

As shown in the literature, Dung's acceptability semantics can also be defined through labellings [Baroni et al., 2018a]. Briefly, a labelling assigns exactly one label to each argument: either **in**, **out**, or **undec**. The arguments labelled **in** constitute an extension E under a given semantics; **out** arguments are defeated by arguments in E, and arguments labelled **undec** are neither in the extension nor defeated by E. For an $EAF = \langle Ar, att, att2\rangle$, since attacks on attacks and the reinstatement of attacks may affect the acceptability of arguments, labels are also assigned to attacks in att, so that if $(x, y) \in att$ is **in** (resp. **out**), then this denotes that the attack (x, y) is successful (resp. unsuccessful). Also, analogously to the labellings for arguments, attacks can be labelled as undec. As a result, whereas the attacks at the argument level (*i.e.* those in the *att* relation) are labelled, second order attacks (*i.e.* those in the *att2* relation) are not. Formally:

Definition 10 (Def. 7 in [Modgil, 2009a]). *A labelling for an EAF $\langle Ar, att, att2 \rangle$ is a pair of total functions $(\mathcal{L}_{Ar}, \mathcal{L}_{att})$ such that:*

1. $\mathcal{L}_{Ar} : Ar \mapsto \{\text{in}, \text{out}, \text{undec}\}$

2. $\mathcal{L}_{att} : att \mapsto \{\text{in}, \text{out}, \text{undec}\}$

For $S \in \{\text{in}, \text{out}, \text{undec}\}$: $S(\mathcal{L}_{Ar}) = \{x \in Ar \mid \mathcal{L}_{Ar}(x) = S\}$; $S(\mathcal{L}_{att}) = \{(x, y) \in att \mid \mathcal{L}_{att}((x, y)) = S\}$

Definition 11 (Def. 8 in [Modgil, 2009a]). *Let $\mathcal{L} = (\mathcal{L}_{Ar}, \mathcal{L}_{att})$ be a labelling for an EAF $\langle Ar, att, att2 \rangle$. $\forall x \in Ar$:*

1. *$x \in \text{out}(\mathcal{L}_{Ar})$ is legally* **out** *iff $\exists (y, x) \in att$ such that $\mathcal{L}_{Ar}(y) = \text{in}$ and $\mathcal{L}_{att}((y, x)) = \text{in}$.*

2. *$x \in \text{in}(\mathcal{L}_{Ar})$ is legally* **in** *iff $\forall (y, x) \in att$, either $\mathcal{L}_{Ar}(y) = \text{out}$ or $\mathcal{L}_{att}((y, x)) = \text{out}$.*

3. *$x \in \text{undec}(\mathcal{L}_{Ar})$ is legally* **undec** *iff :*

    (a) *$\nexists (y, x) \in att$ such that $\mathcal{L}_{Ar}(y) = \text{in}$ and $\mathcal{L}_{att}((y, x)) = \text{in}$; and*

    (b) *it is not the case that: $\forall y \in Ar$, $(y, x) \in att$ implies $\mathcal{L}_{Ar}(y) = \text{out}$ or $\mathcal{L}_{att}((y, x)) = \text{out}$.*

$\forall (x, y) \in att$:

1. *$(x, y) \in \text{out}(\mathcal{L}_{att})$ is legally* **out** *iff $\exists (z, (x, y)) \in att2$ such that $\mathcal{L}_{Ar}(z) = \text{in}$.*

2. *$(x, y) \in \text{in}(\mathcal{L}_{att})$ is legally* **in** *iff $\forall (z, (x, y)) \in att2$, $\mathcal{L}_{Ar}(z) = \text{out}$.*

3. *$(x, y) \in \text{undec}(\mathcal{L}_{att})$ is legally* **undec** *iff :*

    (a) *$\nexists (z, (x, y)) \in att2$ such that $\mathcal{L}_{Ar}(z) = \text{in}$; and*

(b) it is not the case that: $\forall z \in Ar$, $(z,(x,y)) \in att2$ implies $\mathcal{L}_{Ar}(z) = \text{out}$.

For $\text{S} \in \{\text{in}, \text{out}, \text{undec}\}$:

- An argument $x \in Ar$ is said to be illegally S iff $x \in \text{S}(\mathcal{L}_{Ar})$, and it is not legally S.

- An attack (y,x) is said to be illegally S iff $(y,x) \in \text{S}(\mathcal{L}_{att})$, and it is not legally S.

Then, the admissible, preferred and stable EAF labellings are defined in [Modgil, 2009a] as follows:

Definition 12 (Def. 9 in [Modgil, 2009a]). Let $\mathcal{L} = (\mathcal{L}_{Ar}, \mathcal{L}_{att})$ be a labelling for an EAF $\langle Ar, att, att2 \rangle$.

- $\mathcal{L}$ is admissible iff:
  1. no $x \in Ar$ is illegally in or illegally out;
  2. no $(y,x) \in att$ is illegally in or illegally out; and
  3. $\forall x, y \in \text{in}(\mathcal{L}_{att})$, it is not the case that $(y,x) \in att$ and $(x,y) \in att$.

- $\mathcal{L}$ is preferred iff it is admissible and there is no admissible labelling $\mathcal{L}'$ such that $\text{in}(\mathcal{L}_{Ar}) \subset \text{in}(\mathcal{L}'_{Ar})$.

- $\mathcal{L}$ is stable iff it is admissible, $\text{undec}(\mathcal{L}_{Ar}) = \varnothing$ and $\text{undec}(\mathcal{L}_{att}) = \varnothing$.

Then, Modgil shows that the admissible, preferred and stable extensions of an EAF are in one-to-one correspondence with the corresponding labellings of their arguments. Specifically, for $E \in$ {admissible, preferred, stable}, E is a σ-extension of EAF iff there exists a σ-labelling $(\mathcal{L}_{Ar}, \mathcal{L}_{att})$ of EAF such that $\text{in}(\mathcal{L}_{Ar}) = E$.

Example 3 (cont'd) The only preferred and stable labelling corresponding to the EAF of the weather example is $(\mathcal{L}_{Ar}, \mathcal{L}_{att})$, where $\text{in}(\mathcal{L}_{Ar}) = \{e, c', b\}$, $\text{out}(\mathcal{L}_{Ar}) = \{a, c\}$, $\text{in}(\mathcal{L}_{att}) = \{\beta, \eta\}$ and $\text{out}(\mathcal{L}_{att}) = \{\alpha, \epsilon\}$.

The line of work on EAF was extended in different ways. For instance, in [Modgil and Bench-Capon, 2008; Modgil, 2009b], a specific class of EAF has been defined (the *hierarchical EAF*). This kind of framework is stratified so that attacks at some level i are only attacked by arguments that belong to the next level up. For instance, the EAF of Example 3 could be partitioned into 3 levels: level 1 corresponding to $(\{a,b\}, \{(a,b),(b,a)\}, \{(c,(b,a)),(c',(a,b))\})$, level 2 corresponding to $(\{c,c'\}, \{(c,c'),(c',c)\}, \{(e,(c,c'))\})$ and level 3 corresponding to $(\{e\},\{\},\{\})$. Note that there are always two kinds of attacks in these hierarchical EAFs, so second-order attacks exist.

In [Modgil and Bench-Capon, 2008; Modgil, 2009b], the Value-based Argumentation Frameworks introduced in [Bench-Capon, 2003] $(VAF)^2$ are translated into hierarchical EAFs.

Another version of hierarchical EAF which accounts for attacks originating in a set of arguments is also used in [Modgil and Prakken, 2010] in order to establish links with $ASPIC+$ [Prakken, 2010]. These links allow the introduction of *structured EAFs* that satisfy the postulates proposed in [Caminada and Amgoud, 2007].

Moreover, EAFs can be considered as meta-argumentation frameworks, *i.e.* frameworks able to argue about the argumentation process itself. Indeed, the relation *att2* given in EAF can be viewed as a "meta-element" expressing information about the argumentation process (how to take into account the attacks between two arguments when preferences exist). In [Boella *et al.*, 2009a], a study of meta-argumentation is presented with a methodology and some techniques, among them a flattening technique that transforms an EAF into a Dung argumentation framework introducing meta-arguments; in fact, this flattening gives good results in the case of a hierarchical EAF, see [Polberg, 2017]. An application of this technique to the EAF is given in the following definition, which simplifies Def. 10 of [Boella *et al.*, 2009a] in order to avoid some irrelevant meta-arguments and attacks.

Definition 13 (Def. 10 in [Boella *et al.*, 2009a]). *Let $\langle Ar, att, att2 \rangle$ be an EAF. The flattened version of this EAF is the AF defined by:*

- *the set of arguments $= \{acc(a) | a \in Ar\} \cup \{X_{ab}, Y_{ab} | (a,b) \in att\}$*

---

[2] Value-based Argumentation Frameworks have been introduced for persuasion situations. They take into account valued arguments and audiences.

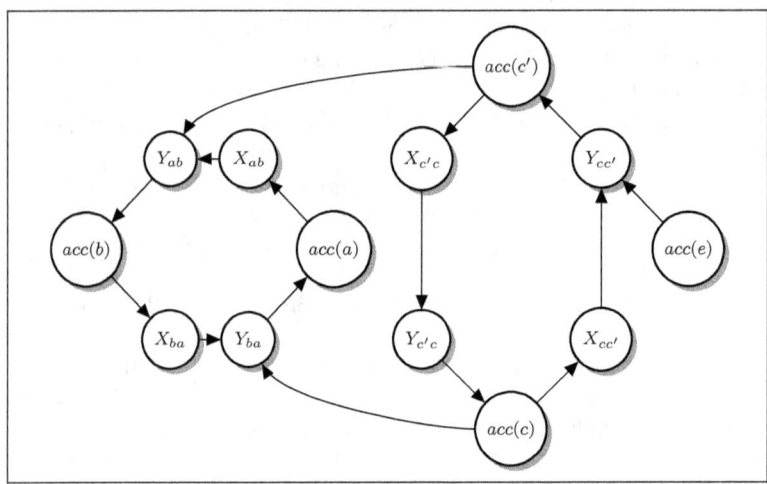

Figure 3: The flattened EAF for Example 3

- *the binary attack relation =*
  $\{(X_{ab}, Y_{ab}) | (a,b) \in att\} \cup$
  $\{(Y_{ab}, acc(b)) | (a,b) \in att\} \cup$
  $\{(acc(a), X_{ab}) | (a,b) \in att\} \cup$
  $\{(acc(c), Y_{ab})) | (c,(a,b)) \in att2\}$

Example 3 (cont'd) *See in Figure 3, the flattening of this EAF.*

Then with Dung semantics, the preferred (and also stable and complete) extension of this flattened EAF is the set that contains $acc(e)$, $acc(c')$, $acc(b)$ and does not contain $acc(a)$, $acc(c)$.

Note that several other flattening processes are proposed in literature:

- In [Boella *et al.*, 2008b], an *EAF* is proposed in order to argue about coalitions of agents. Even though the starting point used in that study is only semi-formal and so not completely abstract (the definition of coalitions is done using agents, goals, ...), the built *EAF* is abstract considering that arguments are coalitions, and attacks represent either attacks between coalitions (for instance because they have the same goal) or the impact of some preferences over these attacks. The flattening process proposed in order to take into account this EAF is very similar to the one defined in [Boella et al., 2009a]: only names for the meta-arguments are different. Thus, the same resulting argumentation framework is produced.

- In [Modgil and Bench-Capon, 2011], another flattening process for EAF is proposed. It is not similar to the previous ones defined in [Boella et al., 2008b; Boella et al., 2009a] in the sense that the structure of the graph is not the same (more nodes and more edges). Nevertheless, it is shown in [Villata et al., 2011] that all of them correspond to the same "argumentation pattern", i.e. to the same behaviour of the second-order attacks.[3]

Note that, in [Modgil and Bench-Capon, 2011, Section 5], some similarities are exhibited between EAF and other approaches, but no formal comparison is done. These approaches are the $AFRA$ (see Section 3.3) and Gabbay's approach (see Section 3.2).

3.2 The Higher-Level Argumentation Frame ($HLAF$)

Gabbay pursued his study of "higher-level networks" introduced in [Barringer et al., 2005] through several papers [Gabbay, 2009a; Gabbay, 2009b], using the idea of meta-argumentation. These networks are more general than EAFs since one can find attacks to attacks at any level (and not only second-order attacks); moreover, other kinds of attacks can be found in these networks. For instance:

- attacks whose source is either a set of elements (joint or conjunctive attacks), or another attack,

- attacks whose target is a set of elements (disjunctive attacks).

---

[3]In [Villata et al., 2011], an argumentation pattern is defined as a multi-labelling of a set of arguments associated to a propositional formula reflecting constraints about this labelling. For instance, the EAF defined by three arguments a, b and c with an attack from a to c and a second-order attack from b to (a, c) is characterized by the constraint:
$[(\mathcal{L}ab(c) = \text{in}) \rightarrow (\mathcal{L}ab(a) = \text{out} \vee \mathcal{L}ab(b) = \text{in})] \wedge$
$[(\mathcal{L}ab(a) = \text{in} \wedge \mathcal{L}ab(b) = \text{out}) \rightarrow (\mathcal{L}ab(c) = \text{out})]$.

Gabbay's aim was to define a framework rich enough to generalize all the existing networks (including the use of a support relation, but supports are not accounted for in those papers). In the argumentation context, the following basic definition for Higher-Level Argumentation Frames ($HLAF$) considers only attacks from one argument to another argument or another attack:

Definition 14 (Def. 1.1 in [Gabbay, 2009b]). *Let Ar be a set of arguments. Level $(0, n)$ argumentation frames are defined as follows:*

1. *A pair $(a, b) \in Ar \times Ar$ is called a level $(0, 0)$ attack.*

2. *If $c \in Ar$ and α is a level $(0, n)$ attack then (c, α) is a level $(0, n+1)$ attack.*

3. *A level $(0, n)$ argumentation frame is the pair $\langle Ar, att \rangle$, where att contains level $(0, m)$ attacks for $0 \leq m \leq n$.*

It is obvious to see that an EAF can be viewed as a particular case of $HLAF$: it is a level $(0, 1)$ argumentation frame with a specific constraint about the sources of level $(0, 1)$ attacks (see Item 4 of Definition 4):

Example 3 (cont'd) *The level $(0, 1)$ argumentation frame corresponding to the weather example is:*

- $Ar = \{a, b, c, c', e\}$

- $att = \{(a, b), (b, a), (c, c'), (c', c), (c, (b, a)), (c', (a, b)), (e, (c, c'))\}$, *the 4 first attacks being level $(0, 0)$ attacks and the 3 last ones being level $(0, 1)$ attacks.*

Then, two kinds of approaches are proposed in order to take into account these networks: labelling-based semantics and flattening processes.[4]

For the first approach, Gabbay proposed the following labelling-based semantics as in Caminada's works [Caminada, 2006; Caminada and Gabbay, 2009]:

---

[4]A third approach is also evoked in Gabbay's works: the translation into logical formalisms (logic programming in [Gabbay, 2009a] and intuitionistic logic in [Gabbay and Gabbay, 2016]). Nevertheless, this approach will not be developed here due lack of space.

1 - HIGHER-ORDER INTERACTIONS IN ABSTRACT ARGUMENTATION

Definition 15 (Def. 2.2 in [Gabbay, 2009b])**.** *Consider $\langle Ar, att \rangle$ a level $(0, n)$ argumentation frame. Let $\mathcal{L}ab : Ar \cup att \to \{\text{in}, \text{out}, \text{undec}\}$. $\mathcal{L}ab$ is a complete labelling if, for every $\beta \in Ar \cup att$, it holds that:*

1. *$\mathcal{L}ab(\beta) = \text{in}$ if there is no a such that $(a, \beta) \in att$.*

2. *$\mathcal{L}ab(\beta) = \text{out}$ if there exists a such that $(a, \beta) \in att$, $\mathcal{L}ab(a) = \text{in}$ and $\mathcal{L}ab((a, \beta)) = \text{in}$.*

3. *$\mathcal{L}ab(\beta) = \text{in}$ if for all a such that $(a, \beta) \in att$, $\mathcal{L}ab(a) = \text{out}$ or $\mathcal{L}ab((a, \beta)) = \text{out}$.*

4. *$\mathcal{L}ab(\beta) = \text{undec}$ if for all a such that $(a, \beta) \in att$, either ($\mathcal{L}ab(a) = \text{out}$ or $\mathcal{L}ab((a, \beta)) = \text{out}$), or ($\mathcal{L}ab(a) = \text{in}$ and $\mathcal{L}ab((a, \beta)) = \text{undec}$), or ($\mathcal{L}ab(a) = \text{undec}$ and $\mathcal{L}ab((a, \beta)) = \text{in}$), or ($\mathcal{L}ab(a) = \mathcal{L}ab((a, \beta)) = \text{undec}$). And moreover, for some a such that $(a, \beta) \in att$, either $\mathcal{L}ab(a) = \text{undec}$ or $\mathcal{L}ab((a, \beta)) = \text{undec}$.*

Example 3 (cont'd) *The set of elements that must be labelled is:*

$$Ar \cup att = \{a, b, c, c', e, (a, b), (b, a), (c, c'), (c', c), (c, (b, a)),$$
$$(c', (a, b)), (e, (c, c'))\}.$$

And the corresponding complete labelling in the sense of Definition 15 is:

- *the elements labelled* **in**: *e, c', b, $(e, (c, c'))$, $(c, (b, a))$, $(c', (a, b))$, (c', c),*

- *the elements labelled* **out**: *c, a, (c, c'), (a, b), (b, a),*

- *the elements labelled* **undec**: *nothing.*

That gives the following "complete extension":
$$\{e, c', b, (e, (c, c')), (c, (b, a)), (c', (a, b)), (c', c)\}$$

It is interesting to note that attacks are also labelled and so can be viewed as belonging to the corresponding "extensions", in contrast to the semantics defined in the EAF approach (see Definition 7).

The translation approach for HLAFs has been already described in the previous section, consisting of a translation into a Dung's *AF*. Indeed, Definition 13 exactly corresponds to Gabbay's flattening process applied to the EAF case. However, Gabbay considers that the translation process described in Definition 13 is not enough in order to represent generalized Higher-Level Argumentation Frames and, in particular, attacks whose source is another attack. Indeed, in such a case, joint attacks must be used in order to capture the real meaning of those attacks (see the discussion in [Gabbay, 2009a, Section 2]). Notwithstanding this, following the translation approach it is easy to see that Gabbay's $HLAF$s can be considered as a conservative generalization of AFs.

In [Gabbay, 2009b] some comparisons are presented between HLAFs and EAF (see Section 3.1), inductive defense semantics proposed by [Hanh et al., 2011] (see Section 3.4) and $AFRA$ (see Section 3.3), but they remain only informal. Nevertheless, using some examples, Gabbay shows that all these approaches do not coincide. Example 4 illustrates this point by comparing $HLAF$ and EAF (the comparison with [Hanh et al., 2011]'s work can be found in Section 3.4, whereas the one with $AFRA$ is given in Section 3.3).

Example 4 (See Figure 3 in [Gabbay, 2009b]). *Consider for instance the following example.*

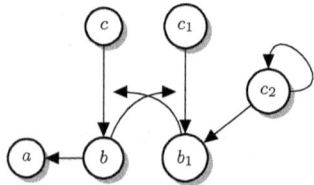

In this example, using the EAF semantics, the complete extension is $\{a, c, c_1\}$. With Gabbay's approach, at least two complete labellings exist, corresponding to the sets of arguments $\{a, c, c_1\}$ and $\{c, c_1\}$. Indeed, in the $HLAF$, argument b_1 can be labelled either out *or* undec.

Note that Gabbay's ideas can also be applied in structured argumentation. Indeed, a recent work [Arisaka et al., 2019] proposes a structured vision of argumentation by "blocks" (an argument being viewed as an argumentation). So, in that work, since each block is an argumentation graph and interactions exist between blocks, a kind of "recursivity" can be identified as it has been done in Gabbay's works.

3.3 Argumentation Frameworks with Recursive Attacks ($AFRA$)

In [Baroni et al., 2009; Baroni et al., 2011] the authors proposed the Argumentation Framework with Recursive Attacks ($AFRA$) as a generalization of the AF, where attacks are allowed to target other attacks as well as arguments. The recursiveness of their approach relies on the fact that these attacks on attacks can appear at any level, thus allowing for higher-order attacks.

As argued by the authors, from a conceptual view, such a generalization supports a straightforward representation of reasoning situations which are not easily accommodated within Dung's framework. In particular, as part of their motivation and similarly to [Modgil, 2009b], the authors propose an example where higher-order attacks are partly used to encode preferences between conflicting arguments. However, as also stated by the authors in [Baroni et al., 2011], further levels of recursive attacks can be considered in the area of modelling decision processes.

Definition 16 (Def. 3 in [Baroni et al., 2011]).
An Argumentation Framework with Recursive Attacks *(AFRA) is a pair $\langle Ar, att \rangle$, where Ar is a set of arguments and $att \subseteq Ar \times (Ar \cup att)$ is an attack relation.*

Given an attack $\alpha = (a, X) \in att$, a is said to be the source of α, denoted as $\mathbf{s}(\alpha) = a$, and X is the target of α, denoted as $\mathbf{t}(\alpha) = X$. Moreover, the authors introduce an abbreviated notation for recursive attacks, avoiding to explicitly show all the recursive steps implied in their definition; for instance, an attack $(a, (b, c))$ can be expressed as (a, α), where $\alpha = (b, c)$. Then, as in Dung's framework, the authors introduce a graph-like notation for the $AFRA$ where nodes correspond to arguments and edges represent attacks that are labelled with their associated Greek letters.

Example 3 (cont'd) *The AFRA corresponding to the weather example can be defined by the sets:*

- $Ar = \{a, b, c, c', e\}$, and
- $att = \{\alpha, \beta, \gamma, \delta, \epsilon, \eta, \theta\}$, where $\alpha = (a, b)$, $\beta = (b, a)$, $\gamma = (c', \alpha)$, $\delta = (c, \beta)$, $\epsilon = (c, c')$, $\eta = (c', c)$, $\theta = (e, \epsilon)$.

The graphical representation for this $AFRA$ is given below, where arguments are in circles and the Greek letters labelling attacks are within squares:

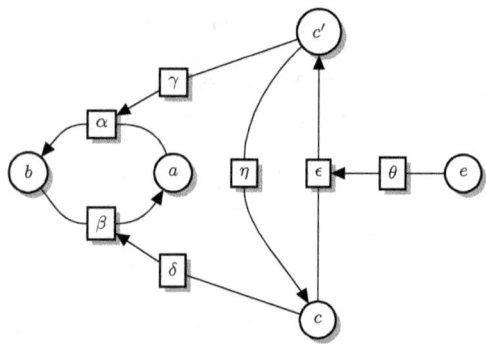

A key difference between the $AFRA$ and the other approaches discussed in the previous subsections is that the authors of [Baroni et al., 2009; Baroni et al., 2011] conceive an attack as an entity able to affect any other entity (be it an argument or an attack) rather than just a by-product of how arguments relate to each other. Consequently, all semantic notions for $AFRA$ are defined following Dung's methodology, except for the fact that attacks are included as first-class elements in those definitions. As a result, similarly to Gabbay's approach where attacks are labelled (see Section 3.2), the extensions of an $AFRA$ may not only include arguments, but also attacks.

As a starting point different types of defeat are defined, which regard attacks (rather than their source arguments) as the subjects able to defeat arguments or other attacks. This is also coherent with the fact that an attack can be made ineffective by attacking the attack itself. Moreover, according to the idea that an attack is strictly related to its source, a defeat over an attack also occurs in a situation where the source of the attack is itself defeated.

Definition 17 (Defs. 4, 5 and 6 in [Baroni et al., 2011]). *Let $\langle Ar, att \rangle$ be an $AFRA$, $\alpha \in att$ and $X \in Ar \cup att$. α defeats X, denoted $\alpha \rightarrow^R X$, if $\mathbf{t}(\alpha) = X$ (direct defeat), or $X = \beta \in att$ and $\mathbf{t}(\alpha) = \mathbf{s}(\beta)$ (indirect defeat).*

Then, based on this notion of defeat, the notions of conflict-freeness, acceptability, admissibility and extensions under different semantics are introduced.

Definition 18 (Defs. 7, 8, 10 in [Baroni et al., 2011]). *Let $\langle Ar, att \rangle$ be an $AFRA$, $S \subseteq Ar \cup att$:*

- *S is conflict-free iff $\nexists \alpha, X \in S$ such that $\alpha \to^R X$.*

- *$X \in S$ is acceptable w.r.t. S iff $\forall \alpha \in att$ such that $\alpha \to^R X$, $\exists \beta \in S$ such that $\beta \to^R \alpha$.*

- *S is admissible iff it is conflict-free and each element of S is acceptable w.r.t. S.*

Note that, whereas [Baroni et al., 2009] just considered the preferred semantics, [Baroni et al., 2011] extended the results to also cover the complete, grounded, stable, semi-stable and ideal semantics. Nonetheless, as mentioned before, in this chapter we will only focus on the four classical semantics since they are the ones covered by most approaches. For the purpose of defining the grounded semantics, [Baroni et al., 2011] defines the characteristic function analogously to [Dung, 1995].

Definition 19 (Def. 9 in [Baroni et al., 2011]). *The characteristic function of $AFRA = \langle Ar, att \rangle$ is defined as follows:*

$$F_{AFRA} : 2^{Ar \cup att} \mapsto 2^{Ar \cup att}$$

$$F_{AFRA}(S) = \{X \in Ar \cup att \mid X \text{ is acceptable w.r.t. } S\}$$

Definition 20 (Defs. 11 to 14 in [Baroni et al., 2011]). *Let $AFRA = \langle Ar, att \rangle$ and $S \subseteq Ar \cup att$:*

- *S is a complete extension of $AFRA$ iff S is admissible and every element of $Ar \cup att$ which is acceptable w.r.t. S belongs to S (i.e. $F_{AFRA} \subseteq S$).*

- *S is the grounded extension of $AFRA$ iff it is the least fixed point of F_{AFRA}.*

- *S is a preferred extension of $AFRA$ iff it is a maximal (w.r.t. set inclusion) admissible set.*

- S is a stable extension of $AFRA$ iff S is conflict-free and $\forall X \in Ar \cup att$, if $X \notin S$ then $\exists \alpha \in S$ such that $\alpha \to^R X$.

Example 3 (cont'd) *The preceding definitions can be illustrated on the weather example as follows. On the one hand, each attack in att originates a direct defeat on its target, namely, $\alpha \to^R b$, $\beta \to^R a$, $\gamma \to^R \alpha$, $\delta \to^R \beta$, $\epsilon \to^R c'$, $\eta \to^R c$ and $\theta \to^R \epsilon$. On the other hand, the indirect defeats are: $\alpha \to^R \beta$, $\beta \to^R \alpha$, $\epsilon \to^R \eta$ and $\eta \to^R \epsilon$. Then, for instance, the set $\{a,b\}$ is conflict-free even though a and b are the source and target of the attack α (also, the target and source of the attack β). As discussed before, this is because defeats can only be originated by attacks; hence, for instance, any set containing just arguments will be conflict-free in the $AFRA$.*

Note that, similarly to what occurs in the EAF, the defeat from θ to ϵ breaks the two-length attack cycle involving arguments c and c'. Hence, The two-length cycle involving arguments a and b is also broken. Consequently, e.g. , $\{e, \theta, c', \eta, \gamma, b, \beta\}$ is an admissible set of this $AFRA$. Moreover, this set is the only complete extension, thus being the grounded extension and the only preferred extension of the framework, which is also stable. Once again, note that this result aligns with the result obtained for the EAF, whose corresponding extension was $\{e, c', b\}$.

In contrast, we can highlight a difference between the result for the $AFRA$ and the one obtained for Gabbay's level $(0, n)$ argumentation frame: whereas the attack $(c, (b, a))$ was labelled as in in the $HLAF$, δ (the corresponding attack in the $AFRA$) does not belong to the extension. This is because the attack η, which is defended against ϵ by the undefeated attack θ, directly defeats c and therefore, indirectly defeats δ. Consequently, b belongs to the $AFRA$ extension while it does not belong to the corresponding extension of the $HLAF$.

Another simpler example allows to compare Gabbay's approach with $AFRA$.

Example 5. *Consider the following very simple framework.*

In this case, with the $AFRA$ semantics, the set $\{\alpha, c\}$ is admissible (the attack from b to c is made ineffective by α), whereas with Gabbay's approach c cannot be labelled **in** without a being labelled **in**.

Another difference appears for the complete semantics: in the $AFRA$, the only complete extension is the set $\{a, \alpha, c\}$ (since α defeats β), whereas in the $HLAF$ the complete labelling is the set $\{a, \alpha, c, \beta\}$ (there is no link between α and β).

In both cases, the difference is due to the notion of defeat adopted by the $AFRA$ (see Definition 17), which accounts for indirect defeats.

In addition to proposing several argumentation semantics, [Baroni et al., 2011] shows that many properties satisfied by Dung's AF also hold for the $AFRA$. First, the characteristic function of the $AFRA$ is shown to be monotonic w.r.t. set inclusion (differently from the EAF's). Then, the authors prove that stable extensions of the $AFRA$ are also preferred extensions but not vice-versa. In addition, they include results showing that every preferred extension is a complete extension but not vice-versa, that the grounded extension is the least complete extension, and that every $AFRA$ possesses at least one preferred extension.

As another set of results, [Baroni et al., 2011] formally shows that when an $AFRA$ coincides with an AF (when no higher-order attacks occur) the generalized notions for the $AFRA$ are compatible with the ones for the AF. There exists a correspondence at the level of acceptability semantics (e.g. grounded, preferred, stable, complete semantics) but there is no correspondence between more basic semantic notions. Consequently, this means that $AFRA$s are not a conservative generalization of AFs since, among other things, the notion of conflict-freeness does not coincide at the $AFRA$ and AF level (see for instance the fact that the set $\{a, b\}$ on Example 5 is conflict-free in an $AFRA$ but not in an AF).

Also, a flattening method is proposed to express an $AFRA$ as an AF, drawing the relevant correspondences concerning the different semantic notions and argumentation semantics. Then, following the flattening technique, the extensions of the $AFRA$ are the extensions of its associated AF.

Definition 21 (Def. 10 in [Baroni et al., 2011]). *Let $AFRA = \langle Ar, att \rangle$, the corresponding argumentation framework is $AF = \langle Ar_{AF}, att_{AF} \rangle$,*

where:
$$Ar_{AF} = Ar \cup att$$
$$att_{AF} = \{(\alpha, X) \mid \alpha \in att, X \in (Ar \cup att), \alpha \rightarrow^R X\}$$

Example 5 (cont'd) *Applying the $AFRA$-AF flattening from Definition 21 we obtain the following AF:*

Here, we have that the only complete extension of the associated AF is $\{a, \alpha, c\}$ (in accordance with the result obtained by directly applying the acceptability semantics on the $AFRA$).

[Baroni et al., 2011] formally shows that the two approaches for determining acceptability in the $AFRA$ (*i.e.* the direct computation approach and the flattening approach) are equivalent. As remarked by the authors, this kind of correspondence is very useful as it allows one to reuse or adapt, in the context of $AFRA$, the large corpus of results and implementations available for Dung's framework. In particular, as will be shown in Section 6, the flattening of an $AFRA$ into an AF is exploited for implementing a reduction-based approach to compute the $AFRA$ extensions.

Finally, [Baroni et al., 2011] draws a detailed comparison between the $AFRA$ and Modgil's EAF, highlighting four points: the fact that EAF only allows for second-order attacks whereas the $AFRA$ allows for higher-order attacks at any level; the differences in the definition of conflict-freeness; the non-monotonicity of Modgil's characteristic function for the general case of EAFs versus the monotonicity of the $AFRA$ characteristic function; and, related to the previous point, the fact that the grounded extension of the EAF is not the least complete extension of an EAF in the general case (whereas this relationship does hold for the $AFRA$).

3.4 The inductive semantics for $HLAF$

In [Hanh et al., 2011] the authors proposed a new inductive semantics for Gabbay's Higher-Level Argumentation Frames ($HLAF$)[5] introduced in Section 3.2. The authors argued that their semantics, based on an inductive defense relation, is sceptical and grounded towards the acceptability of attacks in a sense that an attack is "acceptable" w.r.t. a set of arguments S only if it is inductively defended by S, but could be credulous towards the acceptability of arguments. They motivated their semantics by stating that Gabbay's approach, as well as Modgil's approach, may yield counter-intuitive results in some cases, such as the one illustrated by the example below.

Example 6 (Introduction Ex. in [Hanh et al., 2011]). *Consider a framework like the one depicted below, consisting of attacks $\alpha_1 = (a, a)$ and $\alpha_{i+1} = (a, \alpha_i)$ for $i \geq 1$:*

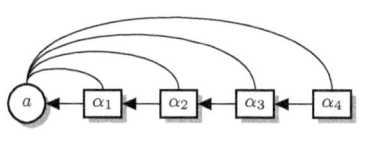

In this figure, each attack α_i is represented by an "arrow" that goes from its source (a) to its target $(a$ or $\alpha_{i-1})$ "across the box" α_i that just gives the name of the attack.

The authors state that they find it rather hard to imagine any practical interpretation of this framework. Then, they state that as a sceptical reasoner one would not want to draw any conclusion (as a result, not accept a). This is because an agent arguing for a has to rely on an infinite line of defense $\alpha_2, \alpha_4, \ldots$. Then, they argue that the semantics for $HLAF$ introduced in Section 3.2, as well as the corresponding $AFRA$ semantics from Section 3.3 will yield a unique preferred extension $\{a, \alpha_2, \alpha_4, \ldots\}$, and they find this result counter-intuitive.

In order to avoid undesired results like the one mentioned above, [Hanh et al., 2011] proposes the inductive semantics of $HLAF$s which,

---

[5] In [Hanh et al., 2011] the authors referred to Gabbay's formalism as the Extended Argumentation Framework (EAF); however, in order not to confuse it with Modgil's EAF (see Section 3.1) here we will keep Gabbay's naming for $HLAF$.

in a situation like the one corresponding to Example 6, will yield the empty set as the only extension. For simplicity, the authors define their semantics for *bounded HLAF*s, but mention that their results could be easily generalized for the case of unbounded $HLAF$. Briefly, a $HLAF$ $\langle Ar, att \rangle$ is said to be bounded if each argument or attack in the framework has a finite number of attacks against it.

They start by defining the notion of *inductive defense*, which captures a sceptical attitude of rational agents towards the acceptance of attacks.

Definition 22 (Def. 3.1 in [Hanh et al., 2011]). *Given $HLAF = \langle Ar, att \rangle$, $S \subseteq Ar$ and $\beta \in att$:*

- *S inductively defends (i-defends) β within 0-steps iff there is no argument $c \in Ar$ such that $(c, \beta) \in att$.*

- *S i-defends β within $(k+1)$-steps iff either: S i-defends β within k-steps; or for each $c \in Ar$, if $(c, \beta) \in att$, then there exists $d \in S$ such that:*

    - *$(d, c) \in att$ and S i-defends (d, c) within k-steps, or*
    - *$(d, (c, \beta)) \in att$ and S i-defends $(d, (c, \beta))$ within k-steps.*

Example 3 (cont'd) *Given the $HLAF$ corresponding to the weather example, it holds that γ, δ, η and θ are i-defended by any set of arguments within 0-steps (thus, within k-steps for $k \geq 1$) since they are not attacked by any argument in the framework.*

Then, it holds that β is i-defended by the set $S = \{c'\}$ within 1-steps because for the only argument c such that $(c, \beta) = \delta \in att$, there exists $c' \in S$ such that $(c', c) = \eta \in att$ and S i-defends η within 0 steps.

In contrast, α and ϵ are not i-defended by any set; moreover, they are respectively attacked by γ and θ, which are i-defended by any set within 0-steps.

Then, accounting for this notion of inductive defense, they characterize the new acceptability semantics of $HLAF$ as follows.

Definition 23 (Defs. 3.2 to 3.5 in [Hanh et al., 2011]). *Given $HLAF = \langle Ar, att \rangle$ and $S \subseteq Ar$:*

1 - HIGHER-ORDER INTERACTIONS IN ABSTRACT ARGUMENTATION

- *S is i-conflict-free iff $\nexists a, b \in S$ such that $(a, b) \in att$ and S i-defends (a, b) (within any number of steps).*

- *An argument $a \in Ar$ is i-acceptable w.r.t. S iff for each $b \in Ar$ such that $(b, a) \in att$, there exists $c \in S$ such that:*
    - *$(c, b) \in att$ and S i-defends (c, b); or*
    - *$(c, (b, a)) \in att$ and S i-defends $(c, (b, a))$.*

- *S is i-admissible iff it is i-conflict-free and every argument in S is i-acceptable w.r.t. S.*

- *The characteristic function based on i-defense is defined as follows:*

    $F_I : 2^{Ar} \mapsto 2^{Ar}$

    $F_I(S) = \{a \in Ar \mid a \text{ is i-acceptable w.r.t. } S\}$

- *S is an i-preferred extension iff it is a maximally (w.r.t. set inclusion) i-admissible set.*

- *S is an i-complete extension iff it is an i-admissible set and each argument that is i-acceptable w.r.t. S belongs to S.*

- *S is the grounded i-extension iff it is the least fixed point of F_I.*

The semantic notions of *HLAF* based on i-defense can be illustrated on the weather example.

Example 3 (cont'd) *Given that e is an unattacked argument, it holds that $F_I(\varnothing) = \{e\}$. Then, $F_I(\{e\}) = \{e, c'\}$ since, as shown before, there exists $c' \in Ar$ such that $(c', c) = \epsilon \in att$ but $(e, \epsilon) = \theta \in att$, where θ is i-defended by $\{e\}$. Finally, $F_I(\{e, c'\}) = \{e, c', b\}$ since the attack $\alpha = (a, b)$ is itself attacked by $\gamma = (c', \alpha)$ and γ is i-defended by $\{e, c'\}$. Moreover, $\{e, c', b\}$ is the least fixed point of the characteristic function and a maximal i-admissible set; thus, it corresponds to both the i-grounded extension and the only i-preferred extension of HLAF. As a result, the outcome in this case coincides with that obtained for Modgil's EAF. In addition, the outcome aligns with that obtained for the AFRA (the extension obtained here is contained in the extension obtained for the AFRA), which was shown to differ from the one obtained with Gabbay's semantics for HLAF.*

Another example illustrates the differences between Gabbay's approach and inductive defense semantics.

Example 4 (cont'd) *Recall that, with Gabbay's approach, at least two complete labellings are possible corresponding to the sets $\{a,c,c_1\}$ and $\{c,c_1\}$. Let us now consider the i-defense semantics for HLAF. Given that c and c_1 are the only unattacked arguments, it holds that $F_I(\varnothing) = \{c,c_1\}$. Then, we have $F_I(\{c,c_1\}) = \{c,c_1\}$. Note that $a \notin F_I(\{c,c_1\})$ because, even though c attacks b (the only attacker of a), a is not acceptable w.r.t. $\{c,c_1\}$ since the attack (c,b) is not i-defended by $\{c,c_1\}$ (within any number of steps); the only attacks i-defended by $\{c,c_1\}$ are (c_2,c_2), (c_2,b_1), $(b,(c_1,b_1))$, $(b_1,(c,b))$ and (b,a) (all of which are, in particular, i-defended within 0 steps). Consequently, $\{a,c,c_1\}$ cannot be an extension using the inductive semantics of [Hanh et al., 2011].*

In [Hanh et al., 2011] the authors formally showed that their inductive semantics preserves the key properties of well-established semantics for abstract argumentation, such as the Fundamental Lemma and the monotonicity of the characteristic function. Furthermore, it is obvious to see that, in the case of a $HLAF$ with no higher-order attacks, all attacks in the framework will be i-defended; thus, i-conflict-freeness turns into Dung's conflict freeness, and the same holds for acceptability, admissibility, etc. And so this approach is a conservative generalisation of Dung's approach.

Moreover, in [Hanh et al., 2011], some links were also established with other higher-order approaches. It was shown that any extension obtained with Modgil's EAF semantics, Gabbay's $HLAF$ semantics or the $AFRA$ semantics contains a sceptical part corresponding to an extension obtained under the i-defense semantics, in addition to a credulous part resulting from the credulousness towards the acceptance of attacks. Formally, that corresponds to: let S be an extension obtained with Modgil's EAF semantics, Gabbay's $HLAF$ semantics or the $AFRA$ semantics, S contains a greatest (w.r.t. set-inclusion) i-extension T (for the homonym semantics), *i.e.* $T \subseteq S$ and $\forall U$ being an i-extension (for the same semantics), if $U \subseteq S$ then $U \subseteq T$. This result is derived differently following the other higher-order approaches that are studied in [Hanh et al., 2011].

- First, the authors stated that inductive defense semantics could be viewed as a sceptical approach to the semantics of Gabbay;

in that way, for instance, the g-grounded[6] extension corresponds to the union of the i-grounded extension and the set of attacks i-defended by it. Then, they state that the truly sceptical part of any g-complete extension can be characterized by an i-complete extension; here, they again highlight that the difference in the complete extensions results from the credulousness of Gabbay's approach w.r.t. the acceptance of attacks.

- Then, regarding the relationship between [Hanh et al., 2011]'s semantics and the $AFRA$ semantics, the authors state that they differ in the conditions imposed over acceptable attacks, and is related to the existence of indirect defeats in the $AFRA$ (see Definition 17). In that way, an attack will be acceptable in the $AFRA$ only if both the attack and its source argument are defensible. Nevertheless, despite this difference, the authors establish a correspondence between i-complete extensions and bcgg-complete[7] extensions: a bcgg-complete extension is equal to the union of an i-complete extension and the set of attacks coming from arguments in the i-complete extension that are i-defended by it.

- And finally, as to the relationship between [Hanh et al., 2011]'s semantics and Modgil's semantics for the EAF, the authors remark the following: whereas Modgil's EAF could be viewed as a special case of general extended frameworks such as the $HLAF$, its semantics are based on the underlying intuition that attacks against attacks represent preferences between conflicting arguments (thus motivating the last clause of Definition 4). Therefore, this insight suggests that different intuitions and applications could lead to different classes and different semantics for extended argumentation frameworks (as also evidenced by the non-monotonicity of Modgil's characteristic function for the general case of EAF). In spite of these differences, the authors in [Hanh et al., 2011] argue that their i-grounded semantics captures the most sceptical

---

[6]The grounded extension according to Gabbay's semantics. In the remainder of this section, we will refer to the extensions obtained under Gabbay's σ semantics as the g-σ extensions.

[7]Similarly to the notation for Gabbay's approach, these denote the complete extensions obtained by the $AFRA$ semantics.

part of Modgil's grounded semantics since, as shown in [Modgil, 2009b], the characteristic function of hierarchical and preference symmetric frameworks is monotonic (where in the former case, the hierarchical restriction essentially ensures that any defense of an attack is inductive). Then, they enforce i-admissible sets by requiring m-conflict-freeness,[8] characterizing the notion of mi-admissibility.[9] As a result, they show that mi-admissibility could be viewed as a sceptical part in the credulous semantics of Modgil. In other words, that any m-preferred[10] extension contains a maximal (w.r.t. set inclusion) mi-admissible set.

To end this section it is worth mentioning that the line of work on inductive defense semantics started by [Hanh et al., 2011] was recently continued in [Li and Wu, 2019]. There, the authors defined a new semantics for $HLAF$ accounting for *infinite* inductive defense, since the notion of i-defense characterized in [Hanh et al., 2011] is only inductively defined for finite steps. For that purpose, they defined a notion of *renovation sets* to recognize "valid attacks", similarly to the "i-defense of an attack" in [Hanh et al., 2011]. Then, they formally showed the relationship between the notion of i-defense and their renovation sets: an attack α is i-defended by a set of arguments S within k-steps iff there exists a finite renovation set of α w.r.t. S, which renovates α within k-steps. In that way, they state that the semantics of [Hanh et al., 2011] can also be expressed with finite renovation sets.

3.5 The Recursive Argumentation Framework (RAF)

In [Cayrol et al., 2017], another framework that allows representing both simple and higher-order attacks (*i.e.* attacks from an argument to either another argument or another attack) is considered.

Definition 24 (Def. 4 in [Cayrol et al., 2017]). *A Recursive Argumentation Framework (RAF) is a tuple $\langle Ar, att, \mathbf{s}, \mathbf{t} \rangle$ where Ar is a finite*

---

[8]That is, conflict-freeness as defined in Definition 5 by Modgil.

[9]"mi-admissibility" means that this notion is defined mixing notions given in Modgil's work and [Hanh et al., 2011].

[10]Like before, m-preferred extension is used to denote a preferred extension obtained with Modgil's semantics for EAF.

and non-empty set of arguments, att is a finite set disjunct from Ar representing attack names, **s** is a function from att to Ar mapping each interaction to its source, and **t** is a function from att to $(Ar \cup att)$ mapping each interaction to its target.

Note that a *RAF* can be graphically represented in the same way as an *AFRA* (see Section 3.3).

Acceptability semantics for argumentation frameworks with higher-order attacks have been defined in a direct way in [Cayrol et al., 2017]. The idea is to specify the conditions under which the arguments are considered as accepted directly on the extended framework, without translating the original framework into an *AF*. Moreover, due to the defeasible nature of attacks (attacks may be affected by other attacks), conditions under which the attacks are accepted must also be specified. Indeed, some attacks may not be "valid", in the sense that they cannot defeat the argument or attack they are targeting. So, acceptability conditions for arguments should be given with respect to valid attacks and, conversely, attacks should be declared valid with respect to other arguments or attacks. For instance, the fact that two arguments may be conflicting depends on the validity of the attack between them. Hence, the traditional notion of extension defined in terms of a set of arguments is replaced by a pair of a set of arguments and a set of attacks, called a "structure".

Definition 25 (Def. 5 in [Cayrol et al., 2017]). *Consider $RAF = \langle Ar, att, \mathbf{s}, \mathbf{t} \rangle$. A structure of RAF is a pair (S, Γ) with $S \subseteq Ar$ and $\Gamma \subseteq att$.*

Intuitively, given a structure $U = (S, \Gamma)$, S contains the arguments that are accepted "owing to" U and Γ contains the attacks which are valid "owing to" U (the meaning of "owing to" depending on the considered semantics).

In the following, we recall the acceptability conditions for structures, and the definitions of the semantics that are given in [Cayrol et al., 2017]. The key notion is the fact that a set of arguments (resp. attacks) can be "defeated" (resp. "inhibited") w.r.t. a given structure.

Definition 26 (Equations (1) – (4) in [Cayrol et al., 2017]). *Consider $RAF = (Ar, att, \mathbf{s}, \mathbf{t})$. Let $U = (S, \Gamma)$ be a structure of RAF, $a \in Ar$ and $\alpha \in att$.*

- a is defeated w.r.t. U iff there is $\beta \in \Gamma$ with $\mathbf{s}(\beta) \in S$ and $\mathbf{t}(\beta) = a$,
- α is inhibited w.r.t. U iff there is $\beta \in \Gamma$ with $\mathbf{s}(\beta) \in S$ and $\mathbf{t}(\beta) = \alpha$.

$Def(U)$ (resp. $Inh(U)$) denotes the set of arguments (resp. attacks) that are defeated (resp. inhibited) w.r.t. U.

Then semantics for RAF are defined as follows:

Definition 27 (Defs. 6, 7 in [Cayrol et al., 2017]). *Consider $RAF = \langle Ar, att, \mathbf{s}, \mathbf{t}\rangle$. Let $U = (S, \Gamma)$ be a structure of RAF.*

- U is conflict-free *iff* $S \cap Def(U) = \varnothing$ *and* $\Gamma \cap Inh(U) = \varnothing$.
- *Let $a \in Ar$ and $\alpha \in att$. a (resp. α) is* acceptable *w.r.t. U iff for each $\beta \in att$ with $\mathbf{t}(\beta) = a$ (resp. $\mathbf{t}(\beta) = \alpha$), either $\beta \in Inh(U)$ or $\mathbf{s}(\beta) \in Def(U)$. $Acc(U)$ denotes the set of all arguments and attacks that are acceptable w.r.t. U.*
- U is admissible *iff it is conflict-free and for each $x \in (S \cup \Gamma)$, x is acceptable w.r.t. U.*
- U is complete *iff it is conflict-free and $Acc(U) = S \cup \Gamma$.*
- U is stable *iff it is conflict-free and satisfies $Ar \setminus S \subseteq Def(U)$ and $att \setminus \Gamma \subseteq Inh(U)$.*
- U is preferred *iff it is a $\subseteq$-maximal admissible structure.*
- U is grounded *iff it is the $\subseteq$-minimal conflict-free structure $U = (S, \Gamma)$ satisfying $Acc(U) \subseteq S \cup \Gamma$.*[11]

Example 3 (cont'd) *The structure $(\{b, c', e\}, \{\beta, \delta, \gamma, \eta, \theta\})$ is a complete, preferred and stable structure and the grounded structure of the RAF corresponding to the weather example. At this point we can remark an important difference with $AFRA$: whereas η defeats δ (because it defeats its source c) in $AFRA$, η does not inhibit δ w.r.t. this structure in RAF. Hence, we obtain different results for the grounded, complete,*

---
[11]The definition for the grounded structure was given in [Cayrol et al., 2020], which is an extended version of [Cayrol et al., 2017].

preferred and stable semantics, where δ is left out of the AFRA extension, but is included in the corresponding RAF structure.

The notion of structure has been strengthened in order to obtain a conservative generalization of Dung's frameworks for the conflict-free, admissible, complete, stable and preferred semantics. It is worth noting that in an AF, each attack is considered as valid, in the sense that it may affect its target. The next definition strengthens the notion of structure by adding a condition on attacks that will force every acceptable attack to be valid.

Definition 28 (Defs. 11 to 13 in [Cayrol et al., 2017]). *Consider RAF = $\langle Ar, att, \mathbf{s}, \mathbf{t} \rangle$.*

1. *A d-structure on RAF is a structure $U = (S, \Gamma)$ with $(Acc(U) \cap att) \subseteq \Gamma$.*

2. *A conflict-free (resp. admissible, complete, preferred, stable) d-structure is a conflict-free (resp. admissible, complete, preferred, stable) structure which is also a d-structure.*

This result has also been extended to the grounded semantics in [Cayrol et al., 2020]. The conservative generalization proved in [Cayrol et al., 2017; Cayrol et al., 2020] relies upon a correspondence between a Dung's framework (and its extensions) and a "non-recursive" RAF (and its d-structures), where a non-recursive RAF is a RAF in which no attack targets another attack.

Another one-to-one correspondence has been proved in [Cayrol et al., 2017]. Indeed the RAF and the $AFRA$ approaches give similar results for complete, preferred and stable semantics but, once again, it is not the case when we consider conflict-freeness and admissibility (see Propositions 2 to 5 in [Cayrol et al., 2017]). Moreover this correspondence needs to apply some constraints on the semantics results (it is not a direct one). Example 5, already used for comparing Gabbay's approach with $AFRA$, can also be used for illustrating these points.

Example 5 (cont'd) *First, the set $\{\alpha, \beta\}$ cannot be conflict-free in AFRA (since α defeats β), whereas the structure $(\varnothing, \{\alpha, \beta\})$ is conflict-free in RAF.*

Moreover, recall that $\{c, \alpha\}$ is an admissible set of the $AFRA$, while the structure $(\{c\}, \{\alpha\})$ is not admissible with the RAF approach. Indeed, in $AFRA$, α defeats β (or b) despite the absence of its source while, in RAF, an attack whose source is not accepted cannot defeat other arguments or attacks.

Consider now the semantics level, for instance for the preferred semantics. With the RAF approach, the preferred structure is $(\{a, c\}, \{\alpha, \beta\})$ whereas with the $AFRA$ approach, the preferred extension is $\{a, c, \alpha\}$. In that case, if we want to obtain a RAF structure from an $AFRA$ extension, we need to add to the structure all those attacks whose only reason for being defeated, according to $AFRA$, is because of the attacks towards their source (here β). Conversely, the $AFRA$ extension is obtained from the RAF structure by the removal of attacks whose source is not in the structure (here β, too).

Note that, on Example 5, RAF produces results similar to Gabbay's approach. This is also the case when we consider Example 4.

Example 4 (cont'd) *In this example, considering the complete labellings obtained with Gabbay's approach and the structures of the RAF approach, the same results are obtained: first a, c, c_1, and all attacks are labelled* in *and are in the same structure; second c, c_1, and all attacks except (c, b) and (c_1, b_1) are labelled* in *and are in the same structure.*

These two examples show some correspondences between RAF and Gabbay's higher-level argumentation frames. Nevertheless, these correspondences remain to be proven, particularly because between labellings and structures a main difference exists: the undec value.

3.6 Comparison between Higher-order approaches: a first and succinct summary

Throughout Section 3 we highlighted many differences and similarities in order to compare the five approaches introduced in this section (EAF–Section 3.1–, $HLAF$–Section 3.2–, $AFRA$–Section 3.3–, i-semantics for $HLAF$–Section 3.4–and RAF–Section 3.5). These comparison points were introduced when pertinent (depending on the definitions and examples that were discussed at that point in the text). So, in order to facilitate the reading and the understanding of this chapter, we just recall here the main comparison points between all these approaches.

- First of all, these approaches have been compared with Dung's framework and generally they are a conservative generalization of the latter when no higher-order attacks are present. Nevertheless, this result does not hold for the $AFRA$ when we consider some basic notions such as conflict-freeness (see Example 5 in Section 3.3).

- An EAF can be viewed as a particular case of $HLAF$ (a level $(0,1)$ argumentation frame with a specific constraint), but EAF and $HLAF$ do not coincide from a semantics point of view (see Example 4 in Section 3.2).

- Examples 3 and 5 in Section 3.3 illustrate the same results between $HLAF$ and $AFRA$: $HLAF$ and $AFRA$ do not coincide from a semantics point of view.

- Moreover, a detailed comparison between $AFRA$ and EAF can be found in [Baroni et al., 2011] highlighting the fact that EAF and $AFRA$ do not coincide from a semantics point of view.

- Another comparison is available concerning $HLAF$, EAF and i-semantics, yielding once again the same result: $HLAF$, EAF and i-semantics do not coincide (see Examples 3 and 4 in Section 3.4 and the text given at the end of Section 3.4).

- No comparison exists between RAF and the other higher-order approaches, except for $AFRA$. In this case, Example 5 in Section 3.5 can be used for illustrating the fact that RAF and $AFRA$ do not coincide from a semantics perspective.

Note that a more complete analysis and comparison of these approaches can be found in Section 7.

4 Different variants of first-order bipolar argumentation frameworks

In this section we will present some bipolar argumentation frameworks, which are amongst the most-widely used in the literature and inspired the approaches from Section 5. Then we will end this section by briefly discussing the links between the developments on bipolar argumentation frameworks and works in structured argumentation that also account for a notion of support, as well as mentioning other works that contemplate the existence of support relations for performing legal reasoning, for mining arguments and relations from debates, and for identifying arguments and their relations in an empirical study.

Bipolar Argumentation Frameworks (BAFs) were firstly introduced in [Karacapilidis and Papadias, 2001; Verheij, 2003; Amgoud et al., 2004] and further developed in [Cayrol and Lagasquie-Schiex, 2005], where the authors discuss the use of bipolarity in argumentation, analyzing how it appears under different forms in each step of the argumentation process. Briefly, a BAF extends Dung's AF by considering two independent interactions between arguments, with diametrically opposed nature: an attack relation and a support relation. Over the years, different *interpretations* for the notion of support were proposed in the literature, leading to the formalization of variants of BAFs.

In this section, we will start by introducing the characterization of BAF given in [Cayrol and Lagasquie-Schiex, 2005], where a *general* notion of support is considered (*i.e.* a support relation that does not impose constraints on the arguments it relates, other than expressing a positive relationship between them). Then, we will introduce the Argumentation Framework with Necessities (AFN) originally proposed in [Nouioua and Risch, 2010], whose support relation is interpreted as *necessity*, meaning that if an argument a supports another argument b, then the acceptance of a is required to get the acceptance of b. Finally, we will present the approach of [Oren and Norman, 2008], where an *evidential* interpretation of support is considered to capture a particular notion: an argument cannot be accepted unless it is supported by evidence. For each of these approaches, we will only provide the basic definitions of the framework and the types of attack they consider, without entering into details about the different methods they propose for determining the accepted arguments of the framework.

As mentioned before, other interpretations for the notion of support such as *deductive* [Boella et al., 2010] or *backing* [Cohen et al., 2012] have been considered in the literature (we refer the reader to [Cayrol and

Lagasquie-Schiex, 2013; Cohen et al., 2014] for a full account of support in abstract argumentation). Note that the deductive approach will be introduced in Section 5.1 since it accounts for higher-order interactions.

4.1 The General Bipolar Argumentation Framework

As briefly mentioned at the beginning of this section, a Bipolar Argumentation Framework (BAF) extends Dung's AF by incorporating a support relation that is defined independently from the attack relation. Formally:

Definition 29 (Def. 1 in [Cayrol and Lagasquie-Schiex, 2005]).
A Bipolar Argumentation Framework (BAF) $\langle Ar, att, sup \rangle$ consists of a set Ar of arguments, a binary relation att called an attack relation, and another binary relation sup called a support relation.

Similarly to Dung's AF, a BAF can also be represented by a directed graph, with two kinds of edges: solid arrows for the attack relation and double arrows for the support relation. The notion of BAF and its graphical representation are illustrated by the following example, taken from [Cohen et al., 2014] (in turn, inspired on [Prakken and Vreeswijk, 2002; Cayrol and Lagasquie-Schiex, 2009]):

Example 7 (Introduction example in [Cohen et al., 2014]). *Consider the following arguments exchanged during the meeting of the editorial board of a newspaper:*

Argument i: *Information I concerning person P should be published.*

Argument p: *Information I is private, so P denies publication.*

Argument s: *I is an important information concerning P's son.*

Argument m: *P is the new prime minister, so everything related to P is public.*

It is clear that some conflicts appear during the discussion. That is the case of the conflict between arguments p and i, and between arguments m and p. On the other hand, there is a relation between arguments

p and s, which is clearly not a conflict. Moreover, s provides a new piece of information enforcing argument p.

This discussion can be represented by a BAF as the one depicted below:

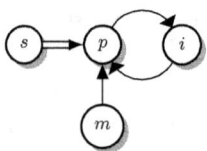

Given the coexistence of the support and attack relations in a BAF, [Cayrol and Lagasquie-Schiex, 2005] introduce the notions of *supported* and *secondary*[12] attack, which combine a sequence of supports with a direct attack.

Definition 30 (Def. 3 in [Cayrol and Lagasquie-Schiex, 2005]). *Given a $BAF = \langle Ar, att, sup \rangle$ and $a, b \in Ar$. A supported attack from a to b exists iff there exists a sequence of arguments $a = a_1, \ldots, a_n = b$ ($n \geq 3$) such that $(a_i, a_{i+1}) \in sup$, with $(1 \leq i \leq n-2)$, and $(a_{n-1}, a_n) \in att$.*

A secondary attack from a to b exists iff there exists a sequence of arguments $a = a_1, \ldots, a_n = b$ ($n \geq 3$) such that $(a_1, a_2) \in att$ and $(a_i, a_{i+1}) \in sup$, with $(2 \leq i \leq n-1)$.

By extension, the authors in [Cayrol and Lagasquie-Schiex, 2005] state that a sequence of two arguments a, b such that $(a, b) \in att$ (*i.e.* a direct attack) is also considered to be a supported attack.

Example 7 (cont'd) *Given the BAF from this example, we have the direct attacks specified by the attack relation (which are also considered to be supported attacks), and a supported attack from s to i. On the other hand, no secondary attacks exist.*

Having established the conflicts that arise from the coexistence of the attack and support relations, the authors of [Cayrol and Lagasquie-Schiex, 2005] turn to establish the conditions under which the acceptable arguments of a BAF can be identified. For that, several alternatives were proposed in [Cayrol and Lagasquie-Schiex, 2005; Cayrol

---

[12]In [Cayrol and Lagasquie-Schiex, 2005] the authors use the terminology *indirect* attack; however, in later works they adopted the terminology *secondary*.

and Lagasquie-Schiex, 2007; Cayrol and Lagasquie-Schiex, 2010], ranging from the direct characterization of the classical semantics for BAF (in particular, considering a wider range of admissible sets and preferred extensions, by imposing additional constraints related to the support relation), to the characterization of a Dung-like AF associated with BAF, called the Coalition Argumentation Framework (CAF), where arguments correspond to *coalitions* of arguments from the BAF that are linked by the support relation. Since the aim of this section is just to introduce the basic formalization of the BAF, focusing on the interpretation of support it adopts, we will not go into further details about these approaches and refer the interested reader to [Cayrol and Lagasquie-Schiex, 2013; Cohen et al., 2014]; this also applies to the approaches to be introduced in the following subsections.

4.2 The Argumentation Framework with Necessities

In [Nouioua and Risch, 2010] the authors firstly introduced the Argumentation Framework with Necessities (AFN), an extension of Dung's AF that incorporates a specialized kind of support relation between arguments: the necessity relation. Briefly, the necessity relation establishes that if an argument a supports another argument b, then a is necessary to obtain b. In that way, "if b is accepted then a is also accepted" and, conversely, "if a is not accepted then b cannot be accepted" either. The authors continued their work on AFNs in [Nouioua and Risch, 2011; Boudhar et al., 2012; Nouioua, 2013]; in particular, in [Nouioua and Risch, 2011] they proposed a generalization of their framework to account for sets of supporting arguments. In this section we will introduce the basic notions surrounding the formalization of AFN as proposed in [Nouioua and Risch, 2011] and [Boudhar et al., 2012], since these are the ones that inspired some of the developments presented later in Section 5.

Definition 31 (Def. 4 in [Boudhar et al., 2012]). *An Argumentation Framework with Necessities (AFN) is defined by $\langle Ar, att, sup \rangle$, where Ar is a set of arguments, $att \subseteq Ar \times Ar$ is a binary attack relation and $sup \subseteq Ar \times Ar$ is a binary irreflexive and transitive relation, called the necessity relation.*

The authors in [Boudhar et al., 2012] state that the irreflexive and transitive nature of *sup* excludes any risk to have a cycle of necessities. In particular, they state that such cycles are undesirable because they correspond to a kind of fallacy (begging the question).

Given the intended meaning of the support relation in AFN, which specializes the general support relation in BAF, positive relationships like the one illustrated on Example 7 might not be well captured by the necessity relation. That is, given arguments s and p such that s supports p, it is neither the case that s is necessary for p, nor that p is necessary for s. Hence, this relation cannot be accommodated within the AFN support relation. The necessity relation of the AFN is illustrated by the following example, partly taken from [Villata et al., 2012].[13]

Example 8 (Ex. from [Villata et al., 2012]). *Consider the following (partial) argument exchange during a degree committee meeting:*

Argument a (**Prof$_1$**): *Student X cannot apply for a PhD on May*

Argument b (**Student** X): *I will graduate on March*

Argument c (**Prof$_2$**): *X is missing a grade in the logics course*

Argument d (**Prof$_3$**): *On the academic transcript, there is no grade in the logics course*

Argument e (**Student** X): *The professor of the logics course said I passed the exam*

This informal exchange could be represented by the AFN depicted below, where attacks are depicted by single arrows and supports are depicted by double arrows:

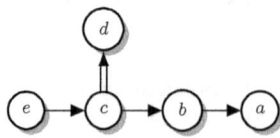

---

[13]The complete example will be introduced later in Section 5.1.

1 - HIGHER-ORDER INTERACTIONS IN ABSTRACT ARGUMENTATION

Here, among other relationships, we can highlight the fact that argument c is necessary for argument d.

In [Nouioua and Risch, 2011] the authors argued that, unlike a general support relation like the one introduced in Section 4.1, the necessity relation has the advantage to ensure that its interaction with the attack relation generates new attacks having exactly the same nature as the direct ones. These *extended attacks* are defined by combining a sequence of supports with a direct attack.

Definition 32 (Def. 2 in [Nouioua and Risch, 2011]). *Let $\langle Ar, att, sup \rangle$ be an AFN and $a, b \in Ar$. There is an extended attack from a to b iff there exists $c \in Ar$ such that either: $(a, c) \in att$ and $(c, b) \in sup$, or $(c, b) \in att$ and $(c, a) \in sup$. The direct attack $(a, b) \in att$ is considered to be a particular case of extended attack.*

Example 8 (cont'd) *Here, in addition to the direct attacks expressed in the attack relation att, there exists an extended attack from e to d, and an extended attack from d to b.*

It should be noted that the first kind of extended attack presented in Definition 32 coincides with the secondary attacks from the BAF (see Definition 30). This kind of extended attack is meant to enforce the acceptability constraint derived from the necessity interpretation of support; specifically, an extended attack of the first kind, where a attacks c and c supports b, is meant to enforce the constraint that if c is not accepted (in particular, in a case where a is accepted), then b should not be accepted either.

On the other hand, the second kind of extended attack is somewhat irrelevant. To illustrate this, let us consider the situation on Example 8. There, there is an extended attack of the second kind from d to b, expressing that if d is accepted, then b must not be accepted. In particular, given the constraint imposed by the necessary support relation, if d is accepted, then c must also be accepted (since c is necessary for d). Then, because of the attack from c to b, in such a case b will not be accepted. Hence, the extended attack from d to b seems a little bit useless. Nevertheless, it can be noted that the second kind of extended attack was introduced in [Nouioua and Risch, 2011] in order to highlight the duality between the necessary interpretation of support and the deductive interpretation of support originally proposed in [Boella et al., 2010][14]. For a full account of the duality between necessary support and deductive support we refer the reader to [Cayrol and Lagasquie-Schiex, 2013; Cohen et al., 2014].

Regarding the acceptability calculus in AFN, different approaches were proposed in the literature, similarly to the case of the BAF. On the one hand, the authors provided a direct characterization of the classical semantics for AFN. On the other hand, they introduced an alternative approach for obtaining the extensions of an AFN by characterizing an associated AF, obtained by considering the arguments of the original AFN and the extended attacks among them (hence, including also the direct attacks).

Finally, [Nouioua and Risch, 2011] also proposed an extension of the AFN in which the necessity relation can express the fact that a given argument requires at least one element among a set of arguments. The resulting framework is called Generalized Argumentation Framework with Necessities ($GAFN$), introduced below.

Definition 33 (Def. 8 in [Nouioua and Risch, 2011]). *A $GAFN$ is defined by a tuple $\langle Ar, att, sup \rangle$ where Ar is a set of arguments, $att \subseteq Ar \times Ar$ is an attack relation and $sup \subseteq ((2^{Ar} \backslash \varnothing) \times Ar)$ is a necessity relation.*

In particular, the support relation in a $GAFN$ encodes the following constraint: given $S \subseteq Ar$ and $a \in Ar$, $(S, a) \in sup$ means that the acceptance of a requires the acceptance of at least one of the arguments in S; in other words, "if a is accepted, then there exists $b \in S$ such that b is also accepted". This generalization of the AFN was then considered in [Nouioua, 2013], where a characterization of additional semantics directly on the $GAFN$ were given following the extension-based approach, in addition to introducing labelling-based semantics for the framework. In particular, as will be shown in Section 5.2, the formalization of the AFN and the $GAFN$ inspired the characterization of different argumentation frameworks with recursive attacks and necessary supports.

---

[14]Recall that the approach to deductive support will be introduced later in Section 5 (specifically, in Section 5.1) since it also accounts for higher-order interactions.

4.3 The Evidential Bipolar Argumentation Framework

In argumentation theory it is usually assumed that the premises (thus, the arguments they belong to) always hold since argumentation frameworks represent a snapshot of the arguments and relations involved on the reasoning process. However, alternative approaches like [Oren and Norman, 2008] consider that arguments should be backed up by evidence. Evidential reasoning involves determining which arguments are applicable based on some evidence. In that way, the approach to evidential support proposed in [Oren and Norman, 2008] intends to capture a particular notion: an argument cannot be accepted unless it is supported by evidence.

The *Evidential Argumentation System* was firstly introduced in [Oren and Norman, 2008], extending Dung's *AF* by incorporating a specialized support relation to capture the notion of *evidential support*; this line of work was later continued in [Polberg and Oren, 2014]. Despite the original naming of their system, for uniformity purposes with other approaches to bipolar abstract argumentation, from hereon we will refer to this system as the *Evidential Bipolar Argumentation Framework* (*EBAF*).

The support relation in the *EBAF* enables to distinguish between *prima-facie* and *standard* arguments. On the one hand, *prima-facie* arguments do not require support from other arguments to stand, whereas standard arguments must be linked to at least one *prima-facie* argument through a chain of supports. Given the evidential interpretation of support, an argument in the *EBAF* will be accepted only if it is supported through a chain of arguments, each of them being itself supported. At the beginning of this chain of supporting arguments there is a special argument η that represents support from the environment (*i.e.* the existence of supporting evidence).

Definition 34 (Def. 3.1 in [Polberg and Oren, 2014]). *An EBAF is a tuple $\langle Ar, att, sup \rangle$, where Ar is a set of arguments, $att \subseteq (2^{Ar}\setminus\varnothing) \times Ar$ is the attack relation, and $sup \subseteq (2^{Ar}\setminus\varnothing) \times Ar$ is the support relation. A special argument $\eta \in Ar$ is distinguished, such that $\nexists (X, y) \in att$ where $\eta \in X$; and $\nexists X$ where $(X, \eta) \in att$ or $(X, \eta) \in sup$.*

Since the environment requires no support, η cannot appear as the second element of a member of *sup*; moreover, it cannot be attacked by any set of arguments. In addition, since any argument attacked by the environment will be unconditionally defeated it makes no sense to include such arguments, therefore prohibiting the environment from appearing in a set originating an attack. Also note that, differently from the previous approaches, and inspired on [Nielsen and Parsons, 2006], the attack relation is not binary. Given $X \subseteq Ar$ and $a \in Ar$, $(X, a) \in att$ reads as follows: "if all the arguments in X are accepted, then a cannot be accepted". Then, for the evidential support relation, $(X, a) \in sup$ reads as: "the acceptance of a requires the acceptance of all the arguments in X"; furthermore, accepted arguments need to trace back to the special argument η.

Since the core idea of the *EBAF* is that valid arguments (in particular, attackers) need to trace back to the environment, the authors define the notion of evidence supported attack (e-supported attack) as follows.

Definition 35 (Def. 3.2 and 3.4 in [Polberg and Oren, 2014]). *Let $\langle Ar, att, sup \rangle$ be an EBAF, $a \in Ar$ and $S \subseteq Ar$:*

- *a has evidential support (e-support) from S iff $a = \eta$ or there is a non-empty $S' \subseteq S$ such that $(S', a) \in sup$ and $\forall b \in S'$, b has e-support from $S \backslash \{a\}$.*

- *a has minimal e-support from S if there is no $S' \subset S$ such that a has e-support from S'.*

- *S carries out an evidence-supported attack (e-attack) on argument a iff $(S', a) \in att$ where $S' \subseteq S$, and for all $s \in S'$, s has e-support from S.*

- *an e-supported attack by S on a is minimal iff there is no $S' \subset S$ that carries out an e-supported attack on a.*

Finally, semantics for *EBAF* have been characterized in [Oren and Norman, 2008] and then reformulated in [Polberg and Oren, 2014]. In addition, [Polberg and Oren, 2014] formally established a correspondence between *EBAF* and *GAFN* and identified correspondences between the properties of both of these systems to the properties obtained in Dung's argumentation framework. Briefly, the translation is such that unsupported arguments in the $GAFN$ will correspond to arguments supported by η in $EBAF$. Then, each attack from a to b in $GAFN$ will be translated into an attack from $\{a\}$ to b in $EBAF$. Then, the generalized support relation of $GAFN$ is translated in a way such that all sets of supporting arguments for a given argument a in $GAFN$ are combined into different sets of supporting arguments for a in $EBAF$. Formally:

Definition 36 (Transl. 1 in [Polberg and Oren, 2014]). *Let $\langle Ar, att, sup \rangle$ be a GAFN. The corresponding EBAF $\langle Ar', att', sup' \rangle$ is created as follows:*

- $Ar' = Ar \cup \{\eta\}$.

- *For every two arguments $a, b \in Ar$ such that $(a, b) \in att$, put $(\{a\}, b)$ in att'.*

- *Let $a \in Ar$ and $Z = \{Z_1, \ldots, Z_n\}$ be a collection of all sets Z_i such that $(Z_i, a) \in sup$. If Z is empty, then put $(\{\eta\}, a)$ in sup'; otherwise, for all $Z' \in (Z_1 \times \ldots \times Z_n)$, add (Z'_S, a) to sup', where Z'_S is the set of all elements in Z'.*

Following the preceding translation we can model the AFN from Example 8 as an $EBAF$.

Example 8 (cont'd) *The arguments and interactions in the degree committee meeting can be represented by the EBAF $\langle Ar, att, sup \rangle$ with*

- $Ar = \{a, b, c, d, \eta\}$

- $att = \{(\{e\}, c), (\{c\}, b), (\{b\}, a)\}$

- $sup = \{(\{\eta\}, e), (\{\eta\}, c), (\{\eta\}, b), (\{\eta\}, a), (\{c\}, d)\}$

This $EBAF$ is depicted below where, for simplicity, the special argument η is omitted. Instead, prima-facie arguments (i.e. arguments supported by η) are represented using solid outlines whereas standard arguments are represented with dashed outlines. So, the only standard argument in this example is d. Also, since every attack and support in the $EBAF$ originates from a singleton set, the solid arrows (resp. the double arrows) directly depart from the argument originating the attack (resp. the support).

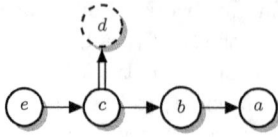

Here, every attack in att corresponds to a minimal e-supported attack.

4.4 Links with Support in Structured Argumentation and Others

Most research on bipolar argumentation systems has been carried out at the abstract level. Notwithstanding this, there exist other works tackling the issue of dealing with the notion of support in other contexts. In this section we will briefly comment on some of them, divided into two groups.

The first group of works addresses the notion of support in three of the major structured argumentation systems: $ASPIC+$ [Modgil and Prakken, 2018], Assumption-Based Argumentation (ABA) [Cyras et al., 2018] and Defeasible Logic Programming ($DeLP$) [García and Simari, 2018]. On the one hand, [Prakken, 2014] and [Cohen et al., 2018] studied different forms of support in $ASPIC+$ and analyzed whether they correspond to any of the existing interpretations of support at the abstract level, showing that $ASPIC+$ sub-argument relation is a special case of necessary support, and can be considered as a special case of evidential support. On the other hand, [Cyras et al., 2017] studied necessary support, deductive support, and the coalitions approach for BAF in the context of ABA. In particular, they proved that the aforementioned interpretations of support in BAFs correspond, under the (respective) admissible and preferred semantics (where defined), to the admissible and preferred semantics of a restricted kind of ABA frameworks, called *bipolar*. Finally, [Cohen et al., 2011] extended $DeLP$ by incorporating a new kind of rules and arguments, corresponding to Toulmin's notion of backing [Toulmin, 1958]. Then, the authors showed that this extended version of $DeLP$ can be used to instantiate the bipolar abstract argumentation framework that adopts the *backing* interpretation of support

1 - HIGHER-ORDER INTERACTIONS IN ABSTRACT ARGUMENTATION

originally proposed in [Cohen et al., 2012].

The second group of works we consider contemplates the existence of support relations in different ways: using BAFs to perform legal reasoning, mining arguments and support relations, and providing an empirical study showing the usefulness of support relations. [Kawasaki and Takahashi, 2018] proposes a transformation from PROLEG [Satoh et al., 2010] to a BAF where the support relation is originated in a set of arguments, and gives a semantics for that BAF in a way that guarantees that a PROLEG answer set coincides with the set of accepted arguments in the BAF; as stated by the authors, their aim is that the meaning of legal reasoning is preserved by their proposed semantics. The work by Cabrio and Villata [Cabrio and Villata, 2013] discusses and evaluates, on a sample of natural language arguments extracted from Debatepedia, the support and attack relations among arguments in BAFs adopting different interpretations of support (general, necessary, deductive) with respect to the more specific notions of textual entailment and contradiction. They investigated the distribution of those attacks in the debates, showing that all these interpretations of support (and the corresponding attacks) are verified in human debates, though with different frequency. Finally, [Polberg and Hunter, 2018] describes the results of an experiment in which participants were asked to judge dialogues in terms of agreement and structure. Among other findings, the data they collected supports the use of BAFs, since the notion of defence does not necessarily account for all of the positive relations between the statements viewed by the participants.

The works accounted for in this subsection serve to establish a connection between the developments on bipolar argumentation at the abstract and structured levels, as well as providing an empirical justification for using BAFs, their application in the legal domain, and the mining of support relations. Nevertheless, it should be noted that none of the approaches discussed above accounts for the existence of higher-order or recursive interactions (neither attack nor support). Consequently, in principle, they would not be suitable to instantiate the approaches to higher-order interactions that will be addressed in this chapter. A deeper study on how to accommodate these structured approaches to fit the existing literature about higher-order interactions in abstract argumentation is certainly of interest. However, since this chapter is meant to focus on recapping the state of the art on higher-order interactions in abstract argumentation, such study is out of scope and will be addressed on future works.

5 Different supports, so different higher-order bipolar approaches

The works presented in this section respect the typology of the support relation: deductive, necessary and finally evidential support.

5.1 Higher-order deductive supports

In [Boella et al., 2010; Villata et al., 2012], the authors pursued their previous work presented in [Boella et al., 2009a] by the introduction of supports in the same meta-argumentation framework. They only considered deductive supports: "a deductively supports b" means that "if a is accepted then b is also accepted". In fact, this is also the first work in which this notion of deductive support is formally defined and used; so, we can consider that Bipolar Argumentation Frameworks with Deductive Support ($BAFD$s) are introduced in [Boella et al., 2010]. Moreover, in order to take into account "defeasible supports" (supports that can be attacked), [Boella et al., 2010; Villata et al., 2012] use second-order interactions with different constraints following the nature of the interaction:

- the attack relation att and the support relation sup are binary relations over the set of arguments (they are called simple attacks and supports);

- a second attack relation $att2$ targets either a simple attack, or a simple support (second-order attack);

- the source of a second-order attack is either an argument or a simple attack.

This "second-order bipolar argumentation framework" can be flattened using Def. 9 in [Boella et al., 2010]. Note that, since the second-order attacks cannot be attacked, this original definition can be simplified as follows:[15]

Definition 37. Let $\langle Ar, att, sup, att2 \rangle$ be a second-order argumentation framework defined with:

- Ar being the set of arguments,
- $att : Ar \times Ar$ being the set of simple attacks,
- $sup : Ar \times Ar$ being the set of simple deductive supports,
- $att2 : (Ar \cup att) \times (att \cup sup)$ being the set of second-order attacks.

The flattened version of this framework is the Dung argumentation framework defined by:

- the set of arguments =
  $\{acc(a) | a \in Ar\} \cup \{X_{ab}, Y_{ab} | (a,b) \in att\} \cup \{Z_{ab} | (a,b) \in sup\}$

- the binary attack relation =
  $\{(X_{ab}, Y_{ab}) | (a,b) \in att\} \cup$
  $\{(Y_{ab}, acc(b)) | (a,b) \in att\} \cup$
  $\{(acc(a), X_{ab}) | (a,b) \in att\} \cup$
  $\{(acc(c), Y_{ab})) | (c,(a,b)) \in att2 \text{ and } (a,b) \in att\} \cup$
  $\{(Z_{ab}, acc(a))) | (a,b) \in sup\} \cup$
  $\{(acc(b), Z_{ab}) | (a,b) \in sup\} \cup$
  $\{(acc(c), Z_{ab})) | (c,(a,b)) \in att2 \text{ and } (a,b) \in sup\} \cup$
  $\{(Y_{cd}, Y_{ab})) | ((c,d),(a,b)) \in att2 \text{ and } (a,b) \in att\}$[16]

As argued by the authors in [Boella et al., 2010], the coexistence of attacks and supports towards arguments in their framework leads to the existence of new attacks, which reinforce the acceptability constraints imposed by the deductive support relation. Specifically, they consider *supported attacks* (like in Definition 30 for the BAF) and *mediated attacks*, defined as follows.

---

[15]Note that Definition 37 extends Definition 13 given in Section 3.1.

[16]In [Boella et al., 2010], the authors consider that the source of an attack that targets a support must always be an argument. Nevertheless, this constraint does not appear in Def. 9 given in [Boella et al., 2010].

Definition 38 (Def. 7 in [Boella et al., 2010]). *Let $\langle Ar, att, sup, att2 \rangle$ be a second-order argumentation framework and $a, b \in Ar$. A mediated attack from a to b exists if there is a sequence of arguments $b = a_1, \ldots, a_n$ ($1 < n$) such that for all $1 \leq i < n$, $(a_i, a_{i+1}) \in sup$ and $(a, a_n) \in att$.*

In [Villata et al., 2012], this approach has also been extended taking into account prioritized supports and has been applied to structured argumentation and to the Abstract Dialectical Framework (ADF) developed by Brewka and Woltran (see [Brewka et al., 2018] for an overview).

The following example illustrates Definition 37.

Example 8 (cont'd) *Consider the following additional arguments exchanged during the degree committee meeting:*

Argument f (Student X): *I was in hospital in the date of the logics exam*

Argument g (Prof$_1$): *There is no record of your stay in the hospital*

Argument h (Student X): *The professor of logics was ill and could not register my exam*

The exchange accounting for every argument and interaction can be represented by the following directed graph in which one can find, among other things, that d supports c (under the deductive interpretation of support, the direction of the arrow previously representing the necessary support from c to d is now reversed), h attacks the support (d, c), and f attacks the attack (c, b):

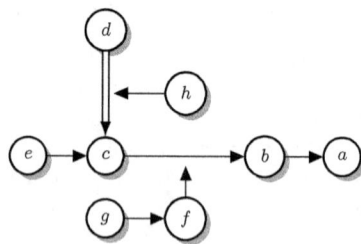

(attacks are represented with solid arrows and supports with double arrows)

Here, there exists a supported attack from d to b and there exists a mediated attack from e to d.

This framework can be flattened into a simple AF (only arguments and simple attacks) as follows:

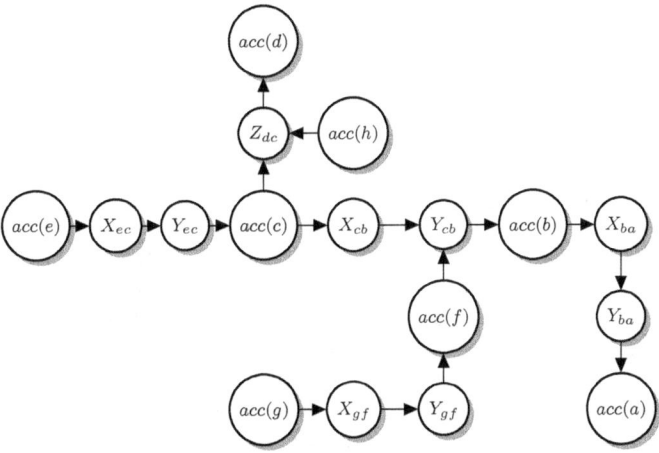

The arguments b, d, e, g and h are acceptable in this AF and belong to any classical extension (grounded, preferred, stable, ...). Moreover, if we consider that the meta-arguments Y (resp. Z) represent the attacks (resp. the support), we can also conclude that the attacks (e, c), (g, f), (b, a) and of course, since they cannot be attacked, (f, (c, b)) and (h, (d, c)) are acceptable in this AF and belong to any classical extension. In contrast, the attack (c, b) and the support (d, c) are not acceptable in this AF.

Note that d is acceptable since the support (d, c) is invalidated by the attack coming from h. Otherwise, in the case the argument h is not considered, the existence of this support and the fact that c is not acceptable would imply that d would also not be acceptable.

Considering the used flattening process and the fact that the semantics in these second-order deductive bipolar frameworks are defined as in AF, it is obvious to see that these frameworks are a conservative generalization of AF, of $BAFD$[17] and of EAF. And so considering the differences between EAF and the other approaches (Gabbay's approach, $AFRA$ and RAF), it is also obvious to see that there is no one-to-one correspondence between second-order deductive bipolar frameworks and these approaches.

5.2 Higher-order necessary supports

Throughout this section, recall that "a necessary supports b" means that "if b is accepted then a is also accepted" (duality between necessary and deductive supports).

5.2.1 $ASAF$ approach

In [Cohen et al., 2015] the authors firstly proposed the *Attack-Support Argumentation Framework* ($ASAF$) taking its basis from the $AFRA$ and the AFN (see Sections 3.3 and 4.2). Specifically, the $ASAF$ features a necessary support relation and an attack relation allowing for attacks and supports between arguments, as well as attacks and supports from an argument to the attack and support relations, at any level. This line of work was further pursued in [Cohen et al., 2016] and [Gottifredi et al., 2018], where the latter consolidates the previous works showing different (and equivalent) alternatives for addressing the acceptability calculus in the $ASAF$, and showing the relationship w.r.t. the frameworks it is inspired on.

As stated in [Gottifredi et al., 2018], the intuition behind the existence of a higher-order support in the $ASAF$ (*i.e.* a support targeting an attack/support) is that the supporting argument provides the context under which the targeted interaction holds. Hence, for instance, given a support β from an argument a to an attack or a support X, argument a should be accepted in order for the interaction X to hold. Similarly, extending the intuition behind the existence of a recursive attack relation (*e.g.* as in the EAF to model preferences), higher-order attacks in an $ASAF$ (*i.e.* attacks targeting an attack/support) capture the intuition that the attacking argument provides a context under which the

---

[17]Indeed, $BAFD$ corresponds to these second-order deductive bipolar frameworks without any second-order attacks.

targeted interaction should not hold.

Definition 39 (Def. 11 in [Gottifredi et al., 2018]). *An* Attack-Support Argumentation Framework *(ASAF) is a tuple $\langle Ar, att, sup \rangle$ where Ar is a set of arguments, $att \subseteq Ar \times (Ar \cup att \cup sup)$ is an attack relation and $sup \subseteq Ar \times (Ar \cup att \cup sup)$ is a necessary support relation. It is assumed that sup is acyclic and $att \cap sup = \varnothing$.*

Since in $ASAF$ attacks and supports can be attacked or supported, abbreviated notations for the interactions are proposed (similarly to what is done in the $AFRA$, see Section 3.3, or in the RAF, see Section 3.5), making use of $\mathbf{s}(\cdot)$ and $\mathbf{t}(\cdot)$ for identifying the source and target of an interaction. Then, for instance, an attack from an argument a to a support from b to X (with X being an argument, an attack or a support) will be represented by a pair $\alpha = (a, \beta)$ in the attack relation att of the $ASAF$, where $\beta = (b, X)$ is a pair belonging to the support relation sup of the $ASAF$; in this case, it holds that $\mathbf{s}(\alpha) = a$, $\mathbf{t}(\alpha) = \beta$, $\mathbf{s}(\beta) = b$ and $\mathbf{t}(\beta) = X$.

Given the duality between the necessary and deductive interpretations of support discussed in Section 4.2, we can represent the discussion held during the degree committee meeting with an $ASAF$, since the only support involves arguments c and d. Nonetheless it should be noted that, for deductive supports targeting an interaction, a necessary support cannot be obtained directly by reversing the support since that would imply that the resulting necessary support originates in an interaction (and this is not allowed in the $ASAF$).

Example 8 (cont'd) *The complete exchange of arguments can be represented by the following ASAF. Similarly to before, arguments are depicted in circles, attacks are depicted using solid arrows, supports are depicted using double arrows, and attacks/supports are labelled with Greek letters in squares:*

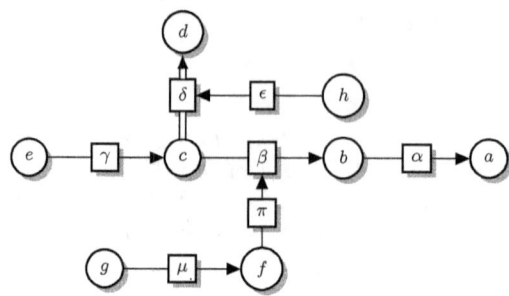

In [Gottifredi et al., 2018] the authors provided a characterization of the $ASAF$ semantics directly on the framework. In order to do that, the authors followed the same methodology applied for the $AFRA$ (see Section 3.3) which consists on first identifying the different kinds of defeat that can occur in the $ASAF$ and then define some basic semantic notions to finally characterize the complete, preferred, stable and grounded semantics of the framework.

Definition 40 (Defs. 12 - 18 in [Gottifredi et al., 2018]). *Let $ASAF = \langle Ar, att, sup \rangle$, $\alpha \in att$, $X \in (Ar \cup att \cup sup)$ and $S \subseteq sup$.*

- *α unconditionally defeats X, denoted α udef X iff either $(\alpha, X) \in att$, or $X \in att$ and $(\alpha, \mathbf{s}(X)) \in att$.*

- *α conditionally defeats X given the set S, denoted α cdef X given S iff there exists a sequence of arguments $[a_1, \ldots, a_n]$ ($n \geq 2$) such that for every a_i ($1 \leq i < n$), $(a_i, a_{i+1}) \in sup$, and it holds that $\mathbf{t}(\alpha) = a_1$ and either: $a_n = X$, or $a_n = \mathbf{s}(X)$ and $X \in att$; the set S is the union of the supports $(a_i, a_{i+1}) \in sup$.*

Note that the preceding definition allows arguments, attacks or supports to be defeated. In the first bullet, if $(\alpha, X) \in att$, a defeat reminiscing the direct defeat of the $AFRA$ would occur; on the other hand, if $(\alpha, \mathbf{s}(X)) \in att$ with $X \in att$, a defeat akin to the indirect defeat of the $AFRA$ takes place. Then, in the second bullet, if $a_n = X$ (and $\mathbf{t}(\alpha) = a_1$), a defeat corresponding to the first kind of extended attack of the AFN occurs; on the other hand, if $a_n = \mathbf{s}(X)$ with $X \in att$ (and again, $\mathbf{t}(\alpha) = a_1$) we have a new kind of defeat, which combines the behavior of the first kind of extended attack from the AFN and the indirect defeat from the $AFRA$.

Then, based on these defeats, the notions of conflict-freeness, acceptability and admissibility for ASAF are defined as follows:

Definition 41 (Defs. 19–21 in [Gottifredi et al., 2018]). *Let $ASAF = \langle Ar, att, sup \rangle$ and $S \subseteq (Ar \cup att \cup sup)$.*

- *S is conflict-free iff $\nexists \alpha, X, \in S$, $\nexists S' \subseteq S$ such that either α udef X or α cdef X given S'.*

- *$X \in (Ar \cup att \cup sup)$ is acceptable w.r.t. S iff it holds that:*
    1. *$\forall \alpha \in att$ such that α udef X: $\exists \beta \in S$, $\exists S' \subseteq S$ such that β udef α or β cdef α given S'.*
    2. *$\forall \alpha \in att$, $\forall T \subseteq sup$ such that α cdef X given T: $\exists \beta \in S$, $\exists S' \subseteq S$, $\exists \gamma \in \{\alpha\} \cup T$ such that β udef γ or β cdef γ given S'.*

From the semantic notions defined in Definition 41, the complete, preferred, stable, and grounded extensions of the $ASAF$ can be defined.

Definition 42 (Def. 22 in [Gottifredi et al., 2018]). *Let $ASAF = \langle Ar, att, sup \rangle$ and $S \subseteq (Ar \cup att \cup sup)$.*

- *S is a complete extension of $ASAF$ iff it is admissible and $\forall X \in (Ar \cup att \cup sup)$, if X is acceptable w.r.t. S, then $X \in S$.*

- *S is a preferred extension of $ASAF$ iff it is a maximal (w.r.t. $\subseteq$) admissible set of $ASAF$.*

- *S is a stable extension of $ASAF$ iff it is conflict-free and $\forall X \in (Ar \cup att \cup sup) \setminus S$, $\exists \alpha \in S$, $\exists S' \subseteq S$ such that α udef X or α cdef X given S'.*

- *S is the grounded extension of $ASAF$ iff it is the smallest (w.r.t. $\subseteq$) complete extension of $ASAF$.*

Example 8 (cont'd) *The only complete, preferred and stable extension of this $ASAF$, which is also its grounded extension, is $\{e, \gamma, d, \epsilon, h, g, \mu,$*

$b, \alpha\}$. In particular note that, even though γ cdef d given $\{\delta\}$, it holds that ϵ udef δ. Consequently, d is acceptable w.r.t. the set $\{\epsilon\}$; moreover, note that the set $\{\gamma, d\}$ is conflict-free because it does not contain δ (the support required for the existence of the conditional defeat of γ on d). More generally, every set of arguments, attacks and supports from the ASAF that does not include all the necessary elements for the existence of a defeat (either unconditional or conditional) is conflict-free; again, this characteristic is inherited from the AFRA.

Recently, [Alfano et al., 2020] proposed labelling-based semantics for the $ASAF$.[18] Briefly, a *labelling* for an $ASAF$ $\langle Ar, att, sup \rangle$ is a total function $\mathcal{L} : (Ar \cup att \cup sup) \mapsto \{\text{in}, \text{out}, \text{undec}\}$. Given a labelling $\mathcal{L}$, we define $\text{in}(\mathcal{L}) = \{X \mid \mathcal{L}(X) = \text{in}\}$, $\text{out}(\mathcal{L}) = \{X \mid \mathcal{L}(X) = \text{out}\}$, and $\text{undec}(\mathcal{L}) = \{X \mid \mathcal{L}(X) = \text{undec}\}$. Also, when convenient, a labelling $\mathcal{L}$ can be represented by the triple $(\text{in}(\mathcal{L}), \text{out}(\mathcal{L}), \text{undec}(\mathcal{L}))$.

Then, the *complete labellings* are defined in [Alfano et al., 2020] as follows. $\mathcal{L}$ is a *complete labelling* of an $ASAF$ $\langle Ar, att, sup \rangle$ iff for every $X \in (Ar \cup att \cup sup)$ it holds that: (1) $\mathcal{L}(X) = \text{in}$ iff $\forall \alpha \in att, \forall S \subseteq sup$ such that α cdef X given S, $\exists Y \in (\{\alpha\} \cup S)$ such that $\mathcal{L}(Y) = \text{out}$; and (2) $\mathcal{L}(X) = \text{out}$ iff $\exists \alpha \in att, \exists S \subseteq sup$ such that α cdef X given S and $\forall Y \in (\{\alpha\} \cup S), \mathcal{L}(Y) = \text{in}$.

In other words, for X to be labelled as in by a complete labelling of an $ASAF$ the following conditions must be satisfied: for every set of elements originating a defeat on X, one of the elements in the set is labelled as out (*i.e.* either the attack or one of the supports, if they exist). Analogously, for X to be labelled as out, it must be the case that there exists a set of elements originating a defeat on X where every element in the set (*i.e.* the attack and every support) is labelled as in. Finally, if X is neither labelled as in nor as out, it is labelled as undec.

[Alfano et al., 2020] mentions that there exists a one-to-one correspondence between complete extensions and complete labellings of an $ASAF$. Specifically, they state that each complete extension E is in one-to-one correspondence with a complete labelling $\mathcal{L} = (E, E^+, (Ar \cup att \cup sup) \setminus (E \cup E^+))$, where $E^+ = \{X \in (Ar \cup att \cup sup) \mid \exists \alpha \in E, \exists S \subseteq E$ such that α cdef X given $S\}$. That is, the complete labelling $\mathcal{L}$ cor-

---

[18]Note that [Alfano et al., 2020] provides all the corresponding definitions in inline text; thus, we maintain inline definitions in this chapter.

responding to a complete extension E of an $ASAF$ is given by the triple $(\text{in}(\mathcal{L}), \text{out}(\mathcal{L}), \text{undec}(\mathcal{L}))$, where $\text{in}(\mathcal{L}) = E$, $\text{out}(\mathcal{L}) = E^+$, and $\text{undec}(\mathcal{L}) = (Ar \cup att \cup sup) \setminus (E \cup E^+))$.

Then, as argued by the authors in [Alfano et al., 2020], the preferred, stable and grounded labellings of an $ASAF$ can be defined in terms of the complete labellings of the framework: $\mathcal{L}$ is a *preferred* (resp. *stable, grounded*) *labelling* of $ASAF$ iff it is a complete labelling such that $\text{in}(\mathcal{L})$ is a preferred (resp. *stable, grounded*) extension of $ASAF$.

Example 8 (cont'd) *The only complete labelling of the $ASAF$ (also, its grounded labelling and its only preferred and stable labelling) is* $(\{e, \gamma, d, \epsilon, h, g, \mu, b, \alpha\}, \{c, \delta, f, \pi, \beta, a\}, \varnothing)$.

Finally, in [Cohen et al., 2015] the authors proposed to translate an $ASAF$ into an AF in order to be able to determine the extensions of the framework. In that way, they first translated the $ASAF$ into its associated AFN and finally, translated the AFN into an AF. The translation given in [Cohen et al., 2015] was later refined in [Gottifredi et al., 2018] and is shown below.

Definition 43 (Defs. 23, 24, 9 and 10 in [Gottifredi et al., 2018]). *Let* $ASAF = \langle Ar, att, sup \rangle$.

The AFN associated with $ASAF$ is $\langle Ar_{AFN}, att_{AFN}, sup_{AFN} \rangle$ *with*
$Ar_{AFN} = Ar \cup att \cup sup \cup \{\beta^+, \beta^- \mid \beta \in sup\}$
$att_{AFN} = \{(\alpha, X) \mid \alpha \in att, \ \mathbf{t}(\alpha) = X\} \cup$
$\quad\quad\quad \{(b, \beta^-), (\beta^-, Y) \mid \beta \in sup, \ \mathbf{s}(\beta) = b, \ \mathbf{t}(\beta) = Y\}$
$sup_{AFN} = \{(a, \alpha) \mid \alpha \in att, \ \mathbf{s}(\alpha) = a\} \cup$
$\quad\quad\quad \{(\beta, \beta^+), (\beta, \beta^-), (b, \beta^+) \mid \beta \in sup, \ \mathbf{s}(\beta) = b\}$
The AF associated with $ASAF$ is $\langle Ar_{AF}, att_{AF} \rangle$*, where:*
$Ar_{AF} = Ar_{AFN}$
$att_{AF} = att_{AFN} \cup \{(a, b) \mid \exists c \in Ar_{AFN} \text{ with}$
$\quad\quad\quad (a, c) \in att_{AFN}, (c, b) \in sup_{AFN}\}$

Note that the second set of attacks added to att_{AF} in Definition 43 exactly corresponds to the first kind of extended attack in the AFN, as described in Def. 32. Then, as mentioned before, extensions of an $ASAF$ can be obtained from extensions of its AF as follows:

Definition 44 (Defs. 25 - 27 in [Gottifredi et al., 2018]). *Let $ASAF = \langle Ar, att, sup \rangle$ and $AF = \langle Ar_{AF}, att_{AF} \rangle$ be its associated argumentation* framework. If S is an extension of AF under the complete, preferred, stable or grounded semantics, then $S' = S\setminus\{\beta^+, \beta^- \mid \beta \in sup\}$ is an extension of $ASAF$ under the same semantics.

Example 8 (cont'd) *The AF associated with the $ASAF$, obtained with Definition 43, is depicted below:*

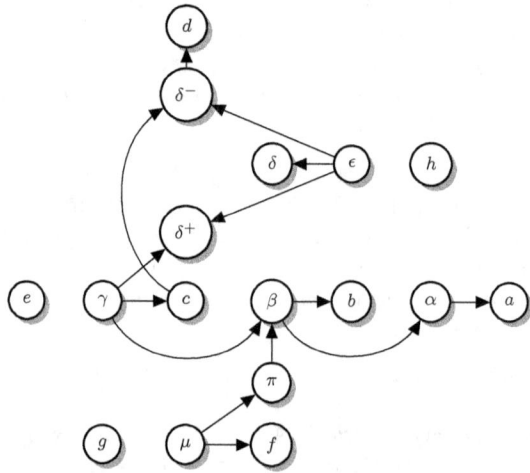

Here, the only complete extension (also, the grounded extension and the only preferred and stable extension) of the associated AF is $\{e, \gamma, g, \mu, b, \alpha, h, \epsilon, d\}$; furtheremore, this is also the only complete, grounded, preferred and stable extension of the $ASAF$. Note that the resulting extension differs from the one obtained in the deductive case: the attack π (corresponding to the attack $(f, (c, b))$ in the deductive approach) does not belong to the $ASAF$ extension because of the indirect defeat coming from μ; this is due to the fact that the $ASAF$ approach takes $AFRA$ as basis.

In [Cayrol et al., 2016] an alternative translation of an $ASAF$ into an AF was proposed with the aim of addressing the acceptability calculus of the framework. This alternative translation also accounts for an intermediate translation into an AFN and is driven by three features that can be identified in interactions involved in a recursion: *groundness*, *validity* and *activation*; following this translation, interactions have to be active in order to be included in the extensions of an *ASAF*. Specifically, as proposed in [Cayrol et al., 2016], an interaction is considered to be grounded if its source is accepted. The validity of an interaction is determined by looking at the interactions that may affect it, that is, interactions attacking and supporting it. Finally, an interaction is considered to be active if it is both grounded and valid; then, for instance, an interaction that is attacked by another interaction that is active will not be considered as valid. The translation of [Cayrol et al., 2016] follows:

Definition 45 (Defs. 4 and 8 in [Cayrol et al., 2016]).
Let $ASAF = \langle Ar, att, sup \rangle$. The AFN associated with $ASAF$ is $\langle Ar_{AFN}, att_{AFN}, sup_{AFN} \rangle$, where:
$Ar_{AFN} = Ar \cup \{\alpha \mid \alpha = (a, X) \in att\} \cup \{\beta \mid \beta = (b, Y) \in sup\}$
$att_{AFN} = \{(\alpha, X) \mid \alpha \in att, \mathbf{t}(\alpha) = X\}$
$sup_{AFN} = \{(a, \alpha) \mid \alpha \in att \cup sup, \mathbf{s}(\alpha) = a\} \cup$
$\{(\alpha, X) \mid \alpha \in sup, \mathbf{t}(\alpha) = X\}$

The AF associated with $ASAF$ is $\langle Ar_{AF}, att_{AF} \rangle$, where:
$Ar_{AF} = Ar_{AFN} \cup \{N_{XY} \mid (X, Y) \in sup_{AFN}\}$
$att_{AF} = \{(\alpha, X) \mid (\alpha, X) \in att_{AFN}\} \cup$
$\{(\alpha, N_{XY}) \mid (\alpha, X) \in att_{AFN}, \alpha \in att, X \in sup, \mathbf{t}(X) = Y\} \cup$
$\{(X, N_{XY}), (N_{XY}, Y) \mid (X, Y) \in sup_{AFN}\} \cup$
$\{(N_{XY}, N_{YZ}) \mid (X, Y) \in sup_{AFN}, X \in sup, Y \in sup, \mathbf{t}(Y) = Z\}$

As proposed in [Cayrol et al., 2016], extensions of the $ASAF$ can be obtained from extensions of its AF obtained through Definition 45 by just filtering out the N_{XY} arguments.[19]

Example 8 (cont'd) *The AF associated with the ASAF, obtained with Definition 45, is depicted in Figure 4. The only complete extension (also, the grounded extension and the only preferred and stable extension) of the associated AF is* $\{e, \gamma, N_{c\beta}, N_{c\delta}, g, \mu, N_{f\pi}, b, \alpha, h, \epsilon, d\}$. *As a result, by filtering out the N-arguments, the only complete, grounded, preferred and stable extension of the ASAF is* $\{e, \gamma, g, \mu, b, \alpha, h, \epsilon, d\}$.

---

[19]Note that [Cayrol et al., 2016] provides no formal definition as to how to obtain the correspondence between extensions of the $ASAF$ and extensions of its associated AF.

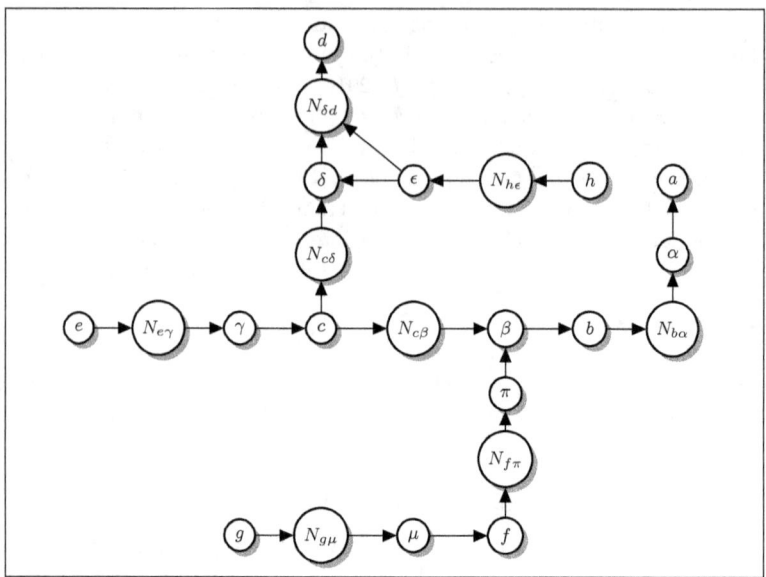

Figure 4: *AF* associated with the *ASAF* of Example 8, following Definition 45

It should be noted that, although the same outcome was obtained for Example 8 when considering the translations of Definition 43 and Definition 45, this does not hold for the general case. The reason for this difference relies on the fact that, differently from [Gottifredi et al., 2018], for a support to be accepted in [Cayrol et al., 2016] it must be the case that its source is also accepted. This difference is illustrated by the following example.

Example 9. *Consider the ASAF depicted below:*

With the translation of Definition 43 we obtain the following associated AF:

1 - HIGHER-ORDER INTERACTIONS IN ABSTRACT ARGUMENTATION

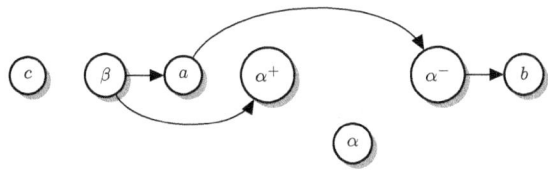

On the other hand, with the translation of Definition 45 we obtain the associated AF depicted below:

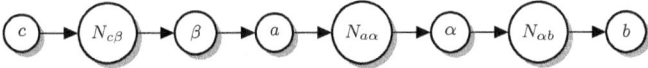

In the former case, the only complete, grounded, preferred and stable extension of the associated AF is $\{c, \beta, \alpha^-, \alpha\}$ and thus, the only complete, grounded, preferred and stable extension of the ASAF would be $\{c, \beta, \alpha\}$. In the latter case, the only complete, grounded, preferred and stable extension of the associated AF is $\{c, \beta, N_{a\alpha}, N_{\alpha b}\}$; consequently, the only complete, grounded, preferred and stable extension of the ASAF would be $\{c, \beta\}$.

Note that a one-to-one correspondence exists between *ASAF without support* and *RAF* (indeed *RAF* and *AFRA* approaches give similar results for semantics level, and *ASAF* are a conservative generalization of *AFRA*). Nevertheless, it is not the case when we consider *ASAF* with support (so *ASAF* that are not only *AFRA*).

5.2.2 *RAFN* approach

In [Cayrol et al., 2018b; Cayrol et al., 2018c], the authors pursued their works about *RAF*, presented in [Cayrol et al., 2017] (see Section 3.5), by the definition and the study of an extension, called Recursive Argumentation Framework with Necessity (*RAFN*), that is able to take into account higher-order necessary supports. The approach presented in [Cayrol et al., 2018b; Cayrol et al., 2018c] is similar to the one used in [Cayrol et al., 2017]: formalization of *RAFN* and direct definition of semantics.

Note that, differently from the $ASAF$ approach, the source of a necessary support in $RAFN$ can be a set of arguments; on the other hand, like in the $ASAF$, this is not the case for an attack.

Definition 46 (Def. 17 in [Cayrol et al., 2018b]). *A Recursive Argumentation Framework with Necessity ($RAFN$) is a tuple $\langle Ar, att, sup, \mathbf{s}, \mathbf{t}\rangle$, where Ar, att and sup are three pairwise disjunct sets respectively representing arguments, attacks and supports names, $\mathbf{s}$ is a function from $(att \cup sup)$ to $(2^{Ar} \setminus \varnothing)$ mapping each interaction to its source, and $\mathbf{t}$ is a function from $(att \cup sup)$ to $(Ar \cup att \cup sup)$ mapping each interaction to its target. It is assumed that $\forall \alpha \in att$, $\mathbf{s}(\alpha)$ is a singleton.*

$RAFN$ semantics are defined using the extension of the notion of "structure" for RAF (see Definition 25 in Section 3.5): A structure of the $RAFN$ is a triple $U = (S, \Gamma, \Delta)$ such that $S \subseteq Ar$, $\Gamma \subseteq att$ and $\Delta \subseteq sup$. Intuitively, the set S represents the set of "acceptable" arguments w.r.t. the structure U, while Γ and Δ respectively represent the set of "valid attacks" and "valid necessary supports" w.r.t. U.

In order to define the structures corresponding to each semantics, some additional notions are introduced. Intuitively, an element x (argument, attack or support) can be defeated w.r.t. U iff there is a "valid attack" w.r.t. U that targets x and whose source is "acceptable" w.r.t. U. Concerning the notion of *supported elements* w.r.t. a structure, elements (arguments, attacks, supports) which receive no necessary support do not require any support, so they are supported w.r.t. any structure; and an element x is supported w.r.t. a given structure U if *for each* support α (which can be regarded as supported), the source of α contains *at least one* argument of U that can be regarded as supported. An element of a $RAFN$ is considered as being still supportable as long as *for each* non-defeated support, *there exists at least one* argument in its source, which is non-defeated and regarded as supportable. And finally, elements that are defeated or that are unsupportable are said to be *unacceptable* (they cannot be accepted). Then an attack $\alpha \in att$ is *unactivable*[20] (such an attack cannot be "activated" in order to defeat the element that it is targeting) iff it is either unacceptable or its source is unacceptable. The

---

[20]This is the word used in [Cayrol et al., 2018b] and a neologism. It expresses the impossibility of activating an attack.

following notation is used in the next definitions: let $E \subseteq (Ar \cup att \cup sup)$, $\overline{E} = (Ar \cup att \cup sup) \setminus E$.

Definition 47 (Defs. 18 to 20 in [Cayrol *et al.*, 2018b]). *Let $RAFN = (Ar, att, sup, \mathbf{s}, \mathbf{t})$. Given a structure $U = (S, \Gamma, \Delta)$:*

1. *For $X \in \{Ar, att, sup\}$, $Def_X(U) = \{x \in X | \exists \alpha \in \Gamma, \mathbf{s}(\alpha) \in S$ and $\mathbf{t}(\alpha) = x\}$.*
   $Def(U) = Def_{Ar}(U) \cup Def_{att}(U) \cup Def_{sup}(U)$ *denotes the set of all defeated elements w.r.t. U.*

2. $Supp(U) = \{x | \forall \alpha \in \Delta$ *such that* $\mathbf{t}(\alpha) = x$, *if* $\alpha \in Supp(U_{-x})$ *then* $\mathbf{s}(\alpha) \cap (S \cap Supp(U_{-x})) \neq \varnothing\}$ *with* $U_{-x} = U \setminus \{x\}$. U *is self-supporting iff* $(S \cup \Gamma \cup \Delta) \subseteq Supp(U)$.

3. $UnSupp(U) = \overline{Supp(U')}$ *denotes the set of* unsupportable *elements w.r.t. U.*

4. $UnAcc(U) = Def(U) \cup UnSupp(U)$ *denotes the set of* unacceptable *elements w.r.t. U.*

5. $UnAct(U) = \{\alpha \in att | \alpha \in UnAcc(U)$ *or* $\mathbf{s}(\alpha) \subseteq UnAcc(U)\}$ *denotes the set of* unactivable *attacks w.r.t. U.*

Note that the set of elements supported by a structure are defined using a self-reference. Indeed one wants to avoid the situation in which an element x would be supported only because x is supported.

Then semantics can be defined as follows.

Definition 48 (Defs. 21 and 22 in [Cayrol *et al.*, 2018b]). *Let $RAFN = (Ar, att, sup, \mathbf{s}, \mathbf{t})$. Given a structure $U = (S, \Gamma, \Delta)$:*

- $x \in Ar \cup att \cup sup$ *is acceptable w.r.t. U iff (i)* $x \in Supp(U)$ *and (ii) for each attack $\alpha \in att$ with $\mathbf{t}(\alpha) = x$, $\alpha \in UnAct(U)$.*
  $Acc(U)$ *denotes the set of all elements that are acceptable w.r.t. U.*

- U *is conflict-free iff* $S \cap Def_{Ar}(U) = \varnothing$, $\Gamma \cap Def_{att}(U) = \varnothing$ *and* $\Delta \cap Def_{sup}(U) = \varnothing$.

- U *is admissible iff it is conflict-free and* $(S \cup \Gamma \cup \Delta) \subseteq Acc(U)$.

- U is complete *iff it is conflict-free and* $(S \cup \Gamma \cup \Delta) = Acc(U)$.

- U is preferred *iff it is a* $\subseteq$-*maximal complete structure,*

- U is stable *iff it is complete and* $\overline{(S \cup \Gamma \cup \Delta)} = UnAcc(U)$.

- U is grounded *iff it is a* $\subseteq$-*minimal complete structure.*

All the definitions can be illustrated using Example 8.

Example 8 (cont'd) *The graphical representation for the RAFN corresponding to this example is the same as the one given for the ASAF in Section 5.2.1.*

In this example, using the previous definitions, there is only one structure that is grounded, preferred and stable: $(\{b,d,e,g,h\}, att, \varnothing)$. *Here the only interaction that is not acceptable is the support* (d,c) *(its attacker ϵ being acceptable). This is one difference between this approach and the approaches presented in Sections 5.1 and 5.2.1. Here, the attack β (i.e. (c,b)) is acceptable since its attacker (π) is unactivable (even if it is acceptable), the source of π being unacceptable.*

The next examples illustrate further differences between the $ASAF$ and the $RAFN$ approach. Indeed several differences can be outlined (even if we exclude cycles of necessary supports, and assume that interactions are binary ones). First, in $ASAF$, attacks and supports are combined to obtain extended (direct or indirect) defeats and these defeats are used in the definition of conflict-freeness. In contrast, in $RAFN$, the notions of support and attack are dealt with separately.

Example 10 (Ex. 16 in [Cayrol *et al.*, 2018b]). *Consider the simple argumentation framework with only 2 necessary supports (so without any higher-order interaction).*

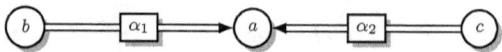

As for acceptability, following the ASAF semantics defined directly over the framework (see Section 5.2.1), an element is acceptable w.r.t. a set of elements whenever it can be defended against each defeat. So, in the particular case when there is no attack, each element of the framework would be acceptable w.r.t. any set, and the sets $\{a, \alpha_1, \alpha_2\}$, $\{a, b, \alpha_1, \alpha_2\}$, $\{a, c, \alpha_1, \alpha_2\}$ *(among others) are admissible.*

In contrast, $RAFN$ acceptability explicitly requires a support. So, the structures $(\{a\}, \varnothing, \{\alpha_1, \alpha_2\})$, $(\{a,b\}, \varnothing, \{\alpha_1, \alpha_2\})$ and $(\{a,c\}, \varnothing, \{\alpha_1, \alpha_2\})$ are not admissible with $RAFN$ semantics.

Another difference was already pointed out in [Cayrol et al., 2017], where correspondences have been provided between a RAF and an $ASAF$ without support. Indeed, in an $ASAF$, an attack is not acceptable whenever its source is not acceptable.

Example 11 (Ex. 15 in [Cayrol et al., 2018b]). *Let $RAFN$ be the following argumentation framework:*

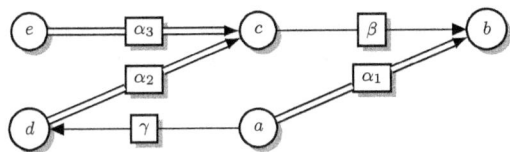

With $RAFN$ semantics, β is not attacked and not supported so β must belong to each complete structure.
With $ASAF$ semantics, if β is acceptable w.r.t. a set S, then c must also be acceptable w.r.t. S. If S is a complete extension, S contains a, γ, α_1, α_2 and α_3. As c is defeated by γ given $\{\alpha_2\}$, it cannot be the case that c is acceptable w.r.t. S. So β cannot belong to any complete extension.

Note also that the $RAFN$ is a conservative generalization of the $GAFN$ (see Section 4.2 in [Cayrol et al., 2018b]).

Moreover, since $RAFN$ are obviously a conservative generalization of RAF, they inherit a one-to-one correspondence with $AFRA$ in the case of the complete, preferred and stable semantics but only when there is *no support* (so when $RAFN$ are reduced to RAF).

5.3 Higher-order evidential supports

In [Cayrol et al., 2018a], the RAF is extended with the introduction of evidential supports. Recall that, as presented in Section 4.3, the evidential understanding of the support relation introduced in [Oren and

Norman, 2008] allows to distinguish between two different kinds of arguments: *prima-facie* and *standard arguments*. *Prima-facie* arguments were already present in [Verheij, 2003] as those that are justified whenever they are not defeated. On the other hand, *standard arguments* are not directly assumed to be justified and must inherit support from prima-facie arguments through a chain of supports.

This extension of *RAF*, called Recursive Evidence-Based Argumentation Framework (*REBAF*), can be defined as follows:

Definition 49 (Def. 13 in [Cayrol et al., 2018a]). *A recursive evidence-based argumentation framework (REBAF) is a sextuple* $\langle Ar, att, sup, \mathbf{s}, \mathbf{t}, PF \rangle$ *where Ar, att and sup are three (possible infinite) pairwise disjunct sets respectively representing arguments, attacks and supports names;* $PF \subseteq Ar \cup att \cup sup$ *is a set representing the* prima-facie *elements that do not need to be supported; functions* $\mathbf{s} : (att \cup sup) \longrightarrow 2^{Ar} \setminus \varnothing$ *and* $\mathbf{t} : (att \cup sup) \longrightarrow (Ar \cup att \cup sup)$ *respectively map each attack and support to its source and its target.*

Then the definition of *REBAF* semantics uses similar notions and techniques to the ones used in [Cayrol et al., 2018b] for the *RAFN*. For instance, the notion of structure in *REBAF* and, given a structure U, the sets $Def(U)$ and $Def_X(U)$ exactly correspond to the equivalent notions in *RAFN*. The other notions are of course adapted to account for the constraints emerging from the evidential interpretation of the support relation:

Definition 50 (Sec. 3.2 in [Cayrol et al., 2018a]). *Let* $\langle Ar, att, sup, \mathbf{s}, \mathbf{t}, PF \rangle$ *be a REBAF. Let* $U = (S, \Gamma, \Delta)$ *be a structure of REBAF.*

- $Supp(U) = PF \cup \{\mathbf{t}(\alpha) | \exists \alpha \in \Delta \cap Supp(U_{-\mathbf{t}(\alpha)}), \mathbf{s}(\alpha) \subseteq S \cap Supp(U_{-\mathbf{t}(\alpha)})\}$ $with^{21}$ $U_{-\mathbf{t}(\alpha)} = U \setminus \{\mathbf{t}(\alpha)\}$.

- $UnAcc(U) = Def(U) \cup \overline{Supp(U')}$ *with* $U' = (\overline{Def_{Ar}(U)}, att, \overline{Def_{sup}(U)})$.

- $UnAct(U) = \{\alpha \in att | \alpha \in UnAcc(U) \text{ or } \mathbf{s}(\alpha) \cap UnAcc(U) \neq \varnothing\}$.

---
[21]By abuse of notation, we write $U \setminus X$ instead of $(S \setminus X, \Gamma \setminus X, \Delta \setminus X)$ with $X \subseteq (Ar \cup att \cup sup)$.

Note that the notion of self-supporting structure in $REBAF$ is the same as the one given for $RAFN$. Then using these notions, the definitions for acceptability, admissibility, conflict-freeness and also for the complete semantics given for $RAFN$ (see Definition 48) can be reused. Some differences appear for the preferred and stable semantics; furthermore, no definition is given in [Cayrol et al., 2018a] for the grounded semantics, but a definition is proposed in [Cayrol and Lagasquie-Schiex, 2020]:

Definition 51. (Defs. 16 in [Cayrol et al., 2018a] and 2.14 in [Cayrol and Lagasquie-Schiex, 2020]) *Let* $REBAF = \langle Ar, att, sup, \mathbf{s}, \mathbf{t}, PF \rangle$. *Let* $U = (S, \Gamma, \Delta)$ *be a structure of* $REBAF$.

- U is preferred *iff it is a* $\subseteq$-*maximal admissible structure,*
- U is stable *iff* $(S \cup \Gamma \cup \Delta) = \overline{UnAcc(U)}$.[22]
- U is grounded *iff it is a* $\subseteq$-*minimal complete structure.*

All these notions can be illustrated on Example 8.

Example 8 (cont'd) *First of all, we must choose the set of prima-facie elements. Indeed, without prima-facie elements, most semantics will yield an empty set. Like for* $EBAF$, *elements that are not the target of a support are assumed to be prima-facie (the prima-facie elements are represented using solid outlines whereas standard elements are represented with dashed outlines). So, the only standard element is the argument d.*

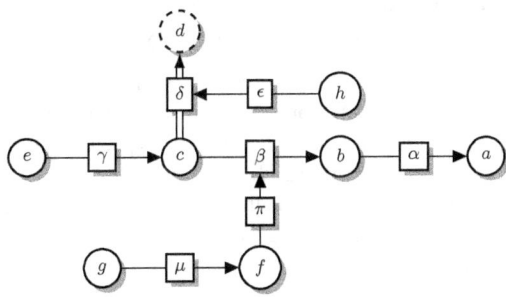

---

[22] Note that this already implies conflict-freeness.

Note that the structure $(\{b, e, g, h\}, att, \varnothing)$ is the only complete, grounded, preferred and stable structure, as in the RAFN case. Here, d is not acceptable since it is not supported (its support being attacked: both δ and its source c are attacked).

In [Cayrol *et al.*, 2018a], several links with other approaches have been proven:

- In Section 4 in [Cayrol *et al.*, 2018a]): $REBAF$ are a conservative generalization of RAF, and so inherit a one-to-one correspondence with $AFRA$ in the case of the complete, preferred and stable semantics but only when there is *no support* and when *each element is prima-facie*.

- In Section 5 in [Cayrol *et al.*, 2018a]): a one-to-one correspondence between $REBAF$ and finite $EBAF$; this correspondence does not work when we consider non-finite $EBAF$.

- In Section 4 in [Cayrol *et al.*, 2018a]): since the type of support used in $ASAF$ (necessary support) is different from the one used in $REBAF$ (evidential support), no correspondence can be established. And the same result occurs with $BAFD$ (deductive support).

- In Section 6 in [Cayrol *et al.*, 2018a]): $REBAF$ are a conservative generalization of AF considering the notion of d-structure (see Definition 28).

5.4 Comparison between Higher-order bipolar approaches: a first and succinct summary

Throughout Section 5, many differences and similarities were highlighted in order to compare the four approaches introduced in this section (the higher-order deductive framework in Section 5.1, $ASAF$ in Section 5.2.1, $RAFN$ in Section 5.2.2, and $REBAF$ in Section 5.3); moreover some links with the higher-order approaches from Section 3 were also given. All these comparison points have been introduced when it was pertinent (depending on the definitions and examples discussed at that point in the text). So in order to facilitate the reading and the understanding of this chapter, the main comparison points are recalled here.

1 - Higher-Order Interactions in Abstract Argumentation

- First of all, as for the higher-order approaches from Section 3, most of the presented higher-order bipolar approaches are a conservative generalization of Dung's framework when neither higher-order nor bipolar interactions exist. The only (partial) exception is the *ASAF* since it is inspired on the *AFRA* and so it inherits the same problem: the generalization holds only at the semantics level but not for the basic semantic notions (such as, for instance, conflict-freeness).

- Second, since each presented higher-order bipolar approach is built upon a specific higher-order approach, the bipolar version is a conservative generalization of the framework it is based on when no supports exist. So this link exists between the higher-order deductive framework and the *EAF*, between the *ASAF* and the *AFRA*, and between the *RAFN* or the *REBAF* and the *RAF*. Of course, the same result holds (with some nuances) when we compare a higher-order bipolar approach and the bipolar framework it is based on, when no higher-order interactions exist.

- Third, because of the three types of support they consider, it is difficult to establish links between all higher-order bipolar approaches. So higher-order deductive frameworks are not comparable with *ASAF*, *RAFN* or *REBAF*; *ASAF* or *RAFN* are not comparable with the higher-order deductive framework or with *REBAF*; and the *REBAF* is not comparable with the three other frameworks.

- And finally, in [Cayrol *et al.*, 2018b], a comparison between *ASAF* and *RAFN* has been carried out, yielding the same results as the ones between *AFRA* and *RAF*: these two approaches do not coincide even if there exists a one-to-one correspondence.

Another point of comparison, not addressed by any of the higher-order bipolar approaches, regards the way in which they treat support cycles: whether they prevent them in the definition of the framework, whether they allow them but reject them in the definition of the semantics, etc.

As it can be noted in Definition 39, the $ASAF$ requires the support relation to be acyclic. As argued by the authors in [Gottifredi et al., 2018], this restriction is inspired on the restrictions placed on the support relation of the AFN (see Def. 31), in which the support relation is required to be irreflexive and transitive. On the one hand, by being acyclic, the support relation of the $ASAF$ is also irreflexive; on the other hand, the transitive nature of necessary support is captured in the $ASAF$ by explicitly considering a sequence of supports in the definition of the conditional defeats. In contrast, we can note that neither the $BAFD$ with second-order attacks, the $RAFN$ nor the $REBAF$ impose restrictions on the support relation of the framework.

Given the deductive interpretation of support adopted by the $BAFD$, we can note that the existence of support cycles would be resolved by the corresponding Dung semantics in the translated AF. If we take the simplest odd-length support cycle, we can consider a self-supporting argument a; in such a case, the translated AF would be such that an even-length attack cycle between a and Z_{aa}. Similarly, a two-length support cycle between two arguments a and b would yield an even-length attack cycle in the translated AF, namely: $b \to Z_{ab}$, $Z_{ab} \to a$, $a \to Z_{ba}$ and $Z_{ba} \to b$. In both cases (odd-length and even-length support cycles), the resulting attack cycle in the AF would be of even length. Consequently, unless the cycle is broken, the arguments involved in the support cycle would be rejected by the grounded semantics, and possibly accepted by the complete, preferred or stable semantics.

At last, the treatment of support cycles in $RAFN$ and $REBAF$ is analogous. Both frameworks allow support relations originating in a set of arguments. Then, they define the set of supported elements by a structure, in which they prevent an element from being supported by itself (by considering U_{-x} in Definition 47 and $U_{-\mathbf{t}(\alpha)}$ in Definition 50). Consequently, since the acceptable elements w.r.t. a structure have to be supported by the structure, this prevents the semantics from accepting an argument that is just supported by itself (either directly or indirectly).

A more complete analysis of the four higher-order bipolar approaches is given in Section 7.

6 Computational issues and some applications

This section starts by introducing computational approaches that implement alternative semantics for some of the frameworks discussed in Sections 3 and 5. Then, we briefly discuss some applications of these frameworks or their underlying ideas to solve problems such as finding solutions to the *liar paradox* [Field, 2008] and the construction of deductive mathematical proofs.

6.1 Computational issues

Several works concern the semantics computation for higher-order frameworks. They describe either logical approaches, or the use of dialectical proofs, or some more direct algorithms.

6.1.1 ASP Encodings for *EAF* and *AFRA*

As discussed in [Dvorak et al., 2015], and also evidenced by the different editions of the *International Competition on Computational Models of Argumentation (ICCMA)*,[23] reduction-based approaches for the implementation of argumentation related problems have become very popular. Among others, reductions to Answer Set Programming (ASP) [Marek and Truszczyński, 1999; Niemelä, 1999] and propositional logic became suitable for the relevant reasoning problems [Thimm and Villata, 2017; Gaggl et al., 2020].

In [Dvorak et al., 2015] the authors proposed an ASP reduction-based approach to compute acceptability in Modgil's *EAF*. For that purpose, they proposed an alternative (but equivalent) characterization for the acceptance of arguments in an *EAF*, which allowed them to design succinct ASP encodings for all standard semantics of the *EAF*. Briefly, the new characterization of acceptability for *EAF* given in [Dvorak et al., 2015] relies on the consideration of a single reinstatement set for the defeats. As shown by the authors, since the union of two reinstatement sets for the same set of arguments S is also a reinstatement set, there exists a unique maximal reinstatement set.

---

[23]http://argumentationcompetition.org

Based on the new definitions, they proposed ASP encodings for EAF. Briefly, the answer-sets of the combination of an encoding for a semantics σ with an ASP representation of an EAF are in one-to-one correspondence to the set of σ-extensions of this EAF. The encoding is partitioned into several modules, and they begin with an input database for a given $EAF = \langle Ar, att, att2 \rangle$. Next we introduce the facts encoding an EAF; for further details and a full description of the encodings, including the definition of modules for each semantics, we refer the reader to [Dvorak et al., 2015]:

$$\widehat{EAF} := \begin{aligned} &\{\mathbf{arg}(x). \mid x \in Ar\} \cup \\ &\{\mathbf{att}(x,y). \mid (x,y) \in att\} \cup \\ &\{\mathbf{d}(x,y,z). \mid (x,(y,z)) \in att2\} \end{aligned}$$

It is worth mentioning that their proposed encodings were incorporated within *ASPARTIX - Answer Set Programming Argumentation Reasoning Tool*,[24] an ASP-based argumentation system for representing and evaluating Dung's AF semantics and some of its extended frameworks, such as Modgil's EAF. In particular, the evaluation of semantics over an EAF is provided in the web-interface *GERD - Genteel Extended argumentation Reasoning Device*.[25] Figure 5 illustrates the use of the *GERD* tool on the EAF of Example 3, where argument c' is denoted as cp.

As mentioned before, different ASP encodings for Dung's framework exist (see *e.g.* [Egly et al., 2010]). Then, based on the encodings for Dung's framework and its semantics, *ASPARTIX* also offers the possibility to evaluate the *AFRA* semantics. In order to be able to use those, an *AFRA* is encoded by an ASP encoding similar to the one provided for the EAF, with the addition of some predicates allowing to translate the *AFRA* into an AF (following the translation described at the end of Section 3.3). The corresponding encoding provided in the *ASPARTIX* website[26] is shown below.

An *AFRA* is encoded by a sequence of statements where each statement either encodes an argument, or an attack between arguments, or

---

[24]http://www.dbai.tuwien.ac.at/research/argumentation/aspartix
[25]http://gerd.dbai.tuwien.ac.at
[26]https://www.dbai.tuwien.ac.at/research/argumentation/aspartix/afra.html

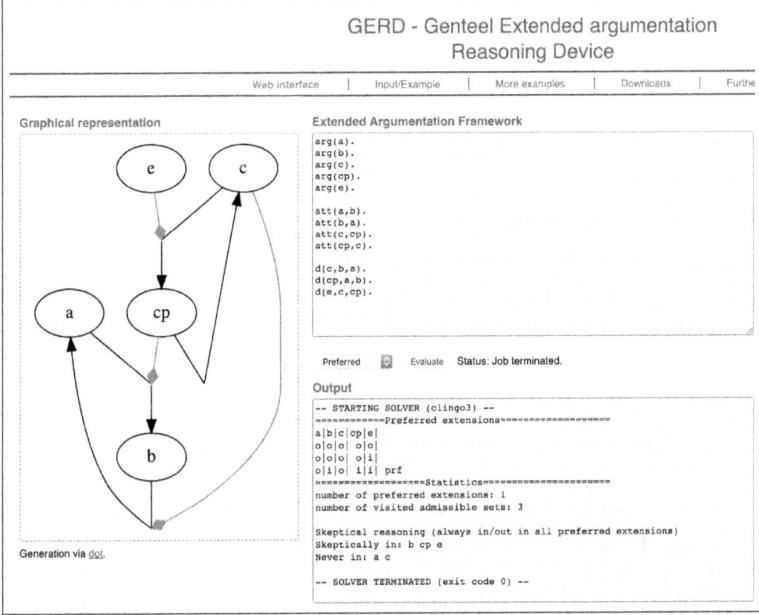

Figure 5: Screenshot of *GERD* illustrating the *EAF* of Example 3 and its preferred extension $\{e, c', b\}$.

% arguments
$\mathbf{arg}(X) \leftarrow \mathbf{afraA}(X)$.
$\mathbf{arg}(R) \leftarrow \mathbf{afraR}(R, X, Y), \mathbf{afraA}(X)$.

% direct defeat
$\mathbf{att}(V, W) \leftarrow \mathbf{afraR}(V, X, W), \mathbf{arg}(W), \mathbf{afraA}(X)$.

% indirect defeat
$\mathbf{att}(V, A) \leftarrow \mathbf{att}(V, W), \mathbf{afraR}(A, W, X), \mathbf{afraA}(W)$.

Figure 6: ASP encoding to translate an *AFRA* into an *AF*.

an attack towards another attack. The facts representing $AFRA = \langle Ar, att \rangle$ are:

$$\widehat{AFRA} := \{\mathbf{afraA}(x). \mid x \in Ar\} \cup \{\mathbf{afraR}(\alpha, x, y). \mid \alpha = (x,y) \in att\}$$

Finally, the ASP implementation of the translation from an $AFRA$ into an AF is shown in Figure 6.

6.1.2 Logical encoding of $REBAF$ and RAF

Another logical approach is presented in [Cayrol and Lagasquie-Schiex, 2018; Cayrol and Lagasquie-Schiex, 2020]. In these works, the authors use a three-sorted logic with equality in order to encode several variants of argumentation frameworks (AF, RAF and $REBAF$). With that work, the authors want to characterize in a logical way the meaning of each type of interaction, to encode the acceptance condition for arguments and interactions and then provide a computational issue for the semantics of these argumentation frameworks.

In this logic, the three sorts are: **arg** a sort for arguments, **att** a sort for attacks and **esup** a sort for evidential supports. Two function symbols s and t can be applied to objects of the sort **att** or **esup** to capture source and target of these interactions. The target can be either of sort **arg** or of sort **att** or of sort **esup** and the source can only be of sort **arg**. Note that this encoding takes into account only the case of interaction sources that are singletons.

Different unary predicates are also used for encoding each element of the argumentation framework: For a node a of the argumentation graph, $Acc(a)$ expresses the status of being accepted, whereas $Nacc(a)$ expresses that a cannot be accepted (implicitly: w.r.t. a given semantics); in other words, the meaning of $Nacc(a)$ is stronger than "a is not accepted". The language also admits atoms of the form $Val(\alpha)$ for attack or support names (intuitively, $Val(\alpha)$ means that the interaction named α is valid w.r.t. a given argumentation semantics). There is also the predicate symbol $PrimaFacie$ for denoting prima-facie elements (so for arguments and interactions).

Since one purpose is to obtain a logical characterization of structures, and so of acceptability, some additional unary predicate symbols are given: $Supp$ for denoting supported elements (arguments, attacks or supports), $UnSupp$ for denoting unsupportable elements and $eAcc$ (resp. $eVal$) for denoting acceptability for arguments (resp. for interactions, i.e. attacks or supports). Note that $eAcc(x)$ ("x is e-accepted") can be understood as "x is accepted and supported" and similarly $eVal(\alpha)$ ("α is e-valid") can be understood as "α is valid and supported".

Using this vocabulary, the formulae describing a given argumentation framework, for instance a $REBAF$, can be partitioned in two sets:

- The first set contains the formulae describing the general behaviour of each interaction, possibly recursive, i.e. how an interaction interacts with arguments and other interactions related to it.

- The second set contains the formulae encoding the specificities of the current framework (enumeration of the arguments and interactions that belong to this framework).

Then, several formulae are introduced for encoding the different principles that govern argumentation semantics. There are formulae for capturing the defence principle, the reinstatement principle and the stability principle.[27] Then extensions under a given semantics (admissible, complete, preferred, grounded, or stable) can be characterized by models of logical theories obtained by combining some of these formulae.

Note that, if we consider finite argumentation frameworks, all the previous formulae can be rewritten in propositional logic and a SAT solver is enough for computing the structures resulting from $REBAF$ semantics.

Next we provide a very simple example in order to illustrate these ideas and notions, and present its complete encoding.

Example 12 (Ex. 1.2 in [Cayrol and Lagasquie-Schiex, 2020]). *Consider the following REBAF.*

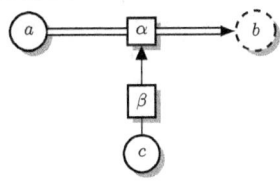

---

[27] Note that the first set describes the conflict-freeness principle.

The set of formulae describing this REBAF is the following:
$\Sigma(REBAF) = \{(eVal(\beta) \land eAcc(c)) \to \neg Val(\alpha),$
  $Supp(a),$
  $Supp(c),$
  $Supp(\alpha),$
  $Supp(\beta),$
  $(eAcc(a) \land eVal(\alpha)) \to Supp(b),$
  $(Supp(a) \land Acc(a)) \leftrightarrow eAcc(a),$
  $(Supp(b) \land Acc(b)) \leftrightarrow eAcc(b),$
  $(Supp(c) \land Acc(c)) \leftrightarrow eAcc(c),$
  $(Supp(\alpha) \land Val(\alpha)) \leftrightarrow eVal(\alpha),$
  $(Supp(\beta) \land Val(\beta)) \leftrightarrow eVal(\beta) \}$

The following ideas are used for obtaining $\Sigma(REBAF)$: first, a prima-facie element is supported. Second, an element is e-accepted if and only if it is accepted and supported. Third, if an attack and its source are e-accepted, then its target cannot be accepted (resp. valid). And finally, if a support and its source are e-accepted, then its target is supported.

Then, the set of formulae $\Sigma_{ss}(REBAF)$ describing the notion of supported/unsupportable element is obtained from $\Sigma(REBAF)$ by adding formulae among which:
  $Supp(b) \to (eAcc(a) \land eVal(\alpha))$
  $\neg UnSupp(a)$
  $\neg UnSupp(c)$
  $\neg UnSupp(\alpha)$
  $\neg UnSupp(\beta)$
  $Unsupp(b) \leftrightarrow \Big((eVal(\beta) \land eAcc(c)) \lor UnSupp(a) \lor UnSupp(\alpha) \Big)$

The first formula in $\Sigma_{ss}(REBAF)$ expresses the fact that if the target of a support is supported, then this support and its source are e-accepted. The other formulae correspond to the unsupported status: first, a prima facie element is not unsupported; second, if the target of a support is unsupported, then this support is not valid, or this support or its source are unsupported.

The set of formulae $\Sigma_d(REBAF)$ describing the principle of defence is obtained from $\Sigma_{ss}(REBAF)$ by adding formulae among which:
  $Val(\alpha) \to (UnSupp(\beta) \lor UnSupp(c))$

The previous formula describes the defense of α: if α is defended (so valid) then its attacker β or the source of β are unsupported (here, this is the only way to invalidate the attack on α since neither β nor its source are in turn attacked).

The principle of reinstatement is expressed using the set of formulae $\Sigma_r(REBAF)$ obtained from $\Sigma_{ss}(REBAF)$ by adding the formulae:
$Acc(a)$
$Acc(b)$
$Acc(c)$
$Val(\beta)$
$(UnSupp(c) \vee UnSupp(\beta)) \rightarrow Val(\alpha)$

The four first formulae correspond to the case of an unattacked element: it does not need a defense for being accepted or valid. The last formula gives the condition for the reinstatement of an element that is the target of an attack (the reverse condition of the one given for the defense).

And finally, $\Sigma_s(REBAF)$ describing the stability principle is obtained from $\Sigma_{ss}(REBAF)$ by adding the formulae:
$Acc(a)$
$Acc(b)$
$Acc(c)$
$Val(\beta)$
$\neg Val(\alpha) \rightarrow eVal(\beta) \wedge eAcc(c)$
$\neg Supp(x) \rightarrow UnSupp(x)$ for $x \in \{a, c, \alpha, \beta\}$

The formulae in $\Sigma_s(REBAF)$ give the impact of either non-accepted/ non-valid elements, or non-supported elements. For instance, if α is not valid, then its attacker β and the source of β are in the extension (so resp. e-valid and e-accepted).

From $\Sigma_d(REBAF)$ it can be deduced that $\neg Val(\alpha)$ then $\neg eVal(\alpha)$, $\neg Supp(b)$ and $\neg eAcc(b)$. That corresponds to the fact that no admissible structure contains b (resp. α, though being supported).

Moreover, there is a model of $\Sigma_d(REBAF)$ satisfying $eAcc(a)$, $eAcc(c)$ and $eVal(\beta)$. That corresponds to the fact that $(\{a,c\}, \varnothing, \{\beta\})$ is an admissible structure. This is also a $\subseteq$-maximal model. That corresponds to the fact that $(\{a,c\}, \{\beta\}, \varnothing)$ is a preferred structure; this is also a complete structure (since it corresponds to a model of $\Sigma_d(REBAF) \cup$

$\Sigma_r(REBAF))$ and a stable structure (since it corresponds to a model of $\Sigma_s(REBAF))$.

6.1.3 Dialectical proof procedure for Modgil's EAF

In addition to characterizing labellings for the EAF, in [Modgil, 2009a] the author defined a dialectical framework for EAF game proof theories, allowing to establish the justified status of an argument to be tested, and providing a basis for algorithmic development of EAF semantics. Analogously to dialectical proof procedures for Dung's AF, such theories consider a dialogue between two players: P (proponent) and O (opponent), each of which are referred to as the other's counterpart. A game begins with P moving an initial argument x to be tested. Then, O and P take turns in moving arguments that attack their counterpart's last move, where attacks can be either on an argument or an attack posed by their counterpart; alternatively, the players can also backtrack to a counterpart's previous move and initiate a new dialogue. In particular, Modgil's approach assumes the use of a finite EAF containing a finite number of arguments (thus, a finite number of attacks).

Then, a legal move function ϕ_{PC} is defined, which places restrictions on the players' moves, for the preferred credulous game (*i.e.* for determining whether an argument belongs to some preferred extension of the corresponding EAF). As argued by the author, since every admissible set of an EAF is a subset of a preferred extension, it suffices to show membership to an admissible set in order to show membership to a preferred extension. Briefly, the ϕ_{PC} game is a tree of ϕ_{PC}-dialogues whose root is P's initial move of an argument. Also, the ϕ_{PC} function is such that it prevents O from moving arguments and attacks that have already been attacked by P in a dialogue d, since P will have already fulfilled its burden of defense with respect to these arguments/attacks. In addition, P can only move an argument x in d if: 1) x does not attack itself; 2) no argument y, and no attack (y, x) or (x, y) has been moved by P; and 3) x does not symmetrically attack some y moved by P.

Next, we illustrate Modgil's approach on the weather example:

Example 3 (cont'd) *Given the EAF representing the weather example, three ϕ_{PC} winning strategies for b are depicted below. The different moves in each strategy are identified by the argument put forward, the* player introducing the argument, and a number indicating the order in which they are played. Also, the notation $a \rightarrow$ means that argument a attacks the previous argument, and the notation $c \twoheadrightarrow$ means that argument c attacks the attack between the two previous arguments:

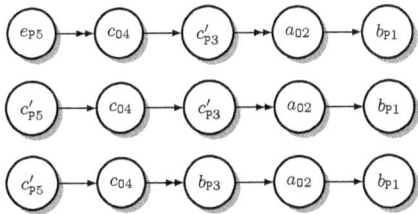

It should be noted that each winning strategy for b consists of a single dialogue. This is because the opponent O has no alternatives to counterattack the arguments/attacks put forward by the opponent P. Also, each strategy corresponds to an admissible set of EAF, from top to bottom: $\{e, c', b\}$ and $\{b, c'\}$ (the admissible set for the last two strategies coincides). Hence, b is (credulously) accepted w.r.t. the only preferred extension $\{e, c', b\}$ of EAF.

6.1.4 Dialectical proof procedure for i-defense semantics of $HLAF$

In [Hanh et al., 2011] the authors introduced a dialectical proof procedure for their inductive defense semantics of $HLAF$ (see Section 3.4) based on [Dung and Thang, 2009; Thang et al., 2009], where two unified frameworks of dialectical proof procedures were proposed.

Similarly to Modgil's approach discussed in the previous section, [Hanh et al., 2011] proposes to evaluate the acceptability of arguments by resolving disputes between two players identified as proponent and opponent. They propose to represent disputes through *dispute derivations*, in which tuples $t_i = \langle P_i, O_i, SP_i, SO_i \rangle$ summarizing the history of the dispute up to step i are successively constructed by expanding the previous one. Given $HLAF = \langle Ar, att \rangle$, the set $P_i \subseteq Ar \cup att$ in each tuple represents the set of arguments and attacks put forward by the proponent (up to step i) that have not been defended by the proponent and hence are open to attacks by the opponent. Also, $SP_i \subseteq Ar$ is the set of all arguments presented by the proponent (up to step i). Consequently, the proponent does not need to re-defend arguments in $SP_i \backslash P_i$. On the other hand, $O_i \subseteq att$ is a set of attacks of the opponent against arguments presented by the proponent in previous steps that are not yet counter-attacked by the proponent. Thus, an attack $\alpha = (a, b) \in O_i$ needs to be counter-attacked by the proponent on either a or α. In addition, $SO_i \subseteq att$ is the set containing attacks by the opponent (up to step i) that have been counter-attacked by the proponent.

Thus, a dispute derivation for an argument a is a sequence of the tuples described above, satisfying the following conditions:

- $P_i \subseteq Ar \cup att$; $SP_i \subseteq Ar$; and $O_i, SO_i \subseteq att$.

- $P_0 = SP_0 = \{a\}$, and $O_0 = SO_0 = P_n = O_n = \varnothing$.

- At step i, an element X is selected from either P_i (i.e. an argument or attack put forward by the proponent that has to be defended) or from O_i (i.e. an attack from the opponent that has to be counter-attacked). The sets corresponding to the next tuple $(i+1)$ are obtained as follows:

    - if $X \in P_i$: $P_{i+1} = P_i \backslash \{X\}$, $O_{i+1} = O_i \cup \{\alpha \mid \alpha = (Y, X) \in att\}$, $SP_{i+1} = SP_i$ and $SO_{i+1} = SO_i$; or
    - if $X \in O_i$: $O_{i+1} = O_i \backslash \{X\}$, $SO_{i+1} = O_i \cup \{X\}$, P_{i+1} augments P_i with an attack α targeting X and with the source of α (as long as the latter does not already belong to SP_i), and SP_{i+1} augments SP_i with the source of attack α (if not already present).

It should be noted that, since at each step the selection can be made from P_i or O_i, the sequence of steps does not necessarily correspond to alternating moves by the different players. Consecutive selections from P_i would correspond to consecutive plays by the opponent (searching to attack the selected proponent's argument or attack), whereas consecutive selections from O_i would correspond to consecutive plays by the proponent (searching to counter-attack the opponent's selected attack).

Then, the authors showed that if $\langle P_0, O_0, SP_0, SO_0 \rangle \ldots \langle P_n, O_n, SP_n, SO_n \rangle$ is a dispute derivation for an argument a, then SP_n is an i-admissible set that contains a. Let us now illustrate the construction of a dispute derivation on the weather example.

Example 3 (cont'd) *The construction of a dispute derivation for b is depicted in Figure 7, where the notation $\underline{X}$ means that X is selected in the corresponding step, and $Attack_X = \{\alpha \in att \mid \alpha = (a, X)\}$. The sequence $\langle P_0, O_0, SP_0, SO_0 \rangle \ldots \langle P_9, O_9, SP_9, SO_9 \rangle$ is a dispute derivation showing that b is acceptable w.r.t. its constructed i-admissible set $SP_9 = \{b, c', e\}$ which, in particular, is the only i-preferred extension of the $HLAF$.*

Finally, the authors in [Hanh et al., 2011] stated that a proof procedure for i-defense semantics can be reduced to a procedure searching for dispute derivations, which could be directly implemented by means of, for instance, base derivations defined in [Thang et al., 2009].

6.1.5 Algorithmic approaches for computing extensions of argumentation frameworks with higher-order interactions

In this section we will briefly discuss different approaches proposed in the literature for computing the extensions of argumentation frameworks that include higher-order interactions.

- In [Nofal et al., 2014] the authors proposed a series of algorithms allowing to enumerate the extensions of frameworks with higher-order attacks, such as the ones discussed in Section 3. In particular, they take the $AFRA$ as a case-study and propose algorithms for enumerating the preferred, stable, complete stage, semi-stable, ideal and grounded semantics of the framework. For illustration purposes, we will next describe the algorithm for obtaining the preferred extensions of an $AFRA = \langle Ar, att \rangle$, and show its application on Example 3.

  Briefly, the algorithm considers five labels: **IN**, **OUT**, **MUST_OUT**, **BLANK** and **UNDEC**. The **BLANK** label is the initial label for all arguments and attacks. In each iteration, a **BLANK** attack $\alpha \in att$ is labelled **IN** to indicate that α might be in a preferred extension. As a selection rule, attacks whose target is the source

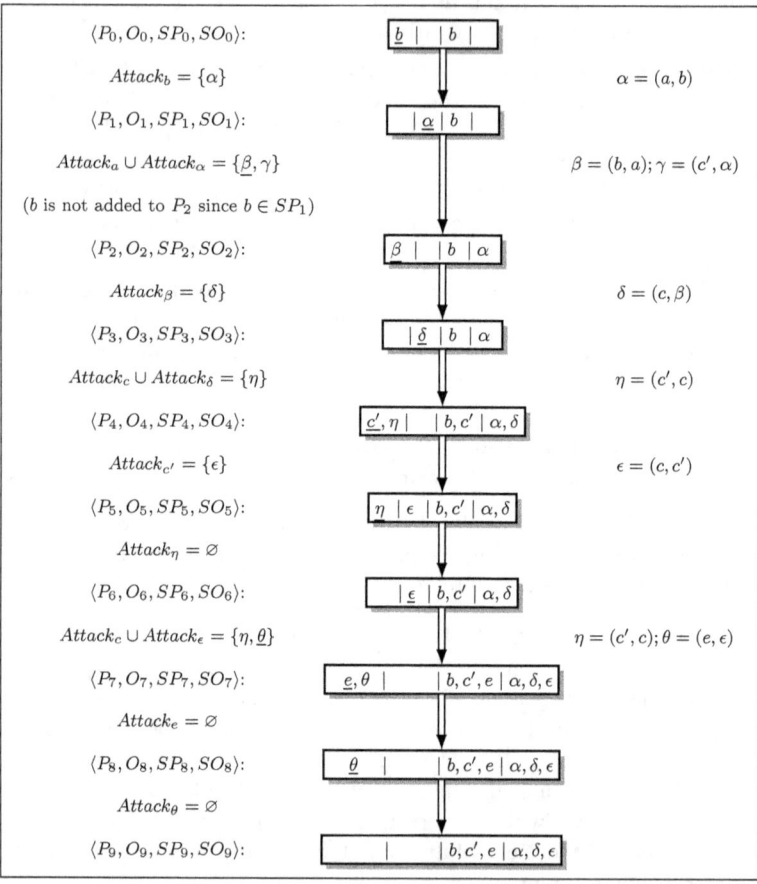

Figure 7: Construction of a dispute derivation for argument b corresponding to the $HLAF$ of Example 3 (arrows represent transitions between steps)

of the larger number of attacks are chosen first. Every time an attack is labelled IN, the labels of some attacks and arguments might change accordingly. An argument $a \in Ar$ is labelled OUT iff there is $\alpha \in att$ with the label IN such that $\mathbf{t}(\alpha) = a$. An attack $\beta \in att$ is labelled OUT iff there is $\alpha \in att$ with the label IN such that $\mathbf{t}(\alpha) \in \{\beta, \mathbf{s}(\beta)\}$. A BLANK argument a is labelled IN, implying that a might be in a preferred extension, iff there is $\alpha \in att$ with the label IN such that $\mathbf{s}(\alpha) = a$, or for each $\beta \in att$ such that $\mathbf{t}(\beta) = a$, the label of β is OUT. Then, each attack $\beta \in att$ with the label BLANK or UNDEC is labelled MUST_OUT iff there is $\alpha \in att$ with the label IN such that $\mathbf{t}(\beta) \in \{\alpha, \mathbf{s}(\alpha)\}$; finally, if some problem arises at this point (inconsistency between the labels assigned), the chosen attack α is labelled UNDEC to try to find a preferred extension excluding it.

Figure 8 exemplifies the algorithm to enumerate the preferred extensions on the $AFRA$ from Example 3, where attacks are selected to be labelled as IN in the following order: η, β, γ, θ.

- In [Alfano et al., 2018] the authors proposed an algorithm for efficiently recomputing the extensions of $BAFD$s with or without second-order attacks (see Section 5.1) after an update on the framework has been performed. Briefly, an update consists of the addition or removal of an argument, an attack or a support; however, as highlighted in [Alfano et al., 2018], updates concerning an argument can be easily performed without requiring to recompute an extension. Their algorithm builds on the incremental approach proposed for Dung's AF in [Alfano et al., 2017] and, given an initial $BAFD$, a semantics, an initial extension for it under the chosen semantics and an update, it computes an extension of the updated $BAFD$. This is achieved by introducing a meta-argumentation translation (analogous to the one proposed in Definition 37) according to which an initial $BAFD$, as well as its extension and an update, are transformed into a Dung's AF with a suitable initial extension and update.

In addition, the authors identify different conditions under which an update over a $BAFD$ is *irrelevant*, in the sense that the original

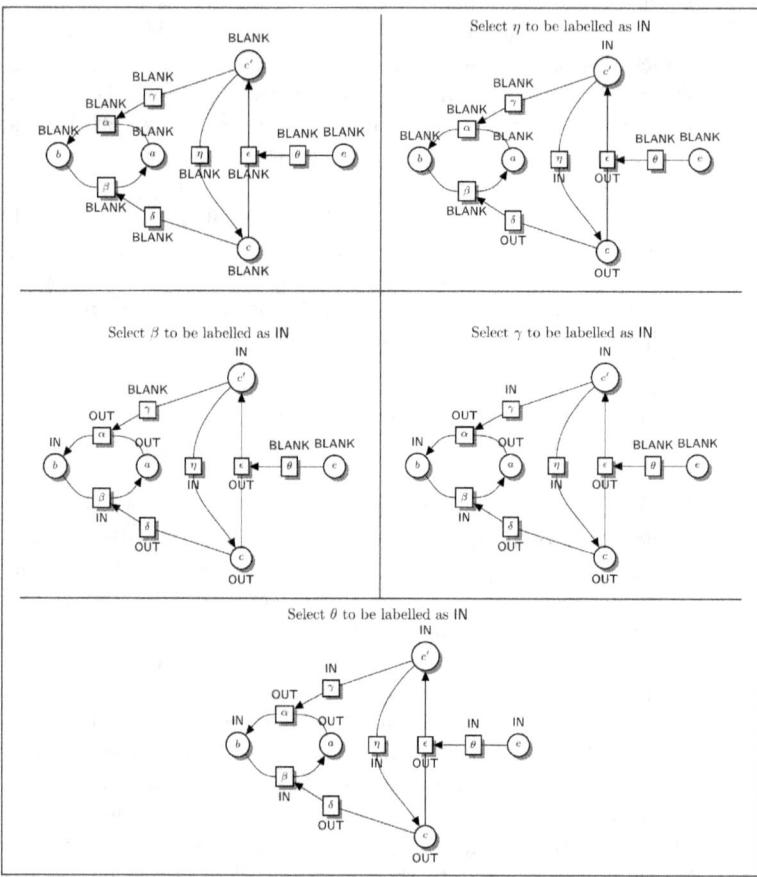

Figure 8: Application of the algorithm to enumerate the preferred extensions of the $AFRA$ corresponding to Example 3. The final labelling corresponds to the only preferred extension $\{e, \theta, c', \eta, \gamma, b, \beta\}$

input extension is still an extension of the updated framework; for this, only the stable and preferred semantics are considered. In other words, irrelevant updates are still applied on the input framework, yielding an updated framework; what occurs in those cases is that the extension of the updated framework does not need to be recomputed. Whereas the conditions characterizing the irrelevant updates are defined in [Alfano et al., 2018] in terms of labellings for the $BAFD$, no formal definition of labellings for $BAFD$ is given; instead, the extensions-labellings correspondence proposed for Dung's AF (see [Caminada and Gabbay, 2009]) is exploited.

Finally note that, even though the algorithm of [Alfano et al., 2018] was envisioned for computing an extension of an updated $BAFD$, it could also be iteratively used for computing an extension of a static $BAFD$ in the following way: start with the $BAFD$ containing all arguments and no attacks nor supports as initial framework, and the set of all arguments as the initial extension; then, add the attacks and supports one-by-one by considering them as updates, with the restriction that the second-order attacks have to be added after adding the interactions they target.

- In line with the work discussed in the previous item, [Alfano et al., 2020] proposed an incremental approach for efficiently computing extensions of an $ASAF$ after performing an update, considering the complete, preferred, stable and grounded semantics. Differently from the previous approach for $BAFD$, labellings for the $ASAF$ were formally characterized in [Alfano et al., 2020] (see Section 5.2.1) and accounted for in the developed algorithm.

The approach of [Alfano et al., 2020] also relies on a transformation of an $ASAF$ into a Dung's AF which, as argued by the authors, improves the one proposed in Definition 43 from two standpoints: i) it is direct, meaning that it does not require the two-step process of [Gottifredi et al., 2018] which first obtains an AFN and then an AF; and ii) the size of the resulting AF is smaller than that of the one obtained by applying Definition 43. Notwithstanding this, as shown in [Alfano et al., 2020], the translation they proposed yields equivalent extensions to those of the corresponding $ASAF$ under the considered semantics.

In addition, the authors formally characterized the irrelevant updates for an $ASAF$, for which an extension E of an updated $ASAF$ can be directly obtained without requiring its overall computation. Note that, like in the case of $BAFD$, irrelevant updates are still applied on the input $ASAF$, yielding an updated $ASAF$. However, differently from the case of $BAFD$, an $ASAF$ extension may also contain attacks and supports; therefore, in the presence of irrelevant updates, the updated extension will not necessarily coincide with the original extension but could easily be obtained without requiring its overall recomputation. On the one hand, for an irrelevant update deleting an attack or a support, an extension of the updated $ASAF$ can be simply obtained by deleting the corresponding interaction from the original extension. On the other hand, for an irrelevant update corresponding to an addition of an attack or a support, the situation depends on the nature of the interaction: whereas a support will always be added to the extension of the updated $ASAF$, an attack will only be added to the extension in cases where its source argument belonged to the original extension.

Finally note that, like in the case of the $BAFD$, the incremental algorithm for the $ASAF$ could be used for obtaining an extension of the framework in the static case. Furthermore, as argued by the authors in [Alfano et al., 2020], their proposed translation from an $ASAF$ into an AF could be used for obtaining $ASAF$'s extensions even in the static case, where updates are not considered (and the same would hold for their translation of a $BAFD$ into an AF).

6.1.6 The GRAFIX tool

Several tools have been developed by the argumentation community, each of them having its specificities (see for instance, the web-interface $GERD$ evoked in Section 6.1.1). Among them, the GRAFIX tool has been proposed for creating and handling enriched abstract argumentation graphs, in particular those with higher-order interactions (RAF,

1 - HIGHER-ORDER INTERACTIONS IN ABSTRACT ARGUMENTATION

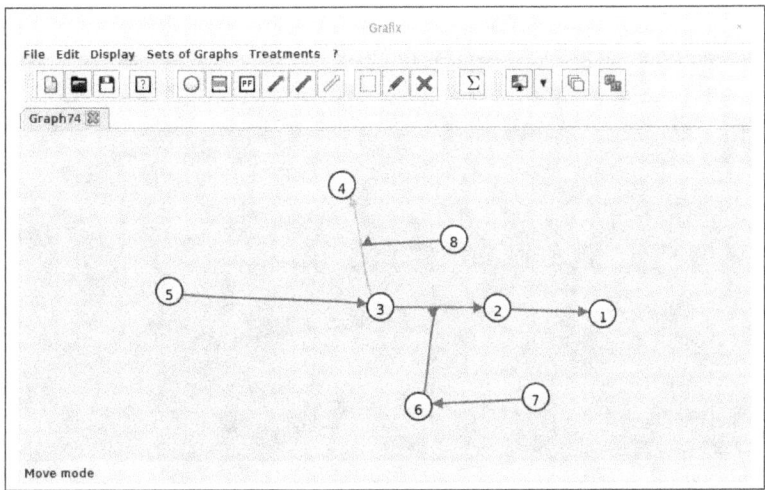

Figure 9: Vizualization of the *REBAF* version of Example 8 with GRAFIX Arguments are numbered as follows: 1 for a, ..., 8 for h. Attacks (resp. supports) are represented with red (resp. green) arrows.

REBAF and *RAFN*), following some of the approaches described in this chapter [Cayrol et al., 2017; Cayrol et al., 2018b; Cayrol et al., 2018c; Cayrol et al., 2018a; Cayrol and Lagasquie-Schiex, 2018].

GRAFIX is a graphical tool[28] encoded in Java language (see [Cayrol and Lagasquie-Schiex, 2008]). It allows for the definition and the visualization of many kinds of argumentation graphs and the execution of some treatments on these graphs. Among these treatments, there is the computation of the well-known acceptability semantics. Another example of treatment is the translation of argumentation graphs into logical bases and the use of these bases for computing some acceptability semantics.

Figure 9 is a screenshot corresponding to the creation of Example 8 with this tool.

Then, Figure 10 shows the corresponding preferred structure com-

---

[28]The visualization part of the tool is realized thanks to the GraphStream library (see [University of Le Havre, 2011]).

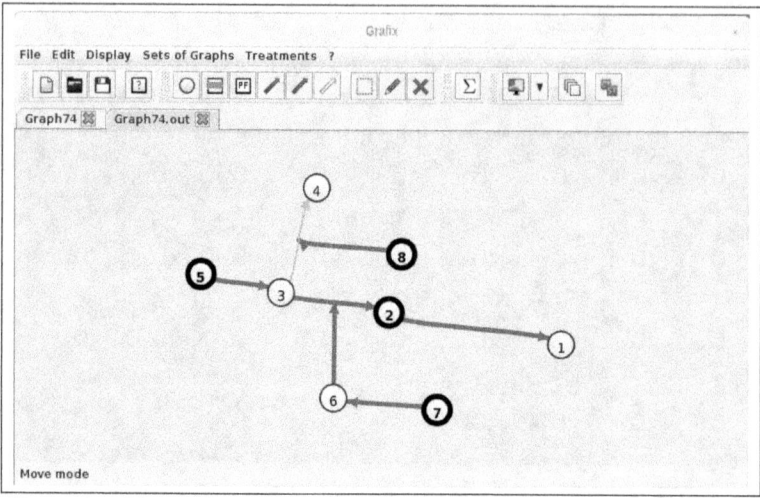

Figure 10: The preferred structure of the $REBAF$ version of Example 8 with GRAFIX (all elements except d, numbered 4 in the figure, are prima-facie). Elements that belong to the preferred structure are in bold.

puted with the GRAFIX tool when we consider that this framework is a $REBAF$ and that all elements except from argument d are prima-facie.

6.2 Applications

In the literature, higher-order frameworks are used for representing and solving different problems. Here, we present two examples of such applications.

- [Dauphin and Cramer, 2017] proposed the *Extended Explanatory Argumentation Framework* ($EEAF$), which extends the *Explanatory Argumentation Framework* of [Seselja and Straßer, 2013] by incorporating recursive attacks, joint attacks and a support relation. As argued by the authors, they apply the meta-argumentation methodology in order to incorporate these elements. The key feature of these frameworks is the existence of a set of *explananda* (scientific phenomenons of which, unlike arguments, the acceptability is not being questioned) and an explanatory relation relating arguments to other arguments or to *explanandum*, suitable for modelling the interaction between explanation and argumentation in scientific debates.

Definition 52 (Def. 18 in [Dauphin and Cramer, 2017]). *An Extended Explanatory Argumentation Framework (EEAF) is a tuple $\langle Ar, X, att, exp, inc, sup \rangle$, where Ar is a set of arguments, X is a set of explananda, $att \subseteq (2^{Ar} \cup exp \cup att) \times (Ar \cup exp \cup att \cup sup)$ is a higher-order attack relation, $exp \subseteq (Ar \times Ar) \cup (Ar \times X)$ is an explanatory relation, $inc \subseteq Ar \times Ar$ is an incompatibility relation, and $sup \subseteq Ar \times Ar$ is a support relation.*

Note that the attack relation *att* not only allows for joint attacks and higher-order attacks, but also for attacks originating in other attacks. On the other hand, the incompatibility relation is used to identify opposing theories, as scientists usually do not accept multiple explanations of a given phenomenon at the same time.

Then, as argued by the authors in [Dauphin and Cramer, 2017], the semantics of their $EEAF$ are defined by flattening their framework into an Explanatory Argumentation Framework. An Explanatory Argumentation Framework is a tuple $\langle Ar', att', X', exp', inc' \rangle$ (*i.e.* it has the same structure as the $EEAF$ minus the support relation), with the restriction that the attack relation att' is defined over pairs of arguments.

This translation is such that the set of arguments of the flattened Explanatory Argumentation Framework is comprised of: meta-arguments $acc(a)$ and $rej(a)$ for each argument in the $EEAF$, meta-arguments $X_{a,b}$ and $Y_{a,b}$ for each attack $(a,b) \in att$, meta-arguments $P_{a,b}$ and $Q_{a,b}$ for each pair $(a,b) \in exp$, a meta-argument $e(S)$ for each joint-attack having S as its set of originating arguments, and a meta-argument $Z_{a,b}$ for each pair of arguments $a, b \in Ar$. Also, the set of *explananda* in the flattened Explanatory Argumentation Framework is the same as the set of the corresponding $EEAF$. Then, the different relations of the $EEAF$ are mapped into the relations of its corresponding Explanatory Argumentation Framework by using the meta-arguments listed above.

Finally, the authors illustrate the applicability of the $EEAF$ on an example which focuses on two groups of solutions to the *liar paradox*. As stated by the authors, the arguments they considered are extracted from the book *Saving Truth from Paradox* [Field, 2008].

Example 13 (Ex. from [Dauphin and Cramer, 2017]). *Given the following arguments:*

ep: *This explanandum represents the paradox.*

a: *The paracomplete, paraconsistent and semi-classical solutions which provide explanations for the paradox by weakening classical logic.*

b: *The underspill and overspill solutions which provide their own explanation of the paradox by suggesting that for some predicates F, F is true of some objects that are not F or vice-versa.*

c: *We did not change logic to hide the defects in other flawed theories such as Ptolemaic astronomy, so why should we change the logic simply to hide these paradoxes?*

d: *There is no known way of saving these flawed theories such as Ptolemaic astronomy and even if there was, there is little benefit to doing so.*

f: *We have worked out the details of the new logics and they allow us to conserve the theory of truth.*

g: *Changing the logic implies changing the meaning.*

h: *Change of meaning is bad.*

i: *The change is mere.*

j: *This is no "mere" relabelling.*

k: *Change of truth schema is a change of the meaning of "true".*

l: *The paradox forces a change of meaning.*

[Dauphin and Cramer, 2017] proposed to model the knowledge in this discussion through the $EEAF$ depicted below, where attacks

(including joint attacks) are depicted using solid arrows, the support is depicted using a double arrow, the explanatory relation is depicted using dashed arrows, and the incompatibility relation is depicted with a dotted line:

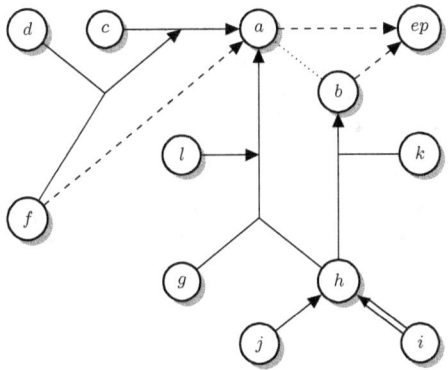

The flattened Explanatory Argumentation Framework corresponding to this EEAF is depicted in Figure 11 taken from [Dauphin and Cramer, 2017]; as argued by the authors, less-relevant auxiliary arguments are omitted in the figure for the sake of visibility (e.g. the $rej(X)$ meta-arguments that do not attack other meta-arguments, and the $Z_{X,Y}$ meta-arguments for which no support $(X, Y) \in sup$ exists).

Then, two argumentative core extensions are identified: $\{a, c, d, f, g, j, k, l\}$ and $\{b, c, d, f, g, j, k, l\}$, each of which corresponds to the two rivaling solutions (because a and b are incompatible). As explained in [Dauphin and Cramer, 2017], this is due to the fact that even though the author in [Field, 2008] might have a preference for one solution or the other, in the excerpt being analyzed, he is merely defending the solutions represented in a from attacks, and making no argument which attacks the solutions represented in b.

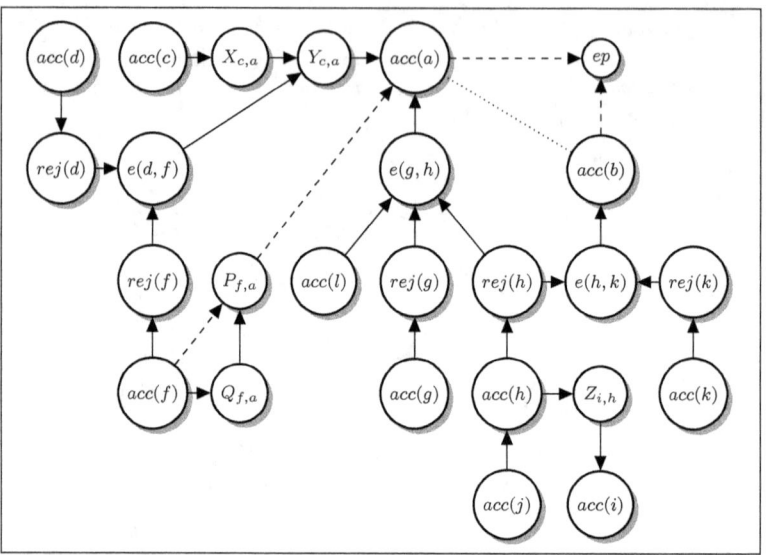

Figure 11: Explanatory Argumentation Framework corresponding to the *EEAF* of Example 13

- Among the existing applications of higher-order frameworks, the works [Boudjani et al., 2018; Boudjani et al., 2020] propose an application to the domain of deductive mathematical proofs. This application has been implemented under the form of a tool, named CLEAR (Constructing and evaLuating dEductive mAthematical pRoofs), designed for students that take mathematics and logics courses. It allows students to build deductive proofs collaboratively using a structured argumentative debate and allows teachers to evaluate these proofs. A light structure is used for modelling the logical arguments: a pair (Δ, α) such that α is a conclusion safely obtained from Δ. Then, the classical notions of rebuttal and undercutting can be used in order to define the attacks (see [Baroni et al., 2018b, Chapter 9]). The higher-order framework used is the one presented in [Cayrol et al., 2016] (see Section 5.2.1), with reversed supports, since the meaning of the support relation used in the tool is the deductive one, whereas [Cayrol et al., 2016] uses the necessary one. Note that the duality between deductive and necessary support can be used in CLEAR since the support relation cannot target another interaction (Definition 12 in [Boudjani et al., 2020]).

So the support relation stands for deduction, and the attack (defeat) relation stands for conflict, this last one being a higher-order relation (targets can be arguments or other relations). Moreover, the tool gives the possibility to aggregate two or more arguments in order to create a "collective support" to another argument.

The following example illustrates the kind of argumentation framework we can build with CLEAR.

Example 14 (from [Boudjani et al., 2020]). *Consider the following theorem that must be proven: "Let ABC be a right triangle in A. Consider that $AB = 4$ and $BC = 5$ and prove that $AC = 3$."*

The following propositions are available in order to build this proof:

ABC is a right triangle in A	$AB^2 = BC^2 + AC^2$
$BC^2 = AC^2 + AB^2$	$AC^2 = BC^2 - AB^2$
$AB^2 = 16$	$BC^2 = 25$
$AC^2 = 9$	$BC = 5$
$AB = 4$	$AC = 3$

The (simplified) debate between students is shown below. On the one hand, some informal arguments are given. On the other hand, the debate reflects the exchanges between students about the building of the proof; hence, some arguments, deductions or attacks are sometimes "surprising":

a_1: *If ABC is a right triangle in A then $BC^2 = AC^2 + AB^2$.*

a_2: *No, if ABC is a right triangle in A then $AB^2 = BC^2 + AC^2$ (and so a_2 attacks a_1).*

a_{inf1}: *(informal argument) Argument a_2 is false (and so a_{inf1} attacks a_2).*

a_{inf3}: *(informal argument) a_2 cannot attack a_1 since a_1 is correct. This relation must be removed. So a_{inf3} attacks the attack from a_2 to a_1.*

a_3: *(deduced from a_1)* If $BC^2 = AC^2 + AB^2$ then $AC^2 = BC^2 - AB^2$.

a_{inf2}: *(informal argument)* Applying the Pythagorean Theorem, $BC^2 = AC^2 + AB^2$ *(that gives another way for deducing a_3).*

a_{inf5}: *(informal argument)* a_{inf2} is redundant with a_1 *(and so a_{inf5} attacks a_{inf2}).*

a_4: If $AC^2 = BC^2 - AB^2$ and $AB^2 = 16$ and $BC^2 = 25$ then $AC^2 = 9$. Moreover a_4 can be deduced from a_2.

a_{inf4}: *(informal argument)* No, a_2 does not allow the deduction of a_4. This relation must be removed. So a_{inf4} attacks the support from a_2 to a_4.

a_5: if $BC = 5$ then $BC^2 = 25$.

a_6: if $AB = 4$ then $AB^2 = 16$. And so the aggregation of a_3, a_5 and a_6 allows the deduction of a_4.

a_7: *(deduced from a_4)* If $AC^2 = 9$ then $AC = 3$.

An additional argument a_8 can be created in order to represent the aggregation of a_3, a_5 and a_6 that must be used together for deducing a_4 (joint support).

And so the corresponding higher-order argumentation framework with deductive supports can be represented as follows:

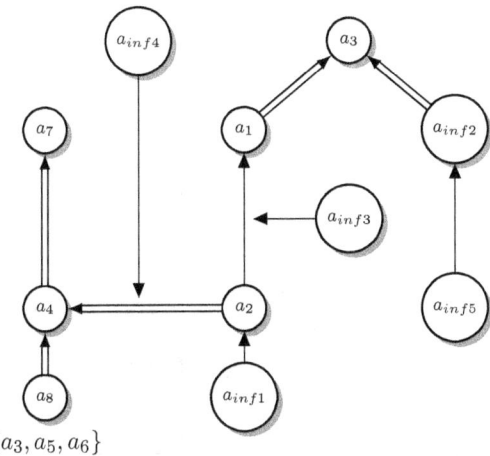

Note that the link between a_3 and a_8 does not really appear in the graph (but it is recorded in the tool and can be used in the final steps). In the same way, a_5 and a_6 are not depicted in the figure; they are isolated and their only role is to be involved in the creation of a_8. The corresponding $ASAF$ can be represented with the same graph in which the direction of the support edges has been reversed. And then, using the $ASAF$ semantics, the preferred extensions can be computed. Of course, we are interested in the extensions that contain argument a_7 that corresponds to the conclusion we try to prove. These extensions will be used for building "proof graphs" for a_7, and then these graphs will be presented to the teachers for evaluation and discussion.

Here, a proof graph for a_7 is: from a_1 one can deduce a_3; from a_3, a_5 and a_6 one can deduce a_4; from a_4 one can deduce a_7.

7 Analysis

In this chapter, at least ten different approaches are presented and, clearly, a synthetic point of view is needed. This is the aim of this section.

Approach	attack order	support order (0 means no support)	support type: deductive (D) necessary (N), evidential (E), undefined (U)	collective attack (source is a set)	collective support (source is a set)	disjunctive interaction (target is a set)	source can be an interaction
[Barringer et al., 2005]	any	any	U	no	no	no	no
EAF	2	0	NC	no	no	no	no
[Hanh et al., 2011]	any	0	NC	yes	yes	yes	yes
[Gabbay, 2009b]	any	0	NC	yes	yes	yes	yes
AFRA	any	0	NC	no	no	no	no
RAF	any	0	NC	no	no	no	no
[Boella et al., 2010]	2	1	D	no	no	no	yes
ASAF	any	any	N	no	no	no	no
RAFN	any	any	N	no	yes	no	no
REBAF	any	any	E	yes	yes	no	no

Table 1: Interactions taken into account (*NC* means "Not Concerned")

First, Table 1 describes the kind of interaction that is taken into account by the different approaches. As we can see in this table, many different possibilities exist regarding the type of interaction (attacks and/or supports, with at least 4 "types" of support), regarding the "order" of these interactions (no interaction, first-order, second-order, and at any level), and regarding the form of their source or their target (one element or a set of elements). No approach is general enough to take into account all these possibilities. But in fact, the question arises whether it would be interesting to have such an approach.

Then, for each approach (except from [Barringer et al., 2005], which

Approach	Directly	After transformation into an AF
[Barringer et al., 2005]	No semantics definition	
EAF	yes	yes (meta-arg+flattening)
[Hanh et al., 2011]	yes	no
[Gabbay, 2009b]	yes	yes (meta-arg+flattening)
$AFRA$	yes	yes (meta-arg+flattening)
RAF	yes	no
[Boella et al., 2010]	no	yes (meta-arg+flattening)
$ASAF$	yes	yes (meta-arg+flattening)
$RAFN$	yes	no
$REBAF$	yes	no

Table 2: Semantics definition

does not propose semantics), Table 2 gives the method followed for defining semantics, whereas Table 3 presents the type of results produced by these semantics. The interesting point here is the fact that almost all approaches have developed semantics in a direct way, some of them also proposing a transformation of their framework into a Dung-like meta-argumentation framework. This transformation facilitates the understanding of the framework and the use of the existing solvers in computational issues. So, for the approaches that do not propose this transformation, it could perhaps be interesting to identify the associated meta-argumentation framework. Concerning the semantics results, four alternatives exist and our personal opinion is that, since interactions can be attacked or supported, they should also appear as outputs of the semantics.

Table 4 synthesizes the links between all approaches answering to the question: Who extends who? Clearly, all the proposed frameworks extend Dung's framework; then, they differ, either on the type of support that is taken into account, or on the way in which they take into account the higher-order attacks. So, three distinct families appear:

- the first one is issued from the seminal work [Barringer et al., 2005] and the EAF,

Approach	Set of			
	arguments	labellings arg+int	arguments +interactions	arguments + meta-arg
[Barringer et al., 2005]	No semantics definition			
EAF	yes	yes	no	yes
[Hanh et al., 2011]	yes	no	no	no
[Gabbay, 2009b]	no	yes	no	yes
$AFRA$	no	no	yes	yes
RAF	no	no	yes	no
[Boella et al., 2010]	no	no	no	yes
$ASAF$	no	yes	yes	yes
$RAFN$	no	no	yes	no
$REBAF$	no	no	yes	no

Table 3: Semantics output, defined in terms of a set of: arguments, labellings of arguments, arguments + interactions or arguments + meta-arguments

- the second family follows Baroni et al's work around the $AFRA$,

- and the last family follows Cayrol et al's work with the notion of RAF.

Note that the first two families also follow ideas of the meta-argumentation approach.

And finally, Table 5 lists the known links between these approaches in terms of their semantics (and the associated properties). Considering two approaches i and j, several links can exist:

1 - Higher-Order Interactions in Abstract Argumentation

	AF	BAF deductive (D), necessary (N), evidential (E), undefined interpretation (U)	[Barringer et al., 2005]	EAF	[Hanh et al., 2011]	[Gabbay, 2009b]	AFRA	RAF	[Boella et al., 2010]	ASAF	RAFN
[Barringer et al., 2005]	✓	✓(U)									
EAF	✓										
[Hanh et al., 2011]	✓			✓							
[Gabbay, 2009b]	✓			✓							
AFRA	✓										
RAF	✓										
[Boella et al., 2010]	✓	✓(D)	✓								
ASAF	✓	✓(N)						✓			
RAFN	✓	✓(N)							✓		
REBAF	✓	✓(E)							✓		

Let i (resp. j) be the approach given on the line (resp. column). $i \checkmark j$ means that i is an extended argumentation framework issued from j.

Table 4: Who extends who?

- First i is a conservative generalization of j (denoted by $i \triangleright j$). That corresponds to the fact that, when the approach i is used on an argumentation framework corresponding to the approach j, all results are strictly identical (for all notions involved in semantics: conflict-freeness, acceptability, admissibility, ...). So this link appears only when i is an extension of j. However, it can be the case that i extends j but i is not a conservative generalization of j (see e.g. the relationship between AFRA and AF).

- Other links can appear between two approaches, allowing to relate frameworks of the same kind (e.g. two approaches using a set of arguments and a set of higher-order attacks) or frameworks with a different structure (e.g. one approach that considers a set of arguments and higher-order attacks and supports, and another considering a set of arguments and first-order attacks and supports). We identify four cases:

- either there is a complete one-to-one correspondence between the approaches i and j (denoted by $i = j$): i and j give exactly the same results (for all notions involved in semantics: conflict-freeness, acceptability, admissibility, ...); in this case, i and j are applied on the same data, whereas in the case of $\triangleright$, the link exists only if i is applied on the more simple data corresponding to the scope of j;
- or there is a partial one-to-one correspondence between the approaches i and j (denoted by $i \approx j$): i and j give the same results (for all notions involved in semantics: conflict-freeness, acceptability, admissibility, ...) if we consider *some constraints either on i, or on j*;
- or there is a partial one-to-one correspondence between the approaches i and j *but only at the semantics level* (denoted by $i \sim j$): i and j give the same results if we only consider semantics such as complete, grounded, preferred or stable (and so the results differ when we consider some other notions as, for instance, conflict-freeness or acceptability); sometimes, some constraints must also to be considered (for instance, if i is applied on the more simple data corresponding to the scope of j);
- or there is no one-to-one correspondence between the approaches i and j (denoted by $i \neq j$): i and j do not give the same results for some semantics, even if some constraints are given on i or j.

Of course, there is also the trivial case that no link can exist between i and j only because they correspond to frameworks of different nature (for instance i takes into account evidential supports, whereas j takes into account necessary supports). This case will be denoted by NC ("Not Concerned") in Table 5. Here the main point is that, even if some links have already been established, a lot of work remains to be done in order to completely compare all these approaches. Note also that no approach is strictly equivalent to another one (there is no i, j such that $i = j$). That means that each approach has its own peculiarities and meets special needs. That also explains why it is difficult to unify these approaches.

	AF	BAF deductive (D), necessary (N), evidential (E)	[Barringer et al., 2005]	EAF	[Hanh et al., 2011]	[Gabbay, 2009b]	AFRA	RAF	[Boella et al., 2010]	ASAF	RAFN
[Barringer et al., 2005]		No semantics									
EAF	▷	NC									
[Hanh et al., 2011]	▷	NC	No semantics	≠							
[Gabbay, 2009b]	▷	NC		≠	≠						
AFRA	~	NC		≠	≠	≠					
RAF	▷	NC					~				
[Boella et al., 2010]	▷	▷ (D)		▷		≠	≠	≠			
ASAF	~	~ (N)					▷	~	NC		
RAFN	▷	▷ (N)					~	▷	NC	≠	
REBAF	▷	≈ (E)					~	▷	NC	NC	NC

Let i (resp. j) be the approach given on the line (resp. column).

- $i \triangleright j$ means that i is a conservative generalization of j (i gives exactly the same results that j when we consider the restriction of i to j).

- $i \approx j$ means that there exists a one-to-one correspondence, but with some constraints (depending of the case).

- $i \sim j$ means that there exists a one-to-one correspondence, but only at the semantics level (sometimes with constraints).

- $i \neq j$ means that no one-to-one correspondence can exist between i and j.

- NC means "Not Concerned".

Table 5: Links between approaches in terms of semantics and associated properties

8 Conclusion

It is now time to conclude this long chapter. Its aim was to propose a state of the art on higher-order abstract bipolar argumentation frameworks, *i.e.* abstract argumentation frameworks that allow interactions targeting other interactions, these interactions being either attacks or supports. This survey is as exhaustive as possible, but, since this topic is currently a very hot topic, it is possible, even probable, that some works are missing.

Nevertheless, considering all the works presented here, we can at least conclude on some points:

- the study of higher-order interactions (bipolar or not) in abstract argumentation is clearly an important topic since it allows an enriched representation of knowledge;

- many distinct approaches addressing this topic were proposed since the seminal work published in [Barringer *et al.*, 2005];

- these approaches can be partitioned into a smaller number of "families";

- even if there exist some links between these families, it is not so simple to unify them into a single general approach because they address different needs and use different methods, or even adopt different interpretations for the notion of support;

- and so, a lot of work remains to be done in this topic: for an eventual unification, but also for computational issues (study of complexity, algorithms);

- in order to boost this last point, the introduction of some dedicated tracks in the ICCMA competition could be of great help.

Acknowledgements

This work was partially supported by Universidad Nacional del Sur under grant 24/N046.

References

[Alfano et al., 2017] Gianvincenzo Alfano, Sergio Greco, and Francesco Parisi. Efficient computation of extensions for dynamic abstract argumentation frameworks: An incremental approach. In *Proceedings of the Twenty-Sixth International Joint Conference on Artificial Intelligence, IJCAI 2017, Melbourne, Australia, August 19-25, 2017*, pages 49–55, 2017.

[Alfano et al., 2018] Gianvincenzo Alfano, Sergio Greco, and Francesco Parisi. A meta-argumentation approach for the efficient computation of stable and preferred extensions in dynamic bipolar argumentation frameworks. *Intelligenza Artificiale*, 12(2):193–211, 2018.

[Alfano et al., 2020] Gianvincenzo Alfano, Andrea Cohen, Sebastian Gottifredi, Sergio Greco, Francesco Parisi, and Guillermo R. Simari. Dynamics in abstract argumentation frameworks with recursive attack and support relations. In *Proc. of ECAI 2020 - 24th European Conference on Artificial Intelligence*, 2020.

[Amgoud and Cayrol, 2002] Leila Amgoud and Claudette Cayrol. A reasoning model based on the production of acceptable arguments. *Annals of Mathematics and Artificial Intelligence*, 34:197–216, 2002.

[Amgoud et al., 2000] Leila Amgoud, Nicolas Maudet, and Simon Parsons. Modelling dialogues using argumentation. In *Proc. of ICMAS*, pages 31–38, 2000.

[Amgoud et al., 2004] Leila Amgoud, Claudette Cayrol, and Marie-Christine Lagasquie-Schiex. On the bipolarity in argumentation frameworks. In *10th International Workshop on Non-Monotonic Reasoning (NMR 2004), Whistler, Canada, June 6-8, 2004, Proceedings*, pages 1–9, 2004.

[Arisaka and Satoh, 2017] Ryuta Arisaka and Ken Satoh. Voluntary manslaughter? a case study with meta-argumentation with supports. In *New Frontiers in Artificial Intelligence. JSAI-isAI 2016. LNCS 10247*, pages 241–252. Springer, 2017.

[Arisaka et al., 2019] Ryuta Arisaka, Francesco Santini, and Stefano Bistarelli. Block argumentation. In Matteo Baldoni, Mehdi Dastani, Beishui Liao, Yuko Sakurai, and Rym Zalila-Wenkstern, editors, *PRIMA 2019: Principles and Practice of Multi-Agent Systems - 22nd International Conference, Turin, Italy, October 28-31, 2019, Proceedings*, volume 11873 of *Lecture Notes in Computer Science*, pages 618–626. Springer, 2019.

[Baroni et al., 2009] Pietro Baroni, Federico Cerutti, Massimiliano Giacomin, and Giovanni Guida. Encompassing attacks to attacks in abstract argumentation frameworks. In *Symbolic and Quantitative Approaches to Reasoning with Uncertainty, 10th European Conference, ECSQARU 2009, Verona,*

Italy, July 1-3, 2009. Proceedings, pages 83–94, 2009.

[Baroni et al., 2011] Pietro Baroni, Federico Cerutti, Massimiliano Giacomin, and Giovanni Guida. AFRA: Argumentation framework with recursive attacks. *International Journal of Approximate Reasoning*, 52:19–37, 2011.

[Baroni et al., 2018a] Pietro Baroni, Martin Caminada, and Massimiliano Giacomin. Abstract argumentation frameworks and their semantics. In Pietro Baroni, Dov Gabbay, Massimiliano Giacomin, and Leendert van der Torre, editors, *Handbook of Formal Argumentation*, chapter 4, pages 159–236. College Publications, 2018.

[Baroni et al., 2018b] Pietro Baroni, Dov Gabbay, and Massimiliano Giacomin. *Handbook of Formal Argumentation*. College Publications, 2018.

[Barringer et al., 2005] Howard Barringer, Dov Gabbay, and John Woods. Temporal dynamics of support and attack networks : From argumentation to zoology. In Dieter Hutter and Werner Stephan, editors, *Mechanizing Mathematical Reasoning, Essays in Honor of Jörg H. Siekmann on the Occasion of His 60th Birthday. LNAI 2605*, pages 59–98. Springer Verlag, 2005.

[Bench-Capon, 2003] Trevor Bench-Capon. Persuasion in practical argument using value-based argumentation frameworks. *Journal of Logic and Computation*, 13(3):429–448, 2003.

[Boella et al., 2008a] Guido Boella, Leendert W. N. van der Torre, and Serena Villata. Attack relations among dynamic coalitions. In *Proceedings of the 20th Belgian-Dutch Artificial Intelligence Conference, BNAIC 2008*, pages 25–32. Universiteit Twente, Enschede, 2008.

[Boella et al., 2008b] Guido Boella, Leendert W. N. van der Torre, and Serena Villata. Social viewpoints for arguing about coalitions. In The Duy Bui, Tuong Vinh Ho, and Quang-Thuy Ha, editors, *Intelligent Agents and Multi-Agent Systems, 11th Pacific Rim International Conference on Multi-Agents, PRIMA 2008, Hanoi, Vietnam, December 15-16, 2008. Proceedings*, volume 5357 of *Lecture Notes in Computer Science*, pages 66–77. Springer, 2008.

[Boella et al., 2009a] Guido Boella, Dov Gabbay, Leendert van der Torre, and Serena Villata. Meta-argumentation modelling I: methodology and techniques. *Studia Logica*, 93(2-3):297–355, 2009.

[Boella et al., 2009b] Guido Boella, Leendert W. N. van der Torre, and Serena Villata. On the acceptability of meta-arguments. In *Proceedings of the 2009 IEEE/WIC/ACM International Conference on Intelligent Agent Technology, IAT 2009, Milan, Italy, 15-18 September 2009*, pages 259–262. IEEE Computer Society, 2009.

[Boella et al., 2010] Guido Boella, Dov Gabbay, Leendert van der Torre, and Serena Villata. Support in abstract argumentation. In Pietro Baroni, Federico Cerutti, Massimiliano Giacomin, and Guillermo Ricardo Simari, editors, *Computational Models of Argument: Proceedings of COMMA 2010, Desenzano del Garda, Italy, September 8-10, 2010.*, volume 216 of *Frontiers in Artificial Intelligence and Applications*, pages 111–122. IOS Press, 2010.

[Boudhar et al., 2012] Imane Boudhar, Farid Nouioua, and Vincent Risch. Handling preferences in argumentation frameworks with necessities. In *ICAART 2012 - Proceedings of the 4th International Conference on Agents and Artificial Intelligence, Volume 1 - Artificial Intelligence, Vilamoura, Algarve, Portugal, 6-8 February, 2012*, pages 340–345, 2012.

[Boudjani et al., 2018] Nadira Boudjani, Abdelkader Gouaïch, and Souhila Kaci. CLEAR: argumentation frameworks for constructing and evaluating deductive mathematical proofs. In Sanjay Modgil, Katarzyna Budzynska, and John Lawrence, editors, *Computational Models of Argument - Proceedings of COMMA 2018, Warsaw, Poland, 12-14 September 2018*, volume 305 of *Frontiers in Artificial Intelligence and Applications*, pages 281–288. IOS Press, 2018.

[Boudjani et al., 2020] Nadira Boudjani, Abdelkader Gouaich, and Souhila Kaci. Argumentation frameworks for constructing, representing and evaluating deductive mathematical proofs. Technical Report lirmm-02458939, LIRMM, 2020.

[Brewka et al., 2018] Gerhard Brewka, Stefan Ellmauthaler, Hannes Strass, Johannes P. Wallner, and Stefan Woltran. Abstract dialectical frameworks. In Pietro Baroni, Dov Gabbay, Massimiliano Giacomin, and Leendert van der Torre, editors, *Handbook of Formal Argumentation*, chapter 5, pages 237–285. College Publications, 2018.

[Cabrio and Villata, 2013] Elena Cabrio and Serena Villata. Detecting bipolar semantic relations among natural language arguments with textual entailment: a study. In *Proceedings of the Joint Symposium on Semantic Processing. Textual Inference and Structures in Corpora, JSSP 2013, Trento, Italy, November 20-22, 2013*, pages 24–32, 2013.

[Caminada and Amgoud, 2007] Martin Caminada and Leila Amgoud. On the evaluation of argumentation formalisms. *Artificial Intelligence*, 171(5-6):286–310, 2007.

[Caminada and Gabbay, 2009] Martin Caminada and Dov Gabbay. A logical account of formal argumentation. *Studia Logica*, 93(2-3):109–145, 2009.

[Caminada, 2006] Martin Caminada. On the issue of reinstatement in argumentation. In *Proc. of the European Conference of Logics in Artificial Intelligence (JELIA)*, pages 111–123, 2006.

[Cayrol and Lagasquie-Schiex, 2005] Claudette Cayrol and Marie-Christine Lagasquie-Schiex. On the acceptability of arguments in bipolar argumentation frameworks. In *Symbolic and Quantitative Approaches to Reasoning* with Uncertainty, 8th European Conference, ECSQARU 2005, Barcelona, Spain, July 6-8, 2005, Proceedings, pages 378–389, 2005.

[Cayrol and Lagasquie-Schiex, 2007] Claudette Cayrol and Marie-Christine Lagasquie-Schiex. Coalitions of arguments in bipolar argumentation frameworks. In *7th International Workshop on Computational Models of Natural Argument*, pages 14–20, 2007.

[Cayrol and Lagasquie-Schiex, 2008] Claudette Cayrol and Marie-Christine Lagasquie-Schiex. The GRAFIX website, 2008. http://www.irit.fr/grafix.

[Cayrol and Lagasquie-Schiex, 2009] Claudette Cayrol and Marie-Christine Lagasquie-Schiex. Bipolar abstract argumentation systems. In Iyad Rahwan and Guillermo Simari, editors, *Argumentation in Artificial Intelligence*, pages 65–84. Springer, 2009.

[Cayrol and Lagasquie-Schiex, 2010] Claudette Cayrol and Marie-Christine Lagasquie-Schiex. Coalitions of arguments: A tool for handling bipolar argumentation frameworks. *International Journal of Intelligent Systems*, 25(1):83–109, 2010.

[Cayrol and Lagasquie-Schiex, 2013] Claudette Cayrol and Marie-Christine Lagasquie-Schiex. Bipolarity in argumentation graphs: Towards a better understanding. *Int. J. Approx. Reasoning*, 54(7):876–899, 2013.

[Cayrol and Lagasquie-Schiex, 2018] Claudette Cayrol and Marie-Christine Lagasquie-Schiex. Logical encoding of argumentation frameworks with higher-order attacks. In *Proc. of the 30th International Conference on Tools with Artificial Intelligence (ICTAI)*. IEEE, 2018.

[Cayrol and Lagasquie-Schiex, 2020] Claudette Cayrol and Marie-Christine Lagasquie-Schiex. Logical Encoding of Argumentation Frameworks with Higher-order Attacks and Evidential Supports. *International Journal on Artificial Intelligence Tools*, 29(03n04):2060003, June 2020.

[Cayrol et al., 2016] Claudette Cayrol, Andrea Cohen, and Marie-Christine Lagasquie-Schiex. Towards a new framework for recursive interactions in abstract bipolar argumentation. In Pietro Baroni, editor, *Proceedings of International Conference on Computational Models of Argument (COMMA)*, pages 191–198. IOS Press, 2016.

[Cayrol et al., 2017] Claudette Cayrol, Jorge Fandinno, Luis Fariñas del Cerro, and Marie-Christine Lagasquie-Schiex. Valid attacks in argumentation frameworks with recursive attacks. In *Proc. of Thirteenth International Symposium on Commonsense Reasoning*, volume 2052. CEUR Workshop Proceedings, 2017.

[Cayrol et al., 2018a] Claudette Cayrol, Jorge Fandinno, Luis Fariñas del Cerro, and Marie-Christine Lagasquie-Schiex. Argumentation frameworks with recursive attacks and evidence-based support. In F. Ferrarotti and

S. Woltran, editors, *Proc. of Tenth International Symposium on Foundations of Information and Knowledge Systems (FoIKS)*, volume LNCS 10833, pages 150–169. Springer-Verlag, 2018.

[Cayrol et al., 2018b] Claudette Cayrol, Jorge Fandinno, Luis Fariñas del Cerro, and Marie-Christine Lagasquie-Schiex. Structure-based Semantics of Argumentation Frameworks with Higher-order Attacks and Supports. In Carlos I. Chesñevar et al, editor, *Argumentation-based Proofs of Endearment. Essays in Honor of Guillermo R. Simari on the Occasion of his 70 th Birthday*, volume 37 of *Tributes*, pages 43–72. College Publications, http://www.collegepublications.co.uk/, 2018.

[Cayrol et al., 2018c] Claudette Cayrol, Jorge Fandinno, Luis Fariñas del Cerro, and Marie-Christine Lagasquie-Schiex. Structure-Based Semantics of Argumentation Frameworks with Higher-Order Attacks and Supports (short paper). In Sanjay Modgil, Katarzyna Budzynska, and John Lawrence, editors, *International Conference on Computational Models of Argument (COMMA)*, pages 29–36. IOS Press, septembre 2018.

[Cayrol et al., 2020] Claudette Cayrol, Jorge Fandinno, Luis Fariñas del Cerro, and Marie-Christine Lagasquie-Schiex. Valid attacks in argumentation frameworks with recursive attacks. *Annals of Mathematics and Artificial Intelligence (Special Issue: Commonsense 2017)*, 2020.

[Cohen et al., 2011] Andrea Cohen, Alejandro Javier García, and Guillermo Ricardo Simari. Backing and undercutting in defeasible logic programming. In *Symbolic and Quantitative Approaches to Reasoning with Uncertainty - 11th European Conference, ECSQARU 2011, Belfast, UK, June 29-July 1, 2011. Proceedings*, pages 50–61, 2011.

[Cohen et al., 2012] Andrea Cohen, Alejandro Javier García, and Guillermo Ricardo Simari. Backing and undercutting in abstract argumentation frameworks. In *Foundations of Information and Knowledge Systems - 7th International Symposium, FoIKS 2012, Kiel, Germany, March 5-9, 2012. Proceedings*, pages 107–123, 2012.

[Cohen et al., 2014] Andrea Cohen, Sebastian Gottifredi, Alejandro Javier García, and Guillermo Ricardo Simari. A survey of different approaches to support in argumentation systems. *Knowledge Eng. Review*, 29(5):513–550, 2014.

[Cohen et al., 2015] Andrea Cohen, Sebastian Gottifredi, Alejandro J García, and Guillermo R Simari. An approach to abstract argumentation with recursive attack and support. *Journal of Applied Logic*, 13(4):509–533, 2015.

[Cohen et al., 2016] Andrea Cohen, Sebastian Gottifredi, Alejandro J. García, and Guillermo R. Simari. On the acceptability semantics of argumentation frameworks with recursive attack and support. In Pietro Baroni, editor, *Pro-

ceedings of International Conference on Computational Models of Argument (COMMA), pages 231–242. IOS Press, 2016.

[Cohen et al., 2018] Andrea Cohen, Simon Parsons, Elizabeth I. Sklar, and Peter McBurney. A characterization of types of support between structured arguments and their relationship with support in abstract argumentation. *Int. J. Approx. Reason.*, 94:76–104, 2018.

[Cyras et al., 2017] Kristijonas Cyras, Claudia Schulz, and Francesca Toni. Capturing bipolar argumentation in non-flat assumption-based argumentation. In *PRIMA 2017: Principles and Practice of Multi-Agent Systems - 20th International Conference, Nice, France, October 30 - November 3, 2017, Proceedings*, pages 386–402, 2017.

[Cyras et al., 2018] Kristijonas Cyras, Xiuyi Fan, Claudia Schulz, and Francesca Toni. Assumption-based argumentation: Disputes, explanations, preferences. In Pietro Baroni, Dov Gabbay, Massimiliano Giacomin, and Leendert van der Torre, editors, *Handbook of Formal Argumentation*, chapter 7, pages 365–408. College Publications, 2018.

[Dauphin and Cramer, 2017] Jérémie Dauphin and Marcos Cramer. Extended explanatory argumentation frameworks. In *Theory and Applications of Formal Argumentation - 4th International Workshop, TAFA 2017, Melbourne, VIC, Australia, August 19-20, 2017, Revised Selected Papers*, pages 86–101, 2017.

[Dung and Thang, 2009] Phan Minh Dung and Phan Minh Thang. A unified framework for representation and development of dialectical proof procedures in argumentation. In *IJCAI 2009, Proceedings of the 21st International Joint Conference on Artificial Intelligence, Pasadena, California, USA, July 11-17, 2009*, pages 746–751, 2009.

[Dung, 1995] Phan Minh Dung. On the acceptability of arguments and its fundamental role in nonmonotonic reasoning, logic programming and n-person games. *Artificial Intelligence*, 77(2):321–358, 1995.

[Dvorak et al., 2015] Wolfgang Dvorak, Sarah A. Gaggl, Thomas Linsbichler, and Johannes P. Wallner. Reduction-based approaches to implement Modgil's extended argumentation frameworks. In Miroslaw Truszczynski Thomas Eiter, Hannes Strass and Stefan Woltran, editors, *Advances in Knowledge Representation, Logic Programming, and Abstract Argumentation. Essays Dedicated to Gerhard Brewka on the Occasion of His 60th Birthday (LNCS, volume 9060)*, pages 249–264. Springer, 2015.

[Egly et al., 2010] Uwe Egly, Sarah Alice Gaggl, and Stefan Woltran. Answer-set programming encodings for argumentation frameworks. *Argument & Computation*, 1(2):147–177, 2010.

[Field, 2008] Hartry Field. *Saving Truth from Paradoxes*. Oxford University

PRess, 2008.

[Gabbay and Gabbay, 2016] Dov Gabbay and Michael Gabbay. The attack as intuitionistic negation. *Logic Journal of the IGPL*, 24(5):807–837, 2016.

[Gabbay, 2009a] Dov Gabbay. Fibring argumentation frames. *Studia Logica*, 93:231–295, 2009.

[Gabbay, 2009b] Dov Gabbay. Semantics for higher level attacks in extended argumentation frames. *Studia Logica*, 93:357–381, 2009.

[Gaggl et al., 2020] Sarah Alice Gaggl, Thomas Linsbichler, Marco Maratea, and Stefan Woltran. Design and results of the second international competition on computational models of argumentation. *Artif. Intell.*, 279, 2020.

[García and Simari, 2018] Alejandro Javier García and Guillermo Ricardo Simari. Argumentation based on logic programming. In Pietro Baroni, Dov Gabbay, Massimiliano Giacomin, and Leendert van der Torre, editors, *Handbook of Formal Argumentation*, chapter 8, pages 409–435. College Publications, 2018.

[Gottifredi et al., 2018] Sebastian Gottifredi, Andrea Cohen, Alejandro Javier García, and Guillermo Ricardo Simari. Characterizing acceptability semantics of argumentation frameworks with recursive attack and support relations. *Artificial Intelligence*, 262:336–368, 2018.

[Hanh et al., 2011] Do Duc Hanh, Phan Minh Dung, Nguyen Duy Hung, and Phan Minh Thang. Inductive defense for sceptical semantics of extended argumentation. *Journal of Logic and Computation*, 21(2):307–349, 2011.

[Karacapilidis and Papadias, 2001] Nikos Karacapilidis and Dimitris Papadias. Computer supported argumentation and collaborative decision making: the HERMES system. *Information systems*, 26(4):259–277, 2001.

[Kawasaki and Takahashi, 2018] Tastuki Kawasaki and Sosuke Moriguchi and Kazuko Takahashi. Transformation from PROLEG to a bipolar argumentation framework. In *2nd International Workshop on Systems and Algorithms for Formal Argumentation (SAFA 2018), Warsaw, Poland, September*, pages 36–47, 2018.

[Li and Wu, 2019] Hengfei Li and Jiachao Wu. Semantics of extended argumentation frameworks defined by renovation sets. In *PRIMA 2019: Principles and Practice of Multi-Agent Systems - 22nd International Conference, Turin, Italy, October 28-31, 2019, Proceedings*, pages 532–540, 2019.

[Marek and Truszczyński, 1999] Victor W. Marek and Mirosław Truszczyński. Stable models and an alternative logic programming paradigm. In *The Logic Programming Paradigm – A 25-Year Perspective*, pages 375–398. Springer, 1999.

[Modgil and Bench-Capon, 2008] Sanjay Modgil and Trevor Bench-Capon. Integrating object and meta-level value based argumentation. In *Proc of International Conference on Computational models of arguments (COMMA)*, pages 240–251, 2008.

[Modgil and Bench-Capon, 2011] Sanjay Modgil and Trevor J. M. Bench-Capon. Metalevel argumentation. *Journal of Logic and Computation*, 21(6):959–1003, 2011.

[Modgil and Prakken, 2010] Sanjay Modgil and Henry Prakken. Reasoning about preferences in structured extended argumentation frameworks. In *Computational Models of Argument: Proceedings of COMMA 2010, Desenzano del Garda, Italy, September 8-10, 2010.*, pages 347–358, 2010.

[Modgil and Prakken, 2018] Sanjay Modgil and Henry Prakken. Abstract rule-based argumentation. In Pietro Baroni, Dov Gabbay, Massimiliano Giacomin, and Leendert van der Torre, editors, *Handbook of Formal Argumentation*, chapter 6, pages 287–364. College Publications, 2018.

[Modgil, 2007] Sanjay Modgil. An abstract theory of argumentation that accommodates defeasible reasoning about preferences. In *Symbolic and Quantitative Approaches to Reasoning with Uncertainty, 9th European Conference, ECSQARU 2007, Hammamet, Tunisia, October 31 - November 2, 2007, Proceedings*, pages 648–659, 2007.

[Modgil, 2009a] Sanjay Modgil. Labellings and games for extended argumentation frameworks. In *IJCAI 2009, Proceedings of the 21st International Joint Conference on Artificial Intelligence, Pasadena, California, USA, July 11-17, 2009*, pages 873–878, 2009.

[Modgil, 2009b] Sanjay Modgil. Reasoning about preferences in argumentation frameworks. *Artificial Intelligence*, 173:901–934, 2009.

[Nielsen and Parsons, 2006] Søren Holbech Nielsen and Simon Parsons. A generalization of dung's abstract framework for argumentation: Arguing with sets of attacking arguments. In *Argumentation in Multi-Agent Systems, Third International Workshop, ArgMAS 2006, Hakodate, Japan, May 8, 2006, Revised Selected and Invited Papers*, pages 54–73, 2006.

[Niemelä, 1999] Ilkka Niemelä. Logic programming with stable model semantics as a constraint programming paradigm. *Ann. Math. Artif. Intell.*, 25(3-4):241–273, 1999.

[Nofal et al., 2014] Samer Nofal, Katie Atkinson, and Paul E. Dunne. Algorithms for argumentation semantics: Labeling attacks as a generalization of labeling arguments. *Journal of Artificial Intelligence Research*, 49:635–668, 2014.

[Nouioua and Risch, 2010] Farid Nouioua and Vincent Risch. Bipolar argumentation frameworks with specialized supports. In *22nd IEEE International Conference on Tools with Artificial Intelligence, ICTAI 2010, Arras,*

France, 27-29 October 2010 - Volume 1, pages 215–218, 2010.

[Nouioua and Risch, 2011] Farid Nouioua and Vincent Risch. Argumentation frameworks with necessities. In *Scalable Uncertainty Management - 5th International Conference, SUM 2011, Dayton, OH, USA, October 10-13, 2011. Proceedings*, pages 163–176, 2011.

[Nouioua, 2013] Farid Nouioua. AFs with necessities: Further semantics and labelling characterization. In *Scalable Uncertainty Management - 7th International Conference, SUM 2013, Washington, DC, USA, September 16-18, 2013. Proceedings*, pages 120–133, 2013.

[Oren and Norman, 2008] Nir Oren and Timothy J. Norman. Semantics for evidence-based argumentation. In P. Besnard, S. Doutre, and A. Hunter, editors, *Proc. of COMMA*, volume 172 of *Frontiers in Artificial Intelligence and Applications*, pages 276–284. IOS Press, 2008.

[Polberg and Hunter, 2018] Sylwia Polberg and Anthony Hunter. Empirical evaluation of abstract argumentation: Supporting the need for bipolar and probabilistic approaches. *Int. J. Approx. Reason.*, 93:487–543, 2018.

[Polberg and Oren, 2014] Sylwia Polberg and Nir Oren. Revisiting support in abstract argumentation systems. In *Computational Models of Argument - Proceedings of COMMA 2014, Atholl Palace Hotel, Scottish Highlands, UK, September 9-12, 2014*, pages 369–376, 2014.

[Polberg, 2016] Sylwia Polberg. Understanding the abstract dialectical framework (preliminary report). *CoRR*, abs/1607.00819, 2016.

[Polberg, 2017] Sylwia Polberg. Intertranslatability of abstract argumentation frameworks. Technical report, Technical Report DBAI-TR-2017–104, Institute for Information Systems, Technical University of Vienna, 2017.

[Prakken and Vreeswijk, 2002] Henry Prakken and Gerard Vreeswijk. Logics for defeasible argumentation. In D. Gabbay and F. Guenthner, editors, *Handbook of Philosophical Logic*, volume 4, pages 218–319. Kluwer Academic Pub., 2002.

[Prakken, 2010] Henry Prakken. An abstract framework for argumentation with structured arguments. *Argument & Computation*, 1(2):93–124, 2010.

[Prakken, 2014] Henry Prakken. On support relations in abstract argumentation as abstractions of inferential relations. In *ECAI 2014 - 21st European Conference on Artificial Intelligence, 18-22 August 2014, Prague, Czech Republic - Including Prestigious Applications of Intelligent Systems (PAIS 2014)*, pages 735–740, 2014.

[Rahwan and Simari, 2009] Iyad Rahwan and Guillermo R. Simari, editors. *Argumentation in Artificial Intelligence*. Springer, 2009.

[Satoh et al., 2010] Ken Satoh, Kento Asai, Takamune Kogawa, Masahiro

Kubota, Megumi Nakamura, Yoshiaki Nishigai, Kei Shirakawa, and Chiaki Takano. PROLEG: an implementation of the presupposed ultimate fact theory of japanese civil code by PROLOG technology. In *New Frontiers in Artificial Intelligence - JSAI-isAI 2010 Workshops, LENLS, JURISIN, AMBN, ISS, Tokyo, Japan, November 18-19, 2010, Revised Selected Papers*, pages 153–164, 2010.

[Seselja and Straßer, 2013] Dunja Seselja and Christian Straßer. Abstract argumentation and explanation applied to scientific debates. *Synthese*, 190(12):2195–2217, 2013.

[Thang et al., 2009] Phan Minh Thang, Phan Minh Dung, and Nguyen Duy Hung. Toward a common framework for dialectical proof procedure in abstract argumentation. *Journal of Logic and Computation*, 19(6):1071–1109, 2009.

[Thimm and Villata, 2017] Matthias Thimm and Serena Villata. The first international competition on computational models of argumentation: Results and analysis. *Artif. Intell.*, 252:267–294, 2017.

[Toulmin, 1958] Stephen E. Toulmin. *The Uses of Argument*. Cambridge University Press, 1958.

[University of Le Havre, 2011] University of Le Havre. GraphStream: A Dynamic Graph Library. http://graphstream-project.org/, 2011.

[Verheij, 2003] Bart Verheij. Deflog: on the logical interpretation of prima facie justified assumptions. *Journal of Logic and Computation*, 13(3):319–346, 2003.

[Villata et al., 2011] Serena Villata, Guido Boella, and Leendert W. N. van der Torre. Argumentation patterns. In Peter McBurney, Simon Parsons, and Iyad Rahwan, editors, *Proceeding of the Eighth International Workshop on Argumentation in Multi-Agent Systems, ArgMAS, Taipei, Taiwan, May 2011*, pages 133–150, 2011.

[Villata et al., 2012] Serena Villata, Guido Boella, Dov Gabbay, and Leendert van der Torre. Modelling defeasible and prioritized support in bipolar argumentation. *Annals of Mathematics and Artificial Intelligence*, 66(1-4):163–197, 2012.

CHAPTER 2

JOINT ATTACKS AND ACCRUAL IN ARGUMENTATION FRAMEWORKS

ANTONIS BIKAKIS
University College London (UCL), UK
a.bikakis@ucl.ac.uk

ANDREA COHEN
Institute for Computer Science and Engineering, CONICET-UNS
Dept. of Computer Science and Engineering, Universidad Nacional del Sur
Bahía Blanca, Argentina
ac@cs.uns.edu.ar

WOLFGANG DVOŘÁK
Institute of Logic and Computation, TU Wien, Austria
dvorak@dbai.tuwien.ac.at

GIORGOS FLOURIS
Foundation for Research and Technology - Hellas (FORTH), Greece
fgeo@ics.forth.gr

SIMON PARSONS
University of Lincoln, UK
sparsons@lincoln.ac.uk

Bikakis, Cohen, Dvořák, Flouris, Parsons

Abstract

While modelling arguments, it is often useful to represent "joint attacks", i.e., cases where multiple arguments jointly attack another (note that this is different from the case where multiple arguments attack another in isolation). Based on this remark, the notion of joint attacks has been proposed as a useful extension of classical Abstract Argumentation Frameworks, and has been shown to constitute a genuine extension in terms of expressive power. In this chapter, we review various works considering the notion of joint attacks from various perspectives, including abstract and structured frameworks. Moreover, we present results detailing the relation among frameworks with joint attacks and classical argumentation frameworks, computational aspects, and applications of joint attacks. Last but not least, we propose a roadmap for future research on the subject, identifying gaps in current research and important research directions.

1 Introduction

As many have already pointed out, the work of Dung [1995] is a cornerstone, arguably *the* cornerstone, of current work on computational argumentation. It was the work that introduced the notion of abstract argumentation and the idea that argumentation could be modelled just as a set of arguments and attacks between them, and it provided an initial set of semantics — complete, grounded, preferred and stable — for the evaluation of a set of arguments and attacks. As such, it is the work on which all subsequent work on abstract argumentation has been built. In addition, because many structured argumentation systems adopt the Dung semantics as a means of establishing which arguments are acceptable, these systems are also built upon [Dung, 1995].

Much of the appeal of [Dung, 1995] lies in its elegant simplicity. The approach relies on just two concepts — arguments and attacks — and yet these simple components can capture a complex range of types of reasoning, reflected in the large set of semantics that have been defined for abstract argumentation systems. However, this very simplicity means that abstract argumentation has limitations in terms of what it can represent. The limitations of representing arguments as atomic entities is widely recognised, and is addressed by work on structured argumentation[1]. However, there are also limitations in the way that [Dung, 1995] handles interactions between arguments. Attacks are binary, so that a given attack is from a single argument to a single argument. Attacks are also atomic in the sense that their impact is assessed independently of other attacks. To use the terminology of [Baroni et al., 2018a], an argument will be out as soon as it is attacked by a single in argument, regardless of any other attacks that may exist. The evaluation of an argument does not, even where arguments have different strengths, take account of whether there are multiple attacks on it. Where strengths are taken into account, it is, effectively, only the strongest attacker that matters.

These limitations, and in particular how they may be overcome, is the subject of this chapter. We are primarily interested in the extension of the [Dung, 1995] model of abstract argumentation to allow non-binary, or "joint" attacks. In particular, we consider the "sets of attacking arguments" (SETAF) approach first suggested in [Nielsen and Parsons, 2007b]. In this approach it is possible to model situations in which two or more arguments jointly attack a single argument, and we explore this approach in depth. This focus also leads us to consider bipolar argumentation frameworks, where joint attacks are a key element, and these frameworks, in turn, lead us to consider joint supports between arguments. We also briefly discuss how joint attacks might be modelled in structured argumentation, and touch on the rather neglected topic of *accrual*, which models situations in which the strength of sets of independent attacking arguments is an aggregate of the strengths of the arguments it contains.

The rest of this chapter is structured as follows. Section 2 motivates the study of joint attacks. Section 3 is perhaps the most central section of the chapter. It introduces the formal model of SETAFs, relates the model to classical abstract argumentation models, considers the computational aspects of SETAFs, and looks at alternative formulations for set-based attacks. Section 4 looks at the uses of joint attacks in bipolar

---

[1] In some systems of structured argumentation, ASPIC+ [Modgil and Prakken, 2018] for example, it is possible to cleanly "lift" a set of abstract arguments from a set of structured arguments in such a way that Dung-style semantics can be applied. In other systems, DeLP [García and Simari, 2018] for example, this is not possible.

argumentation frameworks, and considers the models of joint support that occur in those frameworks as well, while also discussing the use of joint attacks to model higher-order interactions. Section 5 then briefly covers the related topic of accrual, the combination of arguments for or against a given claim. Finally Section 6 looks at future lines of work on joint attacks, and Section 7 provides a brief summary and draws some conclusions.

2 Motivating the need for joint attacks

There are a number of possible motivations for work on joint attacks. One comes from a purely formal consideration of [Dung, 1995]. Therein, Dung considers argumentation frameworks that take the form of a directed graph, with nodes being arguments and edges being attacks between arguments. It is natural to consider a generalisation of these frameworks to ones where the directed graph becomes a directed hypergraph. In its most general form, such a framework would have nodes that represent sets of arguments, and edges that represent attacks between sets of nodes[2]. What we study here is a less general representation in which nodes represent single arguments, and edges represents attacks where the attackers can be a set, but the attackee is constrained to be a singleton. Though less general than the representation just sketched, this is, as we discuss below, a genuine extension of the Dung argumentation framework.

This representation can also be motivated by considering knowledge that is most elegantly represented in a formalism that allows for joint attacks. For example, taken from [Flouris and Bikakis, 2019], consider

---

[2][Nielsen and Parsons, 2007b] briefly considers explicitly representing the most general case of sets of arguments as both attacker and attackee, before settling on the SETAF formalism that we describe below. As [Nielsen and Parsons, 2007b] points out, SETAFs were originally devised during work that allowed arguments about Bayesian networks — work summarised in [Nielsen and Parsons, 2007a] — and not only is the SETAF approach able to capture attacks of sets of arguments on sets of arguments (by attacking each member of the attackee set separately), but it also mirrors the structure of a Bayesian network where conditional probability distributions capture multiple parents affecting a common child, but do not capture a single parent affecting multiple children.

the following aspects of the UK laws governing marriage and civil partnerships[3] (as of early 2020). One is not allowed to enter into a marriage or civil partnership if

(a) you are under 16;

(b) you are closely related to your partner;

(c) you are not single; or

(d) you are under 18 and do not have permission to marry from your parents or guardians.

Much of this can be represented in a standard Dung argumentation framework, with an argument to represent the right to get married or enter a civil partnership (M), which is attacked by arguments that represent being under 16 ($A16$), being closely related (R), and not being single (NS). One might also represent case (d) with a single argument, but this single argument captures both being under 18 *and* not having permission — let's call this argument MWP (for minor without permission). That is fine on its own, but now consider adding additional information about the UK legal system into the framework, [Flouris and Bikakis, 2019] again, this time on voting rights. In the UK you are allowed to vote[4], unless you are under 18, and the natural way to capture this is with an argument (V) representing the right to vote, which is attacked by an argument ($A18$) representing being under 18. How, then, do we capture the relationship between MWP, which incorporates the fact that the individual in question is under 18, and $A18$? We would argue that a natural and elegant way to do this is by replacing MWP by the argument NP, representing the fact that there is no parental permission, and having $A18$ and NP jointly attack M. The resulting SETAF is shown in Figure 1.

Just to make the point that this example of a joint attack is not contrived, Figure 1 contains some other arguments that are found in UK legislation and have a natural representation as a SETAF. (These all reflect the age of majority in the UK, which, as one might expect,

---

[3]https://www.gov.uk/marriages-civil-partnerships
[4]https://www.gov.uk/elections-in-the-uk

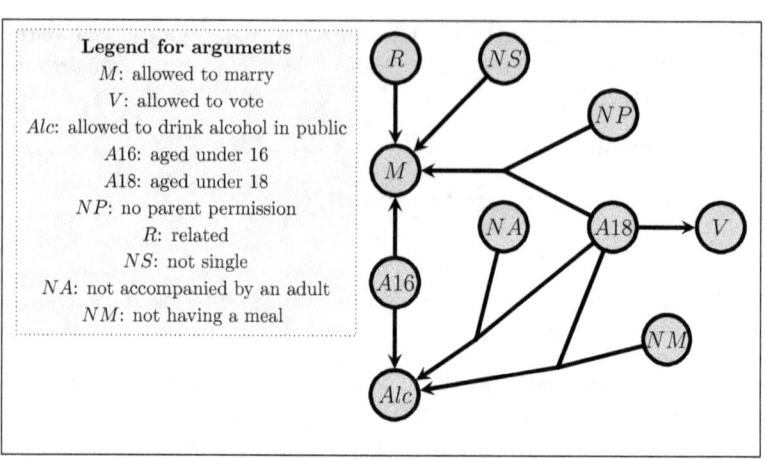

Figure 1: Example of a SETAF, encoding a part of UK legislation.

crops up a lot in the law.) For example, consider the law around alcohol consumption[5]. In the UK, one is allowed to consume alcohol in public (Alc), unless one is under 16, or one is under 18 and not accompanied by an adult (NA), or one is under 18 and not having a meal (NM).

Of course, we are not claiming that using joint attacks is the *only* way to represent the above information. As we mentioned, it is possible to capture all of this in a standard abstract argumentation framework, using what are effectively compound arguments such as "under 18 and not accompanied by an adult". Indeed, [Flouris and Bikakis, 2019] shows that it is always possible to represent a SETAF as a standard abstract argumentation framework, albeit at the cost of a possibly substantial increase in the number of arguments. In addition to this potential cost, a cost both representational and computational, we echo the sentiment expressed in [Nielsen and Parsons, 2007b], that using standard frameworks rather than SETAFs in cases like that of Figure 1 tends to muddle the distinction between arguments and attacks which is the essence of the abstract argumentation approach.

---

[5] https://www.gov.uk/alcohol-young-people-law

3 Modelling joint attacks

In this section, we provide formal considerations associated with the use of collective attacks in argumentation frameworks. These are meant to provide the basic tools towards further formal results on the issue.

More specifically, in Section 3.1, we provide the basic formal definitions associated with SETAFs, as well as their semantics (provided both in terms of extensions, and in terms of labellings). Moreover, a series of formal results on extensions, labellings and their relations are presented, most of which are a direct adaptation of similar results from the standard AF setting.

Section 3.2 studies the relationship among AF and SETAF, and provides answers to the fundamental question of whether SETAFs constitute a more expressive tool than AFs for describing arguments and their relationships.

Section 3.3 provides various computational complexity results related to SETAFs, for different problems pertaining to different semantics. Moreover, algorithms and system implementations that address these problems are considered, including reduction-based approaches.

Further, in Section 3.4 we discuss various alternative models of abstract and structured argumentation accounting for collective attacks, i.e., attacks where a group (i.e., set) of arguments can act either as the attacker, or as the attackee.

3.1 Definitions and semantics

We start our description with the formal definition of SETAFs, including their semantics. In this subsection, we formally describe various types of semantics that have been proposed in the literature, as well as relevant results that should form the formal background and toolbox of anyone aiming to study SETAFs and their properties.

3.1.1 AFs and AF semantics: A brief reminder

An AF was defined in [Dung, 1995] as a pair $AF^{\mathcal{D}} = \langle Ar, att \rangle$ consisting of a (possibly infinite) set of arguments Ar and a binary attack relation att on this set. In principle, an AF is a directed graph, whose nodes correspond to arguments and whose edges correspond to attacks, which essentially represent the fact that a certain argument invalidates another. AFs are given semantics through *extensions*, which are sets of arguments (nodes) that are non-conflicting (i.e., they do not attack each other) and, as a group, "shield" themselves from attacks by other arguments (which are not in the extension). The exact formal meaning given to the term "shield" gives rise to a multitude of different semantics (complete, preferred, stable, etc.) which have been considered in the literature (e.g., see [Baroni et al., 2018a]).

Informally, a set of arguments $S \subseteq Ar$ is: (i) a *conflict-free* extension of $AF^{\mathcal{D}}$ iff it contains no arguments attacking each other; (ii) an *admissible* extension iff it is conflict-free and defends all its elements (i.e., for each argument $a \in Ar$ attacking an argument in S, there is an argument in S attacking a); (iii) a *complete* extension iff it is admissible and contains all the arguments it defends; (iv) a *grounded* extension iff it is minimal (w.r.t. set inclusion) among the complete extensions; (v) a *preferred* extension iff it is maximal among the complete extensions; (vi) a *stable* extension iff it is conflict-free and attacks all the arguments that it does not contain (i.e., all arguments in $Ar \setminus S$); (vii) a *naive* extension iff it is maximal among the conflict-free extensions; $(viii)$ a *semi-stable* extension iff its union with the set of arguments it attacks is maximal among the complete extensions; (ix) an *eager* extension iff it is maximal among the complete extensions that are subsets of every semi-stable extension; (x) an *ideal* extension iff it is maximal among the complete extensions that are subsets of every preferred extension; and (xi) a *stage* extension iff its union with the set of arguments it attacks is maximal among the conflict-free extensions.

3.1.2 A formalism for joint attacks (SETAFs)

To formally represent the notion of joint attacks, Dung's definition for argumentation frameworks was extended in [Nielsen and Parsons, 2007b] for the case where an argument can be attacked by a set of other arguments:

Definition 3.1. *A* Framework with Sets of Attacking Arguments *(SETAF for short) is a pair* $AF^{\mathcal{S}} = \langle Ar, \triangleright \rangle$ *such that* Ar *is a set of arguments and* $\triangleright \subseteq (2^{Ar} \setminus \{\emptyset\}) \times Ar$ *is the* attack *relation.*

It is interesting to note the asymmetry in Definition 3.1: a group of arguments can be the attacker, but not the recipient of an attack. The reason for this asymmetry is justified in [Nielsen and Parsons, 2007b], where it is shown that allowing a set of arguments to be jointly attacked by another does not add to the expressiveness of the proposed model. Indeed, there can be two ways in which a many-to-many attack (say $\{a_1,\ldots,a_n\} \triangleright \{b_1,\ldots,b_m\}$) can be interpreted:

1. The first, called "collective defeat" in [Verheij, 1996a], states that no b_i is accepted whenever all of $a_1,\ldots,a_n$ are accepted. This case can be easily modelled in the setting of Definition 3.1 by creating m attacks of the form $\{a_1,\ldots,a_n\} \triangleright b_i$.

2. The second, called "indeterministic defeat" in [Verheij, 1996a], states that at least one of b_i should not be accepted whenever all of $a_1,\ldots,a_n$ are accepted. This case can also be modelled in the setting of Definition 3.1, by creating m attacks of the form $\{a_1,\ldots,a_n,b_1,\ldots,b_{i-1},b_{i+1},\ldots,b_m\} \triangleright b_i$.

Nevertheless, for simplicity, the attack relationship $\triangleright$ of Definition 3.1 can be extended to apply among sets of arguments. Formally, we say that a set of arguments S attacks another set of arguments T (denoted by $S \blacktriangleright T$) iff there exist $U \subseteq S, a \in T$ such that $U \triangleright a$. Note that we used a different symbol for the extended relation, to avoid confusion. Importantly, $\blacktriangleright$ does not change the semantics of the attack and does not generalise it to attacks among sets of arguments; it is just a syntactic shorthand.

We will write $S \not\triangleright a$ when it is not the case that $S \triangleright a$, and $S \not\blacktriangleright T$ when it is not the case that $S \blacktriangleright T$. For singleton sets, we often write $S \blacktriangleright a$ to denote $S \blacktriangleright \{a\}$. We say that S *defends* an argument a from a set of arguments T that attacks a, iff $S \blacktriangleright T$.

An interesting note for SETAFs, is that certain attacks may be redundant. In particular, if we have that $S \triangleright a$ and $S' \triangleright a$, for $S \subseteq S'$, then the latter attack is implied by the former and is thus redundant (can be removed from the AF^S without change of semantics). This is also evident from the definition of $\blacktriangleright$, which is, in a sense, the "closure" of $\triangleright$.

3.1.3 Semantics (extensions) for SETAFs

With regards to semantics, it is easy to extend the definitions provided for the AF setting (e.g., in [Dung, 1995; Baroni et al., 2018a]) so as to apply for the case of SETAFs (see [Nielsen and Parsons, 2007b; Flouris and Bikakis, 2019]). In all the following definitions, we consider a fixed SETAF $AF^S = \langle Ar, \triangleright \rangle$ and a set of arguments $S \subseteq Ar$.

Definition 3.2. *S is said to be* conflict-free *iff it does not attack itself. Formally, S is conflict-free iff $S \not\blacktriangleright S$.*

Definition 3.3. *An argument $a \in Ar$ is said to be* acceptable *with respect to S, iff S defends a from all attacking sets of arguments in Ar. Formally, a is acceptable with respect to S iff $S \blacktriangleright T$ for all $T \subseteq Ar$ such that $T \blacktriangleright a$. S is said to be* admissible *iff it is conflict-free and each argument in S is acceptable with respect to S. Formally, S is admissible iff $S \not\blacktriangleright S$ and $S \blacktriangleright T$ for all $T \subseteq Ar$ such that $T \blacktriangleright S$.*

In [Dung, 1995], a characteristic function F_{AF^D} was defined to return the arguments acceptable by a set of arguments in an argumentation framework AF^D. This can be easily extended for SETAF (say AF^S) as follows: $F_{AF^S} : 2^{Ar} \mapsto 2^{Ar}$, such that: $F_{AF^S}(S) = \{a \mid a \text{ is acceptable with respect to } S\}$.

Note that admissible extensions can (equivalently) be defined in terms of the characteristic function F_{AF^S} as any conflict-free set such that $S \subseteq F_{AF^S}(S)$.

Definition 3.4. *An admissible set S is called a* complete extension *of AF^S, iff all arguments that are acceptable with respect to S are in S. Formally, S is a complete extension of AF^S iff all the following conditions hold: (a) $S \not\blacktriangleright S$; (b) $S \blacktriangleright T$ for all $T \subseteq Ar$ such that $T \blacktriangleright S$; (c) If, for some $a \in Ar$, $S \blacktriangleright T$ for all $T \subseteq Ar$ such that $T \blacktriangleright a$, then $a \in S$.*

Obviously, complete extensions (of both AFs and SETAFs) can also be equivalently defined using the characteristic function: a conflict-free set S is a complete extension if and only if $F_{AF^S}(S) = S$.

Definition 3.5. *S is called a* preferred extension *of AF^S, iff it is a complete extension and there is no other complete extension T such that $S \subset T$.*

In other words, a preferred extension is a maximal complete extension. In the standard AF setting, it has been shown that preferred extensions can be equivalently defined as maximal admissible extensions (see, e.g., [Baroni et al., 2018a]). It can be easily shown that the same holds true in the SETAF setting [Dvořák et al., 2020c].

Definition 3.6. *S is called a* grounded extension *of AF^S, iff it is a complete extension and there is no other complete extension T such that $T \subset S$.*

Essentially, grounded extensions are minimal complete extensions. Following similar results in the AF setting ([Baroni et al., 2018a]) we can easily show that the following are equivalent also in the SETAF setting:

- S is a grounded extension
- S is the complete extension such that $\{a \in Ar \mid S \triangleright a\}$ is minimal
- S is the complete extension such that $Ar \setminus (S \cup \{a \in Ar \mid S \triangleright a\})$ is maximal

Using the characteristic function, another equivalent characterisation can be formulated, namely that S is a grounded extension if and only if it is the least fixed point of F_{AF^S} (see also [Dung, 1995]).

Definition 3.7. *S is called a* stable extension *of AF^S, iff it is conflict-free and attacks all arguments in $Ar \setminus S$.*

Equivalently, S is stable if and only if $S = \{a \mid S \not\blacktriangleright a\}$. Also, for a stable extension S it holds that $S \cup \{a \mid S \triangleright a\} = Ar$.

Moreover, we can easily show that stable extensions are also preferred, complete and admissible (see also [Flouris and Bikakis, 2019] and Figure 4), thus S is a stable extension if and only if S is a preferred, complete or admissible extension that attacks all arguments in $Ar \setminus S$.

Example 3.8. *Consider the SETAF shown in Figure 2, whose extensions are shown in Table 1. Let us consider in more detail the complete extensions, which are: $\emptyset$, $\{a_1\}$, $\{a_2, a_3, a_5\}$. Note that, for example, $\{a_2, a_3\}$ is admissible and conflict-free but not complete, because it leaves*

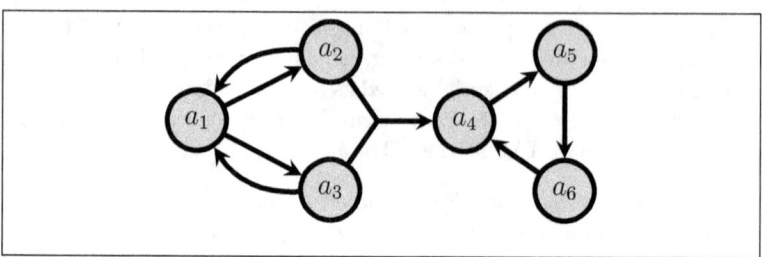

Figure 2: An example SETAF; set attacks are represented as arrows with multiple sources (e.g., $\{a_2, a_3\} \blacktriangleright a_4$); its extensions are shown in Table 1

out a_5, which is acceptable with respect to $\{a_2, a_3\}$. Similarly, $\{a_1, a_2\}$ is not a complete extension because it is not conflict-free, whereas $\{a_5\}$ and $\{a_1, a_5\}$ are not complete extensions because they are not admissible (a_5 is not acceptable with respect to the corresponding set in either case).
The minimal of the complete extensions (namely $\emptyset$) is also grounded, whereas the maximal ones ($\{a_1\}, \{a_2, a_3, a_5\}$) are also preferred. The latter ($\{a_2, a_3, a_5\}$) is also stable, because it attacks all other arguments. Looking at the SETAF illustrated in Figure 3, we note that it also has three complete extensions (namely, $\{a_1\}$, $\{a_1, a_2, a_5\}$, $\{a_1, a_3, a_4\}$), the first of which is also the grounded one ($\{a_1\}$), whereas the other two are the preferred ones $\{a_1, a_2, a_5\}$, $\{a_1, a_3, a_4\}$). However, there is no stable extension, because none of the complete extensions attacks all other arguments in the SETAF.

Definition 3.9. *S is called a* naive extension *of AF^S, iff it is conflict-free and is maximal w.r.t. set inclusion among the conflict-free subsets of Ar.*

Example 3.10. *Returning to the SETAF shown in Figure 2, we note that it has several naive extensions (see Table 1), which are essentially all the maximal subsets of Ar that do not attack themselves. On the other hand, the SETAF of Figure 3 has three naive extensions, namely $\{a_2, a_4\}$, $\{a_1, a_2, a_5\}$, $\{a_1, a_3, a_4\}$.*

Extension type	Extensions
Conflict-free	$\emptyset$, $\{a_1\}$, $\{a_2\}$, $\{a_3\}$, $\{a_4\}$, $\{a_5\}$, $\{a_6\}$, $\{a_1,a_4\}$, $\{a_1,a_5\}$, $\{a_1,a_6\}$, $\{a_2,a_3\}$, $\{a_2,a_3,a_5\}$, $\{a_2,a_3,a_6\}$, $\{a_2,a_4\}$, $\{a_2,a_5\}$, $\{a_2,a_6\}$, $\{a_3,a_4\}$, $\{a_3,a_5\}$, $\{a_3,a_6\}$
Admissible	$\emptyset$, $\{a_1\}$, $\{a_2\}$, $\{a_3\}$, $\{a_2,a_3\}$, $\{a_2,a_3,a_5\}$
Complete	$\emptyset$, $\{a_1\}$, $\{a_2,a_3,a_5\}$
Preferred	$\{a_1\}$, $\{a_2,a_3,a_5\}$
Grounded	$\emptyset$
Stable	$\{a_2,a_3,a_5\}$
Naive	$\{a_1,a_4\}$, $\{a_1,a_5\}$, $\{a_1,a_6\}$, $\{a_2,a_4\}$, $\{a_2,a_6\}$, $\{a_3,a_4\}$, $\{a_3,a_6\}$, $\{a_2,a_3,a_5\}$
Semi-stable	$\{a_2,a_3,a_5\}$
Eager	$\{a_2,a_3,a_5\}$
Ideal	$\emptyset$
Stage	$\{a_2,a_3,a_5\}$

Table 1: Extensions for the SETAF of Figure 2

Definition 3.11. *S is called a* semi-stable extension *of AF^S, iff it is a complete extension and the set $S \cup \{a \in Ar \mid S \blacktriangleright a\}$ is maximal w.r.t. set inclusion among all complete extensions of AF^S.*

Essentially, semi-stable semantics give up the strict requirement of stable semantics that $S \cup \{a \in Ar \mid S \blacktriangleright a\} = Ar$, and require just that $S \cup \{a \in Ar \mid S \blacktriangleright a\}$ is maximal.

Just like in stable extensions, semi-stable extensions are also preferred, complete and admissible (see also [Flouris and Bikakis, 2019] and Figure 4), so the following are equivalent [Dvořák et al., 2020c]:

- S is a semi-stable extension

- S is an admissible extension and $S \cup \{a \in Ar \mid S \blacktriangleright a\}$ is maximal w.r.t. set inclusion among all admissible extensions of AF^S.

- S is a preferred extension and $S \cup \{a \in Ar \mid S \blacktriangleright a\}$ is maximal w.r.t. set inclusion among all preferred extensions of AF^S.

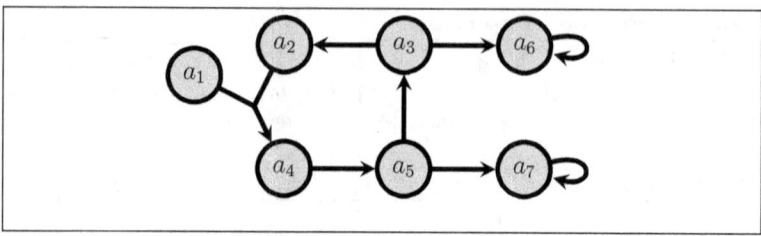

Figure 3: An example SETAF; set attacks are represented as arrows with multiple sources (e.g., $\{a_1, a_2\} \blacktriangleright a_4$); its extensions are shown in Table 2

Example 3.12. *For the SETAF illustrated in Figure 2, where a stable extension exists, this is also the (only) semi-stable extension of the SETAF (see Table 1). However, in the SETAF of Figure 3, where no stable extension exists, one can find two semi-stable extensions, namely: $\{a_1, a_2, a_5\}$, $\{a_1, a_3, a_4\}$. Each of these semi-stable extensions attack (or contain) all arguments except one (a_6 and a_7 respectively).*

Definition 3.13. *S is called an* eager extension *of $AF^{\mathcal{S}}$, iff it is a maximal (with respect to set inclusion) complete extension that is a subset of each semi-stable extension of $AF^{\mathcal{S}}$.*

The maximality requirement implies that we can replace the completeness requirement regarding S with admissibility, i.e., S is an eager extension of $AF^{\mathcal{S}}$, iff it is a maximal (with respect to set inclusion) admissible extension that is a subset of each semi-stable extension of $AF^{\mathcal{S}}$ (see [Dvořák et al., 2020c]).

Example 3.14. *For the SETAF shown in Figure 2, there is only one semi-stable extension, so this is also the eager extension. In the SETAF of Figure 3, where there are two semi-stable extensions, the only eager extension is their intersection, i.e., $\{a_1\}$.*

Definition 3.15. *S is called an* ideal extension *of $AF^{\mathcal{S}}$, iff it is a maximal (with respect to set inclusion) complete extension that is a subset of each preferred extension of $AF^{\mathcal{S}}$.*

Extension type	Extensions
Complete	$\{a_1\}$, $\{a_1, a_2, a_5\}$, $\{a_1, a_3, a_4\}$
Preferred	$\{a_1, a_2, a_5\}$, $\{a_1, a_3, a_4\}$
Grounded	$\{a_1\}$
Stable	(none exists)
Naive	$\{a_2, a_4\}$, $\{a_1, a_2, a_5\}$, $\{a_1, a_3, a_4\}$
Semi-stable	$\{a_1, a_2, a_5\}$, $\{a_1, a_3, a_4\}$
Eager	$\{a_1\}$
Ideal	$\{a_1\}$
Stage	$\{a_1, a_2, a_5\}$, $\{a_1, a_3, a_4\}$

Table 2: Extensions for the SETAF of Figure 3

Again, we can replace the requirement of S being complete, with S being preferred, or admissible (see [Dvořák et al., 2020c]). Moreover, since an ideal extension is a subset of all preferred ones, it is not attacked by any preferred extension, and is in fact the largest complete extension (and admissible set) with this property. Using similar arguments, we can show that an ideal extension is the largest admissible set not attacked by any admissible set, and the largest admissible set not attacked by any complete extension [Dvořák et al., 2020c].

Example 3.16. *For the SETAF shown in Figure 2, the two preferred extensions have an empty intersection, and $\emptyset$ happens to be a complete extension, so the only ideal extension is $\emptyset$. Similarly, for the SETAF of Figure 3, there are two preferred extensions, whose intersection is equal to $\{a_1\}$, and this happens to be a complete extension, so it is also ideal.*

Definition 3.17. *S is called a* stage extension *of AF^S, iff it is conflict-free and $S \cup \{a \in Ar \mid S \blacktriangleright a\}$ is maximal among all conflict-free subsets of Ar.*

Apparently, a stage extension is also naive (see also Figure 4), and, in fact, a stage extension can be equivalently defined as a naive extension such that $S \cup \{a \in Ar \mid S \blacktriangleright a\}$ is maximal among all naive extensions of Ar.

Example 3.18. *For the SETAF shown in Figure 2, which has a stable extension, the (only) stage extension is the stable one, i.e., $\{a_2, a_3, a_5\}$. For the SETAF illustrated in Figure 3, which has no stable extension, there are two stage extensions, which happen to be the same as the semi-stable ones, namely $\{a_1, a_2, a_5\}$, $\{a_1, a_3, a_4\}$. As explained in Example 3.12, each of these semi-stable extensions attack (or contain) all arguments except one (a_6 and a_7 respectively).*

3.1.4 Relationships among extensions

The various extensions are related, in the sense that certain types of extensions are stronger than others (e.g., a preferred extension is also complete, but not vice-versa). Moreover, some types of extensions are guaranteed to exist, others are not, and some extensions are unique. These results have been shown in various works for standard AFs, but [Flouris and Bikakis, 2019] recast them for the SETAF case.

Figure 4 summarises these results. Each arrow in the graph pointing from semantics σ to σ' indicates that every σ-extension of a SETAF is also a σ'-extension of the same SETAF (e.g., every stable extension is also a stage extension). The number (possibly followed by +) that appears next to each semantics indicates the multiplicity of extensions for the specific semantics (e.g., every SETAF has at least one preferred extension). Similarly to Dung-style AFs, for certain semantics, the multiplicity of extensions is different for finite and infinite SETAFs, i.e., SETAFs with finite (respectively infinite) number of arguments. All such arrows are strict, i.e., no semantics is equivalent to another. Note also that in [Flouris and Bikakis, 2019] the relationship among stage and naive semantics is missing.

3.1.5 Labellings

The semantics of AFs can be alternatively defined through labellings, as proposed in [Caminada and Gabbay, 2009]. A labelling is formally defined as a function from arguments to the set {in, out, undec}. Intuitively, an argument belongs to the extension iff it is labelled as in, whereas arguments labelled out are those attacked by the ones labelled in. Finally, the undec labelling is reserved for arguments that are not ac-

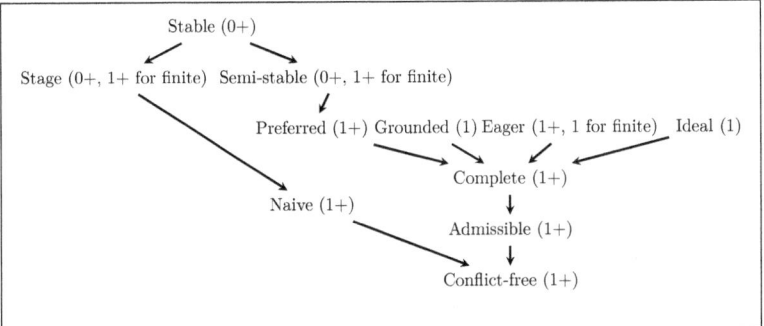

Figure 4: Inclusion relations and multiplicity of extensions for SETAF acceptability semantics

cepted, but are not attacked by an accepted argument either. Although labellings have been originally defined for AFs only [Caminada and Gabbay, 2009], an adaptation for the SETAF case appears in [Flouris and Bikakis, 2019]. Formally, a labelling is a function as follows:

Definition 3.19. *Consider a SETAF $AF^S = \langle Ar, \triangleright \rangle$. A labelling for AF^S is a total function $\mathcal{L}ab : Ar \mapsto \{\text{in}, \text{out}, \text{undec}\}$.*

Note that the labellings of a SETAF are defined over arguments (just like in AFs [Caminada and Gabbay, 2009]), not sets of arguments.

Special classes of labellings can be defined (e.g., conflict-free labellings, admissible labellings, complete labellings, etc) and formally shown to correspond to the respective extensions (conflict-free, admissible, complete, etc). The correspondence is realised through two functions ($Ext2Lab$, $Lab2Ext$), which determine how to generate an extension given a labelling, or vice-versa. It can be shown that, if $\mathcal{L}ab$ is a labelling of a certain type (e.g., complete), then $Lab2Ext(\mathcal{L}ab)$ is an extension of the same type, and, vice-versa, if S is an extension of a certain type (e.g., complete), then $Ext2Lab(S)$ is a labelling of the same type.

In this section, we illustrate these ideas, dealing with complete labellings only, and refer to [Caminada and Gabbay, 2009] and [Flouris and Bikakis, 2019] for further details. We start with the definition of the functions $Lab2Ext$, $Ext2Lab$:

Definition 3.20. *Consider a SETAF $AF^S = \langle Ar, \triangleright \rangle$, and let $\mathcal{E}$ be the set of all possible extensions that can be created over AF^S, $\mathcal{L}$ be the set of all possible labellings that can be created over AF^S. Then:*

Ext2Lab: We define the function $Ext2Lab : \mathcal{E} \mapsto \mathcal{L}$ such that, for $S \in \mathcal{E}$, $\mathcal{L}ab = Ext2Lab(S)$:

- $\mathcal{L}ab(a) = \text{in}$ for all $a \in S$
- $\mathcal{L}ab(a) = \text{out}$ for all $a \notin S$, $S \blacktriangleright a$
- $\mathcal{L}ab(a) = \text{undec}$ for all $a \notin S$, $S \not\blacktriangleright a$

Lab2Ext: We define the function $Lab2Ext : \mathcal{L} \mapsto \mathcal{E}$ such that, for $\mathcal{L}ab \in \mathcal{L}$, $Lab2Ext(\mathcal{L}ab) = \{a \in Ar \mid \mathcal{L}ab(a) = \text{in}\}$.

Clearly, both *Ext2Lab* and *Lab2Ext* are well-defined. Moreover, note that $Ext2Lab(S)$ essentially labels **in** those arguments that are in S, **out** those arguments attacked by S, and **undec** the rest. On the other hand, $Lab2Ext(\mathcal{L}ab)$ contains only the arguments that are labelled **in** by $\mathcal{L}ab$.

Now, we can define complete labellings as follows:

Definition 3.21. *Let $AF^S = \langle Ar, \triangleright \rangle$ be a SETAF. A labelling $\mathcal{L}ab : Ar \mapsto \{\text{in}, \text{out}, \text{undec}\}$ of AF^S is called* complete *iff for all $a \in Ar$:*

1. $\mathcal{L}ab(a) = \text{in}$ *if and only if* $\forall S \blacktriangleright a, \exists b \in S : \mathcal{L}ab(b) = \text{out}$
2. $\mathcal{L}ab(a) = \text{out}$ *if and only if* $\exists S \subseteq Ar$ *such that* $S \blacktriangleright a$ *and* $\mathcal{L}ab(b) = \text{in}$ *for all* $b \in S$

The next step is to prove that complete labellings correspond to complete extensions and vice-versa. The following two theorems prove these points:

Theorem 3.22. *Let $AF^S = \langle Ar, \triangleright \rangle$ be a SETAF and $S \subseteq Ar$ a complete extension of AF^S. Then, $Ext2Lab(S)$ is a complete labelling of AF^S.*

Theorem 3.23. *Let $AF^S = \langle Ar, \triangleright \rangle$ be a SETAF and $\mathcal{L}ab : Ar \mapsto \{\text{in}, \text{out}, \text{undec}\}$ a complete labelling of AF^S. Then, $Lab2Ext(\mathcal{L}ab)$ is a complete extension of AF^S.*

Complete extensions	Complete labellings
$S_1 = \emptyset$	$\mathcal{L}ab_1(a_1)$ = undec, $\mathcal{L}ab_1(a_2)$ = undec, $\mathcal{L}ab_1(a_3)$ = undec, $\mathcal{L}ab_1(a_4)$ = undec, $\mathcal{L}ab_1(a_5)$ = undec, $\mathcal{L}ab_1(a_6)$ = undec
$S_2 = \{a_1\}$	$\mathcal{L}ab_2(a_1)$ = in, $\mathcal{L}ab_2(a_2)$ = out, $\mathcal{L}ab_2(a_3)$ = out, $\mathcal{L}ab_2(a_4)$ = undec, $\mathcal{L}ab_2(a_5)$ = undec, $\mathcal{L}ab_2(a_6)$ = undec
$S_3 = \{a_2, a_3, a_5\}$	$\mathcal{L}ab_3(a_1)$ = out, $\mathcal{L}ab_3(a_2)$ = in, $\mathcal{L}ab_3(a_3)$ = in, $\mathcal{L}ab_3(a_4)$ = out, $\mathcal{L}ab_3(a_5)$ = in, $\mathcal{L}ab_3(a_6)$ = out

Table 3: Complete extensions and complete labellings for the SETAF of Figure 2.

The above theorems show that complete labellings and complete extensions are essentially analogous ways to define the semantics of a SETAF. Similar theorems hold for the other types of extensions/labellings (see [Flouris and Bikakis, 2019], Theorems 5.10, 5.11).

Example 3.24. *Table 3 shows the complete labellings that correspond to the SETAF of Figure 2. Comparing complete extensions with complete labellings, we see that, e.g., the third labelling explicitly rejects a_6 (because it is attacked by a_5, which is accepted), but the second one makes no explicit decision on a_6, as the agent cannot make up its mind on how to resolve the cyclic attack among a_4, a_5, a_6. This distinction cannot be made with the corresponding complete extensions (first column of Table 3). Moreover, we can easily verify that:*

- *the labellings can be generated through the corresponding extensions, using Definition 3.20;*

- *the labellings are all complete labellings (under Definition 3.21);*

- *the extensions could be generated from the labellings, using Definition 3.20.*

Another interesting point to note is that, for complete extensions and labellings, the relationship established by $Ext2Lab$, $Lab2Ext$, is bijective. In other words, for every labelling $\mathcal{L}ab$ and extension S of a SETAF, it holds $Ext2Lab(Lab2Ext(\mathcal{L}ab)) = \mathcal{L}ab$ and $Lab2Ext(Ext2Lab(S)) = S$. This is true for most, but not all, types of labellings; e.g., for admissible labellings, several different labellings may correspond to the same extension through $Lab2Ext$. A complete analysis of this phenomenon can be found in [Flouris and Bikakis, 2019], where the concept of *proper labellings* is introduced to settle this question. Moreover, a rich set of results showing various properties of labellings can be found in [Baroni et al., 2011a; Baroni et al., 2018a]. Although these results have been shown for AFs, recasting them for SETAFs is in most cases easy. Further details on the above are omitted, and the reader is referred to [Baroni et al., 2011a; Baroni et al., 2018a; Flouris and Bikakis, 2019; Caminada and Gabbay, 2009] for more information.

3.2 Relating models for joint attacks with classical AFs

One of the obvious questions regarding SETAFs is whether they constitute a genuine extension of standard AFs (with more expressive power), or whether they are just a shorthand, i.e., syntactic sugar for knowledge that can be anyway represented in the standard Dung setting.

This is a very important question, because, if it turns out that AFs can be used to represent SETAFs, then we would be able to use the more intuitive SETAF formalism for modelling the attacks among arguments, while at the same time exploiting implementations and tools (and complexity results) developed for simple AFs to perform reasoning over the SETAF, by exploiting these translations. In the opposite case, SETAFs should be viewed as a separate, and more expressive branch of computational argumentation, and would require a different set of tools to support reasoning over them.

Interestingly, different works have addressed this problem, and answers have been given from different perspectives. In the rest of this section, we analyse four such works, namely:

- [Dvořák et al., 2019a], who characterise the expressive power of AFs and SETAFs based on the notion of signatures [Dunne et al.,

2015], showing that SETAFs are strictly more expressive than AFs for the most popular semantics.

- [Flouris and Bikakis, 2019], who circumvent the negative result of [Dvořák et al., 2019a] by considering an exponential-sized translation of SETAFs to AFs and appropriate mappings among their semantics, for various semantics.

- [Polberg, 2017], who applies an approach similar to [Flouris and Bikakis, 2019], considers various alternative (and more condensed) translations with similar results (for the most popular semantics).

- [Brewka et al., 2011], who consider the problem of translating Abstract Dialectical Frameworks (ADFs) [Brewka et al., 2018] to AFs; given that SETAFs are a special case of ADFs, this result can be applied for the purposes of this chapter as well, albeit for a limited set of semantics.

3.2.1 Characterising the expressive power using signatures

The approach of [Dvořák et al., 2019a] is based on *signatures* of different semantics (namely complete, grounded, preferred, stable, semi-stable, stage and naive [Dung, 1995; Caminada, 2006; Verheij, 1996b; Bondarenko et al., 1997]) for AFs and SETAFs. Signatures have been originally defined in [Dunne et al., 2015] as a way to characterise the expressive power of an AF, by way of conditions under which a candidate set of subsets of arguments are "realistic", i.e., they correspond to the extensions of some argumentation framework AF for a semantics of interest.

The idea has been extended to other types of argumentation frameworks (e.g., in [Linsbichler et al., 2016; Pührer, 2020; Strass, 2015a; Strass, 2015b] for the ADF case [Brewka et al., 2018]), and employed heavily as a means to compare the expressiveness of different argumentation frameworks with, e.g., normal logic programs and propositional logic [Strass, 2015a; Strass, 2015b].

Formally, given a set of extensions (i.e., a set of sets of arguments) $\mathcal{E}$, $\mathcal{E}$ belongs to the signature Σ_σ^{AF} iff there is an AF framework whose set of extensions, under σ-semantics, is $\mathcal{E}$. Similarly, one can define Σ_σ^k, where

k corresponds to a SETAF that admits only attacks where the attacking set has arity at most k (note that Σ_σ^1 coincides with Σ_σ^{AF} and Σ_σ^∞ coincides with the generic SETAF framework Σ_σ^{SETAF}). By definition, the notion of a signature expresses exactly the sets of extensions that can be constructed given a certain framework type, and for a certain semantics.

The focus of [Dvořák et al., 2019a] is to compute the signatures Σ_σ^k for the considered semantics and for different k. As an example, they define the notion of an incomparable set of sets, where a set of sets $\mathcal{E}$ is incomparable iff all elements of $\mathcal{E}$ are pairwise incomparable, i.e., for $T, U \in \mathcal{E}$, $T \subseteq U$ implies $T = U$. Then, they prove that the set comprising all stable extensions of a SETAF is incomparable, i.e., $\Sigma_{ST}^\infty = \{\mathcal{E} \mid \mathcal{E} \text{ is incomparable}\}$.

Signatures are a powerful tool for determining expressive power. Larger signatures imply that the corresponding framework type is more flexible (and thus more expressive). In particular, if $\mathcal{E} \notin \Sigma_\sigma^1$, then this means that one cannot construct an AF whose σ-extensions are exactly the ones in $\mathcal{E}$. Thus, by comparing Σ_σ^k for various $k \in \{1, 2, ..., \infty\}$, we can determine the relative expressive power of the different framework types.

Using this reasoning, the main conclusion of the paper is that, for all the considered semantics, and for all $k > 0$, SETAFs that allow for collective attacks of $k+1$ arguments are more expressive than SETAFs that only allow for collective attacks of at most k arguments, because $\Sigma_\sigma^k \subset \Sigma_\sigma^{k+1}$. As a corollary, SETAFs are strictly more expressive than AFs, even if restricted to attacks of at most 2 arguments.

It is important however to interpret the above results under the correct lens. In particular, the results of [Dvořák et al., 2019a] tell us that certain sets of extensions that can be constructed using SETAFs, cannot be *directly* constructed through AFs. More specifically, for a given $\mathcal{E} \in \Sigma_\sigma^{SETAF} \setminus \Sigma_\sigma^{AF}$, we know that one can create a SETAF whose set of σ-extensions is exactly $\mathcal{E}$; moreover, there is no AF whose set of σ-extensions is exactly $\mathcal{E}$.

However, if we don't insist on the *direct* construction, one may be able to succeed in constructing $\mathcal{E}$ through some AF, but in another, indirect way. In particular, one could define an appropriate mapping (algorithm) among sets of extensions (say f), and then construct an AF $AF^{\mathcal{D}}$ whose set of extensions is, say, $\mathcal{E}'$, where $f(\mathcal{E}') = \mathcal{E}$. For generality, one should also define a generic way to construct $AF^{\mathcal{D}}$ from the original SETAF $AF^{\mathcal{S}}$, via some other mapping (algorithm), say g. By the results of [Dvořák et al., 2019a], this transformation cannot be a simple rearrangement of the attacks among the existing arguments of the SETAF, but should necessarily involve new, artificial arguments that would somehow encode the "collectivity of attacks".

3.2.2 An exponential translation to encode collectivity of attacks

This approach of "expanding" the SETAF with new arguments in order to get rid of collective attacks (and thus result in an AF) is followed in [Flouris and Bikakis, 2019]. In that paper, a rather straightforward translation is followed, where, for any given SETAF $AF^{\mathcal{S}} = \langle Ar, \triangleright \rangle$, one constructs a so-called *generated AF* $AF^{\mathcal{D}} = \langle Ar', att \rangle$, whose "arguments" are all the non-empty sets of arguments of the original SETAF (i.e., $Ar' = 2^{Ar} \setminus \{\emptyset\}$). The corresponding attack relation att follows in the obvious manner from $\triangleright$. In the above terminology, this is the mapping g.

Then, the authors go on to identify the relationship among the σ-extensions of the $AF^{\mathcal{S}}$ and its corresponding generated $AF^{\mathcal{D}}$, as well as how one can identify the σ-extensions of $AF^{\mathcal{S}}$ through the σ-extensions of $AF^{\mathcal{D}}$, and vice versa (i.e., the mapping f and its inverse).

Various different semantics are considered, including the ones originally defined in [Dung, 1995] (conflict-free, admissible, complete, grounded, preferred, stable), but also naive [Bondarenko et al., 1997], semi-stable [Caminada, 2006], eager [Caminada, 2007], ideal [Dung et al., 2007] and stage [Verheij, 1996b].

The conclusion of the above analysis is that many of the semantics (namely, complete, preferred, grounded, stable and ideal) admit a very simple one-to-one correspondence among the semantics of the SETAF and the generated AF. In particular, a set of arguments $S \subseteq Ar$ of the SETAF ($AF^{\mathcal{S}}$) is a σ-extension if and only if the set $2^S \setminus \{\emptyset\}$ is a σ-extension of the generated AF (recall that an argument in $AF^{\mathcal{D}}$ is a set of arguments from $AF^{\mathcal{S}}$).

For conflict-free and admissible extensions, the situation is similar, except that there are some additional σ-extensions of $AF^\mathcal{D}$ which do not follow this exact pattern. This has effects on the correspondence among naive extensions as well (recall that a naive extension is a maximal conflict-free set). Further, more convoluted correspondences exist for semi-stable, stage, and eager semantics, where the characterisations are complicated by the requirement of maximality (see [Flouris and Bikakis, 2019] for details).

Complexity of characterisations put aside, the work of [Flouris and Bikakis, 2019] shows that one can model a SETAF as an AF in a way that "preserves" the semantics, in the sense that one can determine the σ-extensions of the SETAF by just looking at the AF (and vice-versa). Alas, the proposed transformation for achieving this effect, results to an AF with an exponentially larger number of arguments compared to the SETAF. Note that if we count the size of a SETAF in terms of the number of arguments plus the number of attacks, then we may not get an exponential increase (if a sufficiently large number of attacks exist), although the exponential increase is still true in the worst-case scenario.

3.2.3 Considering more compact translations

A similar, but less extreme "expansion" scheme is followed in [Polberg, 2017], where the problem of translating SETAFs to AFs is considered, among other things. The considered semantics are the standard Dung semantics, i.e., conflict-free, admissible, complete, preferred, grounded and stable [Dung, 1995].

To perform the translation, two translation schemes (and variations thereof) are considered: one is inspired by the so-called coalition approach and the other by the so-called defender approach. Both have a polynomial size compared to the SETAF (assuming that the size of the SETAF is considered to be equal to the number of attacks plus the number of arguments).

The coalition approach is similar to the one proposed in [Flouris and Bikakis, 2019], where an argument in the AF is a set of arguments from the SETAF. However, in [Polberg, 2017] a "condensed" version of the translation is considered, where not all subsets of Ar are included in the generated AF, but only those that are actually the initiators of an attack.

Different ways to translate the attack relation are then considered, with different results with respect to the correspondence among the semantics of the SETAF and the corresponding AF.

The second translation scheme is inspired by [Modgil and Bench-Capon, 2010], and uses arguments in the translated AF that represent "statements" regarding an argument in the SETAF (*e.g.* whether it is accepted, justified, rejected etc). More precisely, for every argument a in the SETAF, two arguments are included in the AF: the argument itself (a), as well as a' which stands for "a is rejected". Moreover, every attack in the SETAF is represented as an argument in the AF (these are called auxiliary arguments).

Then, appropriate attacks are introduced in the new framework. Namely, each argument a attacks its corresponding a', and a' attacks the auxiliary arguments representing an attack involving a as an attacker. The auxiliary arguments representing attacks, attack the corresponding recipient of the attack. In this way, a defends the auxiliary arguments it is involved in, so if a is not accepted, the attack itself (i.e., the auxiliary argument representing it) will not be accepted, and thus the recipient of the attack will be unaffected by the attack. Using this trick, the semantics of the SETAF can be appropriately captured by the AF.

For both translations, the correspondences provided among the σ-extensions of the SETAF and its generated AF are generally elegant, and quite similar to the correspondences of [Flouris and Bikakis, 2019] (note however that the more complex cases of semi-stable, stage and eager semantics are not considered by [Polberg, 2017]).

Despite that, a strong statement is made in [Polberg, 2017] that no full exact SETAF-AF translation can be created. This statement is based on the idea of signatures, and follows similar lines of reasoning as in [Dvořák *et al.*, 2019a]. Therefore, it should be interpreted in the sense of a direct translation, as explained also in our analysis of the results of [Dvořák *et al.*, 2019a].

3.2.4 An indirect translation path, through ADFs

Another interesting translation results as a corollary of the work in [Brewka *et al.*, 2011]. In that paper, the authors do not study SETAFs, but ADFs [Brewka *et al.*, 2018]. An ADF is similar to an AF, except that the acceptance of an argument is determined by an acceptance condition (expressed as a propositional formula) over the acceptance of all its attackers. Thus, for example, one could say that an argument is accepted iff no more than two of its attackers are accepted, or that an argument is accepted iff all of its attackers are accepted.

Note that the expressive power of acceptance conditions allows ADFs to model various different types of relations among arguments, including attack, support, joint attacks or supports, as well as hybrid cases. In particular, it is easy to see that AFs and SETAFs are special cases of ADFs [Polberg, 2016; Linsbichler et al., 2016].

Three different types of semantics have been defined for ADFs in [Brewka et al., 2018], namely models, well-founded models and stable models. In the special case where an ADF is used to describe an AF (or a SETAF), models of the ADF correspond to the stable extensions of the AF (or SETAF) and well-founded models of the ADF correspond to the grounded extensions of the AF (or SETAF). Moreover, for this special case, stable models of the ADF and models of the ADF coincide (see [Dvořák et al., 2020a], Proposition 1), so stable models of the ADF also correspond to stable extensions of the AF (or SETAF). It should be noted here that stable models have been retrospectively redefined in [Brewka et al., 2013], but this redefinition does not break the above correspondences (see Theorem 4 in [Brewka et al., 2013]).

In [Brewka et al., 2011], the authors show that, given an ADF, one can generate an AF such that the stable extensions of the AF correspond (in a formal manner made clear in the paper) to the models of the ADF. A similar correspondence is also shown among the grounded extensions of an AF and the well-founded models of the ADF, as well as among the stable extensions of the AF and the stable models of the ADF. Although SETAFs were not in the scope of the work of [Brewka et al., 2011], the fact that SETAFs are a special case of ADFs, allows us to apply their results to the case considered in this chapter. Moreover, [Brewka et al., 2011] show that the proposed translations are polynomial in size, and can also be computed in polynomial time, where the size of the original ADF (corresponding to a SETAF) is computed as the number of arguments plus the size of the acceptance conditions of the arguments.

3.3 Computational considerations

In this section we give an overview on complexity results of SETAFs and discuss implementation approaches for evaluating SETAFs. As discussed in [Dvořák and Dunne, 2018] understanding the inherent complexity of the reasoning tasks is crucial towards efficient implementations of argumentation systems. In particular, problems on different levels of complexity have different limits concerning scalability and require different techniques to be implemented in a scalable manner. We first introduce the computational tasks we are interested in, then discuss their complexity, and finally discuss algorithms and reduction-based approaches for these tasks.

3.3.1 Computational Problems

The standard problems studied in computational (abstract) argumentation are the tasks of computing extensions of a given semantics and computing the credulous or skeptical consequences under a given semantics [Charwat et al., 2015; Dvořák and Dunne, 2018; Cerutti et al., 2018]. These tasks are investigated in the literature on algorithms, systems, and complexity of abstract argumentation, and are the basis for the different tracks of the International Competition on Computational Models of Argumentation (ICCMA)[6] [Thimm and Villata, 2017; Gaggl et al., 2020]. In the following we provide formal definitions of these computational problems in the context of SETAFs. To this end we will use $\sigma(AF^S)$ to denote the σ-extensions of a SETAF AF^S. We start with the function problems of computing one or all of the extensions of a SETAF w.r.t. a semantics σ:

- *Some Extension* SE_σ: Given SETAF AF^S, compute an extension $E \in \sigma(AF^S)$.

- *Enumerate Extensions* EE_σ: Given SETAF AF^S, compute the extension-set $\sigma(AF^S)$.

Beside these function problems we consider decision problems whose output is either yes or no. These problems are of particular interest

---
[6]http://argumentationcompetition.org/

as they are well-suited for being analysed with the techniques of complexity theory. To this end we consider the skeptical acceptance of an argument, i.e., an argument is skeptically accepted if it is contained in each extension, and credulous acceptance of an argument, i.e., an argument is credulously accepted if it is contained in some extension (for a given semantics σ):

- *Credulous Acceptance* $Cred_\sigma$: Given SETAF $AF^S = \langle Ar, \triangleright \rangle$ and an argument $a \in Ar$, is a contained in some $E \in \sigma(AF^S)$?

- *Skeptical Acceptance* $Skept_\sigma$: Given SETAF $AF^S = \langle Ar, \triangleright \rangle$ and an argument $a \in Ar$, is a contained in each $E \in \sigma(AF^S)$?

Moreover, we consider the frequently-studied problems of verifying a given extension, deciding whether a SETAF has at least one extension, and deciding whether a SETAF has a non-empty extension. These problems are of some interest on their own but are in particular relevant as frequent sub-tasks of reasoning procedures. We next provide the formal definitions of these problems:

- *Verification of an extension* Ver_σ: Given SETAF $AF^S = \langle Ar, \triangleright \rangle$ and a set of arguments $S \subseteq Ar$, is $S \in \sigma(AF^S)$?

- *Existence of an extension* $Exists_\sigma$: Given SETAF $AF^S = \langle Ar, \triangleright \rangle$, is $\sigma(AF^S) \neq \emptyset$?

- *Existence of a non-empty extension* $Exists_\sigma^{\neg \emptyset}$: Given a SETAF $AF^S = \langle Ar, \triangleright \rangle$, does there exist a set $E \neq \emptyset$ such that $E \in \sigma(AF^S)$?

3.3.2 Complexity results for SETAFs

We next discuss the computational complexity of the decision problems introduced in the previous section. The rationale behind the focus on decision problems is that tools of complexity theory are better suited for decision problems than for function problems and that, when chosen carefully, the complexity of the decision problems is also a good indicator for the complexity of the corresponding function problem. In computational argumentation the credulous and skeptical acceptance decision problems together are considered to be a good indicator for the complexity of a semantics.

In this section we assume the reader to have basic knowledge in computational complexity theory.[7] We will consider the following complexity classes: L (logarithmic space), P (polynomial time), NP (non-deterministic polynomial time), coNP (complement of a NP problem), Θ_2^P (polynomial time with non-adaptive NP-oracle calls), Σ_2^P (non-deterministic polynomial time with NP-oracle calls), Π_2^P (complement of a Σ_2^P problem), and D_2^P (intersection of a Σ_2^P and a Π_2^P language).

We have the following relations between these complexity classes:

$$L \subseteq P \subseteq \begin{matrix} NP \\ coNP \end{matrix} \subseteq \Theta_2^P \subseteq \begin{matrix} \Sigma_2^P \\ \Pi_2^P \end{matrix} \subseteq D_2^P$$

We follow [Dvořák et al., 2018] and start our complexity analysis with the observation that SETAFs generalize Dung AFs and thus all the decision problems are at least as hard as the corresponding problem for Dung AFs (cf. [Dvořák and Dunne, 2018, Table 1]). Interestingly, one can also obtain the same upper bounds (see Table 4) as we discuss below. These results for SETAFs show the same complexity as the corresponding Table for Dung AFs (cf. [Dvořák and Dunne, 2018, Table 1][8]). However, there is a subtle difference between the complexity results for Dung AFs and SETAFs. In both cases the complexity is stated w.r.t. the size of the input framework, which in case of Dung AFs is often interpreted as w.r.t. the number of arguments $|Ar|$ in the input framework. This interpretation is not valid for SETAFs where the number of attacks $|\triangleright|$ can be exponentially larger than the number of arguments $|Ar|$ (this even holds for normal forms where redundant attacks are removed). Thus, one has to consider the complexity w.r.t. the number of arguments plus the representation size of the attacks $\triangleright$. The latter is bounded bound by $|Ar| \cdot |\triangleright|$, i.e., is polynomially bounded in $|Ar| + |\triangleright|$.

---

[7] For a gentle introduction into complexity theory in the context of argumentation the reader is referred to [Dvořák and Dunne, 2018]).

[8] Notice that [Dvořák and Dunne, 2018, Table 1] includes $\mathcal{CF}2$ semantics which has not yet been generalised to SETAFs and is thus not included in Table 4. On the other hand, we include eager semantics which has not been considered in [Dvořák and Dunne, 2018, Table 1] (see [Dunne et al., 2013, Table 2] for the complexity results of eager semantics in Dung AFs).

We can thus interpret the complexity results for SETAFs in Table 4 as w.r.t. $|Ar|+|\triangleright|$ [9].

The crucial observation towards the upper bounds is that checking basic properties of a set of arguments, although it is more evolved than in Dung AFs, can still be performed in L. First, to test whether a set S is conflict-free one can iterate over all attacks $(T,a) \in \triangleright$ and check that $T \cup \{a\} \not\subseteq S$. Second, to test $S \blacktriangleright T$ one can iterate over all attacks $(U,b) \in \triangleright$ and test whether $U \subseteq S$ and $b \in T$. Finally, a simple algorithm for testing that a set S defends an argument a iterates over all attacks $(T,a) \in \triangleright$ and for each of these attacks checks that $S \blacktriangleright T$. That is, for all three problems we just need to store a small number of pointers to the input which can be done in logarithmic space.

Proposition 3.25. *Given a SETAF $AF^S = \langle Ar, \triangleright \rangle$, a set of arguments $S \subseteq Ar$, and an argument $a \in Ar$, deciding whether S is conflict-free, deciding whether $S \blacktriangleright a$, and deciding whether $a \in \mathrm{F}_{AF^S}(S)$ are in L.*

Notice that most of the complexity upper bounds for Dung AFs are based on the fact that these three problems can be solved in polynomial-time, and thus these upper bounds also apply to SETAFs (cf. Table 4). We next exemplify this for the credulous acceptance problem of stable semantics.

Proposition 3.26. *$Ver_{ST} \in$ L and $Cred_{ST}$ is NP-complete.*

Proof. First, consider the verification problem Ver_{ST} and an arbitrary SETAF $AF^S = \langle Ar, \triangleright \rangle$. We can verify that a given set S is a stable extension of AF^S by (a) checking that S is conflict-free and (b) checking that for each $a \in Ar \setminus S$ we have $S \blacktriangleright a$. As both can be done in L, we obtain the L membership of Ver_{ST}.

Now consider the credulous acceptance problem $Cred_{ST}$. The NP-hardness is by the corresponding result for AFs. For the upper bound consider an arbitrary SETAF $AF^S = \langle Ar, \triangleright \rangle$ and an argument $a \in Ar$. We can decide the credulous acceptance of a in AF^S by a standard guess & check algorithm. That is, one first uses the non-determinism to guess a set E and then use a deterministic part to verify that E is a stable

---

[9] For a more fine-grained analysis of algorithms for SETAFs one might take into account the actual representation size of the attacks, cf. [Dvořák et al., 2020b].

extension and contains the argument a. This gives an NP procedure for $Cred_{\mathcal{ST}}$. □

Next, let us consider the complexity of ideal semantics, as it is the only case where the upper bound for Dung AFs [Dunne, 2009] does not directly apply to SETAFs. Recall that the ideal extension can be characterised as the maximal admissible set that is not attacked by any other admissible set (Definition 3.15). In order to compute the ideal extension we thus use NP-oracle queries that for each argument ask (a) whether it is credulously accepted w.r.t. preferred semantics and (b) whether it is attacked by some admissible set. We then consider the set E^0 of all arguments that are credulously accepted but not attacked by an admissible set. Notice that E^0 is conflict-free by construction and it is an over-approximation of the ideal extension. We then compute the maximal admissible subset of E^0 by iteratively computing sets E^{i+1} by removing arguments that are not defended by E^i until we reach a fixed-point E. We then have that E is the ideal extension. We have that the NP-oracle queries of the above procedure are independent of each other and thus can be executed in parallel. Moreover, each iteration of the fixed-point computation is in polynomial-time and the fixed-point is reached after at most $n/2$ iterations, i.e., one can compute the fixed-point in polynomial time. Thus the above is a Θ_2^P-algorithm for computing the ideal extension. Hence, we obtain Θ_2^P upper bounds for all reasoning tasks of ideal semantics.

3.3.3 Algorithms for SETAFs

The field of algorithms for SETAFs is rather under-explored with the exception of [Nielsen and Parsons, 2006]. The former studies algorithmic ideas for preferred semantics. We recapitulate their main observations in terms of a simple algorithm (see Algorithm 1) in the style of today's labelling-based algorithms ([Charwat et al., 2015; Cerutti et al., 2018]).

The rough idea of labelling-based algorithms is to start with all arguments unlabelled, in each step pick an argument and then consider two branches: one where we add the argument to the extension, i.e., labelled **in**; and one where we decide that the argument is excluded from the extension, i.e., labelled **out** or **undec** (cf. Section 3.1.5). When all

σ	$Cred_\sigma$	$Skept_\sigma$	Ver_σ	$Exists_\sigma$	$Exists_\sigma^{\neg\emptyset}$
Conflict-free	in L	trivial	in L	trivial	in L
Naive	in L	in L	in L	trivial	in L
Grounded	P-c	P-c	P-c	trivial	in L
Stable	NP-c	coNP-c	in L	NP-c	NP-c
Admissible	NP-c	trivial	in L	trivial	NP-c
Complete	NP-c	P-c	in L	trivial	NP-c
Ideal	Θ_2^P-c	Θ_2^P-c	Θ_2^P-c	trivial	Θ_2^P-c
Eager	Π_2^P-c	Π_2^P-c	D_2^P-c	trivial	Π_2^P-c
Preferred	NP-c	Π_2^P-c	coNP-c	trivial	NP-c
Semi-stable	Σ_2^P-c	Π_2^P-c	coNP-c	trivial	NP-c
Stage	Σ_2^P-c	Π_2^P-c	coNP-c	trivial	in L

Table 4: Complexity of SETAFs ($\mathcal{C}$-c denotes completeness for class $\mathcal{C}$).

arguments are labelled, one tests whether the labelling is valid w.r.t. the considered semantics and, if so, it is added to the output. By that procedure we would consider all possible candidates for valid labellings and thus also obtain all the extensions. In order to design an efficient algorithm one aims to cut off branches that do not lead to valid labellings as soon as possible. One approach are the so-called label propagation rules, i.e., one uses the already fixed labels of the arguments to conclude that other arguments have to obtain a certain label and by that avoids unnecessary branching in the algorithm. For instance, for preferred semantics, given the set of arguments $\mathcal{L}ab_{\text{in}}$ labelled **in** by a partial labelling $\mathcal{L}ab$ we can conclude that all arguments in the set $\mathcal{L}ab_{\text{in}}^+$, i.e., arguments a with $\mathcal{L}ab_{\text{in}} \blacktriangleright a$, must be labelled **out**. Moreover, for attacks that target $\mathcal{L}ab_{\text{in}}$ and have only one argument outside of $\mathcal{L}ab_{\text{in}}^+$ we have that this argument has to be labelled **out**. This is captured by the set $\mathcal{L}ab_{\text{in}}^{\leftarrow}$ defined as $\mathcal{L}ab_{\text{in}}^{\leftarrow} = \{a \in Ar \mid \mathcal{L}ab_{\text{in}} \cup \{a\} \blacktriangleright \mathcal{L}ab_{\text{in}}\}$. This propagation of **out** labels is implemented in Line 8 of Algorithm 1 and triggered whenever a new argument is labelled **in**. Another observation is that we cannot label an argument **in** if this would cause a conflict in the set $\mathcal{L}ab_{\text{in}}$. Many cases where this could happen are already covered by the propagation rules for out labels, but these rules do not cover attacks $(S \cup \{a\}, a) \in \rhd$ with $S \subseteq \mathcal{L}ab_{\text{in}}$. This propagation is implemented by the if condition on Line 4, which prevents the algorithm from starting the branch where the argument a is added to the extension. Finally, when an argument a is already defended by $\mathcal{L}ab_{\text{in}}$ then, due to the maximality of preferred extensions, we know that this argument is in each preferred extension containing $\mathcal{L}ab_{\text{in}}$ and thus we must label a by in. This propagation is implemented by the if condition on Line 12, which prevents the algorithm from starting the branch where the argument a is excluded from the extension.

We obtain that Algorithm 1 returns the preferred labellings of a given SETAF AF^S. Notice that the algorithm can be easily adapted to compute complete labellings, by removing the maximality check on Line 16, or admissible sets, by removing the maximality check on Line 16 and the if condition on Line 12. We can roughly estimate the running time of these algorithms by $O(\exp(|Ar|) \cdot \text{poly}(|Ar|, |\rhd|))$. Notably only the polynomial part depends on the number of attacks while the exponential part solely depends on the number of arguments. Finally, recent work [Dvořák et al., 2020b] suggests to not just label arguments but also label the attacks of a SETAF. It then studies possible label propagation-rules for stable and complete semantics and provides a linear time algorithm (linear w.r.t. the representation of the SETAF) for grounded semantics.

3.3.4 Systems and Reduction-based Approaches

Reduction-based approaches have been successfully applied in the design of argumentation systems, most prominently by systems that are based on modern SAT-solver technology or answer-set programming [Cerutti et al., 2018]. For SETAFs the only system discussed in the literature, i.e., the SETAF module of the ASPARTIX[10] system [Dvořák et al., 2018], is based on answer-set programming.

---

[10] https://www.dbai.tuwien.ac.at/research/argumentation/aspartix

Algorithm 1 pref-lab($AF^\mathcal{S}$)

Require: SETAF $AF^\mathcal{S} = \langle Ar, \rhd \rangle$, global variable $\mathcal{L}$
Ensure: $\mathcal{L}$ is the set of preferred labellings
1: $\mathcal{L} = \emptyset$, $\mathcal{L}ab = \langle \emptyset, \emptyset, \emptyset \rangle$
2: pref-lab($AF^\mathcal{S}, \mathcal{L}ab$)

3: **function** pref-lab($F, \mathcal{L}ab$)
   Require: SETAF $F = \langle A, R \rangle$, partial labelling $\mathcal{L}ab$, global variable $\mathcal{L}$
4: **if** there is an argument $a \in A$ not labeled by $\mathcal{L}ab$ **then**
5:    $a \leftarrow$ pick some unlabeled argument
6:    **if** $\mathcal{L}ab_{\text{in}} \cup \{a\} \in \mathcal{CF}(F)$ **then**
7:      $\mathcal{L}ab'_{\text{in}} = \mathcal{L}ab_{\text{in}} \cup \{a\}$,
8:      $\mathcal{L}ab'_{\text{out}} = \mathcal{L}ab_{\text{out}} \cup \mathcal{L}ab'^{\rightarrow}_{\text{in}} \cup \mathcal{L}ab'^{\leftarrow}_{\text{in}}$
9:      $\mathcal{L}ab'_{\text{undec}} = \mathcal{L}ab_{\text{undec}} \setminus \mathcal{L}ab'_{\text{out}}$
10:     pref-lab($AF^\mathcal{S}, \langle \mathcal{L}ab'_{\text{in}}, \mathcal{L}ab'_{\text{out}}, \mathcal{L}ab'_{\text{undec}} \rangle$)
11:   **end if**
12:   **if** $\{a\} \notin \mathcal{F}_F(\mathcal{L}ab_{\text{in}})$ **then**
13:     pref-lab($AF^\mathcal{S}, \langle \mathcal{L}ab_{\text{in}}, \mathcal{L}ab_{\text{out}}, \mathcal{L}ab_{\text{undec}} \cup \{a\} \rangle$)
14:   **end if**
15: **else**
16:   **if** $\mathcal{L}ab_{\text{in}} \in \mathcal{AD}(F)$ **and** $\mathcal{L}ab_{\text{in}}$ $\subseteq$-max among $\{\mathcal{L}ab_{\text{in}} \mid \mathcal{L}ab \in \mathcal{L}\}$ **then**
17:     $\mathcal{L} = \mathcal{L} \cup \{\mathcal{L}ab\}$
18:   **end if**
19: **end if**
20: **endFunction**

Reduction to Answer-set Programming. Answer-set programming (ASP) [Marek and Truszczyński, 1999; Niemelä, 1999] is a declarative problem solving paradigm with its roots in logic programming and non-monotonic reasoning. Today's answer-set systems [Gebser et al., 2011; Leone et al., 2006] support a rich language and are capable of solving hard problems efficiently. Thus, ASP is a convenient formalism to implement argumentation systems. The ASPARTIX approach [Egly et al., 2010] to argumentation problems relies on a query-based implementation where the argumentation framework is provided as an input database, and one provides fixed queries encoding the different argumentation semantics and reasoning tasks.

Here we briefly highlight the main differences between the ASP encodings of Dung AFs [Egly et al., 2010] and SETAFs [Dvořák et al., 2018]. To this end, we first briefly recall the basic terminology for logic programs (for rigorous definitions see [Cerutti et al., 2018] or [Charwat et al., 2015]). A logic program (under the answer-set semantics) is a set of disjunctive rules r of the form

$$a_1 \vee \cdots \vee a_n \leftarrow b_1, \ldots, b_k, \; not \; b_{k+1}, \ldots, \; not \; b_m$$

where $a_1, \ldots, a_n$ and $b_1, \ldots, b_m$ are atoms, and *not* stands for *default negation*. We refer to a as a positive literal, while we refer to *not a* as a default negated literal. The *head* of r is the set $\{a_1, \ldots, a_n\}$ and the *body* of r is $\{b_1, \ldots, b_m\}$, and a rule r is a *constraint* if $n = 0$. A *fact* is a ground rule without disjunction ($n = 1$) and with an empty body. An *input database* is a set of facts.

In order to evaluate SETAFs with ASP, in a first step, we have to encode SETAFs as an input database for the ASP-program. We introduce three predicates **arg**, **att**, and **mem** to encode a SETAF $AF^S = \langle Ar, \triangleright \rangle$. The predicate **arg** is used to encode arguments, the latter two to encode the set attacks, i.e., **att** encodes which argument is attacked by an attack and **mem** encode which arguments are required to attack that argument. Notice that, this encoding uses a unique identifier for each attack in $\triangleright$. The encoding of a SETAF $AF^S = \langle Ar, \triangleright \rangle$ is then given by $\pi_{setaf}(AF^S) = \{\mathbf{arg}(a). \mid \text{ for } a \in Ar\} \cup \{\mathbf{att}(r, x). \mid \text{ for } r \in \triangleright \text{ and } r = (S, x)\} \cup \{\mathbf{mem}(r, y). \mid \text{ for } r \in \triangleright, r = (S, x), \text{ and } y \in S\}$ (cf. Figure 5). While arguments are represented in the same way as in Dung AFs, Dung AFs allow for a simpler representation of attacks. That is, the encoding of AFs ([Cerutti et al., 2018]) only uses one binary predicate **att** to encode the attacks, containing the attacker and the attacked argument of each attack, and does not use identifiers for attacks.

When it comes to the encoding of semantics one uses predicates **in**(·), **out**(·) to guess whether an argument is in the extension or not (in the same way as for AFs). Notice that the predicate **out**(·) encodes

$\pi_{setaf}(AF^{\mathcal{S}})$

$\mathbf{arg}(a).\ \mathbf{arg}(b).\ \mathbf{arg}(c).$
$\mathbf{att}(r1, c).\ \mathbf{mem}(r1, a).\ \mathbf{mem}(r1, b).$
$\mathbf{att}(r2, b).\ \mathbf{mem}(r2, a).\ \mathbf{mem}(r2, c).$
$\mathbf{att}(r3, a).\ \mathbf{mem}(r3, b).\ \mathbf{mem}(r3, c).$

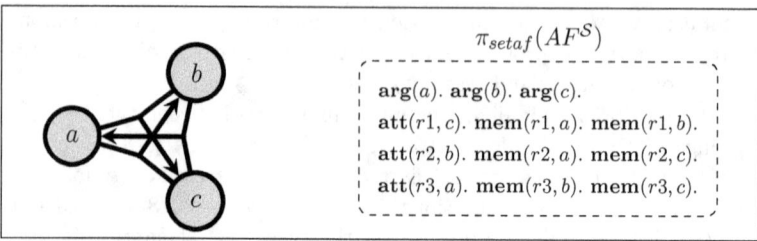

Figure 5: SETAF $AF^{\mathcal{S}} = \langle\{a, b, c\}, \{(\{a, b\}, c), (\{a, c\}, b), (\{b, c\}, a)\}\rangle$ and its ASP-encoding $\pi_{setaf}(AF^{\mathcal{S}})$.

$\pi_{\mathcal{CF}}$

$$
\begin{aligned}
\mathbf{in}(Y) &\leftarrow \mathbf{arg}(Y), \mathit{not}\ \mathbf{out}(Y).\\
\mathbf{out}(Y) &\leftarrow \mathbf{arg}(Y), \mathit{not}\ \mathbf{in}(Y).\\
\mathbf{blocked}(R) &\leftarrow \mathbf{mem}(R, X), \mathbf{out}(X).\\
&\leftarrow \mathbf{in}(X), \mathbf{att}(R, X), \mathit{not}\ \mathbf{blocked}(R).
\end{aligned}
$$

$\pi_{\mathcal{AD}}$

$$
\begin{aligned}
\mathbf{in}(Y) &\leftarrow \mathbf{arg}(Y), \mathit{not}\ \mathbf{out}(Y).\\
\mathbf{out}(Y) &\leftarrow \mathbf{arg}(Y), \mathit{not}\ \mathbf{in}(Y).\\
\mathbf{blocked}(R) &\leftarrow \mathbf{mem}(R, X), \mathbf{out}(X).\\
&\leftarrow \mathbf{in}(X), \mathbf{att}(R, X), \mathit{not}\ \mathbf{blocked}(R).\\
\mathbf{defeated}(R) &\leftarrow \mathbf{att}(R, X), \mathbf{mem}(R, Y),\\
&\qquad \mathbf{att}(R2, Y), \mathit{not}\ \mathbf{blocked}(R2).\\
&\leftarrow \mathbf{in}(X), \mathbf{att}(R, X), \mathit{not}\ \mathbf{defeated}(R).
\end{aligned}
$$

Figure 6: ASP Encodings $\pi_{\mathcal{CF}}$, $\pi_{\mathcal{AD}}$ for $\mathcal{CF}$ and $\mathcal{AD}$ semantics of SETAFs.

that an argument is not in the extension and does not correspond to the label `out`. This guess builds up all possible subsets of arguments which are then filtered by adding constraints that reflect the specific semantics. Here the SETAF encodings differ from the AF encodings as they explicitly define statuses of attacks. First, we call an attack $(T, a) \in \triangleright$ blocked w.r.t. a set $E \subseteq Ar$ if $T \not\subseteq E$. Second, we consider an attack $(T, a) \in \triangleright$ to be defeated by a set E iff $E \blacktriangleright T$. We will exemplary discuss the encodings π_{CF}, π_{AD} for conflict-free sets and admissible sets respectively (cf. Figure 6). In the encoding of the conflict-freeness, with the first two rules one guesses a subset of arguments, the third rule computes the blocked attacks, and the constraint in the fourth line rules out all sets that contain an argument X and have a non-blocked rule attacking X. That is, if we compute the answer-sets of the combined program $\pi_{setaf}(AF^S) \cup \pi_{CF}$ the answer-sets correspond to the conflict-free sets, i.e., the conflict-free sets are given by the $\mathbf{in}(\cdot)$ predicate in the answer-sets. Next, we further extend π_{CF} to an encoding π_{AD} for admissible semantics. That is, we add a rule that computes the defeated attacks and a constraint that rules out sets where an argument of the set is attacked by an undefeated attack. Thus, if we compute the answer-sets of the combined program $\pi_{setaf}(AF^S) \cup \pi_{AD}$ the answer-sets correspond to the admissible sets.

Other Reduction-based Approaches. For Dung AFs and their generalisations, several reduction-based approaches have been studied in the literature and often resulted in argumentation systems [Charwat et al., 2015]. In particular, systems based on modern SAT-solving systems have been successful [Thimm and Villata, 2017; Gaggl et al., 2020]. Beside ASP, none of these approaches have been considered in the literature on SETAFs so far. However, very recently a first version of the SAT-based SETAF system *joukko* appeared online[11]. Thus, one approach towards an efficient SETAF system would be to extend existing approaches that have been successful for AFs to SETAFs. Another approach is to translate SETAFs to AFs or ADFs and use one of the existing systems for these formalisms to evaluate SETAFs. Translations from SETAFs to AFs have been presented in [Polberg, 2017] and [Flouris and Bikakis, 2019](see also Section 3.2 in this chapter). However, when using these translations one is faced with an exponential blow-up in the arguments and thus these translations are not well-suited for computational matters. Recall, that algorithms for SETAFs scale polynomially w.r.t. the number of attacks and exponentially w.r.t. the number of arguments. Thus translating attacks to arguments and using AFs tools can results in a serious computational overhead. Concerning the latter, there are

---

[11]https://bitbucket.org/andreasniskanen/joukko

rather simple translations of SETAFs into ADFs [Linsbichler *et al.*, 2016; Polberg, 2016] (see Section 3.2.4) which do not increase the number of arguments. That is, one can efficiently encode a SETAF as an ADF and then use one of the existing systems for ADFs, e.g., k++ADF[12] [Linsbichler *et al.*, 2018], YADF[13] [Brewka *et al.*, 2017], or DIAMOND[14] [Ellmauthaler and Strass, 2014], to evaluate the SETAF. The attentive reader may argue that the computational complexity of ADFs is higher than that of SETAFs and thus such a reduction might result in significant overheads. However, modern ADF systems are sensitive to the actual complexity of the acceptance conditions in the processed ADF and thus the overheads when processing ADFs with acceptance conditions generated from SETAFs probably will not be as high as one would expect from the worst-case complexity gap.

3.4 Alternative models for attacks involving sets of arguments

SETAFs have not been the only attempt to formalise *collective* attacks[15] in argumentation systems. There have been earlier or more recent related approaches, both in abstract and structured argumentation, each of which captures a slightly different notion of collective attack and with a different aim.

One of the earliest approaches to formalise collective attacks in abstract argumentation was the *collective argumentation theories* proposed by [Bochman, 2003]. These are generalisations of Dung's abstract argumentation frameworks aimed at the representation of the semantics of disjunctive logic programs, but also, more generally, at the description of "reasoning situations in which the conflict between incompatible views or theories is global and cannot be reduced to particular claims made by these theories". [Bochman, 2003] proposes a four-valued semantics, i.e., each argument is assigned a subset of $2^{\{t,f\}}$ and attacks occur among sets of arguments (e.g. $S \hookrightarrow T$) and are interpreted as "at least one of

---

[12]https://www.cs.helsinki.fi/group/coreo/k++adf/
[13]http://www.dbai.tuwien.ac.at/proj/adf/yadf/
[14]http://diamond-adf.sourceforge.net/
[15]We use the term collective to refer to any kind of attack relation that involves sets of arguments, either as attackers or as targets of an attack, or both.

the arguments in the attacked set (T) should be rejected whenever all the arguments from the attacking set (S) are accepted".

[Cayrol and Lagasquie-Schiex, 2010] introduced the notion of coalitions of arguments to represent sets of non-conflicting arguments that are related via the support relation in a bipolar argumentation framework (BAF). Using this notion, a bipolar argumentation framework $AF^{\mathcal{B}}$ can be translated into a Dung-style meta-argumentation framework $C(AF^{\mathcal{B}})$, called "Coalition AF", in which the arguments represent coalitions of arguments of $AF^{\mathcal{B}}$ and the attacks among arguments (called *c-attacks*) correspond to attacks among elements of the corresponding coalitions: S c-attacks T in $C(AF^{\mathcal{B}})$, iff there exist arguments $a, b \in AF^{\mathcal{B}}$ such that $a \in S, b \in T$ and a attacks b in $AF^{\mathcal{B}}$. All arguments belonging to a coalition are then treated in the same way when computing the acceptable arguments: an argument a is acceptable (under the preferred, stable or grounded semantics) in $AF^{\mathcal{B}}$ iff it is a member of a coalition S, which is acceptable (under the same semantics) in $C(AF^{\mathcal{B}})$.

The framework proposed in [Gabbay and Gabbay, 2016] also considers sets of arguments, but as recipients of disjunctive attacks from single arguments. In this framework, the result of an attack from an argument a that is labelled in, to a set of arguments S, is that at least one of the arguments in S must be labelled out. Definition 2.8 and Theorem 2.9 of the same paper show how a finite disjunctive framework can be converted to a Dung-style AF with the same set of extensions, which, combined with the results on the relationship between SETAF and AF that we present in Section 3.2, provide a way to associate SETAF with disjunctive argumentation frameworks. Note also that [Nielsen and Parsons, 2007b] also provides a way to model disjunctive attacks using SETAFs, using the notion of "indeterministic defeat" [Verheij, 1996a] (see Section 3.1 for details).

CumulA [Verheij, 1995; Verheij, 1996a] is an example of a structured argumentation model that supports collective attacks. In this model, arguments are tree-like structures that represent how a conclusion is supported. In order to support situations where a set of arguments should be collectively defeated (*collective defeat*) or at least one of the arguments in a set should be defeated (*indeterministic defeat*), it uses *compound defeaters*, i.e., attack relations where either the source or the target of the attack (or both) are sets of arguments. The meaning of a compound defeater is different than that of joint attacks in SETAFs: if all arguments in the attacking set are undefeated, the arguments in the attacked set are defeated as a group unless one of the arguments in the attacked set has already been defeated by another defeater. In the latter case the compound defeater becomes *inactive*.

Another structured argumentation formalism that incorporates the notion of collective attacks is the Abstract Argumentation System (AAS) from [Vreeswijk, 1997]. An AAS is defined as a triple $(\mathcal{L}, \mathcal{R}, \leq)$, where $\mathcal{L}$ is a language containing the symbol $\bot$, which represents a contradictory proposition, $\mathcal{R}$ is a set of (strict and defeasible) inference rules, and $\leq$ is a preorder on the set of arguments, called *order of conclusive force*, and determining "the relative difference in strength among arguments". Arguments are defined as chains of rules organised as trees. The notion of defeat in AAS is used to capture and resolve conflicts among groupwise incompatible arguments: a set of arguments X defeats an argument a if $X \cup \{a\}$ is incompatible (there is a strict argument b that is based on the conclusions of $X \cup \{a\}$ and has conclusion $\bot$) and X is not undermined by a (there is no $c \in X$ such that $a < c$).

Defeasible Logic [Nute, 1994], which, as shown in [Governatori and Maher, 2000] has an argumentation-theoretic semantics, also supports a type of collective attacks, called *team defeat*. This logic includes a rule priority relation, which is used to resolve conflicts between rules with contradictory conclusions. An attack on a rule r with conclusion p from a rule r' with conclusion $\neg p$ can be invalidated by another rule r'' also with conclusion p that is superior to r'. In this case, we say that r and r'' team defeat r'. Using this feature, we conclude that p is true if for every applicable rule that supports $\neg p$, there is a superior rule for p; in other words, if the rules for $\neg p$ are team defeated by the rules for p[16]. In order to support this feature, the argumentation-theoretic characterisation of Defeasible Logic defines arguments as sets of proof trees supporting the same conclusion and team defeat as a relation between two arguments with opposite conclusions, and requires that an argument team defeats all its attacking arguments to become acceptable. Team defeat is also supported by other rule-based non-

---

[16]For a more detailed discussion on team defeat, see [Billington *et al.*, 2010]

monotonic logics, which use preferences on rules, such as Courteous Logic Programs [Grosof, 1997] and Order Logic [Laenens and Vermeir, 1990]. An interesting problem is to study the possibility of mapping Defeasible Logic, or any of the other rule-based non-monotonic logic that supports team defeat, to SETAF by defining arguments as proof trees and by representing team defeat, between a set of rules R supporting the same conclusion and a rule s supporting the opposite conclusion, as a joint attack from the set of arguments that have a top rule in R to each argument that has s as its top rule.

[Baroni et al., 2018b] recently introduced a semi-structured formalism for argumentation, called LAF-ensembles, capturing a set of essential features of structured arguments, such as their conclusion, their "attackable elements" and their subarguments. They also defined a family of abstract argumentation frameworks, called *set-based* (as their nodes correspond to sets of arguments instead of individual arguments), which are appropriate for representing LAF-ensembles at the abstract level. In set-based argumentation frameworks, the attacks occur at the set level. The main differences between set-based frameworks and SETAFs are that the former allow attacks on sets of arguments and attacks where the source is the empty set; the latter are useful to capture inconsistencies of the theory at the language level (e.g., incompatible subsets of the language in Vreeswijk's AAS [Vreeswijk, 1997]).

Finally, it should be noted that, as also explained in Section 3.2 and shown in [Polberg, 2016], ADFs are generalisations of SETAFs and can therefore model the type of collective attacks used in SETAFs. This is done by setting the following acceptance condition for each argument a: at least one argument from each of the sets of arguments attacking a should be rejected.

4 Applications of joint attacks and models for joint supports

The ideas behind the characterisation of abstract argumentation frameworks with joint attacks, as those described in Section 3, have also been applied in other contexts. In this section we will focus on applications of joint attacks in Bipolar Argumentation Frameworks (BAFs) and argumentation frameworks with higher-order interactions[17].

Briefly, BAFs extend Dung's AF by incorporating a support relation intended to model a positive interaction between the elements it relates. The first works accounting for bipolarity in abstract argumentation conceived the support relation as a binary relation over the set of arguments in the framework (see [Cayrol and Lagasquie-Schiex, 2013; Cohen et al., 2014] for an overview on BAFs). However, later approaches adopted a different view of the support relation, to also account for *joint supports* (i.e., support relations whose source is a set of arguments) or, more generally, *higher-order supports* (i.e., support relations that can target other interactions, either attacks or supports), in addition to arguments.

In this section we will consider approaches to bipolar abstract argumentation that make use of joint attacks, joint supports or both. Finally, we will discuss the possibility of using joint attacks for modelling higher-order attacks and supports (i.e., interactions whose target is another interaction) and the generalised necessary support relation proposed in [Nouioua and Risch, 2011] and also accounted for in [Cayrol et al., 2018b].

4.1 Flat bipolar argumentation frameworks with joint attacks or joint supports

In [Oren and Norman, 2008] the authors used the SETAF as the underlying framework for representing evidence against an argument in order to allow for evidence-based reasoning. They introduced the Evidential Argumentation System (EAS) which further extended the definition of SETAF by incorporating a specialised support relation to capture the notion of *evidential support*. The support relation in the EAS enables to distinguish between *prima-facie* and *standard* arguments; the former arguments do not require support from other arguments to stand, whereas the latter must be linked to at least one *prima-facie* argument through a chain of supports. Moreover, the prima-facie arguments are supported by a special argument η denoting support from the environment or the

---

[17]The latter are the subject of study in Chapter 1 of this handbook [Cayrol et al., 2021].

existence of supporting evidence. Also, analogously to the attack relation, the support relation in an EAS allows for supports to be originated on sets of arguments. Formally:

Definition 4.1. *An EAS is a tuple $\langle Ar, att, sup \rangle$, where Ar is a set of arguments, $att \subseteq (2^{Ar}\backslash\emptyset) \times Ar$ is the attack relation, and $sup \subseteq (2^{Ar}\backslash\emptyset) \times Ar$ is the support relation. A special argument $\eta \in Ar$ is distinguished, such that $\nexists(X,y) \in att$ where $\eta \in X$; and $\nexists X$ where $(X,\eta) \in att$ or $(X,\eta) \in sup$.*

The attack relation in an EAS is interpreted in the same way as the attack relation in the SETAF. Given $X \subseteq Ar$ and $a \in Ar$, $(X,a) \in att$ reads as follows: if all the arguments in X are accepted, then a cannot be accepted. In contrast, the evidential support relation is interpreted as follows. Given $X \subseteq Ar$ and $a \in Ar$, $(X,a) \in sup$ reads as: "the acceptance of a requires the acceptance of every argument in X".

Since the core idea of the EAS is that valid arguments (in particular, those originating attacks) need to trace back to the environment, the authors define the notion of evidence supported attack (e-supported attack). Then, based on this notion, semantics for the EAS have been characterized in [Oren and Norman, 2008] and then reformulated in [Polberg and Oren, 2014], following Dung's methodology.

Generalised Argumentation Frameworks with Necessities (GAFNs) [Nouioua and Risch, 2011] (directly referred to as AFNs in [Nouioua, 2013]) are another kind of bipolar argumentation frameworks that account for interactions between single arguments and sets of arguments but in a different way: a necessity relation between a set of arguments S and an argument a means that the acceptance of a requires the acceptance of *at least one* argument in S.

To illustrate the support relation of GAFNs, let us consider the following example. Suppose that in order to be awarded with a scholarship (s) a student is required to obtain a Bachelor's degree with honours (bh) or justify modest income (mi). In addition, suppose that the student has a bad mark (bm), and that having a bad mark prevents the student from obtaining the honours (regardless of the average of marks). We can represent this scenario by a GAFN with arguments s, bh, mi and bm. On the other hand, there exists an attack from bm to bh, and there exists a necessary support from the set $\{bh, mi\}$ to argument s. It is important to note that, even though the attack from bm to bh will result in bh not being accepted, this does not prevent s from being accepted (in other words, the student will obtain the scholarship). This is because the support towards s is originated in the set $\{bh, mi\}$, where each argument within this set provides an alternative condition for obtaining the scholarship.

Generalised Argumentation Frameworks with Necessities are formally defined as follows:

Definition 4.2. *A Generalised Argumentation Framework with Necessities*
(GAFN) is defined by a tuple $\langle Ar, att, sup \rangle$ where Ar is a set of arguments, $att \subseteq Ar \times Ar$ is an attack relation and $sup \subseteq ((2^{Ar} \setminus \emptyset) \times Ar)$ is a necessity relation.

In [Nouioua, 2013] the author proposed a characterisation of semantics for the GAFN, in addition to those given in [Nouioua and Risch, 2011]. Finally, it should be noted that in [Polberg and Oren, 2014] the authors provided a translation allowing the transformation of a GAFN into an EAS. Briefly, this translation is such that unsupported arguments in the GAFN will be arguments supported by η in the EAS; on the other hand, all sets of supporting arguments in the GAFN are combined into different sets of supporting arguments in the EAS by accounting for their Cartesian product. Finally, for the attack relation it suffices to map the attacking arguments in the GAFN into singleton sets of attacking arguments in the EAS. Then, [Polberg and Oren, 2014] formally established a correspondence between the EAS and the GAFN in terms of their semantics, and identified correspondences between the properties of both frameworks and properties of Dung's AF.

Example 4.3. *Consider the GAFN $\langle Ar, att, sup \rangle$, where:*

- $Ar = \{a, b, c, d, e, f\}$
- $att = \{(b, a), (e, a), (c, d)\}$
- $sup = \{(\{b\}, e), (\{d, f\}, e), (\{a\}, d)\}$

This AFN could be translated into the EAS $\langle Ar', att', sup' \rangle$, where:

- $Ar' = Ar \cup \{\eta\}$
- $att' = \{(\{b\}, a), (\{e\}, a), (\{c\}, d)\}$
- $sup' = \{(\{b,d\}, e), (\{b,f\}, e), (\{a\}, d), (\{\eta\}, a), (\{\eta\}, b), (\{\eta\}, c), (\{\eta\}, f)\}$

For details about the characterisation of semantics for EAS and GAFN we refer the reader to [Oren and Norman, 2008] and [Nouioua and Risch, 2011; Nouioua, 2013], respectively.

4.2 Bipolar argumentation frameworks with joint attacks or joint supports and higher-order interactions

The ideas adopted by the EAS and the GAFN described in the previous section were further exploited in [Cayrol et al., 2018a] and [Cayrol et al., 2018b], where the authors introduced the Recursive Evidence-Based Argumentation Framework (REBAF) and the Recursive Argumentation Framework with Necessity (RAFN). Briefly, these frameworks extend Dung's AF by accounting for attack and support relations that can target not only arguments, but also attacks or supports at any level[18]. As a result, the REBAF adopts the evidential interpretation for the support relation of [Oren and Norman, 2008], whereas the RAFN adopts the generalised necessity interpretation of support proposed in [Nouioua and Risch, 2011]. The formal definitions of these frameworks are included below:

Definition 4.4. *A Recursive Evidence-Based Argumentation Framework (REBAF) is a tuple $\langle Ar, att, sup, \mathbf{s}, \mathbf{t}, PF \rangle$ where Ar, att and sup are pairwise disjoint sets respectively representing the names of arguments, attacks and supports, and $PF \subseteq Ar \cup att \cup sup$ is a set representing the* prima-facie *elements of the framework that do not need to be supported. The functions $\mathbf{s} : (att \cup sup) \mapsto 2^{Ar} \setminus \emptyset$ and $\mathbf{t} : (att \cup sup) \mapsto (Ar \cup att \cup sup)$ respectively map each attack and support to its source and its target.*

---

[18]The REBAF and the RAFN are studied in more detail in Chapter 1 of this handbook [Cayrol et al., 2021].

Definition 4.5. *A Recursive Argumentation Framework with Necessity (RAFN) is a tuple* $\langle Ar, att, sup, \mathbf{s}, \mathbf{t} \rangle$, *where Ar, att and sup are pairwise disjoint sets respectively representing the names of arguments, attacks and supports. The function* $\mathbf{s} : (att \cup sup) \mapsto 2^{Ar} \setminus \emptyset$ *and* $\mathbf{t} : (att \cup sup) \mapsto (Ar \cup att \cup sup)$ *respectively map each attack and support to its source and its target. It is assumed that* $\forall \alpha \in att$, $\mathbf{s}(\alpha)$ *is a singleton.*

Note that, according to Definition 4.4, attacks and supports in a REBAF can have a set of arguments as their source. In contrast, by Definition 4.5, the attack relation of a RAFN is restricted to only allow for arguments as the source of attacks. Then, in both cases, an attack or a support can also be the target of an interaction. Consequently, since these frameworks allow to reason about interactions in addition to arguments, the attacks and supports are also accounted for in the acceptability calculus.[19]

Semantics of REBAF and RAFN are defined using a notion of *structure*, defined as a triple $U = (S, \Gamma, \Delta)$ such that $S \subseteq Ar$, $\Gamma \subseteq att$ and $\Delta \subseteq sup$. Then, the notions of conflict-freeness, acceptability and admissibility as well as the subsequent semantics are defined over these structures, with the idea that the set S represents the set of "acceptable" arguments w.r.t. the structure U, and the sets Γ and Δ respectively represent the sets of "valid attacks" and "valid supports" w.r.t. U. For details about the definition of semantics for REBAF and RAFN, we refer the reader to [Cayrol et al., 2018a] and [Cayrol et al., 2018b], respectively, or to Chapter 1 of this handbook [Cayrol et al., 2021].

4.3 Using joint attacks to model higher-order interactions and generalised necessary supports

Gabbay [2009] proposed Higher-Level Argumentation Frames (HLAFs), which extend Dung's framework by allowing for attacks from arguments targeting not only arguments, but also attacks at any level. A HLAF can be defined as follows:

Definition 4.6. *Let Ar be a set of arguments.* Level $(0, n)$ *argumentation frames are defined as follows:*

---

[19]This feature is also shared by other frameworks such as the AFRA and the ASAF, discussed in Section 4.3.

1. A pair $(a, b) \in Ar \times Ar$ is called a level $(0, 0)$ attack.

2. If $c \in Ar$ and α is a level $(0, n)$ attack then (c, α) is a level $(0, n+1)$ attack.

3. A level $(0, n)$ argumentation frame is the pair $\langle Ar, att \rangle$ where att contains only level $(0, m)$ attacks for $0 \leq m \leq n$.

Note that, although the level of HLAFs is expressed in terms of pairs $(0, n)$ with possibly different values for n, the first component of the pair denoting the level is always 0 (the part of the level associated with the set of arguments). In particular, [Gabbay, 2009] proposed different kinds of approaches in order to define the semantics of HLAFs: the first option consists in translating a HLAF into a Dung's AF; the second alternative corresponds to the characterisation of labellings for HLAF, similarly to the labellings for AFs [Baroni et al., 2018a]; finally, in the third approach Gabbay proposed to translate a HLAF into a logic program. In the following, we will consider the first translation approach, which consists of obtaining a Dung's AF corresponding to a HLAF. Specifically, a HLAF $\langle Ar, att \rangle$ can be translated into an AF $\langle Ar^*, att^* \rangle$, where:

- $Ar^* = Ar \cup \{x_\beta, y_\beta \mid \beta = (a, \alpha) \in att\}$.
- $att^* = \{(a, x_\beta), (x_\beta, y_\beta), (y_\beta, \alpha) \mid \beta = (a, \alpha) \in att\}$.

The new arguments $x_{(a,\alpha)}$ and $y_{(a,\alpha)}$ associated with an attack from a to α respectively represent that the attack is 'live' or 'dead'; moreover, Gabbay argued that the translation of attacks as in the second bullet above is sufficient for attacks which are under attack. This translation is illustrated in Figure 7, where two attacks $\alpha = (a, b) \in att$ and $\beta = (c, \alpha) \in att$ are considered.

Then, given a set of extensions $E_1^+, E_2^+, \ldots, E_n^+$ of the associated AF, the corresponding extensions of the HLAF are $E_i^+ \cap Ar$, $(i = 1, \ldots, n)$.

In spite of proposing the translation described above, [Gabbay, 2009] argued that an attack $(a, b) \in att$ should be viewed as an independent unit, the attack of a on b, which can be itself attacked. In particular, he stated that the preceding translation does not serve its purpose for modelling more general situations, such as attacks originated in other attacks (although the latter are not allowed in the frameworks of Definition 4.6).

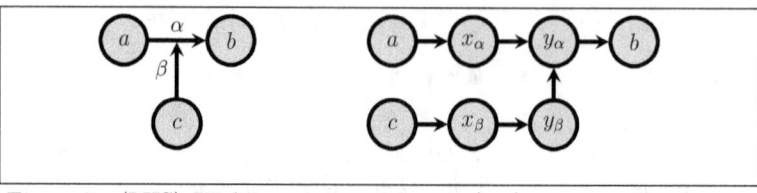

Figure 7: (LHS) HLAF with attacks $\alpha = (a, b)$ and $\beta = (c, \alpha)$; and (RHS) the translation into a series of AF attacks

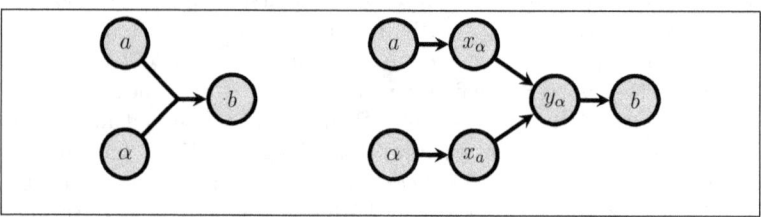

Figure 8: (LHS) Graphical representation of the joint attack by a and α on b, corresponding to an attack $\alpha = (a, b)$; and (RHS) its translation into a sequence of AF attacks.

In that way, the author suggested that an attack $(a, b) \in att$ should be a unit kept 'live' unless attacked itself. Consequently, he proposed an alternative solution making use of *joint attacks*: an attack $\alpha = (a, b)$ is translated in a way such that the argument b is jointly attacked by two arguments a and α Then, both a and α must be 'live' in order for b to be 'dead'. The graphical representation of a joint attack by a and α on b, corresponding to an attack $\alpha = (a, b)$ is shown in Figure 8 on the left.

Given this notion of joint attack, [Gabbay, 2009] proposed a further translation of joint attacks into attacks in a Dung's AF. This translation has some similarities with the one introduced before for directly translating a HLAF into an AF, and is illustrated in Figure 8 on the right for the case of a joint attack by a and α on b.

Alternatively to the translation of joint attacks into attacks at the argument level in a Dung's AF, [Gabbay, 2009] introduced the *frames with joint attacks*:

Definition 4.7. *A frame with joint attacks has the form $\langle Ar, att \rangle$, where*

2 - JOINT ATTACKS AND ACCRUAL

Ar is the set of arguments and $att \subseteq Ar \times Ar \times Ar$ is a ternary relation. We understand $(x, y, z) \in att$ as saying that the two arguments x and y are mounting a joint attack on z.

The author remarked that single attacks can still appear in a frame with joint attacks; these would be attacks of the form $(x, x, y) \in att$. It is important to note that, following Definition 4.7, the frames with joint attacks are a particular case of the SETAFs, where the set of arguments originating an attack is restricted to a maximum of two elements. Consequently, the algorithms and reduction-based approaches for SETAFs discussed in Section 3.3 could also be applied to the frames with joint attacks.

Then, Gabbay introduced definitions analogous to those of [Nielsen and Parsons, 2007b], characterising the extensional semantics of these frameworks. Finally, he proposed a translation from HLAFs into frames with joint attacks, so that extensions of the former correspond to extensions of the latter.

Definition 4.8. Let $\langle Ar, att \rangle$ be a HLAF. The corresponding frame with joint attacks $\langle Ar', att' \rangle$ is defined as follows:

- $Ar' = Ar \cup att$
- $att' = \{(a, \alpha, \beta) \mid \alpha = (a, \beta) \in att\}$

Following this approach, for instance, the HLAF illustrated in Figure 9 on the top can be translated into a frame with joint attacks (or a SETAF) like the one depicted in Figure 9 at the bottom.

Next, we will discuss the possibility of using joint attacks for modelling attacks, including higher-order attacks, through Gabbay's Frames with Joint Attacks (a particular case of SETAFs) in frameworks such as the AFRA [Baroni et al., 2011b] or the ASAF[20] [Gottifredi et al., 2018]. We will start by briefly recalling the definition of these frameworks, as proposed by their authors. As mentioned before, these frameworks are studied in another chapter of this book. Thus, for more details, we refer the interested reader to Chapter 1 of this handbook [Cayrol et al., 2021].

---

[20] In the case of the ASAF, initially, without supports (where such an ASAF would be an AFRA). The means for modelling supports through joint attacks will be discussed later.

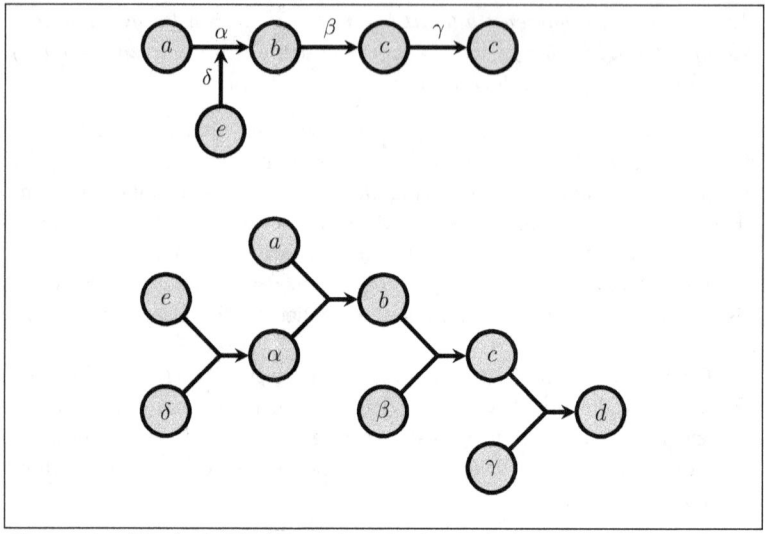

Figure 9: (Top) A HLAF with a higher-order attacks, where Greek letters denote the labels of the attacks; and (Bottom) its corresponding frame with joint attacks

Argumentation Frameworks with Recursive Attacks (AFRAs) [Baroni et al., 2011b] generalise Dung's AFs by incorporating a *recursive attack relation* where attacks are allowed to target other attacks as well as arguments, and the attacks can occur at any level.

Definition 4.9. *An* Argumentation Framework with Recursive Attacks (AFRA) *is a pair* $\langle Ar, att \rangle$ *where:*

- *Ar is a set of arguments;*
- *att is a set of attacks, namely pairs* (a, X) *such that* $a \in Ar$ *and* $(X \in Ar$ *or* $X \in att)$.

Given an attack $\alpha = (a, X) \in att$, a is said to be the source of α, denoted as $\mathbf{s}(\alpha) = a$, and X is the target of α, denoted as $\mathbf{t}(\alpha) = X$. Moreover, similarly to the notation used before for Gabbay's HLAF, [Baroni et al., 2011b] introduces an abbreviated notation for recursive attacks in the AFRA; for instance, an attack $(c, (a, b))$ can be expressed as (c, α), where $\alpha = (a, b)$.

Then, [Baroni et al., 2011b] establishes the different kinds of defeat that can occur between the elements of an AFRA. A key aspect of their formalisation is that they regard attacks (not their source arguments) as the subjects able to defeat arguments and other attacks. Then, an attack can be made ineffective (in other words, defeated) either by attacking the attack itself or by attacking its source. The notions of *direct defeat* and *indirect defeat* are introduced in [Baroni et al., 2011b] as follows:

Definition 4.10. *Let $\langle Ar, att \rangle$ be an AFRA, $\alpha \in att$ and $X \in Ar \cup att$. It is said that α defeats X, denoted $\alpha \to^R X$ if one of the following conditions holds:*

- $\mathbf{t}(\alpha) = X$ *(direct defeat); or*

- $X = \beta \in att$ *and* $\mathbf{t}(\alpha) = \mathbf{s}(\beta)$ *(indirect defeat).*

Then, based on this notion of defeat, the notions of conflict-freeness, acceptability, admissibility and extensions under different semantics are introduced following Dung's methodology. Consequently, the extensions of an AFRA will not only contain the accepted arguments under the corresponding semantics, but also the accepted attacks.

Example 4.11. *The arguments and attacks depicted at the top in Figure 9 correspond to the AFRA $\langle Ar, att \rangle$, where $Ar = \{a, b, c, d, e\}$ and $att = \{\alpha, \beta, \gamma, \delta\}$, with $\mathbf{s}(\alpha) = a$, $\mathbf{t}(\alpha) = b$, $\mathbf{s}(\beta) = b$, $\mathbf{t}(\beta) = c$, $\mathbf{s}(\gamma) = c$, $\mathbf{t}(\gamma) = d$, $\mathbf{s}(\delta) = e$, $\mathbf{t}(\delta) = \alpha$.*

Here, the direct defeats are: $\alpha \to^R b$, $\beta \to^R c$, $\gamma \to^R d$ and $\delta \to^R \alpha$. On the other hand, the indirect defeats are: $\alpha \to^R \beta$ and $\beta \to^R \gamma$. Consequently, β reinstates d, α reinstates c and γ, and δ reinstates α and b. As a result, for instance, the AFRA has only one complete extension (which is also its grounded and only preferred and stable extension), namely $\{a, e, \delta, b, \beta, d\}$. In contrast, if we apply the SETAF semantics on the framework depicted at the top on Figure 9[21], we have that the only complete extension is $\{a, e, \delta, b, \beta, \gamma, d\}$.

---

[21] As stated before, the frames with joint attacks are a particular case of SETAFs.

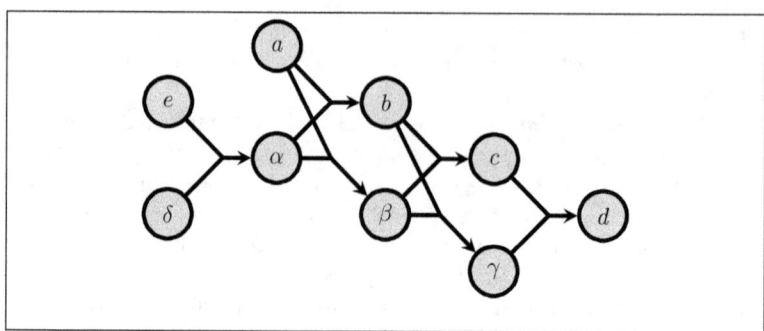

Figure 10: SETAF corresponding to the AFRA from Example 4.11

The difference in the result obtained by applying the AFRA semantics, compared to the one obtained by translating the AFRA into a SETAF and then applying the SETAF semantics, has to do with the fact that the translation proposed in [Gabbay, 2009] does not take into account the indirect defeats. In particular, in the above example, the indirect defeat by β on γ is not captured, leaving γ as an accepted attack[22]. This suggests that we need to establish a different translation of AFRAs into SETAFs, in order to account for the effect of indirect defeats. An alternative translation of an AFRA into a SETAF could be:

Definition 4.12. *Let $\langle Ar, att \rangle$ be an AFRA. The corresponding SETAF $\langle Ar, \triangleright \rangle$ is defined as follows:*

$$
\begin{aligned}
Ar =\ & Ar \cup att \\
\triangleright =\ & \{(\{a, \alpha\}, X) \mid \alpha = (a, X) \in att\} \cup \\
& \{(\{a, \alpha\}, \alpha') \mid \alpha = (a, X) \in att,\ \alpha' \in att,\ \mathbf{s}(\alpha') = X\}
\end{aligned}
$$

The AFRA from Example 4.11, corresponding to the framework depicted at the top of Figure 9, can be translated following Definition 4.12 to obtain the SETAF depicted in Figure 10. Then, applying the SETAF semantics on that framework, the only complete extension coincides with the one obtained with the AFRA semantics in Example 4.11.

---

[22]The indirect defeat by α on β is not captured either; however, since α is not accepted (because it is directly defeated by δ) it does not affect the outcome.

Let us now consider the formalization of the Attack-Support Argumentation Framework (ASAF) [Gottifredi et al., 2018]. Briefly, the ASAF extends Dung's AF by incorporating bipolar higher-order interactions. In that way, the ASAF allows for the representation and reasoning with attack and support relations not only between arguments, but also targeting the attack and support relations themselves. In particular, the support relation of the ASAF is interpreted as necessity [Nouioua and Risch, 2011]. That is, the necessary support relation in the ASAF imposes the following acceptability constraints on the elements it relates: if a supports b, then the acceptance of b implies the acceptance of a; equivalently, the non-acceptance of a implies the non-acceptance of b. Note that the support relation in the ASAF is set to be binary, differently from the necessary support relation of the GAFN introduced in Section 4.1. Some of the following definitions are taken from [Alfano et al., 2020], where the background for the ASAF was succinctly introduced.

Definition 4.13. *An* Attack-Support Argumentation Framework *(ASAF) is a tuple $\langle Ar, att, sup \rangle$ where Ar is a set of arguments, $att \subseteq W$ is the attack relation, and $sup \subseteq W$ is the support relation, with W being the set iteratively defined as follows: $W = Ar \times Ar$ (basic step) and $W = Ar \times W$ (iterative step). It is assumed that sup is acyclic and $att \cap sup = \emptyset$.*

Similarly to the case of the AFRA, an attack $(a, b) \in att$ will be denoted as $\alpha_1 = (a, b)$; analogously, a support $(b, c) \in sup$ will be denoted as $\beta_1 = (b, c)$. Then, for instance, an attack from d to α_1 will be denoted as $\alpha_2 = (d, \alpha_1)$. In general, given an attack $\alpha = (a, X) \in att$, a is called the source of α, denoted $\mathbf{s}(\alpha) = a$, and X is called the target of α, denoted $\mathbf{t}(\alpha) = X$. Analogously, given a support $\beta = (b, Y) \in sup$, b is called the source of β, denoted $\mathbf{s}(\beta) = b$, and Y is called the target of β, denoted $\mathbf{t}(\beta) = Y$.

Like in the AFRA, different kinds of defeat that can occur between the elements of an ASAF. Specifically, they correspond to the two kinds of defeat identified for the AFRA, plus two additional kinds of defeat that arise from the coexistence of the attack and support relations.

Definition 4.14. *Let $\Delta = \langle Ar, att, sup \rangle$ be an ASAF, $\alpha \in att$, $X \in (Ar \cup att \cup sup)$ and $\mathbf{S} \subseteq sup$. We say that α defeats X (given $\mathbf{S}$),*

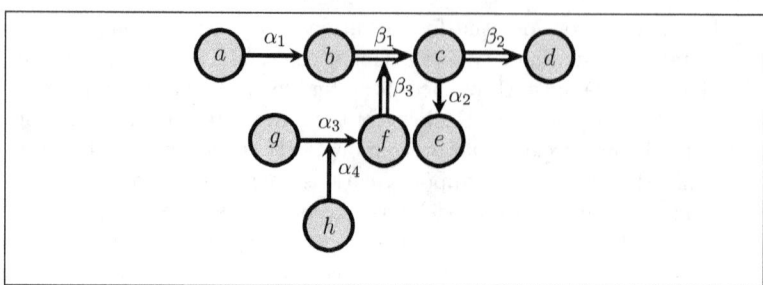

Figure 11: ASAF from Example 4.15

denoted α def X given $\mathbf{S}$ (or simply α def X whenever $\mathbf{S} = \emptyset$) iff one of the following conditions holds:

- there exists a (possibly empty) support path from $\mathbf{t}(\alpha)$ to X, whose corresponding set of supports is $\mathbf{S}$; or

- $X \in att$ and there exists a (possibly empty) support path from $\mathbf{t}(\alpha)$ to $\mathbf{s}(X)$, whose corresponding set of supports is $\mathbf{S}$.

To illustrate these notions, let us consider the following example. Similarly to Dung's AF or the AFRA, an ASAF can be graphically represented using a graph-like notation where two kinds of edges are considered: $\rightarrow$ for the attack relation and $\Rightarrow$ for the support relation. In addition, attacks and supports are labelled with greek letters, following the convention that attacks are labelled with α (possibly with subscripts) and supports are labelled with β (again, possibly with subscripts).

Example 4.15. Consider the ASAF $\langle Ar, att, sup \rangle$, where $Ar = \{a, b, c, d, e, f, g, h\}$, $att = \{\alpha_1, \alpha_2, \alpha_3, \alpha_4\}$ and $sup = \{\beta_1, \beta_2, \beta_3\}$, with $\alpha_1 = (a, b)$, $\alpha_2 = (c, e)$, $\alpha_3 = (g, f)$, $\alpha_4 = (h, \alpha_3)$, $\beta_1 = (b, c)$, $\beta_2 = (c, d)$ and $\beta_3 = (f, \beta_1)$. This framework is depicted in Figure 11, and the following defeats occur: α_1 def b, α_2 def e, α_3 def f, α_4 def α_3, α_1 def c given $\{\beta_1\}$, α_1 def α_2 given $\{\beta_1\}$, α_1 def d given $\{\beta_1, \beta_2\}$, α_3 def β_1 given $\{\beta_3\}$.

The semantics of the ASAF are also defined following Dung's methodology, accounting for the notions of conflict-freeness, acceptability and admissibility, to later characterise the complete, preferred, stable and grounded semantics of the framework. It should be noted that, since the defeats in the ASAF may involve a set of supports, these notions cannot be directly defined by considering the definitions for AFs (see Section 3.1 and Definition 4.14).

Definition 4.16. Let $\Delta = \langle Ar, att, sup \rangle$ be an ASAF and $\mathbf{S} \subseteq (Ar \cup att \cup sup)$.

- $\mathbf{S}$ is conflict-free iff $\nexists \alpha, X \in \mathbf{S}$, $\nexists \mathbf{S}' \subseteq (\mathbf{S} \cap sup)$ such that α def X given $\mathbf{S}'$.

- $X \in (Ar \cup att \cup sup)$ is acceptable w.r.t. $\mathbf{S}$ iff $\forall \alpha \in att$, $\forall \mathbf{T} \subseteq sup$ such that α def X given $\mathbf{T}$: $\exists Y \in (\{\alpha\} \cup \mathbf{T})$, $\exists \alpha' \in \mathbf{S}$, $\exists \mathbf{S}' \subseteq (\mathbf{S} \cap sup)$ such that α' def Y given $\mathbf{S}'$.

- $\mathbf{S}$ is admissible iff it is conflict-free and for all $X \in \mathbf{S}$, X is acceptable w.r.t. $\mathbf{S}$.

Definition 4.17. Let $\Delta = \langle Ar, att, sup \rangle$ be an ASAF and $\mathbf{S} \subseteq (Ar \cup att \cup sup)$.

- $\mathbf{S}$ is a complete extension of Δ iff it is an admissible set and $\forall X \in (Ar \cup att \cup sup)$, if X is acceptable w.r.t. $\mathbf{S}$, then $X \in \mathbf{S}$.

- $\mathbf{S}$ is a preferred extension of Δ iff it is a maximal (w.r.t. $\subseteq$) complete extension of Δ.

- $\mathbf{S}$ is a stable extension of Δ iff it is a complete extension of Δ and $\forall X \in (Ar \cup att \cup sup) \backslash \mathbf{S}$, $\exists \alpha \in \mathbf{S}$, $\exists \mathbf{S}' \subseteq (\mathbf{S} \cap sup)$ such that α def X given $\mathbf{S}'$.

- $\mathbf{S}$ is the grounded extension of Δ iff it is the smallest (w.r.t. $\subseteq$) complete extension of Δ.

The ASAF from Example 4.15 has only one complete extension, which is also the grounded extension and the only preferred and stable extension of the framework: $\{a, e, f, g, h, \alpha_1, \alpha_4, \beta_1, \beta_2, \beta_3\}$. In particular, it can be noted that whereas α_3 def β_1 given $\{\beta_3\}$, the support β_1 is reinstated by α_4, since α_4 def α_3. Then, the defeats from α_1 on c and α_2 given $\{\beta_1\}$ are also reinstated, as well as the defeat from α_1 on d given $\{\beta_1, \beta_2\}$.

As shown in [Gottifredi et al., 2018], an ASAF without support is an AFRA. So, when applying the AFRA semantics on ASAFs without supports we obtain the same outcome as the one obtained under the ASAF semantics. Consequently, the translation of an AFRA into a SETAF could also be applied to the ASAF; nevertheless, some adjustments need to be made in order to account for the defeats involving a set of supports. A possible translation of an ASAF into a SETAF is given below.

Definition 4.18. *Let $\langle Ar, att, sup \rangle$ be an ASAF. The corresponding SETAF $\langle Ar, \triangleright \rangle$ is defined as follows:*

$$Ar = Ar \cup att \cup sup \cup \{\beta^* \mid \beta \in sup\}$$
$$\triangleright = \{(\{a, \alpha\}, X) \mid \alpha = (a, X) \in att\} \cup$$
$$\{(\{a, \alpha\}, \alpha') \mid \alpha = (a, X) \in att, \ \alpha' \in att, \ \mathbf{s}(\alpha') = X\} \cup$$
$$\{(\{a, \alpha\}, X^*) \mid \alpha = (a, X) \in att, \ X \in sup\} \cup$$
$$\{(\{a, \beta\}, \beta^*), (\{\beta^*\}, X) \mid \beta = (a, X) \in sup\} \cup$$
$$\{(\{\beta^*\}, X^*) \mid \beta \in sup, \ X \in sup, \mathbf{t}(\beta) = X\} \cup$$
$$\{(\{\beta^*\}, \alpha) \mid \beta \in sup, \ \alpha \in att, \mathbf{t}(\beta) = \mathbf{s}(\alpha)\}$$

The ASAF from Example 4.15, corresponding to the framework depicted in Figure 11, can be translated following Definition 4.18 to obtain the SETAF depicted in Figure 12. Applying the SETAF semantics on that framework, the only complete extension coincides with the one obtained with the ASAF semantics, namely $\{a, e, f, g, h, \alpha_1, \alpha_4, \beta_1, \beta_2, \beta_3\}$.

Finally, we will briefly discuss the possibility of using joint attacks to model the generalised necessity relation proposed in [Nouioua and Risch, 2011] adopted in frameworks such as the RAFN (see Section 4.2).

Let us recall the example introduced in Section 2, represented using the SETAF from Figure 2. There, we can think of the argument NP as providing a context under which $A18$ attacks M; that is, a person aged under 18 is not allowed to marry whenever parent permission is not provided. So, we could think of this situation as corresponding to the existence of an attack $\alpha_1 = (A18, M)$ and a generalised necessary support $\beta_1 = (\{NP\}, \alpha_1)$. Similarly, given the restriction to drink alcohol, arguments NA and NM can be considered as providing alternative contexts under which $A18$ attacks Alc. Therefore, we could think of representing this situation through an attack $\alpha_2 = (A18, Alc)$ and a

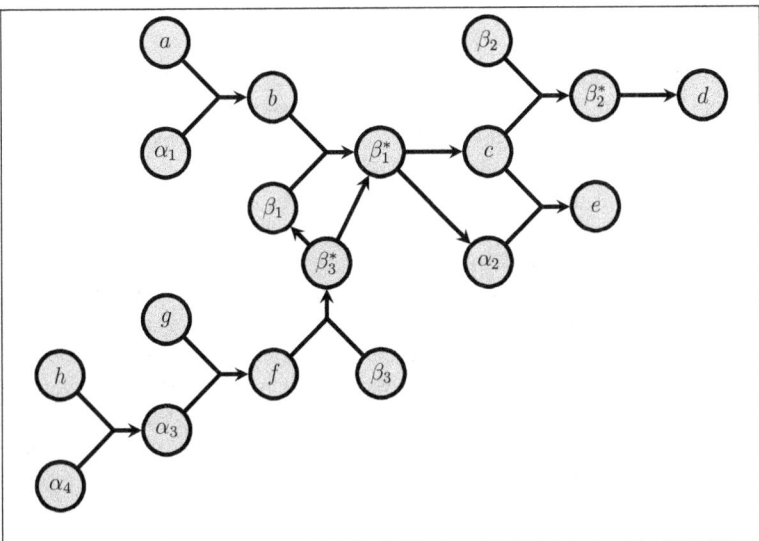

Figure 12: SETAF corresponding to the ASAF from Example 4.15

generalised necessary support $\beta_2 = (\{NA, NM\}, \alpha_2)$. This is because, in this situation, it suffices to have either NA or NM accepted in order to be able to accept α_2 (i.e., in order for the attack from $A18$ to Alc to hold).

The preceding example suggests that joint attacks (as those in the SETAF from Figure 2) may be suitable for modelling the generalised necessary support relation in the case of higher-order supports targeting an attack. However, for instance, if there exists another interaction (say, an attack α_3) targeting β_1, we should be able to model on the SETAF the fact that if α_3 is accepted then β_1 no longer holds and, consequently, that NP does not provide a context under which the attack α_1 from $A18$ to M holds. Nevertheless, if the support β_1 from NP to α_1 is modeled by a joint attack from NP and $A18$ as in Figure 2, we cannot model the attack from α_3 towards β_1 in the SETAF, since β_1 is not made explicit in this representation.

5 Accrual

A parallel line of research in computational argumentation studies the accrual of arguments, i.e., how arguments supporting or refuting the same claim can be combined. The main differences between accrual and joint attacks (at least under the type of joint attacks used in SETAFs) is that, in joint attacks, the strength of an argument or a combination of arguments is not considered when evaluating the effectiveness of attacks, and each argument participating in a joint attack is an essential element of it (in other words if an argument is missing then the attack is ineffective), while in accrual the strength of each argument is taken into account, and adding an argument to an accrual makes the accrual stronger, or more generally it changes its strength and the effectiveness of its attacks (or supports).

A seminal study on the accrual of arguments [Prakken, 2005] set out three principles for accrual:

1. An accrual is sometimes weaker than its accruing elements. This is due to the possibility that the accruing reasons are not independent.

2. An accrual makes its elements inapplicable. More generally, any 'larger' accrual that applies makes all its 'lesser' versions inapplicable. This is because an accrual is meant to consider all available information, while the individual arguments it consists of take only part of the information into account.

3. Flawed reasons or arguments may not accrue. Any treatment of accrual should capture that when an individual reason or argument turns out to be flawed, it does not take part in the accrual.

Prakken also described two general ways to formalise accrual: (*a*) the *knowledge representation* (or else *KR*) approach, which requires formulating a separate rule for each possible combination of the accruing reasons; (*b*) the *inference* approach, where the accrual is part of the inference process, i.e., after all individual reasons have been constructed, those that attack or support the same claim are somehow aggregated and some mechanism is then used to resolve any conflicts between the conflicting sets of reasons. He also proposed a formalisation of accrual using the inference approach, according to which the conclusion of each individual defeasible inference step is labelled with the premises of the applied defeasible inference rule:

$$\phi, \phi \Rightarrow \psi \mathrel{|\!\sim} \psi^{\{\phi, \phi \Rightarrow \psi\}}$$

and a new defeasible inference rule is introduced that takes any set of labelled versions of a certain formula and produces the unlabelled version:

$$\phi^{l_1}, \cdots, \phi^{l_n} \mathrel{|\!\sim} \phi$$

The attack relationships among arguments are adjusted as follows: rebuttal requires that the two arguments support opposite conclusions that are labelled in the same way, while undercut requires that the attacking arguments have unlabelled conclusions. Finally, the following rules ensure that when a set of reasons accrues, any subset of it is inapplicable:

$$\phi^{l_1}, \cdots, \phi^{l_n} \mathrel{|\!\sim} \neg \lceil \phi^{l_1}, \cdots, \phi^{l_{n-1}} \rceil$$

The proposed formalisation satisfies all principles of accrual, but has a computational drawback: it requires considering all possible accruals for every conclusion, which may lead to an exponential increase in the number of arguments.

The idea of combining arguments for and against a claim, albeit under the name "aggregation" rather than "accrual" was studied by argumentation researchers before [Prakken, 2005]. One line of work, that of Fox and colleagues, goes back at least as far as [O'Neil et al., 1989], where the idea of symbolically weighing evidence is formalised in what recognisably is a structured argumentation framework, and arguably as far back as [Fox et al., 1980] where the idea was first applied. The formal development of that work came to a conclusion with [Krause et al., 1995] and [Elvang-Gøransson et al., 1993]. The former paper describes a model that links the simple form of accrual from [O'Neil et al., 1989], which effectively just looks at the numbers or arguments for and against a claim[23], with forms of accrual which connect to probabilistic models. The latter uses the same model to develop a hierarchy of notions of acceptability, coming close to Dung's work at about the same time that work was first published [Dung, 1993].

CumulA [Verheij, 1995; Verheij, 1996a] was another structured argumentation model that dealt with accrual. Additionally to the notion of *compound defeaters*, which we discussed in Section 3.4, it also includes the notions of *coordination* and *narrowings* of arguments. Different arguments supporting the same conclusion can be combined in a *coordinated* argument, while the *narrowing* of a coordinated argument a is an argument b supporting the same conclusion as a but containing a subset of the arguments combined in a (or narrowings of them). CumulA deals with accrual using compound defeaters and the following acceptability condition for arguments: if the narrowing of an argument a is **in** (meaning that the argument is accepted), then a should be **in** too. As shown in [Prakken, 2005], CumulA satisfies all three principles of accrual but the second one (i.e. that an accrual makes its elements inapplicable) is satisfied in a way that is too strong. Because of the acceptability condition described above, which implies that if an accrual is **out** then all its narrowings are **out**, it cannot capture a situation where an accrual is defeated because of subargument defeat so that some of its narrowings can be undefeated.

More recently, [Besnard and Hunter, 2001] developed another account of accrual, based on their logic-based approach to argumentation, though again they do not describe it as such. In [Besnard and Hunter, 2001], lines of discussion about a particular claim — the argument for it, the arguments against it, the arguments against those arguments, and so on — are brought together into an argument tree. Then, all argument trees for or against a claim are assembled into an argument structure. An argument structure thus gathers everything that is relevant to whether or not a claim should be accepted. This, of course, is not much different to what one would get from assembling all of the arguments in a structured framework like ASPIC+ or DeLP that bear on a specific formula into some super-structure. However, whereas most

---

[23]Before dismissing such a simple model, consider how effective such simple models can be [Dawes, 1979].

structured frameworks summarise this higher level structure in a notion of acceptability, [Besnard and Hunter, 2001] defines a "categoriser" which maps a structure to a number, and this can be thought of as the accrued value of the set of arguments in the structure.

[Lucero et al., 2009] proposed an approach for formalising the accrual of arguments in Defeasible Logic Programming using the notion of *a-structure*, a special kind of argument which subsumes different chains of reasoning that provide support for the same conclusion, and *partial attacks* among a-structures, where the attacking a-structure generally affects only the narrowing of the attacked a-structure containing exactly the arguments affected by the conflict. A binary preference relation on a-structures is used to determine the relevant strength of conflicting a-structures and whether an attack succeeds (in which case it constitutes a *defeat*). To deal with combined attacks (situations where two or more a-structures simultaneously attack the same a-structure), they define a process, called *bottom-up sequential degradation*, according to which the defeats are applied in sequence with the "deeper" ones applied first. The described framework satisfies all three principles of accrual. Its main difference with the formalisation proposed in [Prakken, 2005] is that when analysing a theory to determine the accepted (undefeated) a-structures, it only considers maximal accruals (a-structures) and not all possible accruals for a conclusion.

[Gordon, 2018] proposed the use of argument weighing functions as a way to model different types of argument schemes, including some types of argument accrual, in Carneades, a structured argumentation framework. In this framework, an argument is defined as a tuple (s, P, c, u) where s is the scheme that the argument instantiates; P, the premises of the argument, is a finite subset of the underlying logical language $\mathcal{L}$; and c, its conclusion, and u, its undercutter, are elements of $\mathcal{L}$. Its semantics is defined in terms of a labelling, which assigns a value from $\{\text{in}, \text{out}, \text{undec}\}$ to each element of $\mathcal{L}$ and a weighing function, which assigns a value from $[0, 1]$ to each argument and 0 to all arguments such that their undercutter is in. Gordon also provided several examples of weighing functions, some of which are appropriate for modelling different types of accrual. To simulate *convergent arguments*, i.e., arguments that at least one of its premises must be in to support their conclusion, he defined a weighing function that assigns 1 to an argument if at least one of its premises is **in** and its undercutter is not **in**, and 0 otherwise. A weighing function that simulates *cumulative arguments*, i.e., arguments whose strength increases with the number of their acceptable premises, assigns the percentage of the premises of the argument that are **in** to every argument whose undercutter is not **in**. Cumulative arguments are a special type of accrual that does not satisfy Prakken's second principle, since cumulation can only increase the strength of an argument. Another weighing function that simulates accrual takes into account all *factors* (statements) that need to be considered when evaluating the arguments for a certain issue. It does so by assigning to each argument the proportion of its factors that are premises of the argument and are labelled **in**. One limitation of this framework with respect to accrual is that although it handles various forms of accrual at the level of statements, e.g., premises or factors of an argument, it does not provide a way to handle accrual of multiple arguments.

[Prakken, 2019] proposed a formalisation of accrual for ASPIC+, a structured argumentation framework where arguments are tree-like structures constructed from a knowledge base, which is a subset of an underlying logical language $\mathcal{L}$, and a set of inference, strict or defeasible, rules. In this framework, there are two ways to attack an argument a: either at the top inference rule r of a (*undercut*) or at the conclusion of r (*rebuttal*)[24] - in both cases r must be defeasible, otherwise a cannot be attacked. [Prakken, 2019] extended ASPIC+ with the notion of *accrual sets*, which are defined relative to a labelling of the set of arguments S. An accrual set for a literal $\phi \in \mathcal{L}$, denoted as $s_l(\phi)$, is the set of arguments with conclusion ϕ satisfying the following two conditions: (i) for any argument in $s_l(\phi)$ no immediate subargument of a is **out** and no undercutter of a is **in**; (ii) any argument with conclusion ϕ whose undercutters are **out** and its immediate subarguments are **in** must be in $s_l(\phi)$. The extended framework also includes a preference relation $\leq$ on the power set of S, such that any set of arguments containing a strict argument is at least as preferred as every other sub-

---

[24] Note that these definitions are different from the standard ASPIC+ where arguments can also be attacked on their subarguments. As explained in [Prakken, 2019], in the version of ASPIC+ considered in this paper, arguments are constructed recursively and the recursion takes care of subargument attacks.

set of S. It also includes a new defeat relation on arguments, called *l-defeat*, which takes into account accruals: an argument a l-defeats an argument b iff a undercuts b; or a rebuts b, and for some accrual sets for the conclusions of a, $s_l(Conc(a))$, and b, $s_l(Conc(b))$, it holds that $s_l(Conc(a)) \not< s_l(Conc(b))$. A *characteristic function* F is used to compute the labelling of a framework, which satisfies the following conditions: an argument a is **in** iff all arguments that l-defeat a are **out** and all immediate subarguments of a are **in**; a is **out** iff it is defeated by an argument that is labelled **in** or one of its immediate subarguments are **out**. The proposed framework satisfies all principles of accrual and preserves some of the properties of Dung's AFs, such as the existence of complete and preferred labellings and the relations between grounded, complete, stable and preferred semantics.

In the field of abstract argumentation, the most relevant approaches are the frameworks with graded semantics (e.g., see [Baroni *et al.*, 2019] for an overview and a study of their properties) or ranking semantics and social argumentation frameworks (e.g., see [Leite and Martins, 2011; Baroni *et al.*, 2015; Patkos *et al.*, 2016]). Such frameworks provide methods for assessing the strength of an argument based on the aggregate strength of its attackers and the aggregate strength of its supporters (and in some cases the initial valuation of the argument), capturing the main idea of accrual. Some of their general properties are: (i) the larger the set of the attackers on an argument, the lower the strength of the argument under attack; (ii) the larger the set of supporters or defenders of an argument, the higher the strength of the argument they support or defend; and (iii) an argument with 0 strength does not have an effect on the strength of the arguments it attacks or supports. The last property satisfies the third principle of accrual (i.e., that flawed arguments do not accrue), while by considering the aggregate strength of the attackers or supporters of an argument, they essentially satisfy the second principle, i.e., that an accrual makes its elements inapplicable. Properties (i) and (ii), however, violate the second principle, since they imply that an accrual is always stronger than the individual accrued arguments.

6 Proposals for future work on joint attacks

In this section we highlight some emerging topics for future research on joint attacks and accrual.

There are several interesting directions for further research concerning semantics of SETAFs. Standard semantics of AFs have been generalised to SETAFs and their basic properties and relations are settled. However, several prominent semantics have not yet been generalised and analysed on SETAFs, e.g., cf2 [Baroni and Giacomin, 2003], strong admissibility [Baroni and Giacomin, 2007; Caminada and Dunne, 2019] and weak admissibility [Baumann et al., 2020]. Recently, a first approach to transfer also ranking-based semantics to SETAFs has been undertaken [Yun et al., 2020a]. Properties of AF semantics have been studied in versatile aspects [van der Torre and Vesic, 2018; Baumann, 2018] beyond the existing analysis for SETAFs. For instance, generalising the principle-based approach for analysing and comparing semantics to SETAFs would be valuable for the selection of the right argumentation semantics, and understanding the different notions of equivalence also on SETAFs is fundamental for using SETAFs in dynamic settings. Concerning the latter, a first investigation of strong equivalence notions for SETAFs has been done in [Dvořák et al., 2019b].

As another research direction one could consider enhancing the expressiveness of SETAF by extending its basic model with features similar to the ones used in extensions of the AF model, such as the introduction of a joint support relation, weights on (joint) attacks, values promoted by (sets of) arguments, or a preference relation among (sets of) arguments. This would allow associating SETAFs with the corresponding AF extensions, i.e., frameworks for bipolar argumentation [Amgoud et al., 2008; Cayrol and Lagasquie-Schiex, 2005], graded [Grossi and Modgil, 2015] or weighted argumentation [Dunne et al., 2011], value-based [Bench-Capon, 2003], or preference-based argumentation [Amgoud and Vesic, 2014] respectively. A related but somehow orthogonal research direction is the investigation of the relations of SETAFs and other extensions of AFs concerning their expressiveness. Existing investigations in that direction are the embedding of SETAFs in ADFs [Linsbichler et al., 2016] and translations between SETAFs and claim-augmented AFs [Dvořák et al., 2020d].

The translations from SETAF to AFs discussed in Section 3.2 either had the weakness that they might increase the size exponentially or only supported a selection of the semantics. For future research one could investigate alternative translation schemes in order to avoid this pitfall. Ideally, we would like to have a transformation that applies for all semantics and causes a polynomial increase in the size of the framework (in the sense of [Brewka et al., 2011]), while at the same time resulting in elegant correspondences for all the semantics (unlike [Flouris and Bikakis, 2019], where this is true only for some of the semantics). Recall, that [Polberg, 2017] provide a translation that satisfies the latter two properties but only for a selection of the semantics, i.e., the semantics based on complete extensions, with the exception of semi-stable semantics. Also notice that a polynomial increase in the size of the framework can still result in an exponential increase in the number of arguments. Thus, another open question is whether such a potential exponential increase in the number of arguments can be avoided.

On the computational side there are several open challenges. From the theoretical perspective one would be interested in identifying classes of instances that provide milder complexity than general SETAFs. One approach that has been extensively studied for AFs are the so called tractable fragments [Dunne, 2007], i.e., special graph classes like acyclic or bipartite, that allow for efficient reasoning procedures. A more general approach are graph parameters and techniques for parametrised complexity theory that allow for algorithms which are only exponential w.r.t. a graph parameter but polynomial in the size of the AF [Dvořák et al., 2012a; Dvořák et al., 2012b]. From a more practical view one would be interested in efficient labelling-based algorithms [Nofal et al., 2014; Charwat et al., 2015] for SETAFs as well as systems that extend methods that have been successfully applied for AFs [Cerutti et al., 2018]. An important step to boost the development of such systems would be to establish standard formats to share SETAF instances and standard benchmark sets.

Regarding the application of joint attacks, the ideas discussed in Section 4.3 could be further explored. As shown in Section 4.3, the translations from an AFRA [Baroni et al., 2011b] or an ASAF [Gottifredi et al., 2018] into a SETAF yield the same outcomes as those obtained directly by applying the AFRA or ASAF semantics, respectively. However, this was only shown for the examples illustrated in that section. A formal analysis of this correspondence for the general case of an arbitrary ASAF or AFRA is left for future work. In addition, the brief discussion at the end of Section 4.3 can also be the subject of future work, also considering the translations discussed in Section 3.2. Specifically, studying the possibility of using the SETAF for modelling the generalised support relation of frameworks like the RAFN [Cayrol et al., 2018a; Cayrol et al., 2018b], accounting for all cases: first-order supports, higher-order supports targeting attacks and supports, and higher-order supports which can be themselves attacked.

Finally, the potential of collective attacks in structured models of argumentation is rather unexplored. Consider an instantiation scheme like ASPIC+ [Prakken, 2019], or instantiations for logic programs [Caminada et al., 2015] or assumption-based argumentation [Alcântara et al., 2019] that construct AFs from a knowledge base. Using SETAFs instead of AFs as target formalism can significantly reduce the number of arguments and, in certain cases one can even ensure that each statement has a unique argument supporting that statement [Dvořák et al., 2020d]. A first investigation in that direction is [Yun et al., 2019; Yun et al., 2020b] where SETAFs are instantiated from Datalog knowledge bases. There is also scope for relating this kind of approach to work on accrual and the other models that collect related arguments such as the argument trees of [Besnard and Hunter, 2001] and the coalitions of [Cayrol and Lagasquie-Schiex, 2010].

The accrual of arguments is a less studied problem compared to joint attacks. As we discussed in Section 5 most existing approaches are focused on structured argumentation and only few of them satisfy all three principles proposed in [Prakken, 2005]. There are a lot of interesting future research directions in this area such as the systematic comparison and evaluation of the frameworks that support accrual and the development of methods for handling accrual in abstract argumentation. The latter could rely on the recently proposed graded semantics for abstract argumentation or may require the development of a new abstract argumentation framework that explicitly models accrual. Another interesting direction, which could also lead to a solution for this problem, is to study the relation between current approaches for accrual and collective attacks and the mapping between the frameworks that deal with these two different problems.

7 Conclusions

In this chapter we have studied different formalisms that account for joint attacks (or more generally, collective attacks) in abstract argumentation. Also, we discussed the consideration of joint attacks in the literature of structured argumentation as well as the application of joint attacks (and joint supports) in other frameworks such as bipolar argumentation frameworks or argumentation frameworks with higher-order interactions. We also touched upon works on argument accrual which, although not strictly related to the existing models of joint attacks, can be considered as a related topic. In particular, the SETAF [Nielsen and Parsons, 2007b] framework along with its computational complexity, algorithms and applications, was the main subject of study in this chapter.

In Section 3.1 the basic definitions of the framework were provided, followed by the presentation of extension-based semantics of the SETAF and the relationships between them, as well as the introduction of labelling-based semantics for the framework. Then, in Section 3.2 the expressive power of the SETAF was compared against that of Dung's AF [Dung, 1995]. For that, the results and analyses reported in [Dvořák et al., 2019a; Flouris and Bikakis, 2019; Polberg, 2017; Brewka et al., 2011] were accounted for. On the one hand, the characterisation of the expressive power using signatures was discussed, to then consider exponential and compact translations of SETAFs into AFs, and later discuss an indirect translation path consisting of a translation of a SETAF into an ADF [Brewka et al., 2018] and a translation of the latter into an AF. The main conclusion here is that, although SETAFs could be represented by means of Dung's AFs, the SETAF indeed increases the expressive power of the AF. Moreover, as discussed in Section 3.3, the translations from a SETAF into an AF lead to an exponential blow-up in the arguments, making them not well-suited for computational matters.

In Section 3.3 different computational problems for SETAFs were characterised, following the definition of function problems and decision problems for Dung's AFs (cf. [Charwat *et al.*, 2015; Dvořák and Dunne, 2018; Cerutti *et al.*, 2018]). In particular, the decision problems include determining the credulous or skeptical acceptance of an argument under a given semantics, the verification of an extension, or determining the existence of a (non-empty) extension. Then, the computational complexity of these decision problems is addressed, and the results are linked to the existing results for decision problems in a Dung's AF. The conclusion here is that the complexity of decision problems for SETAFs is the same as the complexity of the corresponding problems for Dung's AFs. Notwithstanding this, we should note that although in both cases the complexity is stated w.r.t. the size of the input framework, the size is interpreted differently for AFs and SETAFs. On the one hand, the size of an AF is often interpreted in terms of the number of arguments of the input framework. On the other hand, since the number of attacks in a SETAF may be exponentially larger than the number of arguments in the framework (due to the existence of attacks by sets of arguments), the size of a SETAF should be interpreted in terms of the number of arguments plus the number of attacks.

Also, Section 3.3 briefly discussed the ideas behind labelling-based algorithms for SETAFs, illustrating the algorithm for labelling enumeration under the preferred semantics. Moreover, as mentioned before, different reduction-based approaches for computing the extensions of a SETAF were presented. The former consists of encoding the SETAF and its semantics in Answer-set programming [Marek and Truszczyński, 1999; Niemelä, 1999], whereas the latter rely on translations of a SETAF into a Dung's AF or an ADF. While the drawbacks of the translations into AFs were pointed out above, we should note that the translation into an ADF offers the possibility of using existing systems for ADFs [Linsbichler *et al.*, 2018] without incurring significant overheads.

Section 3.4 recalled alternative models of abstract and structured argumentation which account for attacks involving sets of arguments. While in SETAF, a set of arguments can only be the source of an attack, in other models sets of arguments are only considered as a potential target of an attack (e.g. see [Gabbay and Gabbay, 2016]); or as both potential sources or targets of an attack (e.g. see [Bochman, 2003;

Cayrol and Lagasquie-Schiex, 2010; Verheij, 1995]). The different models also differ in how the arguments within a set are treated. For example, while in the framework of [Cayrol and Lagasquie-Schiex, 2010], all arguments in a coalition are treated in the same way, i.e. they are all either accepted or rejected, in other frameworks, such as the ones proposed in [Bochman, 2003; Gabbay and Gabbay, 2016], a successful attack on a set of arguments has as a result that at least one of the arguments in the attacked set is rejected. Although the different approaches have different aims, an interesting problem is to study the extent to which they can be mapped to each other and whether there is a more general model that captures their different features.

Section 4 addresses the application of models for joint attacks in the context of Bipolar Argumentation Frameworks (BAFs) and argumentation frameworks with higher-order interactions such as those addressed in Chapter 1 [Cayrol et al., 2021]. First, BAFs that make use of joint attacks, joint supports, or both are recalled, highlighting the constraints they impose on the attack and support relations, as well as the adopted interpretations for the notion of support. Then, generalisations of these BAFs are presented, which incorporate higher-order interactions in order to allow for attacks and supports targeting other attacks or supports. Later, an analysis of the possibility of using joint attacks to model higher-order interactions is performed.

On the one hand, Section 4 considered the work by Gabbay on Higher-Level Argumentation Frames [Gabbay, 2009], as well as the proposed translations of HLAFs into Frames with Joint Attacks (a particular case of SETAF). Then, Gabbay's ideas are taken in the context of the AFRA [Baroni et al., 2011b] and the ASAF [Gottifredi et al., 2018], two abstract argumentation frameworks allowing for binary higher-order interactions. Our findings are that, when applying the translation proposed by Gabbay to obtain the SETAF associated with an AFRA or an ASAF without supports, and then applying the SETAF semantics, the corresponding extensions might not be as expected (since the translation does not account for the existence of indirect defeats). Then, translations for obtaining a SETAF corresponding to an AFRA or an ASAF are proposed, and illustrated through examples; their formalisations for the general case of an AFRA or an ASAF are left for future research. Moreover, the possibility of using the SETAF to model generalised necessary supports is briefly analysed in Section 4, leaving an in-depth discussion for future work.

In Section 5 different works addressing the topic of argument accrual were discussed, both at the abstract and structured levels of argumentation. As discussed there, the main difference between the approaches studying argument accrual and those accounting for joint attacks (e.g., as in SETAFs) is that the strength of the arguments combined to originate a joint attack is not accounted for when evaluating the effectiveness of attacks; notwithstanding this, each argument originating a joint attack is an essential element in the sense that the attack becomes ineffective whenever one of its source arguments is missing. In contrast, in accrual, the strength of each argument is taken into account, and adding an argument to an accrual causes changes in the strength and the effectiveness of its attacks (or supports).

Finally, as stated in Section 6, many open challenges remain for research on joint attacks, in addition to those mentioned above.

Acknowledgements

This work was partially supported by EPSRC grant EP/P010105/1 and by Universidad Nacional del Sur grant 24/N046. The opinions expressed in this paper are those of the authors and do not necessarily reflect the opinions of the funders. The authors are grateful to the reviewers of this chapter for their helpful comments.

References

[Alcântara et al., 2019] João F. L. Alcântara, Samy Sá, and Juan Carlos Acosta Guadarrama. On the equivalence between abstract dialectical frameworks and logic programs. *Theory and Practice of Logic Programming*, 19(5-6):941–956, 2019.

[Alfano et al., 2020] Gianvincenzo Alfano, Andrea Cohen, Sebastian Gottifredi, Sergio Greco, Francesco Parisi, and Guillermo R. Simari. Dynamics in abstract argumentation frameworks with recursive attack and support relations. In *Proceedings of the 24th European Conference on Artificial Intelligence*, 2020.

[Amgoud and Vesic, 2014] Leila Amgoud and Srdjan Vesic. Rich preference-based argumentation frameworks. *International Journal of Approximate Reasoning*, 55(2):585–606, 2014.

[Amgoud et al., 2008] Leila Amgoud, Claudette Cayrol, Marie-Christine Lagasquie-Schiex, and Pierre Livet. On bipolarity in argumentation frameworks. *International Journal of Intelligent Systems, Bipolar Representations of Information and Preference (Part 2: reasoning and learning)*, 23(10):1062–1093, 2008.

[Baroni and Giacomin, 2003] Pietro Baroni and Massimiliano Giacomin. Solving semantic problems with odd-length cycles in argumentation. In Thomas D. Nielsen and Nevin Lianwen Zhang, editors, *Proceedings of the 7th European Conference on Symbolic and Quantitative Approaches to Reasoning with Uncertainty*, pages 440–451. Springer, 2003.

[Baroni and Giacomin, 2007] Pietro Baroni and Massimiliano Giacomin. On principle-based evaluation of extension-based argumentation semantics. *Artificial Intelligence*, 171(10-15):675–700, 2007.

[Baroni et al., 2011a] Pietro Baroni, Martin Caminada, and Massimiliano Giacomin. An introduction to argumentation semantics. *The Knowledge Engineering Review*, 26(4):365–410, 2011.

[Baroni et al., 2011b] Pietro Baroni, Federico Cerutti, Massimiliano Giacomin, and Giovanni Guida. AFRA: Argumentation framework with recursive attacks. *International Journal of Approximate Reasoning*, 52:19–37, 2011.

[Baroni et al., 2015] Pietro Baroni, Marco Romano, Francesca Toni, Marco Aurisicchio, and Giorgio Bertanza. Automatic evaluation of design alternatives with quantitative argumentation. *Argument & Computation*, 6(1):24–49, 2015.

[Baroni et al., 2018a] Pietro Baroni, Martin Caminada, and Massimiliano Giacomin. Abstract argumentation frameworks and their semantics. In Pietro Baroni, Dov Gabbay, Massimiliano Giacomin, and Leendert van der Torre, editors, *Handbook of Formal Argumentation*, chapter 4, pages 159–236. College Publications, 2018.

[Baroni et al., 2018b] Pietro Baroni, Massimiliano Giacomin, and Beishui Liao. A general semi-structured formalism for computational argumentation: Definition, properties, and examples of application. *Artificial Intelligence*, 257:158 – 207, 2018.

[Baroni et al., 2019] Pietro Baroni, Antonio Rago, and Francesca Toni. From fine-grained properties to broad principles for gradual argumentation: A principled spectrum. *International Journal of Approximate Reasoning*, 105:252–286, 2019.

[Baumann et al., 2020] Ringo Baumann, Gerhard Brewka, and Markus Ulbricht. Revisiting the foundations of abstract argumentation — semantics based on weak admissibility and weak defense. In *Proceedings of the Thirty-Fourth AAAI Conference on Artificial Intelligence*, pages 2742–2749. AAAI Press, 2020.

[Baumann, 2018] Ringo Baumann. On the nature of argumentation semantics: Existence and uniqueness, expressibility, and replaceability. In Pietro Baroni, Dov Gabbay, Massimiliano Giacomin, and Leendert van der Torre, editors, *Handbook of Formal Argumentation*, chapter 17, pages 839–936. College Publications, 2018. also appears in IfCoLog Journal of Logics and their Applications 4(8):2779–2886.

[Bench-Capon, 2003] Trevor J. M. Bench-Capon. Persuasion in practical argument using value-based argumentation frameworks. *Journal of Logic and Computation*, 13(3):429–448, 2003.

[Besnard and Hunter, 2001] Philippe Besnard and Anthony Hunter. A logic-based theory of deductive arguments. *Artificial Intelligence*, 128:203–235, 2001.

[Billington et al., 2010] David Billington, Grigoris Antoniou, Guido Governatori, and Michael Maher. An inclusion theorem for defeasible logics. *ACM Transactions on Computational Logic*, 12(1), November 2010.

[Bochman, 2003] Alexander Bochman. Collective argumentation and disjunctive logic programming. *Journal of Logic and Computation*, 13(3):405–428, 2003.

[Bondarenko et al., 1997] Andrei Bondarenko, Phan Minh Dung, Robert A. Kowalski, and Francesca Toni. An abstract, argumentation-theoretic approach to default reasoning. *Artificial Intelligence*, 93(1-2):63–101, 1997.

[Brewka et al., 2011] Gerd Brewka, Paul E. Dunne, and Stefan Woltran. Relating the semantics of abstract dialectical frameworks and standard AFs. In *Proceedings of the 22nd International Joint Conference on Articial Intellignce*, 2011.

[Brewka et al., 2013] Gerhard Brewka, Hannes Strass, Stefan Ellmauthaler, Johannes Peter Wallner, and Stefan Woltran. Abstract dialectical frameworks revisited. In *Proceedings of the 23rd International Joint Conference on Artificial Intelligence*, pages 803–809, 2013.

[Brewka et al., 2017] Gerhard Brewka, Martin Diller, Georg Heissenberger, Thomas Linsbichler, and Stefan Woltran. Solving advanced argumentation problems with answer-set programming. In Satinder P. Singh and Shaul Markovitch, editors, *Proceedings of the Thirty-First AAAI Conference on Artificial Intelligence*, pages 1077–1083. AAAI Press, 2017.

[Brewka et al., 2018] Gerhard Brewka, Stefan Ellmauthaler, Hannes Strass, Johannes P. Wallner, and Stefan Woltran. Abstract dialectical frameworks.

In Pietro Baroni, Dov Gabbay, Massimiliano Giacomin, and Leendert van der Torre, editors, *Handbook of Formal Argumentation*, chapter 5, pages 237–285. College Publications, 2018. also appears in IfCoLog Journal of Logics and their Applications 4(8):2263–2318.

[Caminada and Dunne, 2019] Martin Caminada and Paul E. Dunne. Strong admissibility revisited: Theory and applications. *Argument & Computation*, 10(3):277–300, 2019.

[Caminada and Gabbay, 2009] Martin Caminada and Dov M. Gabbay. A logical account of formal argumentation. *Studia Logica*, 93(2/3):109–145, 2009.

[Caminada et al., 2015] Martin Caminada, Samy Sá, João F. L. Alcântara, and Wolfgang Dvořák. On the equivalence between logic programming semantics and argumentation semantics. *International Journal of Approximate Reasoning*, 58:87–111, 2015.

[Caminada, 2006] Martin Caminada. Semi-stable semantics. In *Proceedings of the 1st Conference on Computational Models of Argument*, pages 121–130, 2006.

[Caminada, 2007] Martin Caminada. Comparing two unique extension semantics for formal argumentation: Ideal and eager. In *Proceedings of the 19th Belgian-Dutch Conference on Artificial Intelligence*, pages 81–87, 2007.

[Cayrol and Lagasquie-Schiex, 2005] Claudette Cayrol and Marie-Christine Lagasquie-Schiex. On the acceptability of arguments in bipolar argumentation frameworks. In *Proceedings of the 8th European Conference on Symbolic and Quantitative Approaches to Reasoning with Uncertainty*, pages 378–389, 2005.

[Cayrol and Lagasquie-Schiex, 2010] Claudette Cayrol and Marie-Christine Lagasquie-Schiex. Coalitions of arguments: A tool for handling bipolar argumentation frameworks. *International Journal of Intelligent Systems*, 25(1):83–109, 2010.

[Cayrol and Lagasquie-Schiex, 2013] Claudette Cayrol and Marie-Christine Lagasquie-Schiex. Bipolarity in argumentation graphs: Towards a better understanding. *International Journal of Approximate Reasoning*, 54(7):876–899, 2013.

[Cayrol et al., 2018a] Claudette Cayrol, Jorge Fandinno, Luis Fariñas del Cerro, and Marie-Christine Lagasquie-Schiex. Argumentation frameworks with recursive attacks and evidence-based support. In F. Ferrarotti and S. Woltran, editors, *Proceedings of the 10th International Symposium on Foundations of Information and Knowledge Systems*, pages 150–169. Springer-Verlag, 2018.

[Cayrol et al., 2018b] Claudette Cayrol, Jorge Fandinno, Luis Fariñas del Cerro, and Marie-Christine Lagasquie-Schiex. Structure-based semantics of argumentation frameworks with higher-order attacks and supports. In Sanjay Modgil, Katarzyna Budzynska, and John Lawrence, editors, *International Conference on Computational Models of Argument*, pages 29–36. IOS Press, septembre 2018.

[Cayrol et al., 2021] Claudette Cayrol, Andrea Cohen, and Marie-Christine Lagasquie-Schiex. Higher-order interactions (bipolar or not) in abstract argumentation: A state of the art. In Dov Gabbay, Massimiliano Giacomin, Guillermo R. Simari, and Matthias Thimm, editors, *Handbook of Formal Argumentation*, volume 2, chapter 1. College Publications, 2021.

[Cerutti et al., 2018] Federico Cerutti, Sarah A. Gaggl, Matthias Thimm, and Johannes P. Wallner. Foundations of implementations for formal argumentation. In Pietro Baroni, Dov Gabbay, Massimiliano Giacomin, and Leendert van der Torre, editors, *Handbook of Formal Argumentation*, chapter 15, pages 688–767. College Publications, 2018. also appears in IfCoLog Journal of Logics and their Applications 4(8):2623–2706.

[Charwat et al., 2015] Günther Charwat, Wolfgang Dvořák, Sarah Alice Gaggl, Johannes Peter Wallner, and Stefan Woltran. Methods for solving reasoning problems in abstract argumentation - A survey. *Artificial Intelligence*, 220:28–63, 2015.

[Cohen et al., 2014] Andrea Cohen, Sebastian Gottifredi, Alejandro Javier García, and Guillermo Ricardo Simari. A survey of different approaches to support in argumentation systems. *Knowledge Engineering Review*, 29(5):513–550, 2014.

[Dawes, 1979] Robyn M. Dawes. The robust beauty of improper linear models in decision making. *American Psychologist*, 34(7):571, 1979.

[Dung et al., 2007] Phan Minh Dung, Paolo Mancarella, and Francesca Toni. Computing ideal sceptical argumentation. *Artificial Intelligence*, 171(10-15):642–674, July 2007.

[Dung, 1993] Phan Minh Dung. On the acceptability of arguments and its fundamental role in nonmonotonic reasoning and logic programming. In *Proceedings of the 13th International Joint Conference on Artificial Intelligence*, pages 852–857, Chambéry, France, 1993.

[Dung, 1995] Phan Minh Dung. On the acceptability of arguments and its fundamental role in nonmonotonic reasoning, logic programming and n-person games. *Artificial Intelligence*, 77(2):321–357, September 1995.

[Dunne et al., 2011] Paul E. Dunne, Anthony Hunter, Peter McBurney, Simon Parsons, and Michael Wooldridge. Weighted argument systems: Basic definitions, algorithms, and complexity results. *Artificial Intelligence*, 175(2):457 – 486, 2011.

[Dunne et al., 2013] Paul E. Dunne, Wolfgang Dvořák, and Stefan Woltran.

Parametric properties of ideal semantics. *Artificial Intelligence*, 202:1–28, 2013.

[Dunne et al., 2015] Paul E. Dunne, Wolfgang Dvořák, Thomas Linsbichler, and Stefan Woltran. Characteristics of multiple viewpoints in abstract argumentation. *Artificial Intelligence*, 228:153–178, 2015.

[Dunne, 2007] Paul E. Dunne. Computational properties of argument systems satisfying graph-theoretic constraints. *Artificial Intelligence*, 171(10-15):701–729, 2007.

[Dunne, 2009] Paul E. Dunne. The computational complexity of ideal semantics. *Artificial Intelligence*, 173(18):1559–1591, 2009.

[Dvořák and Dunne, 2018] Wolfgang Dvořák and Paul E. Dunne. Computational problems in formal argumentation and their complexity. In Pietro Baroni, Dov Gabbay, Massimiliano Giacomin, and Leendert van der Torre, editors, *Handbook of Formal Argumentation*, chapter 14, pages 631–687. College Publications, 2018. also appears in IfCoLog Journal of Logics and their Applications 4(8):2557–2622.

[Dvořák et al., 2012a] Wolfgang Dvořák, Sebastian Ordyniak, and Stefan Szeider. Augmenting tractable fragments of abstract argumentation. *Artificial Intelligence*, 186:157–173, 2012.

[Dvořák et al., 2012b] Wolfgang Dvořák, Reinhard Pichler, and Stefan Woltran. Towards fixed-parameter tractable algorithms for abstract argumentation. *Artificial Intelligence*, 186:1–37, 2012.

[Dvořák et al., 2018] Wolfgang Dvořák, Alexander Greßler, and Stefan Woltran. Evaluating SETAFs via Answer-Set Programming. In *Proceedings of the Second International Workshop on Systems and Algorithms for Formal Argumentation*, pages 10–21. CEUR-WS.org, 2018.

[Dvořák et al., 2019a] Wolfgang Dvořák, Jorge Fandinno, and Stefan Woltran. On the expressive power of collective attacks. *Argument & Computation*, 10(2):191–230, 2019.

[Dvořák et al., 2019b] Wolfgang Dvořák, Anna Rapberger, and Stefan Woltran. Strong equivalence for argumentation frameworks with collective attacks. In *Proceedings of the 42nd German Conference on Artificial Intelligence*, pages 131–145. Springer, 2019.

[Dvořák et al., 2020a] Wolfgang Dvořák, Atefeh Keshavarzi Zafarghandi, and Stefan Woltran. Expressiveness of SETAFs and support-free ADFs under 3-valued semantics. In *Proceedings of the 8th Conference on Computational Models of Argument*, pages 191–202. IOS Press, 2020.

[Dvořák et al., 2020b] Wolfgang Dvořák, Anna Rapberger, and Johannes Peter Wallner. Labelling-based algorithms for SETAFs. In *Proceedings of the*

Third International Workshop on Systems and Algorithms for Formal Argumentation, pages 34–46. CEUR-WS.org, 2020.

[Dvořák et al., 2020c] Wolfgang Dvořák, Anna Rapberger, and Stefan Woltran. On the different types of collective attacks in abstract argumentation: equivalence results for SETAFs. *Journal of Logic and Computation*, 30(5):1063–1107, 2020.

[Dvořák et al., 2020d] Wolfgang Dvořák, Anna Rapberger, and Stefan Woltran. On the relation between claim-augmented argumentation frameworks and collective attacks. In *Proceedings of the 24th European Conference on Artificial Intelligence*, pages 721–728. IOS Press, 2020.

[Egly et al., 2010] Uwe Egly, Sarah Alice Gaggl, and Stefan Woltran. Answer-set programming encodings for argumentation frameworks. *Argument & Computation*, 1(2):147–177, 2010.

[Ellmauthaler and Strass, 2014] Stefan Ellmauthaler and Hannes Strass. The DIAMOND system for computing with abstract dialectical frameworks. In *Proceedings of the 5th Conference on Computational Models of Argumen*, volume 266, pages 233–240. IOS Press, 2014.

[Elvang-Gøransson et al., 1993] Morten Elvang-Gøransson, Paul J. Krause, and John Fox. Acceptability of arguments as 'logical uncertainty'. In M. Clarke, R. Kruse, and S. Moral, editors, *Proceedings of the 2nd European Conference on Symbolic and Quantitative Approaches to Reasoning and Uncertainty*, pages 85–90. Springer, 1993.

[Flouris and Bikakis, 2019] Giorgos Flouris and Antonis Bikakis. A comprehensive study of argumentation frameworks with sets of attacking arguments. *International Journal of Approximate Reasoning*, 109, 03 2019.

[Fox et al., 1980] J. Fox, D. Barber, and K. D. Bardhan. Alternatives to Bayes? a quantitative comparison with rule-based diagnostic inference. *Methods of Information in Medicine*, 19:210–215, 1980.

[Gabbay and Gabbay, 2016] Dov M. Gabbay and Michael Gabbay. Theory of disjunctive attacks, Part I. *Logic Journal of the IGPL*, 24(2):186–218, 2016.

[Gabbay, 2009] Dov M. Gabbay. Semantics for higher level attacks in extended argumentation frames. *Studia Logica*, 93:357–381, 2009.

[Gaggl et al., 2020] Sarah Alice Gaggl, Thomas Linsbichler, Marco Maratea, and Stefan Woltran. Design and results of the second international competition on computational models of argumentation. *Artificial Intelligence*, 279, 2020.

[García and Simari, 2018] Alejandro Javier García and Guillermo Ricardo Simari. Argumentation based on logic programming. In Pietro Baroni, Dov Gabbay, Massimiliano Giacomin, and Leendert van der Torre, editors,

Handbook of Formal Argumentation, chapter 8, pages 409–435. College Publications, 2018.

[Gebser et al., 2011] Martin Gebser, Benjamin Kaufmann, Roland Kaminski, Max Ostrowski, Torsten Schaub, and Marius T. Schneider. Potassco: The Potsdam answer set solving collection. *AI Communications*, 24(2):107–124, 2011.

[Gordon, 2018] Thomas F. Gordon. Defining argument weighing functions. *Journal of Applied Logics*, 5(3):747–773, 2018.

[Gottifredi et al., 2018] Sebastian Gottifredi, Andrea Cohen, Alejandro Javier García, and Guillermo Ricardo Simari. Characterizing acceptability semantics of argumentation frameworks with recursive attack and support relations. *Artificial Intelligence*, 262:336–368, 2018.

[Governatori and Maher, 2000] Guido Governatori and Michael J. Maher. An Argumentation-Theoretic Characterization of Defeasible Logic. In *Proceedings of the 14th European Conference on Artificial Intelligence*, pages 469–473, 2000.

[Grosof, 1997] Benjamin N. Grosof. Prioritized conflict handling for logic programs. In Jan Maluszynski, editor, *Proceedings of the 1997 International Symposium on LOgic Programming*, pages 197–211. MIT Press, 1997.

[Grossi and Modgil, 2015] Davide Grossi and Sanjay Modgil. On the graded acceptability of arguments. In *Proceedings of the 24th International Joint Conference on Artificial Intelligence*, pages 868–874. AAAI Press, 2015.

[Krause et al., 1995] Paul J. Krause, Simon Ambler, Morten Elvang-Gøransson, and John Fox. A logic of argumentation for reasoning under uncertainty. *Computational Intelligence*, 11 (1):113–131, 1995.

[Laenens and Vermeir, 1990] E. Laenens and D. Vermeir. A Fixpoint Semantics for Ordered Logic. *Journal of Logic and Computation*, 1(2):159–185, 12 1990.

[Leite and Martins, 2011] João Leite and João G. Martins. Social abstract argumentation. In *Proceedings of the 22nd International Joint Conference on Artificial Intelligence, Barcelona*, pages 2287–2292. IJCAI/AAAI, 2011.

[Leone et al., 2006] Nicola Leone, Gerald Pfeifer, Wolfgang Faber, Thomas Eiter, Georg Gottlob, Simona Perri, and Francesco Scarcello. The DLV system for knowledge representation and reasoning. *ACM Transactions on Computational Logic*, 7(3):499–562, 2006.

[Linsbichler et al., 2016] Thomas Linsbichler, Jörg Pührer, and Hannes Strass. A uniform account of realizability in abstract argumentation. In *Proceedings of the 22nd European Conference on Artificial Intelligence*, pages 252–260. IOS Press, 2016.

[Linsbichler et al., 2018] Thomas Linsbichler, Marco Maratea, Andreas Niskanen, Johannes Peter Wallner, and Stefan Woltran. Novel algorithms for abstract dialectical frameworks based on complexity analysis of subclasses and SAT solving. In Jérôme Lang, editor, *Proceedings of the Twenty-Seventh International Joint Conference on Artificial Intelligence*, pages 1905–1911, 2018.

[Lucero et al., 2009] Mauro J. Gómez Lucero, Carlos I. Chesñevar, and Guillermo R. Simari. On the accrual of arguments in defeasible logic programming. In *Proceedings of the 21st International Joint Conference on Artificial Intelligence*, pages 804–809. Morgan Kaufmann Publishers Inc., 2009.

[Marek and Truszczyński, 1999] Victor W. Marek and Mirosław Truszczyński. Stable models and an alternative logic programming paradigm. In *The Logic Programming Paradigm – A 25-Year Perspective*, pages 375–398. Springer, 1999.

[Modgil and Bench-Capon, 2010] Sanjay Modgil and Trevor J. M. Bench-Capon. Metalevel argumentation. *Journal of Logic and Computation*, 21(6):959–1003, 09 2010.

[Modgil and Prakken, 2018] Sanjay Modgil and Henry Prakken. Abstract rule-based argumentation. In Pietro Baroni, Dov Gabbay, Massimiliano Giacomin, and Leendert van der Torre, editors, *Handbook of Formal Argumentation*, chapter 6, pages 287–364. College Publications, 2018. also appears in IfCoLog Journal of Logics and their Applications 4(8):2319–2406.

[Nielsen and Parsons, 2006] Søren Holbech Nielsen and Simon Parsons. Computing preferred extensions for argumentation systems with sets of attacking arguments. In *Proceedings of the 1st Conference on Computational Models of Argument*, pages 97–108. IOS Press, 2006.

[Nielsen and Parsons, 2007a] Søren Holbech Nielsen and Simon Parsons. An application of formal argumentation: Fusing Bayesian networks in multi-agent systems. *Artificial Intelligence*, 171(10–15):754–775, 2007.

[Nielsen and Parsons, 2007b] Søren Holbech Nielsen and Simon Parsons. A generalization of Dung's abstract framework for argumentation: Arguing with sets of attacking arguments. In *Proceedings of the 3rd International Workshop on Argumentation in Multi-Agent Systems*, pages 54–73, 2007.

[Niemelä, 1999] Ilkka Niemelä. Logic programming with stable model semantics as a constraint programming paradigm. *Annals of Mathematics and Artificial Intelligence*, 25(3–4):241–273, 1999.

[Nofal et al., 2014] Samer Nofal, Katie Atkinson, and Paul E. Dunne. Algorithms for decision problems in argument systems under preferred semantics. *Artificial Intelligence*, 207:23–51, 2014.

[Nouioua and Risch, 2011] Farid Nouioua and Vincent Risch. Argumentation frameworks with necessities. In *Proceedings of the 5th International Conference on Scalable Uncertainty Management*, pages 163–176, 2011.

[Nouioua, 2013] Farid Nouioua. AFs with necessities: Further semantics and labelling characterization. In *7th International Conference on Scalable Uncertainty Management*, pages 120–133, 2013.

[Nute, 1994] Donald Nute. *Defeasible Logic*, page 353?395. Oxford University Press, Inc., USA, 1994.

[O'Neil et al., 1989] Mike O'Neil, Andrzej J. Glowinski, and John Fox. A symbolic theory of decision-making applied to several medical tasks. In *AIME 89: Proceedings of the Second European Conference on Artificial Intelligence in Medicine*, pages 62–71. Springer, 1989.

[Oren and Norman, 2008] Nir Oren and Timothy J. Norman. Semantics for evidence-based argumentation. In *Proceedings of the 2nd Conference on Computational Models of Argument*, pages 276–284, 2008.

[Patkos et al., 2016] Theodore Patkos, Giorgos Flouris, and Antonis Bikakis. Symmetric multi-aspect evaluation of comments - extended abstract. In *Proceedings of the 22nd European Conference on Artificial Intelligence*, pages 1672–1673. IOS Press, 2016.

[Polberg and Oren, 2014] Sylwia Polberg and Nir Oren. Revisiting support in abstract argumentation systems. In *Proceedings of the 5th Conference on Computational Models of Argument*, pages 369–376, 2014.

[Polberg, 2016] Sylwia Polberg. Understanding the abstract dialectical framework. In *Proceedings of the 15th European Conference on Logics in Artificial Intelligence*, pages 430–446, 2016.

[Polberg, 2017] Sylwia Polberg. *Developing the abstract dialectical framework*. PhD thesis, TU Wien, Institute of Information Systems, 2017. available at http://katalog.ub.tuwien.ac.at/AC13773888.

[Prakken, 2005] Henry Prakken. A study of accrual of arguments, with applications to evidential reasoning. In *Proceedings of the 10th International Conference on Artificial Intelligence and Law*, pages 85–94, New York, NY, USA, 2005. ACM.

[Prakken, 2019] Henry Prakken. Modelling accrual of arguments in ASPIC+. In *Proceedings of the Seventeenth International Conference on Artificial Intelligence and Law*, pages 103–112, New York, NY, USA, 2019. Association for Computing Machinery.

[Pührer, 2020] Jörg Pührer. Realizability of three-valued semantics for abstract dialectical frameworks. *Artificial Intelligence*, 278, 2020.

[Strass, 2015a] Hannes Strass. Expressiveness of two-valued semantics for abstract dialectical frameworks. *Journal of Artificial Intelligence Research*, 54:193–231, 2015.

[Strass, 2015b] Hannes Strass. The relative expressiveness of abstract argumentation and logic programming. In *Proceedings of the 29th AAAI Conference on Artificial Intelligence*, pages 1625–1631, 2015.

[Thimm and Villata, 2017] Matthias Thimm and Serena Villata. The first international competition on computational models of argumentation: Results and analysis. *Artificial Intelligence*, 252:267–294, 2017.

[van der Torre and Vesic, 2018] Leendert van der Torre and Srdjan Vesic. The principle-based approach to abstract argumentation semantics. In Pietro Baroni, Dov Gabbay, Massimiliano Giacomin, and Leendert van der Torre, editors, *Handbook of Formal Argumentation*, chapter 16, pages 797–838. College Publications, 2018. also appears in IfCoLog Journal of Logics and their Applications 4(8):2735–2778.

[Verheij, 1995] Bart Verheij. Accrual of arguments in defeasible argumentation. In *Proceedings of the Second Dutch/German Workshop on Nonmonotonic Reasoning*, pages 217–224, 1995.

[Verheij, 1996a] Bart Verheij. *Rules, Reasons, Arguments. Formal studies of argumentation and defeat*. PhD thesis, Universiteit Maastricht, 1996.

[Verheij, 1996b] Bart Verheij. Two approaches to dialectical argumentation: Admissible sets and argumentation stages. In *In Proceedings of the International Conference on Formal and Applied Practical Reasoning*, pages 357–368. Universiteit, 1996.

[Vreeswijk, 1997] Gerard A. W. Vreeswijk. Abstract argumentation systems. *Artificial Intelligence*, 90(1–2):225–279, February 1997.

[Yun et al., 2019] Bruno Yun, Madalina Croitoru, Srdjan Vesic, and Pierre Bisquert. NAKED: n-ary graphs from knowledge bases expressed in Datalog$^\pm$. In *Proceedings of the 18th International Conference on Autonomous Agents and MultiAgent Systems*, pages 2390–2392, 2019.

[Yun et al., 2020a] Bruno Yun, Srdjan Vesic, and Madalina Croitoru. Ranking-based semantics for sets of attacking arguments. In *Proceedings of the 34th Conference on Artificial Intelligence*, pages 3033–3040. AAAI Press, 2020.

[Yun et al., 2020b] Bruno Yun, Srdjan Vesic, and Madalina Croitoru. Sets of attacking arguments for inconsistent Datalog knowledge bases. In *Proceedings of the 8th Conference on Computational Models of Argument*, pages 419–430. IOS Press, 2020.

CHAPTER 3

PREFERENCE IN ABSTRACT ARGUMENTATION

SOUHILA KACI
Université Montpellier 2, LIRMM, France
souhila.kaci@lirmm.fr

LEENDERT VAN DER TORRE
Computer Science and Communication, University of Luxembourg, Luxembourg
leon.vandertorre@uni.lu

SRDJAN VESIC
CNRS, Univ. Artois, CRIL – Centre de Recherche en Informatique de Lens, F-62300 Lens, France
vesic@cril.fr

SERENA VILLATA
Laboratoire I3S, Université Côte d'Azur, CNRS, Inria, France
serena.villata@univ.cotedazur.fr

Abstract

Preference is a key concept in argumentation to represent the comparative strength of arguments. In abstract argumentation, it is represented by an ordinal comparative, or by a numerical function. In this chapter, we study the role of comparative preference in abstract argumentation. This chapter consists of two parts. In the first part, we survey four reductions discussed in the literature to provide semantics to preference-based argumentation frameworks, and we present ten principles for such semantics.

Some of these principles have been mentioned in the literature before, and some of them are new. We provide a complete analysis for the four Dung semantics, based on these four reductions and the ten principles. In the second part of the chapter, we give an outlook of the various open research challenges concerning preference in abstract argumentation. We discuss alternative semantics not based on reductions, principles in the context of symmetric attack, the relation to structured argumentation, and the dynamics of preference and argumentation.

1 Introduction

In general, as witnessed by the first volume of the Handbook of Formal Argumentation [Baroni et al., 2018], research in formal argumentation is evenly balanced between studies of abstract argumentation, and studies in structured argumentation. At the abstract level, there are dialogue games, algorithms and a principle-based analysis of the various semantics [Baroni and Giacomin, 2007; van der Torre and Vesic, 2018], and at the structured level there are argument schemes, dialogues, algorithms, and rationality postulates.

However, the balance changes when we consider the concept of preference among arguments. Preferences are used in argumentation to represent the comparative strength of arguments, a natural and commonly adopted concept in argumentation. Therefore, in structured argumentation, preference plays a central role in the formal theories and results. Maybe surprisingly, there exist very few formal results and studies about the role of preference in *abstract* argumentation. In this chapter, we study the role of comparative preference in abstract argumentation. In the first part, we survey and extend the formal study on preference in abstract argumentation, and in the second part we discuss open challenges.

First, most of the formal results that involve preference in abstract argumentation are concerned with different ways to define the semantics of preference-based argumentation frameworks, based on different relations between the concept of attack and the concept of preference at the abstract level. In the traditional approach [Amgoud and Cayrol, 2002], an argument defeats another one if it attacks it and the attacked argument is not preferred to the attacker. Moreover, Amgoud and Vesic [2014] propose an alternative reduction, in which the attack may be reversed, and Kaci *et al.* [2018] propose two more alternatives. This raises the question how we can choose among these four reductions for a particular application. Moreover, it raises the question whether there are more alternatives, and if so, how we can guide the search for possibly alternative semantics for preference-based argumentation.

In addition, these different semantics are often analyzed using a principle-based approach, analogous to the principle-based analysis of the semantics of argumentation frameworks. Amgoud and Vesic [2014] study so-called *Conflict-freeness* and *Generalisation*. Kaci *et al.* [2018] introduce *Extension Growth*, which says that if we add preferences, then we can infer more, and *Extension Selection*, where the intersection of the extensions grows because there are fewer extensions, but the extensions do not become larger. However, they do not provide a complete analysis. Moreover, compared to the number of principles studied in abstract argumentation in general, the number of principles investigated for preference-based argumentation is quite small. In this chapter, we survey and extend this principle-based analysis of preference-based argumentation frameworks.

Second, we give an overview of future research directions concerning the formal study of the role of preference in abstract argumentation. More precisely, we question the relation between attack and preference by considering alternative semantics not based on reductions, and then we consider the representation of preference-based argumentation frameworks. In particular, we consider principles which hold only for symmetric attack, which constitutes a particularly promising fragment of preference-based argumentation. Third, we consider the relation to structured argumentation, and finally we consider the dynamics of preference and argumentation.

From these challenges, the relation with structured argumentation is subject of debate. Roughly, whereas we handle preference at the abstract level of Dung's Argumentation Framework (AF), assuming a preference order between arguments, some researchers are skeptical about this approach, on the grounds that the assumptions made at the abstract level may not hold when considering preferences at the structured level [Modgil and Prakken, 2012]. In the opinion of these authors, although the abstract approach is very popular in some research groups, it lies on a very shaky ground. In particular, it involves two binary relations on the set of arguments, where usually none of them has any restrictions, and the interaction between them is not restricted either. As arguments are treated in an absolutely abstract way, the proponents of the second position claim, it is no wonder that strange things may happen. We do not disagree with this observation, which in particular impacts on the study of dynamics of argumentation. However, we believe this observation does not discredit the use of a principle-based analysis. Moreover, whereas most structured approaches use the traditional reduction, in our opinion the alternative reductions can be adopted in structured argumentation as well.

The layout of this chapter is as follows. In Section 2, we introduce the semantics of preference-based argumentation frameworks. In Section 3, we introduce our ten principles, and we verify whether the semantics satisfy the ten principles. Section 4 discusses various open challenges concerning preference in abstract argumentation.

2 Preference-based abstract argumentation

In this section, we introduce the semantics of preference-based argumentation frameworks, based on a Dung semantics and one out of four reductions.

2.1 Dung's argumentation theory

Dung introduced an approach where the acceptance of an argument depends only on the defeat relation among arguments and the chosen argumentation semantics, but not on the internal structure of the arguments [Dung, 1995a]. Dung [1995a] refers to an "attack" relation, but in preference-based argumentation "attack" is used for something else, see Definition 2.4 below.

Definition 2.1 (Argumentation framework [Dung, 1995a]). *An argumentation framework (AF) is a tuple $\langle \mathcal{A}, \text{Def} \rangle$ where $\mathcal{A}$ is a set of arguments and $\text{Def} \subseteq \mathcal{A} \times \mathcal{A}$ is a binary defeat relation.*

3 - Preference in Abstract Argumentation

In this chapter we suppose that the set of arguments of each argumentation framework is finite. Given $a, b \in \mathcal{A}$, $(a, b) \in$ Def stands for a defeats b. The outcome of an argumentation framework is a set of sets of arguments, called *extensions* and denoted by $E(\mathcal{A}, \text{Def})$, that are robust against defeats. The extensions rely on two conditions namely *conflict-freeness* and *defense*.

Definition 2.2 (Conflict-freeness, Defense [Dung, 1995a]). *Let $\langle \mathcal{A}, \text{Def} \rangle$ be an AF.*

- $\mathbb{A} \subseteq \mathcal{A}$ *is* conflict-free *if there are no $a, b \in \mathbb{A}$ such that $(a, b) \in$ Def.*

- $\mathbb{A} \subseteq \mathcal{A}$ defends c *if $\forall b \in \mathcal{A}$ with $(b, c) \in$ Def, $\exists a \in \mathbb{A}$ such that $(a, b) \in$ Def.*

We distinguish several definitions of extension, each corresponding to an acceptability semantics that formally rules the argument evaluation process.

Definition 2.3 (Acceptability semantics [Dung, 1995a]). *Let $\langle \mathcal{A}, \text{Def} \rangle$ be an AF.*

- $\mathcal{S} \subseteq \mathcal{A}$ *is* admissible *iff it is conflict-free and defends all its elements.*

- *A conflict-free $\mathcal{S} \subseteq \mathcal{A}$ is a* complete extension *iff $\mathcal{S} = \{A \mid \mathcal{S} \text{ defends } A\}$.*

- $\mathcal{S} \subseteq \mathcal{A}$ *is the* grounded extension *iff it is the smallest (for set inclusion) complete extension.*

- $\mathcal{S} \subseteq \mathcal{A}$ *is a* preferred extension *iff it is a largest (for set inclusion) complete extension.*

- $\mathcal{S} \subseteq \mathcal{A}$ *is a* stable extension *iff it is a preferred extension that defeats all arguments in $\mathcal{A} \backslash \mathcal{S}$.*

2.2 Preference-based argumentation frameworks

A preference-based argumentation framework may be seen as an instance of Dung's framework. It is based on a binary attack relation between arguments and a preference relation over the set of arguments.

Definition 2.4 (Preference-based argumentation framework). **[Simari and Loui, 1992; Amgoud and Cayrol, 2002]** *A Preference-based Argumentation Framework (PAF) is a 3-tuple $\langle \mathcal{A}, \text{Att}, \succeq \rangle$ where $\mathcal{A}$ is a set of arguments, Att is a binary attack relation $\subseteq \mathcal{A} \times \mathcal{A}$ and $\succeq$ is a second binary relation over $\mathcal{A}$, called preference relation. Moreover, we write $a \succ b$ for $a \succeq b$ and not $b \succeq s$.*

Given $a, b \in \mathcal{A}$, $(a, b) \in$ Att stands for a attacks b. By convention it is often assumed that preferences are transitive, or that the preference relation $\succ$ is an order (irreflexive and transitive). However, it is not always the case that preference satisfies transitivity in real world scenarios. Moreover, this is an issue orthogonal to the discussion in this paper. In this chapter, we do not impose the transitivity of preferences. Only Theorem 4.9 depends on the validity of transitivity for the preference relation.

In order to compute the extensions of a preference-based argumentation framework $\langle \mathcal{A}, \text{Att}, \succeq \rangle$ the latter can be reduced to a Dung's AF $\langle \mathcal{A}, \text{Def} \rangle$, as we will see in Section 2.3. The extensions of a preference-based argumentation framework, denoted by $\mathcal{E}(\mathcal{A}, \text{Att}, \succeq)$, are simply the extensions of the argumentation framework it represents.

We end this subsection with an observation further discussed in section 4.1. A preference-based argumentation framework may also be seen as an extension instead of an instance of an argumentation framework, in the sense that if the preference relation is the universal relation, such that every argument is equally preferred, the attack and the defeat relation coincide. In this sense, we can also define semantics for preference-based argumentation frameworks which are not based on a reduction.

2.3 Reductions: from PAF to AF

In this section, we present different ways to reduce a PAF into Dung's AF and we illustrate them through a running example.

Reduction 1 [Amgoud and Cayrol, 2002] has been used widely in preference-based argumentation. The basic idea is that an attack succeeds only when the attacked argument is not preferred to the attacker.

Definition 2.5 (Reduction 1 [Amgoud and Cayrol, 2002]). *Let $\langle \mathcal{A}, \text{Att}, \succeq \rangle$ be a preference-based argumentation framework and $\langle \mathcal{A}, \text{Def} \rangle$ be the AF it represents. Then, $\forall a, b \in \mathcal{A}$,*

$$(a,b) \in \text{Def iff } (a,b) \in \text{Att}, b \not\succ a. \qquad (1)$$

This reduction has been criticised by Amgoud and Vesic [2009; 2010] as it may lead to extensions that are not conflict-free. The problem occurs when there is an attack from an argument to a preferred argument. This attack is called *critical* by Amgoud and Vesic [2010].

Example 2.6. *(Reduction 1) Let $\langle \mathcal{A}, \text{Att}, \succeq \rangle$ with $\mathcal{A} = \{a, b\}$, $\text{Att} = \{(a,b)\}$ and $b \succ a$. We have $\text{Def} = \emptyset$. Both a and b are accepted using any semantics although they are conflicting w.r.t. Att.*

Amgoud and Vesic [2010] propose to *repair* the argumentation framework to avoid this drawback. They extend Reduction 1 by enforcing a defeat from an argument to another when the former is preferred but attacked by the latter. This is Reduction 2.

Definition 2.7 (Reduction 2, from [Amgoud and Vesic, 2010]). *Let $\langle \mathcal{A}, \text{Att}, \succeq \rangle$ be a preference-based argumentation framework and $\langle \mathcal{A}, \text{Def} \rangle$ be the AF it represents. Then, $\forall a, b \in \mathcal{A}$,*

$$(a,b) \in \text{Def iff } (a,b) \in \text{Att}, b \not\succ a, \text{ or } (b,a) \in \text{Att}, (a,b) \notin \text{Att}, a \succ b. \qquad (2)$$

Example 2.8. *(Example 2.6 continued – Reduction 2) We have $(b,a) \in \text{Def}$. b is accepted.*

Reduction 2 solves the shortcoming of Reduction 1. Kaci *et al.* [2018] argue that nevertheless it is based on an implicit strong constraint. That is, an argument never succeeds to attack a preferred argument. This view gives a *power* to preferred arguments which goes against the idea underlying argumentation. Kaci *et al.* consider a parent who refuses that his child watches TV in the evening during the week because he has courses the next day. However, the child says that his courses have been cancelled. Then to maintain the refusal to watch TV the parent should provide another argument attacking his child's argument. The idea is: if an argument is attacked by a less preferred argument (critical attack) then the former should defend itself against its attacker. They formalize this idea in Reduction 3.

Definition 2.9 (Reduction 3 [Kaci et al., 2018]). *Let $\langle \mathcal{A}, \text{Att}, \succeq \rangle$ be a preference-based argumentation framework and $\langle \mathcal{A}, \text{Def} \rangle$ be the AF it represents. $\forall a, b \in \mathcal{A}: (a,b) \in \text{Def}$ iff*

- $(a,b) \in \text{Att}, b \not\succ a$, *or*

- $(a,b) \in \text{Att}, (b,a) \notin \text{Att}$

Example 2.10. *(Example 2.6 continued – Reduction 3) We have $(a,b) \in \text{Def}$. a is accepted.*

Reduction 3 in turn can also be criticized arguing that it would not be natural to make successful an attack from a less preferred argument. Reduction 4 below mixes Reduction 2 and 3.

Definition 2.11 (Reduction 4 [Kaci et al., 2018]). *Let $\langle \mathcal{A}, \text{Att}, \succeq \rangle$ be a preference-based argumentation framework and $\langle \mathcal{A}, \text{Def} \rangle$ be the AF it represents. $\forall a, b \in \mathcal{A}: (a,b) \in \text{Def}$ iff*

- $(a,b) \in \text{Att}, b \not\succ a$ *or*

- $(b,a) \in \text{Att}, (a,b) \notin \text{Att}, a \succ b$ *or*

- $(a,b) \in \text{Att}, (b,a) \notin \text{Att}$

Example 2.12. *(Example 2.6 continued – Reduction 4) We have $(a,b) \in \text{Def}$ and $(b,a) \in \text{Def}$. The grounded extension is empty, complete extensions are $\{\}, \{a\}, \{b\}$, and there are two preferred/stable extensions $\{a\}$ and $\{b\}$.*

Figure 1 illustrates the differences between the four reductions.

3 - PREFERENCE IN ABSTRACT ARGUMENTATION

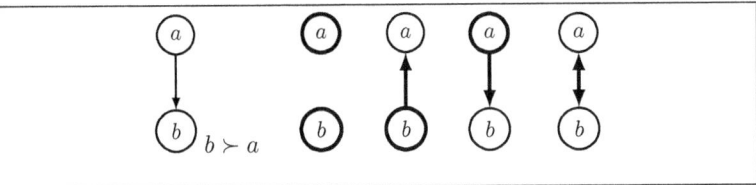

Figure 1: The defeat relation is visualised using thick lines, and if arguments are accepted in grounded semantics (and thus in extensions of all semantics considered in this paper) they are visualised with a thick circle. From left to right: the original argumentation framework, the result after applying Reduction 1, Reduction 2, Reduction 3 and Reduction 4.

3 Principle-based comparison and analysis

We now define the status of an argument. Note that we want each argument to have one and exactly one status. This helps us to simplify the presentation in the rest of the chapter.

Definition 3.1 (Status of an argument). *Let $\mathcal{F} = \langle \mathcal{A}, \text{Att}, \succeq \rangle$ be a PAF. If the set of extensions is empty, all the arguments are declared to be rejected. Otherwise, we say that an argument is*

- *skeptically accepted, if it belongs to all extensions;*
- *credulously accepted, if it is not skeptically accepted and it belongs to at least one extension;*
- *rejected, if it does not belong to any extension.*

We write $\text{Status}(a, \mathcal{F}) = \text{sk}$ (resp. cr, rej) when a is skeptically accepted (resp. credulously accepted, rejected). We define the order $\geq$ on the set of statuses as expected: $\text{sk} > \text{cr} > \text{rej}$. We denote the set of skeptically accepted (resp. credulously accepted, rejected) arguments of a PAF by $\text{Sk}(\mathcal{A}, \text{Att}, \succeq)$ (resp. $\text{Cr}(\mathcal{A}, \text{Att}, \succeq)$, $\text{Rej}(\mathcal{A}, \text{Att}, \succeq)$).

We define the path following the standard definition in the literature, in which we can travel over the directed arcs in both directions.

Definition 3.2 (Path). *For two arguments $a, b \in \mathcal{A}$, we say that there is a path between a and b if there exist $a_1, \ldots, a_n \in \mathcal{A}$ such that (a, a_1), $(a_1, a_2), \ldots (a_n, b) \in \text{Att} \cup \text{Att}^{-1}$.*

3.1 Principles

We now introduce the principles which we use then for our principle-based analysis of preference-based argumentation frameworks. Recall that we do not suppose that the preference relation is transitive. Also, we do not suppose that transitive closure is added in any case unless explicitly stated. The first principle says that an extension cannot contain arguments that attack each other.

Principle 1 (Conflict-freeness).

$$\text{If } (a,b) \in \text{Att then } \nexists E \in \mathcal{E}(\mathcal{F}) \text{ s.t. } \{a,b\} \in E$$

The second principle says that adding preferences helps to keep only some of the extensions.

Principle 2 (Preference selects extensions 1).

$$\mathcal{E}(\mathcal{A}, \text{Att}, \succeq \cup \succeq') \subseteq \mathcal{E}(\mathcal{A}, \text{Att}, \succeq)$$

The third principle is a special case of the second principle.

Principle 3 (Preference selects extensions 2).

$$\mathcal{E}(\mathcal{A}, \text{Att}, \succeq) \subseteq \mathcal{E}(\mathcal{A}, \text{Att}, \emptyset)$$

The fourth principle says, roughly speaking, that extensions grow when preferences are added.

Principle 4 (Extension refinement).

$$\forall E' \in \mathcal{E}(\mathcal{A}, \text{Att}, \succeq \cup \succeq'),\ \exists E \in \mathcal{E}(\mathcal{A}, \text{Att}, \succeq) \mid E \subseteq E'$$

The fifth principle says that each skeptically accepted argument stays skeptically accepted when preferences are added.

Principle 5 (Extension growth).

$$\text{Sk}(\mathcal{A}, \text{Att}, \succeq) \subseteq \text{Sk}(\mathcal{A}, \text{Att}, \succeq \cup \succeq')$$

The sixth principle says that adding preferences cannot increase the number of extensions.

3 - Preference in Abstract Argumentation

Principle 6 (Number of extensions).

$$\mid \mathcal{E}(\mathcal{A}, Att, \succeq \cup \succeq') \mid \ \leq \ \mid \mathcal{E}(\mathcal{A}, Att, \succeq) \mid$$

The seventh principle says that adding a preference $a \succeq b$ cannot worsen the status of argument a.

Principle 7 (Status conservation).

$$\texttt{Status}(a, (\mathcal{A}, Att, \succeq \cup \{(a,b)\})) \geq \texttt{Status}(a, (\mathcal{A}, Att, \succeq))$$

The eighth principle says that if an argument is strictly preferred to all the other arguments, it is not rejected.

Principle 8 (Preference-based immunity).

$$\text{If } (a,a) \notin Att \text{ and } \forall b \in \mathcal{A} \setminus \{a\}, a \succ b$$
$$\text{then } \texttt{Status}(a, (\mathcal{A}, Att, \succeq)) \neq \texttt{rej}$$

The ninth principle says that if there is no path between a and b, then adding preferences between them will not change anything.

Principle 9 (Path preference influence 1). *If there is no path between a and b then:*

$$\mathcal{E}(\mathcal{A}, Att, \succeq) = \mathcal{E}(\mathcal{A}, Att, \succeq \cup \{(a,b)\})$$

The last principle is a variant of the ninth one.

Principle 10 (Path preference influence 2). *If $(a,b) \notin Att$ and $(b,a) \notin Att$, then:*

$$\mathcal{E}(\mathcal{A}, Att, \succeq) = \mathcal{E}(\mathcal{A}, Att, \succeq \cup \{(a,b)\})$$

The first six principles have already been presented in the literature [Kaci et al., 2018], while the other principles are new. It is immediate to see that P2 implies P3, P4 and P6. Also, for semantics that always return at least one extension, P2 implies P5.

Observe that P7 can be equivalently stated as

$$\texttt{Status}(a, (\mathcal{A}, Att, \succeq \cup \{(a,b) \mid b \in B\})) \geq \texttt{Status}(a, (\mathcal{A}, Att, \succeq)).$$

It is immediate to see that the previous formulation implies P7. Conversely, if a semantics satisfies P7, we can show that it satisfies the previous equation by successively applying P7 several times (to be precise, $|B|$ times).

We can also note that under those semantics that always return at least one extension, P4 implies P5.

Proposition 3.3. *Let $\mathcal{E}$ be a semantics that always returns at least one extension. If $\mathcal{E}$ satisfies P4, then it satisfies P5.*

Proof. Let $a \in \mathtt{Sk}(\mathcal{A}, Att, \succeq)$ be an arbitrary skeptically accepted argument of $(\mathcal{A}, Att, \succeq)$. Let $E' \in \mathcal{E}(\mathcal{A}, Att, \succeq \cup \succeq')$. Since $\mathcal{E}$ always returns at least one extension, there is at least one such E'. Let $E \in \mathcal{E}(\mathcal{A}, Att, \succeq)$ such that $E \subseteq E'$. (Such a set exists since P4 is satisfied.) Hence, $a \in E$. Consequently, $a \in E'$. Since E' was arbitrary, we conclude that a belongs to every extension of $(\mathcal{A}, Att, \succeq \cup \succeq')$. Furthermore, $(\mathcal{A}, Att, \succeq \cup \succeq')$ has at least one extension. Hence, $a \in \mathtt{Sk}(\mathcal{A}, Att, \succeq \cup \succeq')$. □

We see from their respective definitions that P9 implies P10.

3.2 Satisfaction of principles

We now study which principles are satisfied by which reductions. The table summarizing all the results is depicted in Figure 1 on page 22. We start by R1. To see that P1 is not satisfied it is sufficient to consider the graph with only two arguments a and b, such that $Att = \{(a,b)\}$ and $b \succeq a$. For all four studied semantics, we have a unique extension $\{a,b\}$.

P3 is not satisfied by any semantics, as shown by the counter-example in figure 2. Since P2 implies P3 and P3 is not satisfied, we conclude that P2 is not satisfied.

P4 and P5 are also violated by all the four semantics. Consider the graph in figure 3. The unique extension of $(\mathcal{A}, Att, \emptyset)$ is the set $\{a,c\}$ and the unique extension of $(\mathcal{A}, Att, \{(b,a)\})$ is the set $\{a,b\}$. Hence, both P4 and P5 are violated.

P6 is trivially satisfied by grounded semantics. It is violated by complete, preferred and stable semantics, as can be seen from the counter-example in Figure 4.

3 - PREFERENCE IN ABSTRACT ARGUMENTATION

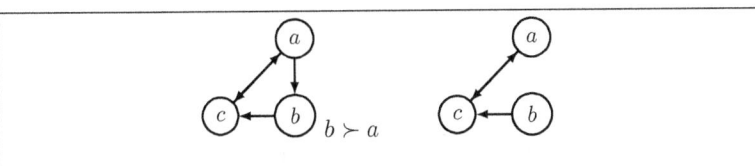

Figure 2: Example showing that P3 is not satisfied by R1 nor by R2 under any of the four semantics. In the rest of the chapter we follow the same way of depicting the argumentation frameworks in the figures, namely, the original framework on the left, and the framework after applying the reduction on the right. In this, as well as in the following examples, we leave the calculation of the extensions to the reader.

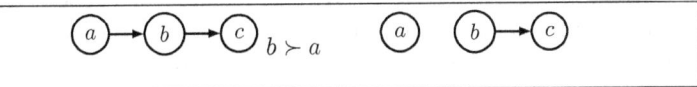

Figure 3: Example showing that P5 is not satisfied by R1 under any of the four semantics

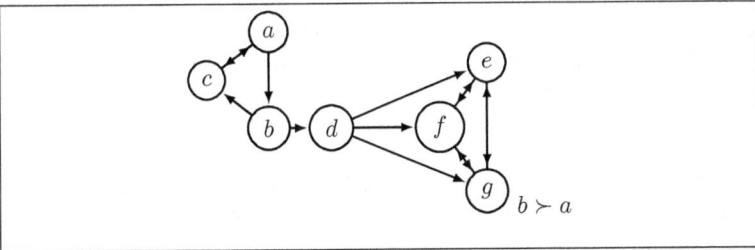

Figure 4: Example showing that P6 is violated by R1 and R2 under complete, preferred and stable semantics. Due to the size of the argumentation framework, in order to keep the chapter within a reasonable page limit, we do not depict its version after applying the reductions.

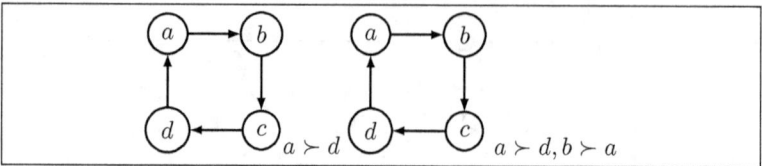

Figure 5: Example showing that P4 and P5 are violated by R2 under all studied semantics

P7 is satisfied by the four semantics. More concretely, Amgoud and Vesic [2020] show that deleting an attack can only improve the status of an argument but never worsen it. Under R1, adding a preference (a,b) will result either in deleting the attack (b,a), if it exists or will induce no change in the argumentation graph. Hence, the status of a cannot be worsen.

P8 is satisfied under all semantics except stable, since a non-attacked argument is in all extensions. The principle is not satisfied under stable semantics, because there might not be any extensions.

P9 and P10 are satisfied since adding a preference between a and b does not influence the graph. Hence, the extensions of the two argumentation graphs coincide trivially, since the graphs coincide.

Let us now study the reduction R2. P1 is satisfied since the reduction preserves conflict-free sets, i.e. for every set of arguments $S \subseteq \mathcal{A}$, S is conflict-free in the original graph if and only if S is conflict-free in the graph obtained after applying reduction R2.

Figure 2 shows that P3 is not satisfied by R2. Consequently, P2 is not satisfied neither.

P4 and P5 are not satisfied, as can be seen from the example in Figure 5.

P6 is trivially satisfied by grounded semantics. The other semantics violate this principle, as illustrated by the example in figure 4.

Proposition 3.4. *R2 satisfies P7 under all the studied semantics.*

Proof. Denote by $\mathcal{F}$ (resp. $\mathcal{F}'$) the graph obtained after applying R2 on $(\mathcal{A}, Att, \succeq)$, resp. $(\mathcal{A}, Att, \succeq \cup\{(a,b)\}))$.

If $(b,a) \notin Att$, $\mathcal{F}$ and $\mathcal{F}'$ coincide. If $(a,b) \in Att$ and $(b,a) \in Att$, the proof follows from the fact that removing an attack on a cannot worsen its status [Amgoud and Vesic, 2020]. In the rest of the proof, we study the case when $(b,a) \in Att$ and $(a,b) \notin Att$.

Stable semantics.

Let a be skeptically accepted in $\mathcal{F}$. By means of contradiction, suppose that a is not skeptically accepted in $\mathcal{F}'$; let E be an extension of $\mathcal{F}'$ such that $a \notin E$. Note that E is conflict-free in $\mathcal{F}$; it also attacks all the arguments in its exterior in $\mathcal{F}$. Hence, E is a stable extension of $\mathcal{F}$. Contradiction.

Let a be credulously accepted in $\mathcal{F}$. Let E be an extension of $\mathcal{F}$ such that $a \in \mathcal{F}$. Since E attacks all arguments in its exterior in $\mathcal{F}$, E also attacks all arguments in its exterior in $\mathcal{F}'$. Hence, E is a stable extension of $\mathcal{F}'$. Consequently, a is not rejected in $\mathcal{F}'$.

Preferred semantics.

Let a be skeptically accepted in $\mathcal{F}$. By means of contradiction, let E be an extension of $\mathcal{F}'$ such that $a \notin E$.

Case 1: $b \in E$. E is conflict-free in $\mathcal{F}$. It is also admissible in $\mathcal{F}$. Thus, there exists E' such that $E \subseteq E'$ and E' is an extension of $\mathcal{F}$. Since b attacks a, $a \notin E'$. Thus, a is not skeptically accepted in $\mathcal{F}$, contradiction.

Case 2: $b \notin E$. Note that E is admissible in $\mathcal{F}$. Let E' be such that $E \subseteq E'$ and E' is an extension of $\mathcal{F}$. Since a is skeptically accepted, $a \in E'$. Note that E' is admissible in $\mathcal{F}'$. Hence, E is not an extension of $\mathcal{F}'$, contradiction.

Let a be credulously accepted in $\mathcal{F}$. Let E be an extension of $\mathcal{F}$ such that $a \in E$. Set E is admissible in $\mathcal{F}'$, hence there exists E' such that $E \subseteq E'$ and E' is an extension of $\mathcal{F}'$. Thus, a is not rejected in $\mathcal{F}'$.

Grounded semantics.

Let f_C be the function such that, for every set of arguments S, $f_C(S)$ is the set of arguments defended by S. Then, the grounded extension is the least fixed point of $f_C(\emptyset)$ [Dung, 1995b]. Denote the grounded extension of $\mathcal{F}$ by GE and the grounded extension of $\mathcal{F}'$ by GE'. Suppose that $a \in GE$ in $\mathcal{F}$.

Let E_0 be the set of non-attacked arguments of $\mathcal{F}$ and Let E_0' be the set of non-attacked arguments of $\mathcal{F}'$.

Since b attacks a in $\mathcal{F}$, $a \notin E_0$. Since $a \in GE$, $b \notin GE$.

For each $i \in \{0, 1, \ldots\}$ let $E_{i+1} = f_C(E_i)$ and $E'_{i+1} = f_C(E'_i)$. Let us

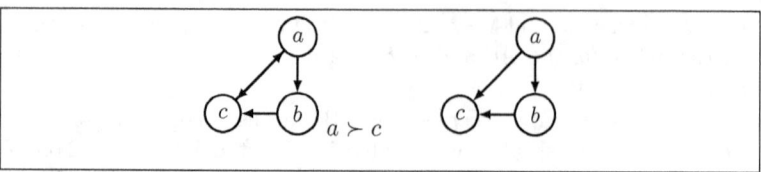

Figure 6: Example showing that P3 is not satisfied by R3 under grounded semantics

prove by induction on i that $E_i \subseteq E'_i$ for every i.

Base: $E_0 \subseteq E'_0$ since $a \notin E$ and $b \notin E$.

Step: Let $E_i \subseteq E'_i$. Let x be an arbitrary argument defended by E_i. For each y such that y attacks x in $\mathcal{F}$, there exists $z \in E_i$ such that z attacks y in $\mathcal{F}$. Note that $z \neq b$, since $b \notin GE$. Hence, since $z \in E'_i$, x is defended by E'_i in $\mathcal{F}'$. Thus, $E_i \subseteq E'_i$.

Consequently, $GE \subseteq GE'$. Thus, a is accepted in $\mathcal{F}'$.

Complete semantics.

Let a be skeptically accepted in $\mathcal{F}$. Skeptical acceptance under complete semantics coincides with acceptance under grounded semantics, and we already showed the property for grounded semantics.

Let a be credulously accepted in $\mathcal{F}$. Let E be an extension of $\mathcal{F}$ such that $a \in E$. Note that E is admissible in $\mathcal{F}'$. Hence, there exists a complete extension E' in $\mathcal{F}'$ such that $E \subseteq E'$. This means that a is not rejected in $\mathcal{F}'$. □

P8 is satisfied by preferred, complete and grounded semantics since a non-attacked argument belongs to all extensions under those semantics. It is not satisfied under stable semantics, since stable extensions might not exist.

P9 and P10 are satisfied since the preference between a and b can only influence the attacks between a and b.

Let us now study R3. P1 is satisfied since the reduction preserves conflict-free sets.

P3 is not satisfied by grounded semantics, as shown by the example in figure 6. Consequently, P2 is not satisfied by grounded semantics neither. Figure 7 shows that P2 and P3 are violated by preferred semantics.

Proposition 3.5. *R3 satisfies P2 under complete and stable semantics.*

3 - Preference in Abstract Argumentation

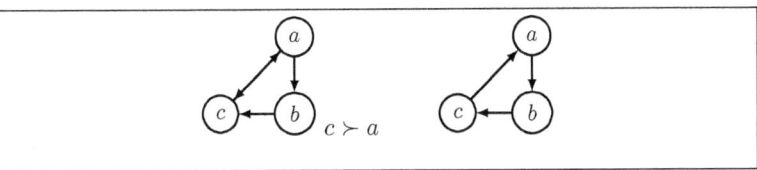

Figure 7: Example showing that P3 and P4 are not satisfied by R3 under preferred semantics and that P5 is not satisfied by R3 under stable and preferred semantics

Proof. Let $\mathcal{F}$, resp. $\mathcal{F}'$ be the argumentation graph obtained after applying R3 on $(\mathcal{A}, Att, \succeq)$, resp. $(\mathcal{A}, Att, \succeq \cup \succeq')$.

Complete semantics.

Suppose E is an extension of $\mathcal{F}'$ and prove it is an extension of $\mathcal{F}$. Note the E is conflict-free in $\mathcal{F}$. Let us show that E is admissible in $\mathcal{F}$. Let $x \in E$ and let y attack x in $\mathcal{F}$. If x attacks y in $\mathcal{F}$, it defends itself. Else, suppose that x does not attack y in $\mathcal{F}$. This means that x does not attack y in $\mathcal{F}'$. Hence, there exists $z \in E$ such that z attacks y in $\mathcal{F}'$. Thus, z attacks y in $\mathcal{F}$ and z defends x. We conclude that E is admissible in $\mathcal{F}$.

Let us prove that E is a complete extension of $\mathcal{F}$. By means of contradiction, suppose that E defends x in $\mathcal{F}$ with $x \notin E$. Let y be an attacker of x in $\mathcal{F}'$ such that no argument defends x against the attack of y in $\mathcal{F}'$. This means that y attacks x in $\mathcal{F}$. Let $z \in E$ be such that z attacks y in $\mathcal{F}$. We know that z does not attack y in $\mathcal{F}'$. It must be that y attacks z in $\mathcal{F}'$. Since we supposed no argument from E attacks y in $\mathcal{F}'$, set E is not admissible in $\mathcal{F}'$, contradiction. So, it must be that E is a complete extension of $\mathcal{F}$.

Stable semantics.

Let E be a stable extension of $\mathcal{F}'$. Then E is conflict-free in $\mathcal{F}$. Let x be an argument in exterior of E. Note that E attacks x in $\mathcal{F}'$. Hence, E attacks x in $\mathcal{F}$. Thus, E is a stable extension of $\mathcal{F}$. □

As a consequence of the previous result, P3, P4, P5 and P6 are satisfied by R3 under complete and stable semantics.

Since P2 is satisfied under complete and stable semantics, this means that each extension of $(\mathcal{A}, Att, \succeq \cup \succeq')$ is an extension of $(\mathcal{A}, Att, \succeq)$, which implies that P4 holds under complete and stable semantics. Let us now show why R3 satisfies P4 under grounded semantics. Let E be the grounded extension of $(\mathcal{A}, Att, \succeq \cup \succeq')$ under reduction R3. Since P2 is satisfied under complete semantics, and E is a complete extension of $\mathcal{F}'$, then E is a complete extension $\mathcal{F}$. Thus, P4 is satisfied since the grounded extension is contained in every complete extension.

Preferred semantics does not satisfy P4 as shown by Figure 7.

P5 is not satisfied by stable and preferred semantics, a counter example is depicted in Figure 7.

P6 is satisfied under complete and stable semantics since it follows from P2. It is trivially satisfied by grounded semantics.

Proposition 3.6. *R3 satisfies P6 under preferred semantics.*

Proof. Let $\mathcal{F}$, resp. $\mathcal{F}'$ be the argumentation graph obtained after applying R3 on $(\mathcal{A}, Att, \succeq)$, resp. $(\mathcal{A}, Att, \succeq \cup \succeq')$. Let us first show that each admissible set E of $\mathcal{F}'$ is admissible in $\mathcal{F}$. Suppose E is admissible in $\mathcal{F}'$. It is immediate to see that E is conflict-free in $\mathcal{F}$. Suppose now that $x \in E$ and that y attacks x in $\mathcal{F}$. If x attacks y in $\mathcal{F}$ the proof is over. Else, suppose that x does not attack y in $\mathcal{F}$. This means that y attacks x in $\mathcal{F}'$ and x does not attack y in $\mathcal{F}'$. Since E is admissible in $\mathcal{F}'$, there exists $z \in E$ such that z attacks y in $\mathcal{F}'$. Consequently, z attacks y in $\mathcal{F}$. Hence, E is admissible in $\mathcal{F}$.

We conclude that the set of admissible sets of $\mathcal{F}'$ is a subset of the set of admissible sets of $\mathcal{F}$. Observe that this means that every extension of $\mathcal{F}$ is a superset of at least one of the extensions of $\mathcal{F}'$, and every extension of $\mathcal{F}'$ has at least one superset that is an extension of $\mathcal{F}$.

Let us now prove the proposition. By means of contradiction, suppose the contrary. From the above observation we deduce that there exists a maximal admissible set E of $\mathcal{F}$ (i.e. an extension of $\mathcal{F}$) such that E is not an extension of $\mathcal{F}'$ and that there exist distinct E_1 and E_2 such that $E_1, E_2 \subseteq E$ and both E_1 and E_2 are extensions of $\mathcal{F}'$.

Define $E' = E_1 \cup E_2$. Note that E' is conflict-free in $\mathcal{F}'$ since

- $E_1 \cup E_2 \subseteq E$,

- E is conflict-free in $\mathcal{F}$

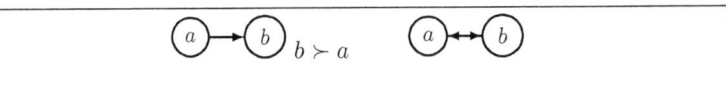

Figure 8: R4 falsifies several principles under all semantics

- for each set S, we have: S is conflict-free in $\mathcal{F}$ if and only if S is conflict-free in $\mathcal{F}'$.

Note that since E_1 defends its arguments in $\mathcal{F}'$ and E_2 defends it arguments $\mathcal{F}'$ and since their union is conflict-free in $\mathcal{F}'$, it must be that E' is admissible in $\mathcal{F}'$. Contradiction, since $E_1 \subsetneq E'$ (this is true since E_2 is not a subset of E_1). □

P7 is satisfied since removing an attack towards an argument can only improve its status [Amgoud and Vesic, 2020].

P8 is not satisfied, the counter example is the argumentation graph with two arguments a and b, where b attacks a and there are no preferences. Adding the preference $a \succ b$ will not improve the status of a.

P9 and P10 are satisfied since the preference between a and b only influences the attacks between those two arguments.

We now study R4. P1 is satisfied since R4 preserves conflict-freeness.

P2, P3, P4 and P5 are not satisfied, as can be seen in figure 8. The same is a counter example for P6 for all the semantics except grounded.

Proposition 3.7. *R4 satisfies P7 under all the studied semantics.*

Proof. Denote by $\mathcal{F}$ (resp. $\mathcal{F}'$) the graph obtained after applying R4 on $(\mathcal{A}, Att, \succeq)$, resp. $(\mathcal{A}, Att, \succeq \cup\{(a,b)\}))$.

If $(b, a) \notin Att$, $\mathcal{F}$ and $\mathcal{F}'$ coincide. If $(a, b) \in Att$ and $(b, a) \in Att$, the proof follows from the fact that removing an attack on a cannot worsen its status [Amgoud and Vesic, 2020]. In the rest of the proof, we study the case when $(b, a) \in Att$ and $(a, b) \notin Att$.

Stable semantics.

Let a be skeptically accepted in $\mathcal{F}$. By means of contradiction, suppose that a is not skeptically accepted in $\mathcal{F}'$; let E be an extension of $\mathcal{F}'$ such that $a \notin E$. Note that E is conflict-free in $\mathcal{F}$; it also attacks all the arguments in its exterior in $\mathcal{F}$. Hence, E is a stable extension of $\mathcal{F}$. Contradiction.

Let a be credulously accepted in $\mathcal{F}$. Let E be an extension of $\mathcal{F}$ such that $a \in \mathcal{F}$. Since E attacks all arguments in its exterior in $\mathcal{F}$, E also attacks all arguments in its exterior in $\mathcal{F}'$. Hence, E is a stable extension of $\mathcal{F}'$. Consequently, a is not rejected in $\mathcal{F}'$.

Preferred semantics.

Let a be skeptically accepted in $\mathcal{F}$. By means of contradiction, let E be an extension of $\mathcal{F}'$ such that $a \notin E$.

Case 1: $b \in E$. E is conflict-free in $\mathcal{F}$. It is also admissible in $\mathcal{F}$. Thus, there exists E' such that $E \subseteq E'$ and E' is an extension of $\mathcal{F}$. Since b attacks a, $a \notin E'$. Thus, a is not skeptically accepted in $\mathcal{F}$, contradiction.

Case 2: $b \notin E$. Note that E is admissible in $\mathcal{F}$. Let E' be such that $E \subseteq E'$ and E' is an extension of $\mathcal{F}$. Since a is skeptically accepted, $a \in E'$. Note that E' is admissible in $\mathcal{F}'$. Hence, E is not an extension of $\mathcal{F}'$, contradiction.

Let a be credulously accepted in $\mathcal{F}$. Let E be an extension of $\mathcal{F}$ such that $a \in E$. Set E is admissible in $\mathcal{F}'$, hence there exists E' such that $E \subseteq E'$ and E' is an extension of $\mathcal{F}'$. Thus, a is not rejected in $\mathcal{F}'$.

Complete semantics.

Let a be skeptically accepted in $\mathcal{F}$. By means of contradiction, suppose that E is an extension of $\mathcal{F}'$ such that $a \notin E$.

Case 1: $b \in E$. Observe that E is admissible in $\mathcal{F}$. Note that E defends exactly same arguments in $\mathcal{F}$ and $\mathcal{F}'$. Thus, E is an extension of $\mathcal{F}$. Contradiction.

Case 2: $b \notin E$. E is admissible in $\mathcal{F}$. Note that E defends b in $\mathcal{F}$ if and only if b is not attacked in $\mathcal{F}$. Case 2.1: b is attacked in $\mathcal{F}$. Then E contains all the arguments it defends in $\mathcal{F}$, hence E is a complete extension of $\mathcal{F}$. Contradiction. Case 2.2: b is not attacked in $\mathcal{F}$. Then, there exists a complete extension E' of $\mathcal{F}$ such that $E \subseteq E'$ and $b \in E'$. This means that $a \notin E'$, contradiction.

Let a be credulously accepted in $\mathcal{F}$. Let E be an extension of $\mathcal{F}$ such that $a \in E$. Note that E is admissible in $\mathcal{F}'$. Hence, there exists a complete extension E' in $\mathcal{F}'$ such that $E \subseteq E'$. This means that a is not rejected in $\mathcal{F}'$.

Grounded semantics.
Let f_C be the function such that, for every set of arguments S, $f_C(S)$ is the set of arguments defended by S. Then, the grounded extension is the least fixed point of $f_C(\emptyset)$ [Dung, 1995b]. Denote the grounded extension of $\mathcal{F}$ by GE and the grounded extension of $\mathcal{F}'$ by GE'. Suppose that $a \in GE$ in $\mathcal{F}$.

Let E_0 be the set of non-attacked arguments of $\mathcal{F}$ and Let E'_0 be the set of non-attacked arguments of $\mathcal{F}'$.

Since b attacks a in $\mathcal{F}$, $a \notin E_0$. Since $a \in GE$, $b \notin GE$.

For each $i \in \{0, 1, \ldots\}$ let $E_{i+1} = f_C(E_i)$ and $E'_{i+1} = f_C(E'_i)$. Let us prove by induction on i that $E_i \subseteq E'_i$ for every i.

Base: $E_0 \subseteq E'_0$ since $a \notin E$ and $b \notin E$.

Step: Let $E_i \subseteq E'_i$. Let x be an arbitrary argument defended by E_i. For each y such that y attacks x in $\mathcal{F}$, there exists $z \in E_i$ such that z attacks y in $\mathcal{F}$. Note that $z \neq b$, since $b \notin GE$. Hence, since $z \in E'_i$, x is defended by E'_i in $\mathcal{F}'$. Thus, $E_i \subseteq E'_i$.

Consequently, $GE \subseteq GE'$. Thus, a is accepted in $\mathcal{F}'$. □

P8 is satisfied by all the semantics except grounded, since after applying R4, set $\{a\}$ is admissible. Hence, it is in at least one stable / preferred / complete extension. The counter-example for grounded semantics is the graph with two arguments a and b, where the only attack is from b to a. The grounded extension is the empty set, hence a is rejected.

Like in the case of other reductions, P9 and P10 are satisfied.
Table 1 summarises how the proposed reductions satisfy these principles with respect to standard Dung's semantics (i.e., complete, grounded, preferred and stable).

4 Open Research Challenges for Preference in Abstract Argumentation

We believe that the concept of preference is important, because it is a natural way to represent and reason about the comparative strength of arguments. Given the limited attention preference has received in the literature of abstract argumentation, we believe that much more work

	R1	R2	R3	R4
P1	×	cgps	cgps	cgps
P2	×	×	cs	×
P3	×	×	cs	×
P4	×	×	cgs	×
P5	×	×	cg	×
P6	g	g	cgps	g
P7	cgps	cgps	cgps	cgps
P8	cgp	cgp	×	cps
P9	cgps	cgps	cgps	cgps
P10	cgps	cgps	cgps	cgps

Table 1: Comparison among the semantics and the principles. We refer to Dung's semantics as follows: complete (c), grounded (g), preferred (p), stable (s). When a principle is never satisfied by a certain reduction for all semantics, we use the × symbol. P1 refers to Principle 1, the same holds for the others.

needs to be done. For example, more principles for preference-based argumentation can be defined and for the four reductions, the principles of this chapter can be verified against other semantics of argumentation frameworks proposed in the literature.

In this section, we discuss four other open challenges we see in the formal study of preference in abstract argumentation. The first challenge concerns the semantics of preference-based argumentation frameworks, the second challenge concerns the kind of preference-based argumentation frameworks considered, the third challenge concerns the relation between the role of preference in abstract argumentation and in structured argumentation, and the fourth challenge concerns the dynamics of preference and argumentation.

4.1 Alternative semantics for preference in abstract argumentation

The first open challenge in the formal study on preference in abstract argumentation concerns the semantics of preference-based argumentation frameworks. In the reduction-based approach, the meaning of preference is local, in the sense that a preference between arguments a and b only affects the attack among arguments a and b. Due to this locality, and the small number of reductions, the number of semantics is just a multiple of the number of semantics for Dung argumentation frameworks.

However, if we consider a more global use of the preferences among arguments, the number of possible semantics for preference-based argumentation frameworks is much larger. For example, consider the use of preference in choice theory, or in non-monotonic logic, or in belief revision. In all these areas, the introduction of preference leads to an explosion of the number of possibilities. Likewise, the use of preference in structured argumentation increases the number and breadth of the formal approaches and results.

In the literature of formal argumentation, we are aware of only one approach by Amgoud and Vesic [2014] introducing the idea of "refining argumentation frameworks by preferences," by which they mean that they "refine the result of the PAF," or more precisely "they allow to choose some extensions among the set of extensions of the repaired framework."

Kaci et al. [2018] call it the selection approach, and they illustrate it by the following example. Assume a set of extensions, and a preference relation over arguments. To get the best extensions, we need to lift the preference relation over arguments to a preference relation over sets of arguments, or extensions. It is well known that there are various ways to make this lifting more precise, see for example the work of Amgoud and Vesic [2014]. Kaci et al. [2018] use the following lifting: if argument a is preferred to argument b, then all extensions containing argument a but not b are either better than all extensions containing b but not a, or the two extensions are incomparable.

Definition 4.1. *Let $E(\mathcal{A}, \text{Def})$ be a set of extensions according to a Dung semantics, and let $\succ$ be an order (irreflexive and transitive) over $\mathcal{A}$, called preference relation. $E \subseteq \mathcal{A}$ is at least as good as $E' \subseteq \mathcal{A}$ if $\forall a, b \in \mathcal{A}$: we do not have that $a \succ b$ and $a \in E' \setminus E$ and $b \in E \setminus E'$. E is better than E' iff E is at least as good as E' and E' not at least as good as E. E is best if there is no E' that is better than E.*

Kaci et al. [2018] define a semantics of preference-based argumentation by a two-step process. First, they select the extensions of the framework according to a regular abstract semantics. Then they use the preference relation to select the best extensions among them. We identify Def=Att in Dung's AF and define $\mathcal{E}(\mathcal{A}, \text{Att}, \succ)$ as a selection of the best extensions of $(\mathcal{A}, \text{Att})$.

Definition 4.2. *The extensions $\mathcal{E}(\mathcal{A}, \text{Att}, \succ)$ are the best extensions among $E(\mathcal{A}, \text{Att})$ based on $\succ$.*

In preference elicitation, the preferences are extracted from a user step by step [Viappiani, 2014]. Assume now that for each step where a user is queried for preference, we consider the arguments the user accepts, given the knowledge about the preference relation thus far. In such a setting, it would be quite useful if the set of accepted arguments is increasing monotonically. Moreover, the other way around, we can even consider scenarios where the interest in the extension guides the order in which the preferences are elicited.

Example 4.3. *Consider the framework below under preferred or stable semantics. The left PAF prefers extension $\{a, c\}$ over extension $\{b, d\}$, but if we add $b \succ a$ then both $\{a, c\}$ and $\{b, d\}$ are best extensions.*

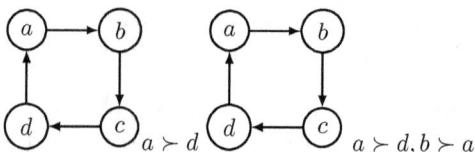

This is just one of the many possibilities, and its simplicity has some drawbacks. For example, if the empty set is an extension, e.g., in complete semantics, then it is always a best extension. Just like in the case of the reductions, this illustrates that more alternatives can be defined, and a principle-based approach is needed to choose among the alternatives.

Moreover, Amgoud and Vesic [2014] consider also the combination of reduction and selection. In particular, they introduce so-called *rich preference-based argumentation frameworks* that allow to take into account the two roles of preferences (reduction and selection) through a two-step process. First, the preferences are used to define a reduction and calculate a set of "preliminary" extensions. Second, the preference is once again used to select the best among those preliminary extensions. The interested user is referred to the corresponding journal publication [Amgoud and Vesic, 2014] for a further discussion.

However, despite the lack of approaches in the existing literature, it is straightforward to generalize semantic approaches for argumentation frameworks to preference-based argumentation. For example, a common idea in abstract argumentation semantics is to use an SCC recursive algorithm [Baroni et al., 2005; Baroni et al., 2014], where SCC stands for the standard graph-theoretic notion of strongly connected component, defined as follows.

Definition 4.4. *An* Att*-path is a sequence* $\langle a_0, \ldots, a_n \rangle$ *of arguments where* $(a_i, a_{i+1}) \in$ Att *for* $0 \leq i < n$ *and where* $a_j \neq a_k$ *for* $0 \leq j < k \leq n$ *with either* $j \neq 0$ *or* $k \neq n$.

Let $F = \langle \mathcal{A}, \text{Att} \rangle$ *be an AF, and let* $a, b \in \mathcal{A}$. *We define* $a \sim b$ *iff either* $a = b$ *or there is an* Att*-path from* a *to* b *and there is an* Att*-path from* b *to* a. *The equivalence classes under the equivalence relation* $\sim$ *are called* strongly connected components *(SCCs) of* F. *We denote the set of SCCs of* F *by* $\text{SCCs}(F)$. *Given* $S \subseteq \mathcal{A}$, *we define* $D_F(S) := \{b \in \mathcal{A} \mid \exists a \in S : (a, b) \in \text{Att} \land a \not\sim b\}$.

Cramer and van der Torre [2019] define the SCC-recursive scheme as a function that maps a semantics σ to another semantics $\text{SCC}(\sigma)$, such that σ is applied only to the strongly connected components. The simplest case is defined as follows.

Definition 4.5. *Let* σ *be an argumentation semantics. The argumentation semantics* $\text{SCC}(\sigma)$ *is defined as follows. Let* $F = \langle \mathcal{A}, \text{Att} \rangle$ *be an AF, and let* $S \subseteq \mathcal{A}$. *Then* S *is an* $\text{SCC}(\sigma)$*-extension of* F *iff either*

- $|\text{SCCs}(F)| = 1$ *and* S *is a* σ*-extension of* F, *or*
- $|\text{SCCs}(F)| > 1$ *and for each* $C \in \text{SCCs}(F)$, $S \cap C$ *is an* $\text{SCC}(\sigma)$*-extension of* $F|_{C \setminus D_F(S)}$.

We refer the interested reader to the extensive literature on SCC recursion [Baroni et al., 2005; Baroni et al., 2014]. What we would like to emphasize here is how this idea can be used as well for preference-based argumentation. For example, we can use the preferences only when we consider the strongly connected components, not when we consider the recursion.

4.2 Preference-based argumentation frameworks

The second open challenge in the formal study on preference in abstract argumentation concerns the representation of preference-based argumentation frameworks. On the one hand, the preference relation can be represented by a numerical function rather than comparative preferences. As this is studied in a accompanying paper in this handbook, we do not consider this possibility in this chapter. On the other hand, we can impose various kinds of constraints on the attack and preference relation in preference-based argumentation frameworks.

For preference, this is well known. Much has been written in the formal literature on preference and preference logic about intransitive preference, totally ordered preference, and so on.

Less is known about constraints on the attack relation. In Dung's argumentation, it is well known that a symmetric attack relation leads to a collapse of many semantics, and in general to a trivialization of the theory. For this reason, symmetric attacks are rarely studied. However, a symmetric attack also has advantages. One of the questions in abstract argumentation is to distinguish the different interpretations of directional attack. A symmetric attack, however, is very similar to negation in logic and thus more easily understood and manipulated.

For example, consider an argumentation theory in which each argument is represented by a propositional formula, and 'argument A attacks argument B' is defined as '$A \wedge B$ is inconsistent'. Unfortunately, such a simple argumentation theory is not very useful, since the attack relation is symmetric, and the various semantics reduce to, roughly, one of the following two statements: 'an argument is acceptable iff it is part of all/some maximal consistent subsets of the set of arguments.' However, such a simple argumentation theory is useful again when we add the additional condition that argument A is at least as preferred as argument B. It may also be worthwhile to consider the generalizations of Dung's framework in propositional argumentation [Boella *et al.*, 2005].

As the notion of symmetric attack is considered to be very important, Kaci *et al.* [2006] introduce a new term for it, and refer to it as the conflict relation. The new preference-based argumentation framework considers a conflict and a preference relation. The conflict relation should not be interpreted as an attack relation, since a conflict relation is symmetric, and an attack relation is usually asymmetric.

Definition 4.6 (Conflict+preference argumentation framework). *A conflict+preference argumentation framework is a triplet $\langle \mathcal{A}, \mathcal{C}, \succeq \rangle$ where $\mathcal{A}$ is a set of arguments, $\mathcal{C}$ is a binary symmetric conflict relation defined on $\mathcal{A} \times \mathcal{A}$ and $\succeq$ is a (total or partial) pre-order (preference relation) defined on $\mathcal{A} \times \mathcal{A}$.*

An acyclic argumentation framework is an argumentation framework in which the attack relation is acyclic, a symmetric argumentation framework is an argumentation framework in which the attack relation is symmetric, *etc.* Kaci et al. [2006] define an acyclic strict attack relation as follows. Assume the attack relation is such that there is an attack path where argument A_1 attacks argument A_2, argument A_2 attacks argument A_3, etc, and argument A_n attacks argument A_1, then we have that all the arguments in the attack path attack the previous one. Consequently, if argument A strictly attacks B if A attacks B and not vice versa, then the strict attack relation is acyclic.

Definition 4.7 (Acyclic argumentation framework). *A strictly acyclic argumentation framework is an argumentation framework $\langle \mathcal{A}, \mathcal{R} \rangle$ where the attack relation $\mathcal{R} \subseteq \mathcal{A} \times \mathcal{A}$ satisfies the following property:*

If there is a set of attacks $A_1 \mathcal{R} A_2$, $A_2 \mathcal{R} A_3$, $\cdots$, $A_n \mathcal{R} A_1$ then we have that $A_2 \mathcal{R} A_1$, $A_3 \mathcal{R} A_2$, $\cdots$, $A_1 \mathcal{R} A_n$.

Kaci et al. [2006] show that strictly acyclic argumentation frameworks are characterized by conflict+preference argumentation frameworks. For this they use an unusual kind of reduction, defined as follows.

Definition 4.8. *Let $\langle \mathcal{A}, \mathcal{R} \rangle$ be an argumentation framework and $\langle \mathcal{A}, \mathcal{C}, \succeq \rangle$ a conflict+preference argumentation framework. We say that $\langle \mathcal{A}, \mathcal{C}, \succeq \rangle$ represents $\langle \mathcal{A}, \mathcal{R} \rangle$ iff for all arguments A and B of $\mathcal{A}$, we have $A \mathcal{R} B$ iff $A \mathcal{C} B$ and $A \succeq B$. We also say that $\mathcal{R}$ is represented by $\mathcal{C}$ and $\succeq$.*

Theorem 4.9. *$\langle \mathcal{A}, \mathcal{R} \rangle$ is a strictly acyclic argumentation framework if and only if there is a conflict+preference argumentation framework $\langle \mathcal{A}, \mathcal{C}, \succeq \rangle$ that represents it.*

The preference-based argumentation theory introduced by Kaci and colleagues is a variant of Reduction 1, their representation theorem does not hold if Reduction 1 is used. As far as we know it is an open problem which properties the attack relation satisfies (if the defeat relation is symmetric). More generally, the question may be raised which principles hold only for preference-based argumentation frameworks with a symmetric attack relation, but not in general.

4.3 Preference in argumentation

The third challenge concerning preference in abstract argumentation is to relate it to the different ways proposed in the literature to compute a preference relation over arguments depending on the internal structure of the arguments.

We are aware only of one formal result, which we present in the first subsection. The other subsections present other ways in the literature in which preference is calculated in structured argumentation, and for which no corresponding result in abstract argumentation has been provided yet.

4.3.1 Structured arguments & Implicit priorities

The introduction of preferences in argumentation theory was first proposed by Simari and Loui [1992]. In their framework, an argument is structured in the form of a tuple $\langle H, h \rangle$, where H is a set of formulas and h is a formula. H is called the support of the argument and h is its conclusion. By abuse of language we say that $\langle H, h \rangle$ is an argument for h. Inspired by non-monotonic reasoning, the authors build the arguments from a set of defeasible knowledge, denoted by $\mathbb{D}$. They also consider a consistent set of knowledge denoted by K. Generally $K \cup \mathbb{D}$ is inconsistent. A pair $\langle H, h \rangle$ is an argument if and only if it satisfies the following properties:

1. $H \subseteq \mathbb{D}$ and h is a formula of the language,

2. $K \cup H \vdash h$,

3. $K \cup H$ is a consistent subset of $K \cup \mathbb{D}$,

4. $\nexists H' \subset H$ such that $K \cup H' \vdash h$.

We say that an argument can be activated if the premise of each rule in its support is true. This notion defines the specificity between two arguments. An argument is more specific than another argument if and only if each time the former can be activated the latter can also be activated, but the reverse is not true. The specificity relation defines the preference relation. Being more specific makes an argument preferred.

4.3.2 Structured arguments & Explicit priorities

Priorities in the above setting are implicitly represented by the specificity principle. Amgoud et al. [1996] consider explicit priorities. More precisely, in their approach, knowledge is represented by a set of weighted propositional logic formulas $\mathcal{K}$. Let $\mathcal{K} = \{(\phi_i, \alpha_i) | i = 1, \cdots, n\}$, where ϕ_i is a propositional logic formula and $\alpha_i \in (0, 1]$ is the certainty/priority degree associated with ϕ_i. An argument is also a pair $\langle H, h \rangle$. Conditions (1)-(4) given above apply in this case too, where $K = \emptyset$ and $\mathbb{D}$ is replaced with $\mathcal{K}^* = \{\phi_i | (\phi_i, \alpha_i) \in \mathcal{K}\}$.

One can then construct a function $w : \mathcal{A} \to [0, 1]$, where $w(\langle H, h \rangle)$ depends on the weight of formulas involved in H [Amgoud et al., 1996]. Thus, an argument is preferred iff it is preferred w.r.t. w.

4.3.3 Abstract arguments and values

In some applications, the arguments need to be compared not on the basis of their internal structure but with respect to the viewpoints or decisions they promote. This may be due to the fact that the internal structure of the arguments is not available or because the values must be considered. This is particularly true in persuasion dialogues when the preference over values induces the preference over arguments promoting those values [Bench-Capon, 2003]. The more important the value, the more preferred the argument is. Thus, if two arguments are conflicting then the argument promoting a preferred value is accepted. For example, suppose that two parents discuss whether their son can watch the soccer game on the TV or whether he should prepare for his exam. Watching the game allows their son to discuss it with his friends, which promotes his sociability. On the other hand, preparing for his exam promotes his education. If the parents consider that sociability is not more important than education, then the child should prepare for his exam.

Bench-Capon [2003] developed an argumentation framework which models the above considerations (see Chapter 5 of this handbook [Atkinson and Bench-Capon, 2021] for a detailed discussion). Like Dung's framework he considers abstract arguments. Moreover he considers (i) a set of values promoted by the arguments and (ii) a set of audiences, following Perelman [1980], where each audience corresponds to a preference relation over values.

Definition 4.10 (Value-based argumentation framework, from [Bench–Capon, 2003]). *A Value-based Argumentation Framework (VAF) is a five-tuple, VAF* $= \langle \mathcal{A}, \text{Att}, V, \text{val}, \mathcal{D} \rangle$, *where* $\mathcal{A}$ *is a set of arguments,* Att *is an attack relation,* val *is a function which maps from elements of* $\mathcal{A}$ *to elements of* V *and* $\mathcal{D}$ *is the set of possible audiences. An audience specific argumentation framework is a five-tuple, VAF*$_{ad} = \langle \mathcal{A}, \text{Att}, V, \text{val}, >_{ad} \rangle$, *where* $ad \in \mathcal{D}$ *is an audience and* $>_{ad}$ *is a partial order over* V.

It is worth noticing that Att is independent of val, in the sense that two arguments promoting the same value may be related with the attack relation.

Definition 4.11 (Audience-specific value-based argumentation framework, from [Bench-Capon, 2003]). $\langle \mathcal{A}, \text{Att}, V, \text{val}, >_{ad} \rangle$ *represents* $\langle \mathcal{A}, \text{Att}, \succ \rangle$ *if and only if* $\forall A, B \in \mathcal{A}$, *we have*

$$a \succ b \text{ if and only if } \text{val}(a) >_{ad} \text{val}(b).$$

Concerning the existence of audience-specific value-based argumentation frameworks representing a preference-based argumentation framework, the situation is the same as between preference-based argumentation frameworks and argumentation frameworks. Every audience-specific value-based argumentation framework represents precisely one preference-based argumentation framework, and each preference-based argumentation framework is represented by an equivalence class of alphabetic variants of audience-specific value-based argumentation frameworks [Kaci and van der Torre, 2008]. The acceptable extensions of an audience-specific value-based argumentation framework are again simply the acceptable extensions of the unique preference-based argumentation framework it represents.

4.3.4 Abstract arguments & NMR on preferences over values

In Bench-Capon's framework an argument promotes at most one value. However, in practice it may be the case that arguments promote multiple values. Moreover, in Bench-Capon's framework, a value v_1 being more important than (or preferred to) a value v_2 is interpreted as any argument promoting v_1 being preferred to any argument promoting v_2. One can also imagine other ways to compare the arguments promoting v_1 and v_2. Kaci and van der Torre [2008] extend Bench-Capon's framework in order to take into account the previous considerations. More specifically, they consider *(i)* arguments promoting multiple values, and *(ii)* various kinds of preferences over values.

4.3.5 Valued Preference-based Argumentation Frameworks with Varied Strength Defeats

Kaci & Labreuche [2014] developed a preference-based argumentation framework in which the preference relation over arguments has varied strengths. The valued preference relation can be computed from a Boolean preference relation or from a valuation of arguments. Associated with an attack relation, the valued preference relation leads to a defeat relation with varied strengths.

4.3.6 Argument-based preferences over arguments

Modgil [2009] developed a preference-based argumentation framework, called extended argumentation framework, in which the arguments are abstract entities and the preference relation over the set of arguments is not defined by external information (e.g., ordered values or information pervaded with implicit or explicit priorities). Preferences over arguments are supported by arguments. More precisely, Modgil uses Reduction 1 in which the condition $not(b \succ a)$ is supported by an argument. In other words, $(a,b) \in$ Def if and only if $(a,b) \in$ Att and there is no argument claiming that b is preferred to a.

Definition 4.12 (Extended argumentation framework [Modgil, 2009]). *An extended argumentation framework is a three-tuple $\langle \mathcal{A}, \text{Att}, \mathcal{H} \rangle$ such that $\mathcal{A}$ is a finite set of arguments and*

- $\text{Att} \subseteq \mathcal{A} \times \mathcal{A}$ *is an attack relation,*
- $\mathcal{H} \subseteq \mathcal{A} \times \text{Att}$: $(a, (b, d))$ *stands for "a claims that d is preferred to b",*
- *if $(a, (b, d)), (a', (d, b)) \in \mathcal{H}$ then $(a, a'), (a', a) \in \text{Att}$.*

4.4 Dynamics of preference and argumentation

The fourth open challenge in preference-based abstract argumentation is the dynamics of argumentation. In the literature, often two kinds of dynamics are distinguished. One kind of dynamics can be found in dialogue, where the speech acts of the agents may affect the knowledge bases of the agents. The other kind of dynamics can be found in some principles, when they compare distinct argumentation frameworks.

Some of the principles discussed in this chapter are dynamic in a special way: the attack relation is considered to be fixed, but only the preference relation changes. Whereas the dynamics of argumentation frameworks may be criticized, we believe that the dynamics of the preference relation occurs naturally in many applications e.g. recommender systems [Jannach et al., 2010].

5 Related research

Amgoud and Vesic [2014] have proposed a new PAF which satisfies two requirements: conflict-freeness of extensions with respect to the attack relation (this is Principle 1 in our analysis), and, in the absence of critical attacks, the extensions of a preference-based argumentation framework coincide with the extensions of Dung's argumentation framework. Moreover the PAF handles critical attacks following Reduction 2. The second requirement can be roughly reformulated as: if the preference relation is empty, extensions of a PAF and Dung's framework coincide.

To compute extensions, Amgoud and Vesic [2014] define a preference relation over the powerset of arguments using the so-called democratic and elitist relations. Maximal conflict-free subsets are extensions of the PAF (grounded, preferred and stable extensions). The Definition 4.1 of Kaci et al. [Kaci *et al.*, 2018] is incomparable with democtratic and elitist relations. Definition 4.1 applies on the PAF with ignoring the associated preference relation while [Amgoud and Vesic, 2014] applies democtratic and elitist relations on the PAF with Reduction 2.

There is a striking similarity at the abstract level between preference based argumentation and support in bipolar argumentation, as both can be seen as reductions [Yu and Torre, 2020].

Other approaches have been proposed in the literature to reason over preferences in argumentation. Among them, in [Prakken and Sartor, 1997], the underlying language is Extended Logic Programming (ELP), which includes strict rules and defeasible rules, while the preference information is provided under the form of an ordering on the defeasible rules. They study two different cases with respect to the nature of this ordering. In the first one, the ordering is fixed and indisputable (i.e., strict priorities), while in the second case priorities are themselves defeasibly derived as conclusions within the system. To support defeasible priorities, they allow stating rules and constructing arguments about priorities.

The argumentation framework proposed in [Governatori and Maher, 2000] uses the language of Defeasible Logic. In this framework, the rule priority relation of Defeasible Logic is used to determine whether an argument is defeated by a counter-argument.

The framework proposed in [Kakas and Moraitis, 2003; Kakas and Moraitis, 2006] also includes the notion of dynamic priorities in the context of negotiating agents. Roles and context define in a natural way dynamic preferences on the decisions implied by the negotiation strategies of the agents as their environment changes. The underlying monotonic logic includes a special type of rules that are used to give priority between competing rules in case of conflict. Based on these rules, they build arguments on priorities, and reason with them to give preference to specific arguments in the system.

Similarly, the framework proposed in [Bikakis and Antoniou, 2010] also adopts Defeasible Logic as the underlying formalism for building arguments. However, it assumes a distributed argumentation system, in which each agent has its own knowledge base and agents create arguments by combining their local knowledge with the beliefs of other agents. The preferences among arguments in this case are derived from the preferences of an agent on the agents it imports information from.

In [Cyras and Toni, 2016], the authors introduce a new approach to handling preferences in Assumption-Based Argumentation (ABA) called ABA+. More precisely, ABA+ assumes preferences on the object level (i.e., over assumptions) and incorporates them directly into the definition of attack, rather than assuming preferences on the meta level (e.g., over arguments). ABA+ also allows for preferences in generic ABA frameworks, as opposed to allowing for preferences only in flat ABA frameworks.

[Brewka et al., 2013] introduce static and dynamic preferences into Abstract Dialectical Frameworks (ADFs). Roughly, they handle dynamic preferences as follows: they first guess some (stable, preferred, grounded) extension M. Some nodes in M carry the preference information. They extract this information and check whether M can be reconstructed under the preference information, thus verifying that the preferences represented in the model itself were taken into account adequately.

There are also works that deal with preferences in the context of ranking-based / gradual argumentation semantics [Amgoud et al., 2017], but we do not study them in this chapter.

6 Summary

In this paper, we presented a principle-based analysis of the semantics of of preference-based argumentation frameworks. We considered four reductions to move from PAFs to a Dung-like abstract argumentation on which standard semantics can be applied to compute the set of accepted arguments, and we proposed a set of ten principles we used to study the considered reductions.

The results of this paper give rise to many new research questions. Many more principles can be defined in our framework (e.g., following [Rienstra et al., 2015]), and used in the analysis. In particular, it is striking that many principles have a dynamic flavor, and we conjecture that many approaches to dynamics of argumentation [Booth et al., 2013] can be used as a source for principles. We are in particular interested in principles that distinguish the various PAF semantics. We believe that there are not many new reductions to be found, but more PAF semantics can be defined not based on reductions. We observe that the resolution-based family of abstract argumentation semantics [Baroni et al., 2011] seems also related to Reduction 3 introduced in this paper.

References

[Amgoud and Cayrol, 2002] Leila Amgoud and Claudette Cayrol. Inferring from inconsistency in preference-based argumentation frameworks. *International Journal of Approximate Reasoning*, 29(2):125–169, 2002.

[Amgoud and Vesic, 2009] Leila Amgoud and Srdjan Vesic. Repairing preference-based argumentation frameworks. In *IJCAI'09*, pages 665–670, 2009.

[Amgoud and Vesic, 2010] Leila Amgoud and Srdjan Vesic. On the role of preferences in argumentation frameworks. In *ICTAI'10*, pages 219–222, 2010.

[Amgoud and Vesic, 2014] Leila Amgoud and Srdjan Vesic. Rich preference-based argumentation frameworks. *Int. J. Approx. Reasoning*, 55(2):585–606, 2014.

[Amgoud and Vesic, 2020] Leila Amgoud and Srdjan Vesic. Dung's semantics satisfy attack removal monotonicity. *CoRR*, abs/2007.04221, 2020.

[Amgoud et al., 1996] Leila Amgoud, Claudette Cayrol, and Daniel LeBerre. Comparing arguments using preference orderings for argument-based reasoning. In *Proc. of the 8th International Conference on Tools with Artificial Intelligence, ICTAI 1996*, pages 400–403, 1996.

[Amgoud et al., 2017] Leila Amgoud, Jonathan Ben-Naim, Dragan Doder, and Srdjan Vesic. Acceptability semantics for weighted argumentation frameworks. In Carles Sierra, editor, *Proc. of the 26th International Joint Conference on Artificial Intelligence, (IJCAI'17)*, pages 56–62, 2017.

[Atkinson and Bench-Capon, 2021] Katie Atkinson and Trevor Bench-Capon. Value-based argumentation. In Dov Gabbay, Massimiliano Giacomin, Guillermo R. Simari, and Matthias Thimm, editors, *Handbook of Formal Argumentation*, volume 2, chapter 5. College Publications, 2021.

[Baroni and Giacomin, 2007] Pietro Baroni and Massimiliano Giacomin. On principle-based evaluation of extension-based argumentation semantics. *Artif. Intell.*, 171(10-15):675–700, 2007.

[Baroni et al., 2005] Pietro Baroni, Massimiliano Giacomin, and Giovanni Guida. Scc-recursiveness: a general schema for argumentation semantics. *Artif. Intell.*, 168(1-2):162–210, 2005.

[Baroni et al., 2011] Pietro Baroni, Paul E. Dunne, and Massimiliano Giacomin. On the resolution-based family of abstract argumentation semantics and its grounded instance. *Artif. Intell.*, 175(3-4):791–813, 2011.

[Baroni et al., 2014] Pietro Baroni, Guido Boella, Federico Cerutti, Massimiliano Giacomin, Leendert W. N. van der Torre, and Serena Villata. On the input/output behavior of argumentation frameworks. *Artif. Intell.*, 217:144–197, 2014.

[Baroni et al., 2018] P. Baroni, D. Gabbay, and M. Giacomin. *Handbook of Formal Argumentation*. College Publications, 2018.

[Bench-Capon, 2003] Trevor J.M. Bench-Capon. Persuasion in practical argument using value-based argumentation frameworks. *Journal of Logic and Computation*, 13(3):429–448, 2003.

[Bikakis and Antoniou, 2010] Antonis Bikakis and Grigoris Antoniou. Defeasible contextual reasoning with arguments in ambient intelligence. *IEEE Trans. Knowl. Data Eng.*, 22(11):1492–1506, 2010.

[Boella et al., 2005] Guido Boella, Joris Hulstijn, and Leendert W. N. van der Torre. A Logic of Abstract Argumentation. In *ArgMAS*, pages 29–41, 2005.

[Booth et al., 2013] Richard Booth, Souhila Kaci, Tjitze Rienstra, and Leendert W. N. van der Torre. A logical theory about dynamics in abstract argumentation. In *Scalable Uncertainty Management - Proc. of the 7th International Conference, SUM 2013*, pages 148–161, 2013.

[Brewka et al., 2013] Gerhard Brewka, Hannes Strass, Stefan Ellmauthaler, Johannes Peter Wallner, and Stefan Woltran. Abstract dialectical frameworks revisited. In Francesca Rossi, editor, *IJCAI 2013, Proceedings of the 23rd International Joint Conference on Artificial Intelligence, Beijing, China, August 3-9, 2013*, pages 803–809. IJCAI/AAAI, 2013.

[Cramer and van der Torre, 2019] Marcos Cramer and Leon van der Torre. SCF2 - an argumentation semantics for rational human judgments on argument acceptability. In Christoph Beierle, Marco Ragni, Frieder Stolzenburg, and Matthias Thimm, editors, *Proceedings of the 8th Workshop on Dynamics of Knowledge and Belief (DKB-2019) and the 7th Workshop KI & Kognition (KIK-2019) co-located with 44nd German Conference on Artificial Intelligence (KI 2019), Kassel, Germany, September 23, 2019*, volume 2445 of *CEUR Workshop Proceedings*, pages 24–35. CEUR-WS.org, 2019.

[Cyras and Toni, 2016] Kristijonas Cyras and Francesca Toni. ABA+: assumption-based argumentation with preferences. In Chitta Baral, James P. Delgrande, and Frank Wolter, editors, *Principles of Knowledge Representa-*

3 - Preference in Abstract Argumentation

tion and Reasoning: Proceedings of the Fifteenth International Conference, KR 2016, Cape Town, South Africa, April 25-29, 2016, pages 553–556. AAAI Press, 2016.

[Dung, 1995a] Phan M. Dung. On the acceptability of arguments and its fundamental role in non-monotonic reasoning, logic programming and n-person games. *Artificial Intelligence*, 77:321–357, 1995.

[Dung, 1995b] Phan M. Dung. On the Acceptability of Arguments and Its Fundamental Role in Nonmonotonic Reasoning, Logic Programming, and n-Person Games. *Artificial Intelligence*, 77(2):321–357, 1995.

[Governatori and Maher, 2000] Guido Governatori and Michael J. Maher. An argumentation-theoretic characterization of defeasible logic. In Werner Horn, editor, *ECAI 2000, Proceedings of the 14th European Conference on Artificial Intelligence, Berlin, Germany, August 20-25, 2000*, pages 469–473. IOS Press, 2000.

[Jannach et al., 2010] D. Jannach, M. Zanker, A. Felfernig, and G. Friedrich. *Recommender Systems - An Introduction*. Cambridge University Press, 2010.

[Kaci and Labreuche, 2014] S. Kaci and Ch. Labreuche. Valued preference-based instantiation of argumentation frameworks with varied strength defeats. *International Journal of Approximate Reasoning*, 55(9):2004–2027, 2014.

[Kaci and van der Torre, 2008] S. Kaci and L. van der Torre. Preference-based argumentation: Arguments supporting multiple values. *IJAR*, 48:730–751, 2008.

[Kaci et al., 2006] Souhila Kaci, Leendert W. N. van der Torre, and Emil Weydert. Acyclic argumentation: Attack = conflict + preference. In *ECAI 2006, Proc. of the 17th European Conference on Artificial Intelligence*, pages 725–726, 2006.

[Kaci et al., 2018] Souhila Kaci, Leendert W. N. van der Torre, and Serena Villata. Preference in abstract argumentation. In *Computational Models of Argument - Proceedings of COMMA 2018, Warsaw, Poland, 12-14 September 2018*, pages 405–412, 2018.

[Kakas and Moraitis, 2003] Antonis C. Kakas and Pavlos Moraitis. Argumentation based decision making for autonomous agents. In *The Second International Joint Conference on Autonomous Agents & Multiagent Systems, AAMAS 2003, July 14-18, 2003, Melbourne, Victoria, Australia, Proceedings*, pages 883–890. ACM, 2003.

[Kakas and Moraitis, 2006] Antonis C. Kakas and Pavlos Moraitis. Adaptive agent negotiation via argumentation. In Hideyuki Nakashima, Michael P. Wellman, Gerhard Weiss, and Peter Stone, editors, *5th International Joint Conference on Autonomous Agents and Multiagent Systems (AAMAS 2006),*

Hakodate, Japan, May 8-12, 2006, pages 384–391. ACM, 2006.

[Modgil and Prakken, 2012] Sanjay Modgil and Henry Prakken. Resolutions in structured argumentation. In Bart Verheij, Stefan Szeider, and Stefan Woltran, editors, *Computational Models of Argument - Proceedings of COMMA 2012, Vienna, Austria, September 10-12, 2012*, volume 245 of *Frontiers in Artificial Intelligence and Applications*, pages 310–321. IOS Press, 2012.

[Modgil, 2009] Sanjay Modgil. Reasoning about preferences in argumentation frameworks. *Artificial Intelligence*, 173(9-10):901–934, 2009.

[Perelman, 1980] Ch. Perelman. Justice, law and argument. *Reidel, Dordrecht*, 1980.

[Prakken and Sartor, 1997] Henry Prakken and Giovanni Sartor. Argument-based extended logic programming with defeasible priorities. *J. Appl. Non Class. Logics*, 7(1):25–75, 1997.

[Rienstra et al., 2015] Tjitze Rienstra, Chiaki Sakama, and Leendert W. N. van der Torre. Persistence and monotony properties of argumentation semantics. In *Theory and Applications of Formal Argumentation - Proc. of the 3rd International Workshop, TAFA 2015*, pages 211–225, 2015.

[Simari and Loui, 1992] G.R. Simari and R.P. Loui. A mathematical treatment of defeasible reasoning and its implementation. *Artificial Intelligence*, 53:125–157, 1992.

[van der Torre and Vesic, 2018] Leendert van der Torre and Srdjan Vesic. The principle-based approach to abstract argumentation. In Pietro Baroni, Dov Gabbay, Massimiliano Giacomin, and Leendert van der Torre, editors, *The Handbook of Formal Argumentation*. College Publications, 2018.

[Viappiani, 2014] Paolo Viappiani. Preference modeling and preference elicitation: An overview. In Mouzhi Ge and Francesco Ricci, editors, *Proceedings of the First International Workshop on Decision Making and Recommender Systems (DMRS2014), Bolzano, Italy, September 18-19, 2014*, pages 19–24, 2014.

[Yu and Torre, 2020] Liuwen Yu and Leendert Van Der Torre. A principle-based approach to bipolar argumentation. In *Proceedings of the 18th International Workshop on Non-Monotonic Reasoning*, 2020.

CHAPTER 4

COLLECTIVE ACCEPTABILITY IN ABSTRACT ARGUMENTATION

DOROTHEA BAUMEISTER
Institut für Informatik, Heinrich-Heine-Universität Düsseldorf,
Germany
d.baumeister@hhu.de

DANIEL NEUGEBAUER
Institut für Informatik, Heinrich-Heine-Universität Düsseldorf,
Germany
daniel.neugebauer@hhu.de

JÖRG ROTHE
Institut für Informatik, Heinrich-Heine-Universität Düsseldorf,
Germany
rothe@hhu.de

Abstract

This chapter highlights the collective acceptability problem in multiagent argumentation, which is related to the problem of collective decision making in the field of computational social choice at the intersection of social choice theory, theoretical computer science, and artificial intelligence. Specifically, the chapter surveys various approaches to collective acceptability and showcases useful methods for structural aggregation of argumentation frameworks and their properties.

1 Introduction

In this chapter, we will be concerned with collective decision making in multiagent scenarios, specifically focusing on how methods from computational social choice (COMSOC) can be employed to solve the problem of collective acceptability in multiagent argumentation. Computational social choice (see the textbooks edited by Brandt *et al.* [2016] and Rothe [2015]) is an interdisciplinary area at the interface of social choice theory, computer science, and (distributed) artificial intelligence. A core research stream in COMSOC is the study of voting. While voters have individual preferences over alternatives among which they seek to determine the best choices, called the *winners* of the election, in argumentation theory we are faced with agents who have individual views on the arguments and on the attack relation among arguments and who seek to determine the best choices of arguments, i.e., *acceptable sets of arguments*, according to certain semantics. Just as preference aggregation through voting rules has been studied in COMSOC, various ways of aggregating argumentation frameworks (AFs) of different types have been proposed.

We will survey the latter through the COMSOC lens, paying particular attention to how methods and concepts originally designed for COMSOC have been applied in the context of argumentation. For example, axiomatic properties of voting mechanisms have been thoroughly studied in social choice theory and also in COMSOC, and in Section 3 we will present what is known about the axiomatic properties of methods used to aggregate AFs. In Section 4, we will turn to specific aggregation methods for AFs. In particular, we will present existing aggregation operators and their properties that make use of partial, incomplete, and value-based AFs, the former two being related to uncertainty and incomplete information about the argumentation at hand and the latter being related to the impact a ranking of values assigned to arguments may have on the argumentation process.

Uncertainty can occur both in voting and in argumentation, even though due to distinct sources of incomplete information. In voting, we may be faced with "noisy" elections [Procaccia *et al.*, 2007; Xia, 2012; Caragiannis *et al.*, 2016], for example due to incorrect vote counts, either by accident or with malicious intent, or with voters who are simply too

4 - COLLECTIVE ACCEPTABILITY IN ABSTRACT ARGUMENTATION

lazy to rank all the alternatives [Baumeister et al., 2012a; Menon and Larson, 2017]. In argumentation, dynamic changes in a given AF [Cayrol et al., 2010], different and changing individual views and beliefs of the involved agents [Cayrol et al., 2008], or uncertainty in the underlying knowledge base used to instantiate the AF [Modgil and Prakken, 2013] all may lead to uncertainty about the status of an AF [Coste-Marquis et al., 2007; Baumeister et al., 2018b; Niskanen et al., 2020; Skiba et al., 2020].

Central concepts used for voting with uncertainty are the notions of *possible* and *necessary winner*, proposed by Konczak and Lang [2005] and studied in more depth by, e.g., Xia and Conitzer [2011], Chevaleyre et al. [2012], and Baumeister et al. [2011, 2012b]. Intuitively, assuming that the voters' preferences over the alternatives are incomplete, a *possible winner* is an alternative that can be made win for *some* complete extension of the voters' preferences, whereas a *necessary winner* is an alternative that must win for *every* complete extension of the voters' preferences. These notions are so important that they could be beneficially applied in other areas as well, such as in fair division by Bouveret et al. [2010], Baumeister et al. [2017], and Kuckuck and Rothe [2019]; in algorithmic game theory by Kerkmann et al. [2020] and Rothe et al. [2018]; and in judgment aggregation by Baumeister et al. [2015a, 2020].

As mentioned earlier, at the intersection of computational social choice and formal argumentation lies the problem of *collective acceptability*. Acceptability in the standard argumentation model means that we are looking at a single representation of an argumentation and try to determine which arguments are acceptable under a certain semantics. Collective acceptability, on the other hand, is concerned with acceptability in a set of several related, but potentially different representations of a single argumentation. Problems of collective acceptability can arise in various applications. When individual agents each create their personal representation of the argumentation from a private knowledge base, different or missing information in these knowledge bases will have the effect that the argumentation representations will be different. Determining argument acceptability in the individual views will have limited significance, while collective acceptability with regard to all individual views can incorporate all information available to the agents and produce more meaningful results. In strategic scenarios, arguments known to all agents (and thus present in all individual views) may represent arguments that were already given in a shared discussion, while arguments known only to some, but not all agents may represent arguments that these agents could still bring forward. Here, collective acceptability may help anticipate which arguments might become acceptable in the further progress of the discussion, and which of the available arguments should be brought forward to improve an agent's position in the argumentation.

There are two fundamentally different approaches to collective acceptability in the literature. In the survey by Bodanza *et al.* [2017], these are called the argument-wise and the framework-wise approach. The *argument-wise* approach determines acceptability in the individual views using standard methods, and then defines *semantic aggregation* methods—e.g., voting procedures—to aggregate the individually accepted arguments into a single collectively acceptable set of arguments. The *framework-based* approach defines *structural aggregation* methods to aggregate individual views into a collective representation first, and then determines acceptability in the collective representation, either by standard methods or by dedicated methods for that representation.

Figure 1 illustrates these two approaches to the problem of collective acceptability. The elements in the top-left corner represent the individual views of m agents. Semantic evaluation on the individual views (left arrows going down) produces individually acceptable outcomes (bottom left), which can be aggregated using semantic aggregation (bottom arrows going right) into a collectively acceptable outcome. Alternatively, structural aggregation on the individual views (top arrows going right) produces a single collective representation of the argumentation (top right), which can be semantically evaluated (right arrow going down) for a collectively acceptable outcome.

Semantic aggregation is suitable for applications where the outcome of the individual views (i.e., the accepted arguments) is considered to be the most important part of the views, overshadowing the importance of the underlying structure of the argumentation. This might be the case when agents have created their individual argumentations with the specific purpose to support the acceptability of certain key arguments. The agents might not be very interested in finding a collective view of the

4 - Collective Acceptability in Abstract Argumentation

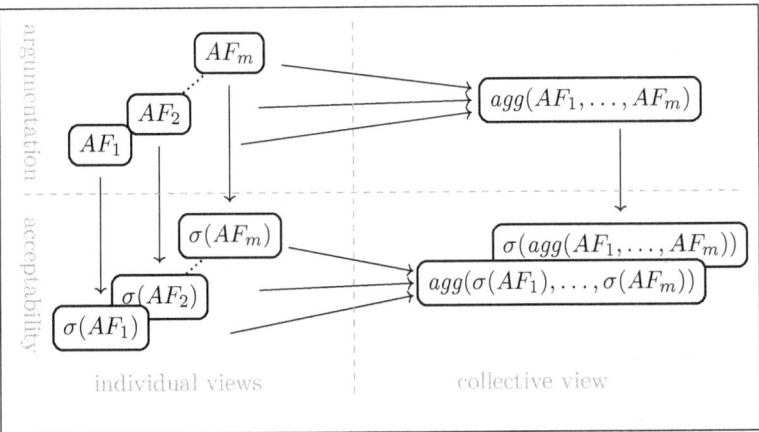

Figure 1: Schematic overview of different approaches to collective acceptability in formal argumentation, where σ is a general placeholder for some kind of acceptability criterion on AFs, and agg is a general placeholder for some aggregation operator on AFs.

argumentation that is close to their individual view, but they are highly interested in having collectively accepted arguments that are close to the accepted arguments in their view. Semantic aggregation is applied by Delobelle et al. [2016] via merging extensions, and by Caminada and Pigozzi [2011] and Awad et al. [2017] via merging labelings. A related research problem to the argument-wise approach to collective acceptability is the problem of *realizability* [Dunne et al., 2015] or *synthesis* [Niskanen et al., 2019], where sets of accepted arguments are given as input, and the goal is to find an AF whose sets of accepted arguments are the same as, or as close as possible to, the given sets. These methods can be used to augment the results of semantic aggregation methods to not only obtain collectively acceptable arguments as a result, but also an AF that produces these. Additional constraints can be used to make sure that the AF created is as close as possible to the input AFs. Relatedly, the *enforcement* problem asks whether a given set of arguments can be enforced as an acceptable set by a finite number of elementary changes to a given AF. The original definition by Baumann and Brewka [2010] allows the addition of single arguments along with incident attacks, while subsequent work allows the addition and deletion of attacks [Wallner et al., 2017], or both simultaneously [Coste-Marquis et al., 2015].

The second approach of structural aggregation is suitable when the structure in the individual views is more important than their accepted arguments. For example, this might be the case when each agent has access to a limited part of the information on a subject matter, so the significance of which arguments are accepted in the individual views is rather limited. Merging the individual fragments of the information into an aggregated representation of the argumentation creates a better basis for semantic evaluation in such situations. In this chapter, we focus on structural aggregation.

2 Preliminaries

We recall the model of abstract argumentation frameworks Dung [1995].

Definition 2.1. *An* argumentation framework (AF) *is a pair* $AF = \langle Ar, att \rangle$ *consisting of a finite set Ar of arguments and a binary attack relation* $att \subseteq Ar \times Ar$.

As an example, consider the set of arguments, $Ar = \{a, b, c, d, e, f, g\}$, displayed in Table 1. These arguments might be given in the context of a virus disease that is spreading in a population, and they make different suggestions on how to react to the disease. We will use this argumentation as a running example throughout this chapter.

Arguments a and b mutually attack each other, and so do d and e, as well as f and g. Further, argument c attacks b, argument f attacks c and e, and argument g attacks b and d. This is formally captured by the attack relation

$$att = \{(a,b), (b,a), (c,b), (d,e), (e,d), (f,e), (f,c), (g,d), (f,g),\\ (g,b), (g,f)\}.$$

Figure 2 gives a graph representation of this AF, $\langle Ar, att \rangle$.

An argumentation *semantics* σ maps a given argumentation framework AF to the set of σ-*extensions* of AF, which are the sets of arguments that are acceptable in AF with respect to σ. A set S of arguments

a	"Everybody needs access to medical supplies for personal protection."
b	"Medical supplies are not sufficient, so hospitals must have priority access to medical supplies."
c	"Medical supplies are sufficient for all."
d	"The disease problem must be solved by government health officials, the population should stay out of it."
e	"Everybody has a right to know how dangerous the disease really is."
f	"Information about the dangers of the disease may cause a panic in the population, leading to hoarding of medical supplies and thus a shortage of these."
g	"If people know enough about the disease, they can effectively protect themselves without the need of medical supplies."

Table 1: Arguments used in the AF shown in Figure 2

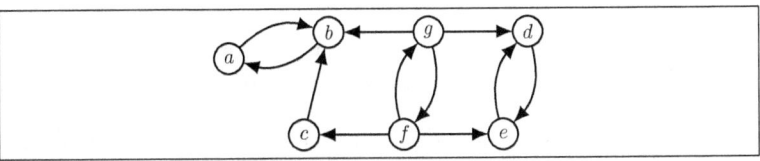

Figure 2: Graph representation of the AF used in our running example, where arguments are displayed as nodes and attacks as directed edges

is called *conflict-free* ($\mathcal{CF}$) if there are no attacks among arguments in S. A conflict-free set S is *admissible* ($\mathcal{AD}$) if it defends each of its arguments, where we say $S \subseteq Ar$ *defends* $a \in Ar$ if for each argument $b \in Ar$ attacking a, there is an argument $c \in S$ attacking b. A maximal (with respect to set inclusion) admissible set S is said to be *preferred* ($\mathcal{PR}$). A conflict-free set S is *stable* ($\mathcal{ST}$) if every argument outside of S is attacked by some argument in S. An admissible set that is closed under defense—i.e., that includes every argument that it defends—is called *complete* ($\mathcal{CO}$). The unique minimal complete set is the *grounded* extension ($\mathcal{GR}$).

In this chapter, we consider aggregation operators on argumentation frameworks, which are mappings that aggregate a set of argumentation frameworks into a collective representation. We use the following general notation. Let $\mathbb{AF}$ denote the universe of all possible argumentation frameworks. An *aggregation operator agg* : $\mathbb{AF}^m \to 2^{\mathbb{AF}}$ for argumentation frameworks is a mapping from a set of m input argumentation frameworks (which may represent the individual views of m agents) to a set of aggregate argumentation frameworks. An aggregation operator is called *resolute* if it always outputs a singleton, i.e., if $|agg(P)| = 1$ for all profiles $P \in \mathbb{AF}^m$, and *irresolute* otherwise. Generalized aggregation operators may use an extended target format that goes beyond standard argumentation frameworks.

3 Axiomatic Properties of Aggregation Methods in Argumentation

In social choice theory, mechanisms for collective decision making are studied with respect to various axioms. Such axioms express desirable behavior of these mechanisms. Unfortunately, there is a number of impossibility results such as Arrow's Theorem (Arrow [1951 revised edition 1963]) and the Gibbard–Satterthwaite Theorem (Gibbard [1973]; Satterthwaite [1975]) showing that it is impossible to fulfill some of the most basic criteria simultaneously. Similar questions arise in collective argumentation, especially when different views of agents should be aggregated. This is, for example, the case when agents have different opinions about the attack relation. An important question is then which properties of the individual attack relations will be preserved by a given aggregation rule. In the case of argumentation, these properties will be specifically related to the various semantics. In this section, we will review results by Chen and Endriss [2019] about the preservation of semantic properties in the aggregation of abstract argumentation frameworks. This work focuses on the case where all agents consider the same set of arguments but have different opinions on the attacks among them. However, a generalization where the argument sets may differ for each agent would also be possible.

The general setting we consider is a common finite set of arguments

Ar and a set $N = \{1,\ldots,n\}$ of agents. The individual view of agent $i \in N$ is represented as an argumentation framework $AF_i = \langle Ar, att_i \rangle$. The profile $P = (att_1,\ldots, att_n)$ consist of all individual attack relations. Additionally, $N_r^P = \{i \in N \mid r \in att_i\}$ is the set of supporters of attack r in profile P. To aggregate these individual views, we use aggregation rules. For a fixed number of n agents, they are formally defined as a function

$$F : (Ar \times Ar)^n \to Ar \times Ar.$$

One class of aggregation rules often used in the context of judgment aggregation are quota rules that have been introduced by Dietrich and List [2007] and further studied by, e.g., Baumeister et al. [2015a, 2020]. The idea is that an element will be included in the aggregated outcome if the agreement exceeds some given quota. According to Chen and Endriss [2019], the definition is as follows in the context of argumentation frameworks.

Definition 3.1 (Quota Rule). *For $q \in \{1,\ldots,n\}$ and a profile P, the quota rule F_q is defined as*

$$F_q(P) = \{r \in Ar \times Ar \mid |N_r^P| \geq q\}.$$

Hence, all attacks that are supported by at least q agents are accepted.

Prominent examples are the *majority rule* F_q with quota $q = \lceil \frac{n}{2} \rceil$ and the *nomination rule* F_1 with quota 1. The latter rule requires only one nomination of every attack that is included in the aggregated argumentation framework, which is a reasonable choice especially in argumentation where conflicts should be taken seriously. Another example of a rule is the *dictatorship* of a specific agent. The outcome for the dictatorship of agent $i \in N$ is always the individual AF of this agent. In contrast to the quota rules defined above, this rule does not take into account the attack relations of all agents.

Example 3.2. *Recall our running example from Table 1 and Figure 2 in Section 2. We consider a profile consisting of the attack relations (as shown in Figures 3a–3c) of the three individual AFs AF_1, AF_2, and AF_3 representing the individual views of three agents over the same set of arguments. When using the majority rule, all attacks with at least two*

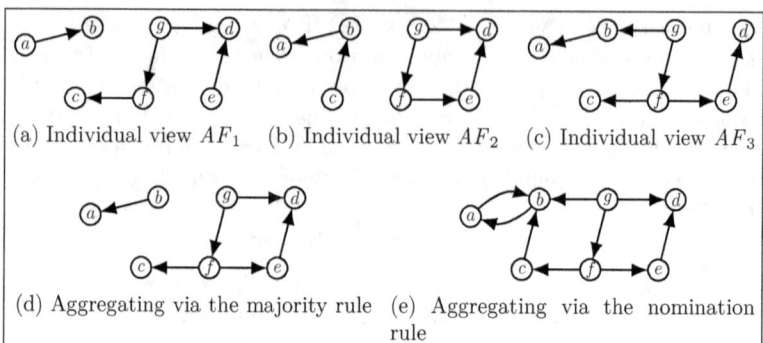

Figure 3: Graph visualizations for Example 3.2

supporters will be included in the aggregated AF. The result is shown in Figure 3d. Under the nomination rule, all attacks that are present in at least one individual AF are contained in the aggregated outcome, as shown in Figure 3e.

Other rules that are used in voting or judgment aggregation may also be transferred to abstract argumentation. For example, rules that minimize the distance to the individual votes or the individual judgments (for one of the common types of distance between preferences or judgment sets), like the Kemeny rule in voting [Kemeny, 1959; Hemaspaandra et al., 2005] and in judgment aggregation [de Haan, 2017; de Haan, 2016], can also be used for the aggregation of argumentation frameworks. They will not be considered here, though.

An important property of aggregation rules is the existence of agents with veto power. These are agents which may not be ignored, and hence only attacks that exist in their individual attack relation may be included in the aggregated outcome.

Definition 3.3 (Veto Power). *Agent $i \in N$ has veto powers under aggregation rule F if for every profile P, it holds that*

$$F(P) \subseteq att_i.$$

It is obvious that under the majority and nomination rules, no agent has veto powers. In a dictatorship, the dictator has veto powers.

4 - Collective Acceptability in Abstract Argumentation

The most basic axioms in social choice are anonymity and neutrality. Their intuitive meaning in the context of voting rules is that all voters and all candidates, respectively, are treated equally, which are quite basic fairness criteria. This can be directly transferred to agents and attacks for AF aggregation rules.

Definition 3.4 (Anonymity and Neutrality).

- *An aggregation rule F is* anonymous *if for all profiles P and all permutations $\pi : N \to N$, it holds that*

$$F(P) = F(att_{\pi(1)}, \ldots, att_{\pi(n)}).$$

- *An aggregation rule F is* neutral *if for all profiles P and all attacks a, a' with $N_a^P = N_{a'}^P$, it holds that*

$$a \in F(P) \Leftrightarrow a' \in F(P).$$

From the definition of quota rules it follows that they are anonymous and neutral. However, there may also be reasons for AF aggregation rules that are not anonymous. This is for example the case when some agents are considered to be experts and their view (maybe on a subset of the arguments) should have more weight in the aggregated outcome. An example for such rules are so-called *qualified majority rules*, where a subset of the agents has veto powers. The acceptance of an attack depends on acceptance by a weak majority and by the agents that have veto powers. Similar reasons can justify aggregation rules that are not neutral, since some arguments may be more important than others. Properties of qualified majority rules, which are not anonymous, have been studied by Tohmé et al. [2008]. A dictatorship is obviously not anonymous and not neutral.

In social choice theory, there are many different formulations for independence axioms. The common idea is that the choice between two alternatives should only depend on their relation and not on other ("irrelevant") alternatives. For the aggregation of AFs, we require that the acceptance of an attack only depends on the supporters of this attack.

Definition 3.5 (Independence). *An aggregation rule F is* independent *if for all profiles P, P' and all attacks a with $N_a^P = N_a^{P'}$, it holds that*

$$a \in F(P) \Leftrightarrow a \in F(P').$$

As the definition of quota rules relies only on the number of supporters for the attacks, they satisfy independence. Again, dictatorships violate independence. As for anonymity and neutrality, there may be reasons to consider AF aggregation rules that are not independent. If additional structural relations between the arguments are considered that have to be respected in the aggregated outcome, it may be useful to consider non-independent aggregation rules. The situation in judgment aggregation is similar, but here the relation between the different issues are present through the given formulas. However, independence is a key axiom for several impossibility results in judgment aggregation. Terzopoulou and Endriss [2020] provide a recent study of alternative formulations for independence in judgment aggregation.

A very intuitive property for aggregation mechanisms is that additional support should not be harmful. This is captured by the monotonicity axiom. Violation of this axiom is considered to be a serious disadvantage, as stated, for example, by Fishburn [1977]. A prominent example of a voting rule violating monotonicity is the Dodgson rule (see Dodgson [1876]), as shown by Fishburn [1977] for five alternatives and by Fishburn [1982] even for four alternatives; see also Brandt [2009].

For the aggregation of argumentation frameworks, monotonicity requires that a selected attack will never be rejected if it receives more support from the agents.

Definition 3.6 (Monotonicity). *An aggregation rule F is* monotonic *if for all profiles P, P' and all attacks a with $N_a^P \subseteq N_a^{P'}$ and $N_{a'}^P = N_{a'}^{P'}$ for all attacks $a' \neq a$, it holds that*

$$a \in F(P) \Rightarrow a \in F(P').$$

Similarly to the other axioms, monotonicity is satisfied by quota rules but violated by dictatorships.

The last two properties we consider are unanimity and groundedness. We will follow Chen and Endriss [2019] who write: "Note that, in line with the existing literature in argumentation theory on the one hand and social choice theory on the other, we use the term 'grounded' in two unrelated ways (grounded extension vs. grounded aggregation rules)." It will always be clear from the context whether we use the term *grounded* in the sense of argumentation theory or social choice theory.

While unanimity requires that an aggregation rule has to follow a unanimous decision on an attack, groundedness requires that at least one supporter must exist for an attack to be selected.

Definition 3.7 (Unanimity and Groundedness).

- An aggregation rule F is unanimous *if for all profiles P, it holds that*
$$F(P) \supseteq att_1 \cap \cdots \cap att_n.$$

- An aggregation rule F is grounded *if for all profiles P, it holds that*
$$F(P) \subseteq att_1 \cup \cdots \cup att_n.$$

The quota lies between 1 and the number of agents. Hence all quota rules are unanimous and grounded. Dictatorships are also grounded, since all attacks are contained in the individual AF of the dictator, but unanimity is obviously violated. To summarize, quota rules satisfy all introduced basic axioms: anonymity, neutrality, independence, monotonicity, unanimity and groundedness.

A very important concept in social choice theory is *collective rationality* (see Arrow [1951 revised edition 1963]) with respect to some given property. In the case of transitive preferences of individuals, collective rationality would imply that the aggregated preference is also transitive. The Condorcet paradox Condorcet [1785] shows that this is not the case for the pairwise majority comparison between alternatives. Collective rationality has also been studied in judgment aggregation by List and Pettit [2002] and in graph aggregation by Endriss and Grandi [2017]. For AF-aggregation, Chen and Endriss [2019] define collective rationality with respect to some *AF-property Prop* $\subseteq 2^{Ar \times Ar}$, which is the set of all attack relations that satisfy *Prop*.

Definition 3.8 (Preserving a Property). *Given an aggregation rule F and some AF-property Prop, we say that F preserves Prop if for every profile P with $att_i \in Prop$, $i \in N$, it holds $F(P) \in Prop$.*

Two basic AF-properties are *acyclicity* and *coherence*. They are very attractive because they reduce the number of possible extensions. If the attack relation of an AF is acyclic, the grounded, stable, preferred, and complete extension coincide, and hence there is exactly one. Coherent AFs are those where preferred and stable semantics coincide. An aggregation rule F *preserves acyclicity (coherence)* if for every profile consisting of acyclic (coherent) individual AFs the outcome of F is also an acyclic (coherent) AF. Obviously, acyclicity is stronger than coherence; however, the results show that preserving acyclicity is easier than preserving coherence.

Theorem 3.9. [Chen and Endriss [2019]]

- Let $|Ar| \geq |N|$. Then for any neutral and independent aggregation rule F that preserves acyclicity, there is at least one agent that has veto powers.

- Let $|Ar| \geq 4$. Then any unanimous, grounded, and independent aggregation rule F that preserves coherence is a dictatorship.

Tohmé et al. [2008] showed that already qualified majority rules preserve acyclicity. Since these rules include an agent with veto powers, this is a special case of the result above. In quota rules, which are independent, no agent has veto powers, hence they do not preserve acyclicity. As an example for the majority rule, consider the simple case of three arguments $\{a, b, c\}$ and three agents with the attack relations $att_1 = \{(a,b), (b,c)\}$, $att_2 = \{(b,c), (c,a)\}$, and $att_3 = \{(c,a), (a,b)\}$. It holds that all individual AFs are acyclic, but the outcome of the majority rule includes the three attacks $\{(a,b), (b,c), (c,a)\}$ which actually form a cycle. Since all quota rules satisfy the basic axioms, they do not preserve coherence.

When considering the basic semantics, the grounded semantics is the only one that always has a unique solution. In this case, an interesting question is also the preservation of the membership property under the grounded semantics, i.e., whether any argument that is contained in the grounded extension of each individual AF is also contained in the grounded extension of the aggregated AF. For the study of semantics with nonunique extensions, there are different ways to formulate argument acceptability appropriately.

Let σ be the stable, preferred, or complete semantics, F an aggregation rule, and $P = (att_1, \ldots, att_n)$ a profile. We say F *preserves credulous argument acceptability under* σ if for all arguments $a \in Ar$ that belong to some extension under σ for every AF_i, $i \in N$, there is some σ-extension of $F(P)$ that contains a. On the other hand, F is said to *preserve sceptical argument acceptability under* σ if for all arguments $a \in Ar$ that belong to all extensions under σ for every AF_i, $i \in N$, all σ-extensions of $F(P)$ contain a. Since the grounded extension is unique, both notions coincide for this extension.

Unfortunately, argument acceptability in either form is not compatible with a set of very basic axioms, unless we allow dictatorships.

Theorem 3.10. [Chen and Endriss [2019]] *Let* $|Ar| \geq 4$ *and* σ *be the stable, preferred, complete, or grounded semantics. Then any unanimous, grounded, and independent aggregation rule F that preserves either credulous or sceptical argument acceptability under σ is a dictatorship.*

The proof of this theorem relies on similar results on graph aggregation obtained by Endriss and Grandi [2017]. This result shows that no quota rule preserves argument acceptability.

Example 3.11. *Consider again the individual attack relations as shown in Figures 3a–3c. The grounded extension for AF_1, AF_2, and AF_3 is $\{a,c,e,g\}$. Thus argument a is contained in the grounded extension for every individual agent. However, for the argumentation framework resulting from the majority rule (see Figure 3d) the grounded extension is $\{b,c,e,g\}$ and thus does not contain argument a. This shows that the majority rule does not preserve argument acceptability under grounded semantics.*

The next properties do no longer focus on single arguments but on sets of arguments. Let P be a profile. An aggregation rule F *preserves conflict-freeness (admissibility)* if for all sets of arguments $A \subseteq Ar$ that are conflict-free (admissible) for all AF_i, $i \in N$, it holds that A is also conflict-free (admissible) in $F(P)$. In contrast to the previous impossibility result (Theorem 3.10), the results here are more positive. For conflict-freeness there is a very general result, whereas for admissibility there is at least one reasonable rule that satisfies some basic criteria.

Theorem 3.12. [Chen and Endriss [2019]]

- *Every aggregation rule that is grounded preserves conflict-freeness.*

- *Let $|Ar| \geq 4$. The nomination rule is the only unanimous, grounded, anonymous, neutral, independent, and monotonic aggregation rule that preserves admissibility.*

A counter-example for the majority rule and the preservation of admissibility rule is given in the following example.

Example 3.13. *Recall the example shown in Figures 3a–3c. It holds that $\{a, c, e, g\}$ is admissible for AF_1, AF_2, and AF_3. However, for the framework that results from a majority aggregation (see Figure 3d) $\{a, c, e, g\}$ is not admissible since a is not defended. This means, the majority rule does not preserve admissibility.*

In addition to preservation of admissibility and conflict-freeness, the question of preservation under a given semantics is interesting. Formally, an aggregation rule F *preserves extensions under the stable (complete, grounded, preferred) semantics* if for all sets $A \subseteq Ar$ that are stable (complete, grounded, preferred) for all AF_i, $i \in N$, it holds that A is also stable (complete, grounded, preferred) in $F(P)$. The results here differ, depending on the semantics considered. For the case of the complete, preferred, and grounded semantics, there is again a negative result that builds up on known results by Endriss and Grandi [2017] on graph aggregation. On the other hand, the result for the stable semantics is more positive, as the nomination rule indeed preserves stable extensions.

Theorem 3.14. [Chen and Endriss [2019]]

- *Let $|Ar| \geq 5$. Then any unanimous, grounded, and independent aggregation rule F that preserves either complete or preferred extensions is a dictatorship.*

- *Let $|Ar| \geq 4$. Then any unanimous, grounded, and independent aggregation rule F that preserves grounded extensions is a dictatorship.*

- *The nomination rule preserves stable extensions.*

The following example shows that the majority rule does not preserve extensions under the grounded, stable, preferred, or complete semantics.

Example 3.15. *Consider again the individual argumentation frameworks shown in Figures 3a–3c. Note that all three attack relations are acyclic, and hence the grounded, stable, preferred, and complete semantics coincide. As mentioned before, the grounded extension for all three argumentation frameworks is $\{a, c, e, g\}$. However this is not preserved under majority aggregation, since in the resulting argumentation framework the grounded extension is $\{b, c, e, g\}$. Since all attack relations are acyclic, the same example shows that the majority rule does not preserve grounded, stable, preferred, or complete extensions.*

A different view on the preservation of extensions has been taken by Dunne *et al.* [2012], who consider different variants of these problems from the point of view of computational complexity. Following their work, Delobelle *et al.* [2018] propose specific AF aggregation rules.

4 Specific Aggregation Methods in Argumentation

In this section, we survey different specific structural aggregation operators for abstract argumentation frameworks from the literature, starting with the pioneering work on AF aggregation by Coste-Marquis *et al.* [2007], who use *partial* argumentation frameworks as a supporting notion for a framework of parameterized aggregation operators. Next, we show how the model of *incomplete* argumentation frameworks due to Baumeister *et al.* [2018b] is used to implement a simple, very general structural aggregation operator. Finally, we present the work of Airiau *et al.* [2017], who employ *value-based* argumentation frameworks due to Bench-Capon [2003] as a target formalism of structural aggregation operators, using different values associated with arguments as a possible explanation for the differences in the input AFs. A less in-detail, but broader overview of aggregation operators in formal argumentation can be found in the survey by Bodanza *et al.* [2017].

4.1 Partial Argumentation Frameworks

Coste-Marquis *et al.* [2007] introduced the notion of partial argumentation framework specifically as an intermediate format for the implementation of aggregation operators on argumentation frameworks.

Definition 4.1 (Partial Argumentation Framework (PAF)). *A partial argumentation framework (PAF) is a triple $PAF = \langle Ar, att, ign \rangle$ and consists of*

- *a set Ar of arguments,*

- *an attack relation $att \subseteq (Ar \times Ar)$ specifying attacks known to exist,*

- *and an ignorance relation $ign \subseteq (Ar \times Ar)$ specifying attacks whose existence is not known.*

It is assumed that $att \cap ign = \emptyset$. A third relation is the non-attack relation $non = (Ar \times Ar) \setminus (att \cup ign)$, which is implicitly given by the other two.

Aggregating several individual AFs via PAFs is a two-step process. First, each AF is expanded to incorporate the information that is present in the other AFs. An expansion must include the union of all arguments from all AFs. On the other hand, an expansion must respect the information that is present in the original AF; in particular, all arguments and all attacks of the original AF must be present in the expanded AF, too, and also every attack that does not exist in the original AF must be non-existent in the expanded AF. However, this still leaves a lot of freedom, allowing for many different expansion operators. Expansion operators use PAFs as their target format in order to be able to represent ignorance about the existence of attacks.

Definition 4.2 (Expansion Operator). *Let $\mathbb{PAF}$ denote the universe of all partial AFs. We consider mappings $exp : \mathbb{AF}^m \to \mathbb{PAF}$ that map an argumentation framework $AF = \langle Ar, att \rangle$ and a profile $P = (AF_1, \ldots, AF_{m-1})$ with $AF_i = \langle Ar_i, att_i \rangle$ of other individual AFs to an expanded PAF representation $exp(AF; P) = \langle Ar', att', ign' \rangle$ of AF, that incorporates the information given by the other individual AFs AF_i. More formally, exp is called an* expansion operator *if it satisfies the following conditions:*

- $Ar' = Ar \cup \bigcup_{i=1}^{m-1} Ar_i$, *i.e., all arguments from all individual views are included;*

- $att' \supseteq att$, i.e., all known attacks from AF are preserved; and

- $non' \supseteq (Ar \times Ar) \setminus att$, i.e., all non-attacks from AF are preserved.

Every agent may choose their own expansion operator, it is not required that all agents use the same one in an aggregation process. In their paper, Coste-Marquis et al. [2007] focus on the *consensual expansion operator*, which is defined as follows.

Definition 4.3 (Consensual Expansion). *The* consensual expansion operator exp_C *maps an* $AF = \langle Ar, att \rangle$ *and a profile* $P = (AF_1, \ldots, AF_{m-1})$ *with* $AF_i = \langle Ar_i, att_i \rangle$ *to a PAF* $\langle Ar', att', ign' \rangle$ *with*

- $Ar' = Ar \cup \bigcup_{i=1}^{m-1} Ar_i$ *(as required)*,

- $att' = att \cup \left(\left(\bigcup_{i=1}^{m-1} att_i\right) \setminus conf(AF, P)\right) \setminus non\right)$, *and*

- $ign' = conf(AF, P) \setminus (att \cup non)$.

The helper function $conf(AF, P) = (att \cup \bigcup_{i=1}^{m-1} att_i) \cap (non \cup \bigcup_{i=1}^{m-1} non_i)$ *identifies those attacks for which there is a conflicting opinion in the input AFs about whether or not they exist.*

An agent that uses the consensual expansion operator is confident to include new attacks in their set att' of attacks known to exist, when all other agents that have both incident arguments in their individual view agree on the existence of the attack. Likewise, when all other agents that know both incident arguments of an attack agree that it does *not* exist, the agent is confident to include it in their set non' of attacks known to not exist. All other new attacks are included in the ignorance relation ign'.

Example 4.4. *Recall our running example from Table 1 and Figure 2 in Section 2. Consider three individual AFs* $(AF_i = \langle Ar_i, att_i \rangle)_{i \in \{1,2,3\}}$ *that each represent the subjective views of a participant in the example discussion, where*

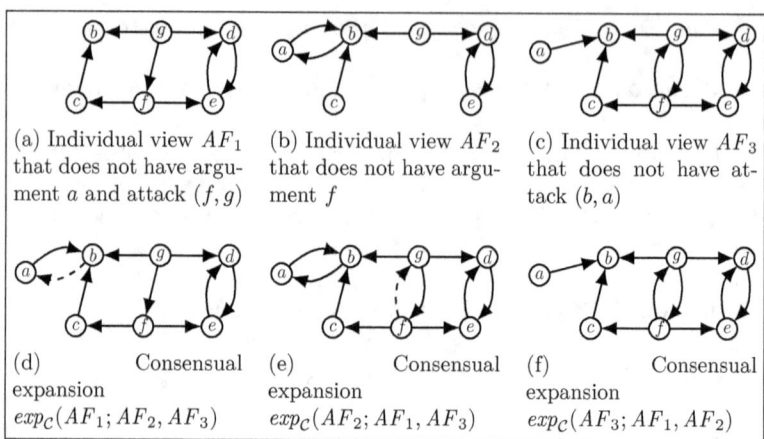

Figure 4: Graph visualizations for the consensual expansion in Example 4.4, where attacks in an ignorance relation ign_i are drawn as dashed arrows

$$Ar_1 = \{b, c, d, e, f, g\}, \quad att_1 = \{(c, b), (d, e), (e, d), (f, e), (f, c),$$
$$(g, d), (g, b), (g, f)\},$$
$$Ar_2 = \{a, b, c, d, e, g\}, \quad att_2 = \{(a, b), (b, a), (c, b), (d, e), (e, d),$$
$$(g, d), (g, b)\},$$
$$Ar_3 = \{a, b, c, d, e, f, g\}, \quad att_3 = \{(a, b), (c, b), (d, e), (e, d), (f, e),$$
$$(f, c), (g, d), (f, g), (g, b), (g, f)\}.$$

These individual AFs are illustrated in Figures 4a, 4b, and 4c, respectively. The stable extensions are $\{c, e, g\}$ for AF_1, $\{a, c, e, g\}$ for AF_2, and $\{a, c, e, g\}$ and $\{a, d, f\}$ for AF_3.

When all individual views are expanded using consensual expansion, we obtain the PAFs $PAF_1 = exp_C(AF_1; AF_2, AF_3)$, $PAF_2 = exp_C(AF_2; AF_1, AF_3)$, and $PAF_3 = exp_C(AF_3; AF_1, AF_2)$ that are visualized in Figures 4d, 4e, and 4f.

A profile of expanded individual views can now be aggregated using the *fusion* step. Fusion operators identify a single AF or a set of AFs that is "as close as possible" to all expanded individual views. To implement the notion of "closeness," a fusion operator is parameterized by a pseudo-distance $d : \mathbb{PAF}^2 \to \mathbb{R}^+$ on PAFs—satisfying $d(PAF_1, PAF_2) = d(PAF_2, PAF_1)$ and $d(PAF_1, PAF_2) = 0 \Leftrightarrow PAF_1 = PAF_2$—and an aggregation function $\otimes : \mathbb{R}^m \to \mathbb{R}$ that is used to aggregate m distances into a single score.

As an example, Coste-Marquis et al. [2007] define an edit distance on PAFs, which penalizes outright disagreement about the existence of a relation $r \in (Ar \times Ar)$ with an increase of the distance by 1, while if one of the two PAFs has r in its ignorance relation and the other does not, this is penalized with an increase of 0.5.

Definition 4.5 (Edit Distance on PAFs). *The* edit distance $d_e : \mathbb{PAF}^2 \to \mathbb{R}^+$ on PAFs *is defined as follows:*

$$d_e(\langle Ar, att_1, ign_1\rangle, \langle Ar, att_2, ign_2\rangle) =$$
$$\sum_{r \in (Ar \times Ar)} 1 \cdot (\mathbb{1}_{att_1 \cap non_2}(r) + \mathbb{1}_{non_1 \cap att_2}(r))$$
$$+ 0.5 \cdot (\mathbb{1}_{ign_1 \cap att_2}(r) + \mathbb{1}_{ign_1 \cap non_2}(r) + \mathbb{1}_{att_1 \cap ign_2}(r) + \mathbb{1}_{non_1 \cap ign_2}(r)),$$

where $\mathbb{1}_X$ denotes the indicator function *for a set X, defined by $\mathbb{1}_X(x) = 1$ if $x \in X$, and $\mathbb{1}_X(x) = 0$ otherwise.*

We are now ready to define fusion operators.

Definition 4.6 (Fusion Operator). *Let $m \in \mathbb{N}$, d be a pseudo-distance on PAFs, and $\otimes$ be an aggregation function on $\mathbb{R}^m$. The* fusion operator $fusion_{d,\otimes} : \mathbb{PAF}^m \to 2^{\mathbb{AF}}$ *maps a profile $(PAF_1, \ldots, PAF_m)$ of partial AFs obtained through expansion to the set of AFs that minimize the aggregated distance (with respect to d and $\otimes$) to all input PAFs, i.e.,*

$$fusion_{d,\otimes}(PAF_1, \ldots, PAF_m) =$$
$$\{\langle Ar^*, att^*\rangle \mid Ar^* = \bigcup_{i=1}^{m} Ar_i, att^* \subseteq (Ar^* \times Ar^*),$$
$$\text{and } \langle Ar^*, att^*\rangle \text{ minimizes } \otimes_{i=1}^{m} (d(\langle Ar^*, att^*\rangle, PAF_i))\}.$$

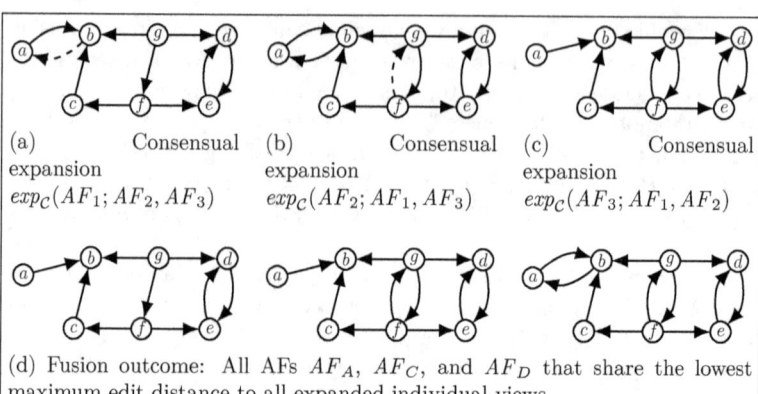

(a) Consensual expansion $exp_C(AF_1; AF_2, AF_3)$

(b) Consensual expansion $exp_C(AF_2; AF_1, AF_3)$

(c) Consensual expansion $exp_C(AF_3; AF_1, AF_2)$

(d) Fusion outcome: All AFs AF_A, AF_C, and AF_D that share the lowest maximum edit distance to all expanded individual views

Figure 5: Graph visualizations for the fusion in Example 4.4, where attacks in an ignorance relation ign_i are drawn as dashed arrows

An aggregation operator is now obtained by chaining m expansion operators and a fusion operator together.

Definition 4.7 (PAF Aggregation Operator). *Let $m \in \mathbb{N}$. For a profile $\exp = (exp_1, \ldots, exp_m)$ of m expansion operators, for a given pseudo-distance d on PAFs, and for a given aggregation function $\otimes$ on $\mathbb{R}^m$, the aggregation operator $agg^{pafs}_{\exp,d,\otimes} : \mathbb{AF}^m \to 2^{\mathbb{AF}}$ is defined as*

$$agg^{pafs}_{\exp,d,\otimes}(AF_1, \ldots, AF_m) = fusion_{d,\otimes}(exp_1(AF_1; AF_2, \ldots, AF_m),$$
$$\ldots, exp_m(AF_m; AF_1, \ldots, AF_{m-1})).$$

In general, PAF aggregation operators are irresolute, i.e., they may return a set of aggregates instead of a single aggregate, since the fusion operator may find several AFs that share the lowest aggregated distance to the expanded individual views.

Example 4.8. *We continue our example. The expanded views PAF_1, PAF_2, and PAF_3 that we obtained through consensual expansion all share the same set $\{a, b, c, d, e, f, g\}$ of arguments, have ignorance relations $ign_1 = \{(b, a)\}$, $ign_2 = \{(f, g)\}$, and $ign_3 = \emptyset$, and the following attack relations:*

$att_1 = \{(a,b),(c,b),(d,e),(e,d),(f,e),(f,c),(g,d),(g,b),(g,f)\}$,
$att_2 = \{(a,b),(b,a),(c,b),(d,e),(e,d),(f,e),(f,c),(g,d),(g,b),(g,f)\}$,
$att_3 = \{(a,b),(c,b),(d,e),(e,d),(f,e),(f,c),(g,d),(f,g),(g,b),(g,f)\}$.

For better readability, we display the expanded individual views again in Figures 5a, 5b, and 5c.

For the fusion operator, we use the edit distance d_e, and as an aggregation function we use the maximum function max. *The operator $fusion_{d_e,\max}$ identifies all argumentation frameworks $\langle \{a,b,c,d,e,f,g\}, att \rangle$ for which the maximum of the edit distances to each expanded individual view PAF_1, PAF_2, or PAF_3 is minimal. Among all candidates $\langle \{a,b,c,d,e,f,g\}, att \rangle$, we only need to consider those that share the attacks and non-attacks that all PAF_i agree on, because any deviation would increase the distance to every individual view. The only relations for which there is disagreement among the PAF_i are (b,a) and (f,g). This leaves four candidates AF_A, AF_B, AF_C, and AF_B with attack relations:*

$att_A = \{(a,b),(c,b),(d,e),(e,d),(f,e),(f,c),(g,d),(g,b),(g,f)\}$,
$att_B = \{(a,b),(b,a),(c,b),(d,e),(e,d),(f,e),(f,c),(g,d),(g,b),(g,f)\}$,
$att_C = \{(a,b),(c,b),(d,e),(e,d),(f,e),(f,c),(g,d),(f,g),(g,b),(g,f)\}$,
$att_D = \{(a,b),(b,a),(c,b),(d,e),(e,d),(f,e),(f,c),(g,d),(f,g),$
$(g,b),(g,f)\}$.

The edit distances are as follows:

- $d_e(AF_A, PAF_1) = 0.5$, $d_e(AF_A, PAF_2) = 1.5$,
  $d_e(AF_A, PAF_3) = 1$,

- $d_e(AF_B, PAF_1) = 0.5$, $d_e(AF_B, PAF_2) = 0.5$,
  $d_e(AF_B, PAF_3) = 2$,

- $d_e(AF_C, PAF_1) = 1.5$, $d_e(AF_C, PAF_2) = 1.5$,
  $d_e(AF_C, PAF_3) = 0$,

- $d_e(AF_D, PAF_1) = 1.5$, $d_e(AF_D, PAF_2) = 0.5$,
  $d_e(AF_D, PAF_3) = 1$,

AF_B has an edit distance $d_e(AF_B, PAF_3) = 2$ to PAF_3, while the other three AFs share a maximum distance of 1.5 to any of the PAFs. Therefore, the result of the fusion step is

$$fusion_{d_e, \max}(PAF_1, PAF_2, PAF_3) = \{AF_A, AF_C, AF_D\}.$$

These AFs are visualized in Figure 5d. AF_A has $\{a, c, e, g\}$ as its only stable extension, AF_C has stable extensions $\{a, c, e, g\}$ and $\{a, d, f\}$, and AF_D has stable extensions $\{a, c, e, g\}$, $\{a, d, f\}$, and $\{b, d, f\}$. In particular, we can observe that the aggregate has an extension where argument b is accepted, although b is not accepted in any of the individual views.

The fact that the aggregation results can have acceptable (sets of) arguments that are not acceptable in any individual view bears some similarities with the discursive dilemma from judgment aggregation theory,[1] where a profile of individually consistent judgment sets can be aggregated into a single judgment via some majority criterion such that this aggregated judgment contains a conclusion that is not accepted in any of the individual judgments. Since we are concerned with framework-based structural aggregation, we regard the information provided by the individual views (i.e., the existing arguments and attacks) as primarily relevant, overshadowing the outcome (i.e., extensions and accepted arguments) of the individual views. Like premise-based aggregation rules in judgment aggregation (see Kornhauser and Sager [1993], Dietrich and Mongin [2010]), structural aggregation operators for AFs allow the outcome of the aggregate to have precedence over the aggregation of the outcomes.

4.2 Incomplete Argumentation Frameworks

Incomplete Argumentation Frameworks (IAFs) further generalize PAFs and can represent uncertainty about the existence of individual arguments [Baumeister et al., 2015c], uncertainty about the existence of individual attacks [Baumeister et al., 2015b], or both simultaneously [Baumeister et al., 2018b].

---

[1] Generalizing the *doctrinal paradox* due to Kornhauser and Sager [1986], Pettit [2001] introduced the *discursive dilemma*; both are discussed by Mongin [2012] in more detail.

Definition 4.9 (Incomplete Argumentation Framework). *An* incomplete argumentation framework (IAF) *is a quadruple* $IAF = \langle Ar, Ar^?, att, att^? \rangle$ *and consists of*

- *a set* Ar *of* definite arguments *known to exist,*
- *a set* $Ar^?$ *of* uncertain arguments *of which each may or may not exist,*
- *a set* $att \subseteq (Ar \cup Ar^?) \times (Ar \cup Ar^?)$ *of* (conditionally) definite attacks *that exist if and only if both incident arguments exist, and*
- *a set* $att^? \subseteq (Ar \cup Ar^?) \times (Ar \cup Ar^?)$ *of* uncertain attacks *of which each may or may not exist, but only if both incident arguments exist.*

We assume that $Ar \cap Ar^? = \emptyset$ *and* $att \cap att^? = \emptyset$.

An IAF is a compact representation of a set of possible worlds—namely, all the standard AFs that can be obtained from it by deciding for each uncertain element whether or not it should be included. Each such AF is called a *completion* of the IAF.

Definition 4.10 (Completion). *Let* $IAF = \langle Ar, Ar^?, att, att^? \rangle$ *be an incomplete AF. An AF* $AF^* = \langle Ar^*, att^* \rangle$ *is called a* completion *of* IAF *if it satisfies*

$$Ar \subseteq Ar^* \subseteq Ar \cup Ar^? \text{ and}$$
$$att|_{Ar^*} \subseteq att^* \subseteq \left(att \cup att^?\right)|_{Ar^*}.$$

That is, every completion of IAF must include each of its definite arguments and may include any of its uncertain arguments. Attacks can only be included if both incident arguments are included (indicated by the restriction operator $att'|_{Ar'} = att' \cap (Ar' \times Ar')$). Every conditionally definite attack that has both incident arguments present in a completion, must be included in that completion. Uncertain attacks may be included, but only if both incident arguments are included.

Incomplete AFs define acceptability criteria for sets of arguments or for individual arguments, which are derived from the standard AF criteria of extensions and credulous/skeptical acceptance in the completions of an IAF.

Definition 4.11 (Acceptability in IAFs)**.** *Let* $IAF = \langle Ar, Ar^?, att, att^? \rangle$ *be an incomplete AF and let σ be a semantics.*

- *A set $S \subseteq Ar \cup Ar^?$ of arguments in IAF is accepted possibly (respectively, necessarily) for semantics σ if S is a σ extension in some completion of IAF (respectively, in all completions of IAF)*[2]*. This notion of acceptability is derived from the verification problem (σ-VER) for standard AFs and formalized via the problems of possible verification (σ-PV) and necessary verification (σ-NV) by Baumeister et al. [2018b] and Fazzinga et al. [2020].*

- *A single argument $a \in Ar \cup Ar^?$ in IAF is possibly credulously accepted (respectively, necessarily credulously accepted) for semantics σ if a is a member of some σ extension S in some completion of IAF (respectively, in all completions of IAF). Similarly to verification, this notion is derived from the credulous acceptance problem (σ-CA) for standard AFs and formalized via the problems of possible credulous acceptance (σ-PCA) and necessary credulous acceptance (σ-NCA) by Baumeister et al. [2018a].*

- *A single argument $a \in Ar \cup Ar^?$ in IAF is possibly skeptically accepted (respectively, necessarily skeptically accepted) for semantics σ if a is a member of all σ extensions S in some completion of IAF (respectively, in all completions of IAF). Again, this notion is derived from the skeptical acceptance problem (σ-SA) for standard AFs and formalized via the problems of possible skeptical acceptance (σ-PSA) and necessary skeptical acceptance (σ-NSA) by Baumeister et al. [2018a].*

Note that the problems of possible/necessary acceptability in IAFs can be restricted to definite targets (i.e., $S \subseteq Ar$ for sets of arguments or $a \in Ar$ for single arguments) without changing their complexity: For "necessary" problem variants, we get a trivial "no" answer if (part of) the target is uncertain, since then it cannot be present in all completions.

---

[2]This is a revised definition due to Fazzinga *et al.* [2020], who point out some problematic behavior of the initial definition by Baumeister *et al.* [2018b], where it was only required that $S \cap Ar^*$ (instead of S) is a σ extension in some completion of IAF in order for S to be accepted in that completion. In this chapter, we focus on the revised definition.

4 - COLLECTIVE ACCEPTABILITY IN ABSTRACT ARGUMENTATION

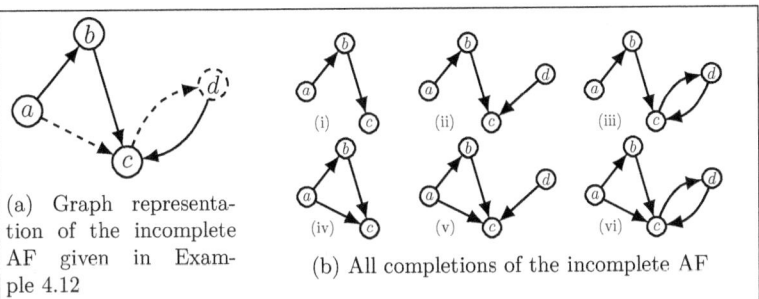

(a) Graph representation of the incomplete AF given in Example 4.12

(b) All completions of the incomplete AF

Figure 6: Incomplete AF and corresponding completions for Example 4.12

For "possible" problem variants, we can disregard all completions that do not contain all target arguments, since they cannot produce "yes" answers. The remaining cases constitute exactly the original problem where all target arguments are definite.

Example 4.12. *Consider an incomplete AF $IAF = \langle Ar, Ar^?, att, att^? \rangle$ with three definite arguments in $Ar = \{a, b, c\}$, an uncertain argument in $Ar^? = \{d\}$, three (conditionally) definite attacks in $att = \{(a,b), (b,c), (d,c)\}$, and two uncertain attacks in $att^? = \{(a,c), (c,d)\}$. Its graph representation is given in Figure 6a, where uncertain elements are distinguished by dashed lines, and all its completions are given in Figure 6b.*

When we determine the stable extensions of the completions, we see that completion (i) has $\{a,c\}$ as its only stable extension, completion (iii) has $\{a,c\}$ and $\{a,d\}$ as its stable extensions, completion (iv) has $\{a\}$ as its only stable extension, and all other completions—namely, (ii), (v), and (vi)—have $\{a,d\}$ as their only stable extension. For the stable semantics, this means that:

- *$\{a\}$, $\{a,c\}$, and $\{a,d\}$ are possibly accepted, while no other set of arguments is possibly accepted, and no set of arguments is necessarily accepted.*

- *Argument a is necessarily skeptically accepted, since it is a member of every stable extension in every completion.*

- *Argument b is not possibly credulously accepted, since it is not a member of any stable extension in any completion.*

- *Argument c is possibly skeptically accepted, since in completion (i), c is a member of all stable extensions. However, c is not necessarily credulously accepted, since there are completions—e.g., (iv)—where c is not included in any stable extension.*

The computational complexity of decision problems that capture these notions of acceptability was analyzed by Baumeister *et al.* [2018b] and Fazzinga *et al.* [2020] for sets of arguments and by Baumeister *et al.* [2018a] and Niskanen *et al.* [2020] for individual arguments.³ Table 2 gives an overview of the complexities compared to the respective complexity of the base problem for standard AFs without uncertainty. Results marked with ♠ are by Dung [1995], marked with ♣ are by Baumeister *et al.* [2018b], marked with • are by Fazzinga *et al.* [2020], marked with ★ are by Dimopoulos and Torres [1996], marked with ♦ are by Baumeister *et al.* [2018a], marked with ▲ are by Coste-Marquis *et al.* [2005], marked with ▼ are by Niskanen *et al.* [2020], marked with † are by Rey [2014], and marked with ‡ are by Dunne and Bench-Capon [2002]. Results marked with an asterisk (*) are straight-forward and are not formally proven. We can observe various cases where the computational hardness is not increased by the introduction of uncertainty—in particular, for necessary verification (σ-NV). Further, experimental results by Niskanen *et al.* [2020] indicate that even the hard cases (i.e., those problems that are complete for NP, coNP, or a class even higher in the polynomial hierarchy) may be tamed in practice through suitable encodings.

Even though incomplete AFs were conceived as a very general representation of various different kinds of structural uncertainty in AFs, in particular they can serve as a target formalism for structural aggregation of different individual AFs. In the following, we define and illustrate the aggregation operator from sets of AFs to a single IAF as introduced by Baumeister *et al.* [2018b]. This aggregation operator is very liberal

---

³We mention in passing that similarly to the way Baumeister *et al.* [2018b] study possible and necessary variants of the *verification problem*, Skiba *et al.* [2020] have done so for the *existence problem*.

and imposes only a single condition: When all individual AFs agree on the existence (respectively, nonexistence) of an element, then this element must exist (respectively, cannot exist) in the aggregate. Arguments or attacks for which there is disagreement are included as uncertain elements, thus allowing to include or exclude them via completions.

Definition 4.13 (IAF Aggregation Operator). *Denote with* $\mathbb{IAF}$ *the universe of all incomplete argumentation frameworks. For every* $m \in \mathbb{N}$, *the aggregation operator* $agg^{inc} : \mathbb{AF}^m \to \mathbb{IAF}$ *maps any set* $\{AF_1, \ldots, AF_m\}$ *with* $AF_i = \langle Ar_i, att_i \rangle$ *of* m *individual AFs to* $IAF = \langle Ar, Ar^?, att, att^? \rangle$ *with*

$$Ar = \bigcap_{i=1}^{m} Ar_i,$$

$$att = \{(a,b) \in (Ar \cup Ar^?) \times (Ar \cup Ar^?) \mid \\ (\forall i \in \{1, ..., m\})\, [a, b \in Ar_i \Rightarrow (a,b) \in att_i]\},$$

$$Ar^? = \left(\bigcup_{i=1}^{m} Ar_i\right) \setminus Ar,$$

$$att^? = \left(\bigcup_{i=1}^{m} att_i\right) \setminus att.$$

That is, every argument that occurs in every individual AF is included as a definite argument, and all other arguments that occur in individual AFs are included as possible arguments. An attack is included as (conditionally) definite attack if it occurs in every individual AF that includes both incident arguments, and as a possible attack otherwise.

Example 4.14. *Again, we use our running example with the same three individual AFs* $(AF_i = \langle Ar_i, att_i \rangle)_{i \in \{1,2,3\}}$ *that each represent the subjective views of a participant in the example discussion, where*

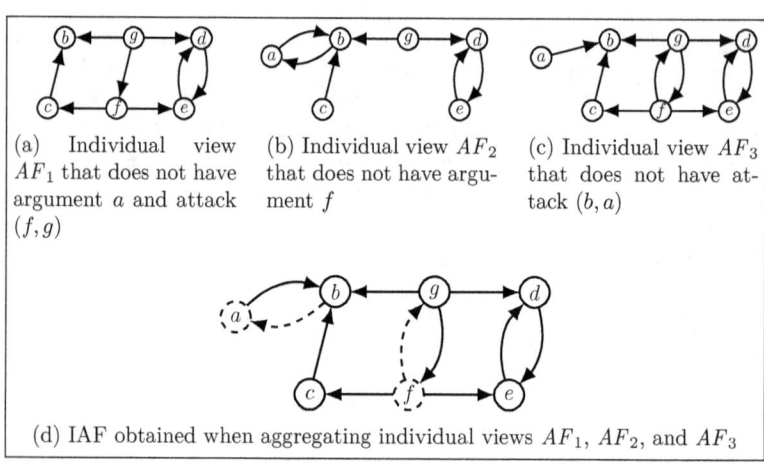

Figure 7: Graph representation of the individual AFs in Example 4.14

$$Ar_1 = \{b, c, d, e, f, g\}, \quad att_1 = \{(c, b), (d, e), (e, d), (f, e), (f, c),$$
$$(g, d), (g, b), (g, f)\},$$
$$Ar_2 = \{a, b, c, d, e, g\}, \quad att_2 = \{(a, b), (b, a), (c, b), (d, e), (e, d),$$
$$(g, d), (g, b)\},$$
$$Ar_3 = \{a, b, c, d, e, f, g\}, \quad att_3 = \{(a, b), (c, b), (d, e), (e, d), (f, e),$$
$$(f, c), (g, d), (f, g), (g, b), (g, f)\}.$$

For better readability, the individual AFs are illustrated again in Figures 7a, 7b, and 7c. The stable extensions are $\{c, e, g\}$ for AF_1, $\{a, c, e, g\}$ for AF_2, and $\{a, c, e, g\}$ and $\{a, d, f\}$ for AF_3.

Aggregating these AFs with the method of Definition 4.13 yields the IAF $IAF = \langle Ar, Ar^?, att, att^? \rangle$ displayed in Figure 7d, with:

4 - COLLECTIVE ACCEPTABILITY IN ABSTRACT ARGUMENTATION

$Ar = \{b, c, d, e, g\}$,
$Ar^? = \{a, f\}$,
$att = \{(a,b), (c,b), (d,e), (e,d), (f,e), (f,c), (g,d), (g,b), (g,f)\}$,
$att^? = \{(b,a), (f,g)\}$.

Arguments a and f do not occur in every individual AF and are therefore uncertain arguments, all other arguments are definite arguments. For the attacks, we have three different cases: Everyone agrees on the existence of the attacks (c,b), (d,e), (e,d), (g,d), and (g,b), so these are definite attacks. For the attacks (a,b), (f,e), (f,c), and (g,f), not all individual AFs include both incident arguments, but those that do, agree on their existence, so these are conditionally definite attacks (which, like definite attacks, are included in the set att, but they might vanish alongside incident uncertain arguments in a completion). For the remaining attacks (b,a) and (f,g), there is disagreement about their existence among individual views who include both incident arguments, so these are uncertain attacks.
IAF has the following nine completions.

- *One includes neither of the arguments a or f and has $\{c, e, g\}$ as its only stable extension.*

- *Two include argument a, but not f, and both have $\{a, c, e, g\}$ as their only stable extension.*

- *Two include argument f, but not a, both have $\{c, e, g\}$ as a stable extension, and one of them has $\{b, d, f\}$ as a second stable extension.*

- *The remaining four completions include bot arguments a and f. All have $\{a, c, e, g\}$ as a stable extension, two of them have $\{a, d, f\}$ as a second stable extension, and one of these two has $\{b, d, f\}$ as a third stable extension.*

As in the example used in the previous section, we can again observe that IAF has completions where argument b is accepted (i.e., b is possibly credulously accepted in IAF), even though b is not accepted in any of the individual views.

Since the IAF aggregation operator produces a single IAF as its output, and every IAF is a compact representation of a set of AFs, the operator can be seen as an irresolute AF aggregation operator. However, the existing acceptability semantics for IAFs make it unnecessary to resolve the remaining ambiguity through some sort of fusion, since collective acceptability can already be determined at this stage. One could say that the IAF aggregation operator circumvents the issues inherent to aggregation by skipping the actual aggregation altogether. Instead of being a single aggregate, an IAF that represents a set of individual AFs can be seen as a kind of convex closure of those individual views. Every reasonable structural aggregate of the individual views will lie within that closure (i.e., will be a completion of the IAF). As such, the acceptability semantics for the IAF provide a way to define a very cautious collective acceptability for the individual views, which avoids the deliberate choice of a specific aggregate AF.

4.3 Value-Based Argumentation Frameworks

Bench-Capon [2003] proposed the notion of (audience-specific) value-based argumentation framework (see also Chapter 5 of this handbook [Atkinson and Bench-Capon, 2021]).

Definition 4.15 (Value-Based Argumentation Framework).
An audience-specific value-based argumentation framework (AVAF) *is a quintuple $VAF_\alpha = \langle Ar, att, V, val, \succ_\alpha \rangle$, where*

- *$\langle Ar, att \rangle$ is an AF (without self-attacks, i.e., att is supposed to be irreflexive),*

- *V is a nonempty set of (social or moral) values,*

- *$val : Ar \to V$ is a mapping assigning values to arguments, and*

- *$\succ_\alpha$ is a transitive and asymmetric (thus, in particular, irreflexive) relation reflecting the value preferences of audience α on V.*

4 - COLLECTIVE ACCEPTABILITY IN ABSTRACT ARGUMENTATION

A value-based argumentation framework (VAF) *is similarly defined as a quintuple* $\langle Ar, att, V, val, \succ \rangle$, *where* Ar, att, V, *and* val *have the same meaning as above but* $\succ$ *is the set of all possible audiences, i.e., of all possible preferences on* V.

If V is a singleton, or if no preferences among the values in V are expressed, an AVAF degenerates to an ordinary AF. In an AVAF, each audience α can be identified with its preference relation $\succ_\alpha$ over values in V, and while the function *val* mapping arguments to values is fixed for everyone, the preferences $\succ_\alpha$ over values are specific to this particular audience (or agent) α. The point is that an attack in an AVAF succeeds only if the attacked argument is not preferred to the attacking argument by the audience: From α's point of view, an argument of lower (social or moral) value cannot defeat an argument of higher value. Note that if both arguments are assigned the same value, or if there is no preference between two values, an attack between such arguments does succeed.

Definition 4.16 (Defeat Relation). *Let* $VAF_\alpha = \langle Ar, att, V, val, \succ_\alpha \rangle$ *be an AVAF. An argument* $a \in Ar$ α*-defeats an argument* $b \in Ar$ *if and only if* a *attacks* b *in* $\langle Ar, att \rangle$ *and not* $val(b) \succ_\alpha val(a)$.

Example 4.17. *For illustration, we extend the AF* $\langle Ar, att \rangle$ *from our running example (recall Table 1 and Figure 2 in Section 2) as follows. Suppose that*

- *the arguments a and g are mapped to the value* PERSONAL SAFETY *(represented by white vertices in Figure 8);*

- *the arguments b, c, d, and f are mapped to the value* COLLECTIVE SAFETY *(represented by gray vertices in Figure 8); and*

- *the argument e is mapped to the value* RIGHT TO BE INFORMED *(represented by a black vertex in Figure 8).*

The six graphs displayed in Figure 8 show the six AVAFs resulting from each possible complete, strict preference ranking of these three values in

$$\succ = \{\succ_\alpha, \succ_\beta, \succ_\gamma, \succ_\delta, \succ_\eta, \succ_\zeta\},$$

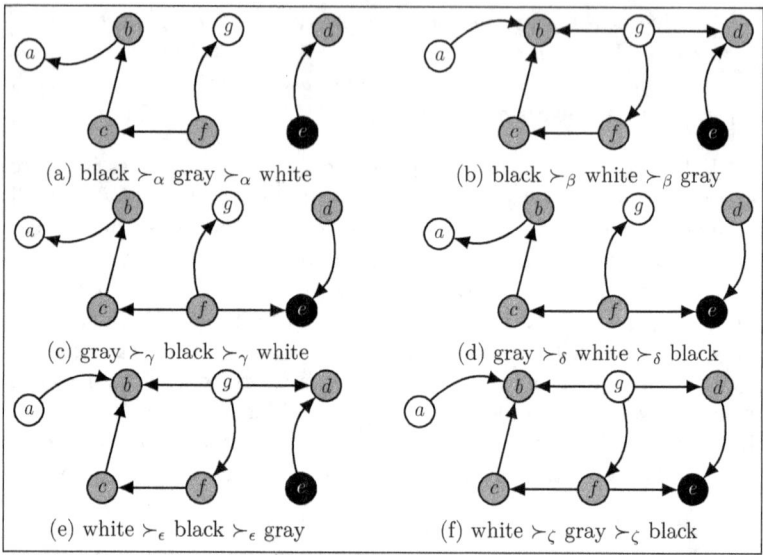

Figure 8: Defeat relation in six AVAFs for Example 4.17.

where the arguments are vertices and a directed edge from x to y in VAF_ψ with $\succ_\psi \in \succ$ means that x ψ-defeats y.

For example, audience α ranks these values by $black \succ_\alpha gray \succ_\alpha white$, so the attacks (a, b), (d, e), (f, e), (g, b), (g, d), and (g, f) do not succeed in VAF_α because the attackers have lower value than the attackees in these cases and thus they do not defeat them.

The α-defeat relation from Definition 4.16 is an irreflexive binary relation on Ar, just as an attack relation on Ar. Therefore, an AVAF $VAF_\alpha = \langle Ar, att, V, val, \succ_\alpha \rangle$ induces another AF $\langle Ar, \alpha\text{-}defeat \rangle$ with α-defeat being a subrelation of att.

We thus can define the same semantics that were introduced for AFs in Section 2 for AVAFs as well. For example, a set S of arguments is said to be *conflict-free for audience* α if no argument in S α-defeats any other argument in S: For any two arguments $a, b \in S$, if a attacks b

4 - COLLECTIVE ACCEPTABILITY IN ABSTRACT ARGUMENTATION

then $val(b) \succ_\alpha val(a)$.[4] Further, S is *admissible for audience* α if it is conflict-free for α and for each argument $a \in S$ and for each argument $b \in Ar$ such that b α-defeats a, there is an argument $c \in S$ that α-defeats b. If S is a maximal (with respect to set inclusion) admissible set for audience α, it is said to be *preferred by audience* α. And S is *stable for audience* α if it is conflict-free for α and every argument outside of S is α-defeated by some argument in S.

Note that for any AVAF VAF_α having no cycle in which all arguments are assigned the same value, the associated AF $\langle Ar, \alpha\text{-}defeat \rangle$ will have no cycle (as any cycle would be broken when an argument of lower value attacks an argument of higher value).[5] In such cases, there exists a unique, nonempty preferred extension for audience α.

Example 4.18. *Continuing Example 4.17, the six AVAFs from Figure 8 have the following preferred extensions for their audiences:*

- *{b, e, f} for Figure 8a with value preference black $\succ_\alpha$ gray $\succ_\alpha$ white;*

- *{a, c, e, g} for Figure 8b with value preference black $\succ_\beta$ white $\succ_\beta$ gray, Figure 8e with value preference white $\succ_\epsilon$ black $\succ_\epsilon$ gray, and Figure 8f with value preference white $\succ_\zeta$ gray $\succ_\zeta$ black; and*

- *{b, d, f} for Figure 8c with value preference gray $\succ_\gamma$ black $\succ_\gamma$ white and Figure 8d with value preference gray $\succ_\delta$ white $\succ_\delta$ black.*

---

[4]This definition of conflict-free sets in preference-based AFs—i.e., AFs which feature a defeat relation derived from the attack relation and some preference over arguments—has been criticized by Amgoud and Vesic [2011] and Modgil and Prakken [2013], who argue that sets of arguments that attack each other should not be considered conflict-free, even when none of the internal attacks succeed, i.e., are defeats. Following the work on rationalizability by Airiau et al. [2017], however, in this chapter we stick to the original definition.

[5]Indeed, Bench-Capon [2003] argues that single-valued cycles in a VAF indicate that the reasoning giving rise to them must be flawed. He further points out that, while in standard AFs even cycles are in fact required (in particular, two-cycles help to deal with uncertain and incomplete information) and odd cycles are at least plausible [Dung, 1995], cycles should be avoided in value-based argumentation: Odd cycles in VAFs are like *paradoxes* indicating that nothing can be believed, and even cycles are like *dilemmas* requiring a choice between alternatives to be made.

Note that none of the arguments in the AVAFs of Example 4.18 is preferred by every audience, yet they all are preferred by some audience. Bench-Capon [2003] refers to these properties as objective and subjective acceptability:

Definition 4.19 (Objective and Subjective Acceptability).
Let $\langle Ar, att, V, val, \succ \rangle$ be a VAF. An argument $a \in Ar$ is

- objectively acceptable *if a is contained in a preferred extension of every audience in $\succ$;*

- subjectively acceptable *if a is contained in a preferred extension of some audience in $\succ$; and*

- indefensible *if a is neither objectively nor subjectively acceptable (e.g., in case a is attacked by an objectively acceptable argument with the same value).*

For AVAFs having no single-valued cycle, the preferred extension can be efficiently computed, as it is unique (which greatly simplifies the situation because, recalling Definition 4.11, there is no difference between skeptical and credulous acceptance). For VAFs, however, with no audience specified, all possible audiences need to be taken into account for the problem of determining whether an argument is objectively acceptable, subjectively acceptable, or indefensible. Bench-Capon [2003] discusses this problem, focusing on certain simple cases, such as when the number of values is two.

Let us have a look at Figure 8 again. We know that all six AVAFs displayed there result from one and the same AF $\langle Ar, att \rangle$, each according to another audience. We started from $\langle Ar, att \rangle$ and then created the six AVAFs by removing certain attacks (namely, those attacks where the attacked argument had a higher value than the attacking argument).

Now, let us ask the converse question: Suppose we are given (or observe) a number of AFs, not necessarily over the same set of arguments and each with its own attack relation. Suppose further that these attack relations are, in fact, defeat relations (i.e., the observed AFs, in fact, are each associated with an AVAF in the way described after Example 4.18). The question is whether these AVAFs can be *rationalized*, i.e., whether they can be explained in terms of a single common master AF, a common set of values, and a common value function, together with a profile of individual preference orders, one for each agent. This *rationalizability problem* has been introduced and studied by Airiau et al. [2017]. We now define their notion more formally; in fact, we define an entire class of rationalizability problems, parameterized by a set of constraints that the solutions are required to fulfill.

Definition 4.20 (Rationalizability).
A profile $\langle \langle Ar_1, 1\text{-}defeat \rangle, \langle Ar_2, 2\text{-}defeat \rangle, \ldots, \langle Ar_n, n\text{-}defeat \rangle \rangle$ of AFs[6] is said to be rationalizable by an AVAF *for a given set of constraints (some to be specified below) if there are*

- *an attack relation att on $Ar = \bigcup_{i=1}^{n} Ar_i$,*

- *a nonempty set V of (social or moral) values,*

- *a mapping $val : Ar \to V$ assigning values to arguments, and*

- *a profile $\langle \succeq_1, \succeq_2, \ldots, \succeq_n \rangle$ of preference orders on V,[7] each meeting the given constraints,*

such that for all agents i, $1 \leq i \leq n$, and for any two arguments $a, b \in Ar_i$, a i-defeats b if and only if a attacks b in $\langle Ar, att \rangle$ and not $val(b) \succ_i val(a)$.

For a given rationalizable profile, $\langle Ar, att \rangle$ is referred to as its master AF *and att as its* master attack relation.

Examples of types of *constraints* considered by Airiau et al. [2017] are:

- the master attack relation *att* may be required to be fixed,

- the set V of values and the value function $val : Ar \to V$ may be required to be fixed,

- the number of values in V may be bounded by some constant, and

---
[6]For notational convenience, we from now on use numbers $1, 2, \ldots, n$ instead of Greek letters to denote agents or audiences.
[7]Unlike Bench-Capon [2003], Airiau et al. [2017] use *preorders* (which are reflexive, transitive binary relations) or *weak orders* (which, in addition, are complete) for the agents' value preferences. The strict part of a preorder is denoted as $\succ$.

- the value preferences $\succeq_i$ may be required to be weak orders.

Airiau et al. [2017] study the question of whether it is possible, given some set of constraints, to characterize the class of profiles of AFs that can be rationalized by an AVAF for these constraints. They also investigate the computational complexity of the rationalizability problem that asks whether a given profile of AFs is rationalizable by an AVAF for a given set of constraints.

They start with the case of a single agent (where profiles contain just a single AF) and first observe that every single AF is rationalizable when there are no constraints, so rationalizability is trivial in this case. Next, they show that the single-agent rationalizability problem can be solved efficiently in many cases.

Theorem 4.21. [**Airiau et al. [2017]**] *For a single agent, the rationalizability problem can be solved in (deterministic) polynomial time whenever any of the following sets of constraints is given:*

- *the master attack relation is fixed;*

- *the master attack relation, the set of values, and the value function are fixed; and*

- *the master attack relation, an upper bound on the number of values is given, and the agent has a weak preference order.*

For multiple agents, however, the situation is more complicated. For example, the easy observation that every single AF is rationalizable in the absence of constraints does not carry over to profiles with more than one agent, as Airiau et al. [2017] demonstrate with the following example.

Example 4.22. *Suppose there are two agents, 1 and 2, discussing three arguments, a, b, and c. While agent 1 thinks they form a cycle, a 1-defeating b, b 1-defeating c, and c 1-defeating a, agent 2 believes that the three arguments are isolated, i.e., the 2-defeat relation is empty. Of course, for this profile to be rationalizable by any AVAF (without any constraints), a master attack relation would have to include at least the attacks from a to b, from b to c, and from c to a. However, for agent 2* to be able to remove these attacks in her 2-defeat relation, 2's preference would have to include at least the comparisons $val(a) \succ_2 val(c)$, $val(c) \succ_2 val(b)$, and $val(b) \succ_2 val(a)$, which by transitivity of $\succ_2$ implies $val(a) \succ_2 val(a)$, a contradiction. Hence, even when there are no constraints whatsoever, this profile is not rationalizable by any AVAF.

While the rationalizability problem is not trivial when there are no constraints in the multiagent case, Airiau et al. [2017] show that it can still be solved efficiently in this case and also when the only constraints are that the master attack relation and/or the value set and value function are fixed. In their proof, they show that any multiagent rationalizability problem can be decomposed into a set of n single-agent rationalizability problems that can be solved independently of each other by Theorem 4.21.

It follows that, among the constraints they consider, a true multi-agent rationalizability problem can be obtained only when an upper bound on the number of values is given. For this problem in general, they have the following result that is obtained by a reduction from the 3-colorability problem, one of the standard NP-complete problems [Garey and Johnson, 1979].

Theorem 4.23. [Airiau et al. [2017]] *For multiple agents, the rationalizability problem is NP-complete whenever a fixed master attack relation and an upper bound of at least three on the number of values are given.*

Whether Theorem 4.23 also holds for an upper bound of at least *two* values, or whether the rationalizability problem becomes tractable in this case, is an interesting open problem. However, when the agents share a *common* set of arguments, we obtain an efficiently solvable special case of the general problem.

Theorem 4.24. [Airiau et al. [2017]] *For multiple agents, the rationalizability problem can be solved in (deterministic) polynomial time whenever a fixed master attack relation is given and there are at most two values.*

As a recommendation for further reading, Lisowski et al. [2019] study two approaches for agents with individual AVAFs to arrive at a common consensus: They may either seek to aggregate their preferences on values (making use of preference aggregation techniques such as voting [Baumeister and Rothe, 2015]), or they may seek to aggregate their defeat graphs (making use of the graph aggregation techniques proposed by Endriss and Grandi [2017]). Exploring the strengths and limitations of both approaches separately, Lisowski et al. [2019] also propose a third option that combines these two approaches and thus avoids some drawbacks that they may have on their own.

5 Conclusion and Outlook

In this chapter, we highlighted the problem of collective acceptability in formal argumentation and its relations to computational social choice theory. We formally defined structural aggregation for abstract argumentation frameworks, surveyed axiomatic evaluation criteria for aggregation operators, and showcased three specific structural aggregation methods from the literature in more detail.

Beyond the structural aggregation operators covered here, there are some that employ numerical weights to implement the aggregation. Delobelle et al. [2018] define an aggregation operator using weighted AFs [Dunne et al., 2011] with weights on attacks as a refinement of the PAF aggregation operator by Coste-Marquis et al. [2007]. Gabbay and Rodrigues [2012] propose an aggregation operator for AFs that uses weights on arguments and attacks to fuel a system of equations that determines the collectively most acceptable arguments. Hunter and Noor [2020] recently initiated the study of aggregation operators for probabilistic AFs [Li et al., 2011].

Wallner [2020] goes beyond the purely abstract perspective of AFs and proposes constraints on aggregation operators that can incorporate information which has been abstracted away in the AF representation.

It is an interesting open task to verify axiomatic properties for structural AF aggregation operators—e.g., the properties presented in Section 3—for the aggregation methods presented in Section 4.

Acknowledgments

This work was supported in part by DFG grants BA 6270/1-1, RO 1202/14-2, and RO 1202/21-1 and by the project "Online Participation" funded by the NRW Ministry for Innovation, Science, and Research.

References

[Airiau et al., 2017] S. Airiau, E. Bonzon, U. Endriss, N. Maudet, and J. Rossit. Rationalisation of profiles of abstract argumentation frameworks: Characterisation and complexity. *Journal of Artificial Intelligence Research*, 60:149–177, 2017.

[Amgoud and Vesic, 2011] L. Amgoud and S. Vesic. A new approach for preference-based argumentation frameworks. *Annals of Mathematics and Artificial Intelligence*, 63(2):149–183, 2011.

[Arrow, 1951 revised edition 1963] K. Arrow. *Social Choice and Individual Values*. John Wiley and Sons, 1951 (revised edition 1963).

[Atkinson and Bench-Capon, 2021] Katie Atkinson and Trevor Bench-Capon. Value-based argumentation. In Dov Gabbay, Massimiliano Giacomin, Guillermo R. Simari, and Matthias Thimm, editors, *Handbook of Formal Argumentation*, volume 2, chapter 5. College Publications, 2021.

[Awad et al., 2017] E. Awad, R. Booth, F. Tohmé, and I. Rahwan. Judgement aggregation in multi-agent argumentation. *Journal of Logic and Computation*, 27(1):227–259, 2017.

[Baumann and Brewka, 2010] R. Baumann and G. Brewka. Expanding argumentation frameworks: Enforcing and monotonicity results. In *Proceedings of the 3rd International Conference on Computational Models of Argument*, volume 216 of *FAIA*, pages 75–86. IOS Press, September 2010.

[Baumeister and Rothe, 2015] D. Baumeister and J. Rothe. Preference aggregation by voting. In J. Rothe, editor, *Economics and Computation. An Introduction to Algorithmic Game Theory, Computational Social Choice, and Fair Division*, Springer Texts in Business and Economics, chapter 4, pages 197–325. Springer-Verlag, 2015.

[Baumeister et al., 2011] D. Baumeister, M. Roos, and J. Rothe. Computational complexity of two variants of the possible winner problem. In *Proceedings of the 10th International Conference on Autonomous Agents and Multiagent Systems*, pages 853–860. IFAAMAS, May 2011.

[Baumeister et al., 2012a] D. Baumeister, P. Faliszewski, J. Lang, and J. Rothe. Campaigns for lazy voters: Truncated ballots. In *Proceedings* of the 11th International Conference on Autonomous Agents and Multiagent Systems, pages 577–584. IFAAMAS, June 2012.

[Baumeister et al., 2012b] D. Baumeister, M. Roos, J. Rothe, L. Schend, and L. Xia. The possible winner problem with uncertain weights. In *Proceedings of the 20th European Conference on Artificial Intelligence*, pages 133–138. IOS Press, August 2012.

[Baumeister et al., 2015a] D. Baumeister, G. Erdélyi, O. Erdélyi, and J. Rothe. Complexity of manipulation and bribery in judgment aggregation for uniform premise-based quota rules. *Mathematical Social Sciences*, 76:19–30, 2015.

[Baumeister et al., 2015b] D. Baumeister, D. Neugebauer, and J. Rothe. Verification in attack-incomplete argumentation frameworks. In *Proceedings of the 4th International Conference on Algorithmic Decision Theory*, pages 341–358. Springer-Verlag *Lecture Notes in Artificial Intelligence #9346*, September 2015.

[Baumeister et al., 2015c] D. Baumeister, J. Rothe, and H. Schadrack. Verification in argument-incomplete argumentation frameworks. In *Proceedings of the 4th International Conference on Algorithmic Decision Theory*, pages 359–376. Springer-Verlag *Lecture Notes in Artificial Intelligence #9346*, September 2015.

[Baumeister et al., 2017] D. Baumeister, S. Bouveret, J. Lang, N. Nguyen, T. Nguyen, J. Rothe, and A. Saffidine. Positional scoring-based allocation of indivisible goods. *Journal of Autonomous Agents and Multi-Agent Systems*, 31(3):628–655, 2017.

[Baumeister et al., 2018a] D. Baumeister, D. Neugebauer, and J. Rothe. Credulous and skeptical acceptance in incomplete argumentation frameworks. In *Proc. COMMA*, volume 305 of *FAIA*, pages 181–192. IOS Press, 2018.

[Baumeister et al., 2018b] D. Baumeister, D. Neugebauer, J. Rothe, and H. Schadrack. Verification in incomplete argumentation frameworks. *Artif. Intell.*, 264:1–26, 2018.

[Baumeister et al., 2020] D. Baumeister, G. Erdélyi, O. Erdélyi, J. Rothe, and A. Selker. Complexity of control in judgment aggregation for uniform premise-based quota rules. *Journal of Computer and System Sciences*, 112:13–33, 2020.

[Bench-Capon, 2003] T. Bench-Capon. Persuasion in practical argument using value-based argumentation frameworks. *Journal of Logic and Computation*, 13(3):429–448, 2003.

[Black, 1958] D. Black. *The Theory of Committees and Elections*. Cambridge University Press, 1958.

[Bodanza et al., 2017] G. Bodanza, F. Tohmé, and M. Auday. Collective argumentation: A survey of aggregation issues around argumentation frameworks. *Argument & Computation*, 8(1):1–34, 2017.

[Bouveret et al., 2010] S. Bouveret, U. Endriss, and J. Lang. Fair division under ordinal preferences: Computing envy-free allocations of indivisible goods. In *Proceedings of the 19th European Conference on Artificial Intelligence*, pages 387–392. IOS Press, August 2010.

[Brandt et al., 2016] F. Brandt, V. Conitzer, U. Endriss, J. Lang, and A. Procaccia, editors. *Handbook of Computational Social Choice*. Cambridge University Press, 2016.

[Brandt, 2009] F. Brandt. Some remarks on Dodgson's voting rule. *Mathematical Logic Quarterly*, 55(4):460–463, 2009.

[Caminada and Pigozzi, 2011] M. Caminada and G. Pigozzi. On judgment aggregation in abstract argumentation. *Journal of Autonomous Agents and Multi-Agent Systems*, 22(1):64–102, 2011.

[Caragiannis et al., 2016] I. Caragiannis, A. Procaccia, and N. Shah. When do noisy votes reveal the truth? *ACM Transactions on Economics and Computation*, 4(3):Article 15, 15:1–15:30, 2016.

[Cayrol et al., 2008] C. Cayrol, F. de Saint-Cyr, and M. Lagasquie-Schiex. Revision of an argumentation system. In *Proceedings of the 11th International Conference on Principles of Knowledge Representation and Reasoning*, pages 124–134. AAAI Press, September 2008.

[Cayrol et al., 2010] C. Cayrol, F. de Saint-Cyr, and M. Lagasquie-Schiex. Change in abstract argumentation frameworks: Adding an argument. *J. Artif. Intell. Res.*, 38:49–84, 2010.

[Chen and Endriss, 2019] W. Chen and U. Endriss. Preservation of semantic properties in collective argumentation: The case of aggregating abstract argumentation frameworks. *Artificial Intelligence*, 269:27–48, 2019.

[Chevaleyre et al., 2012] Y. Chevaleyre, J. Lang, N. Maudet, J. Monnot, and L. Xia. New candidates welcome! Possible winners with respect to the addition of new candidates. *Mathematical Social Sciences*, 64(1):74–88, 2012.

[Condorcet, 1785] J.-A.-N. de Caritat, Marquis de Condorcet. *Essai sur l'application de l'analyse à la probabilité des décisions rendues à la pluralité des voix*, 1785. Facsimile reprint of original published in Paris, 1972, by the Imprimerie Royale. English translation appears in I. McLean and A. Urken, *Classics of Social Choice*, University of Michigan Press, 1995, pages 91–112.

[Coste-Marquis et al., 2005] S. Coste-Marquis, C. Devred, and P. Marquis. Symmetric argumentation frameworks. In *Proceedings of the 8th European Conference on Symbolic and Quantitative Approaches to Reasoning and Uncertainty*, pages 317–328. Springer-Verlag *Lecture Notes in Artificial Intelli-*

gence #3571, July 2005.

[Coste-Marquis et al., 2007] S. Coste-Marquis, C. Devred, S. Konieczny, M. Lagasquie-Schiex, and P. Marquis. On the merging of Dung's argumentation systems. *Artif. Intell.*, 171(10):730–753, 2007.

[Coste-Marquis et al., 2015] S. Coste-Marquis, S. Konieczny, J. Mailly, and P. Marquis. Extension enforcement in abstract argumentation as an optimization problem. In *Proceedings of the 24th International Joint Conference on Artificial Intelligence*, pages 2876–2882. AAAI Press/IJCAI, July 2015.

[de Haan, 2016] R. de Haan. Parameterized complexity results for the Kemeny rule in judgment aggregation. In *Proceedings of the 22nd European Conference on Artificial Intelligence*, pages 1502–1510. IOS Press, August/September 2016.

[de Haan, 2017] R. de Haan. Complexity results for manipulation, bribery and control of the Kemeny judgment aggregation procedure. In *Proceedings of the 16th International Conference on Autonomous Agents and Multiagent Systems*, pages 1151–1159. IFAAMAS, May 2017.

[Delobelle et al., 2016] J. Delobelle, A. Haret, S. Konieczny, J. Mailly, J. Rossit, and S. Woltran. Merging of abstract argumentation frameworks. In *Proceedings of the 15th International Conference on Principles of Knowledge Representation and Reasoning*, pages 33–42. AAAI Press, April 2016.

[Delobelle et al., 2018] J. Delobelle, S. Konieczny, and S. Vesic. On the aggregation of argumentation frameworks: operators and postulates. *Journal of Logic and Computation*, 28(7):1671–1699, 2018.

[Dietrich and List, 2007] F. Dietrich and C. List. Judgment aggregation by quota rules: Majority voting generalized. *Journal of Theoretical Politics*, 19(4):391–424, 2007.

[Dietrich and Mongin, 2010] F. Dietrich and P. Mongin. The premise-based approach to judgment aggregation. *Journal of Economic Theory*, 145:562–582, 2010.

[Dimopoulos and Torres, 1996] Y. Dimopoulos and A. Torres. Graph theoretical structures in logic programs and default theories. *Theoret. Comput. Sci.*, 170(1):209–244, 1996.

[Dodgson, 1876] C. Dodgson. A Method of Taking Votes on more than two Issues. Pamphlet printed by the Clarendon Press, Oxford, and headed "not yet published" (see the discussions in [McLean and Urken, 1995; Black, 1958], both of which reprint this paper), 1876.

[Dung, 1995] P. Dung. On the acceptability of arguments and its fundamental role in nonmonotonic reasoning, logic programming and n-person games. *Artificial Intelligence*, 77(2):321–357, 1995.

[Dunne and Bench-Capon, 2002] P. Dunne and T. Bench-Capon. Coherence in finite argument systems. *Artificial Intelligence*, 141(1):187–203, 2002.

[Dunne et al., 2011] P. Dunne, A. Hunter, P. McBurney, S. Parsons, and M. Wooldridge. Weighted argument systems: Basic definitions, algorithms, and complexity results. *Artificial Intelligence*, 175(2):457–486, 2011.

[Dunne et al., 2012] P. Dunne, P. Marquis, and M. Wooldridge. Argument aggregation: Basic axioms and complexity results. *Proceedings of the 4th International Conference on Computational Models of Argument*, 245:129–140, 2012.

[Dunne et al., 2015] P. Dunne, W. Dvořák, T. Linsbichler, and S. Woltran. Characteristics of multiple viewpoints in abstract argumentation. *Artificial Intelligence*, 228:153–178, 2015.

[Endriss and Grandi, 2017] U. Endriss and U. Grandi. Graph aggregation. *Artificial Intelligence*, 245:86–114, 2017.

[Fazzinga et al., 2020] B. Fazzinga, S. Flesca, and F. Furfaro. Revisiting the notion of extension over incomplete abstract argumentation frameworks. In *Proceedings of the 29th International Joint Conference on Artificial Intelligence*, pages 1712–1718. ijcai.org, July 2020.

[Fishburn, 1977] P. Fishburn. Condorcet Social Choice Functions. *SIAM Journal on Applied Mathematics*, 33(3):469–489, 1977.

[Fishburn, 1982] P. Fishburn. Monotonicity paradoxes in the theory of elections. *Discrete Applied Mathematics*, 4(2):119–134, 1982.

[Gabbay and Rodrigues, 2012] D. Gabbay and O. Rodrigues. A numerical approach to the merging of argumentation networks. In *International Workshop on Computational Logic in Multi-Agent Systems*, pages 195–212. Springer, 2012.

[Garey and Johnson, 1979] M. Garey and D. Johnson. *Computers and Intractability: A Guide to the Theory of NP-Completeness*. W. H. Freeman and Company, 1979.

[Gibbard, 1973] A. Gibbard. Manipulation of Voting Schemes: A General Result. *Econometrica*, 41(4):587–601, 1973.

[Hemaspaandra et al., 2005] E. Hemaspaandra, H. Spakowski, and J. Vogel. The complexity of Kemeny elections. *Theoretical Computer Science*, 349(3):382–391, 2005.

[Hunter and Noor, 2020] A. Hunter and K. Noor. Aggregation of perspectives using the constellations approach to probabilistic argumentation. In *Proceedings of the 34nd AAAI Conference on Artificial Intelligence*, pages 2846–2853. AAAI Press, 2020.

[Kemeny, 1959] J. Kemeny. Mathematics Without Numbers. *Dædalus*, 88:577–

591, 1959.

[Kerkmann et al., 2020] A. Kerkmann, J. Lang, A. Rey, J. Rothe, H. Schadrack, and L. Schend. Hedonic games with ordinal preferences and thresholds. *Journal of Artificial Intelligence Research*, 67:705–756, 2020.

[Konczak and Lang, 2005] K. Konczak and J. Lang. Voting procedures with incomplete preferences. In *Proceedings of the Multidisciplinary IJCAI-05 Workshop on Advances in Preference Handling*, pages 124–129, July/August 2005.

[Kornhauser and Sager, 1986] L. Kornhauser and L. Sager. Unpacking the court. *Yale Law Journal*, 96(1):82–117, 1986.

[Kornhauser and Sager, 1993] L. Kornhauser and L. Sager. The one and the many: Adjudication in collegial courts. *California Law Review*, 349(3):382–391, 1993.

[Kuckuck and Rothe, 2019] B. Kuckuck and J. Rothe. Duplication monotonicity in the allocation of indivisible goods. *AI Communications*, 32(4):253–270, 2019.

[Li et al., 2011] H. Li, N. Oren, and T. Norman. Probabilistic argumentation frameworks. In *Proceedings of the International Workshop on Theorie and Applications of Formal Argumentation*, volume 7132 of *LNAI*, pages 1–16. Springer, July 2011.

[Lisowski et al., 2019] G. Lisowski, S. Doutre, and U. Grandi. Aggregation in value-based argumentation frameworks. *arXiv preprint arXiv:1907.09113*, 2019.

[List and Pettit, 2002] C. List and P. Pettit. Aggregating Sets of Judgments: An Impossibility Result. *Economics and Philosophy*, 18(1):89–110, 2002.

[McLean and Urken, 1995] I. McLean and A. Urken. *Classics of Social Choice*. University of Michigan Press, 1995.

[Menon and Larson, 2017] V. Menon and K. Larson. Computational aspects of strategic behaviour in elections with top-truncated ballots. *Journal of Autonomous Agents and Multi-Agent Systems*, 31(6):1506–1547, 2017.

[Modgil and Prakken, 2013] S. Modgil and H. Prakken. A general account of argumentation with preferences. *Artificial Intelligence*, 195:361–397, 2013.

[Mongin, 2012] P. Mongin. The doctrinal paradox, the discursive dilemma, and logical aggregation theory. *Theory and Decision*, 73(3):315–355, 2012.

[Niskanen et al., 2019] A. Niskanen, J. Wallner, and M. Järvisalo. Synthesizing argumentation frameworks from examples. *Journal of Artificial Intelligence Research*, 66:503–554, 2019.

[Niskanen et al., 2020] A. Niskanen, D. Neugebauer, M. Järvisalo, and J. Rothe. Deciding acceptance in incomplete argumentation frameworks. In

Proceedings of the 34nd AAAI Conference on Artificial Intelligence, pages 2942–2949. AAAI Press, 2020.

[Pettit, 2001] P. Pettit. Deliberative democracy and the discursive dilemma. *Philosophical Issues*, 11(1):268–299, 2001.

[Procaccia et al., 2007] A. Procaccia, J. Rosenschein, and G. Kaminka. On the robustness of preference aggregation in noisy environments. In *Proceedings of the 6th International Conference on Autonomous Agents and Multiagent Systems*, pages 416–422. IFAAMAS, May 2007.

[Rey, 2014] L. Rey. Complexity of abstract argumentation. Master's thesis, Heinrich-Heine-Universität Düsseldorf, Institut für Informatik, Düsseldorf, Germany, August 2014.

[Rothe et al., 2018] J. Rothe, H. Schadrack, and L. Schend. Borda-induced hedonic games with friends, enemies, and neutral players. *Mathematical Social Sciences*, 96:21–36, 2018.

[Rothe, 2015] J. Rothe, editor. *Economics and Computation. An Introduction to Algorithmic Game Theory, Computational Social Choice, and Fair Division*. Springer Texts in Business and Economics. Springer-Verlag, 2015.

[Satterthwaite, 1975] M. Satterthwaite. Strategy-Proofness and Arrow's Conditions: Existence and Correspondence Theorems for Voting Procedures and Social Welfare Functions. *Journal of Economic Theory*, 10(2):187–217, 1975.

[Skiba et al., 2020] K. Skiba, D. Neugebauer, and J. Rothe. Complexity of possible and necessary existence problems in abstract argumentation. In *Proceedings of the 24th European Conference on Artificial Intelligence*, pages 897–904. IOS Press, August/September 2020.

[Terzopoulou and Endriss, 2020] Z. Terzopoulou and U. Endriss. Neutrality and relative acceptability in judgment aggregation. *Social Choice and Welfare*, 2020.

[Tohmé et al., 2008] F. Tohmé, G. Bodanza, and G. Simari. Aggregation of attack relations: A social-choice theoretical analysis of defeasibility criteria. In *Proceedings of the 5th International Symposium on Foundations of Information and Knowledge Systems*, pages 8–23. Springer-Verlag *Lecture Notes in Artificial Intelligence #4932*, February 2008.

[Wallner et al., 2017] J. Wallner, A. Niskanen, and M. Järvisalo. Complexity results and algorithms for extension enforcement in abstract argumentation. *Journal of Artificial Intelligence Research*, 60:1–40, 2017.

[Wallner, 2020] J. Wallner. Structural constraints for dynamic operators in abstract argumentation. *Argument & Computation*, 11:151–190, 2020.

[Xia and Conitzer, 2011] L. Xia and V. Conitzer. Determining possible and necessary winners given partial orders. *Journal of Artificial Intelligence Re-*

search, 41:25–67, 2011.

[Xia, 2012] L. Xia. Computing the margin of victory for various voting rules. In *Proceedings of the 13th ACM Conference on Electronic Commerce*, pages 982–999. ACM Press, June 2012.

σ	σ-VER	σ-PV	σ-NV
$\mathcal{CF}$	$\in$ P ♠	$\in$ P ♣	$\in$ P ♣
$\mathcal{AD}$	$\in$ P ♠	$\in$ P •	$\in$ P ♣
$\mathcal{ST}$	$\in$ P ♠	$\in$ P •	$\in$ P ♣
$\mathcal{CO}$	$\in$ P ♠	$\in$ P •	$\in$ P ♣
$\mathcal{GR}$	$\in$ P ♠	$\in$ P •	$\in$ P ♣
$\mathcal{PR}$	coNP-c. ★	Σ_2^p-c. ♣	coNP-c. ♣

σ	σ-CA	σ-PCA	σ-NCA
$\mathcal{CF}$	$\in$ P *	$\in$ P ♦	$\in$ P ♦
$\mathcal{AD}$	NP-c. ★	NP-c. ♦	Π_2^p-c. ♦
$\mathcal{ST}$	NP-c. ★	NP-c. ♦	Π_2^p-c. ♦
$\mathcal{CO}$	NP-c. ▲	NP-c. ♦	Π_2^p-c. ♦
$\mathcal{GR}$	$\in$ P *	NP-c. ♦	coNP-c. ♦
$\mathcal{PR}$	NP-c. ★	NP-c. ♦	Π_2^p-c. ♦

σ	σ-SA	σ-PSA	σ-NSA
$\mathcal{CF}_{\neq\emptyset}$	$\in$ P ▼	$\in$ P ▼	$\in$ P ▼
$\mathcal{AD}_{\neq\emptyset}$	coNP-c. †	Σ_2^p-c. ▼	coNP-c. †
$\mathcal{ST}$	coNP-c. ★	Σ_2^p-c. ♦	coNP-c. ♦
$\mathcal{CO}$	$\in$ P ▲	NP-c. ♦	coNP-c. ♦
$\mathcal{GR}$	$\in$ P *	NP-c. ♦	coNP-c. ♦
$\mathcal{PR}$	Π_2^p-c. ‡	Σ_3^p-c. ♦	Π_2^p-c. ♦

Table 2: Overview of the computational complexity of decision problems in standard and incomplete argumentation frameworks for the six original semantics. The results are grouped into verification variants (top), credulous acceptance variants (center), and skeptical acceptance variants (bottom). Each block has the semantics in the left-most column, the base problem for standard AFs in the second column, and the "possible" and "necessary" variants of the base problem in the third and fourth column, respectively. For a complexity class $\mathcal{C}$, we write $\mathcal{C}$-c. to denote completeness for $\mathcal{C}$. For the problem σ-SA and its generalizations, only nonempty conflict-free ($\mathcal{CF}_{\neq\emptyset}$) and nonempty admissible ($\mathcal{AD}_{\neq\emptyset}$) sets are considered, since these problems are trivial for general $\mathcal{CF}$ or $\mathcal{AD}$ sets.

CHAPTER 5

VALUE-BASED ARGUMENTATION

KATIE ATKINSON
Department of Computer Science, University of Liverpool, UK
katie@liverpool.ac.uk

TREVOR BENCH-CAPON
Department of Computer Science, University of Liverpool, UK
tbc@liverpool.ac.uk

Abstract

Value-based argumentation is concerned with recognising, accounting for, and reasoning with, the social purposes promoted by agents' beliefs and actions. Value-based argumentation frameworks extend Dung's abstract argumentation frameworks by ascribing an additional property to arguments, representing the values they promote, and recognising audiences. Values are ordered according to the preferences of an audience (different audiences will have different preferences) and an attack is successful only if the value of the attacked argument is not preferred to its attacker by its audience. Arguments can be related to values through the use of an argumentation scheme, thus enabling us to structure value-based argumentation. We describe the motivation of value-based argumentation, its formal description and properties, the argumentation scheme and its associated critical questions and some of the applications to which value-based argumentation has been put.

1 Philosophical motivations for value-based argumentation

The formal models of value-based argumentation that are presented in this chapter are intended to capture various philosophical concepts that are reflected in everyday human reasoning. In this section we explain the key philosophical accounts that motivated the development of the computational models of value-based argument.

1.1 Values and audiences

The inspiration for value-based argumentation originally came from the book *New Rhetoric* of Perelman and Olbrechts-Tyteca [1969]. The key insight of the *New Rhetoric* was that the acceptability of an argument depended not only on the argument itself, or on available counterarguments, but on the audience to which it was addressed. For an argument to be accepted, *its audience has to accept it*. In subsequent work on this topic Perelman says:

> If men [sic] oppose each other concerning a decision to be taken, it is not because they commit some error of logic or calculation. They discuss apropos the applicable rule, the ends to be considered, the meaning to be given to values, the interpretation and characterisation of facts. [Perelman, 1980], p150.

Perelman's academic roots were in jurisprudence and he drew on legal disputes to support his argument:

> Each [party] refers in its argumentation to different values [...] the judge will allow himself to be guided in his reasoning by the spirit of the system: i.e. by the values which the legislative authority seeks to protect and advance. [Perelman, 1980], p152.

Consideration of this had also been noted in AI and Law. In their highly influential work, Berman and Hafner [1993] discussed what should happen in factor-based reasoning with cases [Bench-Capon, 2017] when there were no precedents to allow the case to be decided. They argued that in such cases the decision should be made according to which social purposes would be promoted by deciding for the plaintiff and which would be promoted by deciding for the defendant, and the decision made according to which would better serve the prevalent social values. Note that this means that different arguments can be accepted in different jurisdictions (attitudes to the death penalty in Georgia and Minnesota were very different in the 1970s), and at different times (*"stare decisis* would bow to changing values"[1]).

Thus there seems something missing from a purely logical view: sometimes the logic will fail to compel, and we will need to make a choice on other grounds. Since the situation occurs in important arenas like law, we do not want the choice to be arbitrary: we want to provide rational grounds for such choice. As Perelman puts it:

> Logic underwent a brilliant development during the last century when, abandoning the old formulas, it set out to analyze the methods of proof used effectively by mathematicians. ... One result of this development is to limit its domain, since everything ignored by mathematicians is foreign to it. Logicians owe it to themselves to complete the theory of demonstration obtained in this way by a theory of argumentation. [Perelman and Olbrechts-Tyteca, 1969], p10.

The situation is reflected in Dung's abstract argumentation. Sometimes, the acceptability of an argument will not be unequivocally determined by the framework. Given a dilemma (cycles with even length in standard argumentation frameworks [Bench-Capon, 2014]) the restrictive grounded semantics will allow neither horn to be embraced, whereas the more permissive preferred semantics will allow either proposition to be believed, but offer no reason to opt for one rather than the other. Value-based argumentation attempts to offer reasons for this choice as part of a "theory of argumentation".

---

[1] Justice Marshall in Furman v Georgia 408 U.S. 238 (1972).

1.2 Direction of fit

The other key influence on value-based argumentation was the work of John Searle on practical reasoning and his notion of *direction of fit* [Searle, 2003]. Searle wrote

> Assume universally valid and accepted standards of rationality, assume perfectly rational agents operating with perfect information, and you will find that rational disagreement will still occur; because, for example, the rational agents are likely to have different and inconsistent values and interests, each of which may be rationally acceptable. [Searle, 2003], p. xv.

Searle's idea was that such rational disagreement was possible because of direction of fit. There is only a single actual world, and a single history of that world, and so our beliefs about the present and the past have to match that actual world. Because there is only one actual world, there is a right answer to questions of fact, and while there may be disagreement, this is something that should be capable of resolution, given complete information. Values, interests and aspirations can play no part in such *theoretical* reasoning: that would be to indulge in wishful thinking.

The future is, however, a different matter. There are many possible futures, and we can, through our actions, play a part in determining which will come to pass. In *practical* reasoning, reasoning about what we should do, we attempt to fit the world to our desires, so that our actions will bring about the future that we prefer. But here different values, interests, aspirations and even tastes, may be a legitimate source of rational disagreement. Some may find it strange if someone prefers vanilla ice cream to chocolate, but it is not irrational. Of course, these aspirations can affect deeper matters: in politics a desire for tax rises may exhibit a preference for equality over economic growth. Such a preference is not a matter of rationality, but of the values that one wishes to be expressed in a society.

Thus in practical reasoning, rational disagreement is to be expected [Bench-Capon, 2002c]. The notion of direction of fit, however, applies not only to actions, but to the law. Disagreement is at the heart of law, and even at the highest level judges differ as to the proper outcome of a case. Five-to-four decisions occur in almost a fifth of cases heard by the US Supreme Court[2]. Not only is disagreement common, it is expected: that is why appeal courts typically comprise an odd number of judges, and why the more important the court the more the judges, so that the US Supreme Court has nine[3]. Nor can judicial disagreement be considered irrational: after all, the minority will produce an opinion stating their reasons for their views. To a certain extent the judges are trying to fit their view of the current cases to the existing law: the doctrine of *stare decisis* means that their decision should be consistent with past decisions. For a logical analysis of precedential constraint, see [Horty and Bench-Capon, 2012]. However, it is often the case that the precedents do not fully constrain the decision: it may be that all of them can be distinguished according to some features of the current case. For such cases the judges are free to decide for either party. Here they try to fit the law as it will be after their decision (for the current case will serve as a precedent for future cases) to the way they desire the law should be. That is, they consider which decision will promote the purposes of the law better, as described in [Berman and Hafner, 1993]. Therefore, as in practical reasoning, the values and aspirations of the judges will determine their decision [Atkinson and Bench-Capon, 2005]. The justification is that the majority opinion of a properly appointed court should reflect the prevailing values of its society.

1.3 Value-based argumentation

To reflect the situation where the dispute is about how best to fit the world to our desires it is clear that a basic assumption of Dung's argumentation frameworks, that attacks always succeed in defeating the argument they attack, must be relaxed. As an example, while it is true that Sarah will not be able to go on holiday if she buys a new car, this

---

[2]Between 2000 and 2018, according to the US Supreme Court as reported in *The Washington Post* https://www.washingtonpost.com/news/posteverything/wp/2018/06/28/those-5-4-decisions-on-the-supreme-court-9-0-is-far-more-common/. Last accessed February 2020.

[3]Nine is the traditional number. As we write, in the run up to the 2020 Presidential election, there is speculation that this may be increased.

attack can simply be ignored if Sarah prefers the holiday: she can continue to make do with her current car. For a different person, however, perhaps a petrol-head like Jeremy, the attack will be decisive and the holiday plans abandoned.

Thus to reflect debates where values, aspirations and tastes matter, not only in everyday practical reasoning, but in important areas such law, politics and ethics as well, a method of augmenting Dung's framework with a notion of values was needed. *Values* was the term used to cover these subjective preferences. It is a term widely understood in this sense in popular media, and the notion of a *value premise* is a key part of the Lincoln-Davis debate format used throughout the USA as the basis for competitive debating in a number of leagues[4]. Thus the general notion of values is felt to be widely understood. For example, the French Republic was based on the three values of liberty, equality and fraternity. In value-based reasoning there have been many different sets of values used for different problems. Generally it is held that the identification of the relevant set of values is part of the formulation of the problem to be discussed [Atkinson and Bench-Capon, 2007c]. Some attempts have been made to provide a basis for the identification of values: e.g. van der Weide *et al.* [2009] used Schwartz Value Theory [Schwartz, 1992] and Bench-Capon [2020] used Maslow's hierarchy of needs [Maslow, 1943]. Often, however, it seems that a very general account is not best suited to a particular problem, and the use of problem specific value sets remains common.

2 Values in abstract argumentation frameworks

Value-based argumentation first appeared in the context of an extension to Dung's abstract argumentation framework, first in [Bench-Capon, 2002b] and later in the journal version [Bench-Capon, 2003a]. The basic idea was to extend Dung's notion of an augmentation framework as pair

---

[4]Including the National Speech and Debate Association, or NSDA (formerly known as the National Forensics League, or NFL) competitions, and related debate leagues such as the National Christian Forensics and Communication Association, the National Catholic Forensic League, the National Educational Debate Association, the Texas University Interscholastic League, Texas Forensic Association, Stoa USA and their affiliated regional organizations.

of a set of arguments and a set of the attacks between them, $\langle Ar, att \rangle$ by adding a set of values, V, a function mapping the members of Ar onto V, val, and a set of audiences P, expressed as orderings on V. Note that P might contain all the factorially many possible orderings on V, or only a selection of them. This might be to represent a particular set of agents with specific preferences, or some constraint on the ordering itself. For example, in order to represent facts, theoretical arguments are typically related to the value *truth*. Then to avoid wishful thinking, truth must be the most preferred value for every audience.

2.1 Extending Dung's argumentation frameworks with values

Accordingly, a value-based argumentation framework (VAF) is defined as an extension of a Dung-style argumentation framework (AF).

Definition 2.1 (Value-Based Argumentation Framework (VAF)). *A value-based argumentation framework is a 5-tuple $VAF = \langle Ar, att, V, val, P \rangle$ where Ar is a finite set of arguments, att is an irreflexive binary relation on Ar, V is a nonempty set of values, val is a function which maps from elements of Ar to elements of V and P is the set of possible audiences (represented as orderings on V). We say that an argument $a \in Ar$ relates to a value $v \in V$ if accepting a promotes or defends v: The value in question is given by $val(ar)$. For every $ar \in Ar$, $val(ar) \in V$.*

Note that if there is a single value, (perhaps *truth*), a VAF is equivalent to a standard Dung AF. If every argument maps to its own distinct value, we have a similar situation to the Preference Based Frameworks of Amgoud and Cayrol [Amgoud and Cayrol, 1998] and [Amgoud and Cayrol, 2002], except that Preference Based Argumentation uses only a single ordering so that P has only one member, and there is only a single audience.

In order to evaluate the status of arguments with respect to an audience we produce an audience specific value-based argumentation framework.

Definition 2.2 (Audience-Specific VAF (AVAF)). *An audience-specific VAF is a 5-tuple $AVAF = \langle Ar, att, V, val, Valpref_a \rangle$, where Ar, att, V*

and val are as for a VAF, a is an audience, $a \in P$, and $Valpref_a$ is a preference relation (transitive, irreflexive and asymmetric), $Valpref_a \subseteq V \times V$, reflecting the value preferences of audience a. The AVAF relates to the VAF from which it is derived in that Ar, att, V and val are identical, and $Valpref_a$ is the set of preferences derivable from the ordering $a \in P$ in the VAF.

Our purpose in introducing VAFs is to allow us to distinguish between one argument attacking another, and that attack succeeding, so that the attacked argument may or may not be defeated. Whether the attack succeeds depends on the value order of the audience considering the VAF. We therefore define the notion of defeat for an audience:

Definition 2.3 (Defeat for an Audience). *An argument ar $defeats_a$ an argument br for audience a, $(defeats_a(ar, br))$, in an AVAF $\langle Ar, att, V, val, Valpref_a \rangle$ if and only if $attacks(ar, br) \in att$ and $valpref(br, ar) \notin Valpref_a$.*

We can now define audience specific versions of the notions standardly associated with AFs:

Definition 2.4 (Acceptable to an Audience). *An argument $ar \in Ar$ is acceptable to an audience a with respect to a set of arguments S, $(acceptable_a(ar, S))$ if: $\forall(x)(x \in Ar \land defeats_a(x, ar) \rightarrow \exists(y)(y \in S \land defeats_a(y, x))$*

Definition 2.5 (Conflict Free for an Audience). *A set S of arguments is conflict free for an audience a if:*
$\forall(x)\forall(y)(x \in S \land y \in S) \rightarrow (\neg(attacks(x,y) \in att) \lor (valpref(val(y), val(x)) \in Valpref_a))$

Definition 2.6 (Admissability for an Audience). *A conflict free for an audience a set of arguments S is admissible for the audience a if: $\forall(x)(x \in S \rightarrow acceptable_a(x, S)$*

Definition 2.7 (Preferred Extension for an Audience). *A set of arguments S in a value-based argumentation framework $\langle Ar, att, V, val, Valpref_a \rangle$ is a preferred extension for audience-a, $(preferred_a)$, if it is a maximal (with respect to set inclusion) admissible for audience a subset of Ar.*

A practical way of evaluating the status of arguments in an AVAF is to remove from the VAF all the unsuccessful attacks, those for which $valpref(br, ar) \in Valpref_a$, whereupon it can be treated as a standard AF. Thus for any AVAF, $vaf_a = \langle Ar, att_{avaf}, V, val, valpref_a \rangle$ there is a corresponding AF, $af_a = \langle A, att_{af} \rangle$ such that for $(x, y) \in att_{avaf}$, $(x, y), \in att_{af}$ if and only if $defeats_a(x, y)$. The preferred extension of af_a will contain the same arguments as the preferred extension for audience a of the VAF. Note that if the original VAF does not contain any cycles in which all arguments pertain to the same value, af_a will contain no cycles, since every cycle will be broken at the point at which the attack is from an inferior value to a superior one for audience a. Hence both af_a and vaf_a will have a unique, non-empty, preferred extension for such cases.

Theorem 2.8. *Every AVAF with no single-valued cycles has a unique nonempty preferred extension.*

PROOF. *Let avaf be an AVAF, and let af be the standard argumentation framework resulting from removing all failing attacks. If avaf is cycle-free, then af is cycle free and hence by Theorem 2.6 of Bench-Capon [2003a] it has a unique, not-empty preferred extension. But suppose avaf has a cycle. We know that this contains at least two values. Let v be the least preferred value in the cycle, and arg be the final argument in a chain relating to this value. The attack from arg to the next argument in the cycle will fail. Therefore this attack will not appear in af and the cycle will be broken at this point. This applies to all cycles in avaf. Therefore af is cycle free, and so has a unique, non-empty, preferred extension. QED*

Moreover, since the AF derived from an AVAF contains no cycles, the grounded extension coincides with the preferred extension for this audience, and so there is a straightforward polynomial-time algorithm to compute it, given in [Bench-Capon, 2003a].

For the moment we will restrict consideration to VAFs which do not contain any cycles in a single value. For such VAFs, the notions of sceptical and credulous acceptance are of no relevance, since any given audience will accept only a single preferred extension. These preferred extensions may, and typically will, however, differ from audience to audience. We therefore introduce two useful notions: *objective acceptance*,

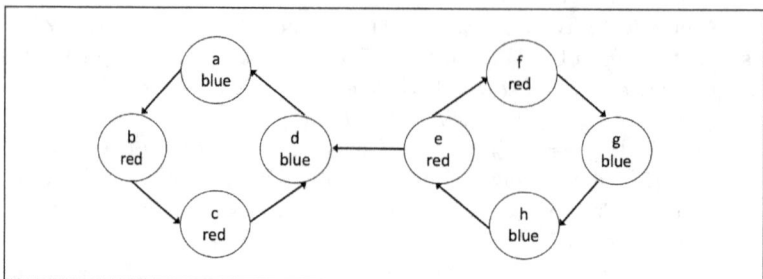

Figure 1: VAF with values red and blue

arguments which are acceptable to all audiences irrespective of their particular value order, and *subjective acceptance*, arguments which can be accepted by audiences with the appropriate value order.

Definition 2.9 (Objective Acceptance.). *Given a VAF $\langle Ar, att, V, val, Valpref \rangle$, an argument $a \in A$ is objectively acceptable if and only if for all valpref $\in Valpref$, a is in every valpref.*

Definition 2.10 (Subjective Acceptance.). *Given a VAF $\langle Ar, att, V, val, Valpref \rangle$, an argument $a \in A$ is subjectively acceptable if and only if for some valpref $\in Valpref$, a is in that valpref.*

An argument which is neither objectively nor subjectively acceptable (such as one attacked by an objectively acceptable argument with the same value) is said to be *indefensible*.

All arguments which are not attacked will, of course, be objectively acceptable. Otherwise, objective acceptance typically arises from cycles in two or more values. For example, consider a three-cycle in two values, say two arguments with V_1 and one with V_2. The argument with V_2 will either resist the attack on it when it is preferred to V_1, or, when V_1 is preferred, fail to defeat the argument it attacks which will, in consequence, be available to defeat its attacker. Thus the argument with V_2 will be objectively acceptable, and both the arguments with V_1 will be subjectively acceptable. A more elaborate example is shown in Figure 1.

There will be two preferred extensions, according to whether $red > blue$, or $blue > red$. If $red > blue$, the preferred extension will be

$\{e, g, a, b\}$, and if $blue > red$, $\{e, g, d, b\}$. Now e, g and b are objectively acceptable, but d, which would have been objectively acceptable if e had not attacked d, is only subjectively acceptable (when $blue > red$), and a, which is indefensible if d is not attacked, is also subjectively acceptable (when $red > blue$). Arguments c, f and h are indefensible. Results characterising the structures which give rise to objective acceptance are given in [6].

The question now arises as to whether it is possible to determine to which audiences an argument is acceptable. This question is fully explored in [Bench-Capon et al., 2007].

2.2 Computational complexity results of value-based argumentation frameworks

Not long after VAFs were first proposed in the literature, a study was conducted on a number of decision problems in VAFs [Dunne and Bench-Capon, 2004]. In that paper it was shown that, for a given audience, those decision questions which are typically computationally hard in the standard Dungian AF setting, actually admit efficient solution methods in the value-based setting. The paper also highlighted a number of questions that arise solely in value-based frameworks that lack efficient decision processes.

The two key questions addressed in the paper concern the decision problems in VAFs of subjective and objective acceptance, as set out in Definitions 2.9 and 2.10 above. Concerning the decision problem of subjective acceptance, it is shown in [Dunne and Bench-Capon, 2004] that the complexity of this problem is NP-complete, and for objective acceptance, the decision problem is shown to be CoNP-complete. The paper also considers decision problems related to determining subjective acceptance by attempting to identify which pair-wise orderings are "critical" in that a given ordering will admit an audience for which an argument is subjectively accepted, whereas reversing this order will yield a situation in which the argument of interest is never accepted. Full results and their proofs are given in [Dunne and Bench-Capon, 2004]. Extrapolating from the results, they demonstrate that the identification of an argument as subjectively or objectively acceptable is just as hard as the corresponding problems of determining credulous and sceptical acceptance in standard argumentation frameworks; see [Dunne and Bench-Capon, 2002] for a full discussion of this point. Further complexity results, especially those concerning which audience can accept a given argument, can be found in [Bench-Capon et al., 2007].

Further studies on computational complexity problems were later reported in [Dunne, 2010]. By considering properties of the directed graph structure defined by taking those values involved in conflicting arguments, Dunne identified an extensive class of argumentation systems for which the subjective and objective decision problems admit polynomial time solutions.

More recently, Nofal et al. [2014] examined specific questions in abstract argumentation frameworks under preferred semantics. They looked at the acceptance problem in standard argumentation frameworks, deciding whether a specific argument is in at least one preferred extension (i.e. it is credulously accepted) or in all such extensions (i.e. it is skeptically accepted). The paper presents an algorithm that enumerates all preferred extensions and builds algorithms that decide the acceptance problem without requiring explicit enumeration of all extensions. The improvements in efficiency brought about by the algorithms are achieved through a number of mechanisms: introduction of new labels for arguments' status, introduction of a new mechanism for pruning the search space so that transitions leading to dead ends are avoided at an early stage, and introduction of a cost-effective heuristic rule that yields earlier identification of arguments for transitions that might reach a goal state designating a preferred extension. The techniques developed for the acceptance problem in AFs are then used analogously to solve decision problems in VAFs, specifically deciding subjective and objective acceptance. Algorithms to solve these problems are defined and full proofs of the soundness and completeness of these algorithms is given in [Nofal et al., 2014].

The studies referenced above set out properties of VAFs with a view to demonstrating their viability for use in domain applications. We now turn to considering how values are captured in accounts of structured argumentation.

3 Values in structured argumentation

In the previous section we showed how abstract value-based argumentation could be used to account for the subjective preferences which come into play when we are reasoning about how to make the world fit our desires. But the question arises: *how do values become attached to arguments?* The discussion in section 1 suggested that arguments for which value preferences are relevant are likely to arise in practical reasoning, reasoning about what to do. We will therefore begin our search for the link between arguments and values by looking at practical reasoning.

3.1 Practical syllogism

Practical reasoning was identified as different from theoretical reasoning by Aristotle in his *Nicomachean Ethics*, The discussion was revived by Anscombe [1978] and Kenny [1978]. Kenny's example of a practical syllogism is

K1: I'm to be in London at 4.15.
   If I catch the 2.30 train, I'll be in London at 4.15.
   So, I'll catch the 2.30 train.

Although Aristotle attempted to present the practical syllogism as a deduction, this position proved difficult to maintain, and Kenny's abductive presentation is now more common. It still has, however, a number of peculiarities.

- The conclusion is not really a prediction. Whether or not I actually catch the train is contingent on a number of things beyond my control. Rather it is a resolution, a *decision* to try to catch the train. The result of practical reasoning should not be a belief, but an action or a plan of action which will realise the desires one has decided to pursue.

- The truth of the premises is not enough to determine the decision. There may be earlier trains, and I may decide to catch one of those to be on the safe side. There may be many other ways of achieving the goal. Like any abduction, its soundness depends on it being the best (for me, in my current circumstance) way to achieve the goal.

- If I do catch the train, there will be many things that I cannot do. If I in fact prefer to do one of these things to being in London, then I may choose one of these other activities.

- There may be a number of other consequences of catching the train which are not desirable. These may be sufficiently undesirable that I decide not to catch the train.

These aspects are somewhat reflected in Searle's formulation in [Searle, 2003]:

S1: I want, all things considered, to achieve E.
The best way, all things considered, to achieve E is to do M.
So, I will do M.

In order to act on the basis of an argument such as K1, therefore, we need to consider alternative actions, alternative goals and any additional consequences, and then choose the best of these alternative goals and actions. Note the element of choice here: we can choose which of our goals we will seek to realise, and which actions to undertake to realise these goals. In order to decide which is best, I need to go beyond the goals themselves, and consider why these states of affairs are wanted. This is where values come in. It is our values that make certain states of affairs goals, because these states of affairs promote our values. In [Atkinson and Bench-Capon, 2014] there was a detailed discussion of how values give rise to a number of types of goal such as maintenance goals, achievement goals, avoidance goals and removal goals.

It is the values associated with these goals that determines which of them should be considered best by a particular person. Which is best will be determined by the preference ordering on values, and so may vary from person to person. Whether I decide to catch the train in K1 depends on the value served by being in London, and the values served by possible alternatives.

In order to assist with the formulation of a computational version of practical reasoning, we decided to propose an argumentation scheme, in the manner of [Walton, 1996].

3.2 Argumentation schemes

Walton's notion of an argumentation scheme is that it is a means of presumptive reasoning: if the premises are true, then we may presumptively draw the conclusion, subject to satisfactorily dealing with critical questions characteristic of the scheme.

Walton [1996] proposes two schemes relating to practical reasoning. The first is the *necessary condition scheme*

W1: G is a goal for agent a.
Doing action A is necessary for agent a to carry out goal G.
Therefore agent a ought to do action A.

The other was quite similar: the *sufficient condition scheme*.

W2: G is a goal for agent a.
Doing action A is sufficient for agent a to carry out goal G.
Therefore agent a ought to do action A.

Walton associates four critical questions with each of these schemes:

- WCQ1: Are there alternative ways of realising goal G?

- WCQ2: Is it possible to do action A?

- WCQ3: Does agent a have goals other than G which should be taken into account?

- WCQ4: Are there other consequences of doing action A which should be taken into account?

Although these arguments are fair reflections of the practical syllogisms K1 and S1, they have no link to values. As we saw above, values are essential for evaluation. Thus if critical question WCQ1 is posed, and it proves that there is an alternative action, say A2, without values we have no reason to say that this is a *better* alternative, and so choose to realise G with A2 rather than A.

For this reason we introduced an argumentation scheme which did have the required link to values. This scheme was first presented in [Atkinson et al., 2004] and was more fully reported in [Atkinson et al., 2006a]. The scheme was stated in [Atkinson, 2005] as:

AS1: In the circumstances R
we should perform action A
to achieve new circumstances S
which will realise some goal G
which will promote some value V.

In [Atkinson, 2005] and [Atkinson et al., 2006a] sixteen critical questions were identified:

- CQ1: Are the believed circumstances true?

- CQ2: Assuming the circumstances, does the action have the stated consequences?

- CQ3: Assuming the circumstances and that the action has the stated consequences, will the action bring about the desired goal?

- CQ4: Does the goal realise the value stated?

- CQ5: Are there alternative ways of realising the same consequences?

- CQ6: Are there alternative ways of realising the same goal?

- CQ7: Are there alternative ways of promoting the same value?

- CQ8: Does doing the action have a side effect which demotes the value?

- CQ9: Does doing the action have a side effect which demotes some other value?

- CQ10: Does doing the action promote some other value?

- CQ11: Does doing the action preclude some other action which would promote some other value?

- CQ12: Are the circumstances as described possible?
- CQ13: Is the action possible?
- CQ14: Are the consequences as described possible?
- CQ15: Can the desired goal be realised?
- CQ16: Is the value indeed a legitimate value?

In [Atkinson and Bench-Capon, 2007c] a seventeenth CQ was added:

- CQ17: Can others act so as to take us to a state other than S?

This scheme allowed arguments for actions to be related to values: instantiating the scheme would give such an argument. Instantiating the critical questions would provide a means of attacking such arguments. This process of reasoning is illustrated in [Bench-Capon et al., 2005] and [Atkinson and Bench-Capon, 2007c].

3.3 Semantics for structured value-based argumentation

In order to provide a semantic underpinning for this argument scheme and critical questions, use was made of the notion of Action Based Alternating Transition Systems (AATS) with values (AATS+V). These were introduced in [Atkinson and Bench-Capon, 2007a] and more fully reported in [Atkinson and Bench-Capon, 2007c]. An AATS is a type of state transition diagram, introduced in [Wooldridge and van der Hoek, 2005], formally based on Alternating-time Temporal Logic [Alur et al., 2002]. In an AATS the states and transitions can be used to represent the current and future situations and the actions which will lead between them. In fact these transitions represent *joint actions*[5], that is, the cumulative effect of every agent relevant to the situation performing one action each. This means that a given action of a particular agent will appear in several transitions, depending on what the other agents do, and an agent may consequently not be in full control of the state that will be reached by using that action.

The definition of an AATS is:

---
[5]No suggestion of cooperation is intended here: the actions are joint solely in the sense that they are performed simultaneously.

Definition 3.1 (AATS ([Wooldridge and van der Hoek, 2005])). .
An Action-based Alternating Transition System *(AATS)* is an $(n+7)$-*tuple* $S = \langle Q, q_0, Ag, Ac_1, \ldots, Ac_n, \rho, \tau, \Phi, \pi \rangle$, *where:*

- Q *is a finite, non-empty set of* states;

- $q_0 \in Q$ *is the* initial state;

- Ag = *{1,...,n} is a finite, non-empty set of* agents;

- Ac_i *is a finite, non-empty set of* actions, *for each* $ag_i \in Ag$ *where* $Ac_i \cap Ac_j = \emptyset$ *for all* $ag_i \neq ag_j \in Ag$;

- $\rho : Ac_{ag} \to 2^Q$ *is an* action pre-condition function, *which for each action* $\alpha \in Ac_{ag}$ *defines the set of states* $\rho(\alpha)$ *from which* α *may be executed;*

- $\tau : Q \times J_{Ag} \to Q$ *is a partial* system transition function, *which defines the state* $\tau(q, j)$ *that would result by the performance of j from state q. This function is partial as not all joint actions are possible in all states;*

- Φ *is a finite, non-empty set of* atomic propositions; *and*

- $\pi : Q \to 2^\Phi$ *is an* interpretation function, *which gives the set of primitive propositions satisfied in each state: if* $p \in \pi(q)$, *then this means that the propositional variable* p *is satisfied (equivalently, true) in state* q.

As presented in [Wooldridge and van der Hoek, 2005], AATS have no values. Therefore they were extended in [Atkinson and Bench-Capon, 2007c] to include values, giving an AATS+V in which the transitions are additionally labelled with the values promoted or demoted by that transition. The additional definitions are:

Definition 3.2 (AATS+V ([Atkinson and Bench-Capon, 2007c])).
Given an AATS, an AATS+V is defined by adding two additional elements as follows:

- V *is a finite, non-empty set of values.*

5 - VALUE-BASED ARGUMENTATION

- $\delta : Q \times Q \times V \rightarrow \{+, -, =\}$ *is a valuation function which defines the status (promoted (+), demoted (−) or neutral (=)) of a value* $v_u \in V$ *ascribed to the transition between two states:* $\delta(q_x, q_y, v_u)$ *labels the transition between* q_x *and* q_y *with one of* $\{+, -, =\}$ *with respect to the value* $v_u \in V$.

With this definition it is possible to describe the practical reasoning argumentation scheme and critical questions in terms of the extended AATS+V. This gives us:

AS2 In the initial state $q_0 = q_x \in Q$,
   Agent $i \in Ag$ should participate in joint action $j_n \in J_{Ag}$ where $j_n{}^i = \alpha_i$,
   Such that $\tau(q_x, j_n)$ is q_y,
   Such that $p_a \in \pi(q_y)$ and $p_a \notin \pi(q_x)$, or $p_a \notin \pi(q_y)$ and $p_a \in \pi(q_x)$,
   Such that for some $v_u \in Av_i$, $\delta(q_x, q_y, v_u)$ is +.

We may now state the critical questions in these terms also.

- CQ1: $q_0 \neq q_x$ and $q_0 \notin \rho(\alpha_i)$.

- CQ2: $\tau(q_x, j_n)$ is not q_y.

- CQ3: $p_a \notin \pi(q_y)$.

- CQ4: $\delta(q_x, q_y, v_u)$ is not +.

- CQ5: Agent $i \in Ag$ can participate in joint action $j_m \in J_{Ag}$, where $j_n \neq j_m$, such that $\tau(q_x, j_m)$ is q_y.

- CQ6: Agent $i \in Ag$ can participate in joint action $j_m \in J_{Ag}$, where $j_n \neq j_m$, such that $\tau(q_x, j_m)$ is q_y, such that $p_a \in \pi(q_y)$ and $p_a \notin \pi(q_x)$ or $p_a \notin \pi(q_y)$ and $p_a \in \pi(q_x)$.

- CQ7: Agent $i \in Ag$ can participate in joint action $j_m \in J_{Ag}$, where $j_n \neq j_m$, such that $\tau(q_x, j_m)$ is q_z, such that $\delta(q_x, q_z, v_u)$ is +.

- CQ8: In the initial state $q_x \in Q$, if agent $i \in Ag$ participates in joint action $j_n \in J_{Ag}$, then $\tau(q_x, j_n)$ is q_y, such that $p_b \in \pi(q_y)$, where $p_a \neq p_b$, such that $\delta(q_x, q_y, v_u)$ is −.

- CQ9: In the initial state $q_x \in Q$, if agent $i \in Ag$ participates in joint action $j_n \in J_{Ag}$, then $\tau(q_x, j_n)$ is q_y, such that $\delta(q_x, q_y, v_w)$ is $-$, where $v_u \neq v_w$.

- CQ10: In the initial state $q_x \in Q$, if agent $i \in Ag$ participates in joint action $j_n \in J_{Ag}$, then $\tau(q_x, j_n)$ is q_y, such that $\delta(q_x, q_y, v_w)$ is $+$, where $v_u \neq v_w$.

- CQ11: In the initial state $q_x \in Q$, if agent $i \in Ag$ participates in joint action $j_n \in J_{Ag}$, then $\tau(q_x, j_n)$ is q_y and $\delta(q_x, q_y, v_u)$ is $+$. There is some other joint action $j_m \in J_{Ag}$, where $j_n \neq j_m$, such that $\tau(q_x, j_m)$ is q_z, such that $\delta(q_x, q_z, v_w)$ is $+$, where $v_u \neq v_w$.

- CQ12: $q_x \notin Q$.

- CQ13: $j_n \notin J_{Ag}$.

- CQ14: $\tau(q_x, j_n) \notin Q$.

- CQ15: $p_a \notin \pi(q)$ for any $q \in Q$.

- CQ16: $v_u \notin V$.

- CQ17: $j_n{}^i = j_m{}^i$, $j_n \neq j_m$ and $\tau(q_x, j_n) \neq \tau(q_x, j_m)$.

This formal account of the practical reasoning argumentation scheme and critical questions enable them to be used in agent systems designed to model practical reasoning scenarios; examples of these are provided in Section 4.

3.4 Dialogue interactions: values in persuasion and deliberation

In the previous sections we have considered reasoning with a specific audience which can determine the value order and evaluate the arguments accordingly. Often, however, values are crucial in dialogues where we have two or more audiences each with their own value order. Two distinct types of such dialogue are *persuasion* and *deliberation* [Walton and Krabbe, 1995].

In persuasion it is the person being persuaded who determines the value order [Bench-Capon, 2002c], since people will accept an argument only if it is acceptable on their own value ordering. Thus the proponents may not be able to use the arguments which convinced them because these will be acceptable on their value order, but perhaps not on the value order of the person they wish to persuade. Thus in a persuasion dialogue it is often necessary to elicit the value order of the other person, so that arguments acceptable to them can be found. Sometimes, however, it will not be possible to find arguments acceptable to the other, in which case the persuader must first try to get them to accept a value ordering and then to accept the argument which is the topic of the dialogue. Such dialogues are modelled in [Bench-Capon and Modgil, 2009]. A strategy for efficient persuasion in dialogues is given in [Atkinson et al., 2012].

Deliberation is different in that while the value orders may well differ, neither party can determine what it should be. Therefore one purpose of a deliberation dialogue is to find a value ordering which will be acceptable to all concerned, so that a solution corresponding to this order can be found, which should be acceptable to all the parties. A set of speech acts to support dialogues based on this view of deliberation is given in [Atkinson et al., 2013] and a tool showing how these speech acts can be used to generate persuasion and deliberation dialogues in agent systems is described in [Kirchev et al., 2019].

4 Key applications of value-based argumentation

In this section we will illustrate the use of value-based argumentation in a number of domains.

4.1 General practical reasoning

We will begin by looking at the use of value-based argumentation in general practical reasoning. Our example will be that used in [Atkinson and Bench-Capon, 2007c], which adapts a well known brain teaser. AI students may be familiar with it as it is often used to illustrate search problems.

The situation is that a farmer is returning from market with a chicken (C), a bag of seeds (S) and his faithful dog (D). He needs to cross a river, and there is a boat (B) but it can only carry the farmer and one of his possessions. He cannot leave the chicken and seeds together because the chicken will eat the seeds. Similarly, he cannot leave the dog and the chicken unattended together because the dog will eat the chicken. His problem is how to organise his crossing.

We will represent the states by two lists, one for the items on the right bank, and one for items on the left. Thus [BCDS, _] will be selected from Q as the initial state, q_0. The complete set of states is shown in Figure 2

The transitions will be formed by various joint actions. We will assume that the animals will eat if they can, and so ignore the possibility of, for example, leaving the dog and chicken alone, and the dog doing nothing. This gives us the following six joint actions.

j_1: All do nothing

j_2: Farmer rows alone, chicken eats seeds if possible, dog eats chicken if possible

j_3: Farmer rows seeds, dog eats chicken if possible

j_4: Farmer rows dog, chicken eats seeds if possible

j_5: Farmer rows chicken, animals do nothing

j_6: All continue their journey home.

We can also identify a number of possible values[6]:

P: Progress - Promoted when farmer moves one of his possessions to the right side of the river. It is demoted when a state is revisited (through the always undesirable "goal" of repetition), and, to a lesser extent, when a possession is rowed from the right bank to the left (Pr). Rowing an item back is preferred to repetition, since

---

[6]Some labels for values are the same as the propositions used in the state description. The context makes it clear which is intended.

repeating a state cannot be progress, whereas reaching a new state by returning an item to the left bank might be on a solution path, even though a *prima facie* backwards step.

S: Farmer has seeds - demoted when farmer loses seeds.

C: Farmer has chicken - demoted when farmer loses chicken.

F: Friendship - promoted when farmer travels with dog (it was for this companionship that he brought the dog with him).

We assume that the farmer values his possessions most, then wishes to make progress, and then have the joys of companionship. His value order is thus

$C, S > P > Pr > F$

We can now apply the joint actions to q_0 and label the transitions according to how they promote or demote the values. Initially five of the six actions are available, since the preconditions for j_6 are not satisfied. The result is shown as the first layer of Figure 2.

We can see that the only action which promotes a value without demoting a preferred value is j_5, and so the farmer will row the chicken, using the following argument:

- Farmer should row the chicken to promote Progress.

From q_2 three actions are possible. But two of them demote progress by reaching previous states, so the argument is

- Farmer should not row the seeds, or do nothing, as that would demote progress. So Farmer should row alone.

Having reached q_6, there are two actions which promote progress, rowing the seeds, and rowing the dog. But rowing the dog additionally promotes friendship, and so that will be chosen. From q_8 the only harmless action is to row the chicken to reach q_{10}. From q_{10} progress can be promoted by rowing the seeds, while all other actions demote a value. From here the only neutral action is to row alone to reach q_{13}. From here the farmer can promote progress by rowing the chicken. Now at

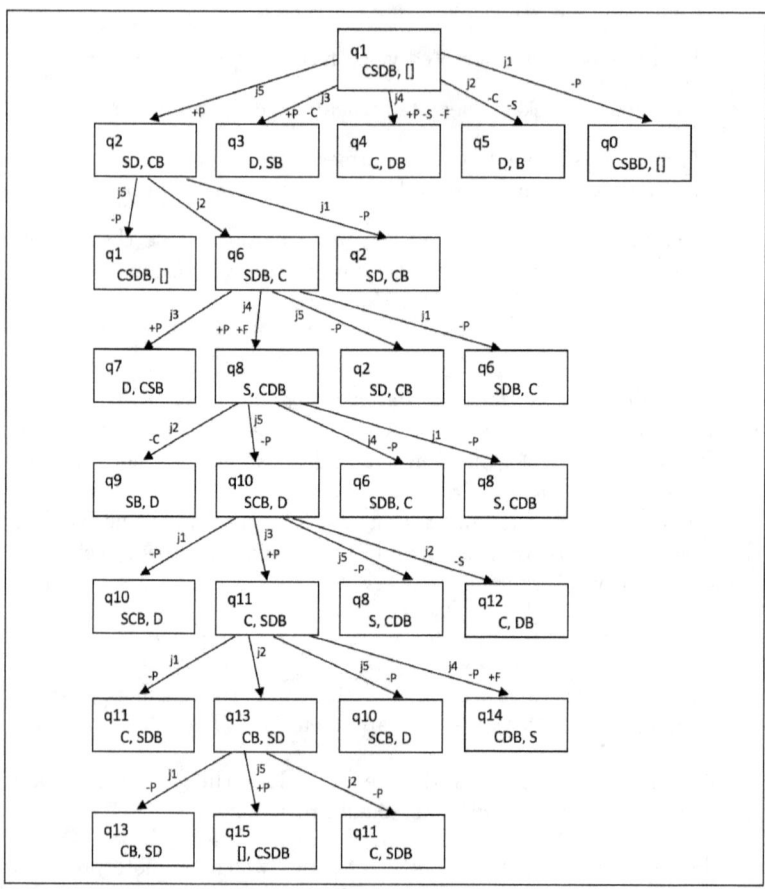

Figure 2: AATS+V for the farmer's river crossing problem. Note that when seeds and chicken are eaten, they no longer appear in the state descriptions.

last everything is on the right bank, and progress can be made by them all proceeding home.

This example shows how the puzzle can be solved by simply considering how to best promote values at every move. No look ahead is needed. In the standard puzzle, heuristic search gives two solutions, since rowing either the dog or the seeds in q_6 will achieve the goal. In the practical reasoning version this is resolved because in q_6 the farmer chooses to row the dog, to promote friendship as well as progress. For another example of practical reasoning, deciding whether to travel by aeroplane or train, see [Bench-Capon and Atkinson, 2009].

4.2 Domain-specific application: law

A domain in which value-based argumentation has been widely used is law, and in that domain arguing with values precedes abstract value-based argumentation and the formal modelling of structured argument with values by over a decade. The notion of values was in introduced to AI and Law by [Berman and Hafner, 1993][7]. In that paper Berman and Hafner noted that when using factor-based reasoning [Bench-Capon, 2017], often there were factor-based arguments for both sides which needed to be chosen between. Factor-based reasoning as proposed in HYPO [Rissland and Ashley, 1987] and CATO [Aleven and Ashley, 1995], however, offered no rationale for choosing between them. The answer given in [Berman and Hafner, 1993] was that the arguments which better served the purposes of the law should be accepted. This idea was elaborated into a more formal theory of reasoning with cases as theory construction, in which value preferences were derived from precedents which were then applied to new cases, in [Bench-Capon and Sartor, 2001] and [Bench-Capon and Sartor, 2003], which was was the basis of the CATE [Chorley and Bench-Capon, 2005b] and AGATHA [Chorley and Bench-Capon, 2005a] systems. In [Greenwood et al., 2003] it was proposed that the argumentation scheme for practical reasoning, described in [Atkinson et al., 2006a] and discussed above, could be used to generate the case based arguments required by factor-based reasoning and link them to values. The wild animals cases of [Berman and Hafner,

---

[7]Berman and Hafner used *purposes* rather than *values*, but they functioned in the same way. We will use *purpose* and *value promoted* as synonymous.

1993] had been modelled as a Dung style argumentation framework in [Bench-Capon, 2002a]. These various strands were brought together in [Bench-Capon et al., 2005], which added values to the AF of [Bench-Capon, 2002a], and evaluated the arguments according to the resulting VAF.

In [Berman and Hafner, 1993] the example cases were some well known property law cases (often used as an introduction to property law in US law schools) concerning wild animals. That paper discussed three cases:

- *Keeble v Hickergill (1707)*. This was an English case in which Keeble rented a duck pond, to which he lured ducks, which he shot and sold for consumption. Hickergill, out of malice, scared the ducks away by firing guns. The court found for Keeble. Two arguments for Keeble are possible: that he was engaged in an economically valuable activity, and that he was operating on his own land. The former reading is adopted in [Berman and Hafner, 1993], but others, e.g. [Bench-Capon and Rissland, 2001], prefer the latter.

- *Pierson v Post (1805)*. In this New York case, Post was hunting a fox with hounds. Pierson intercepted the fox, killed it with a handy fence rail, and carried it off. The court found for Pierson. The argument was that Post had never had possession of the fox. The argument that hunting vermin is a useful activity which needs protection and encouragement formed the basis of the minority opinion. In this case, because of its legal setting, the original complainant, Post, whose role corresponds to the plaintiff in the other cases, is named second. We shall, however, refer to Post as the plaintiff and Pierson as the defendant to maintain consistency of role with the other cases.

- *Young v Hitchens (1844)*. In this English case, Young was a commercial fisherman who spread a net of 140 fathoms in open water. When the net was almost closed, Hitchens went through the gap, spread his net and caught the trapped fish. The case was decided for Hitchens. The basis for this was that Young had never had possession of the fish, and that it was not part of the court's remit to rule as to what constituted unfair competition.

Later work [Bench-Capon and Rissland, 2001] also included four other cases in the discussion:

- *Ghen v Rich (1881)*. In this Massachusetts case, Ghen was a whale hunter who harpooned a whale which subsequently was not reeled in, but was washed ashore. It was found by a man called Ellis, who sold it to Rich. According to local custom, Ellis should have reported his find, whereupon Ghen would have identified his lance and paid Ellis a fee. The court found for Ghen.

- *Conti v ASPCA*[8] *(1974)*. In this New York case, Chester, a parrot owned by the ASPCA, escaped and was recaptured by Conti. The ASPCA found this out and reclaimed Chester from Conti. The court found that they were within their rights to do so.

- *New Mexico vs Morton (1975)* and *Kleepe vs New Mexico (1976)*. These two cases concerned the ownership of unbranded burros normally present on public lands, which had temporarily strayed off them. Both were won by the state.

These seven cases were formalised as a Dung style AF in [Bench-Capon, 2002a] and this was also used in [Bench-Capon et al., 2005]. It is shown in Figure 3.

The twenty six arguments, the arguments they attack and the values associated with them in [Bench-Capon et al., 2005] are shown in Table 1.

The basic approach in [Bench-Capon, 2002a] was to remove the arguments not applicable to a particular case and then consider preferred extensions. Then if argument A was sceptically acceptable, the plaintiff would win, but otherwise the defendant would win (the burden of proof is on the plaintiff). This, however, is not straightforward in the Dung style AF since there are even-length cycles in the AF, and so there will be multiple preferred extensions, some with A and some without.

The cycles in question are:

---

[8]The American Society for the Prevention of Cruelty to Animals.

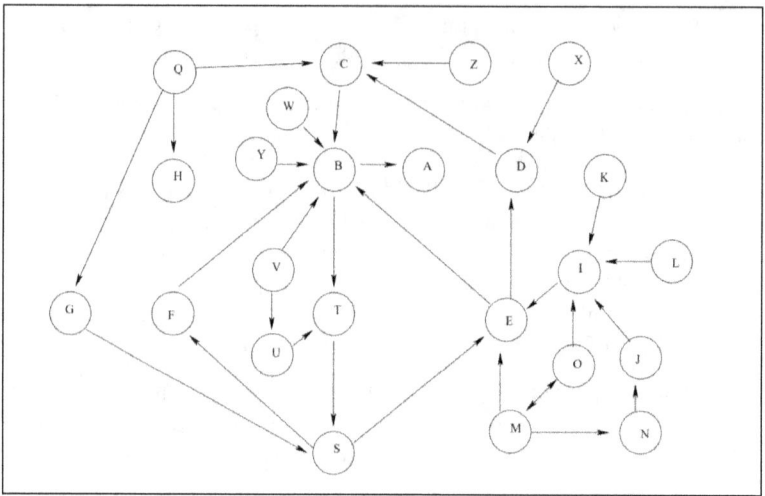

Figure 3: Dung style AF for wild animals cases, as given in [Bench-Capon, 2002a]

- the two-cycle M-O, which arises in *Pierson*
- the four-cycle B-T-S-E, which arises in *Young*
- the four-cycle B-T-S-F, which arises in *Young*

This is precisely the situation for which Berman and Hafner commended the use of values: we need to choose between M, which promotes *clarity*, and O which promotes *useful activity*. In the actual case of *Pierson*, clarity was chosen, so that M was able to resist the attack of O, and so A was not in the preferred extension.

In the case of the two four-cycles that appeared in *Young*, the case was in fact resolved by the acceptance of U, which claimed that deciding what constituted unfair competition was outside the remit of the court. With T defeated, S defeats F, and so defends B. Similarly, S also defeats E and so B is acceptable. Now B defeats A and so the defendant won. Note that V was absent from *Young*. It was, however, present in *Ghen*, which concerned whaling, an industry long governed by clear

5 - VALUE-BASED ARGUMENTATION

CL = Clear law, UA = Useful activity, PR = Protect property rights, EA = Economic activity, CR = The court should not make law

ID	Argument	Attacks	Values
A	Pursuer had right to animal		claim
B	Pursuer not in possession	A, T	CL
C	Owns the land so possesses animals	C	PR
D	Animals not confined by owner	C	
E	Effort promising success made to secure animal made by pursuer	B, D	CL
F	Pursuer has right to pursue livelihood	B	EA
G	Interferer was trespassing	S	PR
H	Pursuer was trespassing	F	PR
I	Pursuit not enough (Justinian)	E	CL
J	Animal was taken (Justinian)	I	CL
K	Animal was mortally wounded (Puffendorf)	I	CL
L	Bodily seizure is not necessary (Barbeyrac), interpreted as animal was brought within certain control (Tompkins)	I	UA
M	Mere pursuit is not enough (Tompkins)	E, O	CL
N	Justinian is too old an authority (Livingston)	J	
O	Bodily seizure is not necessary (Barbeyrac), interpreted as reasonable prospect of capture is enough (Livingston)	I, M	UA
Q	The land was open	G, H, C	PR
S	Defendant in competition with the plaintiff	E, F	EA
T	Competition was unfair	S	EA
U	Not for courts to regulate competition	T	CR
V	The iron holds the whale is an established convention of whaling	B, U	CR
W	Owners of domesticated animals have a right to regain possession	B	PR
X	Unbranded animals living on land belong to owner of land	D	PR
Y	Branding establishes title	B	
Z	Physical presence (straying) insufficient to confer title on owner	C	CL

Table 1: **Arguments in the Wild Animal Cases.**

conventions. Here the courts felt that just as it was not in their remit to determine what was unfair competition, neither could they overturn established conventions on the matter. Thus V was able to defeat U and B and so enable the plaintiff to win. This was forced in the standard AF, but in a VAF the attack from U to T can be resisted by preferring the value of *economic activity* to that promoted by restricting the court's remit, which would enable Young to win, even in the absence of an applicable convention. Such a shift in attitude may well occur (attitudes to regulation of competition swing back and forth), and so *Young* may at some future time be overturned. This illustrates a feature of value-based argumentation in law: because value preferences can change, the outcome of a case may be different at different times and in different jurisdictions. This captures the essence of the role of values in this kind of legal reasoning. A more elaborate discussion in [Bench-Capon *et al.*, 2005] also investigates the role of intermediate concepts [Lindahl and Odelstad, 2004] in moving from facts to legal conclusions.

Further discussions of value-based reasoning in the wild animals cases can be found in [Bench-Capon, 2003b] and [Atkinson and Bench-Capon, 2007b]. In [Wyner *et al.*, 2007] an additional case, *Popov v Hayashi* [Atkinson, 2012] was included. This celebrated case[9], concerned a record breaking home run baseball hit by Barry Bonds of the San Francisco Giants. There was a struggle amongst crowd members over its possession. Popov first laid his glove on the ball, but Hayashi emerged from the ensuing melee with the ball. The wild animals cases were cited in the case, analogies being drawn between the hunted animals and the "fugitive baseball" [Finkelman, 2001]. This case and the wild animals cases were further discussed in [Bench-Capon, 2012].

4.3 Domain-specific application: e-participation

Another domain in which value-based argumentation has proved effective is e-participation. Political disputes often turn on disagreement as to values, and so this is a natural way to model such disputes. In PARMENIDES [Atkinson *et al.*, 2006b], a policy was presented for critique by members of the public through a software tool. The policy was pre-

---

[9]It was the subject of the 2004 comedy documentary film *Up for Grabs* https://www.imdb.com/title/tt0420356/

sented as an instantiation of the practical reasoning scheme AS1 given above. Thus the policy was presented in terms of an understanding of the current situation and what it was meant to achieve in terms of facts, goals and values. The user was then given the opportunity to critique the policy in terms of relevant critical questions characteristic of the scheme[10]. In this way disagreement with the policy could be made precise, and different motives for disagreement identified. For example, different people might doubt whether the current situation was indeed as suggested, others might doubt that the policy would achieve its ends, and yet others might oppose these ends because rejecting the values they promote. PARMENIDES was further developed in [Cartwright and Atkinson, 2009] and later PARMENIDES formed the basis for the development of the Structured Consultation Tool (SCT) [Bench-Capon et al., 2015], produced as part of the IMPACT project[11]. The SCT enabled not only a policy proposal to be critiqued, but also for the users to make proposals of their own, which could be automatically critiqued by instantiating critical questions [Wardeh et al., 2013].

We will base our example in this chapter on that of [Atkinson et al., 2011], which was also used in [Wyner et al., 2012]. The example is an issue in UK Road Traffic policy. The number of fatal road accidents is an obvious cause for concern, and in the UK there are speed restrictions on various types of road, in the belief that excessive speed causes accidents. The policy issue which we will consider is how to reduce road deaths. One option is to introduce speed cameras to discourage speeding.

Following [Atkinson and Bench-Capon, 2007c] the first step is to build a model. In [Atkinson et al., 2011] there was an extensive discussion of how to construct the model on the basis of responses received to a Green Paper[12]. Like [Wyner et al., 2012] we will focus on the final refinement of the model presented in [Atkinson et al., 2011], which in-

---

[10] Not all critical questions were used: for example, those relating to the components of the model were not appropriate.

[11] Integrated Method for Policy Making Using Argument Modelling and Computer Assisted Text Analysis, in the European Framework 7 project (Grant Agreement No 247228) in the ICT for Governance and Policy Modeling theme (ICT-2009.7.3).

[12] A Green Paper is a Government publication issued as part of a consultation process that details specific issues, and then points out possible courses of action in terms of policy and legislation in order to receive feedback from interested parties.

cludes responses from road safety organisations, motoring lobby groups, representations about financial constraints and civil liberties groups.

We now present the AATS+V. States are composed from the propositions considered relevant. In the model of [Atkinson et al., 2011] the propositions that were considered are (given as pairs of positive and negative propositions):

R: The number of road deaths is acceptable; There are more road deaths than there should be.

S: Many motorists break the speed limits; Speed limits are generally obeyed.

P: Privacy is respected; There are additional intrusions on privacy.

These three propositions give rise to, potentially, eight states. We may, if we wish, exclude one or more of these as impossible. For example, if we believe that it is impossible that the number of road deaths is acceptable and yet many motorists break the speed limits, we may introduce constraints on states to filter it out. In [Wyner et al., 2012], we specify all the possible states available for consideration. One state is designated as the current state:

- Many motorists break the speed limits ∧ There are more road deaths than there should be ∧ Privacy is respected.

We consider the impact of changes of state in terms of three values:

L: human life (Life);

B: the financial cost to the Government (Budget); and

F: the impact on civil liberties (Freedom). Here the principal concern is the impact on privacy.

The main agents involved are the *Government* (G), and *Motorists* (M), each considered as a body. In some cases the consequences of action are indeterminate (or at least cannot be determined using the elements we are modelling). To account for this we introduce a third agent, termed *Nature* (N). The action ascribed to Nature determines the outcomes of the actions of the other agents, where these outcomes are uncertain or probabilistic. We take the Government to be the independent agent, the one attempting to fulfill its values, while the actions of the Motorists and Nature are relative to its choices.

The Government has three actions: introducing speed cameras (G_1), educating motorists (G_2), or doing nothing (G_3). Motorists may reduce their speed or do nothing. Nature has two actions according to which fatal accidents are or are not reduced as a result of the Government and motorist actions. Actions are assumed to be always carried out together with other agents, represented as joint actions. The joint actions available are:

- j_0: Government does nothing, motorists do nothing and nature does nothing.

- j_1: Government introduces cameras, motorists do nothing and nature does nothing.

- j_2: Government introduces cameras, motorists reduce speed and nature reduces accidents.

- j_3: Government introduces cameras, motorists reduce speed and nature does nothing.

- j_4: Government educates motorists, motorists reduce speed and nature reduces accidents.

- j_5: Government educates motorists, motorists do nothing and nature reduces accidents. (Being more skilled, the drivers can cope with their excessive speed).

Finally, we have transitions, which relate a source state, a destination state, a joint action, a list of values promoted, and a list of values demoted. The joint action can only be carried out where, in some sense, the conditions for doing the action are met (e.g. where motorists are not speeding, then they cannot reduce speed) and result in a state that also makes sense (e.g. where motorists reduce speed and nature reduces accidents, then motorists are not speeding and accidents are reduced). We can presume that accidents are always reduced when motorists are

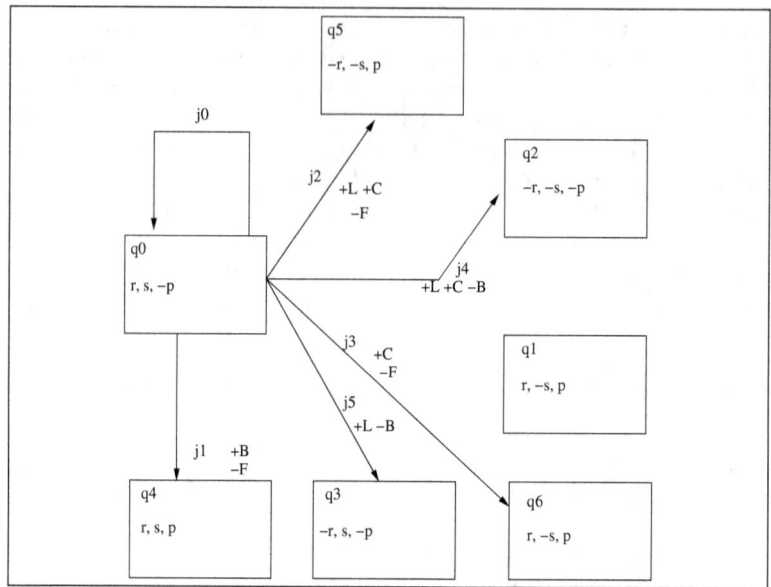

Figure 4: AATS+V for speed camera debate, as given in [Atkinson et al., 2011]

educated since either they do not speed or can control their vehicles better. The transitions from q_0 are shown in Table 2. We are not interested in what happens in subsequent states. The whole AATS+V is shown as Figure 4.

	$j0$	$j1$	$j2$
$q0$	⟨q0,_,_⟩	⟨q0,+B,-F⟩	⟨q5,+L+C,-F⟩
	$j3$	$j4$	$j5$
$q0$	⟨q6,+C,-F⟩	⟨q2,+L+C,-B⟩	⟨q3,+L,-B⟩

Table 2: Final Transition matrix.

On the basis of this model, it seems that introducing speed cameras is a reasonable proposal. The hope is that this would induce motorists to cut their speed, and that the number of accidents would fall, so that

j_2 is performed and q_5 is reached. This can be expressed in the form of AS1:

> The current state is: Many motorists break the speed limits $\wedge$ There are more road deaths than there should be $\wedge$ Privacy is respected.
>
> The action is: The government should introduce speed cameras.
>
> The destination state is: Speed limits are generally obeyed $\wedge$ The number of road deaths is acceptable $\wedge$ There are additional intrusions on privacy.
>
> The values promoted are: Life, Compliance

Note that the Government is expressing a preference for Life and Compliance over Freedom, which is demoted by the action.

This proposal can now be the subject of criticisms. For example,

CQ1 There might be disagreement as to the current situation: it would be possible to deny that many motorists broke the speed limits, or to claim that the number of road deaths was, in fact, acceptable.

CQ2 It might be argued that the action would not have the stated effects. Introducing speed cameras could reach q_4 or q_6 which would fail to promote one or both of our values.

CQ9 The action may demote a value. For example, freedom is demoted by the proposal.

CQ11 Other values can be promoted. There is no ground for this criticism in our example.

CQ13 It might be argued that the model is flawed and the proposed action is not possible. For example, it might be argued that the installation of speed cameras on the scale proposed was simply infeasible.

CQ17 Perhaps one or other of the agents will not perform the hoped for outcome. For example, it might be argued that reducing speed will not in fact reduce accidents and so the joint action will be j_3 leading to q_5 and so failing to promote life.

Using these methods to generate arguments, we can perform a full analysis. There are five arguments to perform an action from instantiating AS1.

PR1 We should perform G_1 to reach q_5 to promote L

PR2 We should perform G_1 to reach q_5 or q_6 to promote C

PR3 We should perform G_1 to reach q_4 to promote B

PR4 We should perform G_2 to reach q_2 or q_3 to promote L

PR5 We should perform G_2 to reach q_2 to promote C

Two arguments to refrain from an action are generated by a contrapositive form of AS1:

NPR1 We should not perform G_1 to avoid q_5 and q_6 since this would demote F

NPR2 We should not perform G_2 to avoid q_2 and q_3 since that would demote B

We accept that q_0 is the current state, and that other features of the model are correct. But we still have CQ17, which gives rise to three objections:

Ob1 Motorists may choose M_0 not M_1: attacks PR1, PR2 and PR5.

Ob2 Reducing speed may not reduce accidents and deaths. Attacks PR1.

Ob3 Motorists may choose M_1 not M_0: attacks PR3.

We now reach the final stage, when we weigh the merits and demerits of competing options, which requires us to identify the attacks between arguments. One source of attack is that a value is demoted: thus NPR1 attacks PR1, PR2 and PR3, and NPR2 attacks PR4 and PR5. Another source of attack, giving rise to symmetric attacks, is an alternative way of promoting the same value: thus PR1 and PR4 mutually attack, and PR2

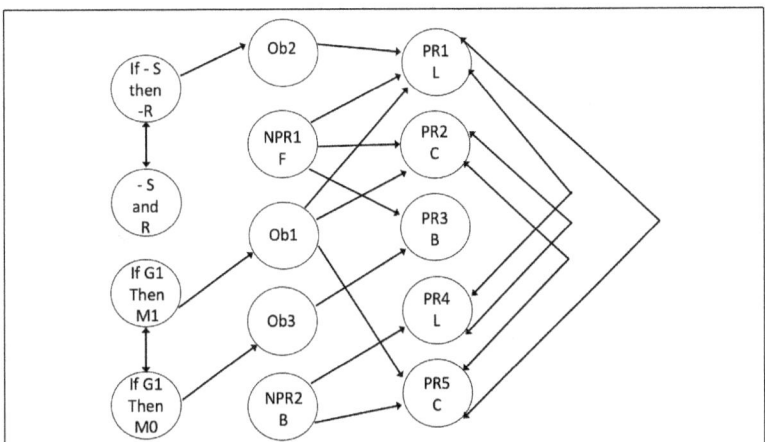

Figure 5: VAF for speed camera debate

and PR5 mutually attack. Finally we have different actions promoting different values: PR1 and PR5 and PR2 and PR4 mutually attack in this way. Finally we can have attacks which question the motive put forward: if PR1 is advanced to justify speed cameras, some may argue that the real expectation is that q_4 will be reached and that the real motive is to save money, rather than lives. This, however, does not challenge the action, but the justification, and we will not include these attacks here. Finally we have arguments representing the actual responses of motorists and nature to the introduction of speed cameras. These will form two two-cycles. We can now evaluate the arguments using a VAF. The VAF is shown in Figure 5.

On the left of the diagram are the two epistemic questions that need to be resolved. In default of anything better let us assume that, on the best evidence available, it is reasonable to expect that motorists will in fact reduce their speed, and that reducing speed will indeed lessen accidents and deaths. Having resolved these two cycles, we have answered the attacks from Ob1 and Ob2, while Ob3 is no longer attacked and will defeat PR3. When arguments are defeated, we can remove them and their attacks (and attacks on them) from the VAF to obtain the

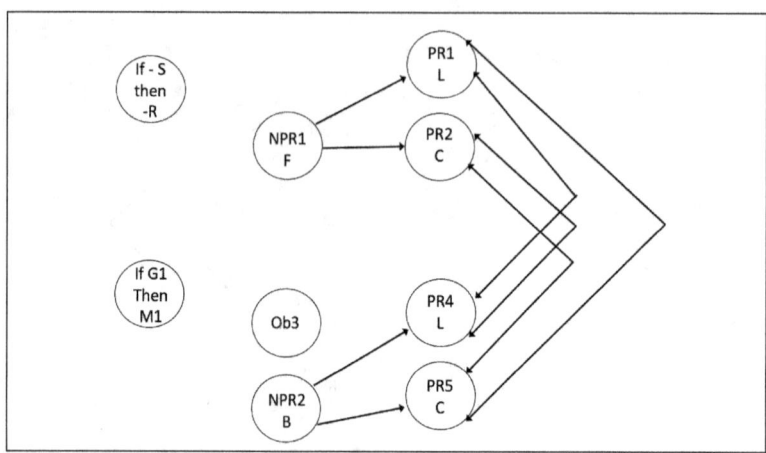

Figure 6: VAF for speed camera debate after epistemic choices

simpler VAF, as shown in Figure 6. Note that if we had made different assumptions about the epistemic questions then a different VAF, and ultimately a different position, would result from this simplification. When an argument is not defeated, but its attack is resisted by a preferred argument, we mark it as *ineffective*. We cannot ignore it, since we have no argument to defeat it, but we will not act upon it. There are no such arguments as yet, since we have not yet exercised preferences, but only chosen between different factual assumptions.

We next consider the two negative arguments based on PRAS2; once we have reached Figure 5 by resolving the epistemic questions, these are unattacked. These arguments will therefore succeed in defeating the arguments they attack unless the value of the attacked argument is preferred to that of the attacker. For NPR1 we must therefore consider Privacy/Freedom against Life to resolve the attack on PR1, and against Compliance to resolve the attack on PR2. A reasonable order would seem to be $L > F > C$: saying that intrusion on privacy is a necessary evil to save lives, but would not be acceptable simply to ensure compliance with speed limits without other gains. NPR1 thus becomes ineffective, which we show in the diagram by shading the argument node. This yields the VAF in Figure 7.

5 - VALUE-BASED ARGUMENTATION

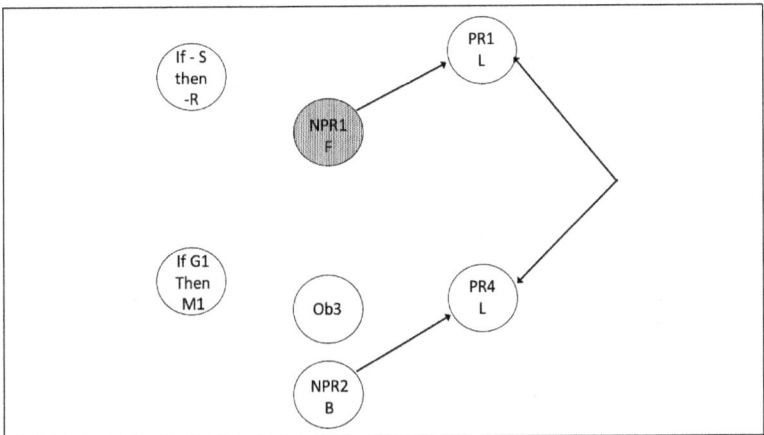

Figure 7: VAF with preferences $L > F > C$

The final question to resolve is whether PR4 can be accepted given NPR2: that is, can we prefer L to B? Unfortunately we are regarding budget as a hard constraint and so we must answer that $B > L$. This means that PR4 falls, leaving a preferred extension for an audience of $B > L > F > C$ comprising: the two factual assumptions, that motorists will reduce their speed, and that less speed means fewer accidents and deaths; the accepted course of action to install cameras to save lives; and two other considerations, that privacy must unfortunately be lessened (represented by the undefeated but ineffective argument), and that budgetary constraints preclude education as an alternative (represented by Obj3). Of course similar reasoning with different assumptions and different value orders would produce different results. If we assumed that motorists would continue to speed with the same value order, we would still install the cameras, but this time on the basis of PR3. If we made the original assumptions but used the value order $B > F > L > C$, we could do nothing, since we would have no way of saving lives without infringing privacy that we could afford, and if we had the value order $F > B > L > C$, we would educate motorists rather than install cameras.

Finally, if we prefer life to freedom, but money is available so that it was possible to prefer L to B, we would have two equally valued arguments, PR1 and PR4, neither attacked except by each other. In this case we should be inclined to choose PR4, since this would mean that the undefeated NPR1 would no longer have to coexist with an argument it attacks, so that it no longer need be regarded as ineffective[13]. In this way we are able to respect the value of privacy, even though F is not preferred to L.

Considerations of these varied alternatives allow us to see how the policy positions favoured depend very critically on how we rank values: the acceptability of a proposal will often depend on whether the public mood has been correctly judged in this respect.

4.4 Domain-specific application: behavioural economics

Value-based reasoning has also been used to explore two "games" used in behavioural economics, the *Dictator Game* [Engel, 2011] and the *Ultimatum Game* [Oosterbeek et al., 2004]. Classical economic theory assumes that people will behave in the manner of "economic man" described as follows by John Stuart Mill [Mill, 1844]:

> [Economics] is concerned with him solely as a being who desires to possess wealth, and who is capable of judging the comparative efficacy of means for obtaining that end.

However experiments performed in behavioural economics cast doubt on this key assumption. In the Dictator Game one player is given a sum of money and is then asked to give the second player as much or as little of it as he wishes. Classical economics would suggest that the player will give nothing, so maximising his own return. Experimentally, however, the results suggest otherwise: most players will give something to the other, sometimes as much as half. No studies report that the canonical model was observed. In one typical study [Forsythe et al., 1994], given $10 to distribute, 79% of participants gave away a positive amount, with 20% giving away half. The mode sum given away in that study was $3. The explanation is that other values come into play here: suggestions

---

[13]One disadvantage of the approach of [Amgoud and Vesic, 2011] in which arguments which resist their attacks also defeat them is that it fails to distinguish between defeated arguments and those which must be acknowledged even though not followed.

include concern for the other, simple generosity, concern for image (no one likes to be thought selfish). This game was thoroughly explored using value-based reasoning in [Bench-Capon *et al.*, 2012]: here we will discuss the more interesting Ultimatum Game.

In the Ultimatum Game the first player is also given a sum of money and asked to decide how much he wishes to offer to the other player. But this time the second player can refuse, in which case both get nothing. Now classical economics suggests that the first player will offer the smallest amount possible and the second player will accept it because, for economic man, anything is better than nothing. As with the Dictator Game, these expectations are not borne out in practice. For example, Nowak and colleagues report that the majority of proposers offer 40–50% and about half of responders reject offers below 30% [Nowak *et al.*, 2000]. These results are robust, and, with some variations, are replicated in all the many studies. Oosterbeek *et al.* [2004] report a meta-analysis of 37 papers with 75 results from Ultimatum Game experiments, which have an average of 40% offered to the responder. The experiments of Henrich *et al.* [2001], carried out over 15 small-scale societies in 12 countries over five continents, report mean offers between 26% and 58%, and note that in some societies there is considerable variation in which offers are rejected: however, again, none suggests that the canonical model is followed by those making and responding to offers. The Ultimatum Game was modelled in [Bench-Capon *et al.*, 2012] and [Atkinson and Bench-Capon, 2018].

First we must model the game as an AATS+V. Obviously the states must include the money held by the two agents. We also wish to represent the reactions of the two players. When the offer is made, it is important whether the second player perceives it as fair, or as insulting. We therefore use a proposition which is true when the second player is annoyed by the offer made. At the end of the game we can consider the reaction of the first player. In particular, if the offer is rejected, a first player who made an ungenerous offer is likely to feel regret that he did not offer more. We therefore use a fourth proposition to record whether the first player feels regret.

Next we turn to actions. Obviously we need that the first player can offer n% of the available sum to the second player and that the second player can accept or reject it. The reception the offer receives will, however, depend critically on the size of n. We will therefore distinguish four cases: where $n > 50$, where $n = 50$, where $n > 0$ but < 50 and where $n = 0$. We should also recognise that the two actions are not chosen simultaneously, and that the choice to accept or reject will depend on how the second player reacts to the offer of the proposer. We therefore introduce a third action, in which the second player chooses a threshold, t, above which he will regard the offer as just, and below which he will feel insulted. We will assume that $t > 0$ and $t < 50$, discounting players who will not be satisfied with even an equal share. While the second player accepts and rejects, the first player can do nothing. This gives the set of joint actions shown in Table 3.

Joint Action	Player 1	Player 2
j1	A1:Offer > 50	B1:Set $t < 50$
j2	A2:Offer 50	B1:Set $t < 50$
j3	A3:Offer $n < 50$ and > 0	B1:Set $t < n$
j4	A3:Offer $n < 50$ and > 0	B1:Set $t > n$
j5	A5:Offer $n = 0$	B1:Set $t > 0$
j6	A4:Do nothing	B2:accept
j7	A4:Do nothing	B3:reject

Table 3: Joint Actions in the Ultimatum Game

Now we must identify some values and the transitions which promote and demote them. First there is economic value, the money, which we shall call M. This can be promoted in respect both of player 1 (M1) and in respect of player 2 (M2). These values are promoted to different degrees according to the size of the player's share. From the literature it appears that some people seem to value fairness, which we shall call E for equality. This is either promoted or not. Third we have the value of generosity (G), which again has been identified as a motivation by various experimenters. Whereas M will be promoted to varying degrees according to the amount of money, E is either promoted or not. What

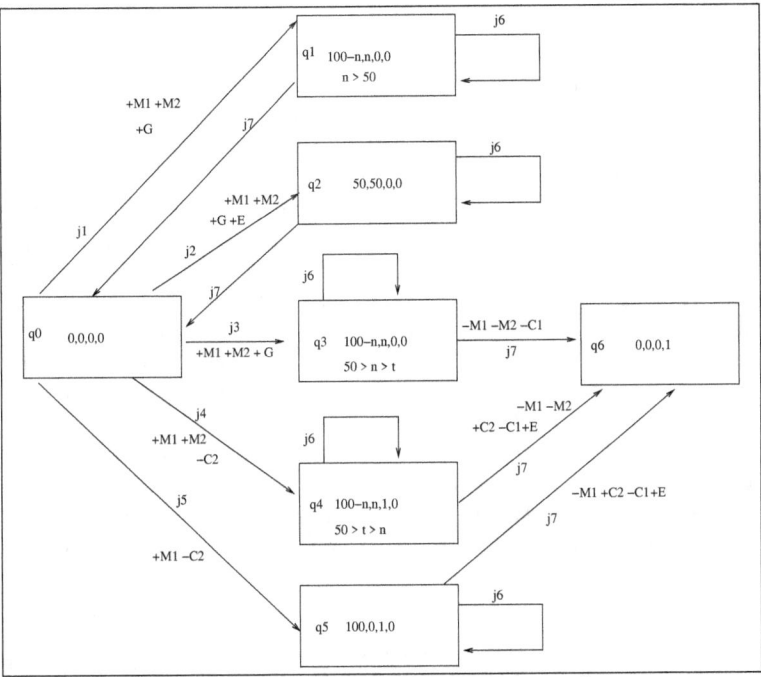

Figure 8: AATS+V for Ultimatum Game, as given in [Bench-Capon et al., 2012]

of G? Experimental evidence suggests that the impact of G does not increase as the amount given increases: we will therefore consider that G, like E, is either satisfied or not, and that any effect of the size of the gift is reflected in M2. Finally either player may be content with the outcome, and we represent this as C1 and C2. Again we will not model degrees of contentment. The AATS+V is shown in Figure 8.

Here will focus on the decision of the second player: the first player needs to think about this since the main aim of an offer is to have it accepted. The VAF for the second player is shown in Figure 9.

What the second player will do will depend on how it orders its values. Thus an offer above 50, or below 50 but above the second player's threshold of acceptability (states q_1 and q_3), will only be rejected if the player prefers equality to both its own and the other player's, money: $E > M1, M2$. Given the set of values we have used, we would expect any player to accept an offer of half the sum, since rejecting in q_2 promotes nothing and demotes money for both players. If the second player is insulted by a non-zero offer and so is in q_4, however, he has a choice of whether to punish the first player and so restore its own pride, or to take the money. Normally we would expect that the player will prefer its own money and its own contentment to the money and contentment of the other agent, and so require $M2 > C2 > M1, C1$ for acceptance, or $C2 > M2 > M1, C1$ for rejection. If E is preferred to both $M2$ and $C2$ the second player will also reject the offer, but here motivated by a desire for equality, rather than the insult. Finally if a zero offer is made we would expect rejection, either because of the insult, or because equality is desired. Indeed a zero offer will only be accepted if the second player prefers the others player's money or contentment to its own contentment: $C1, M1 > C2$. This would be an extreme example of altruism, and we would expect it to be rare. This ordering would also lead to acceptance in q_4.

What the second player will do is crucial. In [Atkinson and Bench-Capon, 2018] the Ultimatum Game was used to explore how an agent can take account of the expected actions of others. There the three actions of our above model were compressed into a set of joint actions as shown in Table 4.

There we say that player one can make a very high offer (vho) of more than half, an equal offer (eo) of half, a fair offer (fo) at the threshold for the second player, or a low offer (lo), below that threshold. All of these may be accepted or rejected by the second player, giving eight joint actions, promoting and demoting values as shown. Note that equality cannot be promoted, since the initial state is one of equality. From this table we can see why most players will make at least a fair offer: only if the first player is desperate to "get one over" the other will a low offer be made, since only a low offer promotes $C1$ but carries with it a high probability of demoting $M1$ and $C1$. How high the offer will go depends on how much the player values the wealth and happiness of the other, and whether it values a feeling of generosity.

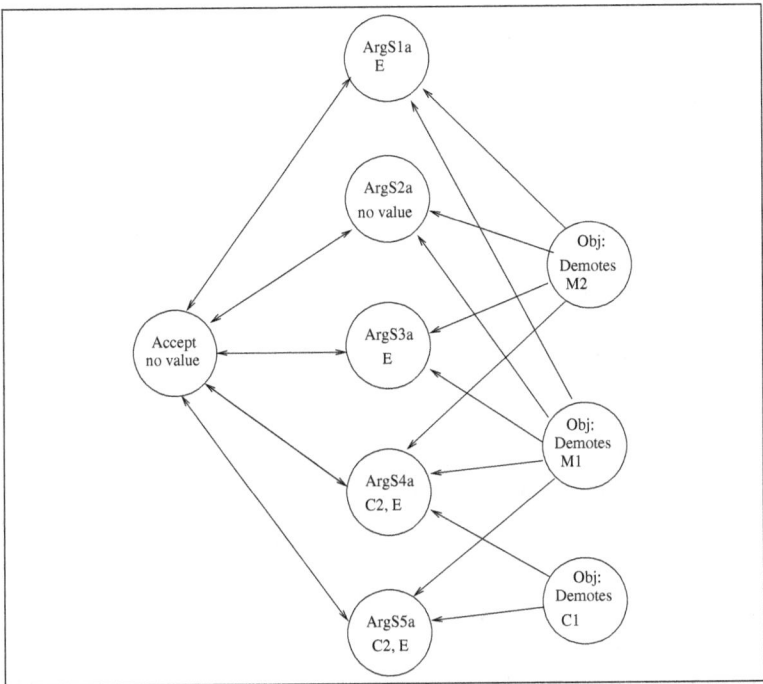

Figure 9: VAF for second player in Ultimatum Game, as given in [Bench-Capon et al., 2012]

In [Henrich et al., 2001] 15 small scale societies from various parts of the world were studied, and it emerged that different groups behave differently. It was suggested that the different societies' actions in the Ultimatum Game could be accounted for in terms of the degree of cooperation and degree of commercial exchange found in daily life. We can relate these characteristics to a value profile. Suppose we associate the value of generosity with the cooperative groups such as the whale hunting Lamelara, and the recognition of C2 (the need to maintain good relations with the other) with commercial exchange. Those who do not engage in cooperative or exchange activities, we term solitary. In [Atkinson and Bench-Capon, 2018] it was found that using value profiles

Joint Action	Proposal	Response	Promoted	Demoted
j1	vho	accept	M1,M2,G, C2	E
j2	vho	reject	G	M1,C1
j3	eo	accept	M1,M2,G,C2	
j4	eo	reject	G	M1,C1
j5	fo	accept	M1,M2	E
j6	fo	reject		M1
j7	lo	accept	M1,M2,C1	E,C2
j8	lo	reject		M1,C1

Table 4: Value promotion and demotion in the Ultimatum Game

representing these three life styles predicted offers and rejections that are very close to the empirical results of [Henrich et al., 2001].

4.5 Other applications

In addition, value-based reasoning has been demonstrated using examples in medicine [Atkinson et al., 2006c], health advice [van der Weide et al., 2009] and [Di Tullio and Grasso, 2011], ontology alignment [Laera et al., 2007] and [Trojahn et al., 2007], an account of the emergence of norms [Bench-Capon and Modgil, 2017] and discussions of ethics [Atkinson and Bench-Capon, 2008]. Most recently in [Bench-Capon, 2020] value-based reasoning has been used as the basis of a novel computational account of virtue ethics in agent systems.

In general, value-based reasoning can be used to model argumentation and reasoning in any domain where the direction of fit is from an agent's desires or needs to the world; any situation in which reasoning about action is required. Such reasoning is pervasive, covering many of the most important aspects of life: from everyday choices such as where to eat or how to travel, to law and politics, and fundamental questions of how we should live.

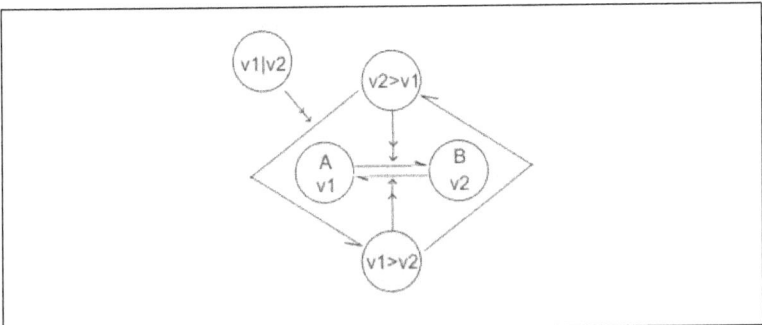

Figure 10: Value-based conflict in extended argumentation framework, as given in [Modgil and Bench-Capon, 2008]

5 Value-based reasoning at the meta-level

Modgil [2009] introduced an elegant and general way of handling preferences: instead of assigning different strengths to arguments, he permitted attacks to themselves be attacked. Such frameworks he termed *Extended Argumentation Frameworks* (EAF). This meant that an attack was unsuccessful not according to whether it was attacking a stronger argument, but according to whether it was itself defeated by some other argument.

The relation between VAFs and EAFs was explored in [Modgil and Bench-Capon, 2008]. A conflict between two arguments is shown as an EAF in Figure 10. There the value preferences are represented as arguments, attacking attacks which require the other preference to succeed. These value preference arguments will, of course, mutually attack. The desired audience represented as an ordering on the values will attack one of these attacks, resolving the framework.

Frameworks of the sort shown in Figure 10 can now be rewritten as standard Dung-style argumentation frameworks using meta level arguments. If we replace arguments by the fact that they are acceptable, e.g. A by $A\ holds$, and introduce arguments that arguments do not hold ($\overline{A}$) and that one argument defeats another ($\overrightarrow{AB}$), we can rewrite Figure 10 as Figure 11.

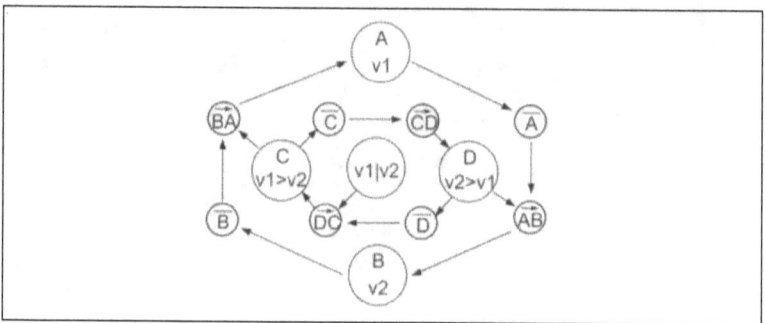

Figure 11: Meta level rewriting of Figure 10 to get a standard AF, as given in [Modgil and Bench-Capon, 2008]

Now an attack $\overrightarrow{AB}$ may fail in two ways: either A may be defeated so that $\overline{A}$ defeats it, or there may be a preference argument that defeats it. There are clear simplifications in this rewriting in that standard AFs can be used instead of the more complicated VAFs and EAFs. The use of EAFs in value-based reasoning was discussed in [Bench-Capon and Atkinson, 2009], and its application to the representation of norms in [Bench-Capon and Modgil, 2019]. A full discussion of meta level argumentation can be found in [Modgil and Bench-Capon, 2011].

6 Concluding remarks

In this chapter we have discussed value-based reasoning. Philosophically it models reasoning where the direction of fit is from an agent's desires to the world: that is where an agent is choosing how to act in order to promote its values, and this covers all domains involving an element of practical reasoning, reasoning about what should be done.

Value-based reasoning was originally presented as a form of abstract argumentation extending Dung's original framework by giving arguments the additional property of *promoting a value*, and evaluating the arguments according to an ordering on those values.

Although there are some theoretical results, the main motivation for value-based reasoning was always applications, especially law where

[Berman and Hafner, 1993] had drawn attention to the role of values in legal decisions, and [Bench-Capon and Sartor, 2003] had incorporated values into theories of case law for particular areas of law. This emphasis on applications was facilitated by the development of a means of doing structured value-based argumentation, based on an argumentation scheme and critical questions semantically underpinned by a form of state transition diagram, AATS+V.

Because of the importance of applications, we have devoted much of this chapter to a detailed discussion of four application domains: general problem solving, law, e-participation and behavioural economics.

Extended argumentation frameworks [Modgil, 2009] offer a means of generalising argumentation involving preferences. Value-based argumentation frameworks fit very well with this framework, since they can be systematically rewritten as standard AFs using meta level arguments describing the status of arguments in the VAF, the value preferences, and the audience concerned. Moving to meta level argumentation, however, does not remove the need for structured value-based argumentation, which is still needed to generate the arguments and attacks. This combination is used in [Bench-Capon and Atkinson, 2009].

The theory of value-based argumentation is fairly well understood, but its potential for modelling applications continues. As a means of representing problems in areas where values are crucial, such as ethics, law and politics, value-based reasoning offers a tried and tested solution.

Acknowledgments

In the many years that we have been working on computational models of value-based argument, we have collaborated with a wide range of people spanning many of the topics covered in this chapter. We are grateful to all colleagues with whom we have worked to progress research in this area: Latifa Al-Abdulkarim, Floris Bex, Elizabeth Black, Dan Cartwright, Alison Chorley, Sylvie Doutre, Paul Dunne, Floriana Grasso, Yanko Kirchev, Loredana Laera, Peter McBurney, Rolando Medellin-Gasque, Sanjay Modgil, Fahd Saud Nawwab, Samer Nofal, Henry Prakken, Giovanni Sartor, Valentina Tamma, Doug Walton, Maya Wardeh, Adam Wyner.

References

[Aleven and Ashley, 1995] Vincent Aleven and Kevin D Ashley. Doing things with factors. In *Proceedings of the 5th International Conference on Artificial Intelligence and Law*, pages 31–41, 1995.

[Alur et al., 2002] Rajeev Alur, Thomas Henzinger, and Orna Kupferman. Alternating-time temporal logic. *Journal of the ACM (JACM)*, 49(5):672–713, 2002.

[Amgoud and Cayrol, 1998] Leila Amgoud and Claudette Cayrol. On the acceptability of arguments in preference-based argumentation. In *Proceedings of the Fourteenth Conference on Uncertainty in Artificial Intelligence*, pages 1–7. Morgan Kaufmann Publishers Inc., 1998.

[Amgoud and Cayrol, 2002] Leila Amgoud and Claudette Cayrol. A reasoning model based on the production of acceptable arguments. *Annals of Mathematics and Artificial Intelligence*, 34(1-3):197–215, 2002.

[Amgoud and Vesic, 2011] Leila Amgoud and Srdjan Vesic. A new approach for preference-based argumentation frameworks. *Annals of Mathematics and Artificial Intelligence*, 63(2):149–183, 2011.

[Anscombe, 1978] G.E.M. Anscombe. On practical reasoning. In Joseph Raz, editor, *Practical Reasoning*, pages 33–45. Oxford University Press, 1978.

[Atkinson and Bench-Capon, 2005] Katie Atkinson and Trevor Bench-Capon. Legal case-based reasoning as practical reasoning. *Artificial Intelligence and Law*, 13(1):93–131, 2005.

[Atkinson and Bench-Capon, 2007a] Katie Atkinson and Trevor Bench-Capon. Action-based alternating transition systems for arguments about action. In *AAAI*, volume 7, pages 24–29, 2007.

[Atkinson and Bench-Capon, 2007b] Katie Atkinson and Trevor Bench-Capon. Argumentation and standards of proof. In *Proceedings of the 11th international conference on Artificial intelligence and law*, pages 107–116, 2007.

[Atkinson and Bench-Capon, 2007c] Katie Atkinson and Trevor Bench-Capon. Practical reasoning as presumptive argumentation using Action based Alternating Transition Systems. *Artificial Intelligence*, 171(10-15):855–874, 2007.

[Atkinson and Bench-Capon, 2008] Katie Atkinson and Trevor Bench-Capon. Addressing moral problems through practical reasoning. *Journal of Applied Logic*, 6(2):135–151, 2008.

[Atkinson and Bench-Capon, 2014] Katie Atkinson and Trevor JM Bench-Capon. Taking the long view: Looking ahead in practical reasoning. In *COMMA*, pages 109–120, 2014.

[Atkinson and Bench-Capon, 2018] Katie Atkinson and Trevor Bench-Capon. Taking account of the actions of others in value-based reasoning. *Artificial Intelligence*, 254:1–20, 2018.

[Atkinson et al., 2004] Katie Atkinson, Trevor Bench-Capon, and Peter McBurney. Justifying practical reasoning. In *Proceedings of the Fourth International Workshop on Computational Models of Natural Argument (CMNA 2004)*, pages 87–90, 2004.

[Atkinson et al., 2006a] Katie Atkinson, Trevor Bench-Capon, and Peter McBurney. Computational representation of practical argument. *Synthese*, 152(2):157–206, 2006.

[Atkinson et al., 2006b] Katie Atkinson, Trevor Bench-Capon, and Peter McBurney. Parmenides: facilitating deliberation in democracies. *Artificial Intelligence and Law*, 14(4):261–275, 2006.

[Atkinson et al., 2006c] Katie Atkinson, Trevor Bench-Capon, and Sanjay Modgil. Argumentation for decision support. In *International Conference on Database and Expert Systems Applications*, pages 822–831. Springer, 2006.

[Atkinson et al., 2011] Katie M Atkinson, Trevor JM Bench-Capon, Dan Cartwright, and Adam Z Wyner. Semantic models for policy deliberation. In *Proceedings of the 13th International Conference on Artificial Intelligence and Law*, pages 81–90, 2011.

[Atkinson et al., 2012] Katie Atkinson, Priscilla Bench-Capon, and Trevor Bench-Capon. A strategy for efficient persuasion dialogues. In *International Conference on Agents and Artificial Intelligence*, pages 332–347. Springer, 2012.

[Atkinson et al., 2013] Katie Atkinson, Trevor Bench-Capon, and Douglas Walton. Distinctive features of persuasion and deliberation dialogues. *Argument & Computation*, 4(2):105–127, 2013.

[Atkinson, 2005] Katie Atkinson. *What Should We Do?* PhD thesis, University of Liverpool, 2005.

[Atkinson, 2012] Katie Atkinson. Introduction to special issue on modelling Popov v. Hayashi. *Artificial Intelligence and Law*, 20(1):1–14, 2012.

[Bench-Capon and Atkinson, 2009] Trevor Bench-Capon and Katie Atkinson. Abstract argumentation and values. In *Argumentation in artificial intelligence*, pages 45–64. Springer, 2009.

[Bench-Capon and Modgil, 2009] Trevor Bench-Capon and Sanjay Modgil. Case law in Extended Argumentation Frameworks. In *Proceedings of the 12th International Conference on Artificial Intelligence and Law*, pages 118–127. ACM, 2009.

[Bench-Capon and Modgil, 2017] Trevor Bench-Capon and Sanjay Modgil.

Norms and value based reasoning: justifying compliance and violation. *Artificial Intelligence and Law*, 25(1):29–64, 2017.

[Bench-Capon and Modgil, 2019] Trevor Bench-Capon and Sanjay Modgil. Norms and extended argumentation frameworks. In *Proceedings of the Seventeenth International Conference on Artificial Intelligence and Law*, pages 174–178, 2019.

[Bench-Capon and Rissland, 2001] Trevor Bench-Capon and Edwina Rissland. Back to the future: Dimensions revisited. In *Legal Knowledge and Information Systems - JURIX 2001*, pages 41–52, 2001.

[Bench-Capon and Sartor, 2001] Trevor Bench-Capon and Giovanni Sartor. Theory based explanation of case law domains. In *Proceedings of the 8th international conference on Artificial intelligence and law*, pages 12–21, 2001.

[Bench-Capon and Sartor, 2003] Trevor Bench-Capon and Giovanni Sartor. A model of legal reasoning with cases incorporating theories and values. *Artificial Intelligence*, 150(1-2):97–143, 2003.

[Bench-Capon et al., 2005] Trevor Bench-Capon, Katie Atkinson, and Alison Chorley. Persuasion and value in legal argument. *Journal of Logic and Computation*, 15(6):1075–1097, 2005.

[Bench-Capon et al., 2007] Trevor Bench-Capon, Sylvie Doutre, and Paul Dunne. Audiences in argumentation frameworks. *Artificial Intelligence*, 171(1):42–71, 2007.

[Bench-Capon et al., 2012] Trevor Bench-Capon, Katie Atkinson, and Peter McBurney. Using argumentation to model agent decision making in economic experiments. *Autonomous Agents and Multi-Agent Systems*, 25(1):183–208, 2012.

[Bench-Capon et al., 2015] Trevor Bench-Capon, Katie Atkinson, and Adam Wyner. Using argumentation to structure e-participation in policy making. In *Transactions on Large-Scale Data-and Knowledge-Centered Systems XVIII*, pages 1–29. Springer, 2015.

[Bench-Capon, 2002a] Trevor Bench-Capon. Representation of case law as an argumentation framework. *Proceedings of JURIX 2002*, pages 103–112, 2002.

[Bench-Capon, 2002b] Trevor Bench-Capon. Value based argumentation frameworks. *arXiv preprint cs/0207059. Originally presented at NMR 2002*, 2002.

[Bench-Capon, 2002c] Trevot Bench-Capon. Agreeing to differ: modelling persuasive dialogue between parties with different values. *Informal Logic*, 22(3), 2002.

[Bench-Capon, 2003a] Trevor Bench-Capon. Persuasion in practical argument using value-based argumentation frameworks. *Journal of Logic and Compu-*

tation, 13(3):429–448, 2003.

[Bench-Capon, 2003b] Trevor Bench-Capon. Try to see it my way: Modelling persuasion in legal discourse. *Artificial Intelligence and Law*, 11(4):271–287, 2003.

[Bench-Capon, 2012] Trevor Bench-Capon. Representing Popov v Hayashi with dimensions and factors. *Artificial Intelligence and Law*, 20(1):15–35, 2012.

[Bench-Capon, 2014] Trevor Bench-Capon. Dilemmas and paradoxes: cycles in argumentation frameworks. *Journal of Logic and Computation*, 26(4):1055–1064, 2014.

[Bench-Capon, 2017] Trevor Bench-Capon. HYPO's legacy: Introduction to the virtual special issue. *Artificial Intelligence and Law*, 25(2):1–46, 2017.

[Bench-Capon, 2020] Trevor Bench-Capon. Ethical approaches and autonomous systems. *Artificial Intelligence*, page 103239, 2020.

[Berman and Hafner, 1993] Donald H Berman and Carole D Hafner. Representing teleological structure in case-based legal reasoning: the missing link. In *Proceedings of the 4th International Conference on Artificial Intelligence and Law*, pages 50–59. ACM, 1993.

[Cartwright and Atkinson, 2009] Dan Cartwright and Katie Atkinson. Using computational argumentation to support e-participation. *IEEE Intelligent Systems*, 24(5):42–52, 2009.

[Chorley and Bench-Capon, 2005a] Alison Chorley and Trevor Bench-Capon. Agatha: Using heuristic search to automate the construction of case law theories. *Artificial Intelligence and Law*, 13(1):9–51, 2005.

[Chorley and Bench-Capon, 2005b] Alison Chorley and Trevor Bench-Capon. An empirical investigation of reasoning with legal cases through theory construction and application. *Artificial Intelligence and Law*, 13(3-4):323–371, 2005.

[Di Tullio and Grasso, 2011] Eugenio Di Tullio and Floriana Grasso. A model for a motivational system grounded on value based abstract argumentation frameworks. In *International Conference on Electronic Healthcare*, pages 43–50. Springer, 2011.

[Dunne and Bench-Capon, 2002] Paul E. Dunne and Trevor J. M. Bench-Capon. Coherence in finite argument systems. *Artif. Intell.*, 141(1/2):187–203, 2002.

[Dunne and Bench-Capon, 2004] Paul E. Dunne and Trevor J. M. Bench-Capon. Complexity in value-based argument systems. In José Júlio Alferes and João Alexandre Leite, editors, *Logics in Artificial Intelligence, 9th European Conference, JELIA 2004, Lisbon, Portugal, September 27-30, 2004,*

Proceedings, volume 3229 of *Lecture Notes in Computer Science*, pages 360–371. Springer, 2004.

[Dunne, 2010] Paul E. Dunne. Tractability in value-based argumentation. In Pietro Baroni, Federico Cerutti, Massimiliano Giacomin, and Guillermo Ricardo Simari, editors, *Computational Models of Argument: Proceedings of COMMA 2010, Desenzano del Garda, Italy, September 8-10, 2010*, volume 216 of *Frontiers in Artificial Intelligence and Applications*, pages 195–206. IOS Press, 2010.

[Engel, 2011] Christoph Engel. Dictator games: a meta study. *Experimental Economics*, 14(4):583–610, 2011.

[Finkelman, 2001] Paul Finkelman. Fugitive baseballs and abandoned property: Who owns the home run ball. *Cardozo L. Rev.*, 23:1609, 2001.

[Forsythe et al., 1994] Robert Forsythe, Joel L Horowitz, Nathan E Savin, and Martin Sefton. Fairness in simple bargaining experiments. *Games and Economic behavior*, 6(3):347–369, 1994.

[Greenwood et al., 2003] Katie Greenwood, Trevor Bench-Capon, and Peter McBurney. Towards a computational account of persuasion in law. In *Proceedings of the 9th International Conference on Artificial Intelligence and Law*, pages 22–31. ACM, 2003.

[Henrich et al., 2001] Joseph Henrich, Robert Boyd, Samuel Bowles, Colin Camerer, Ernst Fehr, Herbert Gintis, and Richard McElreath. In search of homo economicus: behavioral experiments in 15 small-scale societies. *The American Economic Review*, 91(2):73–78, 2001.

[Horty and Bench-Capon, 2012] John Horty and Trevor Bench-Capon. A factor-based definition of precedential constraint. *Artificial Intelligence and Law*, 20(2):181–214, 2012.

[Kenny, 1978] A.J.P. Kenny. Practical reasoning and rational appetite. In Joseph Raz, editor, *Practical Reasoning*, pages 63–80. Oxford University Press, 1978.

[Kirchev et al., 2019] Yanko Kirchev, Katie Atkinson, and Trevor J. M. Bench-Capon. Demonstrating the distinctions between persuasion and deliberation dialogues. In Max Bramer and Miltos Petridis, editors, *Artificial Intelligence XXXVI - 39th SGAI International Conference on Artificial Intelligence, AI 2019, Cambridge, UK, December 17-19, 2019, Proceedings*, volume 11927 of *Lecture Notes in Computer Science*, pages 93–106. Springer, 2019.

[Laera et al., 2007] Loredana Laera, Ian Blacoe, Valentina Tamma, Terry Payne, Jerôme Euzenat, and Trevor Bench-Capon. Argumentation over ontology correspondences in mas. In *Proceedings of the 6th international joint conference on Autonomous agents and multiagent systems*, pages 1–8, 2007.

[Lindahl and Odelstad, 2004] Lars Lindahl and Jan Odelstad. Normative positions within an algebraic approach to normative systems. *Journal of Applied Logic*, 2(1):63–91, 2004.

[Maslow, 1943] Abraham Harold Maslow. A theory of human motivation. *Psychological review*, 50(4):370, 1943.

[Mill, 1844] John Stuart Mill. On the definition of political economy, and on the method of investigation proper to it. *Essays on some unsettled questions of Political Economy*, page 326, 1844.

[Modgil and Bench-Capon, 2008] Sanjay Modgil and Trevor Bench-Capon. Integrating object and meta-level value based argumentation. *Computational Models of Argument: Proceedings of COMMA 2008*, 172:240, 2008.

[Modgil and Bench-Capon, 2011] Sanjay Modgil and Trevor Bench-Capon. Metalevel argumentation. *Journal of Logic and Computation*, 21(6):959–1003, 2011.

[Modgil, 2009] Sanjay Modgil. Reasoning about preferences in argumentation frameworks. *Artificial Intelligence*, 173(9-10):901–934, 2009.

[Nofal et al., 2014] Samer Nofal, Katie Atkinson, and Paul E. Dunne. Algorithms for decision problems in argument systems under preferred semantics. *Artificial Intelligence*, 207:23–51, 2014.

[Nowak et al., 2000] Martin A Nowak, Karen M Page, and Karl Sigmund. Fairness versus reason in the ultimatum game. *Science*, 289(5485):1773–1775, 2000.

[Oosterbeek et al., 2004] Hessel Oosterbeek, Randolph Sloof, and Gijs Van De Kuilen. Cultural differences in ultimatum game experiments: Evidence from a meta-analysis. *Experimental Economics*, 7(2):171–188, 2004.

[Perelman and Olbrechts-Tyteca, 1969] Chaim Perelman and Lucie Olbrechts-Tyteca. The new rhetoric: a treatise on argumentation, trans. *John Wilkinson and Purcell Weaver (Notre Dame, IN: University of Notre Dame Press, 1969)*, 190:411–12, 1969.

[Perelman, 1980] Chaim Perelman. *Justice, law, and argument: Essays on moral and legal reasoning*. Reidal, Dordrecht, 1980.

[Rissland and Ashley, 1987] Edwina L Rissland and Kevin D Ashley. A case-based system for Trade Secrets law. In *Proceedings of the 1st International Conference on Artificial Intelligence and Law*, pages 60–66. ACM, 1987.

[Schwartz, 1992] Shalom H Schwartz. Universals in the content and structure of values: Theoretical advances and empirical tests in 20 countries. *Advances in experimental social psychology*, 25(1):1–65, 1992.

[Searle, 2003] John R Searle. *Rationality in action*. MIT press, 2003.

[Trojahn et al., 2007] Cássia Trojahn, Paulo Quaresma, and Renata Vieira.

An extended value-based argumentation framework for ontology mapping with confidence degrees. In *International Workshop on Argumentation in Multi-Agent Systems*, pages 132–144. Springer, 2007.

[van der Weide et al., 2009] Thomas L van der Weide, Frank Dignum, J-J Ch Meyer, Henry Prakken, and GAW Vreeswijk. Practical reasoning using values. In *International Workshop on Argumentation in Multi-Agent Systems*, pages 79–93. Springer, 2009.

[Walton and Krabbe, 1995] Douglas Walton and Erik Krabbe. *Commitment in dialogue: Basic concepts of interpersonal reasoning.* SUNY press, 1995.

[Walton, 1996] Douglas Walton. *Argumentation schemes for presumptive reasoning.* Lawrence Erlbaum Associates, 1996.

[Wardeh et al., 2013] Maya Wardeh, Adam Wyner, Katie Atkinson, and Trevor Bench-Capon. Argumentation based tools for policy-making. In *Proceedings of the Fourteenth International Conference on Artificial Intelligence and Law*, pages 249–250, 2013.

[Wooldridge and van der Hoek, 2005] M. Wooldridge and W. van der Hoek. On obligations and normative ability: Towards a logical analysis of the social contract. *Journal of Applied Logic*, 3:396–420, 2005.

[Wyner et al., 2007] Adam Wyner, Trevor Bench-Capon, and Katie Atkinson. Arguments, values and baseballs: Representation of Popov v. Hayashi. In *Proceedings of JURIX 2007*, pages 151–160, 2007.

[Wyner et al., 2012] Adam Wyner, Maya Wardeh, Trevor Bench-Capon, and Katie Atkinson. A model-based critique tool for policy deliberation. In *JURIX*, pages 167–176, 2012.

CHAPTER 6

WEIGHTED ARGUMENTATION

STEFANO BISTARELLI

Department of Mathematics and Computer Science, University of Perugia, Italy
`stefano.bistarelli@unipg.it`

FRANCESCO SANTINI

Department of Mathematics and Computer Science, University of Perugia, Italy
`francesco.santini@unipg.it`

Abstract

When dealing with Abstract Argumentation, having preference values on arguments/attacks clearly brings more information to a framework, which can be considered as a directed graph. One of the advantages is the possibility to define a different notion of defence, checking also if the associated preference is stronger than the preference of the considered attack. In the real-world, such values can be represented by "likes" in social-networks, or generic votes in favour of attacks. We focus on qualitative/quantitative preference values on attacks, which indicate their (relative) strength and can measure an argument-pair inconsistency degree. Once assembled, also by moving values from arguments to attacks, it is then possible to redefine semantics, relax the notion of weighted acceptability, and check well-known properties as in Dung's frameworks, e.g., if a framework is well-founded.

1 Introduction and Motivations

In the approach presented in this chapter, we recognise that not all arguments or attacks in an *Abstract Argumentation Framework* (*AF*) [Dung, 1995] are equal in strength. As we will see in the following, considering these strength degrees under the form of "weights" (or values) brings a new perspective when searching for collections of arguments to be considered as collectively acceptable. The reason is that a weighted framework overcomes some natural limitations of the classical frameworks elaborated by P.M. Dung [Dung, 1995], where several extensions for a given semantics may be provided, but with nothing to distinguish between them. If a attacks b and vice-versa, both $\{a\}$ and $\{b\}$ are admissible solutions, but if the attack from a to b is stronger, then $\{b\}$ may not be considered as acceptable anymore. Consider the example where:

a: The car is too expensive for us: we cannot afford it.
b: This is the car I always dreamed of: we should buy it.

In this case, according to Dung's approach, both arguments are credulously accepted, neither is sceptically accepted, and the grounded extension is empty.[1] Hence, the classical analysis is not very useful, and the degree of uncertainty is high: no argument can be always selected without any doubt. However, from what arguments imply in the example above, we can probably derive that attacks are not equal in strength. Reality rules over dreams, and a rational agent cannot go in the direction of buying the desired car because of its price: $\{a\}$ is preferable than $\{b\}$, at least without any additional argument suggesting for a loan. If the attack from a to b is stronger than the inverse attack, then uncertainty can be reduced and $\{a\}$ can be "more acceptable" than $\{b\}$.

In its original formulation [Dung, 1995], a framework *AF* is represented by a pair $\langle Ar, att \rangle$ consisting of respectively a set of arguments and a binary relationship of attack defined among them. Given a framework, it is possible to examine the question on which set(s) of arguments

---

[1] The idea behind the grounded extension is to accept only those arguments that one cannot avoid to accept, to reject only the arguments that one cannot avoid to reject, and abstaining as much as possible. Hence, it represents the most sceptical (or "least committed") semantics among those based on complete extensions [Baroni et al., 2018, Chapter 4].

can be accepted, hence collectively surviving the conflict defined by *att*. Answering this question corresponds to defining an Argumentation semantics. The key idea behind *extension-based* semantics is to identify some sets of arguments (called *extensions*) that survive the conflict "together".

In order to better frame the scope of this chapter, we first clarify the meaning of weights (on attacks) for us. As pointed out in [Dunne et al., 2011], in this chapter we consider three possible interpretations: weights can be seen as *i)* the number of votes in favour of an attack, *ii)* as a measure of the inconsistency of argument-pairs, or *iii)* as rankings of different types of attack. Note that the first interpretation lays a link between Argumentation and Social Choice theory [Dunne et al., 2011]. On the other hand, we will not here survey frameworks where weights represent something different from the above approaches, as for example probabilities: this is for example the topic of Chapter 7 of this handbook [Hunter et al., 2021]. In practice, we here consider weights as a basic strength-value, which may represent various issues like votes provided by users [Egilmez et al., 2013], importance degree of a value it promotes [Bench-Capon, 2003], or trustworthiness of its source [da Costa Pereira et al., 2011]. Therefore, in all such divergent cases, the basic strength may be expressed by a numerical value, leading to *Weighted Argumentation Frameworks*, or simply *WAF*s in the following of the chapter.

The approaches we survey implement both quantitative (e.g., [Dunne et al., 2011], [Kaci and Labreuche, 2011], [Coste-Marquis et al., 2012], and [Bistarelli et al., 2018b]) and qualitative approaches (e.g., [Martínez et al., 2008], [Cayrol et al., 2010], and [Bistarelli et al., 2018b]).[2] Quantitative approaches require to specify a numeric value for each attack, that is 0.7 or 9, for example. Qualitative frameworks express preferences via generic qualitative (usually partial) preference relations over attacks.

There is an established trend in the literature on the formalisation of Argumentation towards considering the strength of arguments/attacks: a summary of the bibliography is given also in Section 2. A shared

---

[2]Note that by using a parametric algebraic structure (i.e., c-semirings, see Section 4), the work in [Bistarelli et al., 2018b] is capable or representing both the preference systems.

motivation among some of these proposals is the observation that not all the arguments are equal, and that the relative strength of the arguments needs to be taken into account somehow. In this chapter, we focus on weighted attacks: if there is an attack $att(a,b)$, then a relation $w(a,b) = s$ returns the weight (s) associated with that directed attack.[3] Hence the definition of an AF needs to include a further relation w. Other works (e.g., Chapters 3 and 5 of this handbook [Kaci et al., 2021; Atkinson and Bench-Capon, 2021]) focus on preference values associated with arguments instead, and offer a different view on a strictly related problem.

The idea of explicitly adding weights to attacks, instead of arguments, was proposed by [Barringer et al., 2005] for the first time. Considering a strength value on attacks allows us to consider a richer and more finer-grained model of frameworks than having weights on arguments. Richer because attacks are usually more than arguments in a framework, i.e., up to $|Ar|^2$ if a framework is a complete directed graph (or *digraph*) with self-attacks: consequently, weights are more than weights on arguments, and such a model can provide richer details. The weight-on-attacks model is also finer-grained, since it is possible to derive a finer definition of acceptability for an extension, which specifies a required level of defence of any argument in an extension: the strength of an argument impacts on all of its neighbours, while the strength on attacks only impacts on the adjacent corresponding argument [Dunne et al., 2011].

Having motivated the use of Weighted Abstract Argumentation as an extension of the model designed by P.M. Dung, and the use of weights on attacks, Figure 1 reports an example of WAF as we intend it in this chapter.

Outline of the Chapter. This chapter is structured as follows: Section 2 briefly surveys the related approaches. Most of these works mainly concern priority-based rules or values associated with arguments. Even if outside the immediate scope of this chapter, these approaches need to be mentioned because they are strictly related to what presented in the

---

[3] With the purpose to lighten the notation, we will use $w(a,b)$ in place of $w(att(a,b))$.

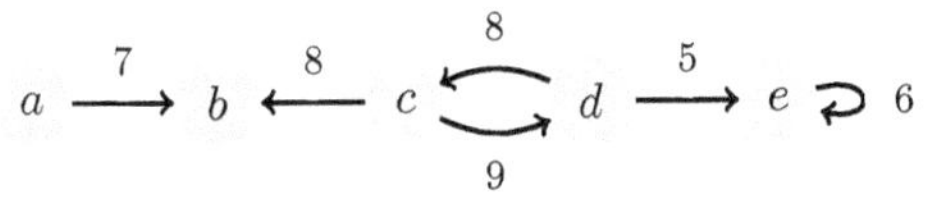

$$a \xrightarrow{7} b \xleftarrow{8} c \overset{8}{\underset{9}{\rightleftarrows}} d \xrightarrow{5} e \rightleftarrows 6$$

Figure 1: An example of WAF.

following.

In Section 3 we present the main techniques in the literature to translate preference values on arguments into weights on attacks: since this chapter is focused on the latter, in this way we suggest a possible bridge towards such two families of frameworks.

Section 4 presents the most important valued-structures where to draw weights from and perform operations. These systems are very specific to single proposals (e.g., weights are in $\mathbb{R}$), or they are more general and can be parametric, such as c-semirings [Bistarelli et al., 1997].[4]

Section 5 introduces how novel definitions of acceptability can be derived when using weights. As advanced in Section 1, the presence of numerical values enriches the model conceived in [Dung, 1995] and allows for reducing uncertainty, simply because more information than just arguments and attacks is embedded in the model. An argument is now defended if the strength of the defence is stronger than the attack strength, by using different aggregation functions on the considered weights. This section also describes the relationships (e.g., implications) among different acceptability notions.

In Section 6 we show how weighted systems can be relaxed; the motivation behind it is dual. On one side, it is related to the notion of defence so that it becomes possible to weaken the condition that defence needs to be stronger than attack: accordingly, weighted defence can be reduced up to the notion of plain defence given by P.M. Dung, which in fact does not consider weights. On the other side, the conflict-freeness required by most semantics can be broken, and an amount of internal inconsistency can be tolerated.

---

[4]C-semirings have been used for the first time to associate a preference value with a *soft constraint*, and find the best solutions of *Soft Constraint Satisfaction Problems* [Bistarelli et al., 1997].

Section 7 shows how classical extension-based semantics in Dung's Argumentation are revised according to *i)* the different notions of weighted acceptability presented in Section 5, *ii)* and also the relaxations in Section 6.

Section 8 summarises how the property of a framework to be well-founded [Dung, 1995] changes in presence of weighted systems.

Section 9 presents tools and real-world applications of weighted Argumentation. All these applications are related to information coming from online social and reviewing platforms, such as *Twitter.com* and *Amazon.com*.

Finally, Section 10 proposes possible future lines of research in the direction outlined in this chapter, and it provides final thoughts and discussion.

2 Related Approaches

This section reports a summary of similar approaches in the literature that concern the use of preference values on arguments (instead of attacks) and the use of priority-based rules, by starting from the second approach.

There have been a number of proposals for extending Dung's framework in order to allow for more sophisticated modelling and analysis of conflicting information. A common theme among some of these proposals is the observation that not all arguments are equal, as we introduced in Section 1. Hence, the relative strength of arguments needs to be taken into account somehow. Such a preference/strength/priority can be modelled in several ways, which we will inspect in the following.

A first well-studied use of preferences in the non-monotonic logic literature is based on the use of priority orderings over formulae in the language or defeasible inference rules. Such methods are usually proposed for structured approaches, instead of the abstract framework of Dung we investigate in this chapter. The strength of arguments is inferred from the strength of the rules from which the arguments are constructed: in this case, priority orderings need to be "lifted" to preferences over arguments. There exist several proposals [Prakken and Sartor, 1997; Modgil and Prakken, 2017; Dung, 2016], and some well-known instantiations are represented for example by $ASPIC^+$ [Modgil and Prakken, 2014] and *Defeasible Logic Programming* [García and Simari, 2004], which comes with *strict* (high priority) and *defeasible* (low priority) rules.

In the literature it is possible to find many proposals where arguments (and not attacks, as in this chapter) are associated with a value or preference. Two of the most well-known proposals are respectively given by [Bench-Capon, 2003] with *Value-based AFs* (*VAFs*), and [Amgoud and Cayrol, 1998] with *Preference-based AFs* (*PAFs*). A *VAF* is a five-tuple $\langle Ar, att, V, val, valpref \rangle$, where Ar is a finite set of arguments, att is an irreflexive binary relation on Ar (i.e., $\langle Ar, att \rangle$ is a standard AF), V is a non-empty set of values, val is a function which maps elements of Ar into elements of V, and $valpref$ is a preference relation (transitive, irreflexive and asymmetric) on $V \times V$. We say that an argument a relates to value v if accepting a promotes or defends v: the value in question is given by $val(a)$, for every $a \in Ar$, $val(a) \in V$. When a VAF is considered by a particular audience, the value ordering is fixed.

A *PAF* is a triplet $\langle Ar, att, Pref \rangle$ where $Pref$ is a partial pre-ordering (reflexive and transitive binary relation) on $Ar \times Ar$. The notion of defence changes accordingly: let a and b be two arguments, b attacks a if-and-only-if $att(b,a)$ and not $a > b$, i.e, a is not (qualitatively) preferred in the partial pre-ordering.

In [Modgil, 2009] the author extends Dung's theory of Argumentation to integrate a meta-level Argumentation concerning preferences. Dung's level of abstraction is preserved, so that arguments expressing preferences are distinguished by being the source of a second attack relation. This abstractly characterises the application of preferences by attacking attacks between the arguments that are subject to preference claims. By proposing a meta-level, the work in [Modgil, 2009] also concerns higher-order models, see also Chapter 3 of this handbook [Kaci *et al.*, 2021].

A quantitative study is proposed in [Leite and Martins, 2011], where the authors define *Social Abstract Argumentation Frameworks*, which basically associate positive and negative votes to each argument. Afterwards, a semantics is essentially given by fix-points of a set of equations that assign, for each argument a, a value that is based on its social support and on how strong the attack a is being subjected to is. This framework has been extend in [Egilmez et al., 2013] by considering weights on attacks as well.

In [Cayrol and Lagasquie-Schiex, 2013] the authors survey the works [Amgoud and Cayrol, 1998; Cayrol et al., 2010; Kaci and Labreuche, 2011], focusing on how to relate preference values and weights, on either arguments or attacks (see Section 3).

One more recently-popular framework is represented by *ranking-based semantics*, whose aim is to elicit a preference score for each argument by considering the structure of a given AF. In case of non-weighted AFs, the scores are extracted by considering properties related to attacks with respect to each argument, as the number of attack/defence paths and their length [Bonzon et al., 2016].

Two other works where preference values are elicited from the framework are [Besnard and Hunter, 2001] and [Cayrol and Lagasquie-Schiex, 2005b]. Furthermore, some works extract a preference from frameworks where arguments are already labelled with a strength score: in this case, such semantics consider both the original weights and the structure of an AF [Amgoud et al., 2017].

3 From Weights on Arguments to Weights on Attacks

As introduced in Section 1, this chapter concerns weighted attacks. Others proposals in the literature deal with values assigned to arguments instead. For this reason, in this section we fill the distance with those works by showing how to pass from values on arguments to values on attacks.

A first possible view is to have numbers associated with arguments: we call a framework like this as WAF_{Ar}, in order to distinguish them from the WAFs in the rest of this chapter, which have weights on attacks instead. A WAF_{Ar} can be described by a triple $\langle Ar, att, w \rangle$, where $w : Ar \to \mathbb{R}$: $w(a)$ is the value associated with an argument in a WAF_{Ar}. Such a framework can be straightforwardly encoded to a PAF following the rule: a is better/equal than b iff $w(a) \geq w(b)$ [Cayrol and Lagasquie-Schiex, 2013].

In [Kaci and Labreuche, 2014] the authors define an *Argumentation Framework with Varied-strength defeat* (*AFV*) as a triplet $\langle Ar, att, Vdef \rangle$, where besides the set of arguments and the attack relationship, *Vdef* is a function from *att* to the interval $[0, 1]$. $Vdef(a, b)$ represents the certainty degree of the statement "*a* attacks *b*". The authors present how to translate WAF_{Ar}s to *PAF*s and *AFV*s. The intuition is that, the larger the preference of argument *a* over argument *b*, the stronger the attack from *a* to *b*. Below we report different alternative approaches (all quantitative) the authors propose for both schemes:

PAF   1. $Vdef(a, b) = 0$ if *b* is better than *a*.

2. $Vdef(a, b) = 0$ if *b* is better than *a*, 1 if *a* is better than *b*, 0 otherwise.

WAF$_{Ar}$ 1. $Vdef(a, b) = 0$ if $w(b) > w(a)$.

2. $Vdef(a, b) = \max(w(a) - w(b), 0)$.

3. $Vdef(a, b) = 1 - \max(w(b) - w(a), 0)$.

Hence, most of the definitions for *Vdef* lead to the suppression of the attacks $att(a, b)$ for which *b* is strictly preferred to *a*.[5] With the third proposal above instead, an attack from *a* to *b* is removed when $w(b) - w(a) = 1$.

The work [Cayrol and Lagasquie-Schiex, 2013] elaborates on [Kaci and Labreuche, 2014]. They consider quantitative as well as qualitative approaches for expressing these preferences. The goal is to translate *PAF*s and WAF_{Ar}s to *Argumentation Frameworks with attacks of Various Strength* (*AF*$_{VS}$), formally $\langle Ar, ATT, \vec{\succeq} \rangle$. Differently from [Kaci and Labreuche, 2014], besides the sets of arguments *Ar*, *ATT* is a finite set of attack relations over *Ar*,[6] and $\vec{\succeq}$ is a binary relation over *ATT*: it expresses a relative strength between the different attack relations in *ATT*. From *PAF*s to *AF*$_{VS}$s, *ATT* and $\vec{\succeq}$ are assembled by respecting some principles considering a qualitative scheme:

P1 The initial set of attacks between arguments in *PAF* must not be modified (no attack appears/disappears).

---

[5] $Vdef(a, b) = 0$ is equivalent to $(a, b) \notin att$.

[6] We use *ATT* for a set of attack relations, and *att* for a single attack relation.

P2 An attack between two equivalent arguments must be strictly stronger than an attack between two incomparable arguments.

P3 An attack from a to b with a strictly preferred to b must be strictly stronger than an attack between two equivalent or incomparable arguments.

P4 Consider an attack from a to c and an attack from b to c of the same class. If a is strictly stronger than b, the attack (a,c) must be strictly stronger than the attack (b,c).

P5 Consider an attack from a to c and an attack from a to b of the same class. If b is strictly stronger than c, the attack (a,c) must be strictly stronger than the attack (a,b).

P2 and **P3** together induce the partitioning of *att* into four classes of attacks $\{\xrightarrow{1}, \xrightarrow{2}, \xrightarrow{3}, \xrightarrow{4}\}$, respectively depending on a better/equivalent/incomparable/worse than b in *PAF*.[7]

The quantitative approach in [Cayrol and Lagasquie-Schiex, 2013] is instead focused on translating a generic WAF_{Ar} with values on arguments to an AF_{VS}. The authors assume a function $w : Ar \to [0,1]$, and a function $f : [0,1] \times [0,1] \to \mathbb{R}$: the strength of $att(a,b)$ is quantified by $f(w(a), w(b))$. As in the aforementioned qualitative scheme from *PAF*s, even in this case some principles impose conditions on the relationship between the weights on arguments and the strength value on derived attacks:

P4' If the weight of a is greater than the weight of b, then the higher the difference of the weights, the stronger the attack from a to b.

P5' If the weight of a is lower than the weight of b, then the higher the difference of the weights, the weaker the attack from a to b.

From **P1**, **P2**, **P4'**, and **P5'** it is possible to derive some conditions on a weighting translation function f:

---

[7] These are exactly the same four classes used in [Martínez et al., 2008], as reported in Section 4.

6 - WEIGHTED ARGUMENTATION

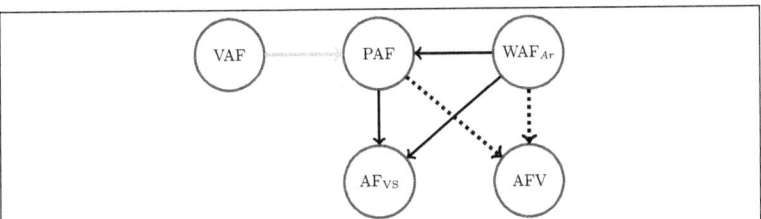

Figure 2: The edges in this graph describe how the different weighting systems proposed in this section (and *VAF* in Section 2) can be translated to others. Plain edges are explained in [Cayrol and Lagasquie-Schiex, 2013], dotted edges in [Kaci and Labreuche, 2014]. The grey edge points to the fact that a unique *PAF* can associated with a *VAF* [Bourguet et al., 2010].

Definition 3.1 (Weighting translation f). *A weighting translation function is a function $f : [0,1] \times [0,1] \to \mathbb{R}$ such that: $\forall x, y, z, t \in [0,1]$,*

- *if $x > y$ and $t > z$ then $f(x,y) > f(y,y) > f(z,t)$.*
- *$f(x,x) = f(y,y)$.*
- *if $x - y > z - t > 0$, then $f(x,y) > f(z,t)$.*
- *if $x - y < z - t < 0$, then $f(x,y) < f(z,t)$.*

An example of a such a function respecting the above constraints is $\forall x, y \in [0,1], f_{\alpha\beta}(x,y) = \alpha(x-y) + \beta$, with $\alpha > 0$ and $\beta > 0$; α amplifies the difference between x and y, while β represents a bias when the difference is 0. For example, if $att(a,b)$, $w(a) = 0.7$, $w(b) = 0.6$, and $f_{11}(x,y) = (x-y) + 1$, then the weight on this attack is 1.1.

To recap, Figure 2 represents how different mapping of weights can be translated one to another (edge direction means "from-to"), with respect to different proposals having a preference on either arguments (WAF_{Ar}, *VAF*, and *PAF*), or attacks (*AFV*, AF_{VS}). Note that a *PAF* can be translated to a class of *VAF*s [Bourguet et al., 2010], and for this reason such an encoding is not represented in Figure 2. Each *VAF* can be however translated into a unique *PAF* instead (grey edge in Figure 2).

Note that in the following we will often refer to the work in [Martínez et al., 2008], which is a variant of the AF_{VS} explained in this section. However, the translations to AF_{VS} described above and in [Cayrol and Lagasquie-Schiex, 2013] are more general, and can lead to more than just four classes of preference among attacks as proposed in [Martínez et al., 2008] instead (see Section 5).[8]

4 Frameworks and Structures to Represent Weights

In the literature, a WAF is a classical AF equipped with a structure to represent weights and some operations to aggregate weights and compare them and prefer one or another. In the following of this section we show how different authors design such a framework of values.

In [Dunne et al., 2011], a weighted argument system is a triple $\langle Ar, att, w\rangle$ where $\langle Ar, att\rangle$ is a Dung-style abstract argument system, and $w : att \longrightarrow \mathbb{R}_>$ is a function assigning real valued weights to attacks. $\mathbb{R}_>$ denotes the real numbers greater than 0, hence attacks are required to have a positive non-zero weight, since an attack could be discarded at no cost otherwise. The aggregation operator is the arithmetic sum, and values are composed in order to compute a relaxation threshold for disregarding attacks (see Section 6). The preference operator is simply $\leq$: for example, 4 is less strong than 5.

A WAF in [Martínez et al., 2008] is a triplet $\langle Ar, ATT, R\rangle$ where Ar is a set of arguments, ATT is in general a set of n binary attack relations $\{\xrightarrow{1}, \xrightarrow{2}, \ldots, \xrightarrow{n}\}$ defined over Ar, and R is a binary relation defined over ATT. The relation $R \subseteq ATT \times ATT$ denotes an order of strength between argument conflicts.[9] The chapter proposes four classes of precedence, that is $\gg$ (and $\ll$), $\approx$, and ?: $att(a, b) \gg att(b, a)$ means that the former attack is stronger than the latter (vice-versa, a weaker attack). Equivalent and incomparable classes are considered as well, i.e, $att(a, b) \approx att(b, a)$ and $att(a, b)?att(b, a)$, respectively. This is accordingly reflected by the definition of defence, where considering

---

[8]For this reason, such translations are not supposed to maintain the defence notion and semantics from PAF or WAF to the work in [Martínez et al., 2008].

[9]A WAF is in this case an AF$_{VS}$ as proposed in Section 3.

$att(a,b)$ and $att(c,a)$ we can have that c is a *strong, weak, normal,* or an *unqualified* defender of b. Therefore, an argument b is defended by T if, and only if, for any argument a such that $att(a,b)$, there is an argument $c \in T$ such that $att(c,a)$, and according to the desired defence strength, $att(c,a) \gg att(a,b)$, $att(c,a) \ll att(a,b)$, $att(c,a) \approx att(a,b)$, and $att(c,a)?att(a,b)$. In such a framework there is an implicit aggregation operator: the attack strength corresponds to the strongest weight among all the counter-attacks. In order to show an example using the notation in [Martínez et al., 2008], given $a \xrightarrow{1} b \xrightarrow{2} c$, a is a strong defender of c if we suppose $\xrightarrow{1} \gg \xrightarrow{2}$; the labels on attacks thus represent a class of attacks.

A *WAF* in [Coste-Marquis et al., 2012] is a triple $\langle Ar, att, w \rangle$ where $\langle Ar, att \rangle$ is a Dung-style Abstract Argumentation Framework, and $w : Ar \times Ar \longrightarrow \mathbb{N}$ is a function assigning a natural number to each attack (i.e. $w(a,b) > 0$ iff $(a,b) \in att$, and a null value otherwise (i.e, $w(a,b) = 0$ iff $(a,b) \notin att$). With respect to [Dunne et al., 2011], the authors of [Coste-Marquis et al., 2012] state that in most situations natural numbers are enough, and this simplifies some of the definitions to come.

In [Coste-Marquis et al., 2012] the authors define $\sigma^{\boxtimes}$-extensions, where σ is one of the given semantics (e.g., admissible), and $\boxtimes$ is an *aggregation function* of weights from $\mathbb{N}^n$ to $\mathbb{N}$, that is the set of natural numbers. To be valid, $\boxtimes$ needs to satisfy three properties: *i) non-decreasingness* (if $x_i \geq x'_i$, then $\boxtimes(x_1, \ldots, x_i, \ldots, x_n) \geq \boxtimes(x_1, \ldots, x'_i, \ldots, x_n)$), *ii) minimality* ($\boxtimes(x_1, \ldots, n_n) = 0$ if $\forall i$, $x_i = 0$), and *iii) identity* ($\boxtimes(x) = x$). In [Coste-Marquis et al., 2012] the authors focus on $+$ and max to simplify the presentation, but several other aggregation functions can be considered as well (e.g., *leximin* or *leximax*). The preference operator, since $\mathbb{N}$ is always used, is simply given by $\leq$: 5 is stronger than 4.

Finally, some other works adopt an algebraic structure named *c-semiring* to represent weights, originally defined in [Bistarelli et al., 1997]; they derive from algebraic semirings and are characterised by two operators as well, that is $\otimes$ and $\oplus$. In practice, c-semirings are *commutative* ($\otimes$ is commutative) and *idempotent* (i.e., $\oplus$ is idempotent) semirings, where $\oplus$ defines a complete lattice: every subset of elements have a *least upper bound*, or *lub*, and a *greatest lower bound*, or *glb*. In fact, c-semirings are semirings where $\oplus$ is used as a preference operator, while $\otimes$ is used to compose preference-values together.

Definition 4.1 (C-semirings [Bistarelli et al., 1997]). *A c-semiring is a tuple* $\mathbb{S} = \langle S, \oplus, \otimes, \bot, \top \rangle$ *such that S is a set, $\top, \bot \in S$, and $\oplus, \otimes : S \times S \to S$ are binary operators making the triples $\langle S, \oplus, \bot \rangle$ and $\langle S, \otimes, \top \rangle$ commutative monoids (semi-groups with identity), satisfying i) $\forall s, t, u \in S, s \otimes (t \oplus u) = (s \otimes t) \oplus (s \otimes u)$ (distributivity), and ii) $\forall s \in S, s \otimes \bot = \bot$ (annihilator). If $\forall s, t \in S, s \oplus (s \otimes t) = s$, the c-semiring is said to be absorptive. In short, c-semirings are commutative and absorptive semirings.*

The idempotency of $\oplus$ leads to the definition of a partial ordering $\leq_\mathbb{S}$ over the set S (S is a poset). Such partial order is defined as $s \leq_\mathbb{S} t$ if and only if $s \oplus t = t$, and $\oplus$ returns the *lub* of s and t (defined also as $\sqcup$, while the *glb* is defined by $\sqcap$). This means that t is "better" than s.

Some more properties can be derived on c-semirings [Bistarelli et al., 1997]: *i)* both $\oplus$ and $\otimes$ are monotone over $\leq_\mathbb{S}$, *ii)* $\otimes$ is intensive (i.e., $s \otimes t \leq_\mathbb{S} s$), and *iii)* $\langle S, \leq_\mathbb{S} \rangle$ is a complete lattice. $\bot$ and $\top$ respectively are the bottom and top elements of such a lattice. When also $\otimes$ is idempotent, *i)* $\oplus$ distributes over $\otimes$, *ii)* $\otimes$ returns the *glb* of two values in S, and *iii)* $\langle S, \leq_\mathbb{S} \rangle$ is a distributive lattice.

Well-known instances of c-semirings are: $\mathbb{S}_{boolean} = \langle \{false, true\}, \vee, \wedge, false, true \rangle$, $\mathbb{S}_{fuzzy} = \langle [0,1], \max, \min, 0, 1 \rangle$, $\mathbb{S}_{bottleneck} = \langle \mathbb{R}^+ \cup \{+\infty\}, \max, \min, 0, \infty \rangle$, $\mathbb{S}_{probabilistic} = \langle [0,1], \max, \times, 0, 1 \rangle$ (also called *Viterbi* semiring), $\mathbb{S}_{weighted} = \langle \mathbb{R}^+ \cup \{+\infty\}, \min, +, +\infty, 0 \rangle$.[10] Note that c-semiring can also deal with non-numeric preference values: for instance, using a c-semiring set $S = \{bad, fair, good\}$, a total ordering as $bad \leq_\mathbb{S} fair \leq_\mathbb{S} good$, and an aggregation operator for which $bad \otimes fair = bad$.

A c-semiring based WAF, i.e., a $WAF_\mathbb{S}$, is a quadruple $\langle Ar, att, w, \mathbb{S} \rangle$, where $\mathbb{S}$ is a c-semiring $\langle S, \oplus, \otimes, \bot, \top \rangle$, Ar is a set of arguments, att the attack binary-relation on Ar, and $w : Ar \times Ar \longrightarrow S$ is a binary function: given $a, b \in Ar$ and $att(a, b)$, then $w(a, b) = s$ means that a attacks b with a weight $s \in S$. Moreover, it is required that $att(a, b)$ iff $w(a, b) <_\mathbb{S} \top$. Note that the Boolean c-semiring can be used to model classical Dung's Argumentation [Bistarelli et al., 2018b].

---

[10]Note that when considering the Weighted c-semiring, it happens that $7 \leq_\mathbb{S} 3$ even if $7 \geq 3$, i.e., lesser means better.

Differently from all the other proposals in this section, a c-semiring is parametric with respect to both the aggregation operator (i.e., $\otimes$) and the preference operator (i.e., $\oplus$), and it is not bound to a single set of values as $\mathbb{R}$ or $\mathbb{N}$. One more advantage is that a Cartesian product of c-semirings (which is still a c-semiring) can model multi-criteria weights, and in general, c-semirings can model partially-ordered preference values besides to totally-ordered ones.

5 Weighted Acceptability

A *weighted acceptability* notion extends the original one given in [Dung, 1995] by considering the two strength levels of attack and defence. For example, only if the defence weight is stronger, then an argument is successfully defended. For this reason, the notion of defence becomes more constrained than the unweighted one, which only considers edges in a digraph. We start by recalling the notion *acceptability* of an argument b in [Dung, 1995], with respect to a set of arguments T.

Definition 5.1 ($\mathbb{D}_{Dung}$). *An argument b is acceptable w.r.t. $T \subseteq Ar$ (or T defends b) iff for any argument $a \in Ar$ s.t. $att(a,b)$, then $\exists c \in T$ s.t. $att(c,a)$.*

Three different definitions of weighted defence are presented in [Martínez et al., 2008], [Coste-Marquis et al., 2012], and [Bistarelli et al., 2018b]. In the following, we condense their main features and we show how they differ.

We start by presenting the notion of acceptability given in [Martínez et al., 2008]. When requiring a preference level $[\gg, \approx]$ (see Section 4), for each attacker a of b there must be either a strong or a normal defender $c \in T$. In Definition 5.2 we report the defence in [Martínez et al., 2008] by using $[\gg, \approx]$.

Definition 5.2 ($\mathbb{D}_{Martínez\ et\ al.}$). *Given a $WAF \langle Ar, ATT, R \rangle$ as formalised in [Martínez et al., 2008], with $R = [\gg, \approx]$, $a, b, c \in Ar$, $T \subseteq Ar$, then b is acceptable w.r.t. T iff $\forall att(a,b), \exists c \in T$ s.t. $att(c,a) \gg att(a,b)$ or $att(c,a) \approx att(a,b)$.*

In practice, only one argument $c \in T$ counter-attacking a with an equal or stronger level is needed to defend an argument b. This defence is typical of *Argumentation Frameworks with attacks of Various Strength*, i.e, the AF$_{VS}$ (see Section 3). A different approach along the same line is [Cayrol et al., 2010]. Hence the notions of defence of these two works are strictly connected.

Note also that the intuition for extending the notion of defence is the same in both AF$_{VS}$s and AFVs (Section 3 and [Kaci and Labreuche, 2011]). The only difference is due to the preference relation over attacks: an AF$_{VS}$ uses a pre-ordering (thus allowing for incomparable attacks), whereas in an AFV the preference relation is based on a function with values on a linearly ordered scale, e.g., the interval $[0, ..1]$, providing a total ordering over attacks.

On the contrary, the idea in [Coste-Marquis et al., 2012] is to aggregate all the weights of counter-attacks, and to check if they are stronger than the considered attack:

Definition 5.3 ($\mathbb{D}_{Coste-Marquis\ et\ al.}$). *Given a WAF $\langle Ar, att, w \rangle$ as defined in [Coste-Marquis et al., 2012], an argument b is acceptable w.r.t. a subset of arguments T iff $\forall a \in Ar$ s.t. $att(a, b)$, we have that $w(T, a) \geq w(a, b)$, where $w(T, a)$ is a shortcut for $\boxtimes_{c \in T} w(c, a)$.*

Thus, an argument b is acceptable if for each attack from $a \in Ar$ against b, the aggregated weight of the collective defence of b is greater than $w(a, b)$.[11]

Finally, we report the definition of acceptability in [Bistarelli et al., 2018b], parametrically given for a c-semiring $\mathbb{S} = \langle S, \oplus, \otimes, \bot, \top \rangle$:

Definition 5.4 ($\mathbb{D}_{Bistarelli\ et\ al.}$). *Given a WAF$_\mathbb{S}$ $\langle Ar, att, w, \mathbb{S} \rangle$ as defined in the work [Bistarelli et al., 2018b], $T \subseteq Ar$ defends $b \in Ar$ (or b is w-acceptable) iff $\forall a \in Ar$ such that $att(a, b)$, we have that $w(a, T \cup \{b\}) \geq_\mathbb{S} w(T \cup \{b\}, a)$, where $w(a, T \cup \{b\})$ is a shortcut for $\otimes_{c \in (T \cup \{b\})}(a, c)$ and $w(T \cup \{b\}, a)$ is a shortcut for $\otimes_{d \in T \cup \{b\}}(d, a)$.*

Besides aggregating the weights of all counter-attacks as in [Coste-Marquis et al., 2012], Definition 5.4 also aggregates the weights of all the attacks from a to $d \cup \{b\}$: in this case, all the attacks from any

---

[11] Such a phrasing of defence is also equivalent to the work in [Bistarelli and Santini, 2010].

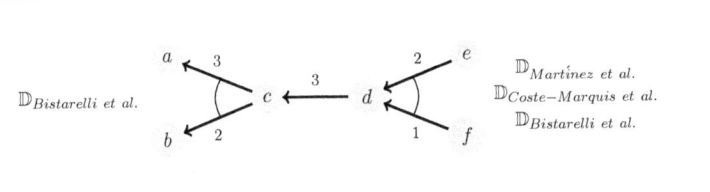

Figure 3: The three notions of defence on the right of this figure aggregate attack weights to check if c is defended from d, while only $\mathbb{D}_{Bistarelli\ et\ al.}$ also aggregates all the weights on the attacks from the same attacker to a set of arguments to be defended (i.e., in this case $\{a,c\}$). Using the Weighted c-semiring, $\{e,f\}$ defends c from d according to $\mathbb{D}_{Coste-Marquis\ et\ al.}$ and $\mathbb{D}_{Bistarelli\ et\ al.}$, since $(2+1) \leq_{Weighted} 3$, but not according to $\mathbb{D}_{Martínez\ et\ al.}$, since $2 \not\leq_{Weighted} 3$ and $1 \not\leq_{Weighted} 3$. Moreover, d defends $\{a,b\}$ from c according to $\mathbb{D}_{Martínez\ et\ al.}$ and $\mathbb{D}_{Coste-Marquis\ et\ al.}$, but not according to $\mathbb{D}_{Bistarelli\ et\ al.}$ since $3 \not\leq_{Weighted} (3+2)$.

argument in Ar towards a set T plus the argument to be accepted need to be considered.

In [Bistarelli *et al.*, 2018b] the authors use c-semirings to represent the other two proposals reported in this section, that is [Martínez *et al.*, 2008] and [Coste-Marquis *et al.*, 2012], as $WAF_\mathbb{S}$. This allows for discovering the relationships among such defences, as described in Section 5, and to easily show an example on how they differ, as highlighted in Figure 3. All three of them aggregate weights (in different ways) towards the same argument d to check if $\{e,f\}$ defends c from it: that is, c is defended if $w(\{e,f\},d) \leq_\mathbb{S} w(d,\{c\})$. Only $\mathbb{D}_{Bistarelli\ et\ al.}$ also aggregates all the weights on the attacks from the same attacker to the set of arguments to be defended: to check if d defends $\{a,b\}$ from c, we need that $w(\{d\},c) \leq_\mathbb{S} w(c,\{a,b\})$.

Properties of Defence. By comparing all the notions of acceptability in the same c-semiring based framework [Bistarelli *et al.*, 2018b], it is possible to catch their relationships as, for example, the implications among them:

- $\mathbb{D}_{Martínez\ et\ al.}, \mathbb{D}_{Coste-Marquis\ et\ al.}, \mathbb{D}_{Bistarelli\ et\ al.} \Rightarrow \mathbb{D}_{Dung}$
- $\mathbb{D}_{Bistarelli\ et\ al.} \Rightarrow \mathbb{D}_{Coste-Marquis\ et\ al.}$
- $\mathbb{D}_{Martínez\ et\ al.} \Rightarrow \mathbb{D}_{Coste-Marquis\ et\ al.}$

Moreover, if we replace the original structure to represent weights in [Coste-Marquis *et al.*, 2012] and [Martínez *et al.*, 2008] with different c-semirings, we obtain further interesting relationships. We recall from Section 4 that $\langle [0,1], \max, \min, 0, 1 \rangle$ is the *Fuzzy* c-semiring and $\langle \{true, false\}, \vee, \wedge, false, true \rangle$ is the *Boolean* c-semiring:

- If $\mathbb{S} = \langle [0,1], \max, \min, 0, 1 \rangle$, then

$$\mathbb{D}_{Martínez\ et\ al.} \Leftrightarrow \mathbb{D}_{Coste-Marquis\ et\ al.}$$

- If $\mathbb{S} = \langle [0,1], \max, \min, 0, 1 \rangle$, then

$$\mathbb{D}_{Bistarelli\ et\ al.} \Rightarrow \mathbb{D}_{Martínez\ et\ al.}$$

- If $\mathbb{S} = \langle \{true, false\}, \vee, \wedge, false, true \rangle$, then

$$\mathbb{D}_{Dung} \Leftrightarrow \mathbb{D}_{Bistarelli\ et\ al.} \Leftrightarrow \mathbb{D}_{Martínez\ et\ al.} \Leftrightarrow \mathbb{D}_{Coste-Marquis\ et\ al.}$$

The last item states that, when dropping weights, the way these defences combine attacks and counter-attacks is irrelevant: they all flatten to the classical acceptability in [Dung, 1995].

6 Relaxations

The two main works that deal with internal relaxations are [Dunne *et al.*, 2011] and [Bistarelli *et al.*, 2018b]. In both of them, the goal is to relax constraints represented by attacks: by tolerating some of them, it is possible to allow some inconsistency level with respect to the original considered framework, where all the attacks are always considered instead.

The main motivation is to obtain progressively more solutions, as one increases the inconsistency level to be tolerated. In this way, it is possible to return non-trivial solutions (e.g., stable extensions) in case conventional (unweighted) AFs have none. Clearly, the cost of such a relaxation represents a preference ordering over the obtained sets of arguments: indeed it is better to prefer extensions that are found with a milder relaxation than others, i.e., extensions on AFs that are closer to the original one.

The key idea in [Dunne et al., 2011] is to consider an *inconsistency budget*, $\beta \in \mathbb{R}_{\geq}$, which is used to characterise how much inconsistency one is prepared to tolerate. The intended interpretation is that, given an inconsistency budget β, it is possible to disregard attacks up to a total weight of β. By doing so, one obtains several AFs on which to compute the desired semantics (or other typical problems in Argumentation). Classical AFs implicitly assume an inconsistency budget of 0, since it is not possible to disregard any attack. By allowing larger inconsistency budgets and consequently dropping more attacks, one can obtain progressively more frameworks and then solutions. The proposed disregard approach applies to all the classical semantics, hence obtaining e.g., β-admissible or β-complete extensions. For instance, the set of β-complete extensions is given by the union of all β-complete extensions on all the AFs obtained by disregarding attacks up to a total of β.

However, the main goal in [Dunne et al., 2011] is to find alternatives when the single most sceptical (or least committed) semantics among all, i.e., the grounded one, returns an empty-set solution: different approaches to relax scepticism correspond to the design of different (unweighted) semantics, as the ideal and eager ones [Baroni et al., 2018, Ch. 4].

Note that the work in [Coste-Marquis et al., 2012] is inspired by the same relaxation as in [Dunne et al., 2011]. Extensions are named as $\sigma_{\boxtimes}^{\beta}$-extensions: β is exactly the same inconsistency budget, but instead of only arithmetic sum, the aggregation of weights is via a parametric operator $\boxtimes$, as explained in Section 5.

In the second relaxation approach that we consider, i.e., [Bistarelli et al., 2018b], given a $WAF_{\mathbb{S}} = \langle Ar, att, w, \mathbb{S} \rangle$, a subset of arguments $T \subseteq Ar$ is α-conflict-free iff $w(T,T) \geq_{\mathbb{S}} \alpha$, where

$$w(T,T) = \bigotimes_{a \in T, b \in T} w(a,b)$$

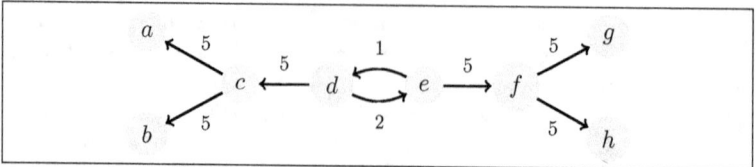

Figure 4: An example of weighted framework taken from [Dunne et al., 2011].

α/β	$\mathcal{PR}$ in [Dunne et al., 2011]	$\mathcal{PR}$ in [Bistarelli et al., 2018b]
0	$\{\{a,b,d,f\},\{c,e,g,h\}\}$	$\{\{a,b,d,f\},\{c,e,g,h\}\}$
1	$\{\{a,b,d,f\},\{c,e,g,h\}\}$	$\{\{a,b,d,f\},\{c,e,g,h\}\}$
2	$\{\{a,b,d,f\},\{c,e,g,h\}\}$	$\{\{a,b,d,f\},\{c,e,g,h\}\}$
3	$\{\{a,b,d,f\},\{c,e,g,h\},\{a,b,d,e,g,h\}\}$	$\{\{a,b,d,f\},\{c,e,g,h\},\{a,b,d,e,g,h\}\}$

Table 1: Considering the WAF in Figure 4, the sets of β-preferred and α-preferred extensions correspond while increasing α/β.

means that we aggregate (using $\otimes$) all the weights associated with attacks in T. This approach is different from [Dunne et al., 2011]: only the original framework is considered, hence no further AF or WAF is derived by disregarding attacks. The relaxation is obtained by tolerating an amount of attacks by breaking the conflict-free condition: an extension may contain attacks up to a threshold of α on the aggregation (i.e., $\otimes$) of their weights.

Figure 4 reports an example of WAF taken from [Dunne et al., 2011]. Table 1 shows β-preferred and α-preferred extensions (using the Weighted c-semiring) on that WAF. On this example, both approaches return the same extensions when using the same α/β.

Table 2 shows the difference between β-grounded and α-grounded extensions. The approach in [Dunne et al., 2011] aims to find several results, while the approach in [Bistarelli et al., 2018b] is more adherent to the original proposal in [Dung, 1995], and always returns a single grounded extension (which in this specific example is always the empty-set).

Relaxing Defence. The two works [Dunne et al., 2011] and [Bistarelli et al., 2018b] can be used to drop attacks in a given framework. All the

α/β	$\mathcal{GR}$ in [Dunne et al., 2011]	$\mathcal{GR}$ in [Bistarelli et al., 2018b]
0	$\{\emptyset\}$	$\emptyset$
1	$\{\emptyset, \{c,e,g,h\}\}$	$\emptyset$
2	$\{\emptyset, \{c,e,g,h\}, \{a,b,d,f\}\}$	$\emptyset$
3	$\{\emptyset, \{c,e,g,h\}, \{a,b,d,f\}, \{a,b,d,e,g,h\}\}$	$\emptyset$

Table 2: Considering the WAF in Figure 4, β-grounded and α-grounded extensions differ on the same α/β.

weighted acceptability notions presented in Section 5 aggregate weights to understand if a defence is stronger than an attack: in general, if the defence strength is higher, then a set of arguments is effectively defended, otherwise the attack is predominant. In this section, we survey two methods that relax the concept of defence by tolerating arguments that are defended at a "milder" level.

A first approach can be found in [Martínez et al., 2008]. From this work, the attack scenario $[T \subseteq arguments, \mathcal{P} = \{\gg\}]$ includes only strongly-defended arguments, since the defence condition is $\gg$. A scenario with defence condition $\mathcal{P}$ can be expanded into another scenario using a different defence condition $\mathcal{Q}$, through the concept of *defence upgrade* and an expansion operator $\uplus$.

Definition 6.1 (Defence upgrade). *Let $p = [T, \mathcal{P}]$ be an attack scenario, and let $\mathcal{Q}$ be a set of defence conditions. The expansion of p according to $\mathcal{Q}$ is defined as $p \uplus \mathcal{Q} = [T' \cup T, \mathcal{P} \cup \mathcal{Q}]$, where $T' = \{a \in Ar \mid a \text{ is acceptable with respect to } [T, \mathcal{Q}]\}$.*

Hence, if $\mathcal{Q}$ is $[\gg, \approx]$, new arguments can be accepted in T', thus extending the set of accepted arguments $T' \cup T$.

The second approach is in [Bistarelli et al., 2018b]. There, the notion of weighted defence can be relaxed in order to be equal to [Coste-Marquis et al., 2012] (see Section 5), and also to the classical defence given by [Dung, 1995]: this leads to completely ignoring weights. The γ-defence (or also $\mathbb{D}_\gamma$) in [Bistarelli et al., 2018b] is parametrised on a given c-semiring (see Section 4) and on a threshold-value γ: such a γ is used to consider arguments that are not "fully" defended according to $\mathbb{D}_{Bistarelli\ et\ al.}$, i.e., for which $w(a, T \cup \{b\}) \not\geq_S w(T, a)$:

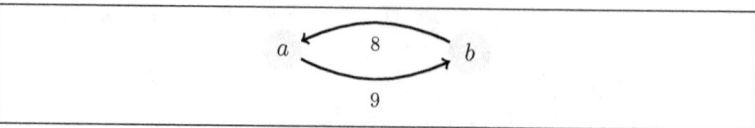

Figure 5: a defends itself according to $\mathbb{D}_{Bistarelli\ et\ al.}$ because the attack weight (9) is stronger than the defence weight (8), but for the same reason b only 1-defends itself: the defence is one unit less than what is really needed for a proper weighted defence (i.e., $9-8$), that is considering w-defence in Section 5.

Definition 6.2 (γ-defence). *Given $\langle Ar, att, w, \mathbb{S} = \langle S, \oplus, \otimes, \bot, \top \rangle \rangle$ and $\gamma \in S$, $T \subseteq Ar$ γ-defends $b \in Ar$ iff $\forall a \in Ar$ such that $att(a,b)$ we have that $w(T,a) \neq \top$ and*

$$\Big(w(a, T \cup \{b\}) \oslash w(T, a)\Big) \geq_\mathbb{S} \gamma$$

Definition 6.2 states that T defends b from a if the difference between the aggregation of the attack weights from a to T (union b) and the one from T to a is better than γ. The $\oslash$ operator is the inverse of $\otimes$ (e.g., the arithmetic $-$ in case $\otimes$ is the arithmetic plus). A simple example is given in Figure 5: a defends itself, while b only 1-defends itself, since $9 - 8 \geq_\mathbb{S} 1$, if we consider $\mathbb{S}$ as the Weighted c-semiring.

Clearly, by progressively increasing γ we consequently relax the constraint that defence needs to be equal or stronger than attack, and we obtain more and more extensions. This motivation is a leitmotif of the whole section.

Note that, if γ is chosen as the $\bot$ of the considered c-semiring, γ-defence is equivalent to the original definition of defence given by P.M. Dung [Bistarelli et al., 2018b]: $\mathbb{D}_\gamma \Leftrightarrow \mathbb{D}_{Dung}$. On the other hand, if $\gamma = \top$, then we have $\mathbb{D}_\gamma \Leftrightarrow \mathbb{D}_{Bistarelli\ et\ al.}$.

When all $a \in Ar$ attack one argument only, γ-defence is equivalent to the defence defined in [Coste-Marquis et al., 2012], that is $\mathbb{D}_{Coste-Marquis\ et\ al.} \Leftrightarrow \mathbb{D}_\gamma$. Finally, by properly choosing γ, it is always possible to obtain a defence that is implied by the one in [Coste-Marquis et al., 2012]: $\mathbb{D}_{Coste-Marquis\ et\ al.} \Rightarrow \mathbb{D}_\gamma$.[12]

---

[12]We remind that $\mathbb{D}_\gamma$ is a relaxation of $\mathbb{D}_{Bistarelli\ et\ al.}$, which on the contrary

7 Semantics in WAFs

Clearly, the concepts of weighted defence and relaxation advanced in Section 5 and Section 6 are not stand-alone, but they are the basis to elaborate on classical extension-based semantics [Dung, 1995].

As introduced in Section 5, in [Dunne et al., 2011] the authors mainly focus on the β-grounded semantics, with the purpose to find alternatives when the unweighted one ([Dung, 1995]) is equal to empty-set, which is not very informative. Nevertheless, also the other β-semantics are briefly presented: they correspond to the union of the extension sets found on all the AFs obtained by disregarding an attack amount up to β.[13]

In [Martínez et al., 2008], only the preferred and grounded semantics (or better, *scenarios*) are explicitly defined; they can be both captured by the use of the expansion operator $\uplus$. For example, Let $p = [T, \mathcal{P}]$ be an admissible scenario. If $p \uplus \mathcal{Q} = [T, \mathcal{P} \cup \mathcal{Q}]$ for any condition $\mathcal{Q}$, then T is a (classic) preferred extension: $T \subseteq Ar$ cannot be expanded. The grounded scenario is instead defined as the least fix-point of $\uplus$ using a defence condition $\mathcal{P}$.

The semantics defined in [Coste-Marquis et al., 2012] follow two distinct paths, whose definitions we report below from the original chapter. Definition 7.1 is used to consider an inconsistency budget β exactly as in [Dunne et al., 2011]; in this case, however, $\boxtimes$ is a more general operator to aggregate weights to find β than the arithmetic sum adopted in [Dunne et al., 2011].

Definition 7.1 ($\sigma_{\boxtimes}^{\beta}$-extensions). *Given a $WAF = \langle Ar, att, w \rangle$, a semantics σ, an aggregation function $\boxtimes$, and a budget β, the set of $\sigma_{\boxtimes}^{\beta}$-extensions, denoted as $\mathcal{E}_{\sigma}^{\boxtimes,\beta}(\langle Ar, att, w \rangle)$, is defined as $\mathcal{E}_{\sigma}^{\boxtimes,\beta}(\langle Ar, att, w \rangle) = \{E \in \mathcal{E}_{\sigma}(Ar, att \setminus att') \mid att' \in Sub(att, w, \beta)\}$, where the function $Sub(att, w, \beta)$ returns the set of subsets of att whose total aggregated weight does not exceed β.*

The $\sigma_{\boxtimes}^{\beta}$-grounded semantics may return several extensions, as in [Dunne et al., 2011]. Afterwards, the same authors propose how to

---

implies the notion of defence in [Coste-Marquis et al., 2012] (see Section 5).

[13]An example of β-preferred extensions is reported in Table 1.

refine them by removing the empty-set from such a set of extensions, with the purpose to not trivialise the sceptical acceptance of arguments: the presence of an empty-set among the results impedes an argument to be sceptically accepted.

The second path in [Coste-Marquis et al., 2012] completely drops the β-budget and only focuses on defining semantics that use a weighted notion of acceptability based on $\otimes$, that is $\otimes$-acceptability. Semantics are simply named as $\otimes$-*semantics*, which are straightforwardly derived from their counterpart in [Dung, 1995]: for example, T is a $\otimes$-complete extension iff T is conflict-free and every argument $a \in Ar$, which is $\otimes$-acceptable with respect to T, belongs to T. In this case, the $\otimes$-grounded semantics returns a single extension.

The approach in [Bistarelli et al., 2018b] is the only one that allows for contemporarily disregarding an "amount" of attacks (also accomplished in [Dunne et al., 2011] and [Coste-Marquis et al., 2012]) *and* requiring defence to be stronger than an attack (also proposed in [Coste-Marquis et al., 2012] and [Martínez et al., 2008]). Moreover, it also permits to relax this latter constraint on defence, up to not considering weights at all.

Therefore, semantics are equipped with two thresholds: α^γ-semantics come with an internal inconsistency budget α (see Section 5), and a threshold γ on the defence relaxation (see Section 6).

Their definition is summarised in the following, given a $WAF_\mathbb{S} = \langle Ar, att, w, \mathbb{S} = \langle S, \oplus, \otimes, \bot, \top \rangle \rangle$ and $\alpha, \gamma \in S$.

- A subset of arguments $T \subseteq Ar$ is α-conflict-free iff $w(T,T) \geq_\mathbb{S} \alpha$. Note that γ is not considered in this semantics, since only the conflict internal to an extension is measured.

- An α-conflict-free set $T \subseteq Ar$ is α^γ-admissible iff the arguments in T are γ-defended by T from the arguments in $Ar \backslash T$.

- An α^γ-admissible $T \subseteq Ar$ is α^γ-complete iff each argument $b \in Ar$ that is γ-defended by T and s.t. $w(T \cup \{b\}, T \cup \{b\}) \geq_\mathbb{S} \alpha$ is in T (i.e., $b \in T$).

- An α^γ-preferred extension is a maximal (with respect to set inclusion) α^γ-admissible subset of Ar.

- An α^γ-admissible set T is also an α^γ-stable extension iff $\forall a \notin T, \exists b \in T$ then $w(b,a) \neq \top$, and $T \cup \{a\}$ is not α^γ-admissible.

Both approaches in [Bistarelli *et al.*, 2018b] and [Coste-Marquis *et al.*, 2012], that is α^γ-semantics and $\boxtimes$-*semantics* respectively, preserve some of the formal properties of the unweighted semantics in [Dung, 1995]: for example, the implications *stable* $\Rightarrow$ *preferred* $\Rightarrow$ *complete* $\Rightarrow$ *admissible* $\Rightarrow$ *conflict-free*.

8 Well-founded WAFs

The goal of this section is to show how the well-foundedness property of frameworks [Dung, 1995] extend to WAFs. We commence by revising the notion of well-foundedness, then we show how it can be adapted to WAFs and how also well-founded WAFs have a single complete/preferred/stable/grounded extension. In this case, the considered framework always provides the same single solution in any these scenarios, consequently eliminating all the possible uncertainty. Since in [Bistarelli *et al.*, 2018b] the authors showed how acceptability in [Martínez *et al.*, 2008] and [Coste-Marquis *et al.*, 2012] can be represented as a $WAF_\mathbb{S}$, we will consider such frameworks in the following of this section in order to propose a single and unitary point of view.

P.M. Dung defines the *sufficient* conditions behind well-foundedness in *AF*s in his pioneering work [Dung, 1995]. A well-founded *AF* is an *AF* without an infinite defeating sequence of arguments.

Definition 8.1 (Well-founded$_{Dung}$ AFs). *An AF is well-founded iff there exists no infinite sequence $a_1, a_2, \ldots, a_n, \ldots$ (with $a_i \in Ar$) such that for each i, $att(a_{i+1}, a_i)$.*

In case of a finite number of arguments, a framework is well-founded if it is acyclic. However, the notions of weighted defence presented in Section 5 consider sets of arguments to check the aggregation strength of attacks and counter-attacks: Definition 8.1 is not enough anymore to capture the aggregation of weights from/to sets, since it is based on plain sequences of arguments.

Because of these reasons, in Definition 8.2 we redefine the notion of sequence of arguments into a sequence of sets, or better, *set-maximal*

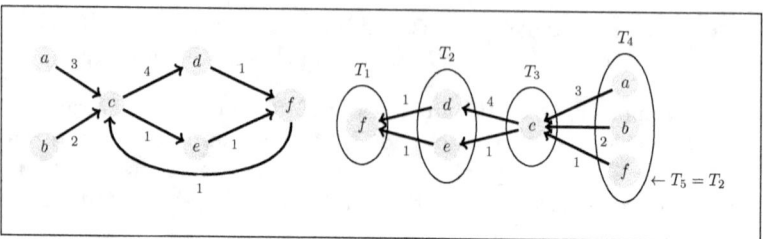

Figure 6: Left: an example of WAF; right: a fragment of an infinite sequence of SMA sets on the framework on the left. T_5 is identical to T_2 and the sequence infinitely continues from it.

attack (SMA) sets. All the remainder of this section is inspired by the work in [Bistarelli and Santini, 2019], definitions and theorems included.

Definition 8.2 (Set-maximal attack (SMA) sets). *Given $WAF_\mathbb{S} = \langle Ar, att, w, \mathbb{S} \rangle$ and $T, U \subseteq Ar$, then T is a set-maximal attack on U, iff*

i) T is conflict-free;

ii) $\forall b \in T, \exists c \in U$ s.t. $att(b,c)$;

iii) there exists no T' s.t. conditions i) and ii) hold and $T \subsetneq T'$.

Example 8.3. *Figure 6 (right) shows a fragment of an infinite sequence of SMA sets, obtained on the WAF represented in Figure 6 (left); it starts from set $\{f\}$. The sequence is: T_1-T_2-T_3-T_4-... (it continues as T_2-T_3-T_4 for an infinite number of times).*

Definition 8.4 generalises the well-foundedness property on different notions of (weighted) defence: for example, the ones in Section 5. Therefore, well-foundedness becomes parametric with respect to the chosen c-semiring *and* the selected defence. Other weighted defences may directly inherit from the definition to check the conditions under which they allow for a well-founded framework.

Definition 8.4 (Generalisation in $WAF_\mathbb{S}$). *Given a $WAF_\mathbb{S} = \langle Ar, att, w, \mathbb{S} \rangle$, if there does not exist an infinite sequence ω of SMA sets $T_1, T_2, \ldots,$ such that for every $T_{i+1}, T_i,$ and T_{i-1} we have that T_{i+1} defends T_{i-1} from each $a \in T_i$ according to a generic weighted defence $\mathbb{D}$, then $WAF_\mathbb{S}$ is well-founded w.r.t. $\mathbb{D}$.*

Hence, besides looking at infinite cycles as in [Dung, 1995] (in this case of sets, and not just of arguments), in WAFs there is one more constraint: to be infinite, this chain of SMA sets needs to respect the constraint imposed by weighted defences: T_{i+1} has to defend T_{i-1} from each $a \in T_i$. Therefore, a framework has less chances to be well-founded when considering weighted defences, since the absence of infinite sequences of SMA sets is not sufficient.

Example 8.5. *Still supposing the WAF in Figure 6 (left), we see that by considering it as a classical framework (dropping weights), the framework is not well-founded$_{Dung}$ because there exists an infinite sequence of arguments given by the cycle $f \leftarrow e \leftarrow c \leftarrow f \leftarrow \ldots$.*

Moreover, considering it as a WAF_S, the related infinite sequence in Figure 6 does not respect the defence conditions of $\mathbb{D}_{Martínez\ et\ al.}$: T_4 cannot defend T_2 from T_3 because there is no attack from T_4 to c (whose values are 3, 2, 1), which is at least as strong as the attack from c to d in T_2 (i.e., 4 is stronger). Consequently, the framework in Figure 6 (left) is not well-founded when using $\mathbb{D}_{Martínez\ et\ al.}$, in accordance with Definition 8.4.

As accomplished for acceptability notions, we can relate the property of a framework to be well-founded by considering the three proposals in Section 5. Since Definition 8.4 is parametrically based on a generic definition of weighted defence, we can plug the three notions directly in: e.g. $wfd_{Martínez\ et\ al.}$ is defined on $\mathbb{D}_{Martínez\ et\ al.}$. We also consider the classical well-founded property in [Dung, 1995], that is wfd_{Dung}, by simply removing weights from a WAF_S, while keeping the same Ar and att.

Theorem 8.6 (Implications and well-foundedness). *Given a WAF_S, the following implications hold (where wfd_* is the well-founded property as derived from the defence notion described in the chapter indicated by $*$):*

1. $wfd_{Dung} \Leftarrow wfd_{Bistarelli\ et\ al.}$; $wfd_{Dung} \Leftarrow wfd_{Martínez\ et\ al.}$; $wfd_{Dung} \Leftarrow wfd_{Coste-Marquis\ et\ al.}$.

2. $wfd_{Coste-Marquis\ et\ al.} \Leftarrow wfd_{Bistarelli\ et\ al.}$.

3. $wfd_{Coste-Marquis\ et\ al.} \Leftarrow wfd_{Martínez\ et\ al.}$.

4. With the Fuzzy c-semiring, $wfd_{Martínez\ et\ al.} \Leftrightarrow wfd_{Coste-Marquis\ et\ al.}$; $wfd_{Martínez\ et\ al.} \Rightarrow wfd_w$.

5. With the Boolean c-semiring,

$$wfd_{Bistarelli\ et\ al.} \Leftrightarrow wfd_{Martínez\ et\ al.} \Leftrightarrow wfd_{Coste-Marquis\ et\ al.}$$

The well-foundedness property is interesting because it points to a framework where there exists only one set of arguments that is worth to be considered under any semantics. According to [Dung, 1995], every well-founded *AF* has exactly one complete extension, which is also grounded, preferred, and stable. The same result is preserved also in each of the weighted approaches in Section 5. Theorem 8.7 formalises this result.

Theorem 8.7 (Uniqueness of extensions). *Given a notion of weighted acceptability, any well-founded WAF where the grounded extension is also complete, has exactly one complete extension, which is also grounded, preferred, and stable.*

Note there is an additional condition with respect to [Dung, 1995]: it is related to the (weighted) grounded extension, which needs to be also complete according to Theorem 8.7. This condition is required by the fact that *WAF*s may have several grounded extensions, as in fact it may happen in [Dunne *et al.*, 2011], [Martínez *et al.*, 2008], [Coste-Marquis *et al.*, 2012], and [Bistarelli *et al.*, 2018b]. This is in general not desirable, since the grounded extension should represent the most sceptical unique point of view. The multiplication of the grounded extensions is due to having several derived frameworks by disaggregating attacks ([Dunne *et al.*, 2011] and [Martínez *et al.*, 2008]), or to imposing additional constraints on weighted acceptability ([Martínez *et al.*, 2008] and [Bistarelli *et al.*, 2018b]).

Uniqueness of the Grounded Extension. It could be interesting to relax the grounded extension in case it corresponds to the empty-set because it does not bring so much information. Nevertheless, it is also to maintain its unicity. In fact, the idea behind the grounded extension is to represent a core set of "least questionable" arguments, being composed of only non-attacked arguments and arguments defended by them (directly and indirectly). For this reason, having several of these least questionable positions makes them not so least questionable anymore.

For example, let us consider a $WAF_\mathbb{S}$ with the set of arguments $Ar = \{a, b, c, d\}$, and $att(a, b)$, $att(b, c)$, $att(b, d)$, all with a weight of 1 (using the *Weighted* c-semiring): $w(a, b) = 1, w(b, c) = 1, w(b, d) = 1$. The set of weighted complete-extensions (*wcom* for short) in [Bistarelli et al., 2018b] is $\{\{a, c\}, \{a, d\}\}$, and then there is no single least element with respect to set inclusion: in a classical formulation of this semantics, we would have two *wgrd* extensions.[14]

Concerning the approach in [Bistarelli et al., 2018b], the authors of [Bistarelli and Santini, 2017] re-obtain a single *wgrd* extension, which however is not always also *wcom*.[15] The *wgrd* extension is there defined as the maximal, w.r.t. set inclusion, *wadm* extension included in the intersection of *wcom* extensions. They follow the same approach used in [Baumann and Spanring, 2015] to alternatively define the ideal and eager semantics.

Definition 8.8 (Weighted grounded). *Given a $WAF_\mathbb{S}$, $F = \langle Ar, att, w, \mathbb{S}\rangle$, an extension $T = wgrd(F)$, iff $T \in wadm(F)$, and $T \subseteq \bigcap wcom(F)$, and $\nexists T' \in wadm(F)$ satisfying $T' \subseteq \bigcap wcom(F)$ s.t. $T \subsetneq T'$.*

Some of the derived properties in [Bistarelli and Santini, 2017] are also reported in the three propositions below.

Proposition 8.9 (Existence and unicity). *$wgrd(F)$ always exists and is unique.*

Proposition 8.10. *If $\mathbb{S}$ is the Boolean c-semiring then $wgrd(F)$ is equivalent to the classical grounded extension [Dung, 1995].*

Proposition 8.11. *$wgrd(F)$ corresponds to the set of sceptically accepted arguments in $wcom(F)$.*

---

[14] According to [Dung, 1995] instead, the grounded extension is $\{a, c, d\}$ in this case. In order to differentiate them from their original formulation, in this paragraph we will use *wcom*, *wadm*, and *wgrd* to point to weighted complete, admissible, and grounded respectively.

[15] This possibility motivates the additional condition in Theorem 8.7.

Considering the initial example in this paragraph, $wgrd(F) = \{a\}$ is not *wcom* but only *wadm*.

9 Tools and Applications

We split this section in two parts. First we describe software tools that can be used as solvers of the weighted problems described in this chapter, and then we show two applications on online platforms, Twitter.com and Amazon.com.

9.1 Tools

In the following we focus on ConArg and ConArgLib, respectively a solver and library based on the former stand-alone solver. Such tools can be used to compute both classical and weighted extensions, and write C++ programs that can implement decision-making procedure using them, for instance.

ConArg[16] [Bistarelli and Santini, 2011; Bistarelli *et al.*, 2016] is an Argumentation reasoner based on *Gecode*[17], which is an open, free, and efficient C++ library where to develop constraint-based applications. ConArg is able to find all the classical extensions on a given classical *AF* [Dung, 1995] and using one of the following semantics: conflict-free, admissible, complete, stable, grounded, preferred, semi-stable, eager, stage, and ideal. In addition, it can check the credulous or sceptical acceptance of a given argument with respect to semantics. Besides classical (unweighted) problems, ConArg also deals with *WAF*s, being able to solve α^γ-conflict-free, α^γ-admissible, α^γ-complete, α^γ-preferred, α^γ-stable, and α^γ-grounded extensions (as presented in Section 7). The solver is offered to users as a command-line executable, or through a Web-interface. ConArg has been extended to deal with probabilities [Bistarelli *et al.*, 2018a] and ranking-based semantics [Bistarelli *et al.*, 2019].

ConArgLib is a C++ library implemented to help developers solve some problems related to extension-based Abstract Argumentation. It

---

[16] http://www.dmi.unipg.it/conarg/.
[17] http://www.gecode.org.

represents one of the first attempts to provide a fast implementation of a library to support the solution of problems in Abstract Argumentation. A developer can use it as the basic brick to directly develop her own applications, instead of interfacing to an external solver: as an example, solving the existence of a non-empty extension, and the credulous/sceptical acceptance of arguments can be used to set-up a decision-making procedure by ranking arguments, and then selecting the decision supported by the strongest ones. ConArgLib solves all the problems solved by ConArg. Moreover, it extends it in two different ways: *i)* a developer may now choose different branching strategies (branching defines the shape of the search tree), and *ii)* can start a parallel search, using more than one thread at the same time.

9.2 Applications

The most popular applications of *WAF*s are addressed to online social platforms, as social networks and e-commerce/reviewing portals. In this case weights can be extracted from different sources: e.g., the number of repetitions the same (or similar) argument has been proposed by different users, the number of likes, the number of shares, etc. In the following sections we show two applications, both using ConArg as the underlying engine.

A further application, using VAFs to help the analysis of Twitter.com discussions, is presented in [Alsinet *et al.*, 2017].

9.2.1 Microdebates

Microdebates [Gabbriellini and Torroni, 2016] are inspired by Twitter's microblogging. A microdebate is a stream of tweets where users annotate their messages by using some special tags. Posts contain terms called hashtags, i.e. a # symbol followed by a text string, representing the stream of news the tweet belongs to. There may be more than one hashtag per post (in case the same post is related to multiple streams).

In such an application, a hashtag identifies the discussion (e.g., *#debateName*): as customary, this ensures that the tweet will appear in the right stream (a microdebate). Moreover, a user may take advantage of two additional special tags: $ and !$. *$opinionName* specifies the opinion a tweet supports, while *!$opinionName* specifies the opinion a tweet counters.

The *Microdebates App* is distributed via Google Play [Yaglikci and Torroni, 2014]. The application has a client-server architecture: the server runs a background process that manages the interaction with Twitte.comr. It maintains a MySQL database containing all the information that is shown on the client side: tweets, topics, word clouds, and other data extracted from the tweets, such as attacks and weights, needed to compute the extensions. In the Microdebates application, weights are determined by counting the tweets that express a given attack. The server retrieves from Twitter.com all new tweets about the topics listed in the database using a Java library for the Twitter.com API. To compute the extensions, the server process invokes ConArg methods for α-preferred extensions.

On the client side, the purpose is to provide the user with an interface to enter new tweets and, mainly, to navigate through microdebates. Microdebates uses computational (weighted) Argumentation to rank opinions and drive the visualisation. The result is a visual summary of the debate that takes into account semantic information such as explicit attack relations that link arguments together. If an argument a belongs to an α-preferred extension T (see Section 7), we know two things: *i)* a can peacefully coexist with the other arguments in T, the inconsistency within T being at most α, and *ii)* for every tweet attacking it, there exists at least another tweet that counters the attack. So, it is reasonable to give arguments in T the status of "popular" argument.

The authors of [Yaglikci and Torroni, 2014] conducted microdebates experiments with ten participants in the 25-34 age group. Each group was given a topic, and a 40-minute time frame to discuss the topic by using Microdebates App. At the end of 40 minutes a two-hour break was given, and then a different topic was proposed, for a total of four different topics The first two topics were *"Are occupy protest movements justified?"* (*#mdoccupy*) and *"Is nuclear energy justified and should it be expanded?"* (*#mdnuke*), whose related WAFs are represented in Figure 7.

The first microdebate has three 1-sceptically preferred arguments (contained in all the extensions) displayed in white, and five 1-credulous-

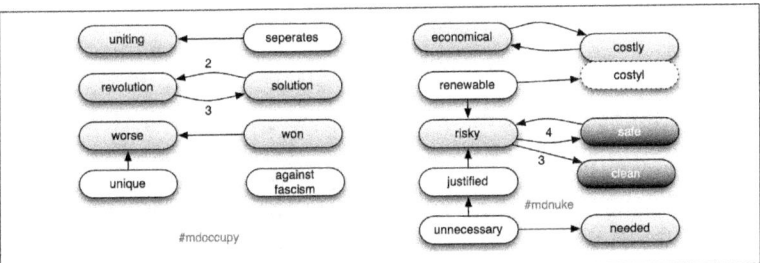

Figure 7: Two WAFs obtained with the Microdebates App, assembled through an experiment with real users (figure taken from [Yaglikci and Torroni, 2014]).

ly preferred arguments (contained in at least one extension) displayed in grey. The second microdebate has two 3-sceptically preferred arguments (white), four 3-credulously preferred arguments (grey), and two losing arguments (black).

9.2.2 Amazon.com Reviews

In [Gabbriellini and Santini, 2015] and [Gabbriellini and Santini, 2016] the authors consider 253 reviews of a selected product (a ballet tutu for kids), extracted from the "Clothing, Shoes and Jeweller" section of Amazon.com. They manually extract abstract arguments from such reviews, and they study how their characteristics, e.g., the distribution of positive (in favour of purchase) and negative ones (against purchase), change through a period of four years. Among other results, they discover that negative arguments tend to permeate also positive reviews. As a second step, by using such observations and distributions, we successfully replicate the reviewers' behaviour by simulating the review-posting process from their basic components, i.e., the arguments themselves.

Reviews are in the period between January 2009 and July 2014. A total of 24 positive and 20 negative claims are manually identified. Arguments with the same claim are grouped together, and the number in each group represents the weight associated with them: for example, "The kid loved it" is a positive claim with a strength of 78, while "The tutu has a bad quality" is a negative claim with a weight of 18. Also attacks were manually extracted and the resulting AF is represented in Figure 8.

The set of classical stable (and semi-stable) extensions counts 256 different instances, which exactly correspond also to the set of preferred extensions. Complete extensions are $6,651$. Since all such information is complicated to be somehow analysed and interpreted (attacks are almost all symmetric), the authors switch to using the weighted Argumentation approach and the relaxation described in Section 6. In this case, it is necessary to relax defence to $\gamma = 22$ in order to obtain more than zero 0^γ-stable extensions, i.e., 16 in this case. Hence, we obtain a small subset of stable extensions with no internal conflict, but able to counter-attack better than the attack they receive. This set represents a refined subset of the original 256 stable extensions obtained when considering the graph as unweighted.

10 Summary and Future Research

In this chapter we have surveyed Weighted Argumentation Frameworks, or WAFs for short. More in particular, we summarised different keypoints of their formalisation, relaxation, semantics, and applications. Such frameworks can be useful when a more fine-grained level of detail is necessary to measure the acceptability of arguments by looking at the strength of attacks.

Preference systems, both qualitative and quantitative, are used in this chapter to refine the notion of defence: different preference systems are used in the literature, from simple values in $\mathbb{R}$ to more general and parametric structures as c-semirings, which also deal with partially-ordered attacks. Having weights as labels allows for tolerating a certain amount of attack among arguments: some works name this inconsistency budget α or β, which is the "sum" of the disregarded weights. Besides such an internal (with respect to extensions) relaxation, also defence can be mitigated by thus allowing more arguments to be defended by the same set: a complete relaxation brings to totally ignore the difference between attack and defence weights, consequently leading to classical defence [Dung, 1995]. In addition, extension-based semantics can be revised according to previous concepts, and it is then possible

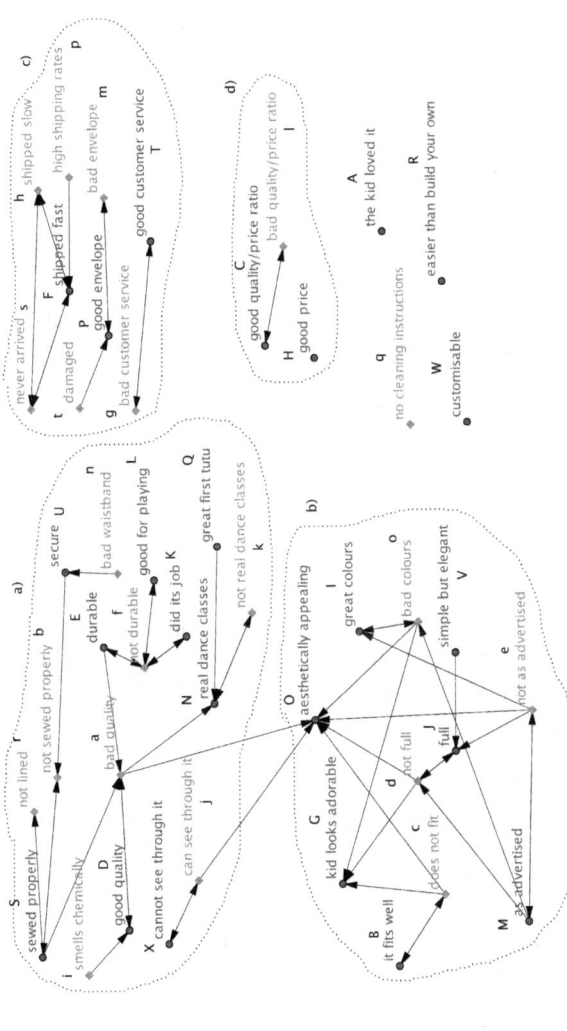

Figure 8: The *AF* manually extracted from 253 reviews about a ballet tutu for kids sold in Amazon.com. Abstract arguments represented by (blue) circles claim in favour of purchasing the tutu, while the ones represented by (red) rhombuses are against purchasing. Claims are grouped in four clusters concerning the product quality (a), aspect (b), shipping (c), and price (d). Figure taken from [Gabbriellini and Santini, 2015].

to define weighted complete, preferred-extensions, and so on. Having redefined *AF*s to *WAF*s, also classical properties of frameworks need to be revised, as for example whether a framework is well-founded or not: in well-founded frameworks, the grounded extension as well as the complete/preferred/stable extensions coincide. Finally, we have presented how *WAF*s naturally find an application to the world of social-media, where information is rated by users, and such scores can be easily converted to preference values for what posted.

Future research. Despite the quite large number of proposals in the literature, there is still a considerable number of open research lines concerning *WAF*s. For example, a few works go beyond the definition of weighted semantics and investigate if well-established properties that hold for classical frameworks in [Dung, 1995] also hold in *WAF*s. Section 8 explores well-founded frameworks, but other properties, such as for example, the existence of (weighted) stable extensions, or the uniqueness of a (weighted) preferred extension, need to be generalised as well. First steps towards formalising these results initiated in [Bistarelli and Santini, 2019].

In addition, considering Section 3, the different translation methods of weights from arguments to attacks are not proven to maintain the semantics between the original framework and the destination one. The relations among these approaches need to be examined further.

Moreover, complexity results are only clearly stated in [Dunne et al., 2011]: the β versions of the decision problems are in fact no harder (although of course no easier) than the corresponding unweighted decision problems, except for the grounded extension: in [Dunne et al., 2011] there are several grounded extensions, and consequently the computation of e.g. the sceptical acceptance of an argument is coNP-complete, instead of being trivial as in [Dung, 1995]. More general results along this direction are needed, considering also the other presented approaches.

One additional path to investigate concerns *Weighted Bipolar Argumentation Frameworks* [Cayrol and Lagasquie-Schiex, 2005a], where the set of edges in a graph is bipartite into support and attack relationships. For example, the proposal in [Amgoud and Ben-Naim, 2018] adopts weights on arguments: it will be interesting to check how compensating defence and attack techniques can be adapted to support and attack edges, and how the relaxation presented in Section 6 can be used in bipolar settings. Along the same line, we can study *weighted Higher-order Argumentation Frameworks*, where attacks may target attacks, and they are in turn associated with a weight.

Finally, existing applications prove that weighted frameworks are really useful when paired to social platforms as Twitter.com or Amazon.com, because such systems natively encourage users to rate posts or reviews. We reckon that the development of *Argument Mining* techniques on social platforms and the analysis of threads via *WAF*s can foster the use of such platforms (or dedicated "rooms" in them) as debating systems, undermining the effect of fake news and summarising how the discussion is structured.

References

[Alsinet et al., 2017] Teresa Alsinet, Josep Argelich, Ramón Béjar, Cèsar Fernández, Carles Mateu, and Jordi Planes. Weighted argumentation for analysis of discussions in twitter. *Int. J. Approx. Reason.*, 85:21–35, 2017.

[Amgoud and Ben-Naim, 2018] Leila Amgoud and Jonathan Ben-Naim. Weighted bipolar argumentation graphs: Axioms and semantics. In *Proceedings of the Twenty-Seventh International Joint Conference on Artificial Intelligence, IJCAI*, pages 5194–5198. ijcai.org, 2018.

[Amgoud and Cayrol, 1998] Leila Amgoud and Claudette Cayrol. On the acceptability of arguments in preference-based argumentation. In *UAI '98: Proceedings of the Fourteenth Conference on Uncertainty in Artificial Intelligence*, pages 1–7. Morgan Kaufmann, 1998.

[Amgoud et al., 2017] Leila Amgoud, Jonathan Ben-Naim, Dragan Doder, and Srdjan Vesic. Acceptability semantics for weighted argumentation frameworks. In *Proceedings of the Twenty-Sixth International Joint Conference on Artificial Intelligence, IJCAI*, pages 56–62. ijcai.org, 2017.

[Atkinson and Bench-Capon, 2021] Katie Atkinson and Trevor Bench-Capon. Value-based argumentation. In Dov Gabbay, Massimiliano Giacomin, Guillermo R. Simari, and Matthias Thimm, editors, *Handbook of Formal Argumentation*, volume 2, chapter 5. College Publications, 2021.

[Baroni et al., 2018] Pietro Baroni, Dov M Gabbay, Massimiliano Giacomin, and Leendert van der Torre. *Handbook of formal argumentation*. College Publications, 2018.

[Barringer et al., 2005] Howard Barringer, Dov M. Gabbay, and John Woods. Temporal dynamics of support and attack networks: From argumentation to zoology. In *Mechanizing Mathematical Reasoning, Essays in Honor of Jörg H. Siekmann on the Occasion of His 60th Birthday*, volume 2605 of *Lecture Notes in Computer Science*, pages 59–98. Springer, 2005.

[Baumann and Spanring, 2015] Ringo Baumann and Christof Spanring. Infinite argumentation frameworks - on the existence and uniqueness of extensions. In *Advances in Knowledge Representation, Logic Programming, and Abstract Argumentation*, volume 9060 of *LNCS*, pages 281–295. Springer, 2015.

[Bench-Capon, 2003] Trevor J. M. Bench-Capon. Persuasion in practical argument using value-based argumentation frameworks. *J. Log. Comput.*, 13(3):429–448, 2003.

[Besnard and Hunter, 2001] Philippe Besnard and Anthony Hunter. A logic-based theory of deductive arguments. *Artif. Intell.*, 128(1-2):203–235, 2001.

[Bistarelli and Santini, 2010] Stefano Bistarelli and Francesco Santini. A common computational framework for semiring-based argumentation systems. In *ECAI 2010 - 19th European Conference on Artificial Intelligence*, volume 215 of *Frontiers in Artificial Intelligence and Applications*, pages 131–136. IOS Press, 2010.

[Bistarelli and Santini, 2011] Stefano Bistarelli and Francesco Santini. Conarg: A constraint-based computational framework for argumentation systems. In *IEEE 23rd International Conference on Tools with Artificial Intelligence, ICTAI*, pages 605–612. IEEE Computer Society, 2011.

[Bistarelli and Santini, 2017] Stefano Bistarelli and Francesco Santini. A hasse diagram for weighted sceptical semantics with a unique-status grounded semantics. In *Logic Programming and Nonmonotonic Reasoning - 14th International Conference, LPNMR 2017*, volume 10377 of *Lecture Notes in Computer Science*, pages 49–56. Springer, 2017.

[Bistarelli and Santini, 2019] Stefano Bistarelli and Francesco Santini. Well-foundedness in weighted argumentation frameworks. In *Logics in Artificial Intelligence - 16th European Conference, JELIA*, volume 11468 of *Lecture Notes in Computer Science*, pages 69–84. Springer, 2019.

[Bistarelli et al., 1997] Stefano Bistarelli, Ugo Montanari, and Francesca Rossi. Semiring-based constraint satisfaction and optimization. *J. ACM*, 44(2):201–236, 1997.

[Bistarelli et al., 2016] Stefano Bistarelli, Fabio Rossi, and Francesco Santini. Conarg: A tool for classical and weighted argumentation. In *Computational Models of Argument - Proceedings of COMMA*, volume 287 of *FAIA*, pages 463–464. IOS Press, 2016.

[Bistarelli et al., 2018a] Stefano Bistarelli, Theofrastos Mantadelis, Francesco Santini, and Carlo Taticchi. Probabilistic argumentation frameworks with metaproblog and conarg. In *IEEE 30th International Conference on Tools with Artificial Intelligence, ICTAI*, pages 675–679. IEEE, 2018.

[Bistarelli et al., 2018b] Stefano Bistarelli, Fabio Rossi, and Francesco Santini. A novel weighted defence and its relaxation in abstract argumentation. *Int. J. Approx. Reason.*, 92:66–86, 2018.

[Bistarelli et al., 2019] Stefano Bistarelli, Francesco Faloci, and Carlo Taticchi. Implementing ranking-based semantics in conarg. In *31st IEEE International Conference on Tools with Artificial Intelligence, ICTAI*, pages 1180–1187. IEEE, 2019.

[Bonzon et al., 2016] Elise Bonzon, Jérôme Delobelle, Sébastien Konieczny, and Nicolas Maudet. A comparative study of ranking-based semantics for abstract argumentation. In *Proceedings of the Thirtieth AAAI Conference on Artificial Intelligence*, pages 914–920. AAAI Press, 2016.

[Bourguet et al., 2010] Jean-Rémi Bourguet, Leila Amgoud, and Rallou Thomopoulos. Towards a unified model of preference-based argumentation. In *Foundations of Information and Knowledge Systems, 6th International Symposium, FoIKS*, volume 5956 of *Lecture Notes in Computer Science*, pages 326–344. Springer, 2010.

[Cayrol and Lagasquie-Schiex, 2005a] Claudette Cayrol and Marie-Christine Lagasquie-Schiex. Gradual valuation for bipolar argumentation frameworks. In Lluís Godo, editor, *Symbolic and Quantitative Approaches to Reasoning with Uncertainty, 8th European Conference, ECSQARU*, volume 3571 of *Lecture Notes in Computer Science*, pages 366–377. Springer, 2005.

[Cayrol and Lagasquie-Schiex, 2005b] Claudette Cayrol and Marie-Christine Lagasquie-Schiex. Graduality in argumentation. *J. Artif. Intell. Res.*, 23:245–297, 2005.

[Cayrol and Lagasquie-Schiex, 2013] Claudette Cayrol and Marie-Christine Lagasquie-Schiex. From preferences over arguments to preferences over attacks in abstract argumentation: A comparative study. In *25th IEEE International Conference on Tools with Artificial Intelligence, ICTAI*, pages 588–595. IEEE Computer Society, 2013.

[Cayrol et al., 2010] Claudette Cayrol, Caroline Devred, and Marie-Christine Lagasquie-Schiex. Acceptability semantics accounting for strength of attacks in argumentation. In *ECAI 2010 - 19th European Conference on Artificial Intelligence*, volume 215 of *Frontiers in Artificial Intelligence and Applications*, pages 995–996. IOS Press, 2010.

[Coste-Marquis et al., 2012] Sylvie Coste-Marquis, Sébastien Konieczny, Pierre Marquis, and Mohand Akli Ouali. Weighted attacks in argumentation frameworks. In *Principles of Knowledge Representation and Reasoning: Proceedings of the Thirteenth International Conference, KR*, pages 593–597. AAAI Press, 2012.

[da Costa Pereira *et al.*, 2011] Célia da Costa Pereira, Andrea Tettamanzi, and Serena Villata. Changing one's mind: Erase or rewind? In *IJCAI 2011, Proceedings of the 22nd International Joint Conference on Artificial Intelligence*, pages 164–171. IJCAI/AAAI, 2011.

[Dung, 1995] Phan Minh Dung. On the acceptability of arguments and its fundamental role in nonmonotonic reasoning, logic programming and n-person games. *Artificial Intelligence*, 77:321–357, 1995.

[Dung, 2016] Phan Minh Dung. An axiomatic analysis of structured argumentation with priorities. *Artif. Intell.*, 231:107–150, 2016.

[Dunne *et al.*, 2011] Paul E. Dunne, Anthony Hunter, Peter McBurney, Simon Parsons, and Michael Wooldridge. Weighted argument systems: Basic definitions, algorithms, and complexity results. *Artificial Intelligence*, 175(2):457–486, 2011.

[Egilmez *et al.*, 2013] Sinan Egilmez, João Martins, and João Leite. Extending social abstract argumentation with votes on attacks. In *Theory and Applications of Formal Argumentation - Second International Workshop, TAFA*, volume 8306, pages 16–31. Springer, 2013.

[Gabbriellini and Santini, 2015] Simone Gabbriellini and Francesco Santini. A micro study on the evolution of arguments in amazon.com's reviews. In *PRIMA 2015: Principles and Practice of Multi-Agent Systems - 18th International Conference*, volume 9387 of *LNCS*, pages 284–300. Springer, 2015.

[Gabbriellini and Santini, 2016] Simone Gabbriellini and Francesco Santini. From reviews to arguments and from arguments back to reviewers' behaviour. In *Agents and Artificial Intelligence - 8th International Conference, ICAART*, volume 10162 of *LNCS*, pages 56–72. Springer, 2016.

[Gabbriellini and Torroni, 2016] Simone Gabbriellini and Paolo Torroni. Microdebates: Structuring debates without a structuring tool. *AI Commun.*, 29(1):31–51, 2016.

[García and Simari, 2004] Alejandro Javier García and Guillermo Ricardo Simari. Defeasible logic programming: An argumentative approach. *TPLP*, 4(1-2):95–138, 2004.

[Hunter *et al.*, 2021] Anthony Hunter, Sylwia Polberg, Nico Potyka, Tjitze Rienstra, and Matthias Thimm. Probabilistic argumentation: A survey. In Dov Gabbay, Massimiliano Giacomin, Guillermo R. Simari, and Matthias Thimm, editors, *Handbook of Formal Argumentation*, volume 2, chapter 7. College Publications, 2021.

6 - WEIGHTED ARGUMENTATION

[Kaci and Labreuche, 2011] Souhila Kaci and Cristophe Labreuche. Arguing with valued preference relations. In *Symbolic and Quantitative Approaches to Reasoning with Uncertainty - 11th European Conference, ECSQARU*, volume 6717 of *Lecture Notes in Computer Science*, pages 62–73. Springer, 2011.

[Kaci and Labreuche, 2014] Souhila Kaci and Christophe Labreuche. Valued preference-based instantiation of argumentation frameworks with varied strength defeats. *Int. J. Approx. Reason.*, 55(9):2004–2027, 2014.

[Kaci et al., 2021] Souhila Kaci, Leendert van der Torre, Srdjan Vesic, and Serena Villata. Preference in abstract argumentation. In Dov Gabbay, Massimiliano Giacomin, Guillermo R. Simari, and Matthias Thimm, editors, *Handbook of Formal Argumentation*, volume 2, chapter 3. College Publications, 2021.

[Leite and Martins, 2011] João Leite and João Martins. Social abstract argumentation. In *IJCAI 2011, Proceedings of the 22nd International Joint Conference on Artificial Intelligence*, pages 2287–2292. IJCAI/AAAI, 2011.

[Martínez et al., 2008] Diego C. Martínez, Alejandro Javier García, and Guillermo Ricardo Simari. An abstract argumentation framework with varied-strength attacks. In *Principles of Knowledge Representation and Reasoning: Proceedings of the Eleventh International Conference*, pages 135–144. AAAI Press, 2008.

[Modgil and Prakken, 2014] Sanjay Modgil and Henry Prakken. The $ASPIC^+$ framework for structured argumentation: a tutorial. *Argument & Computation*, 5(1):31–62, 2014.

[Modgil and Prakken, 2017] Sanjay Modgil and Henry Prakken. Abstract rule-based argumentation. *FLAP*, 4(8), 2017.

[Modgil, 2009] Sanjay Modgil. Reasoning about preferences in argumentation frameworks. *Artif. Intell.*, 173(9-10):901–934, 2009.

[Prakken and Sartor, 1997] Henry Prakken and Giovanni Sartor. Argument-based extended logic programming with defeasible priorities. *Journal of Applied Non-Classical Logics*, 7(1):25–75, 1997.

[Yaglikci and Torroni, 2014] Nefise Yaglikci and Paolo Torroni. Microdebates app for android: A tool for participating in argumentative online debates using a handheld device. In *26th IEEE International Conference on Tools with Artificial Intelligence, ICTAI*, pages 792–799. IEEE Computer Society, 2014.

CHAPTER 7

PROBABILISTIC ARGUMENTATION: A SURVEY

ANTHONY HUNTER
Department of Computer Science, University College London, United Kingdom
anthony.hunter@ucl.ac.uk

SYLWIA POLBERG
School of Computer Science and Informatics, Cardiff University, United Kingdom
polbergs@cardiff.ac.uk

NICO POTYKA
Institute for Parallel and Distributed Systems, University of Stuttgart, Germany
nico.potyka@ipvs.uni-stuttgart.de

TJITZE RIENSTRA
Institute for Web Science and Technologies, University of Koblenz-Landau, Germany
rienstra@uni-koblenz.de

MATTHIAS THIMM
Institute for Web Science and Technologies, University of Koblenz-Landau, Germany
thimm@uni-koblenz.de

Abstract

Argumentation is inherently pervaded by uncertainty, which can arise as a result of the context in which argumentation is used, the kinds of agents that are involved in a given situation, the types of arguments that are used, and more. One of the prominent approaches for handling uncertainty in argumentation is probabilistic argumentation, which offers means of quantifying the level of uncertainty we are dealing with. This chapter offers an overview of the state-of-the-art research in this area.

1 Introduction

Argumentation is inherently pervaded by uncertainty. One of the core concepts of defeasible reasoning and therefore argumentation is the fallibility of human perception, which forces us to be able to reason even with incomplete information and to be prepared to retract our conclusions in the face of new data. This is further compounded by applying argumentation in real-life situations, where uncertainty can arise as a result of the context in which argumentation is used, the kinds of agents that are involved in a given situation, the types of arguments that are used, and more. Thus, just like there are multiple sources of uncertainty in argumentation, there are multiple proposals in the literature towards modelling this kind of reasoning. One of the prominent approaches is *probabilistic argumentation*, which often offers means of quantifying the level of uncertainty we are dealing with. While using probability theory as a means to model uncertain aspects of argumentation has been questioned even by Pollock [1995], recent developments in the field showed its adequacy in practical matters and from the perspective of artificial discussion, see also [Verheij, 2014] for a discussion on this. This chapter offers an overview of the state-of-the-art research in this area, and we start by considering various examples that can benefit from incorporating probabilities.

Example 1.1 (Taken from [Hunter, 2014]). *Suppose that there are two witnesses to a criminal escaping in a car. The first witness says that the getaway car is red, and the other witness says that the getaway car is orange. If we take a strict interpretation of the colours, then we have two arguments a and b below, where each argument attacks the other.*

- a = "The getaway car is red".

- b = "The getaway car is orange".

For these arguments, it may be inappropriate to treat "red" and "orange" as contradictory. There is some ambiguity, and hence some imprecision, in the use of these terms. And so, it may be possible to regard these two terms as consistent together. So if we consider the argument graph, there is some uncertainty as to whether a attacks b and vice versa.

The above example highlights that uncertainty may arise when there is ambiguity, a form of imprecision in the language used in the arguments. Another important reason for uncertainty is that real-world arguments presented in natural language are normally enthymemes [Walton, 1989], i.e. arguments with a support that is insufficient for the claim to be entailed and/or a claim that is incomplete. This means that a given enthymeme could be completed into a full argument in more ways than one, and every interpretation may have a certain probability of being the intended one. We consider this in the next example.

Example 1.2 (Taken from [Hunter, 2014]). *Consider the following arguments:*

- a = "The sun is shining now, we should organize a BBQ for this evening".

- b = "The weather report predicts rain this evening".

Here, a is an argument that has incomplete premises for obtaining the claim (i.e. the premise "The sun is shining now" is insufficient for entailing the claim "We should organize a BBQ for this evening"). And b is an argument that has the premise "The weather report predicts rain this evening", but lacks a claim. Implicitly, the claim of b should negate the premises or the claim of argument a.

When a counterargument is an enthymeme, there may be uncertainty as to whether the argument being attacked is attacked because a premise is being contradicted or because the claim is being contradicted. In the above example, it could be that the implicit claim of b is negating the premise "The sun is shining now", or negating the claim "We should organize a BBQ for this evening". Also, because a is an enthymeme with incomplete premises, b could have a claim that contradicts a missing premise of a. For instance, suppose we make the premises of argument a explicit as follows:

- a_1 = "The sun is shining now".
- a_2 = "If the sun is shining now, it will be warm and dry this evening".
- a_3 = "If it is warm and dry this evening, then we should organize a BBQ for this evening".

So if the argument is explicit, it will have premises a_1, a_2 and a_3, and the claim "We should organize a BBQ for this evening". Then the claim of b can negate some combination of the explicit premises and/or the claim of a. Similarly, we can make the argument b explicit.

- b_1 = "If the weather report predicts rain this evening, then it will not be warm and dry this evening".
- b_2 = "It will not be warm and dry this evening".

Here b_1 is used to make the premises explicit, and b_2 is used to make the claim explicit. This claim then explicitly contradicts the premises a_1 and a_2. In other words, if we regard the premises of a as being represented by a_1, a_2 and a_3, and the premises of b being represented by b and b_1, and the claim of b being b_2, then we get the argument graph visible in Figure 1a.

However, there may be other interpretations of a and/or b, such that the interpretation of b does not attack the interpretation of a, then we get the argument graph in Figure 1b. In this way, there is doubt about whether a does indeed attack b, and what the precise structure of a and b is. All interpretations of the arguments and the resulting graphs are only given with some likelihood.

Yet another kind of uncertainty concerns the degree to which a given agent believes or disbelieves an argument or its premises or claims. In real-life, people often tend to trust or agree with certain things only up to

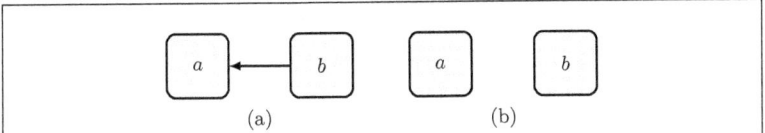

Figure 1: Possible argument graphs created from enthymemes.

a given degree, and being presented with counterarguments can weaken this belief rather than lead to outright rejection of a given argument. For instance, we might be inclined to have some doubt in the weather forecasts or in the stories of a gossiping neighbour. The way a given piece of information is presented to us also affects how we react to it, e.g., framing things a certain way or using particular language are classical strategies used in many areas, from marketing to healthcare. Finally, there is also the issue of imperfection. Human agents do not need to be perfect reasoners and can reach decisions in a biased or flawed manner; at the same time, we may accidentally deem them imperfect or not following certain rules of reasoning because they do not disclose all of their arguments and knowledge that would justify their position. Below we provide examples based on the empirical study from [Polberg and Hunter, 2018] that highlight some of the mentioned behaviours.

Example 1.3 (Taken from [Polberg and Hunter, 2018]). *Consider two listeners to the following discussion on flu shots between agents a and b.*

- $a_1 =$ *Hospital staff members do not need to receive flu shots.*

- $b_2 =$ *Hospital staff members are exposed to the flu virus a lot. Therefore, it would be good for them to receive flu shots in order to stay healthy.*

- $a_3 =$ *The virus is only airborne and it is sufficient to wear a mask in order to protect yourself. Therefore, a vaccination is not necessary.*

- $b_4 =$ *The flu virus is not just airborne, it can be transmitted through touch as well. Hence, a mask is insufficient to protect yourself against the virus.*

- a_5 = *The flu vaccine causes flu in order to gain immunity. Making people sick, who otherwise might have stayed healthy, is unreasonable.*

- b_6 = *The flu vaccine does not cause flu. It only has some side effects, such as headaches, that can be mistaken for flu symptoms.*

When asked, the first listener identifies certain conflicts between what the agents are uttering, as visible in Figure 2a. She admits to strongly agreeing with everything that agent b has said, however, she does not think that what agent a says is entirely wrong. She still, though to a small degree, agrees with a_1, and a_3 prompts only slight disagreement (a_5 is completely disbelieved). She explains that while it is indeed a great idea to get vaccinated and it would be very beneficial to do so, she does not think that "needing" to do that is appropriate.

In addition to the conflicts identified by the first listener, the second listener believes that a_5 attacks b_4 and b_6 attacks a_3 (see Figure 2b). However, she holds a clear anti-vaccine stance and agrees (resp. disagrees) with everything agent a (resp. b) says. The listener states that agent b is expressing lies and inaccuracies and rejects the agent's views without providing any particular kind of evidence to the contrary, thus exhibiting behaviour that could be deemed as biased or not rational.

The aforementioned examples highlight various kinds of uncertainty that can arise concerning the structure of the argument graph or the degree to which a given argument is accepted. While they are fairly straightforward, they are also quite common and reflect the ambiguity and uncertainty that is present in daily life and communication. They also represent challenges that any application of argumentation, particularly one involving human agents, will need to tackle.

The kinds of scenarios we have considered can be conveniently modelled using probabilistic argumentation, particularly with the constellations and the epistemic approaches [Hunter, 2013]. In the *constellations approach*, the uncertainty is in the topology of the graph, by which we understand that certain arguments or relations appear in the graph only with a given probability. This approach is useful when one agent is not sure what arguments and attacks another agent is aware of, or if ambiguity or imprecision of the arguments causes uncertainty in the structure

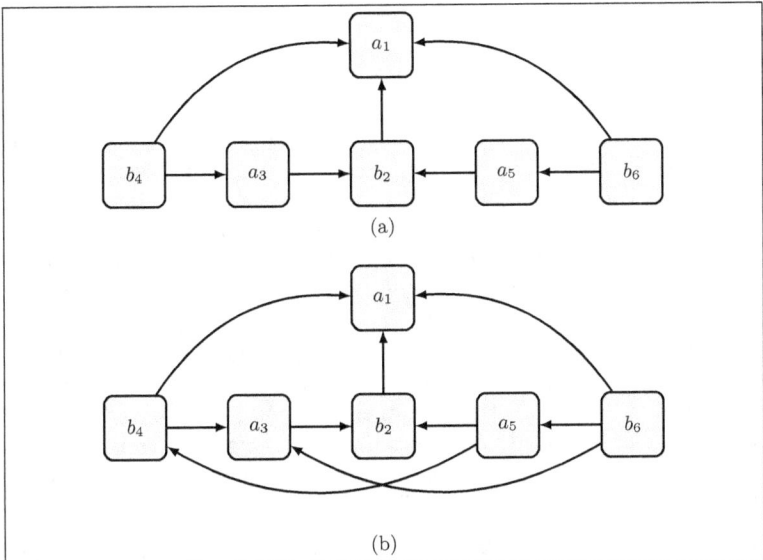

Figure 2: Example of attacks between arguments acquired by two listeners of a vaccine discussion.

of the graph. In the *epistemic approach*, the topology of the argument graph is fixed, but there is uncertainty as to the degree to which each argument is believed. This method can be harnessed when an agent is not certain of their own or another agent's opinion on the arguments, if a given situation calls for a fine-grained way to judge arguments, or if the perceived reasoning of a given agent escapes classical argumentation semantics. In this chapter we will discuss these and further kinds of probabilistic argumentation in more detail and explain how our examples can be modelled.

In Section 2 we recall certain basic notions that we will use throughout this chapter. Sections 3 and 4 investigate the epistemic and constellation approach to probabilistic argumentation, respectively, and Section 5 is dedicated to discussing other kinds of probabilistic approaches. We close with additional discussion and conclusions in Section 6.

2 Preliminaries

Within argumentation we frequently distinguish between the *structured* (or logic-based) and *abstract* approaches. The former define the relations between arguments in terms of the assumed internal structure of these arguments, and often offer means of constructing them from an underlying knowledge base or data source [Besnard and Hunter, 2008; Modgil and Prakken, 2014; Toni, 2014; Garcia and Simari, 2004]. In contrast, *abstract argumentation* [Dung, 1995; Brewka et al., 2014] takes a very simple view on argumentation as it does not presuppose any internal structure of an argument. The arguments and relations between them are assumed to have been constructed, and the focus is put on determining what arguments can be deemed acceptable or not based on how they interact with each other. Various negative and positive kinds of relations have been studied in the literature [Brewka *et al.*, 2014], however, we will focus on the original argumentation framework by Dung [1995] that considers only a binary attack relation.

Definition 2.1. *An abstract argumentation framework AF is a tuple $AF = (Ar, att)$ where Ar is a set of arguments and att is a relation $att \subseteq Ar \times Ar$.*

Let $\mathbb{A}$ denote the set of all abstract argumentation frameworks. For two arguments $a, b \in Ar$ the relation $(a, b) \in att$ means that argument a attacks argument b. With $b^- = \{a \mid (a, b) \in att\}$ we denote the set of attackers of an argument b. For convenience, with $S^- = \bigcup_{a \in S} a^-$ we will denote the set of all attackers of a set of arguments $S \subseteq Ar$.

Abstract argumentation frameworks can be concisely represented by directed graphs, where arguments are represented as nodes and edges model the attack relation (see Figure 2)[1]. The status of a given argument is determined through the appropriate *argumentation semantics*, which often produce answers in the form of extensions [Dung, 1995] or labellings [Wu and Caminada, 2010]. In this work, we use the latter, though we note that for Dung's frameworks, they can be used interchangeably.

---

[1] Note that we only consider finite argumentation frameworks here, i.e., argumentation frameworks with a finite number of arguments.

Definition 2.2. *A labelling $\mathcal{L}ab$ for an abstract argumentation framework $AF = (Ar, att)$ is a function $\mathcal{L}ab : Ar \to \{\text{in}, \text{out}, \text{undec}\}$.*

A labeling $\mathcal{L}ab$ assigns to each argument $a \in Ar$ either the value in, meaning that the argument is accepted, out, meaning that the argument is rejected, or undec, meaning that the status of the argument is undecided. Let $\text{in}(\mathcal{L}ab) = \{a \mid \mathcal{L}ab(a) = \text{in}\}$ and $\text{out}(\mathcal{L}ab)$ resp. $\text{undec}(\mathcal{L}ab)$ be defined analogously. The set $\text{in}(\mathcal{L}ab)$ for a labelling $\mathcal{L}ab$ is also called *extension* [Dung, 1995].

We now recall the three basic semantics of argumentation frameworks: the conflict-free, admissible, and complete semantics[2]. Conflict-freeness represents a certain notion of consistency, i.e., we can jointly accept only those arguments that are not in conflict. The intuition behind admissibility is that an argument can only be accepted if all attackers are rejected and if an argument is rejected then there has to be some reasonable grounds. The idea behind the completeness property is that the status of an argument is only undec if it cannot be classified as neither in nor out.

Definition 2.3. *Let $\mathcal{L}ab : Ar \to \{\text{in}, \text{out}, \text{undec}\}$ be a labelling for $AF = (Ar, att)$. We say that $\mathcal{L}ab$ is*

- *conflict-free ($\mathcal{CF}$) if for no $a, b \in \text{in}(L)$ we have that $a \in b^-$,*

- *admissible ($\mathcal{AD}$) if and only if it is conflict-free and for every $a \in Ar$*

  - *if $\mathcal{L}ab(a) = \text{out}$ then there is $b \in a^-$ with $\mathcal{L}ab(b) = \text{in}$, and*
  - *if $\mathcal{L}ab(a) = \text{in}$ then $\mathcal{L}ab(b) = \text{out}$ for all $b \in a^-$,*

- *complete ($\mathcal{CO}$) if and only if it is admissible and for every $a \in Ar$, if $\mathcal{L}ab(a) = \text{undec}$ then*

  - *there is no $b \in a^-$ s.t. $\mathcal{L}ab(b) = \text{in}$, and*
  - *there exists $c \in a^-$ s.t. $\mathcal{L}ab(c) \neq \text{out}$.*

---

[2] We observe that in the literature, admissibility and conflict-freeness are sometimes viewed as semantics, and sometimes as properties. Given the increased importance of these notions due to the development of various new kinds of semantics and frameworks, we choose to treat them as semantics in this work.

Different additional types of classical semantics [Dung, 1995; Caminada, 2006; Baroni et al., 2011] can be phrased by imposing further constraints such as minimality or maximality.

Definition 2.4. *Let* $Lab : Ar \to \{\text{in}, \text{out}, \text{undec}\}$ *be a complete labelling of* $AF = (Ar, att)$. *Then* Lab *is:*

- grounded *(GR) if and only if* $\text{in}(Lab)$ *is minimal,*
- preferred *(PR) if and only if* $\text{in}(Lab)$ *is maximal, and*
- stable *(ST) if and only if* $\text{undec}(Lab) = \emptyset$.

All statements on minimality/maximality are meant to be w.r.t. $\subseteq$.

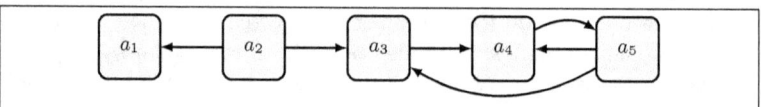

Figure 3: The argumentation framework AF from Example 2.5

Example 2.5. *Consider the abstract argumentation framework* $AF = (Ar, att)$ *where* $Ar = \{a_1, a_2, a_3, a_4, a_5\}$ *and* $att = \{(a_2, a_1), (a_2, a_3), (a_3, a_4), (a_4, a_5), (a_5, a_4), (a_5, a_3)\}$ *(see also Figure 3). The possible labellings under the admissible, complete, grounded, preferred and stable semantics for this framework are listed in Table 1.*

Labeling	a_1	a_2	a_3	a_4	a_5	AD	CO	GR	PR	ST
Lab_1	undec	undec	undec	undec	undec	✓	✗	✗	✗	✗
Lab_2	out	in	out	undec	undec	✓	✓	✓	✗	✗
Lab_3	undec	undec	out	out	in	✓	✗	✗	✗	✗
Lab_4	out	in	out	out	in	✓	✓	✗	✓	✓
Lab_5	out	in	out	in	out	✓	✓	✗	✓	✓

Table 1: Possible labellings of the framework from Figure 3.

We note that a grounded labelling is uniquely determined and always exists [Dung, 1995] and that not every framework is guaranteed to produce a stable labeling.

One of the most common reasoning problems in argumentation concerns the skeptical and credulous acceptance of arguments, by which we understand that they are accepted in all (resp. some) labellings of a given framework:

Definition 2.6. *Let $\sigma \in \{\mathcal{CF}, \mathcal{AD}, \mathcal{CO}, \mathcal{PR}, \mathcal{GR}, \mathcal{ST}\}$ be any semantics. An argument $a \in Ar$ is credulously accepted in AF wrt. σ, written $AF \mid\!\sim_\sigma^c a$, iff $a \in \text{in}(\mathcal{L}ab)$ for some σ-labeling $\mathcal{L}ab$. An argument $a \in Ar$ is skeptically accepted in AF wrt. σ, written $AF \mid\!\sim_\sigma^s a$, iff $a \in \text{in}(\mathcal{L}ab)$ for all σ-labelings $\mathcal{L}ab$. We use $\circ \in \{s, c\}$ as a symbol to refer to any inference mode.*

In Example 2.5, argument a_2 would be skeptically accepted under the complete, grounded, preferred and stable semantics. a_4 and a_5 would also be credulously accepted under the complete, preferred and stable semantics, but not under grounded.

We observe that skeptical reasoning is not considered in the case of conflict-free and admissible semantics since a labelling mapping all arguments to undec is always conflict–free and admissible. Every argument that does not attack itself will be credulously accepted under the conflict-free semantics.

3 The epistemic approach to probabilistic argumentation

We now go beyond classical three-valued semantics of abstract argumentation and turn to the epistemic approach to probabilistic argumentation [Thimm, 2012; Hunter and Thimm, 2014; Baroni et al., 2014; Hunter and Thimm, 2017]. Instead of evaluating abstract argumentation frameworks by labellings, we rely on probability functions. For this section, we define a probability function as follows.

Definition 3.1. *Let X be some finite set and 2^X its power set. A probability function P on X is a function $P : 2^X \to [0, 1]$ that satisfies*

$$\sum_{Y \in 2^X} P(Y) = 1 \quad .$$

Here, a probability function is a function on the set of subsets of some (finite) set which is *normalized*, i.e., the sum of the probabilities of all subsets is one. Let $\mathcal{P}$ be the set of all probability functions.

We use the concept of *subjective probability* [Paris, 1994] for interpreting probabilities. That is, a probability $P(Y)$ for some $Y \subseteq X$ denotes the *degree of belief* we put into Y. Then a probability function P can be seen as an epistemic state of some agent that has uncertain beliefs with respect to X. In probabilistic reasoning [Pearl, 1988; Paris, 1994], this interpretation of probability is widely used to represent and reason over uncertain knowledge.

In the following, we consider probability functions on sets of arguments of an abstract argumentation framework.

Definition 3.2. *Let $AF = (Ar, att)$ be a fixed abstract argumentation framework and let $\mathcal{P}(AF)$ be the set of probability functions of the form $P : 2^{Ar} \to [0,1]$. For $P \in \mathcal{P}(AF)$ and $a \in Ar$, the probability of a is defined as*

$$P(a) = \sum_{a \in Y \subseteq Ar} P(Y) \ .$$

Given some probability function P, the probability $P(a)$ represents the degree of belief that a is *acceptable* wrt. P. In order to bridge the gap between probability functions and labellings, consider the following definition from [Hunter, 2013].

Definition 3.3. *Let $AF = (Ar, att)$ and $P : 2^{Ar} \to [0,1]$ a probability function on Ar. The labelling $\mathcal{L}ab_P : Ar \to \{\text{in}, \text{out}, \text{undec}\}$ defined via the following constraints is called the* epistemic labelling *of P*

- $\mathcal{L}ab_P(a) = \text{in}$ *iff* $P(a) > 0.5$,
- $\mathcal{L}ab_P(a) = \text{out}$ *iff* $P(a) < 0.5$,
- $\mathcal{L}ab_P(a) = \text{undec}$ *iff* $P(a) = 0.5$.

In other words, an argument a is labelled in in $\mathcal{L}ab_P$ when it is believed to some degree (which we identify as $P(a) > 0.5$), it is labelled out when it is disbelieved to some degree (which we identify as $P(a) < 0.5$), and it is labelled undec when it is neither believed nor

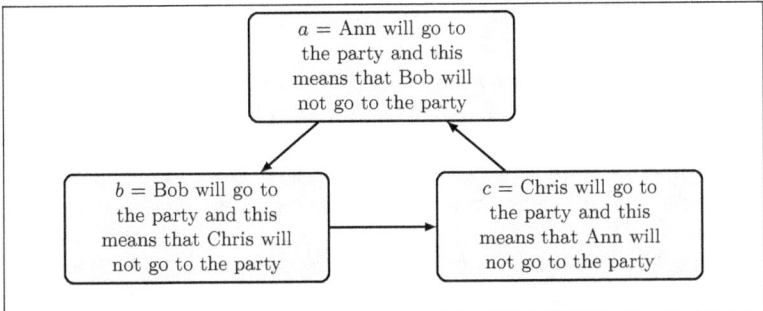

Figure 4: Example of three arguments in a simple cycle.

disbelieved (which we identify as $P(a) = 0.5$). Furthermore, the *epistemic extension* of P is the set of arguments that are labelled in by the epistemic labelling, i.e., X is an epistemic extension iff $X = \text{in}(\mathcal{L}ab_P)$. We say that a labelling $\mathcal{L}ab$ and a probability function P are *congruent*, denoted by $\mathcal{L}ab \sim P$, if for all $a \in Ar$ we have $\mathcal{L}ab(a) = \text{in} \Leftrightarrow P(a) = 1$, $\mathcal{L}ab(a) = \text{out} \Leftrightarrow P(a) = 0$, and $\mathcal{L}ab(a) = \text{undec} \Leftrightarrow P(a) = 0.5$. Note that if $\mathcal{L}ab \sim P$ then $\mathcal{L}ab = \mathcal{L}ab_P$, i.e., if a labelling $\mathcal{L}ab$ and a probability function P are congruent then $\mathcal{L}ab$ is also the epistemic labelling of P.

Example 3.4. *To further illustrate epistemic labelings and extensions, consider the graph given in Figure 4. Here, we may believe that, say, a is valid and that b and c are not valid. In which case, with this extra epistemic information about the arguments, we can resolve the conflict and so take the set $\{a\}$ as the "epistemic" extension. In contrast, there is only one admissible set which is the empty set. So by having this extra epistemic information, we get a more informed extension (in the sense that it has harnessed the extra information in a sensible way).*

In general, we want epistemic extensions to allow us to better model the audience of argumentation. In particular, we can deviate from the principles of classical labeling-based semantics for argumentation frameworks. Consider, for example, when a member of the audience of a TV debate listens to the debate at home, she can produce the abstract argument graph based on the arguments and counterarguments exchanged. Then she can identify a probability function to represent the belief she has in each of the arguments. So she may disbelieve some of the arguments based on what she knows about the topic. Furthermore, she may disbelieve some of the arguments that are unattacked. As an extreme, she is at liberty of completely disbelieving all of the arguments (so to assign probability 0 to all of them). If we want to model audiences, where the audience either does not want to or is unable to add counterarguments to an argument graph being constructed in a given argumentation scenario, we need to take the beliefs of the audience into account. Thus, we need to consider which arguments they believe or disbelieve, which may not correspond to how classical argumentation semantics would evaluate the arguments.

Nevertheless, a completely arbitrary probability function is not very informative, and there are various conditions we can impose on probability functions in order to be able to reason with argumentation frameworks in the probabilistic setting. Recall that in the classical case, semantical conditions such as *conflict-freeness* and *admissibility* (see Definition 2.3) play such a role to constrain the set of labellings. In the literature on the epistemic approach, several *rationality postulates* have been proposed to lift these conditions to the probabilistic setting, see also [Hunter and Thimm, 2017] for an overview.

The first property we consider here is a generalization of the *conflict-freeness* property. Using a probability function to interpret an abstract argumentation framework, means that—in order to reason rationally—we cannot have both high degree of belief in an argument and its attacker at the same time. Let $AF = (Ar, att)$ be an abstract argumentation framework and $P : 2^{Ar} \to [0, 1]$ a probability function over Ar. There are two variants how the notion of *conflict-freeness* can be lifted to the probabilistic case:

COH P is *coherent* wrt. AF if for every $a, b \in Ar$, if $(a, b) \in att$ then $P(a) \leq 1 - P(b)$.

RAT P is *rational* wrt. AF if for every $a, b \in Ar$, if $(a, b) \in att$ then $P(a) > 0.5$ implies $P(b) \leq 0.5$.

Both postulates model the general requirement that, if belief in an argument is high, then the belief in an argument attacked by it should be low. While RAT models a rather crisp version of the requirement, COH is a continuous interpretation.

The other important requirement of argumentative reasoning is that, if all attackers of an argument are believed to a rather low belief, then the argument should be believed to a rather high belief. In the classical setting this is the general *reinstatement principle* and implemented through the completeness property [Baroni et al., 2011]. Aspects of this requirement can be modelled through the following postulates:

FOU P is *founded* wrt. AF if $P(a) = 1$ for every $a \in Ar$ with $a^- = \emptyset$.

TRU P is *trusting* wrt. AF if for every a s.t. for every $b \in a^-$, $P(b) < 0.5$, then $P(a) > 0.5$.

OPT P is *optimistic* wrt. AF if $P(a) \geq 1 - \sum_{b \in a^-} P(b)$ for every $a \in Ar$.

The property FOU states that unattacked arguments should receive maximal degree of belief. The property TRU states that arguments whose attackers are disbelieved, should be believed. The property OPT is a generalisation of that idea and states the degree of belief in an argument should be bounded from below by one minus the sum of the beliefs in the attackers. Note that in the special case of having zero belief in all attackers this property requires to have maximal belief in a, just as in the classical case where an argument is accepted if all its attackers are defeated.

A series of further rationality postulates have been proposed in the literature. We refer to [Baroni et al., 2014; Hunter and Thimm, 2017] for a deeper discussion of this topic.

Example 3.5. *Consider the abstract argumentation framework $AF = (Ar, att)$ depicted in Figure 5 and the probability functions depicted in Table 2 (note that these functions are only partially defined by giving the probabilities of arguments). The following observations can be made:*

- *P_1 is founded and trusting, but neither rational, coherent, nor optimistic.*

- *P_2 is coherent and rational, but neither founded, trusting nor optimistic.*

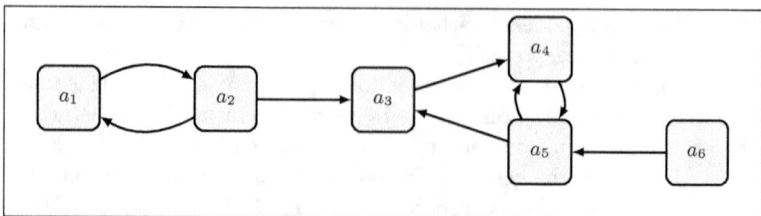

Figure 5: A simple argumentation framework

	a_1	a_2	a_3	a_4	a_5	a_6
P_1	0.2	0.7	0.6	0.3	0.6	1
P_2	0.7	0.3	0.5	0.5	0.2	0.4
P_3	0.7	0.3	0.7	0.3	0	1
P_4	0.7	0.8	0.9	0.8	0.7	1

Table 2: Some probability functions for Example 3.5

- P_3 is coherent, rational, founded, trusting and optimistic.

- P_4 is founded, trusting and optimistic but neither coherent nor rational.

There are correspondences between probability functions satisfying certain rationality postulates and labellings satisfying certain semantics [Hunter and Thimm, 2017]. For example, if $\mathcal{L}ab$ is an admissible labelling in AF then there is a probability function P satisfying COH and OPT with $\mathcal{L}ab \sim P$ [Thimm, 2012]. Moreover, if $\mathcal{L}ab$ is the grounded labelling and P the one probability function satisfying COH and OPT and having maximal *entropy*[3] then $\mathcal{L}ab \sim P$ [Thimm, 2012].

The framework developed so far already allows for reasoning with abstract argumentation frameworks in a way that incorporates probabilistic interpretations of argumentative principles. For example, fixing a set of rationality postulates, one can determine upper and lower bounds for probabilities of arguments by considering the set of a probability

---

[3] Define the *entropy* $H(P)$ of P as $H(P) = -\sum_{E \subseteq Ar} P(E) \log P(E)$.

functions satisfying these rationality postulates [Hunter and Thimm, 2017]. If evidence, i.e., bounds or correct degrees of beliefs of some of the arguments, is available, then this can be incorporated in the model and new bounds for the remaining arguments can be calculated [Hunter and Thimm, 2014; Hunter and Thimm, 2017]. In [Hunter and Thimm, 2017], also a method is proposed to allow for probabilistic reasoning if the available evidence is contradictory.

The framework discussed in this section has been extended and analysed in a number of ways. For example, [Gabbay and Rodrigues, 2015] interprets the epistemic approach within the equational approach and [Baroni et al., 2014] extends the epistemic approach to imprecise probabilities. Moreover, Prakken [Prakken, 2017; Prakken, 2018] provides some first thoughts on how probabilistic reasoning can be applied to structured argumentation approaches and relates his ideas with the epistemic approach. In the remainder of this section, we briefly discuss some further extensions.

3.1 Beliefs in Attacks

A natural extension of the epistemic approach is to not only consider degrees of beliefs of arguments but also of attacks [Polberg et al., 2017; Thimm et al., 2018b]. The intuition here is that a high belief in an attack makes the attack *effective* while low belief in attack means that the attack could almost be ignored. This allows for handling both situations in which the belief in an attack is strongly coupled with the belief one has in the argument carrying it out, as well as those where this dependency is much weaker [Polberg and Hunter, 2018].

From the technical side, this generalization can be achieved either by introducing an additional probability function over the attacks of the framework [Polberg et al., 2017] or considering a single probability function of the union of arguments and attacks [Thimm et al., 2018b]. Rationality postulates for the extended setting are then defined by taking uncertainty of attacks into account and new relationships with the classical abstract argumentation can be found [Polberg et al., 2017]. Furthermore, there are also strong relationships with the approach of [Villata et al., 2011] that introduced the notion of acceptability of attacks, see [Thimm et al., 2018b] for a detailed discussion.

3.2 Dynamics

The epistemic approach provides a static perspective on the abstract argumentation framework under consideration. However, as in classical abstract argumentation semantics, we can also consider a more dynamic setting where agents are situated in a particular environment and exchange arguments. There, agents possess opponent models of the other agents and need to update these models as the dialogue progresses. Epistemic probability distributions can be taken as opponent models, causing non-trivial issues to arise when updates have to be performed. In [Hunter and Potyka, 2017; Hunter, 2016] this setting is discussed in more detail and several approaches to updating such models are presented. Section 5.3 of this chapter offers additional discussion on dynamics and probabilistic argumentation.

3.3 Epistemic Graphs

Abstract argumentation frameworks provide a limited expressivity when it comes to the relationships between arguments. The only relationship that can be modelled is the attack, modelling the intuition that an argument cannot be accepted if one of its attackers is accepted. As the epistemic approach is built on top of abstract argumentation it suffers from the issue. However, for abstract argumentation several extensions have been defined to allow for further relationships such as support (see Chapter 1 in this handbook [Cayrol et al., 2021]). To allow for maximal flexibility, *abstract dialectical frameworks* (ADFs) [Brewka et al., 2018; Polberg, 2016] were proposed as a generalisation of abstract argumentation frameworks that allow to define the acceptability function for any argument.

We can find epistemic approaches that aim to combine probabilities with more expressive frameworks. For instance, [Potyka, 2019] introduces several epistemic postulates for modelling support. This can be further generalized to *epistemic graphs* [Hunter et al., 2018b; Hunter et al., 2020; Hunter et al., 2018a; Hunter et al., 2019] which were introduced as a general means for bringing argumentative and probabilistic reasoning capabilities together. Inspired by the freedom of ADFs, epistemic graphs allow the argument graph to be accompanied by a collection of arbitrary probabilistic constraints which impose restrictions between the degrees of belief of different arguments. The existing analysis considers both modelling and technical aspects of epistemic graphs [Hunter et al., 2018b; Hunter et al., 2020; Hunter and Polberg, 2019] as well as their use in a dynamic setting [Hunter et al., 2018a; Hunter et al., 2019], in particular with respect to opponent modelling in dialogues.

4 The constellation approach to probabilistic argumentation

Another natural approach to introduce probabilities into abstract argumentation is to consider a probability distribution over the possible graph structures of the argumentation framework. Every graph structure can then be seen as the true structure in one possible world. In this way, every possible world can again be associated with a set of accepted arguments. Namely, those arguments that are accepted in the particular graph structure under a given classical semantics. The probability of an argument can then again be computed as the probability of being in a world where the argument is accepted. This approach is called the *constellation approach* to probabilistic argumentation. Here we review the constellation approach to probabilistic argumentation from [Hunter, 2012; Hunter and Thimm, 2016], which extends the methods from [Dung and Thang, 2010] and [Li et al., 2011].

4.1 Motivating Examples

To begin with, we illustrate the usefulness of the constellation approach by means of some examples.

Example 4.1. *Consider the following arguments in a legal case:*

- $a=$ *"John says he was not in town when the robbery took place, and therefore denies being involved in the robbery."*

- $b=$ *"Peter says he was at home watching TV when the robbery took place, and therefore denies being involved in the robbery."*

- $c=$ "Harry says that he is certain that he saw John outside the bank just before the robbery took place, and he also thinks that possibly he saw Peter there too."

The arguments a and b are arguments that each claim that the speaker is not involved in the robbery, and the argument c is by a potential witness casting doubt on the premises of arguments a and b by undercutting their premises. Also, we see that argument c attacks a with explicit certainty and c attacks b with explicit uncertainty.

If we consider both attacks made by argument c, then we get the argument graph given in Figure 6a. However, if we also take into account the doubt in the attack by c on b, then we get the argument graph given in Figure 6b. This means that there is uncertainty over whether the actual argument graph should be Figure 6a or Figure 6b. We can deal with this uncertainty by regarding the set of spanning subgraphs of Figure 6a (i.e. the four subgraphs given in Figure 6) as a sample space, and assigning a probability to each of them such that the sum is 1. For instance, if Harry has only weak confidence in c attacking b, then the probabilities might be 0.2 for Figure 6a and 0.8 for Figure 6b (i.e. $P(AF_1) = 0.2$ and $P(AF_2) = 0.8$).

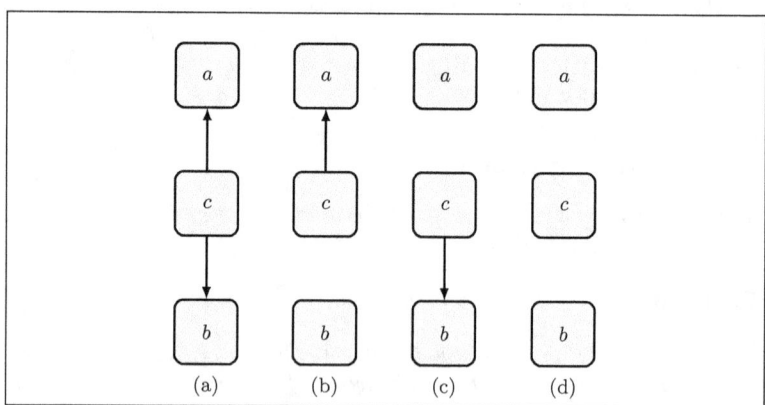

Figure 6: For argument graph AF_1, the subgraphs are (a) AF_1, (b) AF_2, (c) AF_3 and (d) AF_4.

In the example given above, there is explicit uncertainty expressed qualitatively in the attacks made by the arguments. Other situations where uncertainty arises is when there is ambiguity, a form of imprecision in the language used in the arguments, as illustrated in Example 1.1

Another reason for uncertainty in attacks is that real-world arguments presented in natural language are normally enthymemes [Walton, 1989]. An enthymeme is an argument with a support that is insufficient for the claim to be entailed and/or a claim that is incomplete. We see this in Example 1.2 (though Example 1.1 and Example 4.1 also contain enthymemes).

So to summarize, we see at least three kinds of uncertainty that arise in argumentation that we want to capture by quantifying the probability of attack.

Explicit uncertainty of attack Arguments may include some explicit qualification of the attacks made on other arguments. This explicit qualification is usually qualitative (as in Example 4.1), but sometimes it can involve quantitative qualification (such as "I am 99% sure that what John said is a lie").

Implicit imprecision of argument Many arguments have a degree of imprecision in the terminology used (as in Example 1.1). Unless all the language is formally defined, and all participants use the same definitions, it is difficult to avoid some imprecision. This means that when considering two arguments it is not always certain whether or not one attacks the other. For instance, it is possible that "red" and "orange" are consistent together as the description of the same object.

Incomplete premises/claims Most arguments in natural language are enthymemes, which means that they do not explicitly present all their premises and/or claims (as in Example 1.2). With this incompleteness, it is difficult to be certain whether one argument attacks another. If a counterargument has an explicit claim, there may be uncertainty as to whether the attacked argument has the premise that the attacker has contradicted. And if a counterargument has an implicit claim, there may be further uncertainty as to what is being contradicted.

Argumentation often involves multiple agents. This further increases uncertainty in various ways. Consider a typical argumentation scenario where one or more agents are presenting arguments in front of an audience, with the aim of each participant being to persuade the audience to adopt a certain statement. Each participant and the audience have some arguments and counterarguments in mind and may be willing to assimilate further arguments and counterargument.

From an audience's perspective, there may be uncertainty as to what arguments or attacks are in play. The audience may hear various comments in a debate, for example, but they are not sure about the exact set of arguments and attacks that are being put forward. For instance, there may be uncertainty about whether someone has put forward a complex multifaceted argument, or a number of smaller more focused arguments or there may doubt about whether some arguments are just rephrasings of previous arguments. Also, there may be uncertainty about which arguments are meant to be attacked by some argument, which occurs frequently when enthymemes are presented.

From a participant's perspective (i.e., from the perspective of an agent who is about to present arguments and/or attacks during some monological or dialogical argumentation scenario), there may be uncertainty about the audience's opinions, knowledge, values, etc. So when a participant (such as a politician) considers presenting arguments to an audience, the participant might not know for sure which arguments and attacks the audience has in mind. In other words, even before a participant has started, the audience may already have specific arguments and counterarguments in mind and the participant will be adding to those. To handle this, the participant may work with a collection of arguments and counterarguments which he/she assumes will subsume the possibilities for what is held by the audience.

So in general, whether in monological or dialogical situations, we see that there is potentially uncertainty about both arguments and attacks. To address these kinds of uncertainty, we can identify all the possible arguments and attacks that need to be considered. This creates an argument graph, and from this we can identify a probability distribution over the subgraphs of this argument graph. From this distribution, we can then determine the probability that a set of arguments is admissible, or an extension, and the probability that an argument is in an extension.

4.2 Basic definitions

The constellation approach allows us to represent the uncertainty over the topology of the graph. Each subgraph of the original graph is assigned a probability which is understood as the chances of it being the actual argument graph of the agent. It can be used to model what arguments and attacks an agent is aware of. Two important classes of subgraphs are *full subgraphs* that remove arguments, but keep all edges associated with the remaining arguments, and *spanning subgraphs* that keep all arguments, but may remove some of the edges.

Definition 4.2. *Let $AF = (Ar, att)$ and $AF' = (Ar', att')$ be two argument graphs. AF' is a subgraph of AF, denoted $AF' \sqsubseteq AF$, iff $Ar' \subseteq Ar$ and $att' \subseteq (Ar' \times Ar') \cap att$.*

- *$\wp(AF) = \{AF' \mid AF' \sqsubseteq AF\}$ denotes the set of all subgraphs of AF.*

- *A subgraph (Ar', att') is called* **full** *iff $att' = (Ar' \times Ar') \cap att$.*

- *A subgraph (Ar', att') is called* **spanning** *iff $Ar' = Ar$ and $att' \subseteq att$.*

Dependent on the application, we may want to restrict our attention to particular subgraphs. If our uncertainty is about which arguments appear in the graph, then only the full (induced) subgraphs of the argument graph have a non–zero probability. If we are only uncertain about which attacks appear, then it is only the spanning subgraphs of the argument graph that can have a non–zero probability.

Definition 4.3. *A* **subgraph distribution** *P^c is a function $P^c : \wp(AF) \to [0,1]$ with $\sum_{AF' \in \wp(AF)} P^c(AF') = 1$.*

- *P^c is a* **full subgraph distribution** *iff $P^c(AF') = 0$ whenever AF' is not a full subgraph.*

- P^c is a **spanning subgraph distribution** iff $P^c(AF') = 0$ whenever AF' is not a spanning subgraph.

Note that the above definition follows the notation of [Hunter, 2014], a slightly different notation (but equivalent formalisation) can be found in [Fazzinga et al., 2019].

The constellation approach can be applied for different purposes. One application is to define the probability that a set of arguments or a labeling follows the semantics of a particular type (e.g. grounded, preferred, etc.). This can be done by collecting the probabilities of the subgraphs producing the desired extensions or labelings.

Definition 4.4. *For $W \subseteq Ar$ and $\sigma \in \{\mathcal{CF}, \mathcal{AD}, \mathcal{CO}, \mathcal{PR}, \mathcal{GR}, \mathcal{ST}\}$, the probability $P_\sigma(\mathcal{L}ab)$ that a labeling $\mathcal{L}ab : W \to \{\text{in}, \text{out}, \text{undec}\}$ is a σ-labeling is defined as:*

$$P_\sigma(\mathcal{L}ab) = \sum_{AF' \in \wp(AF) \ s.t. \ \mathcal{L}ab \in \mathcal{L}_\sigma(AF')} P^c(AF'),$$

where $\mathcal{L}_\sigma(AF')$ is the set of all σ-labelings of AF'.

Another natural application is to define the probability that an argument is accepted under a given semantics.

Definition 4.5. *Given a semantics $\sigma \in \{\mathcal{AD}, \mathcal{CO}, \mathcal{PR}, \mathcal{GR}, \mathcal{ST}\}$, the probability that $a \in Ar$ is assigned an in status in a σ-labeling is*

$$P_\sigma(a) = \sum_{AF' \in \wp(AF) \ s.t. \ \mathcal{L}ab \in \mathcal{L}_\sigma(AF') \ and \ a \in \text{in}(\mathcal{L}ab)} P^c(AF'),$$

where $\mathcal{L}_\sigma(AF')$ is the set of all σ-labelings of AF'.

We can also define the probability that a set of arguments is an extension.

Definition 4.6. *Given a semantics $\sigma \in \{\mathcal{AD}, \mathcal{CO}, \mathcal{PR}, \mathcal{GR}, \mathcal{ST}\}$, the probability $P_\sigma(W)$ that a set $W \subseteq Ar$ is a σ-extension is:*

$$P_\sigma(W) = \sum_{AF' \in \wp(AF) \ s.t. \ \exists \mathcal{L}ab \in \mathcal{L}_\sigma(AF'): W = \text{in}(\mathcal{L}ab)} P^c(AF'),$$

where $\mathcal{L}_\sigma(AF')$ is the set of all σ-labelings of AF'.

7 - PROBABILISTIC ARGUMENTATION: A SURVEY

Example 4.7. *Consider the graph $AF = (\{a,b\}, \{(a,b)\})$. Its subgraphs are $AF_1 = (\{a,b\}, \{(a,b)\})$, $AF_2 = (\{a,b\}, \emptyset)$, $AF_3 = (\{a\}, \emptyset)$, $AF_4 = (\{b\}, \emptyset)$ and $AF_5 = (\emptyset, \emptyset)$ (see Figure 7). Out of them, AF_1, AF_3, AF_4 and AF_5 are full, and AF_1 and AF_2 are spanning. Consider the following subgraph distribution P^c: $P^c(AF_1) = 0.09$, $P^c(AF_2) = 0.81$, $P^c(AF_3) = 0.01$ and $P^c(AF_4) = 0.09$ and $P^c(AF_5) = 0$. The probability of a given set being a grounded extension is as follows: $P_{\mathcal{GR}}(\{a,b\}) = P^c(AF_2) = 0.81$; $P_{\mathcal{GR}}(\{a\}) = P^c(AF_1) + P^c(AF_3) = 0.1$; $P_{\mathcal{GR}}(\{b\}) = P^c(AF_4) = 0.09$; and $P_{\mathcal{GR}}(\{\}) = P^c(AF_5) = 0$. Therefore $P_{\mathcal{GR}}(a) = 0.91$ and $P_{\mathcal{GR}}(b) = 0.9$.*

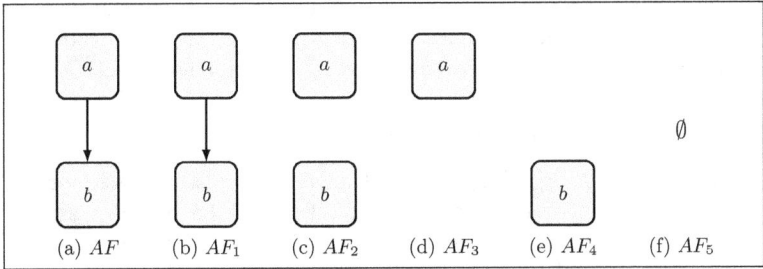

Figure 7: Argument graph AF and its subgraphs.

The proposal for the probabilistic assumption-based argumentation approach of [Dung and Thang, 2010] was the first to consider uncertainty in the topology of the argument graph. It incorporated a probability distribution over sets of arguments, where each set was effectively inducing a subgraph with a probability assignment. It also introduced that the probability of extensions for grounded extensions. Then in a proposal that explicitly considered abstract argumentation, a probability assignment to each argument and to each attack was introduced [Li et al., 2011]. From these two probability functions, a probability distribution over the subgraphs of the argument graph can be obtained. We consider this proposal in detail in the next subsection, and as we will see, it relies on independence between the arguments and between the attacks. This independence assumption does have shortcomings as we will show.

The probability of extensions (first defined by [Dung and Thang, 2010] for grounded semantics) was presented for all semantics in [Li et *al.*, 2011]. The probability of labellings was first proposed by [Riveret et al., 2017].

Probabilistic argumentation that is based on explicitly specifying a probability distribution over full subgraphs was first proposed by [Hunter, 2012; Rienstra, 2012]. Subsequently, it was extended to spanning subgraphs [Hunter and Thimm, 2014], and to all subgraphs [Hunter and Thimm, 2016].

For the probabilistic assumption-based argumentation approach of [Dung and Thang, 2010] a dialectical proof procedure was proposed for grounded semantics in [Thang, 2016] and for credulous and ideal semantics in [Hung, 2016b], and a proof procedure based on Bayesian network algorithms was proposed in [Hung, 2016a; Hung, 2017b], as well as algorithms for approximate calculations [Hung, 2018]. In the remainder of this section, we briefly discuss some further extensions.

4.3 Assuming independence

As mentioned above, in [Li et al., 2011] there is a probability assignment to each argument and to each attack. This can be regarded as the uncertainty as to whether the argument or attack should appear in the argument graph. From these two probability functions, a probability distribution over the subgraphs of the argument graph can be obtained. In this subsection, we will review this proposal.

We will consider a simplified version of the proposal by [Li et al., 2011]. In our simplification, we only consider probabilities of arguments (in contrast to the general case of [Li et al., 2011]) where probabilities of attacks are allowed as well).

Definition 4.8. *A probabilistic argumentation framework* PAF *is a triple* PAF $= (Ar, att, P)$ *where* (Ar, att) *is an abstract argumentation framework and* P *is a function* $P : Ar \to [0, 1]$.

For every argument $a \in Ar$ of a probabilistic argumentation framework PAF the value $P(a)$ is the probability that a is actually present in the argumentation framework. By assuming probabilistic independence between the presence of different arguments, we obtain a probability distribution over sets of arguments. By abuse of notation we denote this probability distribution P as well, which is defined as

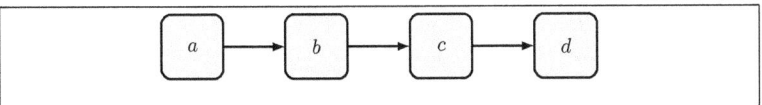

Figure 8: The argumentation framework from Example 4.9

$$P(X) = \prod_{a \in X} P(a) \prod_{a \notin X} (1 - P(a))$$

for all $X \subseteq Ar$. It can be easily shown that $\sum_{X \subseteq Ar} P(X) = 1$, so P is indeed a probability distribution. Given a set $X \subseteq Ar$ of arguments, we denote by $AF{\downarrow}_X$ the induced subgraph of X, i.e. $AF{\downarrow}_X = (X, att \cap (X \times X))$.

Let now $\sigma \in \{\mathcal{CO}, \mathcal{GR}, \mathcal{PR}\}$ be a semantics and $\circ \in \{s, c\}$ be an inference mode. The *probability of acceptance of* a, denoted by $P^{\mathsf{PAF}}_{\circ,\sigma}(a)$, is then defined via

$$P^{\mathsf{PAF}}_{\circ,\sigma}(a) = \sum_{a \in X \subseteq Ar, AF{\downarrow}_X \vdash^{\circ}_{\sigma} a} P(X) \ .$$

In other words, $P^{\mathsf{PAF}}_{\circ,\sigma}(a)$ is the sum of the probabilities of the subgraphs of (Ar, att) where a is accepted wrt. σ and $\circ$.

Example 4.9. *Let $AF = (Ar, att)$ be the abstract argumentation frameworks shown in Figure 8 and consider credulous reasoning wrt. grounded semantics. Let $\mathsf{PAF} = (Ar, att, P)$ be a probabilistic argumentation framework with $P(x) = 0.5$ for all $x \in Ar$. Table 3 lists each subset of $X \subseteq Ar$, together with the set of arguments $x \in Ar$ such that $AF{\downarrow}_X \vdash^{c}_{\mathcal{GR}} x$. For each $X \subseteq Ar$ we have $P(X) = 0.5^4 = 0.0625$. Thus, for each $x \in Ar$ we can calculate the probability $P^{\mathsf{PAF}}_{c,\mathcal{GR}}(x)$ by multiplying the number of subsets of Ar that make x accepted by 0.0625. This yields*

$$P^{\mathsf{PAF}}_{c,\mathcal{GR}}(a) = 0.5, \qquad P^{\mathsf{PAF}}_{c,\mathcal{GR}}(b) = 0.25,$$
$$P^{\mathsf{PAF}}_{c,\mathcal{GR}}(c) = 0.375, \qquad P^{\mathsf{PAF}}_{c,\mathcal{GR}}(d) = 0.3125.$$

Whilst assuming independence brings some advantages, there are situations where it is not appropriate as it does not capture the uncertainty correctly. To illustrate this, we consider the following example.

X	Accepted
$\emptyset$	$\emptyset$
a	a
b	b
a,b	a
c	c
a,c	a,c
b,c	b
a,b,c	a,c
d	d
a,d	a,d
b,d	b,d
a,b,d	a,d
c,d	c
a,c,d	a,c
b,c,d	b,d
a,b,c,d	a,c

Table 3: Choices of X and corresponding accepted arguments in Example 8

Example 4.10. *Consider the argument graph AF_1 in Figure 9 where the meaning for the arguments is as follows.*

- *a = According to John, Peter hit John first.*

- *b = According to Peter, John hit Peter first.*

Assuming that John and Peter are two children having a playground contratemps, and each is accusing the other of having thrown the first punch, we may regard the argument graph AF_1 as a good reflection of the situation. However, if we also suppose that Peter is regularly getting into fights with other children in the playground, then we might want to use probabilistic argumentation. For instance, we may regard AF_2 as the most likely scenario, AF_3 as less likely but not impossible, and AF_6 as impossible because there has clearly been punching and someone is lying. This can easily be captured in the constellations approach with

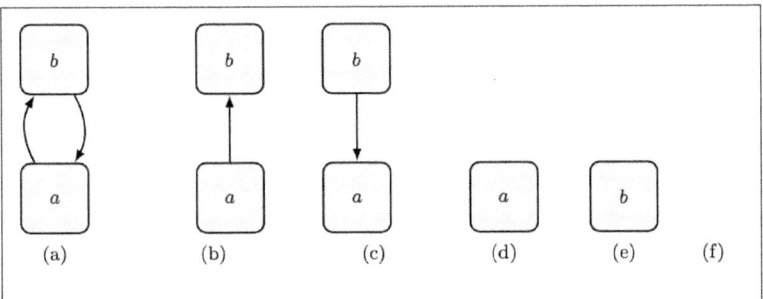

Figure 9: For argument graph AF_1, the subgraphs are (a) AF_1, (b) AF_2, (c) AF_3, (d) AF_4, (e) AF_5, and (f) AF_6 (which is the empty graph).

a probability distribution over the graphs where for instance, argument graph AF_2 has probability 0.9 and argument graph AF_3 has probability 0.1. But using Definition 4.8, we see we cannot obtain this distribution. We can only get a probability distribution over AF_1, AF_4, AF_5, and AF_6. This is because the definition forces us to drop arguments and the attacks that involve them, but does not allow us to just drop attacks.

As we said earlier, in this review, we only provided a simplified version of the proposal by [Li et al., 2011]. In our simplification, we only consider probabilities of arguments (in contrast to the general case of [Li et al., 2011]) where probabilities of attacks are allowed as well). Using the original version, we still would be unable to get the probability distribution of AF_2 having probability 0.9 and argument graph $AF3$ having probability 0.1. The following example, is a further illustration of the shortcoming of the independence assumption over attacks.

Example 4.11. *Consider argument graph AF_1 in Figure 10a where the meaning for the arguments is as follows.*

- *a = CheapAir is going bust.*

- *b = The CheapAir tickets to Paris are a bargain; We should buy them for our holiday.*

- *c = The CheapAir tickets to New York are a bargain; We should buy them for our conference trip.*

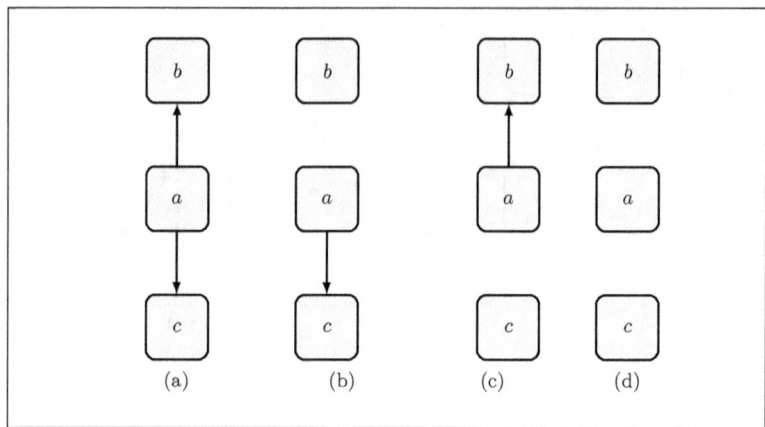

Figure 10: For argument graph AF_1, the subgraphs are (a) AF_1, (b) AF_2, (c) AF_3 and (d) AF_4.

Here the attack by a on b and the attack by a on c are not independent. For instance, if the attack by a on b is shown to be true, then there is a raised probability that the attack by a on c is true.

So the above example illustrates how assuming independence over attacks means that the example is not correctly modelled. In contrast, if we assume a probability distribution over the subgraphs, then the example can be correctly modelled as illustrated in the following example.

Example 4.12. *Continuing Example 4.11, if we start with graph AF_1 in Figure 10a, then there are four subgraphs to consider. Suppose we let the probability function over subgraphs be the following:*

$$P(AF_1) = 0.7, \qquad P(AF_2) = 0, \qquad P(AF_3) = 0, \qquad P(AF_4) = 0.3.$$

The marginals for the attacks are as follows:

$$P(a,b) = P(AF_1) + P(AF_3) = 0.7,$$
$$P(a,c) = P(AF_1) + P(AF_2) = 0.7.$$

So in some settings, it is reasonable to start with a probability distribution over arguments (and attacks), and assume independence between the arguments (and attacks) in order to construct a probability distribution over the subgraphs. In general, it may not be justified to assume independence.

4.4 Further work

For further reading concerning the constellation approach to probabilistic argumentation we refer the readers to [Hunter, 2012; Dung and Thang, 2010; Li et al., 2011; Hunter, 2013; Dondio, 2014a; Hunter, 2014]. Computational results can be found in [Dondio, 2014a; Dondio, 2014b; Fazzinga et al., 2013; Fazzinga et al., 2015; Fazzinga et al., 2019]. In [Doder and Woltran, 2014], we can find a characterization of one of the versions of the constellation approach in terms of probabilistic logic.

Whilst using the constellations approach is computationally expensive [Fazzinga et al., 2015], developments in approximation [Fazzinga et al., 2013; Fazzinga et al., 2018a; Fazzinga et al., 2016b; Fazzinga et al., 2016a; Alfano et al., 2020] and automated reasoning could be harnessed [Bistarelli et al., 2018; Mantadelis and Bistarelli, 2020].

The constellations approach has been generalized to bipolar argumentation [Fazzinga et al., 2018c; Fazzinga et al., 2018b], and to abstract dialectical frameworks [Polberg and Doder, 2014], and it has been used as the basis of a proposal for graded semantics [Thimm et al., 2018a]. In order to deal with uncertainty in the probability assignment, the constellations approach has been extended to support upper and lower bounds on the probability assignments by using credal sets [Morveli-Espinoza et al., 2019], and extended to support Dempster-Shafer theory [Hung, 2017a; Hung, 2017c].

Applications of the constellations approach include modelling the argument graph held by opponents in argumentative discussions [Hunter and Thimm, 2016], aggregating different perspectives on an argument graph [Hunter and Noor, 2020], and evaluating plans [Hung et al., 2015].

5 Other Topics in Probabilistic Argumentation

In the previous sections, we reviewed the two main approaches to probabilistic argumentation, namely the epistemic approach and the constellation approach. In this section, we will give an overview of some other topics in probabilistic argumentation. To begin with, we review some ideas for combining the epistemic and the constellation approach. Subsequently, we will look at some ideas to apply probabilistic techniques in order to learn argumentation frameworks from data. Probabilistic ideas have also been applied to model changing beliefs in argumentation. We will review some ideas in the third subsection. Finally, we sketch some ideas about how argumentation technology can be applied to enhance other probabilistic models.

5.1 Combinations of the Epistemic and the Constellations approach

The *Labelling Framework for Probabilistic Argumentation* [Riveret et al., 2018] is a framework that subsumes and combines a number of approaches considered before. The framework, based on a defeasible logic instantiation of abstract argumentation, deals with three representations of uncertainty, where each representation induces the next. These are probabilistic theory frames (PTFs), probabilistic graph frames (PGFs), and probabilistic labelling frames (PLFs). Roughly speaking, a PTF is a probability distribution over subsets of rules of a given knowledge base. A PTF induces a PGF, which is a probability distribution over subgraphs consisting of arguments constructed on the basis of a set of rules. An argument is either *on* (included) or *off* (not included) in a subgraph. Given a labelling-based semantics, a PGF induces a PLF, which is a probability distribution over labellings. Finally, a PLF can be used to calculate the probability that a statement is accepted. The notion of PGF subsumes the constellations approach but does not enforce independence of different arguments being on or off. Furthermore it is shown that there exists a correspondence between the notion of PLF and a probability distribution over extensions as used in the epistemic approach described in Section 3. Some of the postulates for the epistemic approach are shown to hold under various assumptions about the labelling-based semantics that is used. The presented framework represents a combination of both the constellations and epistemic approach, as it deals with both uncertainty about the topology of an argumentation framework, which arises from uncertainty about elements of the knowledge base, as well as the resulting uncertainty about whether argument is accepted. A similar combination of the constellations and epistemic approach appears in [Rienstra, 2012].

5.2 Learning approaches

The work [Riveret and Governatori, 2016] proposed an anytime algorithm for learning the structure of argumentation graphs from a data stream of labellings. The authors restrict their discussion to grounded labellings, but many ideas can be extended easily to other semantics. The basic idea is to start from a complete graph over the arguments and to collect information about the plausibility of attack edges as labellings arrive. For example, if two arguments are labelled in simultaneously, the attack edge between them can be eliminated. The authors present an algorithm based on 4 rules that allow increasing or decreasing the credibility of edges online while receiving labelling information. If the labellings in the stream are sampled from an unknown argumentation graph such that every labelling has a positive probability of being sampled, then the algorithm will eventually find an argumentation graph that is equivalent to the original graph (in the sense that they have the same complete extensions) with probability 1.

In a similar spirit, [Kido and Okamoto, 2017] consider the problem of estimating the attack relation between arguments from the acceptance statuses given by different users. That is, given sets of arguments accepted by different users, find an attack relation that explains the accepted arguments. To this end, the authors propose a Bayesian network model that defines the probability of extensions and acceptability statues based on the attack relation and the acceptability semantics that can be preferred, stable, grounded or complete. The authors propose definitions for the local probability models in their Bayesian network. Their main assumptions are that larger attack relations are less likely, that all semantics are equally likely and that all extensions under a given semantics and attack relation are equally likely. Probabilistic inference methods can be used to compute the most likely attack relation under the Bayesian network model given the sets of accepted arguments and an acceptability semantics. The authors demonstrate the applicability in an online discussion scenario and prove some analytical guarantees. In general, the true attack relation may not be identifiable from data alone, but in some special cases, it is.

[Riveret et al., 2017] train Boltzman machines on sets of grounded labellings and demonstrate that the trained models can be used to generate grounded labellings with a similar distribution. The sampling algorithm is sound in the sense that it is guaranteed to return only grounded labellings and complete in the sense that every grounded labelling can be generated.

5.3 Dynamics of argumentation and dialogues

The work [Hunter, 2015] proposed epistemic probabilistic argumentation as a model for the belief state of agents in persuasion dialogues. Roughly speaking, the current belief state of the persuadee is represented as a probability distribution over extensions. The persuader can posit arguments that the persuadee may accept or reject. Dependent on the persuadee's judgement, the current belief state can be updated. For this purpose, different update functions have been proposed in [Hunter, 2015] that basically redistribute probability mass such that the persuadee's feedback is taken into account. These update functions can be employed to simulate the outcome of possible persuasion dialogues in order to find an optimal persuasion strategy. Another family of update operators based on the idea of minimizing the distance to the prior beliefs has been introduced in [Hunter and Potyka, 2017] and compared to the redistribution update operators. The language for updates has been extended in [Hunter et al., 2018a] and a fragment that guarantees polynomial runtime guarantees can be found in [Potyka et al., 2019].

In [Shakarian et al., 2016], Belief Revision has been considered for structured probabilistic argumentation. The approach combines Nilsson's probabilistic logic [Nilsson, 1986] and defeasible logic programming with presumptions [Martinez et al., 2012]. Probabilistic logic is used to define an *environmental model* that captures case-specific knowledge which may be uncertain. Defeasible logic programming with presumptions is used to define an *analytical model* that contains general background knowledge which may be inconsistent when combined naively. Both models are combined via an *annotation function* that connects formulas in the environmental and analytical model. The need for belief revision results from the fact that probabilistic logic requires consistent assumptions even if defeasible logic programming does not. The authors study different types of inconsistencies and ways to perform belief revision on the environmental model, the analytical model and the annotation function. In particular, postulates for revising the analytical model and the annotation function are proposed similar to the more general belief revision literature [Hansson, 1997; Falappa et al., 2012] and representation theorems are provided. The usefulness of the approach is demonstrated by an apllication in cyber security [Shakarian et al., 2016]. The approach has been refined in [Simari et al., 2016] by taking quantitative aspects into account like how the probabilities change due to revision.

5.4 Using argumentation for probabilistic models

While this survey deals mainly with probabilistic approaches in modelling uncertainty in argumentation, there are also approaches which use argumentation as a tool to reason about probabilistic models.

[Timmer et al., 2017] proposed an argumentation-based method for explaining inferences in Bayesian networks. They note that Bayesian networks, while widely used and well-understood, are hard to interpret for non-statistical experts. Thus, argumentation may help in explaining Bayesian networks in a way that corresponds to everyday reasoning. Starting with a Bayesian network together with a chosen variable of interest, their method is based on constructing a *support graph*, which consists of reasoning chains that start with potential pieces of evidence and end with the variable of interest. The support graph is then used to construct arguments which are represented within the ASPIC+ argumentation framework [Modgil and Prakken, 2014]. However, the use of argumentation in this setting is somewhat limited when compared with typical argumentation-based methods, as one argument attacks another only if the latter is weaker. As a result, the grounded extension of the resulting argumentation framework coincides with the set of arguments that receive no attacks. This grounded extension acts as an intuitive explanation for the calculation of the likelihood ratio or posterior probability of the variable of interest.

The work [Nielsen and Parsons, 2007] developed an argumentation-based method for Bayesian network fusion. The setting they consider is of a set of agents where each agent is equipped with a Bayesian network representing the agent's domain model. They then address the problem of having the agents compromise and agree on a single Bayesian network. Their model builds on an extension of Dung's [1995] abstract argumentation model supporting set-based attacks. An argumentation framework is defined such that the preferred extensions correspond to possible compromises, as well as a formal multi-agent debate model where a debate ends with finding one of these compromises. The debate model enjoys the property that a debate is guaranteed to end up with the best possible compromise according to a given compromise score function. Note that the method developed here is concerned purely with obtaining a compromise on the *structure* of the agent's Bayesian networks. It does not specify how probabilistic parameterisations of the different Bayesian networks are be fused.

6 Summary and Conclusion

The two main approaches to probabilistic (abstract) argumentation are the constellations and the epistemic approaches. In the constellations approach, there is uncertainty over the topology of the argument graph, whereas in the epistemic approach, the topology of the argument graph is fixed, but there is uncertainty about whether an argument is believed. The epistemic approach has been extended to also allow a probability distribution over subsets of attacks, and thereby represent belief in each attack. A further approach is based on labellings for arguments using in, out, and undecided, augmented with off for denoting that the argument does not occur in the graph. A probability distribution over labellings gives a form of probabilistic argumentation that overlaps with the constellations and epistemic approaches.

There are many other ideas of how to combine probability theory and argumentation, some of which we sketched in the previous section. Others, in particular those relating to probabilistic reasoning with structured argumentation such as [Haenni, 2009] had to left out of this survey.

One particular natural application is learning argumentation graphs from data. Since data is usually noisy and uncertain, probability theory is a natural tool in this area. The additional expressiveness of probabilistic argumentation also raises new questions when it comes to dynamics of argumentation and novel ideas have been studied to model changing beliefs in or by means of probabilistic models. Conversely, argumentation technology has also been used to enhance other probabilistic models, for example, by adding explanation capabilities.

Some research has investigated relationships between Bayesian networks and argumentation. Bayesian networks can be used to model argumentative reasoning with arguments and counterarguments [Vreeswijk, 2004]. In a similar vein, Bayesian networks can be used to capture aspects of argumentation in the Carneades model where the propagation of argument applicability and statement acceptability can be expressed through conditional probability tables [Grabmair et al., 2010]. Argumentation can also be used to help construct Bayesian networks [Bex and Renooij, 2016; Wieten et al., 2019]. Going the other way, arguments can be generated from a Bayesian network, and this can be used to explain the Bayesian network [Timmer et al., 2015]. This involves constructing arguments involving a rule-based language in ASPIC+ for reflecting the network structure. Finally, argumentation can be used to combine multiple Bayesian networks [Nielsen and Parsons, 2007].

Looking forward, we envisage some important developments that will harness key advantages of probabilistic argumentation. These include explaining knowledge learned from data using probabilistic argumentation (see for example [Hunter, 2020]), aggregating and analysing potentially conflicting information from multiple sources (see for example [Hunter and Noor, 2020]), using probabilistic argumentation to dynamically model participants during dialogues so that strategically good moves are made (see for example [Hunter, 2015; Hunter and Thimm, 2016; Hunter et al., 2019]), and using probabilistic argumentation for modelling non-normative behaviour by participants (see for example [Polberg and Hunter, 2018]).

Acknowledgments

The work reported in this chapter has been partially supported by the Deutsche Forschungsgemeinschaft (grants 375588274 and 432308570).

References

[Alfano et al., 2020] G. Alfano, M. Calautti, S. Greco, F. Parisi, and I. Trubitsyna. Explainable acceptance in probabilistic abstract argumentation: Complexity and approximation. In *Proceedings of the 17th International Conference on Principles of Knowledge Representation and Reasoning*, pages 33–43, 2020.

[Baroni et al., 2011] P. Baroni, M. Caminada, and M. Giacomin. An introduction to argumentation semantics. *The Knowledge Engineering Review*, 26(4):365–410, 2011.

[Baroni et al., 2014] P. Baroni, M. Giacomin, and P. Vicig. On rationality conditions for epistemic probabilities in abstract argumentation. In Simon Parsons, Nir Oren, Chris Reed, and Federico Cerutti, editors, *Proceedings of COMMA'14*, volume 266 of *FAIA*, pages 121–132. IOS Press, 2014.

[Besnard and Hunter, 2008] P. Besnard and A. Hunter. *Elements of Argumentation*. The MIT Press, 2008.

[Bex and Renooij, 2016] F. Bex and S. Renooij. From arguments to constraints on a Bayesian network. In *Proceedings of the Sixth International Conference on Computational Models of Argumentation (COMMA'16)*, 2016.

[Bistarelli et al., 2018] S. Bistarelli, T. Mantadelis, F. Santini, and C. Taticchi. Probabilistic argumentation frameworks with metaproblog and conarg. In Lefteri H. Tsoukalas, Éric Grégoire, and Miltiadis Alamaniotis, editors, *Proceedings of ICTAI'18*, pages 675–679. IEEE, 2018.

[Brewka et al., 2014] G. Brewka, S. Polberg, and S. Woltran. Generalizations of Dung frameworks and their role in formal argumentation. *IEEE Intelligent Systems*, 29(1):30–38, Jan 2014.

[Brewka et al., 2018] G. Brewka, S. Ellmauthaler, H. Strass, J. P. Wallner, and S. Woltran. Abstract dialectical frameworks. In Pietro Baroni, Dov Gabbay, Massimiliano Giacomin, and Leendert van der Torre, editors, *Handbook of Formal Argumentation*, chapter 5. College Publications, February 2018.

[Caminada, 2006] M. Caminada. Semi-stable semantics. In *Proceedings of the First International Conference on Computational Models of Argumentation (COMMA'06)*, pages 121–130, 2006.

[Cayrol et al., 2021] Claudette Cayrol, Andrea Cohen, and Marie-Christine Lagasquie-Schiex. Higher-order interactions (bipolar or not) in abstract argumentation: A state of the art. In Dov Gabbay, Massimiliano Giacomin, Guillermo R. Simari, and Matthias Thimm, editors, *Handbook of Formal Argumentation*, volume 2, chapter 1. College Publications, 2021.

[Doder and Woltran, 2014] D. Doder and S. Woltran. Probabilistic argumentation frameworks – a logical approach. In Umberto Straccia and Andrea

Calì, editors, *Proceedings of SUM'14*, volume 8720 of *LNCS*, pages 134–147. Springer, 2014.

[Dondio, 2014a] P. Dondio. Multi-valued and probabilistic argumentation frameworks. In *Proceedings of the Fifth International Conference on Computational Models of Argumentation (COMMA'14)*, 2014.

[Dondio, 2014b] P. Dondio. Toward a computational analysis of probabilistic argumentation frameworks. *Cybernetics and Systems*, 45(3):254–278, 2014.

[Dung and Thang, 2010] P. M. Dung and P. M. Thang. Towards (probabilistic) argumentation for jury-based dispute resolution. In Pietro Baroni, Federico Cerutti, Massimiliano Giacomin, and Guillermo Ricardo Simari, editors, *Proceedings of COMMA'10*, volume 216 of *FAIA*, pages 171–182. IOS Press, 2010.

[Dung, 1995] P. M. Dung. On the Acceptability of Arguments and its Fundamental Role in Nonmonotonic Reasoning, Logic Programming and n-Person Games. *Artificial Intelligence*, 77(2):321–358, 1995.

[Falappa et al., 2012] M. A Falappa, G. Kern-Isberner, . DL Reis, and G. R Simari. Prioritized and non-prioritized multiple change on belief bases. *Journal of Philosophical Logic*, 41(1):77–113, 2012.

[Fazzinga et al., 2013] B. Fazzinga, S. Flesca, and F. Parisi. On the complexity of probabilistic abstract argumentation. In *Proceedings of the 23rd International Joint Conference on Artificial Intelligence (IJCAI'13)*, 2013.

[Fazzinga et al., 2015] B. Fazzinga, S. Flesca, and F. Parisi. On the complexity of probabilistic abstract argumentation frameworks. *ACM Transactions on Computational Logic*, 16(3), May 2015.

[Fazzinga et al., 2016a] B. Fazzinga, S. Flesca, and F. Parisi. On efficiently estimating the probability of extensions in abstract argumentation frameworks. *International Journal of Approximate Reasoning*, 69:106–132, 2016.

[Fazzinga et al., 2016b] B. Fazzinga, S. Flesca, F. Parisi, and A. Pietramala. Computing or estimating extensions' probabilities over structured probabilistic argumentation frameworks. *IfCoLog Journal of Logics and their Applications*, 3(2):177–200, 2016.

[Fazzinga et al., 2018a] B. Fazzinga, S. Flesca, and F. Furfaro. Credulous and skeptical acceptability in probabilistic abstract argumentation: complexity results. *Intelligenza Artificiale*, 12(2):181–191, 2018.

[Fazzinga et al., 2018b] B. Fazzinga, S. Flesca, and F. Furfaro. Probabilistic bipolar abstract argumentation frameworks: complexity results. In Jérôme Lang, editor, *Proceedings of IJCAI'18*, pages 1803–1809. International Joint Conferences on Artificial Intelligence, 2018.

[Fazzinga et al., 2018c] B. Fazzinga, S. Flesca, F. Furfaro, and F. Scala. Computing extensions' probabilities over probabilistic bipolar abstract argumentation frameworks. In Pierpaolo Dondio and Luca Longo, editors, *Proceedings of AI³@AI*IA'18*, volume 2296 of *CEUR Workshop Proceedings*, pages 57–70. CEUR-WS.org, 2018.

[Fazzinga et al., 2019] B. Fazzinga, S. Flesca, and F. Furfaro. Complexity of fundamental problems in probabilistic abstract argumentation: Beyond independence. *Artificial Intelligence*, 268:1–29, 2019.

[Gabbay and Rodrigues, 2015] D. Gabbay and O. Rodrigues. Probabilistic argumentation: An equational approach. *Logica Universalis*, 9(3):345–382, 2015.

[Garcia and Simari, 2004] A. Garcia and Guillermo R. Simari. Defeasible logic programming: an argumentative approach. *Theory and Practice of Logic Programming*, 4(1–2):95–138, 2004.

[Grabmair et al., 2010] M. Grabmair, T. Gordon, and D. Walton. Probabilistic semantics for the carneades argument model using Bayesian networks. In *Proceedings of the Third International Conference on Computational Models of Argumentation (COMMA'10)*, pages 255–266, 2010.

[Haenni, 2009] R. Haenni. Probabilistic argumentation. *Journal of Applied Logic*, 7(2):155 – 176, 2009. Special issue: Combining Probability and Logic.

[Hansson, 1997] S. Hansson. Semi-revision. *Journal of Applied Non-Classical Logics*, 7(1-2):151–175, 1997.

[Hung et al., 2015] N. Duy Hung, S. Kaisaard, and S. Hnin Pwint Oo. Probabilistic argumentation for service restoration in power distribution networks. In T. Theeramunkong, A. M. J. Skulimowski, T. Yuizono, and S. Kunifuji, editors, *Recent Advances and Future Prospects in Knowledge, Information and Creativity Support Systems - Selected Revised Papers from the Tenth International Conference on Knowledge, Information and Creativity Support Systems (KICSS 2015), 12-14 November 2015, Phuket, Thailand*, volume 685 of *Advances in Intelligent Systems and Computing*, pages 54–68. Springer, 2015.

[Hung, 2016a] N. Duy Hung. Computing probabilistic assumption-based argumentation. In Richard Booth and Min-Ling Zhang, editors, *Proceedings of PRICAI'16*, volume 9810 of *LNCS*, pages 152–166. Springer, 2016.

[Hung, 2016b] N. Duy Hung. A probabilistic argumentation engine. In Nikolaos Bourbakis, Anna Esposito, Amol Mali, and Miltos Alamaniotis, editors, *Proceedings of ICTAI'16*, pages 311–318. IEEE, 2016.

[Hung, 2017a] N. Duy Hung. A generalization of probabilistic argumentation with dempster-shafer theory. In Gabriele Kern-Isberner, Johannes Fürnkranz, and Matthias Thimm, editors, *Proceedings of KI'17*, volume 10505 of *LNCS*, pages 155–169. Springer, 2017.

[Hung, 2017b] N. Duy Hung. Inference procedures and engine for probabilistic argumentation. *International Journal of Approximate Reasoning*, 90:163–191, 2017.

[Hung, 2017c] N. Duy Hung. Probabilistic assumption-based argumentation with DST evidence. In *Proceedings of IFSA-SCIS'17*, pages 1–6. IEEE, 2017.

[Hung, 2018] N. Duy Hung. Progressive inference algorithms for probabilistic argumentation. In Tim Miller, Nir Oren, Yuko Sakurai, Itsuki Noda, Bastin Tony Roy Savarimuthu, and Tran Cao Son, editors, *Proceedings of PRIMA'18*, volume 11224 of *LNCS*, pages 371–386. Springer, 2018.

[Hunter and Noor, 2020] A. Hunter and K. Noor. Aggregation of perspectives using the constellations approach to probabilistic argumentation. In *Proceedings of AAAI'20*, 2020.

[Hunter and Polberg, 2019] A. Hunter and S. Polberg. A model-based theorem prover for epistemic graphs for argumentation. In Gabriele Kern-Isberner and Zoran Ognjanović, editors, *Symbolic and Quantitative Approaches to Reasoning with Uncertainty*, pages 50–61, Cham, 2019. Springer International Publishing.

[Hunter and Potyka, 2017] A. Hunter and N. Potyka. Updating probabilistic epistemic states in persuasion dialogues. In Alessandro Antonucci, Laurence Cholvy, and Odile Papini, editors, *Proceedings of ECSQARU'17*, volume 10369 of *LNCS*, pages 46–56. Springer, 2017.

[Hunter and Thimm, 2014] A. Hunter and M. Thimm. Probabilistic argumentation with incomplete information. In Torsten Schaub, Gerhard Friedrich, and Barry O'Sullivan, editors, *Proceedings of ECAI'14*, volume 263 of *FAIA*, pages 1033–1034. IOS Press, 2014.

[Hunter and Thimm, 2016] A. Hunter and M. Thimm. Optimization of dialectical outcomes in dialogical argumentation. *International Journal of Approximate Reasoning*, 78:73–102, 2016.

[Hunter and Thimm, 2017] A. Hunter and M. Thimm. Probabilistic reasoning with abstract argumentation frameworks. *Journal of Artificial Intelligence Research*, 59:565–611, August 2017.

[Hunter et al., 2018a] A. Hunter, S. Polberg, and N. Potyka. Updating Belief in Arguments in Epistemic Graphs. In Michael Thielscher, Francesca Toni, and Frank Wolter, editors, *Proceedings of KR'18*, pages 138–147. AAAI Press, 2018.

[Hunter et al., 2018b] A. Hunter, S. Polberg, and M. Thimm. Epistemic graphs for representing and reasoning with positive and negative influences of arguments. *ArXiv*, 2018.

[Hunter et al., 2019] A. Hunter, S. Polberg, and N. Potyka. Delegated updates in epistemic graphs for opponent modelling. *International Journal of Approximate Reasoning*, 113:207 – 244, 2019.

[Hunter et al., 2020] A. Hunter, S. Polberg, and M. Thimm. Epistemic graphs for representing and reasoning with positive and negative influences of arguments. *Artificial Intelligence*, 281:103236, 2020.

[Hunter, 2012] A. Hunter. Some foundations for probabilistic abstract argumentation. In Bart Verheij, Stefan Szeider, and Stefan Woltran, editors, *Proceedings of COMMA'12*, volume 245 of *FAIA*, pages 117–128. IOS Press, 2012.

[Hunter, 2013] A. Hunter. A probabilistic approach to modelling uncertain logical arguments. *International Journal of Approximate Reasoning*, 54(1):47–81, 2013.

[Hunter, 2014] A. Hunter. Probabilistic qualification of attack in abstract argumentation. *International Journal of Approximate Reasoning*, 55:607–638, 2014.

[Hunter, 2015] A. Hunter. Modelling the persuadee in asymmetric argumentation dialogues for persuasion. In *Proceedings of IJCAI'15*, 2015.

[Hunter, 2016] A. Hunter. Persuasion dialogues via restricted interfaces using probabilistic argumentation. In Steven Schockaert and Pierre Senellart, editors, *Proceedings of SUM'16*, volume 9858 of *LNCS*, pages 184–198. Springer, 2016.

[Hunter, 2020] A. Hunter. Generating instantiated argument graphs from probabilistic information. In *Proceedings of ECAI'20*, 2020.

[Kido and Okamoto, 2017] H. Kido and K. Okamoto. A bayesian approach to argument-based reasoning for attack estimation. In Carles Sierra, editor, *Proceedings of IJCAI'17*, pages 249–255. International Joint Conferences on Artificial Intelligence, 2017.

[Li et al., 2011] H. Li, N. Oren, and T.J. Norman. Probabilistic argumentation frameworks. In Sanjay Modgil, Nir Oren, and Francesca Toni, editors, *Proceedings of TAFA'11*, volume 7132 of *LNCS*, pages 1–16. Springer, 2011.

[Mantadelis and Bistarelli, 2020] T. Mantadelis and S. Bistarelli. Probabilistic abstract argumentation frameworks, a possible world view. *Int. J. Approx. Reason.*, 119:204–219, 2020.

[Martinez et al., 2012] M. V. Martinez, A. J. García, and G. R. Simari. On the use of presumptions in structured defeasible reasoning. In Bart Verheij, Stefan Szeider, and Stefan Woltran, editors, *Computational Models of Argument - Proceedings of COMMA 2012*, volume 245 of *Frontiers in Artificial Intelligence and Applications*, pages 185–196. IOS Press, 2012.

[Modgil and Prakken, 2014] S. Modgil and H. Prakken. The ASPIC+ framework for structured argumentation: a tutorial. *Argument and Computation*, 5:31–62, 2014.

[Morveli-Espinoza et al., 2019] M. Morveli-Espinoza, J. C. Nieves, and C. A. Tacla. An imprecise probability approach for abstract argumentation based on credal sets. In Gabriele Kern-Isberner and Zoran Ognjanović, editors, *Proceedings of ECSQARU'19*, volume 11726 of *LNCS*, pages 39–49. Springer, 2019.

[Nielsen and Parsons, 2007] S. Holbech Nielsen and S. Parsons. An application of formal argumentation: Fusing bayesian networks in multi-agent systems. *Artificial Intelligence*, 171(10-15):754–775, 2007.

[Nilsson, 1986] N. J Nilsson. Probabilistic logic. *Artificial intelligence*, 28(1):71–87, 1986.

[Paris, 1994] J. B. Paris. *The Uncertain Reasoner's Companion – A Mathematical Perspective*. Cambridge University Press, 1994.

[Pearl, 1988] J. Pearl. *Probabilistic reasoning in intelligent systems: networks of plausible inference*. Springer, 1988.

[Polberg and Doder, 2014] S. Polberg and D. Doder. Probabilistic abstract dialectical frameworks. In Eduardo Fermé and João Leite, editors, *Proceedings of JELIA'14*, volume 8761 of *LNCS*, pages 591–599. Springer, 2014.

[Polberg and Hunter, 2018] S. Polberg and A. Hunter. Empirical evaluation of abstract argumentation: Supporting the need for bipolar and probabilistic approaches. *International Journal of Approximate Reasoning*, 93:487–543, 2018.

[Polberg et al., 2017] S. Polberg, A. Hunter, and M. Thimm. Belief in attacks in epistemic probabilistic argumentation. In Serafín Moral, Olivier Pivert, Daniel Sánchez, and Nicolás Marín, editors, *Proceedings of SUM'17*, volume 10564 of *LNCS*, pages 223–236. Springer, 2017.

[Polberg, 2016] S. Polberg. Understanding the Abstract Dialectical Framework. In Loizos Michael and Antonis Kakas, editors, *Proceedings of JELIA'16*, volume 10021 of *LNCS*, pages 430–446. Springer, 2016.

[Pollock, 1995] J. L. Pollock. *Cognitive Carpentry: A Blueprint for How to Build a Person*. The MIT Press, Cambridge, Massachusetts, 1995.

[Potyka et al., 2019] N. Potyka, S. Polberg, and A. Hunter. Polynomial-time updates of epistemic states in a fragment of probabilistic epistemic argumentation. In *Proceedings of ECSQARU'19*, 2019.

[Potyka, 2019] N. Potyka. A polynomial-time fragment of epistemic probabilistic argumentation. *International Journal of Approximate Reasoning*, 115:265 – 289, 2019.

[Prakken, 2017] H. Prakken. On relating abstract and structured probabilistic argumentation: A case study. In Alessandro Antonucci, Laurence Cholvy, and Odile Papini, editors, *Proceedings of ECSQARU'17*, volume 10369 of *LNCS*, pages 69–79. Springer, 2017.

[Prakken, 2018] H. Prakken. Probabilistic strength of arguments with structure. In Michael Thielscher, Francesca Toni, and Frank Wolter, editors, *Proceedings of KR'18*, pages 158–167. AAAI Press, 2018.

[Rienstra, 2012] T. Rienstra. Towards a probabilistic dung-style argumentation system. In *Proceedings of AT'12*, 2012.

[Riveret and Governatori, 2016] R. Riveret and G. Governatori. On learning attacks in probabilistic abstract argumentation. In Catholijn M. Jonker, Stacy Marsella, John Thangarajah, and Karl Tuyls, editors, *Proceedings of AAMAS'16*, pages 653–661. IFAAMAS, 2016.

[Riveret et al., 2017] R. Riveret, D. Korkinof, M. Draief, and J. Pitt. Probabilistic abstract argumentation: An investigation with boltzmann machines. *Argument & Computation*, 8(1):89, 2017.

[Riveret et al., 2018] R. Riveret, P. Baroni, Y. Gao, G. Governatori, A. Rotolo, and G. Sartor. A labelling framework for probabilistic argumentation. *Annals of Mathematics and Artificial Intelligence*, 83(1):21–71, 2018.

[Shakarian et al., 2016] P. Shakarian, G.I. Simari, G. Moores, D. Paulo, S. Parsons, M. A Falappa, and A. Aleali. Belief revision in structured probabilistic argumentation. *Annals of Mathematics and Artificial Intelligence*, 78(3-4):259–301, 2016.

[Simari et al., 2016] G. I. Simari, P. Shakarian, and M. A. Falappa. A quantitative approach to belief revision in structured probabilistic argumentation. *Annals of Mathematics and Artificial Intelligence*, 76(3-4):375–408, 2016.

[Thang, 2016] P. M. Thang. Dialectical proof procedures for probabilistic abstract argumentation. In Matteo Baldoni, Amit K. Chopra, Tran Cao Son, Katsutoshi Hirayama, and Paolo Torroni, editors, *Proceedings of PRIMA'16*, volume 9862 of *LNCS*, pages 397–406. Springer, 2016.

[Thimm et al., 2018a] M. Thimm, F. Cerutti, and T. Rienstra. Probabilistic graded semantics. In Sanjay Modgil, Katarzyna Budzynska, and John Lawrence, editors, *Proceedings of COMMA'18*, volume 305 of *FAIA*, pages 369–380. IOS Press, 2018.

[Thimm et al., 2018b] M. Thimm, S. Polberg, and A. Hunter. Epistemic attack semantics. In Sanjay Modgil, Katarzyna Budzynska, and John Lawrence, editors, *Proceedings of the Seventh International Conference on Computational Models of Argumentation (COMMA'18)*, volume 305 of *Frontiers in Artificial Intelligence and Applications*, pages 37–48, Warsaw, Poland, September 2018.

[Thimm, 2012] M. Thimm. A probabilistic semantics for abstract argumentation. In Luc De Raedt, Christian Bessiere, Didier Dubois, Patrick Doherty, Paolo Frasconi, Fredrik Heintz, and Peter Lucas, editors, *Proceedings of ECAI'12*, volume 242 of *FAIA*, pages 750–755. IOS Press, 2012.

[Timmer et al., 2015] S. T. Timmer, J.-J. Ch. Meyer, H. Prakken, S. Renooij, and B. Verheij. Explaining Bayesian networks using argumentation. In *Proceedings of the 13th European Conference on Symbolic and Quantitative Approaches to Reasoning with Uncertainty (ECSQARU'15)*, 2015.

[Timmer et al., 2017] S.T. Timmer, J.Ch. Meyer, H. Prakken, S. Renooij, and B. Verheij. A two-phase method for extracting explanatory arguments from bayesian networks. *International Journal of Approximate Reasoning*, 80:475 – 494, 2017.

[Toni, 2014] F. Toni. A tutorial on assumption-based argumentation. *Argument & Computation*, 5(1):89–117, 2014.

[Verheij, 2014] B. Verheij. Arguments and their strength: Revisiting pollock's anti-probabilistic starting points. In Simon Parsons, Nir Oren, Chris Reed, and Federico Cerutti, editors, *Proceedings of COMMA'14*, volume 266 of *FAIA*, pages 433–444. IOS Press, 2014.

[Villata et al., 2011] S. Villata, G. Boella, and L. van der Torre. Attack Semantics for Abstract Argumentation. In *Proceedings of the Twenty-Second International Joint Conference on Artificial Intelligence (IJCAI'11)*, pages 406–413, 2011.

[Vreeswijk, 2004] G. A. W. Vreeswijk. Argumentation in bayesian belief networks. In Iyad Rahwan, Pavlos Moraïtis, and Chris Reed, editors, *Proceedings of ArgMAS'04*, volume 3366 of *LNCS*, pages 111–129. Springer, 2004.

[Walton, 1989] D. Walton. *Informal Logic: A Handbook for Critical Argumentation*. Cambridge University Press, 1989.

[Wieten et al., 2019] R. Wieten, F. Bex, H. Prakken, and S. Renooij. Supporting discussions about forensic bayesian networks using argumentation. In *Proceedings of ICAIL'19*, pages 143–152, 2019.

[Wu and Caminada, 2010] Y. Wu and M. Caminada. A labelling-based justification status of arguments. *Studies in Logic*, 3(4):12–29, 2010.

Part II

Dynamics and Dialogues

CHAPTER 8

ENFORCEMENT IN FORMAL ARGUMENTATION

RINGO BAUMANN
Department of Computer Science, Leipzig University, Germany
baumann@informatik.uni-leipzig.de

SYLVIE DOUTRE
Institut de Recherche en Informatique de Toulouse, Université Toulouse Capitole, France
doutre@irit.fr

JEAN-GUY MAILLY
Laboratoire d'Informatique Paris Descartes, Université de Paris, France
jean-guy.mailly@u-paris.fr

JOHANNES P. WALLNER
Institute of Software Technology, Graz University of Technology, Austria
wallner@ist.tugraz.at

BAUMANN, DOUTRE, MAILLY, WALLNER

Abstract

Within argumentation dynamics, a major strand of research is concerned with how changing an argumentation framework affects the acceptability of arguments, and how to modify an argumentation framework in order to guarantee that some arguments have a given acceptance status. In this chapter, we overview the main approaches for enforcement in formal argumentation. We mainly focus on extension enforcement, i.e., on how to modify an argumentation framework to ensure that a given set of arguments becomes (part of) an extension. We present different forms of extension enforcement defined in the literature, as well as several possibility and impossibility results. The question of minimal change is also considered, i.e., what is the minimal number of modifications that must be made to the argumentation framework for enforcing an extension. Computational complexity and algorithms based on a declarative approach are discussed. Finally, we briefly describe several notions that do not directly fit our definition of extension enforcement, but are closely related.

1 Introduction

At the beginning of the 2010s several problems regarding *dynamic aspects* of abstract argumentation have been addressed in the literature [Boella et al., 2009a; Cayrol et al., 2010; Bisquert et al., 2011; Liao et al., 2011]. One much cited problem among these is the so-called *enforcing problem* dealing with changing the acceptability of certain arguments [Baumann and Brewka, 2010]. Over the years, the problem gained more and more attention which finally leads to the writing of this chapter. In its very first version the problem can be briefly summarized as the question whether it is possible, given a specific type of syntactic changes, to modify a given AF such that a desired set of arguments becomes (a subset of) an extension. Consider the following snapshot of a dialogue among agents A and B depicted in Figure 1. Assume it is A's turn and her desired set of arguments is $E = \{a_1, a_2, a_3\}$. Furthermore, A and B are discussing under preferred semantics, which selects maximal conflict-free and self-defending sets of arguments.

8 - Enforcement in Formal Argumentation

Figure 1: Snapshot of a dialogue

In order to enforce E, agent A may come up with new arguments which interact with the old ones (for example through introducing an argument d which attacks b_2 and b_3) and/or question old arguments or attacks between them, respectively (for example through questioning the self-attack of c). Please note that first, in this scenario, enforcing is possible and second, that there are at least two different possibilities to achieve that. This insight leads to a further well-studied issue, namely the so-called *minimal change problem* firstly introduced in [Baumann, 2012b]. This problem is defined as a generalization of the classical enforcing problem since one is not only interested in whether enforcements are *possible*, but also in the *effort needed* to enforce a set of arguments. One numerical measure which is frequently used for this effort corresponds to the number of additions or removals of attacks to reach such an enforcement. The main motivation behind this measure is that adding or removing an isolated argument does not contribute at all to solving or increasing a given conflict, i.e. the conflicting information remains the same. This means, the decrease or increase of a conflict is directly linked to upcoming or disappearing attacks and thus, counting attacks only is a reasonable approach. Regarding the introductory example we obtain a minimal effort of 1 if allowing arbitrary modifications.

In this chapter we give an overview over main variants of enforcement studied in the literature. We give a particular focus on strict and non-strict extension enforcement, whose aim is to modify an AF such that a desired set of arguments becomes exactly (or part of) an extension, under a semantics. A main distinguishing factor among the family of operators for extension enforcement is how an AF may be modified. We highlight here changes corresponding to expansions, i.e., additions of arguments and attacks such as the addition of argument d above, or local updates, i.e., modifying only the attack structure such as questioning the self-attack of c, but also discuss modifications to AFs more broadly, as well. Additionally, we consider as an instance of a change that does not affect the structure of the framework, modifications of the chosen semantics, in order to enforce a set of arguments.

We present main formal properties of extension enforcement derived in the literature, e.g., for impossibility and possibility results, and results for the minimal change problem of extension enforcement. We further survey results regarding the complexity of reasoning on enforcement and present algorithms based on declarative approaches to implement enforcement.

The chapter starts off with recalling formal preliminaries of AFs (Section 2) including types of modifications on AFs. The main section on extension enforcement is Section 3, which first introduces enforcement as a general problem, and focuses on the extension enforcement variant. In this section, we present expansion-based extension enforcement and extension enforcement based on locally updating an attack structure without modifying the set of arguments. Further, minimal change, semantics change, complexity results, and algorithms, are presented. In Section 4 we survey related notions to enforcement, and we close with a discussion of related works (Section 4.5) and with conclusions (Section 5).

2 Formal Preliminaries

In order to keep the chapter self-contained we review all relevant definitions. We start with the basic notions of Dung's abstract argumentation theory [Dung, 1995].

2.1 Argumentation Frameworks and Semantics

An *abstract argumentation framework (AF)* is just a directed graph $F = (A, R)$ where a node $a \in A$ is called an *argument* and a pair $(a, b) \in R \subseteq A \times A$ is interpreted as an *attack* from argument a to argument b. We require that any AF $F = (A, R)$ possesses arguments

8 - ENFORCEMENT IN FORMAL ARGUMENTATION

from a fixed reference set $\mathcal{U}$, i.e. $A \subseteq \mathcal{U}$. Moreover, in this chapter we restrict ourselves to finite AFs, i.e. any AF consists of finitely many arguments and attacks only. Note that this is a common restriction in the literature although actual and potential infinite AFs play an important role in practical applications as well as theoretical considerations (cf. [Baroni et al., 2013; Baumann and Spanring, 2017; Baumann, 2019] for more information). At the heart of Dung's abstract argumentation theory are *argumentation semantics* which formalize intuition of what should be acceptable in the light of conflicts. Two main approaches to argumentation semantics can be found, namely so-called *extension-based* and *labelling-based* versions (cf. [Baroni et al., 2018a] for an introduction and [Baumann, 2018, Sections 2.2, 4.4] for further relations). In this chapter we concentrate on the former only. Consider the following generic definition. The set $\mathcal{F}$ refers to all considered AFs.

Definition 2.1. *A semantics is a total function*

$$\sigma : \mathcal{F} \to 2^{2^{\mathcal{U}}} \quad F = (A, R) \mapsto \sigma(F) \subseteq 2^A.$$

A set of arguments $E \in \sigma(F)$ is called a *σ-extension*. Moreover, we say that a semantics σ is *universally defined* if each AF admits at least one extension with respect to this semantics, i.e. for any $F \in \mathcal{F}$, $|\sigma(F)| \geq 1$. Furthermore, a semantics σ is said to be *uniquely defined* if always exactly one set of arguments is returned, i.e. $|\sigma(F)| = 1$ for any $F \in \mathcal{F}$.

Before presenting the relevant semantics for this chapter we have to introduce some further notation. Given an AF $F = (A, R)$ and a set $E \subseteq A$. We use E_F^+, or simply E^+, for $\{b \mid (a,b) \in R, a \in E\}$. Moreover, $E_F^\oplus$, or simply $E^\oplus$, is called the *range* of E and stands for $E^+ \cup E$. Analogously, E_F^- (or simply E^-) stands for $\{b \mid (b,a) \in R, a \in E\}$, and $E_F^\ominus$ (or simply $E^\ominus$) corresponds to $E^- \cup E$. An argument a is *defended* by E (in F) if for each $b \in A$ with $(b,a) \in R$, b is attacked by some $c \in E$. Finally, $\Gamma_F : 2^A \to 2^A$ with $I \mapsto \{a \in A \mid a \text{ is defended by } I\}$ denotes the so-called *characteristic function* (of F).

Besides conflict-free and admissible sets (abbreviated by *cf* and *ad*) we consider a large number of well-known semantics, namely naive, stage, stable, semi-stable, complete, preferred, grounded, ideal, and eager semantics (abbreviated by $na, stg, stb, sst, co, pr, gr, id, eg$, respectively).

Definition 2.2. *Let $F = (A, R)$ be an AF and $E \subseteq A$.*

1. *$E \in cf(F)$ iff for no $a, b \in E$, $(a, b) \in R$,*
2. *$E \in na(F)$ iff E is $\subseteq$-maximal in $cf(F)$,*
3. *$E \in stg(F)$ iff $E \in cf(F)$ and $E^\oplus$ is $\subseteq$-maximal in $\{I^\oplus \mid I \in cf(F)\}$,*
4. *$E \in stb(F)$ iff $E \in cf(F)$ and $E^\oplus = A$,*
5. *$E \in ad(F)$ iff $E \in cf(F)$ and $E \subseteq \Gamma_F(E)$,*
6. *$E \in sst(F)$ iff $E \in ad(F)$ and $E^\oplus$ is $\subseteq$-maximal in $\{I^\oplus \mid I \in ad(F)\}$,*
7. *$E \in co(F)$ iff $E \in cf(F)$ and $E = \Gamma_F(E)$,*
8. *$E \in pr(F)$ iff E is $\subseteq$-maximal in $co(F)$,*
9. *$E \in gr(F)$ iff E is $\subseteq$-minimal in $co(F)$,*
10. *$E \in id(F)$ iff E is $\subseteq$-maximal in $\{I \mid I \in ad(F), I \subseteq \bigcap pr(F)\}$,*
11. *$E \in eg(F)$ iff E is $\subseteq$-maximal in $\{I \mid I \in ad(F), I \subseteq \bigcap sst(F)\}$.*

It has been shown that any of the introduced semantics is universally defined except the stable one and moreover, grounded, ideal and eager semantics are even uniquely defined (cf. [Baumann and Spanring, 2015] for an overview). In order to get familiar with the introduced definitions consider the following example taken from [Baumann, 2014b].

Example 2.3. *Consider the AF $F = (A, R)$ with $A = \{a, b, c, d, e, f\}$ and $R = \{(a, b), (a, d), (b, c), (c, a), (d, d), (e, d), (e, f), (f, e)\}$. The graphical representation of F is given below.*

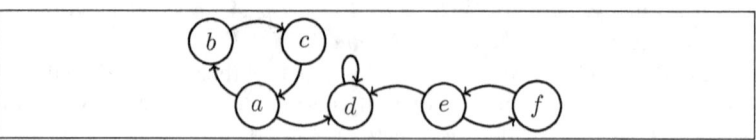

The evaluation of F w.r.t. the introduced semantics is given in the following table. The entry "✓" in row "σ" and line "E" stands for $E \in \sigma(F)$.

	stb	sst	stg	pr	ad	co	gr	id	eg	na
∅				✓	✓	✓	✓			
{e}	✓			✓	✓	✓		✓		
{f}				✓	✓	✓				
{a,e}		✓								✓
{b,e}		✓								✓
{c,e}		✓								✓
{a,f}		✓								✓
{b,f}										✓
{c,f}										✓

Table 1: Evaluation table of F

The AF F is an example for a collapse of stable semantics, i.e. $stb(F) = \emptyset$. The non-existence of stable extensions in F implies the occurrence of odd-length cycles like the 3-cycle $[a, b, c, a]$ or the self-loop $[d, d]$. More precisely, in case of finite AFs we have that being odd-cycle free is sufficient for warranting at least one stable extension [Dung, 1995; Spanring, 2015].

As already indicated in Table 1 there are several well-known subset relations between the considered semantics. For instance, for any AF F we have, $stb(F) \subseteq sst(F) \subseteq pr(F) \subseteq co(F) \subseteq ad(F)$ and $stb(F) \subseteq$

$stg(F) \subseteq na(F)$.

2.2 Acceptance Modes and Structural Changes

In the following we present several acceptance modes and structural changes, that is, changes on the structure (addition or removal of arguments and attacks) of the AF, which can be used to specify a certain type of enforcement.

So-called *credulous* and *sceptical acceptance* are the most common reasoning types for abstract argumentation semantics. They are usually defined for single arguments only. We present their definitions for sets of arguments where the classical single argument acceptance can be obtained by considering the singleton of the argument in question. Moreover, since a non-universally defined semantics σ may return no σ-extension for a given AF F we consider so-called *non-empty sceptical reasoning* which avoids the (possibly) unintended situation that every argument is sceptically accepted due to the emptiness of $\sigma(F)$. A further frequently used acceptance mode is the requirement to be contained in at least one extension, so-called *covered acceptance*[1]. This notion plays a central role in the field of enforcement and is located in-between non-empty sceptical and credulous acceptance.

Definition 2.4. *Given a semantics σ, an AF $F = (A, R)$ and a set $E \subseteq A$. We say that E is*

1. *credulously accepted w.r.t. σ if $E \subseteq \bigcup \sigma(F)$,*

2. *sceptically accepted w.r.t. σ if $E \subseteq \bigcap \sigma(F)$,*

3. *non-empty sceptically accepted w.r.t. σ if $E \subseteq \bigcap \sigma(F)$ and $\sigma(F) \neq \emptyset$,*

4. *covered accepted w.r.t. σ if there is an $E' \in \sigma(F)$, s.t. $E \subseteq E'$.*

For convenience we introduce the following unified notation. We write $E \in cred(F, \sigma)$, $E \in scep(F, \sigma)$, $E \in scep^{\neq \emptyset}(F, \sigma)$ or $E \in cov(F, \sigma)$

---

[1] We mention that this notion is sometimes called credulous acceptance [Dunne, 2007, p. 704]. This is due to the fact that that there are at least two options if generalizing credulous acceptance from arguments to sets of arguments.

for E is credulously, sceptically, non-empty sceptically or covered accepted, respectively. Moreover, for any given reasoning type r we use $E \in r_s(\sigma, F)$ to indicate that there is an equality instead of a subset relation only, e.g. there is an $E' \in \sigma(F)$, s.t. $E = E'$ in the case of covered acceptance (or, said otherwise, $E \in \sigma(F)$). In this case we say that the considered set E is *strictly* accepted. If E is non-empty sceptically accepted w.r.t. σ then E is covered accepted w.r.t. σ (since E must be part of all σ-extensions and there is at least one), and the latter implies that E is credulously accepted w.r.t. σ (since the witness for being covered accepted is a witness for credulous acceptance).

Let us proceed with the running AF exemplifying several acceptance modes.

Example 2.5 (Example 2.3 cont.). *Let $\sigma = stb$. Since $stb(F) = \emptyset$ we obtain $\bigcup stb(F) = \emptyset$ and $\bigcap stb(F) = \mathcal{U}$. Hence, any set $E \subseteq \mathcal{U}$ is sceptically, but not non-empty sceptically accepted, i.e. $E \in scep(F, stb)$ and $E \notin scep^{\neq \emptyset}(F, stb)$. Moreover, E is neither credulously, nor covered accepted, i.e. $E \notin cred(F, stb)$ and $E \notin cov(F, stb)$.*
Consider now $\sigma = pr$. Since $pr(F) = \{\{e\}, \{f\}\}$ we have $\bigcup pr(F) = \{e, f\}$ and $\bigcap pr(F) = \emptyset$. Thus, $\{e, f\}$ is credulously strict but neither sceptically nor non-empty sceptically accepted, i.e. $\{e, f\} \in cred_s(F, pr)$, $\{e, f\} \notin scep(F, pr)$ and $\{e, f\} \notin scep^{\neq \emptyset}(F, pr)$. Moreover, $\{e, f\}$ is not covered accepted whereas $\{e\}$ and $\{f\}$ are and this acceptance is even strict, i.e. $\{e, f\} \notin cov(F, pr)$ and $\{e\}, \{f\} \in cov_s(F, pr)$.

We now introduce typical structural changes. The most general form of dynamic scenarios are so-called *updates* where arguments and attacks can be deleted and added. If we do not delete any information we call the structural change an *expansion* [Baumann and Brewka, 2010; Oikarinen and Woltran, 2011; Baumann, 2012a]. The following kinds of expansions have received particular attention in the literature. *Normal expansions* add new arguments and possibly new attacks which concern at least one of the fresh arguments. Moreover, *local expansions* do not introduce any new arguments but possibly new attacks among the old arguments. Both types of expansions naturally occur in the context of instantiation-based argumentation [Besnard and Hunter, 2008; Caminada and Wu, 2011]. For instance, adding a new piece of information to the underlying knowledge base corresponds to a normal expansion on the AF level. Furthermore, changing the considered notion of attack left the constructed arguments untouched and results in a local expansion. Two further subconcepts of normal expansions are usually considered, so-called *strong* and *weak expansions*. Their names refer to properties of the additional arguments, namely arguments which are never attacked by former arguments (*strong* arguments) and arguments which do not attack former arguments (*weak* arguments). The former type typically occurs in a debate if one tries to strengthen the own point of view via rebutting the opponents arguments. Note that weak expansions seem to be more an academic exercise than a task with practical relevance with regard to real-world argumentation. However, they do play a decisive role in the context of *splittings* [Baumann, 2011; Baumann et al., 2011; Baroni et al., 2018b].

Consider the formal definition of the discussed types of expansions.

Definition 2.6. *An AF G is an expansion of AF $F = (A, R)$ (for short, $F \preceq_E G$) iff $G = (A \cup B, R \cup S)$ for some (maybe empty) sets B and S, s.t. $A \cap B = R \cap S = \emptyset$. An expansion is called*

1. *normal ($F \preceq_N G$) iff $\forall ab\ ((a,b) \in S \rightarrow a \in B \vee b \in B)$,*

2. *strong ($F \preceq_S G$) iff*

    $$F \preceq_N G \text{ and } \forall ab\ ((a,b) \in S \rightarrow \neg(a \in A \wedge b \in B)),$$

3. *weak ($F \preceq_W G$) iff*

    $$F \preceq_N G \text{ and } \forall ab\ ((a,b) \in S \rightarrow \neg(a \in B \wedge b \in A)),$$

4. *local ($F \preceq_L G$) iff $B = \emptyset$.*

Example 2.7. *Consider the following simple AF F. The presented AFs F_X represent examples for $F \preceq_X F_X$. This means, F_N is a normal expansion of F. Note that grey-highlighted arguments or attacks represent added information.*

8 - Enforcement in Formal Argumentation

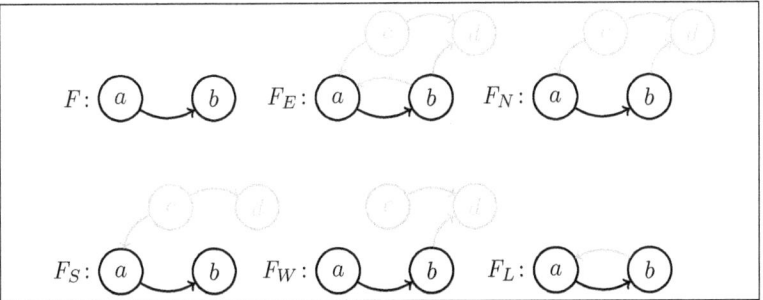

Figure 2: Different kinds of expansions

The natural counter-parts to expansions are so-called *deletions* where no further arguments and attacks are added [Boella et al., 2009a; Bisquert et al., 2011; Baumann, 2014a]. We consider two sub-classes of *deletions* representing the inverse operations to normal and local expansions, namely *normal* and *local deletions*. Normal deletions retract arguments and their corresponding attacks. Such a kind of structural change occurs in the instantiation-based context if we delete information from the underlying knowledge base. Changing to a more restrictive notion of attack corresponds to a local deletion where only attacks are discarded.

In order to present the precise formal meaning of deletions we have to introduce some operations on directed graphs first. First, we use $F \sqcup H$ for the pointwise union of two AFs. In Definition 2.8, such an union is used in order to represent the addition of information (encoded in H) to an initial AF (F). Secondly, the restriction of $F = (A, R)$ to a set $B \subseteq A$ abbreviated as $F|_B$ is given via $(B, R \cap (B \times B))$.

Definition 2.8. *Given an AF $F = (A, R)$, a set of arguments B and a set of attacks S as well as a further AF H. The AF*

$$G = (F \setminus [B, S]) \sqcup H := \left((A, R \setminus S)|_{A \setminus B}\right) \sqcup H$$

is called an update of F (for short, $F \asymp_U G$). An update is called a

1. *deletion ($F \succeq_D G$) iff $H = (\emptyset, \emptyset)$,*

2. *normal deletion ($F \succeq_{ND} G$) iff $F \succeq_D G$ and $S = \emptyset$,*

3. *local deletion* ($F \succeq_{LD} G$) iff $F \succeq_D G$ and $B = \emptyset$.

Let us take a closer look at the definition of $G = (F \setminus [B,S]) \sqcup H$. The AF H plays the role of added information, i.e. it contains new arguments and attacks. Consequently, for all kind of deletions we have $H = (\emptyset, \emptyset)$ which leaves us with $G = F \setminus [B,S]$. The set B contains arguments which have to be deleted. Since attacks depend on arguments, we have to delete the attacks which involve arguments from B too. This operation is formally captured by the restriction of F to $A \setminus B$. Furthermore, the set S contains particular attacks which have to be deleted. This means, the pair $[B,S]$ does not necessarily have to be an AF. Therefore we use $[B,S]$ instead of (B,S). If clear from context we use B and S instead of $[B,\emptyset]$ or $[\emptyset,S]$, i.e. we simply write $F \setminus B$ as well as $F \setminus S$ for normal or local deletions, respectively. Note that the different kinds of expansion presented in Definition 2.6 can be captured by setting $B = S = \emptyset$. Deletions and expansions are dual concepts: $F \preceq_E G$ if and only iff $G \succeq_D F$, and similarly for the normal or local versions.

Example 2.9. *The AF F represents the initial situation. An update as well as arbitrary, normal or local deletion of it are given by F_U, F_D, F_{ND} and F_{LD}. Grey-highlighted arguments or attacks represent added information in contrast to dotted arguments and attacks which represent deleted objects.[2] More formally, in accordance with Definition 2.8 we have that $F_U = (F \setminus [B,S]) \sqcup H$, $F_D = F \setminus [B,S]$, $F_{ND} = F \setminus B$, $F_{LD} = F \setminus S$ where the set of arguments $B = \{c\}$, the set of attacks $S = \{(b,a)\}$ and the AF $H = (\{b,d,e,f\}, \{(d,b),(e,f),(f,d)\})$.*

---

[2]This convention will be used throughout the whole chapter.

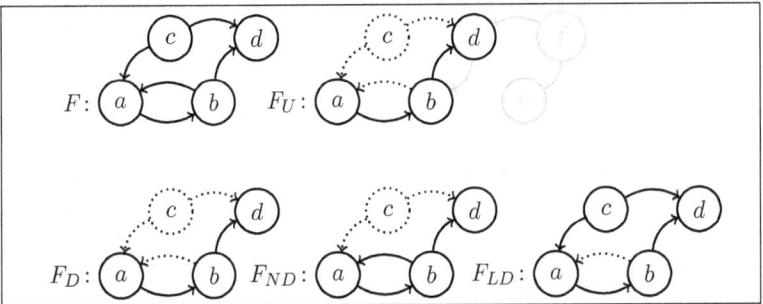

Figure 3: An update and different kinds of deletions

3 Enforcement

3.1 The General Setup

The starting point of any extension enforcement case is:

- an AF F,

- a semantics σ,

- a certain desired set of arguments E, together with

- a reasoning, acceptance mode r, e.g. credulous, sceptical, non-empty sceptical, covered, with a strict or non-strict goal achievement (cf. Section 2.2).

In addition, parameters indicating the way of achieving the enforcement can be specified, namely:

- allowed types of structural changes like update, expansion and deletion (cf. Section 2.2),

- allowed types of semantic changes, if any (cf. Section 3.2.4), and

- whether these changes would have to be minimal, and in which sense (cf. Section 3.2.3).

For illustrative purposes let us assume that r stands for credulous acceptance. Consequently, enforcement is needed if and only if E is not credulously accepted w.r.t. σ in F, i.e. $E \notin cred(F, \sigma)$. This is why we often speak of the *desired* set of arguments E since we want to fix the defect of non-acceptance. In order to achieve this goal we have two main options, namely structural changes and/or semantic changes. More precisely, we are looking for changes of AFs, from F to G, and/or semantics, from σ to τ, s.t. E is credulously accepted w.r.t. τ in G, i.e. $E \in cred(G, \tau)$. The way of how to perform the structural change is fixed in advance. For instance, one may require that only local expansions of F are allowed, i.e. $F \preceq_L G$. The same applies to the semantic change. One may allow changes to any kind of semantics or to admissibility-based ones only. Another option would be to completely forbid semantic changes, i.e. $\tau = \sigma$. In the following definition, we call a *modification type* $M \subseteq \mathcal{F} \times \mathcal{F}$ a relation such that $(F, G) \in M$ iff, when F is an initial AF, then G is a possible result of modifying F. For instance, $M = \preceq_L$ means that only local expansions are authorized.

Consider the following formal definition of an enforcement.

Definition 3.1. *Given two AFs F and G, two semantics σ and τ, a modification type $M \subseteq \mathcal{F} \times \mathcal{F}$, a set of argument E, and a reasoning mode r. A pair (G, τ) is called an (F, σ, M, r)-enforcement of E if*

1. $(F, G) \in M$ *and*

2. $E \in r(G, \tau)$.

Moreover, we call G the τ-enforcing AF *and we say that E is τ-enforced by G.*

The different kinds of expansions and deletions presented in Definitions 2.6, 2.8 are captured by setting $M \in \{\preceq_E, \preceq_N, \preceq_S, \preceq_W, \preceq_L, \asymp_U, \succeq_D, \succeq_{ND}, \succeq_{LD}\}$. Note that $F \preceq_N G$ can be equivalently rewritten as $(F, G) \in \preceq_N$ since $\preceq_N$ is formally a binary relation over $\mathcal{F}$, i.e. $\preceq_N \subseteq \mathcal{F} \times \mathcal{F}$. Whenever F, σ, M and r are clear from context we simply speak of enforcements of E. If the set in question is strictly accepted we speak about a *strict* enforcement (for instance, $r = cov_s$), otherwise *non-strict* (for instance, $r = cov$). Moreover, we distinguish between *conservative* ($\sigma = \tau$) and *liberal* enforcements ($\sigma \neq \tau$). The latter may be interpreted as a change of proof standard or paradigm shift. Imagine a judicial proceeding. Here it is vitally important whether you are accused on the base of criminal or civil law. The required evidence is different and hence the acceptable sets of arguments differ.

Consider the following two examples from [Baumann and Brewka, 2010].

Example 3.2 (liberal, strict). *Given F as presented below, $\sigma = stb$, $M = \asymp_U$, $r = cov_s$ and the desired set $E = \{a_1, a_3\}$.*

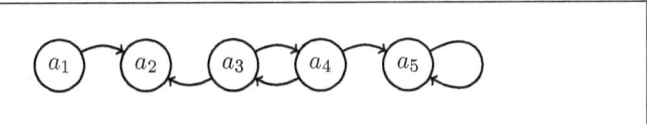

Since $stb(F) = \{\{a_1, a_4\}\}$ we have $E \notin cov_s(stb, F)$. How to enforce E? Define an enforcement (G, τ) of E with $F = G$ and $\tau = pr$. Note that $pr(G) = \{\{a_1, a_3\}, \{a_1, a_4\}\}$ justifies the claim because $E \in cov_s(G, pr)$ holds. The considered enforcement is strict and liberal and F is the pr-enforcing AF.

Example 3.3 (conservative, non-strict). *Given $\sigma = gr$, $M = \preceq_S$, $r = cov$, $E = \{a_2\}$ and $F = (\{a_1, a_2, a_3\}, \{(a_1, a_2), (a_2, a_1), (a_2, a_3)\})$ as presented below.*

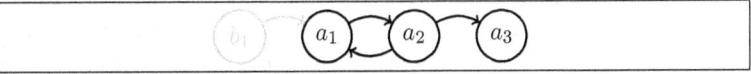

Note that $gr(F) = \{\emptyset\}$. Hence, $E \notin cov(gr, F)$. In this example we allow strong expansions only. Is it possible to enforce E? The answer is "yes". Consider the enforcement (G, τ) of E with G defined as depicted above and $\tau = \sigma$. Since $gr(G) = \{\{b_1, a_2\}\}$ we deduce $E \in cov(G, gr)$. The considered enforcement is non-strict and conservative and G is the gr-enforcing AF.

3.2 Extension Enforcement with Structural Change

We start with a review of one of the most prominent enforcement operators in the literature, named extension enforcement [Baumann and Brewka, 2010; Baumann, 2012b; Coste-Marquis et al., 2015; Doutre and Mailly, 2017b; Wallner et al., 2017; Haret et al., 2018]. Extension enforcement refers to a family of enforcement operators that all deal with covered acceptance, i.e., the enforcement goal is to modify a given AF such that a desired set of arguments becomes an extension, or becomes part of an extension, under a semantics. Both strict and non-strict variants were studied.

The main distinguishing aspect of various extension enforcement operations is what kind of modification type is permitted. Concretely, we look at extension enforcement allowing only expansions (Section 3.2.1), permitting only local modifications (Section 3.2.2), i.e., changing the attack structure, restricting change to be minimal (Section 3.2.3), and changing semantics (Section 3.2.4).

3.2.1 Expansion-based enforcement

In this section we consider conservative (non-)strict enforcements w.r.t. covered reasoning mode under different forms of expansions. More precisely, for a given AF $F = (A, R)$, a semantics σ and a desired set of arguments $E \subseteq A$ we look at pairs (G, σ) being (F, σ, M, cov) enforcements of E. We allow $M \in \{\preceq_E, \preceq_N, \preceq_S, \preceq_W\}$, i.e. arbitrary, normal, strong, and weak expansions are considered. In the following, for the sake of brevity, we do not explicitly mention the covered acceptance mode as well as the conservativeness.

We have already seen a case of non-strict extension enforcement under strong expansions in Example 3.3. We now exemplify some properties of extension enforcement under expansions.

Example 3.4. *Let us consider an AF $F = (A, R)$ with $A = \{a, b, c, d\}$ and an attack relation as shown in Figure 4. Say we want to enforce $E = \{b, d\}$ to be part of an admissible extension in a non-strict manner, and allowing arbitrary expansions. An AF G that ad-enforces these constraints is shown, as well, in Figure 4. That is, expanding by two arguments e and f and adding attacks (b, f), (f, d), and (e, c) results in*

$\{e, b, d\} \in ad(G)$, and thus E is non-strictly enforced to be part of an admissible extension by G. Note that adding the single attack (d, c) only wouldn't do the job since we are interested in non-strict enforcements. However, there are many more ways to non-strictly enforce the desired set E. We encourage the reader to find other witnessing ad-enforcing expansions.

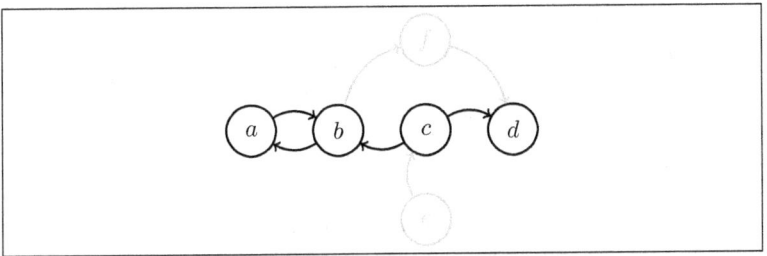

Figure 4: AF and expanded AF from Example 3.4

The preceding example illustrates the existence of enforcements. However, in general, desired enforcements might not exist. Consider the following example.

Example 3.5. Consider again the AF $F = (A, R)$ of Figure 4. We illustrate now three different sources for the impossibility of enforcements.

1. Assume we aim to strictly ad-enforce $E = \{b, d\}$ under normal expansions. While non-strict enforcement of E was possible (cf. Example 3.4), strict enforcement is impossible under normal expansions. The intuition is that $\{b, d\}$ is not admissible in the original AF F (the attack from c to d is not defended) and this fact remains true in any normal expansion G of F. The reason is simply that any new attack in G involves at least one new argument and thus, E can not defend d in G. However, E can be strictly enforced when allowing arbitrary expansions, e.g. adding a defending attack (b, c) is an option.

2. Another reason for impossibility of enforcement occurs when considering enforcement of sets like $\{a, b\}$ under any semantics σ that

preserves conflict-freeness, i.e. $\sigma \subseteq cf$. The reason is that $\{a,b\}$ is conflicting in F and thus, it remains conflicting regardless the considered type of expansion.

3. Even if the set E to be enforced is conflict-free and defends all its elements, enforcement is, under specific semantics, not always possible. Consider the aim to strictly co-enforce $E = \{c\}$ under weak expansions. In F the singleton E is not complete since it defends a and $a \notin E$. Now, weak expansions do not raise new attacks onto existing arguments which implies that former defense relations survive. Thus, for any weak expansion G of F we have E still defends a preventing it from being complete in G.

The previous observations have been firstly formalized in [Baumann and Brewka, 2010, Proposition 1] and later considered further in [Coste-Marquis et al., 2015, Proposition 1]. In the following we recall some results and generalize them to other semantics considered in this article.

Proposition 3.6. *Given an AF $F = (A, R)$ and $E \subseteq A$.*

- *If $E \notin ad(F)$ and $\sigma \subseteq ad$, then there is no AF G strictly σ-enforcing E under normal expansions.*

- *If $E \notin cf(F)$ and $\sigma \subseteq cf$, then there is no AF G (non-)strictly σ-enforcing E under arbitrary expansions.*

- *If E does not contain all defended arguments in F and $\sigma \subseteq co$, then there is no AF G that strictly σ-enforcing E under weak expansions.*

- *If $\sigma \in \{ad, cf, na, stb\}$ and $E \notin \sigma(F)$, then there is no AF G strictly σ-enforcing E under normal expansions.*

Despite several cases being impossible to enforce, there are interesting conditions under which an enforcement is always possible. As an illustration, consider the following example.

Example 3.7. *Say we desire to non-strictly ad-enforce $E = \{b, d\}$ under strong expansions. This means, we want E to be a strict subset of an admissible extension of the expanded framework. An example AF G ad-enforcing $\{b, d\}$ is shown in Figure 5. Here the new argument e is added* which defends both b and d. Since $\{e\}$ is admissible in G we obtain via the famous Fundamental Lemma [Dung, 1995, Lemma 10] that $\{e, b, d\}$ is admissible as desired.

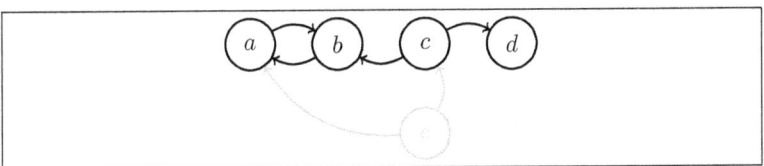

Figure 5: AF from Example 3.7

The observation from the example was generalized to further semantics [Baumann and Brewka, 2010, Theorem 4]. The main construction method is to extend the initial framework with a new argument which attacks all *undesired* arguments. We extend the already proven theorem to all semantics considered in this article.

Theorem 3.8. *Given an AF F, a desired set $E \in cf(F)$ and a semantics $\sigma \in \{ad, stb, pr, co, gr, id\, sst, eg, na, stg\}$. There is a strong expansion G of F non-strictly σ-enforcing E.*

Since strong expansions are particular cases of normal expansions as well as arbitrary expansions, we may state the following corollary.

Corollary 3.9. *Given an AF F, a desired set $E \in cf(F)$ and a semantics $\sigma \in \{ad, stb, pr, co, gr, id\, sst, eg, na, stg\}$. There are arbitrary as well as normal expansions G of F non-strictly σ-enforcing E.*

What about local expansions? Is it possible to (non-)strictly enforce a desired set E with local manipulations only? For most of the existing semantics we may act as follows: given the conflict-freeness of E we attack all remaining arguments first (this is sufficient for $\sigma \in \{ad, stb, pr, co, sst, na, stg\}$) and secondly, add self-loops to the remaining arguments (we additionally cover $\sigma \in \{id, eg\}$).

Theorem 3.10. *Given an AF F, a desired set $E \in cf(F)$ and a semantics $\sigma \in \{ad, stb, pr, co, id\, sst, eg, na, stg\}$. There is a local expansion G of F strictly σ-enforcing E.*

Note that grounded semantics is not included since it requires unattacked arguments which can not be "produced" with the help of local expansions. However, if there is an unattacked argument in the desired set E, then this unattacked argument can be used to attack all the arguments outside the directed set, leading to the strict gr-enforcement of E. Any unattacked argument in the AF can have a similar role for non-strict enforcement.

Theorem 3.11. *Given an AF F and a desired set $E \in cf(F)$, if there is an unattacked argument $a \in E$ (respectively $a \in A$), then there is a local expansion G of F strictly (respectively non-strictly) enforcing E under the grounded semantics.*

Let us turn now to a different aspect of enforcing, namely *how* exactly existing σ-extensions may change when expanding an AF. In general, the change is very much non-monotone: this means, arguments accepted earlier may become unaccepted, others become accepted; the number of extensions may shrink or increase, depending on the new arguments. For instance, it is easy to verify that we obtain a total collapse of stable extensions if we revise an AF by adding a self-defeating argument. Nevertheless, there are a few exceptions as illustrated in the following example taken from [Baumann, 2014b, Example 3.11]

Example 3.12. *Consider the weak expansion G of F as depicted below. In Example 2.3 we already observed that $pr(F) = \{\{e\}, \{f\}\} = \{E_1, E_2\}$.*

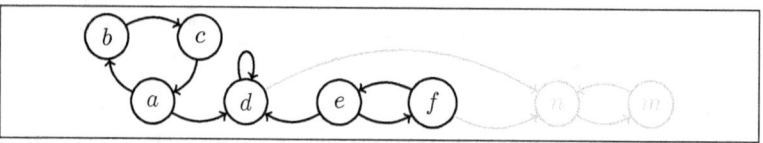

For the weak expansion G we find $pr(G) = \{E_1 \cup \{n\}, E_1 \cup \{m\}, E_2 \cup \{m\}\}$. Consequently, the following interrelations hold:

1. *the number of extensions increased*

2. *every old belief set is contained in a new one*

3. *every new belief set is the union of an old one and a new argument*

The previous example contrasts with the general observation that adding new arguments and attacks may change the outcome of an AF in a nonmonotonic fashion. Such a behaviour allows for reusing already computed extensions and has useful implications w.r.t. justification states. The following theorem [Baumann, 2014b, Theorem 3.2] shows that the class of weak expansions and semantics satisfying the directionality principle guarantee monotonic evolvements. Roughly speaking, the directionality criterion captures the idea that the evaluation of a certain argument should only be affected by its attackers and the attackers of its attackers and so on [Baroni and Giacomin, 2007].

Theorem 3.13. *Given an AF $F = (A, R)$ and a semantics σ satisfying directionality, then for all weak expansions $G = (B, S)$ of F we have:*

1. $|\sigma(F)| \leq |\sigma(G)|$, *(cardinality)*

2. $\forall E \in \sigma(F)\ \exists E' \in \sigma(G)\ \exists C \subseteq B \setminus A,\ s.t.\ E' = E \cup C$ *and (subset)*

3. $\forall E' \in \sigma(G)\ \exists E \in \sigma(F)\ \exists C \subseteq B \setminus A,\ s.t.\ E' = E \cup C$. *(representation)*

It is well-known that admissible, complete, preferred, grounded and ideal semantics satisfy directionality (cf. [van der Torre and Vesic, 2017] for an overview). Having the above theorem at hand we obtain the following relations regarding acceptance modes stating that credulously, sceptically as well as covered accepted sets persist.

Proposition 3.14. *Given an AF $F = (A, R)$ and $\sigma \in \{ad, co, pr, gr, id\}$. For any weak expansions G of F we have:*

- $cred(F, \sigma) \subseteq cred(G, \sigma)$,

- $scep(F, \sigma) \subseteq scep(G, \sigma)$ and

- $cov(F, ad) \subseteq cov(G, ad)$

3.2.2 Attack-based enforcement: Argument-fixed and Local Expansion-based Enforcement

We now turn to extension enforcement under a different kind of modifications to a given AF. In contrast to the previous section on expansion-based enforcement where expansion of the set of arguments and attacks, under certain conditions, was presented, we here look at changes that do not modify the set of arguments, but exclusively focus on updates of the attack structure.

Definition 3.15. *Let $F = (A, R)$ be an AF. We say that G is a local update of F, denoted by $F \asymp_L G$, if there is an AF G' such that $F \preceq_L G'$ and $G' \succeq_{LD} G$.*

In words, an AF $G = (A_G, R_G)$ is a local update of $F = (A_F, R_F)$ if there is an intermediate AF $G' = (A_{G'}, R_{G'})$ that is a local expansion of F (i.e., $A_{G'} = A_F$ and $R_F \subseteq R_{G'}$) and G is a local deletion of G' (i.e., $A_{G'} = A_G$ and $R_{G'} \supseteq R_G$). Put differently, G is a local update of F if the set of arguments stays the same, i.e., $A_G = A_F$, and the attack structure was changed arbitrarily: $R_G = (R_F \setminus R) \cup R'$ for some $R, R' \subseteq A_F \times A_F$.

In this section we consider extension enforcement under local updates [Coste-Marquis et al., 2015]. An intuition of a local update is that the arguments are unmodified, but some new attacks are revealed (e.g., in presence of new information), and some attacks are disputed and discarded (e.g., due to the defeasibility of attacks). Modifying the attacks between existing arguments can also be seen as an update of the preferences between arguments [Amgoud and Cayrol, 2002].[3]

Example 3.16. *Let us look at the same AF F from the preceding section that we used to exemplify expansion-based enforcement. We recall this AF in Figure 6a.*

We begin with looking at enforcement of the set $\{b, d\}$. Say, we desire to have this set of arguments being part of an admissible extension. In F the set $\{b, d\}$ is conflict-free but not admissible: the attack from c onto both b and d is not countered. A local update, in fact a local expansion, that enforces $\{b, d\}$ to be part of an admissible extension is shown in Figure 6b. An attack from b to c suffices to have $\{b, d\}$ defend both b and d.

A different case is exhibited by aiming to have $\{a, b\}$ being admissible: this set neither is conflict-free nor defends its arguments. A possible local update is shown in Figure 6c that enforces $\{a, b\}$ to be exactly an

---

[3]Recall that in preference-based argumentation, the "success" of an attack (a, b) depends on the fact that b is not preferred to a.

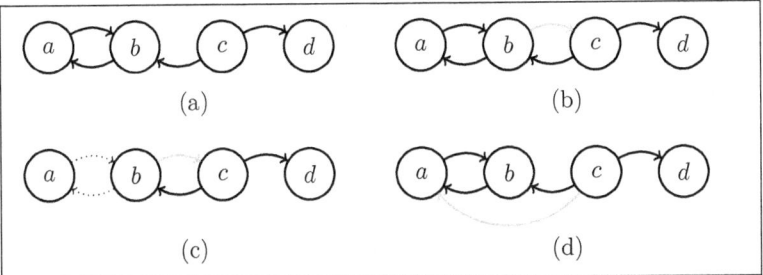

Figure 6: Enforcement by local updates

admissible extension, i.e., realizes strict extension enforcement under local updates and admissibility. Here, the conflicts between a and b are removed, to ensure conflict-freeness, and the attack from b to c is added, to ensure defense.

Finally, consider strict enforcement of $\{c\}$ under complete semantics. The set $\{c\}$ is admissible, yet defends a in F. A possible local update (local expansion) is shown in Figure 6d. Here one attack from c to a ensures that $\{c\}$ does not defend a.

Inspection of the preceding example reveals that several impossible cases, when requiring certain expansions (see previous section), are, in fact, possible under local updates. This is no coincidence: enforcement under local updates is possible for all main semantics of AFs: if $E \neq \emptyset$ is to be enforced, for a given AF $F = (A, R)$ there is the (trivial) local update $G = (A, R')$ with $R' = \{(a,b) \mid a \in E, b \in A \setminus E\}$ (i.e., in G, every argument in E is non-attacked, and every argument in $A \setminus E$ is attacked by all arguments in E). We have $E \in gr(G)$, and since the graph structure of G is acyclic[4], most semantics coincide with the grounded semantics. This observation is formalized next [Coste-Marquis et al., 2015, Proposition 4].

Proposition 3.17. *Let $F = (A, R)$ be an AF and $E \subseteq A$ be a non-*

---

[4]In the case of finite AFs, acyclicity corresponds to the well-foundedness property defined by [Dung, 1995], which implies the coincidence of grounded, stable, preferred and complete semantics. We also refer the reader to [Baroni et al., 2018a] for more details on this topic.

empty set of arguments. There exists a local update G that enforces E (non-)strictly to be (part of) a σ-extension, for all σ considered in this chapter.

Obviously, when $E = \emptyset$, it can always be non-strictly enforced with local updates, since E is included in any set of arguments. It is also the case that E can be strictly enforced with local update.[5] Indeed, for a given AF $F = (A, R)$ we can define the (trivial) local update $G = (A, R')$ with $R' = \{(a, a) \mid a \in A\}$ (i.e., in G, every argument is self-attacking). In this case, the empty set is the only conflict-free set, and thus the only extension for most semantics.

We have seen that enforcing a set of arguments with local updates is possible in general. Both the addition and the removal of attacks are necessary for this results. Indeed, if only local expansions are possible (i.e. removing attacks is not permitted), then a conflicting set E cannot be enforced under any semantics that requires conflict-freeness. Similarly, local deletions are not sufficient for strictly enforcing a set of arguments in all cases. As a matter of example, let us consider again the AF $F = (A, R)$ given at Figure 6a. The set $\{c\}$ cannot be enforced as a stable extensions by only deleting attacks: initially $\{c\}^\oplus = \{b, c, d\} \neq A$, and removing attacks cannot add arguments to the range of $\{c\}$.

3.2.3 Extension Enforcement and Minimal Change

Minimal change is an important topic in other domains of artificial intelligence, like belief change [Alchourrón et al., 1985; Katsuno and Mendelzon, 1992]. In the context of extension enforcement, the question asked is "how much effort will it cost to perform the enforcement?". This effort is defined by [Baumann, 2012b] as the number of attacks that are modified (i.e. either added or removed). Formally,

Definition 3.18. *Given $F = (A, R)$ and $G = (A', R')$, the distance between F and G is $d(F, G) = |(R \setminus R') \cup (R' \setminus R)|$.*

In general, there may be several ways to enforce an extension, even for a fixed type of modification. In that case, minimal change enforce-

---
[5]Except for the stable semantics, since the empty set can never be a stable extension of a non-empty AF.

ment consists in choosing one result that minimizes the distance d between the initial AF and the new one.

Example 3.19. *Figure 7 presents two examples of strong expansions of an AF $F = (A, R)$, with $A = \{a, b, c, d, e\}$ and $R = \{(b, a), (c, a), (d, b), (d, c), (e, d)\}$. This AF has a single stable extension: $stb(F) = \{\{b, c, e\}\}$. Both expansions succeed in non-strictly enforcing the set $\{a\}$ as a stable extension. However, we observe a difference in the number of attacks that have been added. The first one, F_1 (on the left side), adds two attacks, one from the new argument f_1 to c, and another one from f_2 to b; it has a single stable extension $stb(F_1) = \{\{a, e, f_1, f_2\}\}$. The second expansion, F_2 (on the right side), adds a single attack (f_3, e), and it also has a single stable extension: $stb(F_2) = \{\{a, d, f_3\}\}$. With $d(F, F_1) = 2$ and $d(F, F_2) = 1$, F_2 seems to be a more desirable result.*

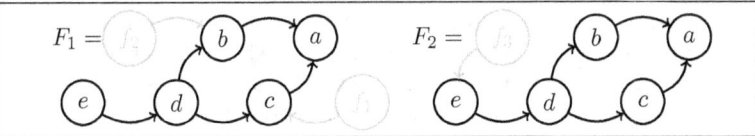

Figure 7: An example of (non-)minimal change

The question of minimal change in enforcement is studied in [Baumann, 2012b]. More specifically, it concerns the minimal change in non-strict enforcement based on normal expansions, as well as the special cases of strong and weak expansions. To do so, he defines the notion of characteristic of a set of arguments S, with respect to an AF F and a modification type $M \subseteq \mathcal{F} \times \mathcal{F}$. This characteristic corresponds to the minimal distance between F and an AF G such that S is included in an extension of G, and G is a possible result for the enforcement (i.e. $(F, G) \in M$). Strict enforcement can be considered as well [Doutre and Mailly, 2017b].

Definition 3.20. *Given a semantics σ, a modification type $M \subseteq \mathcal{F} \times \mathcal{F}$, $x \in \{s, ns\}$ meaning strict or non-strict, and an AF $F = (A, R)$, the*

(σ, M, x)-characteristic of a set $S \subseteq A$ is:

$$N_{\sigma,M}^{F,x}(S) = \begin{cases} 0 & \text{if } x = s, S \in \sigma(F) \\ 0 & \text{if } x = ns, \exists S' \in \sigma(F) \text{ s.t. } S \subseteq S' \\ k & \text{if } k = \min(\{d(F, G) \mid (F, G) \in M, N_{\sigma,M}^{G,x}(S) = 0\}) \\ +\infty & \text{otherwise} \end{cases}$$

Intuitively, the characteristic of a set of arguments S is 0 if this set is already (included in) an extension, k if k is the minimal distance between the initial AF and some AF that enforces S, and $+\infty$ if S cannot be enforced (under the the specified semantics and modification type).

Then, [Baumann, 2012b] introduces the notion of *value function*, that gives a constructive definition of how to compute the characteristic in a finite number of steps, based on properties of the initial AF. We use $V_{\sigma,M}^{F,x}(S)$ to denote this value function.

We start with the case of non-strict enforcement under weak expansion. Baumann shows that for most semantics, either the set S is already included in an extension, or it is impossible to enforce it with a weak expansion [Baumann, 2012b, Theorem 6]. Formally,

Proposition 3.21. *For $\sigma \in \{stb, ad\}$ a semantics, $F = (A, R)$ and AF and $S \subseteq A$ a set of arguments, the value function for non-strict enforcement under weak expansion and the semantics σ is*

$$V_{\sigma,\preceq_W}^{F,ns}(S) = \begin{cases} 0 & \text{if } \exists S' \in \sigma(F) \text{ s.t. } S \subseteq S' \\ +\infty & \text{otherwise} \end{cases}$$

Then, $N_{stb,\preceq_W}^{F,ns}(S) = V_{stb,\preceq_W}^{F,ns}(S)$ and $N_{\sigma,\preceq_W}^{F,ns}(S) = V_{ad,\preceq_W}^{F,ns}(S)$ for $\sigma \in \{ad, co, pr\}$.

Now, we turn to (non-strict) enforcement under strong expansion., i.e. we focus on defining $V_{\sigma,\preceq_S}^{F,ns}(S)$. This case is slightly more involved than the previous one, and it requires additional definitions.

Definition 3.22. *Given $F = (A, R)$ an AF and $X \in cf(F)$,*

- $ad(F, X) = X^{\ominus} \setminus X^{\oplus}$;

- $stb(F, X) = A \setminus X^{\oplus}$.

Intuitively, these sets correspond to the arguments that should be defeated in order to make X an admissible (respectively stable) extension of F. They can be used to define the value function for enforcement under strong expansion, for $\sigma \in \{stb, ad\}$. Interestingly, these value functions can be used also for enforcing a set of arguments under normal expansion or general expansions, as stated by [Baumann, 2012b, Theorem 9].

Proposition 3.23. *For $\sigma \in \{stb, ad\}$ a semantics, $F = (A, R)$ and AF and $S \subseteq A$ a set of arguments, the value function for non-strict enforcement under strong expansion and the semantics σ is*

$$V_{\sigma, \preceq_S}^{F,ns}(S) = \min(\{|\sigma(F, S')| \mid S \subseteq S' \text{ and } S' \in cf(F)\})$$

Then, $N_{stb,M}^{F,ns}(S) = V_{stb,\preceq_S}^{F,ns}(S)$ and $N_{\sigma,M}^{F,ns}(S) = V_{\sigma,\preceq_S}^{F,ns}(S)$ hold for $\sigma \in \{ad, co, pr\}$ and $M \in \{\preceq_E, \preceq_N, \preceq_S\}$.

This means that authorizing more kinds of modifications than the addition of strong arguments is useless regarding the issue of minimal change.

Then, an interesting result [Baumann, 2012b, Proposition 11] states that enforcement is always possible if arbitrary updates are permitted, i.e. attacks can also be deleted (contrary to expansions, where attacks can only be added).

Proposition 3.24. *For $\sigma \in \{stb, sst, pr, co, ad\}$ and any $F = (A, R)$,*

$$N_{\sigma,U}^{F,ns}(S) \leq |R \cap (S \times S)| + |A \setminus S|$$

Intuitively, it says that we can enforce S as (a subset of) an extension by making it conflict-free (i.e. removing the attacks in $R \cap (S \times S)$) and attacking every argument that is not in S (i.e. adding attacks from fresh arguments to arguments in $A \setminus S$). This finite upper bound guarantees that non-strict enforcement under arbitrary updates is always possible. But a more precise evaluation of the characteristics is given by this value function [Baumann, 2012b, Theorem 12]:

Proposition 3.25. *For $\sigma \in \{stb, ad\}$ a semantics, $F = (A, R)$ and AF and $S \subseteq A$ a set of arguments, the value function for non-strict enforcement under arbitrary updates and the semantics σ is*

$$V_{\sigma,U}^{F,ns}(S) = \min(\{|R \cap (S' \times S')| + |\sigma(F, S')| \mid S \subseteq S' \subseteq A\}$$

with $ad(F, S')$ and $stb(F, S')$ as in Definition 3.22. Then, $N_{stb,U}^{F,ns}(S) = V_{stb,U}^{F,ns}(S)$ and $N_{\sigma,U}^{F,ns}(S) = V_{\sigma,U}^{F,ns}(S)$ hold for $\sigma \in \{pr, co, ad\}$.

Finally, [Doutre and Mailly, 2017b] presents characteristics for enforcement under local updates, i.e. when the set of arguments has to remain the same, but attacks between them can be added or deleted. The results are reminiscent of the ones described in this section.

3.2.4 Semantics-based Enforcement

Extension enforcement is usually defined as an operation where the target semantics is given as an input. We call it conservative enforcement when the target semantics is the same as the initial semantics, and liberal enforcement otherwise. On the contrary, [Doutre and Mailly, 2017b] proposes to generalize enforcement, by enhancing operators with a set Σ of possible target semantics. Then, the chosen semantics is the one that allows to enforce the set of arguments with minimal change on the graph. More formally:

Definition 3.26. *For $F = (A, R)$ an AF, $S \subseteq A$ the set of arguments to be enforced and Σ a set of semantics, a strict (resp. non-strict) enforcement of S in F under a given modification type $M \subseteq \mathcal{F} \times \mathcal{F}$, is a pair (G, σ') such that*

1. $(F, G) \in M$;
2. $\sigma' \in \Sigma$ and $S \in \sigma'(G)$ (resp. $S \subseteq S' \in \sigma'(G)$);
3. $\forall \sigma'' \in \Sigma$, $V_{\sigma',M}^{F,x}(S) \leq V_{\sigma'',M}^{F,x}(S)$ (with $x \in \{s, ns\}$).

This means that the new semantics is chosen in a way that guarantees that the change on the graph is minimal. Since the characteristics can be the same for several semantics σ', additional criteria can be used in order to select the new semantics, like the distance between σ' and the initial semantics σ [Doutre and Mailly, 2017a].

Finally, we already mentioned that [Baumann and Brewka, 2010, Section 3.1] discusses the tool of changing semantics in order to enforce a desired set. The authors presented two involved impossibility theorems specifying properties of initial extensions and desired sets, initial and target semantics as well as the considered type of structural change. Regarding the semantic change we have that possible target semantics were restricted to semantics satisfying well-known abstract criteria like admissibility or reinstatement (cf. [van der Torre and Vesic, 2017] for an exhaustive overview). The mentioned theorems show either limitations for exchanging accepted arguments with formerly unaccepted ones (under normal expansions) or limitations for eliminating arguments of existing extensions (under weak expansions).

3.3 Complexity and Algorithms

We review complexity of enforcement problems, in particular expansion-based enforcement, and enforcement based on local updates [Wallner et al., 2017; Coste-Marquis et al., 2015].

In several cases enforcement is, computationally speaking, straightforward if the task consists in checking whether there exists a modified AF that enforces a set of arguments under certain parameters. For instance, extension enforcement under normal expansions for admissible semantics is always possible if the set E to enforce is conflict-free in the given AF (see Section 3.2.1). That is why we look at extension enforcement that aims at minimizing the change induced by an enforcing AF. Concretely, given an AF $F = (A, R)$ we aim at finding an enforcing AF $G = (A', R')$ such that the distance $d(F, G)$ between them is minimal (see Definition 3.18).

Another important aspect for expansion-based enforcement is how many arguments shall be added. That is, if $G = (A', R')$ is an expansion of $F = (A, R)$, how to confine $|A'| - |A|$? This is important from a computational perspective, since allowing for unbounded expansions may complicate computation. We consider here only bounded expansions.

We define the computational problems next, for extension enforcement under bounded expansions and local updates. For local updates no bound is needed, since if the number of arguments $|A|$ does not change, the number of modifications to R is bounded quadratically by $|A|$.

For studying complexity of problems that are inherently optimization problems, such as enforcement when the goal is to find an enforcing AF with a minimum number of modifications to the attack structure, there are several ways to formally approach such problems. One standard way to reveal inherent complexities of optimization problems is to consider a natural decision variant: for a given integer $k \geq 0$, we ask whether there is an enforcing AF with at most k many modifications. We note that another way to study complexity of optimization problems is to consider functional problems instead of decision problems, which is an approach that may give more detailed complexity results (see, e.g., [Krentel, 1988]). However currently no such analysis was carried out for enforcement.

First, we define a decision problem for extension enforcement under bounded expansions.

Extension enforcement under bounded expansions
INSTANCE: an AF $F = (A, R)$, $E \subseteq A$, set A', integer $k \geq 0$, and a semantics σ.
QUESTION: Does there exist an expansion $G = (A \cup A', R')$ of F such that $\exists E' \in \sigma(G)$ with $E \subseteq E'$ and $d(F, G) \leq k$?

In more words, given an AF F, a set $E \subseteq A$ of arguments to enforce, a set of arguments A', an integer $k \geq 0$ and a semantics σ, the task is to decide whether there exists an expansion G of F that enforces E non-strictly under σ, and, moreover, makes at most k many modifications to the attack structure. Note that the expansion G is bounded in the sense that the expanded arguments are already given beforehand, i.e., G has $A \cup A'$ as its arguments. The above definition gives a decision problem for non-strict enforcement. As before, we define strict enforcement analogously by replacing $\exists E' \in \sigma(G)$ and $E \subseteq E'$ with $E \in \sigma(G)$.

Next, we look at a decision problem variant for extension enforcement under local updates.

Extension enforcement under local updates
INSTANCE: an AF $F = (A, R)$, $E \subseteq A$, integer $k \geq 0$, and a semantics σ.
QUESTION: Does there exist a local update $G = (A, R')$ of F such that $\exists E' \in \sigma(G)$ with $E \subseteq E'$ and $d(F, G) \leq k$?

Strict enforcement is again defined as above.

We consider as fragments of these two enforcement problems those sub problems where a semantics is fixed, i.e., extension enforcement under bounded expansions (local updates) under a specific semantics σ, instead of having σ as part of the instance.

Finally, before delving into complexity results from the literature, we provide background on complexity classes used here, and related problems useful to understanding complexity of enforcement. For thorough introductions to computational complexity see, e.g., [Arora and Barak, 2009; Papadimitriou, 2007]. We assume that the reader is familiar with concepts like complexity classes, reductions, completeness, and oracles. Complexity class P is composed of all decision problems which can be decided in polynomial time by a deterministic algorithm. Class NP contains all decision problems that can be decided by a non-deterministic polynomial time algorithm. Class coNP contains all problems that are complementary to a problem in NP. Class Σ_2^P contains all decision problems which can be decided via a non-deterministic polynomial time algorithm that can access an NP oracle. Class Π_2^P contains all problems that are complementary to some problem in Σ_2^P.

Two reasoning tasks on AFs in a static, i.e., non-dynamic setting, are useful to understand the complexity of enforcement. The first one is usually referred to as the Verification problem.

Verification
INSTANCE: an AF $F = (A, R)$, $E \subseteq A$, and a semantics σ.
QUESTION: Does $E \in \sigma(F)$ hold?

That is, the task is to check whether a given set E is a σ-extension. Another useful problem is credulous acceptance of arguments in AFs.

Credulous acceptance
INSTANCE: an AF $F = (A, R)$, $a \in A$, and a semantics σ.
QUESTION: Does $\{a\} \in cred(F, \sigma)$ hold?

In words, an argument is credulously accepted in case there is a σ-extension of a given AF that contains the queried argument.

Complexity of verification and credulous acceptance was established; we summarize complexity results for the main semantics in Table 2.

semantics σ	verification	credulous acceptance
cf	in P	in P
ad	in P	NP-complete
co	in P	NP-complete
stb	in P	NP-complete
pr	coNP-complete	NP-complete

Table 2: Complexity of verification and credulous reasoning in AFs (for an overview see [Dvořák and Dunne, 2018])

3.3.1 Complexity of Enforcement

We illustrate two ways of showing complexity bounds that turn out to be tight in many, but not all, cases.

For an upper bound (i.e. membership in a complexity class), consider the following non-deterministic algorithm (sketch) given an AF F, a set E to enforce, and a semantics σ:

1. non-deterministically guess an AF $G = (A', R')$ that is a bounded expansion (or local update) of F;

2. non-deterministically guess an $E' \subseteq A'$ (for non-strict enforcement only); and

3. verify whether $E' \in \sigma(G)$ and $E \subseteq E'$ (for non-strict enforcement) or $E \in \sigma(G)$ (for strict enforcement).

In case the last step succeeds, then it holds that G enforces E to be a σ-extensions (non-)strictly. As can be seen from this algorithm sketch, a complexity upper bound can be derived from the complexity of the verification problem under σ. Take $\sigma = ad$, i.e., the verification problem under admissibility which is polynomial-time decidable. It follows that extension enforcement under bounded expansions (resp. local updates) is in NP under admissibility. The reason is that the above algorithm sketch directly witnesses membership in NP: one (resp. two) non-deterministic construction(s) and a check in polynomial time show membership for $\sigma = ad$. In the non-deterministic construction of the above algorithm the bound on the expansion is crucial, otherwise a non-bounded, and thus potentially non-polynomially bounded, structure would be constructed. However, this does not imply that enforcement under non-bounded expansions requires large expansions.

There is a similar approach to show lower bounds. Here we distinguish more between strict and non-strict variants. In particular, extension enforcement under bounded expansions (resp. local updates) and σ is C-hard

- if verification under σ is C-hard and the enforcement variant is strict; or

- if credulous acceptance under σ is C-hard and the enforcement variant is non-strict.

The underlying reason is as follows. One can reduce the verification problem to strict extension enforcement and the credulous acceptance problem to non-strict extension enforcement.

For the verification problem under σ, i.e., given an AF F and a set E, consider the extension strict enforcement problem under σ with F, E, and $k = 0$ as input (and $A' = \emptyset$ for expansion-based). Then we are not allowed to make modifications to F, and, therefore, F enforces E to be a σ-extension iff $E \in \sigma(F)$ iff this is a positive instance of the verification problem.

Similarly, the credulous acceptance problem under σ, with F and an argument a as input, is reduced to an instance of non-strict extension enforcement with input F, $E = \{a\}$, and $k = 0$ (and again $A' = \emptyset$). It follows that F enforces E non-strictly if there is an $E' \supseteq E$ with $E' \in \sigma(F)$, implying a positive instance of the credulous acceptance problem.

In several cases the two approaches to show upper and lower bounds yield tight bounds. However, there are notable exceptions.

Let us look first at results obtained for enforcement under bounded expansions, see Table 3. In this case only the non-strict variant was studied [Wallner et al., 2017]. It can be observed that the complexity of this enforcement variant matches complexity of credulous reasoning in static AFs, i.e., the above approaches to show complexity bounds directly result in tight bounds. We remark that complexity of enforcement under conflict-free sets was not presented in [Wallner et al., 2017], however it can be straightforwardly obtained: if the set is conflict-free then enforcement is trivial (and can be checked in polynomial time by scanning the input AF), otherwise, if the given set to enforce is conflicting, no expansion can remove such conflicts, and enforcement is impossible. Since (by definition) any conflict-free set is included in some naive extension, this result also holds for $\sigma = na$.

semantics σ	non-strict
cf	in P
na	in P
ad	NP-c
co	NP-c
stb	NP-c
pr	NP-c

Table 3: Complexity of non-strict extension enforcement under bounded expansions [Wallner et al., 2017]

Let us turn to complexity of enforcement under local update [Wallner et al., 2017; Coste-Marquis et al., 2015], summarized in Table 4. We see that complexity of non-strict enforcement, again, has the same complexity as credulous reasoning in static AFs, except for grounded semantics. Before discussing grounded semantics, let us turn to strict enforcement first.

To some extend surprising are the results for strict extension enforcement, which diverge from non-strict enforcement. For instance, for both admissible and stable semantics strict extension enforcement under local expansions is decidable in polynomial time. The underlying reason is that if E is to be an admissible set (a stable extension), then all conflicts inside the set have to be removed, and each attack from outside countered (each argument outside attacked). The latter means that one can choose an argument inside E to counter non-attacked attackers

semantics σ	strict	non-strict
cf	in P	in P
na	in P	in P
ad	in P	NP-c
co	NP-c	NP-c
gr	NP-c	NP-c
stb	in P	NP-c
pr	Σ_2^P-c	NP-c
sst	Σ_2^P-c	Σ_2^P-c
stg	coNP-hard and in Σ_2^P	Σ_2^P-c

Table 4: Complexity of extension enforcement under local updates [Wallner et al., 2017; Coste-Marquis et al., 2015]

(to achieve defense) or remove the incoming attack. In both cases, it is sufficient to make at least one modification, however one modification is sufficient: adding an attack to counter an attacker (removing an incoming attack might not be sufficient if there are more incoming attacks). For stable semantics, similarly, one attack on unattacked arguments outside E is both necessary and sufficient, only the origin in E is flexible. Overall, this procedure sketches a polynomial-time deterministic algorithm (one can impose an ordering on arguments to make the choice of attacking arguments deterministic).

Finally, let us look at grounded semantics, for which NP-completeness was established for both non-strict and strict extension enforcement under local updates and complete semantics for the strict variant. Recall that both verification and credulous acceptance under grounded semantics is in P, and also verification for complete semantics is in P (Table 2). This means, the lower bounds established by the algorithms above do not result in tight bounds. The intuition behind this "complexity jump" for the strict variant under complete and grounded semantics is that when enforcing some set of arguments E to be complete, one has to be careful about what E defends. That is, enforcing E to be admissible is not the underlying reason for NP hardness, but to avoid having arguments defended that one desires to avoid being defended (as specified by strict enforcement, nothing outside the set E may be defended by

E). In brief, addition or removal of attacks can make E admissible, but implying further arguments being defended. Finding an optimal assignment that accomplishes both having E admissible and nothing outside E being defended by E faces non-deterministic choices. However, the hardness construction to show NP-hardness is somewhat involved.

Finally, for grounded semantics and non-strict enforcement, the intuition for NP-completeness is a bit more direct: there could be a place in the AF to modify such that the grounded extension is significantly expanded and includes the desired E. However, choosing an adequate place in such a way is not direct to find.

Further semantics have been analyzed in [Wallner et al., 2017].

3.3.2 Declarative Algorithms

Main approaches to compute optimal enforcing for AFs rely on declarative programming paradigms based on constraints, in particular maximum Satisfiability (MaxSAT) [Morgado et al., 2013], answer set programming (ASP) [Niemelä, 1999; Gelfond and Lifschitz, 1988], and pseudo Boolean optimization (particularly integer linear programming [Sierksma and Zwols, 2015]).

We present here some of the main ideas for algorithmic approaches to extension enforcement, focusing on the MaxSAT approach [Wallner et al., 2017]. Enforcement via pseudo Boolean optimization is presented in [Coste-Marquis et al., 2015], and via ASP in [Niskanen et al., 2018]. Systems using the MaxSAT approach are presented in [Niskanen et al., 2016b; Coste-Marquis et al., 2015]. We present here encodings and an algorithm for some semantics, for further semantics and details we refer to the original papers.

We briefly recall background on MaxSAT. A literal is either a positive Boolean variable x or a negated Boolean variable $\neg x$. A clause is a disjunction of literals $l_1 \vee \cdots \vee l_n$ and a propositional formula is in conjunctive normal form (CNF) if the formula $\pi = c_1 \wedge \cdots \wedge c_m$ is a conjunction of clauses. Whenever convenient, we will view clauses as a set of literals and a formula in CNF as a set of clauses.

A truth assignment τ assigns either true (1) or false (0) to the Boolean variables. As usual, a truth assignment τ satisfies a variable x if $\tau(x) = 1$. Satisfaction is extended in the usual way to compound formulas, e.g., τ satisfies a literal l if $\tau(x) = 1$ and $l = x$ or $\tau(x) = 0$ and $l = \neg x$. A clause is satisfied by τ if at least one literal of the clause is satisfied, and a formula in CNF is satisfied if each clause is satisfied.

An instance of the partial MaxSAT problem is a pair $\phi = (\phi_h, \phi_s)$ with both ϕ_h and ϕ_s Boolean formulas in CNF (sets of clauses). The former is the set of hard clauses, while the latter is the set of soft clauses. A truth assignment τ is a solution to the partial MaxSAT instance if τ satisfies ϕ_h (the hard clauses). The cost of τ w.r.t. the instance ϕ is $\text{cost}(\phi, \tau) = \sum_{c \in \phi_s} 1 - \tau(c)$, i.e., the number of soft clauses not satisfied. A solution τ to ϕ is optimal if there is no solution τ' to ϕ with $\text{cost}(\phi, \tau') < \text{cost}(\phi, \tau)$. We refer to partial MaxSAT simply as MaxSAT.

We focus on an illustration of a MaxSAT approach to extension enforcement on the variant with local updates. Encoding extension enforcement under local updates can be achieved by encoding an AF, possible modifications, and semantics in MaxSAT. Before delving into encoding this enforcement variant, we recall an encoding of admissible semantics of static AFs similar as in [Besnard and Doutre, 2004]. Given an AF $F = (A, R)$ we define

$$\phi_{cf}(F) = \bigwedge_{(a,b) \in R} \neg a \vee \neg b.$$

We use here arguments as Boolean variables and vice versa. Satisfying truth assignments of this formula correspond directly to conflict-free sets of F in the way that $E \in cf(F)$ iff $\tau(x) = 1$ for $x \in E$ and $\tau(y) = 0$ if $y \notin E$ satisfies ϕ_{cf}. Admissibility can be encoded as follows:

$$\phi_{ad}(F) = \phi_{cf}(F) \wedge \bigwedge_{a \in A} (a \to (\bigwedge_{(b,a) \in R} (\bigvee_{(c,b) \in R} c)))$$

In words, if an argument a is in an admissible set (true in a satisfying assignment) then for each attacker b at least one defender c must be part of the admissible set (true in the assignment), as well, which directly captures the definition of admissibility.

Non-strict extension enforcement under local updates can be encoded by including variables for attacks. We first focus on how to encode constraints for the semantics, which we encode as hard clauses ϕ_h. For notation, for the encodings of conflict-free sets and admissible extensions above we used ϕ, for enforcement formulas we use ψ.

$$\psi_{cf}(F) = \bigwedge_{a,b \in A} (r_{a,b} \to (\neg a \vee \neg b))$$

In words, a new variable $r_{a,b}$ for each pair of arguments a, b is introduced denoting whether there is an attack from a to b. That is, a truth assignment includes now an assignment on the attacks, as well.

Moving on to enforcement under admissibility, which we encode as

$$\psi_{ad}(F) = \psi_{cf}(F) \wedge \bigwedge_{a,b \in A} ((a \wedge r_{b,a}) \to \bigvee_{c \in A} (c \wedge r_{c,b})).$$

That is, if a is assigned to be true and there is an attack (b, a), then this attack has to be defended against, by some c and the corresponding attack (c, b).

Let $F = (A, R)$, $E \subseteq A$ be given as an instance of the non-strict extension enforcement problem under local updates and admissibility. Defining a MaxSAT instance

$$\phi = (\psi_{ad} \wedge \bigwedge_{a \in E} a, \phi_s(F))$$

with

$$\phi_s(F) = \bigwedge_{(a,b) \in R} r_{a,b} \wedge \bigwedge_{(a,b) \notin R, a,b \in A} \neg r_{a,b}$$

results in optimal truth assignments τ to ϕ corresponding to AFs locally updated from an original AF $F = (A, R)$ with a minimum number of modifications that enforce S to be part of an admissible set. To see this, any solution to ϕ satisfies the hard clauses, implying that a truth assignment satisfying $\psi_{ad} \wedge \bigwedge_{a \in E} a$ assigns to true all variables (arguments) in E and possibly more arguments, and further assigns some of the attacks to be true (present in a modified AF) such that the argument variables assigned to true form an admissible set, thus enforcing E.

For strict extension enforcement under local updates, more variables can be fixed, since the set E to be enforced must be exactly a σ-extension, not just be part of one. That is, we can focus on variables only for attacks, since the other variables can be fixed (true for argument variables in E and false otherwise, i.e. $\bigwedge_{a \notin E} \neg a$ can be added to the hard part of the MaxSAT instance).

For instance, strict enforcement under conflict-free sets can be encoded as follows.

$$\psi_{cf}^s(F) = \bigwedge_{a,b \in E} \neg r_{a,b}$$

In more words, there cannot be any attack in the set S to be enforced, all other attacks remain unconstrained for conflict-free sets. Admissible extensions can be encoded by

$$\psi_{ad}^s(F) = \psi_{cf}^s(F) \wedge \bigwedge_{a \in E} \bigwedge_{b \in A \setminus E} (r_{a,b} \to \bigvee_{c \in E} r_{c,b}).$$

The MaxSAT instance is complete by setting

$$\phi = (\psi_{ad}^s(F) \wedge \bigwedge_{a \in E} a \wedge \bigwedge_{a \in A \setminus E} \neg a, \phi_s(F)).$$

We proceed to algorithmic approaches for problems "beyond NP", e.g., strict extension enforcement under local updates and preferred semantics. One approach to such complex problems is to develop an algorithm that uses SAT solvers as subprocedures, and possibly calls a SAT solver multiple times (i.e., an iterative SAT-based procedure). We present an approach based (inspired by) the well-known CEGAR approach [Clarke et al., 2004; Clarke et al., 2003] approach, where CEGAR stands for counterexample guided abstraction refinement. We remark that the term CEGAR is not used unambigously in the literature, and in some communities the term may refer to different concepts. Here, a CEGAR based algorithm works on an abstraction (approximation) of a solution space from which iteratively candidates are drawn. Importantly, due to the approximation, some solutions may be spurious. A SAT call determines whether a candidate is a solution or a spurious solution. In the latter case, the spurious solution is a "counterexample" which is used to refine the approximated solution space (removing as many as possible of the spurious solutions from the space) and a next candidate is produced, until a solution is reached.

For strict extension enforcement under local updates and preferred semantics, the solution space is approximated by considering initially strict extension enforcement under local updates and admissible or complete semantics. It holds that if an AF G enforces a set S strictly under preferred semantics, then G also enforces S strictly under admissible or complete semantics (since a preferred extension is complete and admissible). However, importantly, optimality is not guaranteed this way: an optimal solution to strict extension enforcement under local updates and preferred semantics might not be an optimal solution AF for admissible or complete semantics (since less modifications might be sufficient for admissible or complete semantics, but not for preferred semantics). Nevertheless, strict extension enforcement under complete or admissible semantics can act as an approximation. We focus for illustration on admissible semantics here.

The CEGAR-style algorithm is presented as Algorithm 1. When the loop is entered the first time, the MaxSAT call returns an optimal solution for strict extension enforcement under local updates and admissible semantics. To check whether the AF extractable from the truth assignment τ is a solution also under preferred semantics, we call a SAT solver to determine whether S is a preferred extension in the candidate AF. If so, we return this AF. Otherwise, we found a counterexample and the abstraction is refined. For correctness of the overall algorithm, it is important that a refinement does not remove (all) optimal solutions AFs for strict enforcement under local updates and preferred semantics. We refine here by removing the solution found in the MaxSAT call, i.e., by removing exactly τ from consideration when looking for the next candidate. This is a straightforward refinement. More sophisticated refinements are possible, but require care when designing them (e.g., in order not to violate correctness) [Niskanen and Järvisalo, 2020]. For instance, in some cases refinements can be based on foundational results whether changes on AFs induce changes on semantics [Boella et al., 2009b; Boella et al., 2009a].

The definitions for Algorithm 1 are as follows. From a truth assignment τ we can extract an AF by $\text{EXTRACT}(A, \tau) = (A, R)$ with $R = \{(a, b) \mid \tau(r_{a,b}) = 1\}$. The formula

$$\text{CHECK}(\tau, E) = \phi_{ad}(F') \wedge \bigwedge_{a \in S} a \wedge \bigvee_{b \in A \setminus S} b$$

can be used for checking whether an admissible extension E is a preferred

Algorithm 1 Strict extension enforcement under local updates and preferred semantics

1: $\varphi_h \leftarrow \psi'_{ad}(F)$
2: $\varphi_s \leftarrow \bigwedge_{(a,b)\in R} r_{a,b} \wedge \bigwedge_{(a,b)\notin R, a,b\in A} \neg r_{a,b}$
3: **while** true **do**
4: $\quad \tau \leftarrow \text{MAXSAT}(\varphi_h, \varphi_s)$
5: $\quad result \leftarrow \text{SAT}(\text{CHECK}(\tau, S))$
6: $\quad$ **if** $result = unsatisfiable$ **then return** τ
7: $\quad$ **else** $\phi_h \leftarrow \phi_h \wedge \text{REFINE}(\tau)$
8: **end while**

extension of the modified AF: we guess a superset and check admissibility by the above sub formulas. Refinement is specified via

$$\text{REFINE}(\tau) = \bigvee_{(a,b)\in R'} \neg r_{a,b} \vee \bigvee_{(a,b)\in (A\times A)\setminus R'} r_{a,b}.$$

In words, we exclude in a subsequent search for a solution candidate exactly the currently found candidate AF.

Example 3.27. Consider the AF F from Figure 8(a). That is, we have a chain of attacked arguments from a to b to c, and c attacks both d and e. Further, f is unattacked and does not attack an argument. The unique preferred extension of F is $\{a, c, f\}$. Say we want to strictly enforce $\{b, f\}$ to be exactly a preferred extension, and use Algorithm 1 in order to achieve that. Initially, we solve, via MaxSAT, strict enforcement under admissible semantics to have $\{b, f\}$ being admissible. Say the result is as shown in Figure 8(b), i.e., an attack from f to a is added, resulting in $\{b, d, e, f\}$ being the unique preferred extension, and $\{b, f\}$ being admissible. As the SAT solver call in the algorithm verifies, this AF candidate is not a solution, since, e.g., $\{b, d, e, f\}$ is an admissible superset of $\{b, f\}$, implying that $\{b, f\}$ is not preferred. We exclude, via the refinement step, this candidate AF, and call the MaxSAT solver again. Note that the hard clauses refute the previous candidate AF (i.e., any truth assignment simulating that AF does not satisfy the hard clauses).

In the next steps of the algorithm, all AFs that enforce $\{b, f\}$ to be admissible are checked which make at most one modification (in the

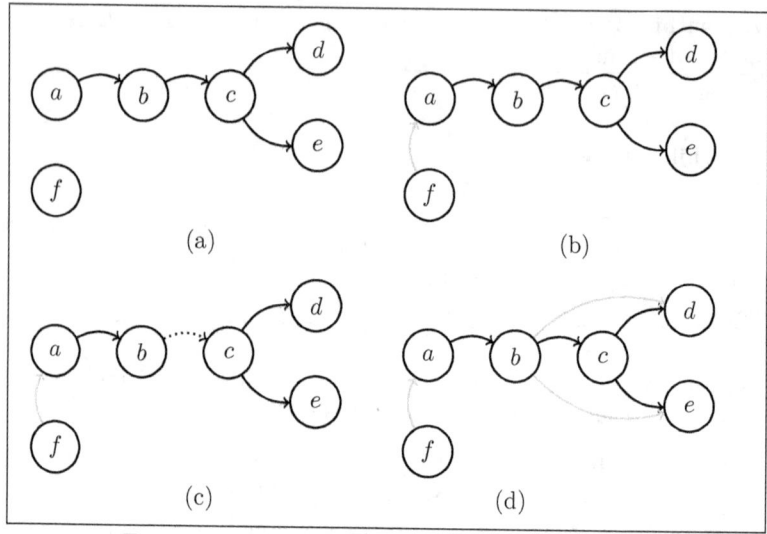

Figure 8: Example candidate AFs for Algorithm 1

above simple refinement). After that it is verified that no modified AF with at most one modification (local update) achieves the strict enforcement under preferred semantics.

For two possible modifications, say the MaxSAT call returns an assignment corresponding to the AF in Figure 8(c). Due to the definition of the MaxSAT instance, we know that $\{b, f\}$ is admissible in this candidate AF, which is like the previous one, except for removal of the attack from b to c. Here $\{b, f\}$ is again admissible, and the unique preferred extension is $\{b, c, f\}$, which is again verified not to be a solution. After checking all AFs that enforce $\{b, f\}$ to be admissible with at most one modification, the algorithm proceeds to at most three modifications, where a possible solution can be found, as illustrated in Figure 8(d).

4 Related Notions to Enforcement

In this section, we overview several notions that are closely related to the enforcement setting described previously.

4.1 Update Using Logical Translations

YALLA (Yet Another Logic Language for Argumentation) [de Saint-Cyr et al., 2016] is a first-order logical language that allows to describe argumentation frameworks and their semantics. Then, operations related to enforcement can be defined through belief change theory, especially belief update [Katsuno and Mendelzon, 1991].

Let us briefly describe the syntax and semantics of YALLA formulas. It is assumed that argumentation frameworks are built from a given *universe* $F_U = (A_U, R_U)$. This means that for any AF $F = (A, R)$, $A \subseteq A_U$ and $R \subseteq R_U \cap (A \times A)$. We write $k = |A_U|$ the number of arguments in the universe. A YALLA formula (or more precisely, YALLA$_U$) is a well-formed first order logic formula such that:

- the set of constant symbols is $V_{const} = \{c_\bot, c_1, \ldots, c_p\}$ where $p = 2^k - 1$;

- the set of function symbols is $V_{func} = \{\text{union}^2\}$;

- the set of predicate symbols is $V_{pred} = \{\text{on}^1, \triangleright^2, \subseteq^2\}$.

The semantics of YALLA is defined through a structure associated with an AF $F = (A, R)$ built on the universe F_U. The domain of this structure is $D = 2^{A_U}$, and it is associated with an interpretation such that:

- the constant symbol $c_\bot$ is associated with the empty set; each constant symbol c_i ($i \in \{1, \ldots, 2^k - 1\}$) is associated with a different non-empty element of D;

- the union function symbol is associated with the binary set-theoretic union over D;

- the on predicate symbol is associated to the characterization function of subsets of A, i.e. on(S) is true if and only if $S \subseteq A$;

- the predicate symbol $\triangleright$ is associated with the set-attack relation induced by R, i.e. $S_1 \triangleright S_2$ if and only if $S_1 \subseteq A$, $S_2 \subseteq A$, and $\exists a_1 \in S_1, a_2 \in S_2$ such that $(a_1, a_2) \in R$;

- the predicate symbol $\subseteq$ is associated with the classical inclusion relation over D.

Some axioms are added to the theory in order to guarantee the meaning of the YALLA formulas. For instance, if a set S_1 is included in A, then any subset of S_1 is included in A as well: this is formalized by $\forall x, y, (\text{on}(x) \wedge y \subseteq x) \Rightarrow \text{on}(y)$. A full description of the YALLA axioms is out of the scope of this chapter; we refer the interested reader to [de Saint-Cyr et al., 2016] for more details on this topic.

An argumentation framework $F = (A, R)$ can be described with the formula

$$\Phi_F = \text{on}(A) \wedge \bigwedge_{x \in A_U \setminus A} \neg\text{on}(\{x\}) \wedge \bigwedge_{(x,y) \in R} (\{x\} \triangleright \{y\}) \wedge \bigwedge_{(x,y) \in R_U \setminus R} \neg(\{x\} \triangleright \{y\})$$

Then, the principles underlying extension-based semantics can also be encoded as YALLA formulas. Given the structure associated with an AF $F = (A, R)$,

- the term t is conflict-free if the formula $\Phi_t^{cf} = \text{on}(t) \wedge \neg(\{t\} \triangleright \{t\})$ is valid;

- the term t_1 defends the term t_2, denoted by $t_1 \blacktriangleright t_2$, if the formula $\forall t_3, ((singl(t_3) \wedge t_3 \triangleright t_2) \rightarrow (t_1 \triangleright t_3))$ is valid, where $singl(t)$ is a formula that is valid if t is a singleton.

The combination of these formulas allows to characterize the admissible sets (i.e. the terms that satisfy of $\Phi_t^{ad} = \Phi_t^{cf} \wedge (t \blacktriangleright t)$). This is the basics of YALLA encoding for the classical Dung's semantics. Additional constraints in the formulas yield encodings Φ_t^σ for the other semantics.

Then, belief update rationality postulates and operators [Katsuno and Mendelzon, 1991] are adapted to take into account the universe $F_U = (A_U, R_U)$. A set of authorized transitions (corresponding to what we call a modification type) is $T \subseteq \Gamma_U \times \Gamma_U$, where Γ_U is the set of all AFs built on the universe F_U. Then, roughly speaking, an update operator $\diamond_T$ is such that, if ϕ is a YALLA formula characterizing an AF F, then for any formula α, $\phi \diamond_T \alpha$ characterizes an AF G such that $(F, G) \in T$.[6]

---
[6]This is actually slightly more subtle than that, since YALLA formulas can characterize sets of AFs.

Finally, enforcing an extension in an AF F can be achieved by updating the formula Φ_F:

$$\Phi_F \diamond_T \Phi_{c_i}^{\sigma}$$

characterizes the AFs that enforce S_i in F, under the modification type T and the semantics σ, where c_i is the YALLA constant symbol that corresponds to the set of arguments S_i, and Φ_t^{σ} is valid if and only if the term t corresponds to a σ-extension (similarly to the way Φ_t^{cf}, described previously, characterizes conflict-free sets).

Another logic-based approach is that of [Doutre et al., 2014], which proposes to translate the argumentation framework and the semantics into logic, to perform the enforcement. In this case, the Dynamic Logic of Propositional Assignments (DL-PA) by [Balbiani et al., 2013], is used to represent update operators as executable programs. The piece of information which causes the update is a formula about acceptance statuses, which should be satisfied by at least one extension of the result (credulous enforcement of the formula) or by each extension of the result (sceptical enforcement of the formula). Forbus' update operator is used to change minimally the attack relation such that the extensions of the new argumentation framework comply with the expected enforcement. An extension of [Doutre et al., 2014] is proposed by [Doutre et al., 2017], which considers also addition and removal of arguments, and by applying the framework to an access control case. Then, [Doutre et al., 2019] generalizes the previous two approaches.

Let us mention that these kinds of approaches based on a belief update operation allow richer forms of enforcement, since complex information about the sets of arguments and the attacks in the AFs can be described. Also, other kinds of belief change operations (e.g. belief revision [Katsuno and Mendelzon, 1992] or belief contraction [Caridroit et al., 2017]) could be defined in these contexts. We refer the interested reader to [Doutre and Mailly, 2018] for more details on the relation between belief change and argumentation.

4.2 Status Enforcement

Status enforcement [Niskanen et al., 2016a] is defined as an operator where two sets of arguments are provided as input, that must be respectively positively and negatively enforced. This operation does not fit the framework described previously, since it is supposed that there is only one set of arguments given in input, that must have exactly one acceptance status with respect to some reasoning mode (see Definition 3.1).

Formally, given an AF $F = (A, R)$, P and N two subsets of A such that $P \cap N = \emptyset$, and σ a semantics,

- the AF $G = (A, R')$ is a credulous status enforcement of (P, N) in F with respect to σ if $P \subseteq \bigcup \sigma(G)$ and $N \cap \bigcup \sigma(G) = \emptyset$;
- the AF $G = (A, R')$ is a sceptical status enforcement of (P, N) in F with respect to σ if $P \subseteq \bigcap \sigma(G)$ and $N \cap \bigcap \sigma(G) = \emptyset$.

In words, status enforcement consists in finding G such that every argument in P is credulously (respectively sceptically) accepted in G, and every argument in N is not credulously (respectively sceptically) accepted in G.

Complexity issues for optimal status enforcement, i.e. finding G such that $d(F, G) = |(R \setminus R') \cup (R' \setminus R)|$ is minimal, have been investigated by [Niskanen et al., 2016a]. Similarly to complexity for optimal extension enforcement (Section 3.3), the complexity results concern a decision problem related to the optimization problem under consideration.

Credulous status enforcement
INSTANCE: an AF $F = (A, R)$, $P \subseteq A$ and $N \subseteq A$ s.t. $P \cap N = \emptyset$, integer $k \geq 0$, and a semantics σ.
QUESTION: Does there exist an AF $G = (A, R')$ such that $P \subseteq \bigcup \sigma(G)$ and $N \cap \bigcup \sigma(G) = \emptyset$ and $d(F, G) \leq k$?

Sceptical status enforcement
INSTANCE: an AF $F = (A, R)$, $P \subseteq A$ and $N \subseteq A$ s.t. $P \cap N = \emptyset$, integer $k \geq 0$, and a semantics σ.
QUESTION: Does there exist an AF $G = (A, R')$ such that $P \subseteq \bigcap \sigma(G)$ and $N \cap \bigcap \sigma(G) = \emptyset$ and $d(F, G) \leq k$?

Two cases are considered: the general case, and the restricted case where $N = \emptyset$ (i.e. only positive arguments must be enforced). Table 5 presents the complexity of these problems for various semantics.

MaxSAT and CEGAR based algorithms in the same spirit as algorithms for extension enforcement (Section 3.3.2) are also provided.

semantics σ	$N = \emptyset$ credulous	$N = \emptyset$ sceptical	N unrestricted credulous	N unrestricted sceptical
cf	in P	trivial	in P	trivial
ad	NP-c	trivial	Σ_2^P-c	trivial
co	NP-c	NP-c	Σ_2^P-c	NP-c
gr	NP-c	NP-c	NP-c	NP-c
stb	NP-c	Σ_2^P-c	Σ_2^P-c	Σ_2^P-c
pr	NP-c	Σ_3^P-c	Σ_2^P-c	Σ_3^P-c

Table 5: Complexity of status enforcement

4.3 Control Argumentation Frameworks

Now we introduce a concept that can be interpreted as a variant of enforcement under uncertain information. Control Argumentation Frameworks (CAFs) [Dimopoulos et al., 2018] are AFs where arguments and attacks are split in three distinct parts:

- the fixed part is made of arguments and attacks that are unquestionably in the system;

- the uncertain part is made of arguments and attacks that may belong to the system, as well as "undirected" attacks: in this case there is for sure a conflict between arguments, but the actual direction is uncertain;

- the control part is made of arguments and attacks that may be used by the agent.

The sets of fixed, uncertain and control arguments are disjoint, as well as the various sets of attacks. Roughly speaking, the fixed part corresponds to certain knowledge, i.e., elements that cannot be influenced neither by the agent nor by its environment (we use "environment" in a wide sense, it also includes other agents). On the contrary, the uncertain part models the agent's knowledge (and beliefs) about the environment (and the other agents); in realistic scenarios, this knowledge is by nature uncertain. Finally, the control part corresponds to the agent's possible actions. When the agent selects a subset of the control arguments and attacks (called a configuration), then it defines a configured CAF, that is the same CAF where the control arguments (and the associated attacks) that have not been selected have been removed. The uncertain part of the CAF induces a set of completions, i.e. classical AFs that are compatible with the knowledge encoded in the CAF. This notion is borrowed from Incomplete Argumentation Frameworks [Coste-Marquis et al., 2007; Baumeister et al., 2018a; Baumeister et al., 2018b].

The notion of controllability of a CAF, with respect to a given target set of arguments, is directly related to enforcement. This target is defined as a subset of the fixed arguments, that is expected to belong to each (or some) extension of each completion. The agent needs to find a configuration that reaches this target. Let us exemplify these concepts.

Example 4.1. *Figure 9 describes a CAF, where the set of fixed arguments is $\{f_1, f_2, f_3, f_4, f_5\}$, the only uncertain argument is u (dashed square argument), and the control arguments are $\{c_1, c_2, c_3\}$ (bold square arguments). The plain arrows represent fixed attacks (e.g. (f_2, f_1) is fixed); the dotted arrow (f_5, f_1) means that it is uncertain whether f_5 actually attacks f_1 or not; the symmetric dashed arrow (u, f_4) means that there is for sure a conflict between u and f_4, but the actual direction is uncertain (it could be (u, f_4), or (f_4, u), or both at the same time). Finally, the bold arrows represent control attacks, they are related to the control arguments that can be selected by the agent.*

We suppose that the target $T = \{f_1\}$ must belong to each stable extension. Without control arguments, this is not possible: there are, for instance, completions where f_5 attacks f_1, and in this case f_1 is not defended. However, with the configuration $\{c_1, c_3\}$, f_1 will be defended against every possible threat coming from the uncertain part: c_1 defends f_1 against f_5, and c_3 defends f_1 against u (that is an undirect threat, since u may defeat f_4, making then f_2 and f_3 acceptable). Similarly, $\{c_2\}$ is a valid configuration, since it allows to guarantee that $\{f_1\}$ is included in every stable extension of every completion.

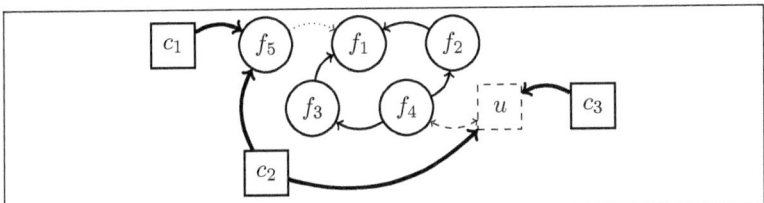

Figure 9: An example of control argumentation framework

Controlling a CAF can be seen as enforcing (non-strictly) an extension in presence of uncertainty. Intuitively, for an AF $F = (A, R)$ and a set of arguments E to be enforced through strong expansion, we can define a CAF that is controllable with respect to E if and only if it is possible to enforce E in F. Indeed, the arguments A and attacks R correspond to the fixed part of the CAF, while the uncertain part is empty. Then, for each $a \in A$, a control argument c_a with a control attack (c_a, a) is added. If E can be enforced in F, then the CAF is controllable (where the configuration to be chosen consists in the set of control arguments that do not attack E). On the opposite, if E cannot be enforced through a strong expansion, then the CAF is not controllable: indeed, the CAF configured by a control configuration is a strong expansion of F, thus E cannot be accepted in this configured CAF. We give a simple example of this transformation.

Example 4.2. Let $F = (A, R)$ be the AF given at Figure 10a. We consider the grounded semantics: $gr(F) = \{\emptyset\}$. Let $E_1 = \{a\}$ be a set of arguments to be (non-strictly) enforced through a strong expansion. This enforcement is possible: for instance, the AF G that is a strong expansion of F where a new argument attacks b yields the expected result. Such an AF G corresponds to the CAF (given at Figure 10b) after it has been configured by $\{c_b\}$ (i.e. the argument c_a and the attack (c_a, a) are removed). So we observe that this CAF is controllable with respect to E_1 and the grounded semantics. On the opposite, $E_2 = \{a, b\}$ cannot be enforced in F with strong enforcement (since it is not conflict-free), and similarly there is no way to configure the CAF with respect to E_2 and the grounded semantics.

In the previous example, we show how non-strict enforcement under

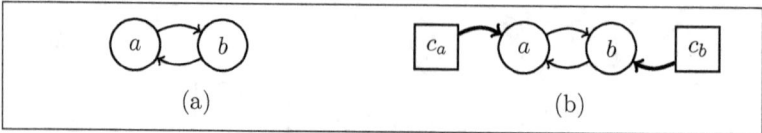

Figure 10: Transforming an AF into a CAF

strong expansion can seen as controlling a CAF. But more generally, since control arguments can attack each others, non-strict enforcement under a normal expansion can also be "translated" in controlling a CAF. On the opposite, configuring a CAF for controlling a target T can be interpreted as enforcing T in all the completions of the CAF with the same normal expansion (where the added arguments and attacks are chosen in the control part).

Let us also briefly mention that detailed complexity results and algorithms for reasoning with CAFs have been provided in [Niskanen et al., 2020], and [Mailly, 2020] defines a weaker form of controllability, that relies on one completion instead of the whole set.

Applying CAFs to Automated Negotiation Let us briefly described how enforcement (or more precisely, CAFs) has been used in a context of automated negotiation [Dimopoulos et al., 2019]. The idea is to represent the (uncertain) knowledge of an agent about her opponent with a CAF. Indeed, negotiation has more chance to reach an agreement if agents have some knowledge about each other; however it is unrealistic to consider that opponent modelling can be done without incomplete or uncertain information. The theory of a negotiating agent is thus made of two parts: a classical AF that represents the agent's personal knowledge, and a CAF that represents her knowledge about her opponent.

It is supposed that agents negotiate about a set of (mutually exclusive) offers $\mathcal{O}$. Each offer may be supported, in AF_1 (the personal knowledge of agent 1), by 0, 1 or several *practical arguments*, i.e. arguments whose conclusions correspond to actions or decisions. The other arguments are *epistemic arguments*, they support knowledge and beliefs. The knowledge of agent 1 about agent 2 is represented in CAF_1^2. The fixed and uncertain parts are supposed to be built from the actual AF of agent 2: the assumption is made that there can be uncertainty (represented in the CAF), ignorance (some arguments or attacks of agent 2 may not appear in the CAF at all), but no mistake (there is no attacks or arguments that appear in CAF_1^2 but not in the personal AF of agent 2). Finally, the control part of CAF_1^2 is made of arguments and attacks chosen in AF_1, that are supposed to be used by agent 1 in order to make its target accepted. Similarly, AF_2 is the personal knowledge of agent 2, and CAF_2^1 represents the (uncertain) knowledge of agent 2 about agent 1.

Each agent selects its preferred offer $o \in \mathcal{O}$ according to its personal knowledge: o has to be supported by a practical argument that is accepted in AF_1; if several offers can be chosen, an assumption is made that the agent has a preference ranking over offers. When the preferred offer o of agent 1 is chosen, she uses her knowledge about agent 2 in order to persuade her to accept o: agent 1 searches for a practical argument in CAF_1^2 that supports o. If such an argument a exists, then three options are possible:

- if a is accepted in each completion of CAF_1^2 without using any control arguments, then agent 1 makes an offer to agent 2 (offer o, supported by argument a);

- otherwise, if a is accepted with the use of some control arguments $c_1, \ldots,$
  c_k, then agent 1 can again make an offer (offer o, supported by argument a, that is accepted because of $c_1, \ldots, c_k$);

- in the last case, a is not accepted even with control arguments, then agent 1 searches for another argument that supports offer o in CAF_1^2.

In the first two cases, if agent 2 accepts the argument a (with, or without control arguments), then the negotiation is a success: offer o is accepted. Otherwise, agent 2 gives to agent 1 the reasons why she rejects a (for instance, she knows some arguments that agent 1 does not know). If agent 1 knows other arguments that support o in CAF_1^2, the process is repeated. Otherwise, this is the end of the round: agents switch their roles, and now agent 2 will choose her preferred offer o', and use her CAF in order to persuade agent 1 to accept o'.

The whole process goes on, until either the agents agree on some offer (in that case, the negotiation is a success), or they do not have available offers (the negotiation fails).

Let us illustrate the process, with an example borrowed from [Dimopoulos et al., 2019].

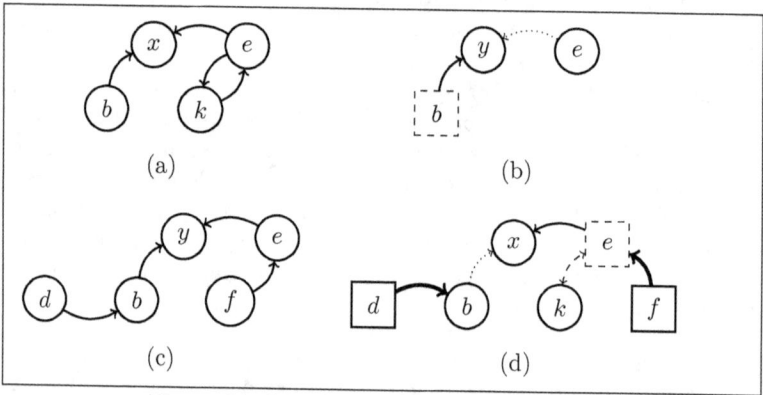

Figure 11: Initial theories of agents 1 and 2

Example 4.3. *Figure 11 shows the negotiation theories of two agents. More precisely, the AFs AF_1 and AF_2 (respectively Figure 11a and Figure 11c) correspond to the personal knowledge of (respectively) agent 1 and agent 2, while CAF_1^2 (Figure 11b) represents the (uncertain) knowledge of agent 1 about agent 2, and vice-versa for CAF_2^1 (Figure 11d). We suppose that both agents use the stable semantics for reasoning, and that there is one offer o, that is supported by arguments x and y. Before starting the negotiation, agent 1 has no reason to accept the offer o (since its supporting argument x is rejected in AF_1), while agent 2 accepts o since y is accepted in AF_2. If agent 1 starts the negotiation, she has no offer to propose (since there is no accepted argument in AF_1 that supports some offer), so the token has to go to agent 2.*

Agent 2 can make an offer. The goal of agent 2 is to persuade agent 1 to accept the offer o, using arguments that agent 1 already knows. This means that she needs to make agent 1 modify her AF in order to accept x (since x is the only argument that supports the offer o in

CAF_2^1). *This persuasion phase goes first through a step that do not use the control part of the CAF: if x is accepted in the CAF with no control argument, agent 2 can send to agent 1 the message "offer o, supported by the accepted argument x". In the present example, this is not the case: there are completions where x is rejected (for instance, the ones where the attack (b,x) exists). So, in the next step, agent 2 searches for a control configuration that allows to make x accepted in each completion. Here, the configuration is the full set of control arguments $\{d,f\}$. Agent 2 can then send the message "offer o, supported by the argument x, that is accepted because d attacks b and f attacks e".*

Receiving this message triggers some updates in agent 1's knowledge. First, she can add the arguments d and f (as well as the attacks (d,b) and (f,e)) in CAF_1^2. Moreover, while argent b was initially uncertain in the CAF, it can now become a fixed argument: since agent 2 sends a message about the argument b, it certainly means that agent 2 knows this argument. Then, agent 1 can also add these arguments and attacks in her AF. The updated AF_1 and CAF_1^2 are shown at Figure 12. Since in the update AF_1, the argument x is accepted, agent 1 can stop the negotiation by accepting the offer o.

Let us suppose that agent 1 has, e.g., some argument i attacking d, then instead of accepting the offer o, she sends the message "reject the offer, because i attacks d". Then agent 2 updates her CAF, and the process continues as illustrated previously until reaching the negotiation success (if some offer can be accepted by both agents) or failure (if no offer can be accepted by both agents, even when exchanging arguments for defending them).

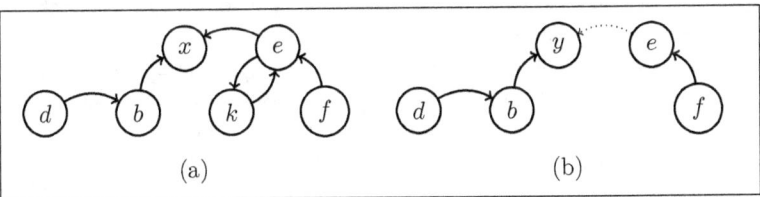

Figure 12: The updated theory of agent 1

Experiments [Dimopoulos et al., 2019] have shown that control arguments and attacks help to increase the agreement rate, even when the percentage of uncertainty in CAFs is high.

4.4 Enforcement under Constraints

We have already seen several approaches to enforcement, and variants to enforcement. Common to many approaches are restrictions on the allowed modifications, such as allowing only expansions, local updates, other types of modifications, or defining allowed modifications via formulas, such as in YALLA.

More broadly, one can impose *constraints* on enforcements. Such constraints can have many different shapes or forms (expansions, deletions, specification as a formula, etc.). In general, constraints can be very useful for applications of enforcement operators. Going into a slightly different direction from before in this chapter, consider the following example.

Example 4.4. *Say we have knowledge about two arguments a and b, and we wish to enforce non-acceptability of a, e.g., because argument a counters a desirable argument. An expansion by c and attack (c, a) does the trick. However, say, in addition, that a is a sub argument of argument b, when inspecting the contents of the arguments. In such a case it seems adequate to require that c attacks the super argument of a, as well, i.e., we want also to add the attack (c, b).*

As suggested by the example, in some situations we may be faced with circumstances that may require specific expansions, or rather ruling out certain expansions. For instance, only considering those expansions that satisfy the condition that if an attack from some argument onto a is added, so must the same attacker also attack b.

Such conditions are not directly captured by the main types of modifications represented in this chapter, but can be incorporated into enforcement, as well. In [Wallner, 2020], several families of constraints are considered, and the survey [Doutre and Mailly, 2018] discusses constraints of dynamics in argumentation, in general. We refer the reader for details to these papers, but highlight a particular type of constraint: implications of presence of arguments and attacks. By allowing constraints that take the form of implications, e.g., of the form mentioned above for attacks, one can specify that attacks on sub arguments must "propagate" to super arguments, which is present in many instances of formal approaches to structured argumentation [Cyras et al., 2018; Modgil and Prakken, 2018].

4.5 Other related works

Beside extension enforcement, adding or removing arguments or attacks to an AF can be seen as another form of enforcement, on the structure of the AF. This relates to dynamic aspects of argumentation. As already mentioned, [Boella et al., 2009a; Cayrol et al., 2010] are among the first approaches that studied the changes implied by such structural enforcements. Other similar approaches are detailed by [Doutre and Mailly, 2018].

In [Ulbricht and Baumann, 2019] the authors studied the question of how to *repair* an AF if nothing is credulously/sceptically accepted. More precisely, the main aim is to restore consistency via removing certain (minimal) sets of arguments or attacks. Note that enforcing a certain non-empty set can be seen as a special kind of repairing given that we are faced with no credulously accepted arguments. The notion of C-restricted semantics [Baumann et al., 2019] is related to enforcement too. It can be shown that a set of arguments is a C-restricted extension if and only if it can be (non-strictly) enforced with a restricted form of expansion.

Normal expansions of AFs have been used for other purposes related to enforcement. For instance, [Booth et al., 2013] describes a framework where an agent's knowledge is represented by an AF F and a propositional formula ϕ that represents an integrity constraint about the complete labellings of the AF. The agent's knowledge is said to be inconsistent if none of the complete labellings satisfies the constraint. Two approaches are proposed for restoring consistency. The first one is a direct use of a normal expansion: the authors have proven that there exists a normal expansion of F that is consistent with ϕ (under some minimal assumption about the consistency of ϕ). The second approach also uses normal expansion, but only after a first step that consists in revising [Katsuno and Mendelzon, 1992] the complete labellings of F by ϕ, in order to compute the so-called fallback beliefs of the agent. Then, a normal expansion allows to obtain a new AF that is consistent with the fallback beliefs. Contrary to the first approach (based only on an expansion), this one guarantees that the agent's complete labellings are as close as possible to the initial complete labellings.

Recently, the inverse problem to extension enforcement was studied, namely the problem of *extension removal* [Baumann and Brewka, 2019]. That is: given an AF F and a set of extensions $\mathcal{E}$, identify an AF H that is as close as possible to F but has none of the extensions in $\mathcal{E}$. In the same way as enforcement shifts revision to the level of extensions, extension removal shifts contraction to the level of extensions.

The approach by [Dunne et al., 2015] aims at checking if a set E of sets of arguments can be the set of extensions of any argumentation framework F with respect to a given semantics σ. This property is named *realizability* of E with respect to σ. Realizability can be seen as a form of enforcement, where a set of extensions has to be enforced, and all the necessary structural changes on the argumentation framework (nothing is known about beforehand) can be done to achieve this.

A further related approach to enforcement is that of learning AFs or synthesis of AFs [Riveret and Governatori, 2016; Riveret, 2016; Niskanen et al., 2019]. In brief, the aim is to construct an AF from certain information available. Different from deterministic logical approaches that construct an AF from a knowledge base (see structured argumentation approaches, e.g., in [Baroni et al., 2018b]), in AF learning or synthesis the information available might not uniquely determine an AF. In the AF synthesis problem [Niskanen et al., 2019], for instance, information about the semantics is given, and the task is to construct an AF that as best as possible matches the given semantic information. In this way, AF synthesis is related to realizability (see above), as well.

5 Conclusion

This chapter has offered an overview of the notion of enforcement in abstract, formal argumentation. A focus has been done on extension enforcement, on its general characterization, and on how it can be achieved: the various changes that can be applied to the structure of the argumentation framework, and/or to the semantics, considering that these changes should be minimal. Results about the complexity of enforcement, and algorithms, showing the feasibility of this approach, have also been presented.

If a general context and a number of specific approaches have been described, many additional proposals exist and keep on being proposed, showing the liveliness of the field. Applications of these formal approaches have also been outlined, and they should be developed in the future.

Regarding future work, several lines of research appear intriguing. Regarding formal foundations, we surveyed the state of the art, yet several directions are open, such as considering further argumentation semantics and their effect on possibility, impossibility, or (computational) cost of (optimal) enforcement. Moreover, different types of modifications can be considered as well, reflecting different updates on the given argumentation.

Beyond Dung's classical argumentation framework, the notion of enforcement can be defined and applied to any enriched argumentation framework, such as value-based argumentation frameworks (see Chapter 5 [Atkinson and Bench-Capon, 2021]), or frameworks with higher-order bipolar interactions (see Chapter 1 [Cayrol et al., 2021]), or with quantitative additions like probabilistic argumentation (see Chapter 7 [Hunter et al., 2021]). The notion can also be extended to semantics other than extension-based, for instance their labelling-based counterparts [Caminada, 2006], or ranking-based semantics [Bonzon et al., 2016].

Chapter 4 [Baumeister et al., 2021] studies (among other notions) Incomplete Argumentation Frameworks (IAFs), that are strongly related to CAFs described in this chapter. Moreover, the possibility of enforcing a set of arguments can be intuitively associated with the notion of possible acceptance in IAFs.

Enforcement is also related to the notion of dialogue (see Chapter 9 [Black et al., 2021]), where it can be put in practice, and to that of strategic argumentation (see Chapter 10 [Governatori et al., 2021]). To go further, an empirical cognitive study of enforcement might be conducted, as it has been done for other argumentation notions (see Chapter 14 [Cerutti et al., 2021]).

Acknowledgments

This work was supported by the Austrian Science Fund (FWF, P30168-N31). Furthermore, this work was supported by the German Federal Ministry of Education and Research (BMBF, 01/S18026A-F) by funding the competence center for Big Data and AI "ScaDS.AI Dresden/Leipzig".

References

[Alchourrón et al., 1985] Alchourrón, C. E., Gärdenfors, P., and Makinson, D. (1985). On the logic of theory change: Partial meet contraction and revision functions. *Journal of Symbolic Logic*, 50(2):510–530.

[Amgoud and Cayrol, 2002] Amgoud, L. and Cayrol, C. (2002). A reasoning model based on the production of acceptable arguments. *Annals of Mathematics and Artificial Intelligence*, 34(1-3):197–215.

[Arora and Barak, 2009] Arora, S. and Barak, B. (2009). *Computational Complexity - A Modern Approach*. Cambridge University Press.

[Atkinson and Bench-Capon, 2021] Atkinson, K. and Bench-Capon, T. (2021). Value-based argumentation. In Gabbay, D., Giacomin, M., Simari, G. R., and Thimm, M., editors, *Handbook of Formal Argumentation*, volume 2, chapter 5. College Publications.

[Balbiani et al., 2013] Balbiani, P., Herzig, A., and Troquard, N. (2013). Dynamic logic of propositional assignments: a well-behaved variant of pdl. In *2013 28th Annual ACM/IEEE Symposium on Logic in Computer Science*, pages 143–152. IEEE.

[Baroni et al., 2018a] Baroni, P., Caminada, M., and Giacomin, M. (2018a). Abstract argumentation frameworks and their semantics. In *Handbook of Formal Argumentation*, chapter 4. College Publications.

[Baroni et al., 2013] Baroni, P., Cerutti, F., Dunne, P. E., and Giacomin, M. (2013). Automata for infinite argumentation structures. *Artificial Intelligence*, 203:104–150.

[Baroni and Giacomin, 2007] Baroni, P. and Giacomin, M. (2007). On principle-based evaluation of extension-based argumentation semantics. *Artificial Intelligence*, 171(10-15):675–700.

[Baroni et al., 2018b] Baroni, P., Giacomin, M., and Liao, B. (2018b). Locality and modularity in abstract argumentation. In Baroni, P., Gabbay, D., Giacomin, M., and van der Torre, L., editors, *Handbook of Formal Argumentation*, chapter 19. College Publications.

[Baumann, 2011] Baumann, R. (2011). Splitting an argumentation framework. In Delgrande, J. P. and Faber, W., editors, *Proc. LPNMR*, volume 6645 of *Lecture Notes in Computer Science*, pages 40–53. Springer.

[Baumann, 2012a] Baumann, R. (2012a). Normal and strong expansion equivalence for argumentation frameworks. *Artificial Intelligence*, 193:18–44.

[Baumann, 2012b] Baumann, R. (2012b). What does it take to enforce an argument? minimal change in abstract argumentation. In Raedt, L. D., Bessiere, C., Dubois, D., Doherty, P., Frasconi, P., Heintz, F., and Lucas, P. J. F., editors, *Proc. ECAI*, volume 242, pages 127–132. IOS Press.

[Baumann, 2014a] Baumann, R. (2014a). Context-free and context-sensitive kernels: Update and deletion equivalence in abstract argumentation. In Schaub, T., Friedrich, G., and O'Sullivan, B., editors, *Proc. ECAI*, volume 263 of *Frontiers in Artificial Intelligence and Applications*, pages 63–68. IOS Press.

[Baumann, 2014b] Baumann, R. (2014b). *Metalogical Contributions to the Nonmonotonic Theory of Abstract Argumentation*. College Publications.

[Baumann, 2018] Baumann, R. (2018). On the nature of argumentation semantics: Existence and uniqueness, expressibility, and replaceability. In Baroni, P., Gabbay, D., Giacomin, M., and van der Torre, L., editors, *Handbook of Formal Argumentation*, chapter 14. College Publications. also appears in IfCoLog Journal of Logics and their Applications 4(8):2779-2886.

[Baumann, 2019] Baumann, R. (2019). *On the Existence of Characterization Logics and Fundamental Properties of Argumentation Semantics*. habilitation treatise, Leipzig University.

[Baumann and Brewka, 2010] Baumann, R. and Brewka, G. (2010). Expanding argumentation frameworks: Enforcing and monotonicity results. In Baroni, P., Cerutti, F., Giacomin, M., and Simari, G. R., editors, *Proc. COMMA*, volume 216 of *Frontiers in Artificial Intelligence and Applications*, pages 75–86. IOS Press.

[Baumann and Brewka, 2019] Baumann, R. and Brewka, G. (2019). Extension removal in abstract argumentation - an axiomatic approach. In *The Thirty-Third AAAI Conference on Artificial Intelligence, AAAI 2019, The Thirty-First Innovative Applications of Artificial Intelligence Conference, IAAI 2019, The Ninth AAAI Symposium on Educational Advances in Artificial Intelligence, EAAI 2019, Honolulu, Hawaii, USA, January 27 - February 1, 2019*, pages 2670–2677.

[Baumann et al., 2011] Baumann, R., Brewka, G., and Wong, R. (2011). Splitting argumentation frameworks: An empirical evaluation. In Modgil, S., Oren, N., and Toni, F., editors, *Proc. TAFA, Revised Selected Papers*, volume 7132 of *Lecture Notes in Computer Science*, pages 17–31. Springer.

[Baumann et al., 2019] Baumann, R., Dvorák, W., Linsbichler, T., and Woltran, S. (2019). A general notion of equivalence for abstract argumentation. *Artif. Intell.*, 275:379–410.

[Baumann and Spanring, 2015] Baumann, R. and Spanring, C. (2015). Infinite argumentation frameworks - On the existence and uniqueness of extensions. In *Advances in Knowledge Representation, Logic Programming, and Abstract Argumentation - Essays Dedicated to Gerhard Brewka on the Occasion of His 60th Birthday*, pages 281–295.

[Baumann and Spanring, 2017] Baumann, R. and Spanring, C. (2017). A study of unrestricted abstract argumentation frameworks. In Sierra, C., editor, *Proc. IJCAI*, pages 807–813. ijcai.org.

[Baumeister et al., 2018a] Baumeister, D., Neugebauer, D., and Rothe, J. (2018a). Credulous and skeptical acceptance in incomplete argumentation frameworks. In Modgil, S., Budzynska, K., and Lawrence, J., editors, *Proc. COMMA*, volume 305 of *Frontiers in Artificial Intelligence and Applications*, pages 181–192. IOS Press.

[Baumeister et al., 2021] Baumeister, D., Neugebauer, D., and Rothe, J. (2021). Collective acceptability in abstract argumentation. In Gabbay, D., Giacomin, M., Simari, G. R., and Thimm, M., editors, *Handbook of Formal Argumentation*, volume 2, chapter 4. College Publications.

[Baumeister et al., 2018b] Baumeister, D., Neugebauer, D., Rothe, J., and Schadrack, H. (2018b). Verification in incomplete argumentation frameworks. *Artificial Intelligence*, 264:1–26.

[Besnard and Doutre, 2004] Besnard, P. and Doutre, S. (2004). Checking the acceptability of a set of arguments. In *Proc. NMR*, pages 59–64.

[Besnard and Hunter, 2008] Besnard, P. and Hunter, A. (2008). *Elements of Argumentation*. The MIT Press.

[Bisquert et al., 2011] Bisquert, P., Cayrol, C., de Saint-Cyr, F. D., and Lagasquie-Schiex, M. (2011). Change in argumentation systems: Exploring the interest of removing an argument. In Benferhat, S. and Grant, J., editors, *Proc. SUM*, volume 6929 of *Lecture Notes in Computer Science*, pages 275–288. Springer.

[Black et al., 2021] Black, E., Maudet, N., and Parsons, S. (2021). Argumentation-based dialogue. In Gabbay, D., Giacomin, M., Simari, G. R., and Thimm, M., editors, *Handbook of Formal Argumentation*, volume 2, chapter 9. College Publications.

[Boella et al., 2009a] Boella, G., Kaci, S., and van der Torre, L. W. N. (2009a). Dynamics in argumentation with single extensions: Abstraction principles and the grounded extension. In Sossai, C. and Chemello, G., editors, *Proc. ECSQARU*, volume 5590 of *Lecture Notes in Computer Science*, pages

107–118. Springer.

[Boella et al., 2009b] Boella, G., Kaci, S., and van der Torre, L. W. N. (2009b). Dynamics in argumentation with single extensions: Attack refinement and the grounded extension (extended version). In McBurney, P., Rahwan, I., Parsons, S., and Maudet, N., editors, *ArgMAS Revised Selected and Invited Papers*, volume 6057 of *Lecture Notes in Computer Science*, pages 150–159. Springer.

[Bonzon et al., 2016] Bonzon, E., Delobelle, J., Konieczny, S., and Maudet, N. (2016). A comparative study of ranking-based semantics for abstract argumentation. In *Proceedings of the 30th AAAI Conference on Artificial Intelligence (AAAI'16)*.

[Booth et al., 2013] Booth, R., Kaci, S., Rienstra, T., and van der Torre, L. W. N. (2013). A logical theory about dynamics in abstract argumentation. In Liu, W., Subrahmanian, V. S., and Wijsen, J., editors, *Proc. SUM*, volume 8078 of *Lecture Notes in Computer Science*, pages 148–161. Springer.

[Caminada, 2006] Caminada, M. (2006). On the issue of reinstatement in argumentation. In Fisher, M., van der Hoek, W., Konev, B., and Lisitsa, A., editors, *Logics in Artificial Intelligence, 10th European Conference, JELIA 2006, Liverpool, UK, September 13-15, 2006, Proceedings*, volume 4160 of *Lecture Notes in Computer Science*, pages 111–123. Springer.

[Caminada and Wu, 2011] Caminada, M. and Wu, Y. (2011). On the limitations of abstract argumentation. In *Benelux Conference on Artificial Intelligence*.

[Caridroit et al., 2017] Caridroit, T., Konieczny, S., and Marquis, P. (2017). Contraction in propositional logic. *International Journal of Approximate Reasoning*, 80:428–442.

[Cayrol et al., 2021] Cayrol, C., Cohen, A., and Lagasquie-Schiex, M.-C. (2021). Higher-order interactions (bipolar or not) in abstract argumentation: A state of the art. In Gabbay, D., Giacomin, M., Simari, G. R., and Thimm, M., editors, *Handbook of Formal Argumentation*, volume 2, chapter 1. College Publications.

[Cayrol et al., 2010] Cayrol, C., de Saint-Cyr, F. D., and Lagasquie-Schiex, M. (2010). Change in abstract argumentation frameworks: Adding an argument. *Journal of Artificial Intelligence Research*, 38:49–84.

[Cerutti et al., 2021] Cerutti, F., Cramer, M., Guillaume, M., Hadoux, E., Hunter, A., and Polberg, S. (2021). Empirical cognitive studies about formal argumentation. In Gabbay, D., Giacomin, M., Simari, G. R., and Thimm, M., editors, *Handbook of Formal Argumentation*, volume 2, chapter 13. College Publications.

[Clarke et al., 2003] Clarke, E., Grumberg, O., Jha, S., Lu, Y., and Veith, H.

(2003). Counterexample-guided abstraction refinement for symbolic model checking. *Journal of the ACM*, 50(5):752–794.

[Clarke et al., 2004] Clarke, E. M., Gupta, A., and Strichman, O. (2004). SAT-based counterexample-guided abstraction refinement. *IEEE Transactions on Computer-Aided Design of Integrated Circuits and Systems*, 23(7):1113–1123.

[Coste-Marquis et al., 2007] Coste-Marquis, S., Devred, C., Konieczny, S., Lagasquie-Schiex, M., and Marquis, P. (2007). On the merging of dung's argumentation systems. *Artificial Intelligence*, 171(10-15):730–753.

[Coste-Marquis et al., 2015] Coste-Marquis, S., Konieczny, S., Mailly, J.-G., and Marquis, P. (2015). Extension enforcement in abstract argumentation as an optimization problem. In Yang, Q. and Wooldridge, M. J., editors, *Proc. IJCAI*, pages 2876–2882. AAAI Press.

[Cyras et al., 2018] Cyras, K., Fan, X., Schulz, C., and Toni, F. (2018). Assumption-based argumentation: Disputes, explanations, preferences. In Baroni, P., Gabbay, D., Giacomin, M., and van der Torre, L., editors, *Handbook of Formal Argumentation*, chapter 7, pages 365–408. College Publications.

[de Saint-Cyr et al., 2016] de Saint-Cyr, F. D., Bisquert, P., Cayrol, C., and Lagasquie-Schiex, M. (2016). Argumentation update in YALLA (yet another logic language for argumentation). *International Journal of Approximate Reasoning*, 75:57–92.

[Dimopoulos et al., 2018] Dimopoulos, Y., Mailly, J.-G., and Moraitis, P. (2018). Control argumentation frameworks. In McIlraith, S. A. and Weinberger, K. Q., editors, *Proc. AAAI*, pages 4678–4685. AAAI Press.

[Dimopoulos et al., 2019] Dimopoulos, Y., Mailly, J.-G., and Moraitis, P. (2019). Argumentation-based negotiation with incomplete opponent profiles. In Elkind, E., Veloso, M., Agmon, N., and Taylor, M. E., editors, *Proc. AAMAS*, pages 1252–1260. International Foundation for Autonomous Agents and Multiagent Systems.

[Doutre et al., 2014] Doutre, S., Herzig, A., and Perrussel, L. (2014). A dynamic logic framework for abstract argumentation. In Baral, C., Giacomo, G. D., and Eiter, T., editors, *Proc. KR*, pages 62–71. AAAI Press.

[Doutre et al., 2019] Doutre, S., Herzig, A., and Perrussel, L. (2019). Abstract argumentation in dynamic logic: Representation, reasoning and change. In *Dynamics, Uncertainty and Reasoning*, pages 153–185. Springer.

[Doutre et al., 2017] Doutre, S., Maffre, F., and McBurney, P. (2017). A Dynamic Logic Framework for Abstract Argumentation: Adding and Removing Arguments (regular paper). In Benferhat, S., Tabia, K., and Ali, M., editors, *Advances in Artificial Intelligence: From Theory to Practice - In-*

ternational Conference on Industrial Engineering and Other Applications of Applied Intelligent Systems (IEA/AIE 2017), volume 10351 of Lecture Notes in Computer Science, pages 295–305. Springer-Verlag.

[Doutre and Mailly, 2017a] Doutre, S. and Mailly, J.-G. (2017a). Comparison criteria for argumentation semantics. In Belardinelli, F. and Argente, E., editors, *Proc. EUMAS/AT, Revised Selected Papers*, volume 10767 of *Lecture Notes in Computer Science*, pages 219–234. Springer.

[Doutre and Mailly, 2017b] Doutre, S. and Mailly, J.-G. (2017b). Semantic change and extension enforcement in abstract argumentation. In Moral, S., Pivert, O., Sánchez, D., and Marín, N., editors, *Proc. SUM*, volume 10564 of *Lecture Notes in Computer Science*, pages 194–207. Springer.

[Doutre and Mailly, 2018] Doutre, S. and Mailly, J.-G. (2018). Constraints and changes: A survey of abstract argumentation dynamics. *Argument & Computation*, 9(3):223–248.

[Dung, 1995] Dung, P. M. (1995). On the acceptability of arguments and its fundamental role in nonmonotonic reasoning, logic programming and n-person games. *Artificial Intelligence*, 77:321–357.

[Dunne, 2007] Dunne, P. E. (2007). Computational properties of argument systems satisfying graph-theoretic constraints. *Artif. Intell.*, 171(10-15):701–729.

[Dunne et al., 2015] Dunne, P. E., Dvořák, W., Linsbichler, T., and Woltran, S. (2015). Characteristics of multiple viewpoints in abstract argumentation. *Artificial Intelligence*, 228:153–178.

[Dvořák and Dunne, 2018] Dvořák, W. and Dunne, P. E. (2018). Computational problems in formal argumentation and their complexity. In Baroni, P., Gabbay, D., Giacomin, M., and van der Torre, L., editors, *Handbook of Formal Argumentation*, chapter 13, pages 631–688. College Publications.

[Gelfond and Lifschitz, 1988] Gelfond, M. and Lifschitz, V. (1988). The stable model semantics for logic programming. In *Proc. ICLP/SLP*, pages 1070–1080. MIT Press.

[Governatori et al., 2021] Governatori, G., Maher, M. J., and Olivieri, F. (2021). Strategic argumentation. In Gabbay, D., Giacomin, M., Simari, G. R., and Thimm, M., editors, *Handbook of Formal Argumentation*, volume 2, chapter 10. College Publications.

[Haret et al., 2018] Haret, A., Wallner, J. P., and Woltran, S. (2018). Two sides of the same coin: Belief revision and enforcing arguments. In Lang, J., editor, *Proc. IJCAI*, pages 1854–1860. ijcai.org.

[Hunter et al., 2021] Hunter, A., Polberg, S., Potyka, N., Rienstra, T., and Thimm, M. (2021). Probabilistic argumentation: A survey. In Gabbay, D.,

Giacomin, M., Simari, G. R., and Thimm, M., editors, *Handbook of Formal Argumentation*, volume 2, chapter 7. College Publications.

[Katsuno and Mendelzon, 1991] Katsuno, H. and Mendelzon, A. O. (1991). On the difference between updating a knowledge base and revising it. In Allen, J. F., Fikes, R., and Sandewall, E., editors, *Proc. KR*, pages 387–394. Morgan Kaufmann.

[Katsuno and Mendelzon, 1992] Katsuno, H. and Mendelzon, A. O. (1992). Propositional knowledge base revision and minimal change. *Artificial Intelligence*, 52(3):263–294.

[Krentel, 1988] Krentel, M. W. (1988). The complexity of optimization problems. *Journal of Computer and System Sciences*, 36(3):490–509.

[Liao et al., 2011] Liao, B. S., Jin, L., and Koons, R. C. (2011). Dynamics of argumentation systems: A division-based method. *Artificial Intelligence*, 175(11):1790–1814.

[Mailly, 2020] Mailly, J. (2020). Possible controllability of control argumentation frameworks. In *Computational Models of Argument - Proceedings of COMMA 2020, Perugia, Italy, September 4-11, 2020*, pages 283–294.

[Modgil and Prakken, 2018] Modgil, S. and Prakken, H. (2018). Abstract rule-based argumentation. In Baroni, P., Gabbay, D., Giacomin, M., and van der Torre, L., editors, *Handbook of Formal Argumentation*, chapter 6, pages 287–364. College Publications.

[Morgado et al., 2013] Morgado, A., Heras, F., Liffiton, M. H., Planes, J., and Marques-Silva, J. (2013). Iterative and core-guided MaxSAT solving: A survey and assessment. *Constraints*, 18(4):478–534.

[Niemelä, 1999] Niemelä, I. (1999). Logic programs with stable model semantics as a constraint programming paradigm. *Annals of Mathematics and Artificial Intelligence*, 25(3-4):241–273.

[Niskanen and Järvisalo, 2020] Niskanen, A. and Järvisalo, M. (2020). Strong refinements for hard problems in argumentation dynamics. In *Proceedings of the 24th European Conference on Artificial Intelligence (ECAI 2020)*, pages 841–848.

[Niskanen et al., 2020] Niskanen, A., Neugebauer, D., and Järvisalo, M. (2020). Controllability of control argumentation frameworks. In *Proceedings of the Twenty-Ninth International Joint Conference on Artificial Intelligence, IJCAI 2020*, pages 1855–1861.

[Niskanen et al., 2016a] Niskanen, A., Wallner, J. P., and Järvisalo, M. (2016a). Optimal status enforcement in abstract argumentation. In Kambhampati, S., editor, *Proc. IJCAI*, pages 1216–1222. IJCAI/AAAI Press.

[Niskanen et al., 2016b] Niskanen, A., Wallner, J. P., and Järvisalo, M.

(2016b). Pakota: A system for enforcement in abstract argumentation. In Michael, L. and Kakas, A. C., editors, *Proc. JELIA*, volume 10021 of *Lecture Notes in Computer Science*, pages 385–400. Springer.

[Niskanen et al., 2018] Niskanen, A., Wallner, J. P., and Järvisalo, M. (2018). Extension enforcement under grounded semantics in abstract argumentation. In Thielscher, M., Toni, F., and Wolter, F., editors, *Proc. KR*, pages 178–183. AAAI Press.

[Niskanen et al., 2019] Niskanen, A., Wallner, J. P., and Järvisalo, M. (2019). Synthesizing argumentation frameworks from examples. *Journal of Artificial Intelligence Research*, 66:503–554.

[Oikarinen and Woltran, 2011] Oikarinen, E. and Woltran, S. (2011). Characterizing Strong Equivalence for Argumentation Frameworks. *Artificial Intelligence*, 175(14-15):1985–2009.

[Papadimitriou, 2007] Papadimitriou, C. H. (2007). *Computational complexity*. Academic Internet Publ.

[Riveret, 2016] Riveret, R. (2016). On learning abstract argumentation graphs from bivalent statement labellings. In *Proc. ICTAI*, pages 190–195. IEEE Computer Society.

[Riveret and Governatori, 2016] Riveret, R. and Governatori, G. (2016). On learning attacks in probabilistic abstract argumentation. In Jonker, C. M., Marsella, S., Thangarajah, J., and Tuyls, K., editors, *Proc. AAMAS*, pages 653–661. ACM.

[Sierksma and Zwols, 2015] Sierksma, G. and Zwols, Y. (2015). *Linear and integer optimization: theory and practice*. CRC Press.

[Spanring, 2015] Spanring, C. (2015). Hunt for the collapse of semantics in infinite abstract argumentation frameworks. In Schulz, C. and Liew, D., editors, *Proceedings of the 2015 Imperial College Computing Student Workshop, ICCSW 2015*, volume 49 of *OASICS*, pages 70–77. Schloss Dagstuhl - Leibniz-Zentrum für Informatik.

[Ulbricht and Baumann, 2019] Ulbricht, M. and Baumann, R. (2019). If nothing is accepted - repairing argumentation frameworks. *J. Artif. Intell. Res.*, 66:1099–1145.

[van der Torre and Vesic, 2017] van der Torre, L. and Vesic, S. (2017). The principle-based approach to abstract argumentation semantics. *The IfCoLog Journal of Logics and their Applications*, 4(8):2727–2780.

[Wallner, 2020] Wallner, J. P. (2020). Structural constraints for dynamic operators in abstract argumentation. *Argument & Computation*, 11(1-2):151–190.

[Wallner et al., 2017] Wallner, J. P., Niskanen, A., and Järvisalo, M. (2017).

Complexity results and algorithms for extension enforcement in abstract argumentation. *Journal of Artificial Intelligence Research*, 60:1–40.

CHAPTER 9

ARGUMENTATION-BASED DIALOGUE

ELIZABETH BLACK
King's College London, UK
elizabeth.black@kcl.ac.uk

NICOLAS MAUDET
Sorbonne University, France
nicolas.maudet@lip6.fr

SIMON PARSONS
University of Lincoln, UK
sparsons@lincoln.ac.uk

Abstract

Dialogue is fundamental to argumentation, providing a dialectical basis for establishing which arguments are acceptable. Argumentation can also be used as the basis for dialogue. In such "argumentation-based" dialogues, participants take part in an exchange of arguments, and the mechanisms of argumentation are used to establish what participants take to be acceptable at the end of the exchange. This chapter considers such dialogues, discussing the elements that are required in order to carry out argumentation-based dialogues, giving examples, and discussing open issues.

1 Introduction

Anyone these days who develops an interest in formal models of argumentation is likely to encounter [Dung, 1995] early in their studies. For many readers the key elements of this paper will be the abstract argumentation framework that it introduces, or its account of the formal process for establishing the acceptability of arguments, work which has led to much subsequent work on argumentation semantics. However, there is another aspect to [Dung, 1995] that interests us here. That is the notion of dialogue. More precisely, it is the idea of dialogue as an exchange between two or more individuals, an exchange which captures features of what would be informally called an "argument". That is, dialogue as the exchange of reasons for or against some matter. This idea crops up right at the start of [Dung, 1995], where readers are asked to imagine the following exchange between two individuals:

A: My government can not negotiate with your government because your government doesn't even recognize my government.

B: Your government doesn't recognize my government either.

A: But your government is a terrorist government.

Later in the introduction, [Dung, 1995] motivates the work in the paper by saying that:

> understanding of the structure and acceptability of arguments is essential for a computer system to be able to engage in exchanges of arguments

and it is this latter idea, or a slight modification of it, that interests us in this chapter. We are interested in how computer *systems* might be able to engage in exchanges of reasons for or against some matter (where such reasons are themselves often referred to as "arguments"[1]).

In keeping with current terminology in artificial intelligence [Russell and Norvig, 2020], we will refer to these computer systems as "agents", considering them to be embedded in a multiagent system [Wooldridge, 2009]. While in this chapter we will only be interested in providing formal models for exchanges between artificial agents, by which we

---

[1] Of course, this double use of the term "argument" is common in natural language, where it is used both to refer to a reason for or against a conclusion, and an exchange of such reasons.

mean that we do not consider aspects of natural language understanding or generation, we acknowledge that some of the agents may be human. That is, the multiagent system may be a combination of software agents and human agents. Indeed, the long-term vision for work on argumentation-based dialogue, as this area of study has become known, is for seamless interaction between humans and artificial agents. Further, the motivation for considering that it makes sense to ground such communication on models of argumentation is exactly that human reasoning and communication appears to naturally align with the process of constructing and evaluating arguments [Mercier and Sperber, 2011].

While the formal developments in [Dung, 1995], and in the large body of work on abstract argumentation that has been built on it, do not directly address exchanges between agents, there is a clear connection to such exchanges. The exchange between A and B that we quoted from [Dung, 1995] can be written as a *dialogue*, that is, as a conversation between multiple individuals. It can also be viewed as part of a *dialectical* process — the process of investigating the truth of an opinion. In this reading, the process is one in which an argument is put forward, followed by a counter-argument, followed by a counter to the counter-argument, and so on. This sequence can then be analysed to determine which of the arguments are acceptable, and variations on the process provide an "argument game" proof theory for different approaches to establishing acceptability [Modgil and Caminada, 2009]. Investigating such proof theories is not our topic here, though, as we will see, there are clear parallels between this work and the kinds of inter-agent dialogue that we are interested in.

Having said what we are not directly concerned with in this chapter, we should say in detail what we are directly concerned with. We have already sketched this, describing a focus on dialogues between agents in a multiagent system. The following example, from [Prakken, 2006], provides more detail on the kind of dialogue we mean:

(1) Paul : My car is very safe. (making a claim)

(2) Olga : Why is your car safe? (asking grounds for a claim)

(3) Paul : Since it has an airbag. (offering grounds for a claim)

(4) Olga : That is true. (conceding a claim) but I disagree that
this makes your car safe: the newspapers recently re-

ported on airbags expanding without cause. (stating a counter-argument)

(5) Paul : Yes, that is what the newspapers say (conceding a claim) but that does not prove anything, since newspaper reports are very unreliable sources of technological information. (undercutting a counter-argument)

(6) Olga : Still your car is not safe, since its maximum speed is very high. (alternative counter-argument)

Thus, it is clear that a focus of this line of work is the development of software systems that can engage in interactions which involve the exchange — and hence the generation and interpretation of — arguments. And we have also suggested that humans might also be part of these interactions. However, this description gives a rather narrow view of the full scope of the work.

For a start, the use of the word "dialogue" can be misleading, suggesting that the focus is only on interactions between just two entities[2], a misunderstanding that can be exacerbated by the focus of much of the literature on dialogues that deal with just two participants. (Sometimes this restriction is explicit, sometimes merely implicit or implied by the examples used.) However, there is no such restriction in general. In other words, the aim of work on argumentation-based dialogue is to support interactions between arbitrarily large numbers of agents.

Second, agents may play a number of different roles in interactions, and these may have an effect on aspects of the dialogue. In a lot of work, agents take part in dialogues as equal participants. That means that interactions are, in some sense, symmetrical, though there is asymmetry that arises because of what utterances are made, not because of the nature of the participants. For example, in the simple persuasion model of [Parsons *et al.*, 2003], which we will discuss below, the first agent to make an utterance is the one that accepts the burden of doing the persuading — making the case — and is also insulated from being persuaded in the sense that the rules of the dialogue do not allow it to change its position. In other cases, one distinguishes between the

---

[2]Perhaps because of a misunderstanding of the dichotomy between monologue and dialogue, or a misreading of the prefix "dia" as meaning "two".

agents engaged in a dialogue, and those observing it. For example, in [Prakken, 2001a] a distinction is made between the two agents who are arguing their positions, and the arbitrator which observes them, decides on the legality of the arguments that they make, and ultimately reaches a decision about which party has won the dispute. In this case, the two arguing agents are not attempting to persuade each other, but to convince the arbitrator. The arbitrator thus plays the role of the judge in some legal systems, settling the dispute based not on its own view, but on the arguments that have been made. In this kind of legal case, in front of a judge, the agents will be looking to make arguments that have legal impact — that are, in some sense the strongest arguments from a legal perspective. In contrast, in front of a jury, the arguers might instead aim to come up with the arguments that they think will best influence jury members, for example seeking to make arguments with emotional force. In either case, the arguers take their audience into account, and evaluate what they might say, choosing between possible arguments, by using some model of the audience. (Section 5.2 discusses how an agent might use such a model to determine which argument to assert.)

In the two cases just described, the audience to the dialogue is sitting in judgement. There is another case in which the audience is not judging, but has the aim of helping the parties in the dialogue reach a solution. Thus they act as a mediator rather than judge. We can also imagine other scenarios, for example where the audience will be reporting back on the dialogue to some external party. In such a situation the participants in the dialogue may be aware of the reporter, or they may be unaware. And if they are aware, they may shade their utterances to influence the external party (that is their aim may be less about their overt goal of "winning" the dialogue and more about securing a favorable report)[3].

All of these facets will influence the detail of the strategies that agents employ in a dialogue, and they may influence the design and structure of the dialogue mechanism. However, all these different dialogues and scenarios point to the same underlying questions, for example;

- What are the main components of a dialogue?

---

[3][Black and Sklar, 2016] considers further some of the issues that arise when there are multiple dialogue participants of different types.

In other words, when we design a dialogue, what are the major elements that we have to consider?

- How are dialogues represented?

  If we want to formally reason about dialogues, how do we represent the various components?

- What makes a good dialogue?

  Once we can represent and reason about (or implement and experiment with) dialogues, we need to establish what makes a good (or better) dialogue.

- How do we design dialogue strategies?

  The questions so far have looked at dialogues as opposed to the participants in the dialogue. When we look at dialogues from the participants' perspective, how do we design the elements that allow dialogue participants to participate?

The remainder of this chapter examines these questions, and discusses work that has been done on them to date. To do so, we have drawn heavily on our own work to illustrate the study of argumentation-based dialogue, making this chapter much more of a position piece than an exhaustive survey. This does not, of course, mean that we are the only researchers to have contributed to this line of work. We have attempted to identify other work that is relevant, and thus to present a relatively balanced picture. Where we have failed to mention work that we should have included, we apologise.

In detail, the contents of the rest of the chapter are as follows. Section 2 provides an introduction to the basic ideas in argumentation-based dialogue, and gives an example of such a dialogue. Then, Section 3 discusses a couple of approaches to representing dialogues, and Section 4 provides an in-depth description of one specific formal dialogue system, illustrating many of the ideas that have been introduced. Section 5 expands on several of the key unresolved issues in argumentation-based dialogue, including how to handle multiparty dialogues, the development of dialogue strategies, and how to handle enthymemes[4] in the context of dialogues. Section 6 concludes.

---

[4] An enthymeme is an incomplete argument, where some of the premises and/or

2 Basic notions

Since work on argumentation-based dialogue pulls together a number of different strands of work, there are several different ways that one might reasonably organise this section. We choose to start with [Walton and Krabbe, 1996], primarily because it provides a widely used classification of different kinds of dialogue.

2.1 Types of dialogue

In particular, [Walton and Krabbe, 1996] suggests that dialogues can be classified by what the participants know, what the participants seek to get from the dialogue, and what the dialogue rules are intended to bring about[5]. The basic classification is often written as follows:

Information-seeking Dialogues are dialogues in which one participant seeks the answer to some question or questions, and looks to obtain answers from another participant, who is believed by the first to know what these answers are.

Inquiry Dialogues are dialogues in which the participants collaborate to answer some question or questions whose answers are not known to either participant.

Persuasion Dialogues involve one participant seeking to persuade another to accept a proposition they do not hold at the beginning of the dialogue. This can mean that the persuadee is agnostic about the position, or it can mean that the persuadee holds the opposite position.

Negotiation Dialogues are dialogues in which the participants bargain over the division of some scarce resource. If a negotiation dialogue terminates with an agreement, then the resource has been

---

the claim of the intended complete argument are omitted. One of the mismatches between formal argumentation, and argumentation as used by people, is that while formal arguments are complete, people usually deal in enthymemes, eliding the "obvious" bits of the arguments that they present.

[5][Walton and Krabbe, 1996] makes use of the notion of the goal of the dialogue, which we find a problematic concept since it suggests intentionality on the part of the rules that specify how the dialogue proceeds.

divided in a manner acceptable to all participants. Though this may seem to be a rather specific notion of negotiation, it can be made quite broad by a suitably careful choice of the scarce resource in question.

Deliberation Dialogues are dialogues in which the participants collaborate to decide what action or course of action should be adopted in some situation. Here, participants share a responsibility to decide the course of action, or, at least, they share a willingness to discuss whether they have such a shared responsibility. Participants may have only partial or conflicting information, and conflicting preferences. As with negotiation dialogues, if a deliberation dialogue terminates with an agreement, then the participants have decided on a mutually-acceptable course of action.

Eristic Dialogues are dialogues in which participants quarrel verbally as a substitute for physical fighting, aiming to vent perceived grievances. At the time of writing, eristic dialogues have not been widely studied in computer science.

A number of these forms of dialogue have been studied in detail by the argumentation community, for example: information-seeking [Fan and Toni, 2012], inquiry [Black and Hunter, 2009; Hitchcock, 1991; Hulstijn, 2000], persuasion[6] [Gordon, 1994; Prakken, 2006; Walton and Krabbe, 1996], negotiation [Rahwan et al., 2003] and deliberation [McBurney et al., 2007].

[Walton and Krabbe, 1996] makes no claim that this set of dialogues is exhaustive, and plenty of other examples can be found in the literature. [Walton and Krabbe, 1996] also suggests that dialogues can be formed from combinations of these basic kinds of dialogue, and this is a point that we will return to below. Since the connection was made between this work — which has its roots in the philosophical study of argumentation — and the work that had been developed in within the AI and multiagent systems communities, these definitions have become widely adopted.

---

[6]Persuasion is perhaps the most widely studied form of dialogue, and these are just a few of the papers to have studied it.

2.2 Hamblin-style dialectical games

In addition to the typology, [Walton and Krabbe, 1996] is also notable for describing a persuasion dialogue. This is done in the form of a *dialogue game* between two or more players, the participants in the dialogue. The game specifies the moves that the players can make, and the rules that dictate which moves are possible when. This is in line with the tradition of Hamblin's *dialectical games* [1970; 1971]. Under this model, each agent holds a *commitment store*, accessible to both agents, which contains statements made during the dialogue. Since commitment stores evolve over time, we index them both by agent and by the step in the dialogue game. Thus we write the commitment stores for x and y at step k as $\mathsf{CS}_{x,k}$ and $\mathsf{CS}_{y,k}$ respectively[7]. The function of commitment stores varies between dialogue systems. The name stems from the idea that agents only put forward statements that they believe are in some sense true, and so making the statements is a commitment to defend that position if challenged. In the dialogue system we shall give as an illustration here, an agent's commitment store is just a subset of what can be inferred from its knowledge base, a subset that it chooses to make public. In other work, [Amgoud et al., 2000b] for example, one agent is allowed to place commitments on another by placing statements into that agent's commitment store. It is perhaps the work of Mackenzie [1979b; 1979a; 1990] which made such models popular outside the formal dialectics community. His game DC is perfect representative of such dialectical games, whose primary objective was to study and avoid fallacies in dialogues. The game is defined through several rules, eg.:

> "No statement may occur if it is a commitment of both speaker and hearer at that stage"

or

> "After a statement p, unless the preceding event was a challenge, p is included in both participants' commitments."

A bit more formally, this means that we can specify for the dialogue move $\langle x, \mathsf{assert}, p \rangle$, meaning agent x asserts formula p, the following conditions and updates:

---

[7]Note, we refer to agents using the variables x and y, such that $x \neq y$.

$\langle x, \mathsf{assert}, p \rangle$	where p is a formula.	
conditions	$p \notin \mathsf{CS}_{x,k}$ or $p \notin \mathsf{CS}_{y,k}$	
update	$\mathsf{CS}_{x,k+1} = \mathsf{CS}_{x,k} \cup \{p\}$ and $\mathsf{CS}_{y,k+1} = \mathsf{CS}_{y,k} \cup \{p\}$	

Other dialogue rules specify what moves can follow what other moves. However, these specifications often conflate different components which should be distinguished.

2.3 Components of a dialogue

We describe in turn the different components of dialogue systems: locutions, dialogues, protocols, and finally agents' strategies.

Locutions and dialogue moves. We start with the set of *moves* available to the agents. We can think of these as being made up of a set of *locutions*, capturing the kinds of things that agents are allowed to say, and which can be instantiated with certain propositions. The instantiated locutions make up the possible dialogue moves. There is no restriction on the locutions that can be used, however a common set of locutions found in such games is:

$$\{\mathsf{assert}, \mathsf{accept}, \mathsf{challenge}, \mathsf{question}, \mathsf{retract}, \mathsf{argue}\}$$

We call the set of locutions L. They allow for claims and their supporting arguments to be stated (or indeed retracted), arguments to be requested and questions to be asked. The set of locutions L can be instantiated to create a set of moves M. Assuming only two agents, we simply write dialogue moves as

$$\langle x, \mathsf{locution}, \phi \rangle$$

where x is the speaker, locution $\in$ L is a locution, and ϕ is a formula built from some content language. For instance, these can be instantiated with propositions (and, in the case of **assert** and **accept**, may be instantiated with sets of propositions). But note that agents are not completely free to combine locutions and propositions — for example rules like that above will specify what is possible.

Dialogue history. A *dialogue* can be seen as a non-empty sequence of dialogue moves. We index the set of moves by the step of the dialogue,

move	possible replies
$\langle x, \textsf{assert}, \phi \rangle$	$\langle y, \textsf{accept}, \phi \rangle, \langle y, \textsf{assert}, \neg\phi \rangle, \langle y, \textsf{challenge}, \phi \rangle$
$\langle x, \textsf{challenge}, \phi \rangle$	$\langle y, \textsf{argue}, (\Phi, \phi) \rangle, \langle y, \textsf{retract}, \phi \rangle$
$\langle x, \textsf{argue}, (\Phi, \phi) \rangle$	$\langle y, \textsf{accept}, \psi \rangle (\psi \in \Phi), \langle y, \textsf{challenge}, \psi \rangle (\psi \in \Phi)$
$\langle x, \textsf{retract}, \phi \rangle$	
$\langle x, \textsf{accept}, \phi \rangle$	

Table 1: Possible replies in persuasion games (where (Φ, ϕ) is an argument with support Φ and claim ϕ)

and denote by d_k a dialogue history of length k.

$$d_k = [m_1, m_2, \ldots, m_k]$$

where each m_k is a well-formed dialogue moves as previously defined. We will write the set of all sequences of dialogue moves as D.

Dialogue protocols. Now, as already discussed, not all sequences of dialogues are *legal*. In general, it could be possible to specify, for any dialogue history, the set of dialogue moves that can be uttered. That is, we can think of a protocol as a function π that defines a subset of M that can be uttered at a particular point in a dialogue.

$$\pi : \mathsf{D} \mapsto 2^{\mathsf{M}}$$

Typically, it is infeasible to specify the allowed moves for every possible dialogue state that can occur. Often, protocols will instead be specified thanks to simple rules: turn-taking, moves allowed after another move, etc. For instance, for persuasion dialogues we may specify the set of possible responses given in Table 1 (adapted from [Prakken, 2006]).

Also, as we saw earlier, it is possible to specify commitment rules stating conditions upon which moves can be made, thus using commitment stores as (part of) the "dialogue state". This approach should be familiar to the AI-oriented reader: such locutions can be considered to be a set of speech acts in the sense of [Austin, 1975; Searle, 1969]. What this means is that although these locutions are merely utterances, they have the properties of actions, in that they change the state of the world. The "state of the world" in this case is the state of the dialogue, and it is changed because the dialogue moves change the state of the commitment stores. Note that the definition of the locutions in terms of conditions and update rules also provides them with an operational semantics [Plotkin, 1981] — conditions and update rules give pre- and post-conditions for the instantiated locutions, thus describing the transitions in world state that they enable. The public nature of the commitment stores also provides the locutions with a social semantics [Singh, 2000].

At this stage it is useful to inspect a bit more closely, following [Fernández and Endriss, 2007], how these various types of dialogue protocols compare. Do they have different expressive power, for instance (in the sense that they allow more dialogues to be captured)? At the basic level lies *deterministic finite-state automaton* (DFA) (aka. *finite state machines*). After each move, a given set of moves is specified, as in our example in Table 1. It is sufficient to store the last dialogue move in a dialogue state to operate. Now, adding commitment stores (sets of propositions agents are committed to, as in Hamblin-like dialogue games) to this state does not make protocols more expressive than DFA. But they can be much more concise and convenient to specify. On the contrary, using some *stacks* (like questions under discussions, which allows embedded sub-questions) would improve the expressivity of the protocol. Often, in practice, protocols are thus specified based on a combination of the last dialogue move and some commitment stores.

A general remark that can be made about the resulting protocols is that they remain mostly "syntactically regulated". An alternative approach, advocated in particular by [Prakken, 2005], is to also base the legality of moves on the current dialectical status of some propositions. In particular, in the context of persuasion dialogues, we may allow moves as long as they modify the dialectical status of the main topic of the discussion. We call such protocols *relevance-based*. This requires of course a reasoning machinery to evaluate this dialectical status of argument.

Agents' strategy. The last remaining component differs a bit, in the sense that it is not part of the dialogue game *per se*. However, it is certainly required to understand or automate agents playing such games. How does an agent decide which dialogue move to play? An agent might make its decision with an overall aim in mind. It might want to comply to rationality principles. It might want to end the dialogue quickly. It might want to drag the dialogue out as long as possible [Gabbay and Woods, 2001b; Gabbay and Woods, 2001a]. It might want to "win", however that is defined for the type of dialogue, if it possibly can. It might want to avoid giving away information [Oren et al., 2006]. The mechanism for making a choice of move that takes this overall aim (or indeed a combination of these aims) into account is a *strategy*.

We can think of the strategy for an agent x as being a function $\mathcal{S}_x$ which takes a dialogue history, the agent's private knowledge base Σ_x, and which returns a subset of dialogue moves. Each agent has its own strategy:

$$\mathcal{S}_x : \mathsf{D} \times \Sigma_x \mapsto 2^{\mathsf{M}}$$

More generally, it would be possible to define *probabilistic strategies*, which return probability distributions over the possible dialogue moves of agents. But the strategies we discuss in this chapter are mostly not probabilistic, i.e. they are *deterministic*.

Finally, we say that a strategy is *decisive* if it returns a single move. We note also that nothing at this stage guarantees that the agent's strategy complies with a given protocol. This is known as the *conformance problem*. One simple way to guarantee this is to filter out illegal moves returned by the strategy. We shall denote $\mathcal{S}_x^\pi$ the strategy of agent x made compliant for protocol π.

As for protocols, we see that specifying a strategy can be tedious, as it may require specifying decisions for each possible dialogue history. When the strategy remains the same at each stage of the dialogue, we talk of a *stationary strategy*, and can simply drop the reference to the dialogue history. More often, strategies may only depend on the previous move of the other agent, and thus can be specified for each possible locution uttered by the other agent.

In the rest of this chapter, we will sometimes use simple *ordered strategies*, that is, a simple preference ranking of the dialogue moves of the agents, assuming those moves comply with rationality constraints. For example, the following simplistic strategy corresponds to an agent

x which would, in any circumstance during the dialogue, prefer (if it is allowed by the protocol π): first to challenge a proposition of the other agent, then to assert some proposition, then concede, and finally to retract some proposition.

$$S_x^\pi = \begin{cases} 1 : \langle x, \text{challenge}, \phi \rangle \\ 2 : \langle x, \text{assert}, \phi \rangle \\ 3 : \langle x, \text{accept}, \phi \rangle \\ 4 : \langle x, \text{retract}, \phi \rangle \end{cases}$$

Suppose the previous move was an assert by the other agent, and that only challenging or accepting are possible replies. In that case the agent would prefer to challenge. Of course, there could be many dialogue moves satisfying a given priority level (e.g. the agent could have many propositions to possibly assert). In that case, further selection functions would be required to make the strategy decisive. There is a lot more to say about strategies, and in particular we may want to take advantage of some information we may have on the other agent(s), in other words consider an *opponent model* on top of the sole dialogue history as discussed. We devote a full section to this recent and lively research question (Section 5.2).

2.4 Evaluation of dialogues

Now comes a difficult question. How can we evaluate the quality of the resulting dialogues? We start with the most obvious properties that one might require.

- *termination*: requires that the dialogue does not run forever. A stronger guarantee is that the dialogue is *deadlock-free*, that is, that there is always a legal move to play at any state of the dialogue. Even these seemingly very basic properties are sometimes difficult to guarantee.

- *successfulness*: whether the goal of the dialogue, or the goal of the participants is fully/partially attained. This criterion can be a simple boolean test, but much finer-grained approaches are possible— think of outcomes in a negotiation for instance.

- *efficiency*: what was the length of the dialogue? Would it have been possible to reach an equally good outcome in fewer communication steps?

- *comparison to full merging of knowledge, or variants of the protocol*: it is often very useful to compare the protocol with (even idealized) alternative solutions, like simply merging all the knowledge of agents (that is, the outcome an omniscient agent would obtain). Or, on the contrary, to a less expressive version of the protocol (for instance, a protocol not allowing the exchange of arguments).

- *relevance of dialogue moves*: was the dialogue coherent overall, were all the moves relevant? These are important aspects to consider when humans are involved in the dialogue system. It is possible to design heuristics evaluating how "close" to the topic under discussion is each move, see for example the work of Rosenfeld and Kraus introducing several heuristics for that purpose in the context of dialogues for persuasion [Rosenfeld and Kraus, 2016a].

There are also criteria that are more difficult to pin down formally, but may still be very important in some contexts–for instance, how cognitively difficult are the reasoning tasks involved in the dialogue? Depending on the application, we may thus require expert feedback to evaluate some of these criteria, which may be difficult to assess automatically.

2.5 Artificial agents playing dialectical games

We now give an example of a specific dialogue game, showing how the components presented above can come together. The game we present is taken from [Parsons *et al.*, 2003], which contains a set of basic dialogue games that are instantiations of the Walton and Krabbe concepts of information-seeking, inquiry and persuasion, dialogue games that are refinements[8] of the more general dialogues introduced in [Amgoud *et al.*, 2000a; Amgoud *et al.*, 2000b], and so are in the Hamblin/MacKenzie

---

[8]In the sense of being refined down to the minimal dialogue games that could be considered instances of the relevant type.

tradition. In fact, this dialogue is a further refinement of the persuasion dialogue from [Parsons et al., 2003][9].

As before, we assume that dialogues take place between two agents, and we refer to these with the variables x and y (where $x \neq y$). Each agent has a private knowledge base, Σ_x and Σ_y respectively, containing their beliefs. We assume that agents reason using some form of structured argumentation system, and so the knowledge base is a collection of formulae in some logical language. We aren't going to impose any conditions on the knowledge base other than that the underlying language contains at least a set of ground literals, or propositions, and the only condition that we will place on the argumentation system is that it can construct arguments from the elements of the knowledge base, and can apply the grounded semantics. (The choice of the grounded semantics is somewhat arbitrary, though were we going to examine the properties of this dialogue system, adopting a semantics that is guaranteed to have an extension, and to only ever have one extension, would be helpful.) Thus, an agent x can construct arguments using[10]:

$$\Sigma_x \cup \mathsf{CS}_{x,k} \cup \mathsf{CS}_{y,k}$$

Here we will assume that there are no degrees of belief to be accounted for. That is, we assume that both agents believe all the elements of Σ_x and Σ_y, and hence $\mathsf{CS}_{x,k}$ and $\mathsf{CS}_{y,k}$ equally. This is, of course, a big assumption, and it is easy to incorporate different levels of belief [Amgoud et al., 2000a; Amgoud et al., 2000b], to relate belief an agent

---

[9]The dialogue games in [Parsons et al., 2003] tied the game to a specific underlying argumentation system, so the version here is more general, and allowed for a range of agent *attitudes* that specified the conditions under which agents could make utterances. The game defined here allows just one attitude and ties that to the use of the grounded semantics, so in that sense it is more specific than that in [Parsons et al., 2003]. The description here also separates protocol and strategy, providing a much more declarative version of the dialogue game, and adopts the **argue** locution.

[10]Since, as we will see, what gets placed into an agent's commitment store are the conclusions of arguments, these will include elements that are not present in the relevant knowledge base. As a result, we can think of a commitment store as being an extension of the corresponding knowledge base rather than a subset, and so consider an agent to have access to the union of the commitment store and that knowledge base though, anything in the commitment store can be inferred from the corresponding knowledge base.

move	conditions
$\langle x, \text{assert}, p \rangle$	$p = \text{Conc}(A)$, s. t. $A \in E_{\mathcal{GR}}(Ar(\Sigma_x \cup \text{CS}_{x,k} \cup \text{CS}_{y,k}))$
$\langle x, \text{assert}, S \rangle$	$\forall s \in S, s = \text{Conc}(A)$, s. t. $A \in E_{\mathcal{GR}}(Ar(\Sigma_x \cup \text{CS}_{x,k} \cup \text{CS}_{y,k}))$
$\langle x, \text{accept}, p \rangle$	$p = \text{Conc}(A)$, s. t. $A \in E_{\mathcal{GR}}(Ar(\Sigma_x \cup \text{CS}_{x,k} \cup \text{CS}_{y,k}))$
$\langle x, \text{accept}, S \rangle$	$\forall s \in S, s = \text{Conc}(A)$, s. t. $A \in E_{\mathcal{GR}}(Ar(\Sigma_x \cup \text{CS}_{x,k} \cup \text{CS}_{y,k}))$
$\langle x, \text{challenge}, p \rangle$	

Table 2: Conditions in persuasion games

has in a proposition to the level of trust it has in the agent who utters it [Parsons et al., 2011], or to allow agents to discuss the level of belief of some proposition that is being discussed [Amgoud and Parsons, 2001].

The protocol used will be similar to the one described in Table 1, at the sole exception that no retract moves are considered. In addition, moves affect the state of the commitment stores, by basically committing agents to the content of the moves they assert or accept (other moves have no consequences on the commitment stores).

The strategy of agents is first based on *rationality conditions*. Agents abiding by these rationality conditions can assert any proposition that is the conclusion of an argument that is in the grounded extension of that agent's argumentation framework[11]. In other words, if agent x can construct an argument from $\Sigma_x \cup \text{CS}_{x,k} \cup \text{CS}_{y,k}$, and that argument is acceptable under the grounded semantics, then its conclusion can be asserted. Similarly, an agent can accept any proposition asserted by another agent, provided that the proposition is the conclusion of an argument that is in the grounded extension of its argumentation framework. As [Parsons et al., 2003] discusses, varying the rationality conditions varies the way that agents behave in a given dialogue game.

Table 2 describes each move as being uttered by x and addressed to y; since x and y are variables that can refer to either participant (where $x \neq y$) this is sufficient to define the rationality and update rules for both of the participating agents. We use the notation $Ar(\Sigma)$ to denote the set of all arguments constructed from the set Σ, $\text{Conc}(A)$ to denote the conclusion of the argument A, and $E_{\mathcal{GR}}(Ar)$ to denote the grounded extension of the set of arguments Ar. We frame each move as taking

---
[11]This is same condition as adopted in [Amgoud et al., 2000a; Amgoud et al., 2000b], restated in terms of the Dung semantics.

place at step k of the dialogue.

Now in terms of protocol, each move has an *update rule* which specifies how the commitment stores of the agents are modified by the move. (An **assert** places the formula with which it is instantiated into the agent's commitment store, other locutions do not change the commitment store.) There are also conditions regarding what types of moves can follow other moves, which are in line with those given in Table 1.

How about strategy? As discussed above, in a persuasion dialogue, one party seeks to persuade another party to adopt a belief or point-of-view he or she does not currently hold. We slightly depart from the description given in [Parsons et al., 2003] and describe a possible strategy, which — when combined with its rationality conditions as discussed above — allows one agent, x to try to persuade another, y, to accept a proposition p:

$$\mathcal{S}_x^\pi = \begin{cases} 1 : \langle x, \text{accept}, \phi \rangle \\ 2 : \langle x, \text{challenge}, \phi \rangle \\ 3 : \langle x, \text{assert}, \neg \phi \rangle \end{cases}$$

An interesting case emerges with the challenge move. Indeed, as procedurally described in [Parsons et al., 2003], the behaviour of the agent should be that "If y has **challenged** p, then x **asserts** S, the premises of its argument for p". Then agent y should inspect in turn each premise. But as the premises of an argument can in turn be challenged, we see that embedded dialogue structures (as in the Paul and Olga example) may occur. To account for this, there are two options:

(i) either the language allows agent x to assert a *collection* of premises in a single move. In that case, the protocol should allow to accept or challenge any proposition in the commitment store of the other agent (thus departing from the simple last move move-reply protocol);

(ii) or the language only allows **assert** moves involving a single proposition at-a-time. In that case, the protocol should be equipped to keep track of propositions challenged, so as to make sure that agents are enforced to respond appropriately to those challenges.

9 - ARGUMENTATION-BASED DIALOGUE

Note also that in both cases, enforcing agents to address the latest **challenge** or **assert** raised requires the use of stack structures, instead of mere sets.

If at any point an agent cannot make the indicated move, it has to concede the dialogue game. An agent also concedes the game if at any point there are no propositions made by the other agent that it hasn't accepted. These rules give two conditions under which the dialogue will terminate with an agent conceding. However, on their own, they do not mean that the protocol will terminate. One agent could, for example, keep making the same assertion repeatedly[12]. However, termination can be guaranteed by adding a rule specifying that **assert** is not allowed if the proposition is already present in the CS, which is the case in the Hamblin's protocol, or, by adding a rule specifying that no move may be repeated, as in [Parsons et al., 2003][13].

While this makes a useful example, it is clearly extremely simple. Agent x's method for persuasion is to assert p, the conclusion that it wants y to accept. Agent y accepts if it can, which it can only do if has no argument against it. If there is a counter-argument, y can assert that, and x may find that more persuasive than its initial argument (particularly if we allow for preferences over beliefs, and hence arguments). If an argument is *challenge*d, the response is to put forward the premises for p, and these are considered in turn. If this does not convince the recipient (as it well might not if the recipient does not have the necessary rules in its knowledge base[14]), then the dialogue will end inconclusively.

In [Parsons et al., 2003] the reason for looking at such a simple

---

[12]A behaviour that any parent will recognise.

[13]With this protocol, if we rely on Hamblin's rule, then, together with the rule about conceding if there is no legal move, it is the case that an agent can win by challenging the same proposition multiple times. When its opponent runs out of alternative justifications the opponent will have nothing new to **assert** and will have to concede. If we adopt the rule that repeating the same move is not possible in place of Hamblin's rule, then the first agent to run out of new moves will have to concede, and that seems a more equitable solution in this case.

[14]The limitation depends on the argumentation system used by the agents. In [Parsons et al., 2003] the underlying system made no distinction between facts (premises or axioms) and rules, so p could be a rule, in the form of a material implication. However, in systems like ASPIC+ [Modgil and Prakken, 2013], where rules cannot be the conclusion of an argument, what can be asserted is more limited.

protocol was to be able to establish some baseline results for dialogues on which more complex dialogue systems could be built. As above, we can show termination, and it is straightforward to obtain results about the conditions under which one participant will be persuaded by another. However, we don't have to look far to find examples of persuasion dialogues that cannot be captured by it. Indeed, we have already seen one — the Paul and Olga example from Section 1. The first three first moves of the Paul and Olga dialogue can be readily captured by our protocol, the assertion of a statement about Paul's car, a question from Olga, and the assertion of the supporting argument that the car has an airbag[15]. However, the fourth move sees Olga both conceding, and making a counter-argument, and in the fifth, Paul does the same. Then in the sixth move, Olga returns to an earlier choice and proposes a new alternative counter-argument. As Prakken [2006] points out, in the sixth move Olga also postpones replying to Paul's counter-argument (by providing another attack on his argument for safety). This kind of flexibility requires different techniques to maintain dialogue coherence, and the approaches in [Prakken, 2001b; Prakken, 2005] aim to provide exactly this.

2.6 A final word on the basics

As mentioned earlier, the work discussed in the chapter as a whole only scratches the surface of existing work on argumentation-based dialogue. The same goes for this section. It has covered the core ideas of argumentation-based dialogue, and it has illustrated many of them using a simple dialogue game. But there are many other aspects of argumentation-based dialogue that it has not covered. Some of these will be discussed later in the chapter, but there are a couple of specific elements that we'll mention here.

First, the argumentation game described here is monotonic in terms of assertions. Once a proposition has been asserted, it remains so. Taken

---

[15] Note that this argument is an enthymeme — it does not include the connection between airbag and safety that we might write as "a car that has an airbag is safe". Such an enthymeme can be asserted in our example protocol whereas the full argument in an ASPIC+-like system would not because the rule "a car that has an airbag is safe" can't be the conclusion of an argument.

together with the idea that a proposition that has been asserted, and hence is placed in a commitment store, will be defended by the relevant agent, this montonicity does not make sense for many dialogue games. Even in the simple persuasion dialogue above, x can assert p and then end up accepting $\neg p$, provided that there is a suitably persuasive argument for it. In such a case it makes no sense to continue to commit to p. As a result, many argumentation-based dialogue systems allow for retraction of prior assertions.

Second, both our discussion and our example, has been for a single kind of dialogue. As we noted when discussing [Walton and Krabbe, 1996], it is possible to imagine dialogues that are combinations of other types. A canonical example is that of a negotiation, for example a negotiation around the purchase of an item, in which one participant transitions into a persuasion dialogue aimed at convincing the other participant that it really needs some feature of the item that justifies a higher price. Now, one could define such combined dialogues — the locutions, protocols, strategies and so on — in addition to the dialogues of which the are composed. One might also look at how dialogues can be combined. Both lines of work have been pursued, with [McBurney and Parsons, 2002] being the first work to explicitly consider the requirements of dialogues which are themselves made up of other dialogues.

Finally, we should mention some seminal work that does not fit directly into the Hamlin/Mackenzie line that we have been describing above. A major contribution here is Prakken's general framework for dialogue [Prakken, 2005], which we mentioned above. An early version of this general approach can be found in [Prakken, 2000], and the basic mechanism, albeit focused on a specific legal dialogue game is in [Prakken and Sartor, 1998]. The key characteristic of this approach is that it is more general than the specific dialogue games that we have been discussing, making it possible to produce results that can evaluate families of dialogue games. [Prakken, 2006] starts from this viewpoint, and indeed in some sense extends it, by identifying and contrasting the key features of many different formal persuasion dialogues. Two of the dialogue systems reviewed in [Prakken, 2006] should also be mentioned. [Walton and Krabbe, 1996] not only classifies different kinds of dialogue, but defines and discusses a dialogue system PPD that is an early example of a general model of persuasion. Finally, Gordon's "pleadings game" [Gordon, 1994], is an early example of a formal dialogue game, one which captures features of US civil procedure, and which includes many of the elements which have now become standard.

3 Representing dialogues

Given the diversity of dialogue or dialectical games proposed in the literature (as briefly discussed in Section 2), it may seem challenging to come up with a common representation scheme. However, such a scheme would in principle bring many benefits, the first one being to facilitate the design of dialogue games, as well as making it possible to compare them and to easily share them among different users. But another potential interest is to make these games amenable to formal verification and analysis of properties. In this section we discuss two proposals which illustrate these two aspects: the dialogue specification language of [Wells and Reed, 2012], and the use of *executable logic* as a formal language, as advocated in [Black and Hunter, 2012a].

3.1 Dialogue specification language

Before setting up their framework, [Wells and Reed, 2012] describe in great detail the landscape of extant dialectal games. They exhibit relations between games, and provide an historical perspective which is interesting in its own right. Based on this analysis, they identified a set of features (See Table 3 for details) which form the basis of their *Dialogue Game Description Language* (DGDL). This is a specific instance of a *Domain Specific Language*, that is "a small language developed to be both concise and powerful" [Wells and Reed, 2012].

They provide a formal specification, under an Extended-Backus Normal Form, and show how several games can be captured in this setting, including DC [Mackenzie, 1979b], which we have already mentioned. Once a game is described, it then becomes possible to proceed to verify whether the description is syntactically correct, for instance.

As dialogue games are also games, a natural question is whether

9 - ARGUMENTATION-BASED DIALOGUE

Moves per turn	Either a single, multiple, or a defined maximum number of moves per turn
Turn Organization	Strict progression
Dialogue magnitude	The maximum number of turns that the dialogue allows
Move types	Range of available locutions or performatives
Move content	The statements, propositions, variables, tokens, or collections thereof that are moved
Openers	A description of the locutional form of the move, eg. "Is it the case that... "
Stores	Collections of in-game artifacts, organized as sets, stacks, or queues
Store contents	In addition to move content, stores can contain locutions or arguments
Store visibility	Whether a given store is public or private
Move legality	The formulation of conditions that must be satisfied for the move to be legal. Inspection of previous moves; store contents; role occupation; store magnitude; store comparisons; the length of the dialogue; correspondence to a given scheme; relation between content elements; form in which the content is presented
Move effects	The effect of a successfully played move upon the dialogue and its components. Prescription of mandated responses; operations on stores; update to status of a game or system; assignment of role
Participants	The number and identity of players that the game supports
Roles	The roles that players can occupy at various stages of a game
Rules	Non-move specific rules that can alter the game as a function of game state rather than as a result of the particular move that has been played

Table 3: Features of dialogue games (from [Wells and Reed, 2012])

the more general *Game Description Language*[16] could also be used to describe dialogue games. Until recently, this language was rather restricted in the sense that incomplete information games could not be handled, but recent developments [Thielscher, 2010] mean that it could in principle be possible to use the Game Description Language to reason about the strategical aspects of agents' involved in dialogue games, even in situations where knowledge about the opponent is incomplete (see Section 5.2 for a more detailed treatment of these questions). It would

---
[16] http://games.stanford.edu/games/gdl.html

thus be interesting to investigate using this general language to capture and reason about dialogue games, although there is the risk that we might lose the domain specific facilities DGDL offers.

3.2 Representing dialogues with logic

In the early 2000s, a stream of papers [Sadri *et al.*, 2001; Torroni, 2002; Endriss *et al.*, 2004] studied communication among agents in the context of *computational logic*. In general, protocols were interpreted as *constraints* that were integrated in the agent's reasoning.

A related but different approach was taken by [Black and Hunter, 2012a], who built a logic-based language aiming at representing dialogical argumentation systems. More precisely, they proposed viewing these systems as operating on a dialogue state, and make use of different modal operators to account for the modification, either of the private or public state: $\oplus \alpha$ (resp. $\ominus \alpha$) adds the literal to (resp. removes the literal from) the *private* state, while $\boxminus \alpha$ (resp. $\boxplus \alpha$) adds the literal α to (resp. removes the literal from) the public state. The designer is then free to define and use predicates which can be used either in the private or public state. For instance, assuming a predicate $b(\phi)$ to represent that the agent believes ϕ and a predicate $t(x)$ to represent that it is x's turn in the dialogue, a simple rule like the following could be stated:

$$b(\phi) \land t(x) \Rightarrow \boxplus claim(\phi) \land \boxminus t(x) \land \boxplus t(y)$$

In words, if the agent believes ϕ and if it is her turn to speak, then she might claim ϕ, which would also release the turn token to the other agent. Following the discussion in Section 2, we see that such a rule integrates both protocol and strategical (under the form of a rationality condition) aspects.

With a few more simple rules (in addition to the rule above) we can capture the first part of the dialogue between Paul and Olga (Section 1). We assume the following predicates: $public(\psi)$ means that ψ is part of the public state; $argue(a, \phi)$ means that the agent can construct an argument a whose claim is ϕ.

$$public(claim(\phi)) \land \neg argue(a, \phi) \land t(x) \Rightarrow \boxplus why(\phi) \land \boxminus t(x) \land \boxplus t(y)$$

$$public(why(\phi)) \land argue(a, \phi) \land t(x) \Rightarrow \boxplus assert(a) \land \boxminus t(x) \land \boxplus t(y)$$

Referring back to the Paul and Olga example (Section 1) (assuming that Paul can, from his private beliefs, construct an argument for why the car is safe based on the fact that it has an airbag, and that Olga cannot construct any argument for why the car is safe) these rules will generate the following behaviour.

n	x	action on dialogue state
0	Paul	$\boxplus claim(safe)$, $\boxplus(t(Olga))$, $\boxminus(t(Paul))$
1	Olga	$\boxplus why(safe)$, $\boxplus(t(Paul))$, $\boxminus(t(Olga))$
2	Paul	$\boxplus assert((\{airbag \implies safe, airbag\}, safe))$, $\boxplus(t(Olga))$, $\boxminus(t(Paul))$

When dialogical systems are *finite* (that is, when for each action rule, there is a finite number of groundings of the rule), then a translation to finite state machines can be obtained, thus allowing to prove properties (like termination, for instance).

4 Example: Value-based deliberation

In this section, we present Black and Atkinson's dialogue system for deliberation [Black and Atkinson, 2011a; Black and Atkinson, 2011b], which allows two agents to agree on an action to perform to achieve some shared goal and formally delineates the public aspects of the dialogue system (such as protocol and moves) from the individual participants' strategies. The participating agents each aim to reach an agreement, but individually they may wish to influence the outcome of the dialogue to better suit their preferences. This type of dialogue therefore sits somewhere between persuasion dialogues (which may be zero sum) and entirely cooperative inquiry dialogues (where agents seek to establish whether a claim can be justified), as is typically the case with real life encounters. Consider, for example, a situation where colleagues at a conference would all like to go to a restaurant for dinner together. Each individual has a set of preferences that they wish to be satisfied by the group's decision about which restaurant to visit. For example, some may require vegetarian options, some may not want to travel far, and some may want to avoid expensive restaurants. In such a scenario, the participants each have their own preferences but are committed to finding an outcome they can all agree to.

4.1 Practical reasoning

The participating agents use the popular *argumentation scheme and critical question* approach [Walton and Krabbe, 1996; Macagno et al., 2018] in order to reason about how to act. Argumentation schemes are stereotypical patterns of reasoning, which are used as presumptive justification for generating arguments. Each scheme has associated with it a set of critical questions, which allow one to identify potential attacks on an argument generated by the scheme. Specifically, the agents make use of the *practical reasoning argument scheme* [Atkinson and Bench-Capon, 2007]:

> In the current circumstances R,
> we should perform action A,
> which will result in new circumstances S,
> which will achieve goal G,
> which will promote value V.

The set of characteristic critical questions that are associated with the scheme can be used to identify challenges to proposals for action that instantiate the scheme. An unfavourable answer to a critical question will identify a potential flaw in the argument. Since the scheme makes use of what are termed as 'values', this caters for arguments based on subjective preferences as well as more objective facts. Such values represent qualitative interests that an agent wishes (or does not wish) to uphold by realising the goal stated.

Each agent has knowledge about the state of the world, the preconditions and effects of actions, and about which values are promoted or demoted by the transition from one state to another (as caused by an action). This knowledge (which is represented as a *value-based transition system* [Black and Atkinson, 2011b], the definition of which we omit here for brevity) can be used to instantiate the practical reasoning argument scheme in order to construct arguments for (or against) actions to achieve a particular goal because they promote (or demote) a particular value.

Definition 4.1. *An* **argument** *constructed by an agent x is a 4-tuple $a = \langle act, p, v, s \rangle$ where:*

– *act is an* **action**;
– *p is a* **goal**;
– *v is a* **value**;
– $s \in \{+, -\}$;
– $s = +$ *iff act is an action that will achieve goal p and will promote value v;*
– $s = -$ *iff act is an action that will achieve goal p but will demote value v.*

For any argument $a = \langle act, p, v, s \rangle$: $\mathsf{Act}(a) = act$; $\mathsf{Goal}(a) = p$; $\mathsf{Val}(a) = v$; $\mathsf{Sign}(a) = s$.
If $\mathsf{Sign}(a) = +$, *then a is a* **positive** *argument* **for** *action* $\mathsf{Act}(a)$.
If $\mathsf{Sign}(a) = -$, *then a is a* **negative** *argument* **against** *action* $\mathsf{Act}(a)$.
The set of all arguments an agent x **can construct** *is denoted* $Args^x$; **the set of all arguments for a particular goal** p **that an agent can construct** *is* $Args^x_p = \{a \in Args^x \mid \mathsf{Goal}(a) = p\}$.
The set of values *for a set of arguments* S *is* $\mathsf{Vals}(S) = \{v \mid a \in S \text{ and } \mathsf{Val}(a) = v\}$.

Given a particular argument for an action, one can then (as mentioned above) identify attacks on that action by posing the various critical questions associated with the practical reasoning argument scheme. This reasoning (through posing of the different critical questions) is split into three stages [Atkinson and Bench-Capon, 2007]: *problem formulation*, where the agents decide on the facts and values relevant to the particular situation under consideration; *epistemic reasoning*, where the agents determine the current situation with respect to the structure formed at the previous stage; and *action selection*, where the agents develop, and evaluate, arguments and counter arguments about what to do. The dialogue system we present here deals only with the action selection stage. It assumes that the agents' problem formulation and epistemic reasoning are sound and that they are in agreement on the output of these stages (this agreement could itself be the product of another dialogue, see for example [Black and Atkinson, 2009]). The critical questions associated with the action selection stage (numbered as in [Atkinson and Bench-Capon, 2007]) are:

CQ5: Are there alternative ways of realising the same consequences?

CQ6: Are there alternative ways of realising the same goal?

CQ7: Are there alternative ways of promoting the same value?

CQ8: Does doing the action have a side effect which demotes the value?

CQ9: Does doing the action have a side effect which demotes some other value?

CQ10: Does doing the action promote some other value?

CQ11: Does doing the action preclude some other action which would promote some other value?

As the focus of the dialogue is to agree to an action that achieves the *goal*, the incidental consequences (CQ5) and other potentially precluded actions (CQ11) are not relevant. The participating agents thus use CQ6–CQ10 to identify attacks on proposed arguments for action. Answers to these questions provide arguments for and against different actions to achieve a particular goal, where each argument is associated with a motivating value. These arguments and the attacks between them (determined by the critical questions) can be represented in a value-based argumentation framework [Bench-Capon, 2002] (see also Chapter 5 of this handbook [Atkinson and Bench-Capon, 2021]), an extension of the abstract argumentation frameworks of Dung [1995]. In a Dung argumentation framework, an argument is admissible with respect to a set of arguments S if its attackers are attacked by some argument in S, and no argument in S attacks an argument in S. In a value-based argumentation framework, an argument succeeds in defeating an argument it attacks only if its value is ranked higher than (if the attack is symmetric) or at least as high as (if the attack is asymmetric) the value of the argument attacked. A particular ordering of the values is characterised as an **audience**; this represents an individual agent's preferences over the values. Arguments in a value-based argumentation framework are admissible with respect to an audience A and a set of arguments S if they are admissible with respect to S in the Dung argumentation framework that results from removing all the attacks that are unsuccessful given the audience A. A maximal admissible set of a value-based argumentation framework is a **preferred extension**.

9 - ARGUMENTATION-BASED DIALOGUE

Value-based argumentation frameworks are commonly considered at an abstract level. The following definition gives a particular instantiation of value-based argumentation frameworks that captures the attack relation between arguments that is generated from the critical questions. Condition (1) of the following attack relation allows for CQ8 and CQ9; condition (2) allows for CQ10[17]; condition (3) allows for CQ6 and CQ7. Note that attacks generated by condition (1) are not symmetrical, whilst those generated by conditions (2) and (3) are.

Definition 4.2. *An* **instantiated value-based argumentation framework** *(i****VAF****) is defined by a tuple* $\langle Ar, att \rangle$ *such that Ar is a finite set of arguments and* $att \subset Ar \times Ar$ *is the* **attack relation**. *A pair* $(a, b) \in att$ *is referred to as "a attacks b" or "b is attacked by a". For two arguments* $a = \langle act, p, v, s \rangle$, $a' = \langle act', p', v', s' \rangle \in Ar$, $(a, a') \in att$ *iff* $p = p'$ *and either:*

(1): $act = act'$, $s = -$ *and* $s' = +$; *or*
(2): $act = act'$, $v \neq v'$ *and* $s = s' = +$; *or*
(3): $act \neq act'$ *and* $s = s' = +$.

An **audience** *for an agent x over the set of values* V *is a binary relation* $\mathcal{R}^x \subset V \times V$ *that defines a total order over* V. *An argument a is* **preferred to** *the argument b under the audience* $\mathcal{R}^x$, *denoted* $a \succ_x b$, *iff* $(\mathsf{Val}(a), \mathsf{Val}(b)) \in \mathcal{R}^x$. *If* $\mathcal{R}^x$ *is an audience over the values* V *for the iVAF* $\langle Ar, att \rangle$, *then* $\mathsf{Vals}(Ar) \subseteq V$.

Given an iVAF and a particular agent's audience, acceptability of an argument is determined as follows. Note that if an attack is symmetric, then an attack only succeeds in defeat if the attacker is more preferred than the argument being attacked; however, as in [Bench-Capon, 2002], if an attack is asymmetric, then an attack succeeds in defeat if the attacker is at least as preferred at the argument being attacked.

Definition 4.3. *Let* $\mathcal{R}^x$ *be an audience and let* $\langle Ar, att \rangle$ *be an iVAF.*

---

[17]It may seem counter intuitive that CQ10 generates attacks between arguments for the same action. Although such arguments do not dispute the action that should be performed, they do dispute the reasons as to why. Where there are two arguments proposed for the same action but each is based upon different values, an agent may only accept the argument based on one of the values. Hence such arguments are seen to be in conflict.

For $(a,b) \in att$ such that $(b,a) \notin att$, a **defeats** b under $\mathcal{R}^x$ iff $b \not\succ_x a$.
For $(a,b) \in att$ such that $(b,a) \in att$, a **defeats** b under $\mathcal{R}^x$ iff $a \succ_x b$.
An argument $a \in Ar$ is **acceptable** w.r.t S under $\mathcal{R}^x$ $(S \subseteq Ar)$ iff: for every $b \in Ar$ that defeats a under $\mathcal{R}^x$, there is some $A_k \in S$ that defeats b under $\mathcal{R}^x$.
A subset S of Ar is **conflict-free** under $\mathcal{R}^x$ iff: no argument $a \in S$ defeats another argument $b \in S$ under $\mathcal{R}^x$.
A subset S of Ar is **admissible** under $\mathcal{R}^x$ iff: S is conflict-free under $\mathcal{R}^x$ and every $a \in S$ is acceptable w.r.t S under $\mathcal{R}^x$.
A subset S of Ar is a **preferred extension** under $\mathcal{R}^x$ iff it is a maximal admissible set under $\mathcal{R}^x$.
An argument A is **acceptable** in the $iVAF$ $\langle Ar, att \rangle$ under audience $\mathcal{R}^x$ iff there is some preferred extension containing it.

The mechanism defined in this section allows an agent to determine attacks between arguments for and against actions. It can then use its preference ordering over the motivating values (i.e., its audience) to determine the acceptability of those arguments.

4.2 Dialogue system

The system we present here [Black and Atkinson, 2011a; Black and Atkinson, 2011b] assumes exactly two participating agents, each with its own identifier taken from the set $I = \{1, 2\}$. Each participant takes it in turn to make a move to the other participant. Participants are referred to using the variables x and y such that: x is 1 if and only if y is 2; x is 2 if and only if y is 1.

The format for moves used in deliberation dialogues is shown in Table 4, and the set of all moves meeting the format defined in Table 4 is denoted M. Also, Sender : $M \mapsto I$ is a function such that Sender($\langle Agent, Type, Content \rangle$) = $Agent$.

An *open* move $\langle x, \text{open}, \gamma \rangle$ opens a dialogue to agree on an action to achieve the goal γ; an *assert* move $\langle x, \text{assert}, a \rangle$ asserts an argument a for or against an action to achieve a goal that is the topic of the dialogue; an *agree* move $\langle x, \text{agree}, act \rangle$ indicates that x agrees to performing action act to achieve the topic; a *close* move $\langle x, \text{close}, \gamma \rangle$ indicates that x wishes to end the dialogue. All dialogues start with an open move. In

9 - Argumentation-based Dialogue

Move	Format
open	$\langle x, \text{open}, \gamma \rangle$
assert	$\langle x, \text{assert}, a \rangle$
agree	$\langle x, \text{agree}, act \rangle$
close	$\langle x, \text{close}, p \rangle$

Table 4: Format for moves used in deliberation dialogues: p is a goal; act is an action; a is an argument; $x \in \mathsf{I}$ is an agent identifier.

order to terminate a dialogue, either: two close moves must appear one immediately after the other in the sequence, in which case the dialogue is unsuccessful and no agreement is reached; or two moves agreeing to the same action must appear one immediately after the other in the sequence, in which case the dialogue is successful and the agents have found an action they can both agree on.

Definition 4.4. A **dialogue**, denoted d_t, is a sequence of moves $[m_1, \ldots, m_t]$ where the following conditions hold:
- m_1 is a move of the form $\langle x, \text{open}, \gamma \rangle$;
- for all $1 \leq s \leq t$, $\text{Sender}(m_s) \in \mathsf{I}$;
- for all $1 \leq s < t$, $\text{Sender}(m_s) \neq \text{Sender}(m_{s+1})$ for all $1 \leq s < t$;
- for all $1 < s \leq t$, if $m_{s-1} = \langle x, \text{close}, \gamma \rangle$ and $m_s = \langle y, \text{close}, \gamma \rangle$, then $s = t$ and d_t is **unsuccessfully terminated**;
- for all $1 < s \leq t$, if $m_{s-1} = \langle x, \text{agree}, act \rangle$ and $m_s = \langle y, \text{agree}, act \rangle$, then $s = t$ and d_t is **successfully terminated with outcome** act.

The **topic** of the dialogue d_t is returned by $\text{Topic}(d_t) = \gamma$.

In order to determine the actions it finds agreeable, an agent considers both the private knowledge it has about actions (their preconditions, effects and the values they promote or demote) and the arguments put forward by the other agent during the dialogue. This is represented in what is called the agent's *dialogue iVAF*, which is the iVAF constructed from the union of the arguments the agent can construct from its own knowledge about actions and the arguments that have been asserted by the other agent[18].

---

[18] While this system does not use explicit commitment stores, one could consider

Definition 4.5. *A dialogue iVAF for an agent x participating in a dialogue d_t is denoted* $\mathsf{dVAF}(x, d_t)$. *If $d_t = [m_1, \ldots, m_t]$, then $\mathsf{dVAF}(x, d_t)$ is the iVAF $\langle Ar, att \rangle$ where $Ar = Args^x_{\mathsf{Topic}(d_t)} \cup \{a \mid \exists m_k = \langle y, assert, a \rangle \ (1 \leq k \leq t)\}$.*

An action is *agreeable* to an agent x if and only if there is some argument *for* that action that is acceptable in x's dialogue iVAF under the audience that represents x's preference over values. Note that the set of actions that are agreeable to an agent may change over the course of the dialogue, as new arguments are asserted by the other agent.

Definition 4.6. *An action act is* **agreeable** *in the iVAF $\langle Ar, att \rangle$ under the audience $\mathcal{R}^x$ iff $\exists a = \langle act, \gamma, v, + \rangle \in Ar$ such that a is acceptable in $\langle Ar, att \rangle$ under $\mathcal{R}^x$. The set of* **all actions that are agreeable to agent x participating in a dialogue** d_t is $\mathsf{AgActs}(x, d_t) = \{a \mid a \text{ is agreeable in } \mathsf{dVAF}(x, d_t) \text{ under } \mathcal{R}^x\}$.

As already discussed, a protocol specifies the moves that an agent is permitted to make at any point in the dialogue. In this dialogue, it is permissible to assert an argument for an action to achieve the goal that is the topic of the dialogue as long as that argument has not previously been asserted in the dialogue[19]. An agent can agree to an action that has been agreed to by the other agent in the preceding move. An agent can also agree to an action that has been proposed by the other participant, unless the agent has previously agreed to that same action and has not asserted any further arguments since then. This is to avoid the situation where an agent keeps repeatedly agreeing to an action that the other agent will not agree to: if an agent makes a move agreeing to an action and the other agent does not wish to also agree to that action, then the first agent must introduce some new argument that may convince the second agent to agree before being able to repeat its agree move. Agents may always make a close move. Note, it is possible to check conformance with the protocol as it only refers to public elements of the dialogue.

Definition 4.7. *The* **deliberation protocol** *is a function* $\pi : \mathsf{D} \mapsto 2^\mathsf{M}$.

---

that an agent's dialogue iVAF is the iVAF constructed from the union of its own knowledge with the other agent's commitment store.

[19] A condition that is similar to those we have seen before.

Let d_t be a dialogue $(1 \leq t)$ such that $\mathsf{Sender}(m_t) = y$, $x' \in \mathsf{I}$, and $\mathsf{Topic}(d_t) = \gamma$.

$$\pi(d_t) = P^{\mathsf{ass}}(d_t) \cup P^{\mathsf{ag}}(d_t) \cup \{\langle x, \mathsf{close}, \gamma\rangle\}$$

such that

$m \in P^{\mathsf{ass}}(d_t)$ iff: $m = \langle x, \mathsf{assert}, a\rangle$, $\mathsf{Goal}(a) = \gamma$ and
$\neg \exists m_{t'} = \langle x', \mathsf{assert}, a\rangle$ where $1 < t' \leq t$

$m \in P^{\mathsf{ag}}(d_t)$ iff: $m = \langle x, \mathsf{agree}, act\rangle$ and either:
$m_t = \langle y, \mathsf{agree}, act\rangle$; otherwise
$\exists m_{t'} = \langle x', \mathsf{assert}, a\rangle$ where $\mathsf{Act}(a) = act$,
$1 < t' \leq t$ and $\forall t''$ such that $1 \leq t'' < t$,
if $\exists m_{t''} = \langle x, \mathsf{agree}, act\rangle$, then
$\exists m_{t'''} = \langle x, \mathsf{assert}, a\rangle$ where $t'' < t''' \leq t$.

Black and Atkinson [Black and Atkinson, 2011a] provide an ordered strategy that agents can use to determine which of the permissible moves (returned by the protocol) to make. According to this strategy: if it is permissible to make a move agreeing to an agreeable action, then make such an agree move; else, if it is permissible to assert an argument *for* an *agreeable* action, then assert some such argument; else, if it is permissible to assert an argument *against* an action that is *not agreeable*, then assert some such argument; else make a close move. When the strategy results in a choice of more than one agree or assert move, a Pick function is used to select a specific move; this is not specified but in its simplest form may return an arbitrary move from the input set. By specifying the Pick function to return exactly one move, we ensure that the strategy is decisive.

Definition 4.8. *The (ordered)* **strategy** *for an agent x is a function* $\mathcal{S}_x : \mathsf{D} \mapsto \mathsf{M}$ *as follows.*

$$\mathcal{S}_x(d_t) = \begin{cases} 1 : \mathsf{Pick}(\mathcal{S}_x^{\mathsf{ag}})(d_t) \\ 2 : \mathsf{Pick}(\mathcal{S}_x^{\mathsf{prop}})(d_t) \\ 3 : \mathsf{Pick}(\mathcal{S}_x^{\mathsf{att}})(d_t) \\ 4 : \langle x, \mathsf{close}, \mathsf{Topic}(d_t)\rangle \end{cases}$$

where the choices for the moves are given by the following subsidiary functions (where $\mathsf{Topic}(d_t) = \gamma$):

$$\begin{aligned}
S_x^{\mathsf{ag}}(d_t) &= \{\langle x, \mathsf{agree}, act\rangle \in P^{\mathsf{ag}}(d_t) \mid act \in \mathsf{AgActs}(x, d_t)\} \\
S_x^{\mathsf{prop}}(d_t) &= \{\langle x, \mathsf{assert}, a\rangle \in P^{\mathsf{ass}}(d_t) \mid a \in Args_\gamma^x, \mathsf{Act}(a) = act, \mathsf{Sign}(a) = +\text{ and} \\
&\qquad act \in \mathsf{AgActs}(x, d_t)\} \\
S_x^{\mathsf{att}}(d_t) &= \{\langle x, \mathsf{assert}, a\rangle \in P^{\mathsf{ass}}(d_t) \mid a \in Args_\gamma^x, \mathsf{Act}(a) = act, \mathsf{Sign}(a) = -, \\
&\qquad act \notin \mathsf{AgActs}(x, d_t) \text{ and} \\
&\qquad \exists m_{t'} = \langle x', \mathsf{assert}, a'\rangle \text{ such that } x' \in \mathsf{I}, \\
&\qquad 1 \leq t' \leq t, \mathsf{Act}(a') = act \text{ and } \mathsf{Sign}(a') = +\}
\end{aligned}$$

The following section gives an example dialogue generated by two agents each using the strategy defined here.

4.3 Example dialogue

This example (first presented in [Black and Atkinson, 2011a]) involves two agents who share the goal to have dinner together (denoted din) and are considering two possible actions that will achieve this goal: go to an Italian restaurant (denoted it); go to a Chinese restaurant (denoted ch). The relevant values are: d, distance to travel; $e1$, agent 1's enjoyment; $e2$, agent 2's enjoyment; and c, cost. The agents' audiences (i.e., their preferences over the values) are:

$$d \succ_1 e1 \succ_1 c \succ_1 e2$$
$$c \succ_2 e2 \succ_2 e1 \succ_2 d$$

Agent 1 starts the dialogue.

$$m_1 = \langle 1, \mathsf{open}, din\rangle$$

At this opening stage of the dialogue, the agents have only their private knowledge about actions, their preconditions, effects and the values they promote or demote. This knowledge is shown in Figures 1 and 2, where the nodes represent arguments and are labelled with the action that they are for (or the negation of the action that they are against) and the value that they are motivated by. If a node is labelled with the negation of an action, this denotes that performing that action will demote the value that labels the node; if a node is labelled with an action, this denotes that performing that action will promote the value that labels the node. The arcs represent the attack relation between arguments, and a double circle round a node means that the argument it represents is acceptable to that agent under its audience. For example, we can see from Figure 1 that agent 1 has two arguments it finds acceptable: $\langle \neg ch, din, e1, -\rangle$ says we should not go for a Chinese as this demotes value $e1$; $\langle it, din, d, +\rangle$ says we should go for an Italian as this promotes value d.

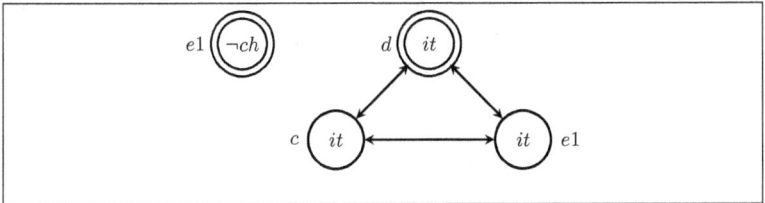

Figure 1: Agent 1's dialogue iVAF at t = 1, dVAF$(1, d_1)$.

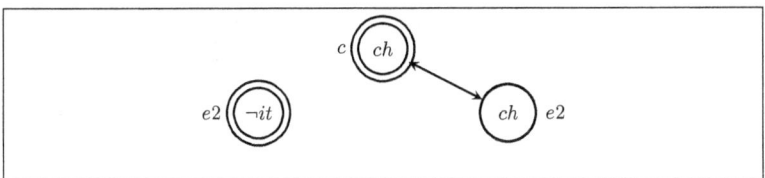

Figure 2: Agent 2's dialogue iVAF at t = 1, dVAF$(2, d_1)$.

It is now agent 2's turn to make a move. Recall that, according to the strategy, if the protocol permits the agent to agree to an action that it finds acceptable, then it will do so. However, the agent is not permitted (according to the protocol) to make an agree move at this stage, since the previous move was not an agree move and no assert moves have yet been made. If no agree moves can be made, the strategy states that (if the protocol allows it and some such argument exists) the agent should make a proposing assert move and assert a positive argument for an action that it finds agreeable. At this point in the dialogue, there is only one argument *for* an action that is acceptable to agent 2 ($\langle ch, din, c, +\rangle$, see Figure 2), hence ch is the only action that is agreeable to agent 2. Agent 2 must therefore, if it can, assert an argument for going to the Chinese restaurant. There are two such arguments that the Pick function could select: $\langle ch, din, c, +\rangle$, $\langle ch, din, e2, +\rangle$. Let us assume that $\langle ch, din, c, +\rangle$ is selected.

$$m_2 = \langle 2, \mathsf{assert}, \langle ch, din, c, +\rangle\rangle$$

This new argument is added to agent 1's dialogue iVAF, to give dVAF$(1, d_2)$ (Figure 3). Note that there is no change to the arguments that agent 1 finds acceptable, since it does not prefer the value that motivates this new argument to those that motivate the arguments that were already present in its dialogue iVAF.

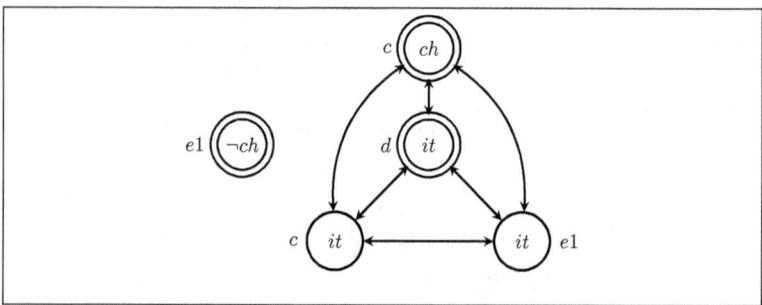

Figure 3: Agent 1's dialogue iVAF at t = 2, dVAF$(1, d_2)$.

Although agent 2 has proposed going to the Chinese restaurant, this action is not agreeable to agent 1 at this point in the dialogue (as there is no argument for this action that is acceptable in Figure 3). There is, however, an argument for the action it ($\langle it, din, d, +\rangle$) that is acceptable in agent 1's dialogue iVAF (Figure 3), and so going to the Italian restaurant is agreeable to agent 1. Hence, agent 1 must make an assert move proposing an argument for the action it, and there are three such arguments that the Pick function can select from: $\langle it, din, d, +\rangle$, $\langle it, din, c, +\rangle$, $\langle it, din, e1, +\rangle$. Let us assume that $\langle it, din, c, +\rangle$ is selected.

$$m_3 = \langle 1, \mathsf{assert}, \langle it, din, c, +\rangle\rangle$$

This new argument is added to 2's dialogue iVAF, to give dVAF$(2, d_3)$ (Figure 4).

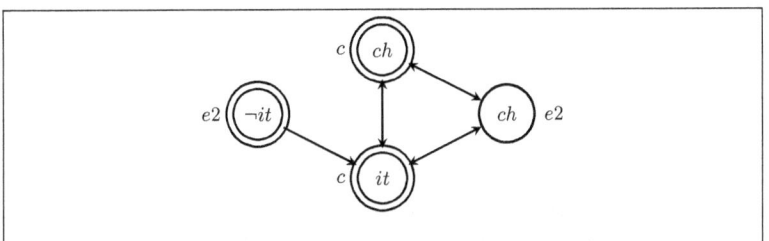

Figure 4: Agent 2's dialogue iVAF at t = 3, dVAF$(2, d_3)$.

Going to the Italian restaurant is now agreeable to agent 2 since the new argument introduced promotes the value ranked most highly for agent 2, i.e. cost, and so this argument is acceptable. So, agent 2 agrees to this action.

$$m_4 = \langle 2, \text{agree}, it \rangle$$

Going to the Italian restaurant is also agreeable to agent 1 (as the argument $\langle it, din, d, + \rangle$ is acceptable in its dialogue iVAF, which is still the same as that shown in Fig. 3 as agent 2 has not asserted any new arguments), hence agent 1 also agrees to this action.

$$m_5 = \langle 1, \text{agree}, it \rangle$$

The dialogue has thus terminated successfully and the agents are each happy to agree to go to the Italian restaurant. Note, however, that this action is agreeable to each agent for a different reason. Agent 1 is happy to go to the Italian restaurant as it promotes the value of distance to travel (the Italian restaurant is close by), whereas agent 2 is happy to go to the Italian restaurant as it will promote the value of cost (as it is a cheap restaurant). The agents need not be aware of one another's audience in order to come to an agreement that they are both happy with[20].

---

[20] Note that the use of values — and the agents' differing preferences over values — adds an extra dimension to the dialogue in comparison with the example given in Section 2, where it is assumed that the agents agree on the defeat relation between arguments. In this case, the agents could also consider a dialogue where they try to reach agreement on a particular audience over the values.

4.4 Evaluation of the dialogue system

Black and Atkinson [2011a] show through theoretical analysis that in their dialogue system:

- all dialogues terminate;
- if a dialogue terminates successfully, then the outcome is agreeable to both participants;
- if a dialogue terminates and there is some action that is agreeable to both participants, then the dialogue will have a successful outcome.

However, they also show the following undesirable property.

- In the case where, if we take the union of the agents' knowledge about actions, there is some action that is agreeable to both of the agents, it is possible that the dialogue will not reach an agreement.

Hence, even if the arguments exist that will enable the agents to reach an agreement, the particular arguments selected for assertion by the Pick function may not allow agreement to be reached. Black and Atkinson [2011b] later explore how an agent may use a model of what it believes to be the other participating agent's preferences over values as a parameter of the Pick function, and how this can lead to better dialogue outcomes.

The behaviour of this dialogue system has also been explored experimentally. [Black and Bentley, 2012] compares the performance of the dialogue system with a simple consensus forming approach, where the agents do not share any knowledge and only agree if there is an action that each finds agreeable given their private knowledge about actions. In this work, random example scenarios (that initialise the agents' private knowledge) are generated, varying the number of arguments, actions and values available to the agents, and the outcome of the dialogue is compared with the outcome of consensus forming. The results show that:

- dialogues are significantly more likely to be successful than consensus forming;

- successful dialogues are more likely with higher numbers of actions and values;

- dialogues produce better quality outcomes than consensus forming (the quality of an outcome is determined by examining whether the action agreed to would be agreeable to one, both or neither of the agents given the union of the knowledge available to each); and

- dialogue length grows exponentially with the number of arguments available.

The first of these results fits with the prediction in [Loui and Moore, 1997] and the results in [Rahwan et al., 2009] — dialogues can expand the space of possible agreements by making participants aware of tradeoffs.

5 Key challenges

In this section, we discuss what we believe to be some of the key challenges that must be addressed if we are to realise the full potential of argumentation dialogue systems.

5.1 Multiparty dialogues

Relaxing the assumption that there are only two agents in the dialogue poses several challenges. [Dignum and Vreeswijk, 2004] listed a number of issues which are still relevant: the system may be open or closed, the variety of roles that agents may take is much larger, the dialogue may be mediated or not, the turn-taking policy becomes more difficult to define, and termination conditions also can be defined in different ways (simply think of what consensus means when more than two agents are involved). There is thus much less work addressing multiparty dialogues in the literature, the exception being negotiation, where it is not uncommon to have settings involving a large number of agents.

In persuasion this is less common. In what follows, we describe a proposal due to [Bonzon and Maudet, 2011]. We assume that agents share the same set of arguments but may have different opinions regarding attack relations among them (as in Section 4, this may result from an underlying value-based argumentation systems, with agents holding

---

1. Agents report their individual view on the issue to the central authority, which then assigns (privately) each agent to PRO or CON.

2. The first round starts with the issue on the gameboard and the turn given to CON.

3. Until a group of agents cannot move:

    (a) agents independently propose moves to the central authority;

    (b) the central authority picks the first (or at random) relevant move from the group of agents whose turn is active, update the gameboard, and passes the turn to the other group

---

Figure 5: A multiparty persuasion protocol

different preference ordering over values). Agents may make claims regarding attacks among arguments.

The protocol is *mediated*, focused on a single *issue*, based on *roles* that agents endorse at the beginning of the dialogue (depending on their stance regarding the issue), and it builds upon the relevance-based protocol idea put forward in [Prakken, 2005]. It is described in Figure 5.

We see that the turn-taking issue is solved here by letting agents of two different "groups" alternate, with moves being picked by a mediator. Termination occurs when a group is left without any legal move to make. Example 5.1 illustrates, for a given initial situation, the different possible executions of the protocol. It is interesting to observe that the outcome may be different depending on the sequence of moves, and that it may differ from the outcome which we would be obtained from merging the agents' argumentation systems (recall that a similar issue occurred in Section 4).

Example 5.1. *Let three agents with their argumentation systems, and the following merged argumentation framework:*

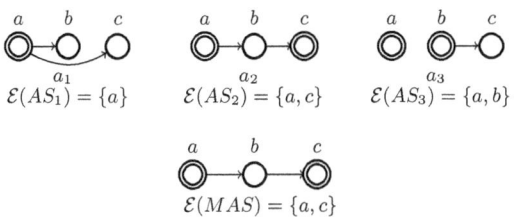

The issue of the dialogue is the argument c. We have $CON = \{a_1, a_3\}$, $PRO = \{a_2\}$. We store in RP_x^t the moves played by agent x until turn t. At the beginning, we have $RP_1^0 = RP_2^0 = RP_3^0 = \emptyset$, $AS^0(GB) = \langle \{c\}, \emptyset \rangle$ and $\mathcal{E}(AS^0(GB)) = \{c\}$.

All the possible sequences of moves allowed by the protocol are represented on tree depicted in Figure 6. As it can be seen on Figure 6, any sequence of the protocol stops with a stable gameboard where $\mathcal{E}(AS(GB)) = \{a, b\}$. Note also that some sequences take more moves than others.

5.2 Dialogue strategies

As discussed earlier in this chapter, a dialogue protocol identifies the space of permissible dialogues. At any point in a dialogue, the protocol will typically identify multiple permissible moves that the agent may make. The choice of which permissible move to make is determined by the agent's *strategy*. In order to improve the likelihood of achieving their dialogue goal, an agent needs to be able to select an effective strategy to apply. Note this section is not intended as an exhaustive review of dialogue strategies, but rather intends to introduce the reader to some of the different approaches that have been taken in the literature and to highlight what we consider to be some of the key challenges for the area. For a different perspective on this, see [Thimm, 2014].

Identifying effective strategies is a challenging problem[21], as an agent must consider not only its own moves but also the possible responses of the other agent(s) participating in the dialogue, often with only limited

---

[21]Chapter 10 of this handbook [Governatori et al., 2021] discusses in depth the complexity of *strategic argumentation problems*, a sub-class of persuasion dialogues where the agents are only able to assert arguments to one another.

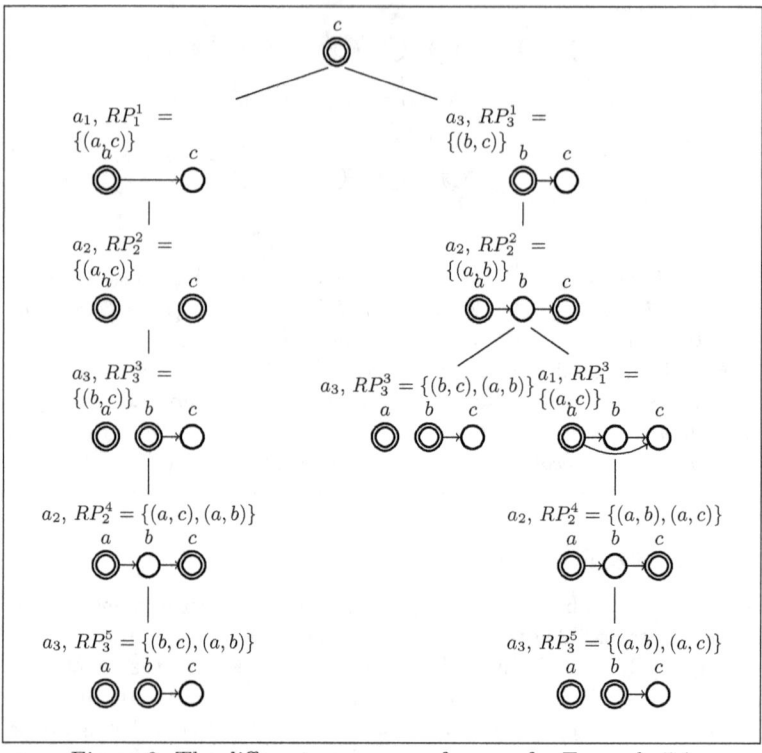

Figure 6: The different sequences of moves for Example 5.1

or uncertain knowledge of the other agents. Further, in adversarial dialogues, one must take care not to divulge information that the other participant(s) may use to gain an advantage. Consider for example the following dialogue, where a child (C) is trying to persuade their parent (P) that they have done their homework (which requires internet access).

P: Have you finished all your homework?

C: Yes, I finished it this afternoon.

P: Are you sure? I thought I heard you chatting online with your friend all afternoon.

C: I wasn't chatting online! The internet hasn't been working all afternoon.

P: If the internet hasn't been working then you can't have finished your homework!

By divulging that the internet has not been working, the child provides new information to their parent, which their parent then uses to construct an argument that counters their child's claim.

Although challenging, identification of effective strategies is crucial if we are to realise the potential of argument dialogue systems, and there has been significant progress towards this aim in recent years. We can categorise approaches to dialogue strategies into two classes: those where the strategising agent uses a model of some aspect of its interlocutor's (or interlocutors') private state to determine a strategy that is optimised for that particular interlocutor, and those that do not use any such model. We will refer to these two classes as *model-based* and *model-free* respectively.

5.2.1 Model-free approaches

Model-free approaches define a strategy — usually for a particular type of dialogue — that can be applied directly by the strategising agent. Typically, these are *ordered strategies* (to use the terminology from Section 2). For example, the dialogue system [Black and Atkinson, 2011a; Black and Atkinson, 2011b] presented in Section 4 provides a strategy for deliberation dialogues as follows.

- If it is permissible to **agree** to an action that the agent finds to be agreeable, then make some such **agree** move; else

- if it is permissible to **assert** a positive argument *for* an action that the agent finds agreeable, then **assert** some such argument; else

- if it is permissible to **assert** a negative argument *against* an action that the agent does not find agreeable, then **assert** some such argument; else

- make a **pass** move.

Model-free strategies can be straightforward to apply and do not require any knowledge of the other agents participating in the dialogue. While they are not specific to a particular interlocutor, they can be made context-dependent so that priorities may vary depending on the current circumstances of the dialogue, as for instance the topic under discussion (see [Kakas et al., 2005] for a proposal along these lines). A disadvantage of this approach is that it is difficult to know how well the strategy will perform, as the complexity of argument dialogue systems makes it hard to theoretically analyse their general behaviour. Table 5 gives some examples of model-free strategies, noting the type of dialogue for which the strategy is intended, the principle embodied by the strategy, and key properties of the strategy.

5.2.2 Model-based approaches

In contrast to model-free approaches, model-based approaches require a strategy to be computed, taking into account the beliefs the strategiser has about the other specific dialogue participant(s). These beliefs are often referred to as an *opponent model*, and we will also use this term. Model-based approaches typically involve mapping the problem of finding a strategy that will be effective against one's opponent(s) to an optimisation problem, so that existing techniques can be exploited to identify such a strategy. Techniques employed include automated planning [Black et al., 2014; Black et al., 2017], decision trees [Hadoux and Hunter, 2017], mixed observability Markov decision processes [Hadoux et al., 2015], evolutionary search [Murphy et al., 2018], partially observable Markov decision processes [Rosenfeld and Kraus, 2016b], game theory [Kacpraz et al., 2014], and Markov games [Rach et al., 2018]. To date, these optimising model-based approaches have primarily focused on a restricted class of persuasion dialogues (hence the term *opponent* model) that only involve the exchange of arguments through assertions (in contrast to dialogues that allow a range of moves such as *accept*, *challenge*, *question*, etc.); this speaks to the challenge of efficiently mapping more flexible dialogue systems to a tractable optimisation problem.

Citation	Type of dialogue	Strategic principle	Properties
[Amgoud and Parsons, 2001]	Persuasion	Ask questions where possible	Proven that if both agents use this strategy then dialogues won't terminate, but if only one does then termination can be achieved.
[Atkinson et al., 2013]	Persuasion over preferences	Prioritise inquiring about attributes that are (subjectively) more important.	Proven to produce dialogues that terminate in fewest number of steps.
[Black and Hunter, 2009]	Inquiry	Exhaustively assert all relevant knowledge.	Proven to be sound and complete in relation to reasoning with union of agents' knowledge.
[Black and Atkinson, 2011a; Black and Bentley, 2012]	Deliberation	Agree if possible. Prioritise positive arguments for an action to negative arguments against an action.	If at termination there is some action agreeable to each agent, successful outcome guaranteed. Experimental evaluation shows benefits over consensus forming with no share of information.
[Kontarinis et al., 2014]	Persuasion	Computation of target sets	Prioritises moves which will affect the issue of the dialogue.
[Medellin-Gasque et al., 2013]	Joint planning	Specifies the order in which the elements of the plan are questioned.	Experimental evaluation shows dialogues produced shorter than those produced by random strategy.
[Murphy et al., 2016]	Uni-directional persuasion	Prioritise supporting arguments that are closer to the topic (as defined by the attack relation).	Experimental evaluation shows success more likely than with random strategy.
[Oren et al., 2006]	Persuasion	Prioritise arguments that reveal less information.	Maintains focus of dialogue. Reasoning is not complete.
[Parsons et al., 2003]	Information-seeking, inquiry and persuasion	Agree if possible. This is not stated explicitly, but is implicit in the protocol.	Proven that dialogues terminate.

Table 5: Examples of model-free strategies from the literature

Representing and updating an opponent model. Model-based approaches make different assumptions about the knowledge available to the strategiser in its opponent model. For example: [Black et al., 2017] assumes that the strategising agent has an uncertain model of the knowledge available to its opponent, captured as a probability distribution over sets of knowledge, but no information about the strategy the opponent employs (seeking strategies that will perform well no matter which moves the opponent plays); [Dimopoulos et al., 2019] builds on their framework of control argumentation systems to distinguish those arguments that are certainly known by the opponent and those which the strategiser does not know whether the opponent knows; [Hadoux and Hunter, 2017] requires a prediction to be made about the decision rule used by the opponent to determine its strategy, and explore how errors in identifying this decision rule affect performance of the system; [Rosenfeld and Kraus, 2016b] and [Hadoux et al., 2015] each assumes the strategising agent has a stochastic model of the opponent's expected dialogue behaviour; while game theoretic approaches such as that of [Kacpraz et al., 2014] require knowledge of the opponent's preferences over dialogue outcomes and assume that the opponent plays optimally to maximise those preferences.

Of course, the effectiveness of model-based strategies depends on how accurate the underlying model's representation of the opponent is. However, the important question of how to develop an accurate opponent model is, as yet, somewhat under-explored in the computational argumentation community. Some works consider how one can develop an opponent model from historical data. For example, [Hadjinikolis et al., 2013] explores how an agent can use its knowledge of past dialogue interactions to predict what is likely to be believed by a new opponent, while [Rosenfeld and Kraus, 2016b] and [Hunter and Polberg, 2017] each show how machine learning techniques can be applied to argumentative data collected from humans in order to make predictions about an opponent ([Rosenfeld and Kraus, 2016b] seeks to predict a stochastic model of an opponent's dialogue behaviour; [Hunter and Polberg, 2017] aims to predict the believability, convincingness and appeal of specific arguments to an opponent). Other works explore how one can update an opponent model either during or after a dialogue with that opponent, based on their dialogue behaviour. For example, [Rach et al., 2019] shows how emotion recognition can be used to determine a model of a human opponent's preferences over arguments, by examining their emotional responses to presented arguments, and [Hunter et al., 2018] explores how an epistemic graph (which uses probabilities to express the degree to which an argument is believed or disbelieved by an agent) can be updated based on the opponent's dialogue behaviour (see Chapter 7 of this handbook [Hunter et al., 2021] for more details).

Rienstra et al. [2013] and Black and Hunter [2016] each look at how to update an uncertain opponent model in the case the opponent makes a move that is inconsistent with some of the possibilities represented by the model. In these works, the opponent model can — in its simplest form — be represented as a set $OM = \{S_1, \ldots, S_n\} \subset 2^{Args}$, where it is believed that the arguments available to the opponent are those in some $S_i \in OM$. If the opponent asserts an argument a that is not part of one of these possible sets, Black and Hunter [2016] — who assume an accurate opponent model in the sense that the opponent's actual arguments are represented by one of the possible sets S_i — remove from the model each S_i that does not contain a. In contrast, Rienstra et al. [2013] add the newly asserted argument to every possible set from the model (and so $S_i := S_i \cup \{a\}$ for each i).

Scalability of model-based approaches. Another key challenge for model-based dialogue strategies is scalability. These approaches typically assume an uncertain model of the opponent. Thus, to evaluate a potential strategy, one may need to consider all the different sets of arguments that could be available to the opponent, and all the possible ways the opponent may behave during the dialogue given each set of arguments that may be available to them. In the worst case, where one has complete uncertainty over which arguments are available to the opponent, there are 2^n — where n is the number of arguments in the problem domain — sets of arguments that may be available to the opponent. For each of these sets, the number of ways the opponent may behave in the dialogue grows factorially with the number of arguments (since, in the general case, agents can choose to assert any number of those arguments, in any order). Similarly, the number of potential strategies to consider for the strategising agent grows factorially with the number of arguments in the problem domain. This scalability challenge is compounded if one is considering multiple opponents. It can be possible to exploit the structured nature of argumentation in order to prune the problem instance and improve scalability. For example, in persuasion dialogues one may be able to remove dominated arguments (arguments that are known to be attacked by an argument that is itself not attacked by anything) from the problem [Hadoux et al., 2015].

Another approach to improving scalability can be to restrict the potential strategies for consideration. The automated planning approach of Black et al. [2017] considers only what they refer to as *simple* strategies (a sequence of moves to be made by the agent regardless of how the opponent behaves in the dialogue, in contrast to a policy where the move to be made by the strategising agent depends on the opponent's moves) returning a simple strategy that maximises — according to the opponent model — the probability of guaranteed success no matter which strategy the opponent employs. One may also consider approaches that, rather than guaranteeing a strategy that is optimal given the opponent model (such as [Hadoux et al., 2015; Hadoux and Hunter, 2017][22]) aim to return a near optimal strategy. For example, Murphy et al. [2018] show how the problem of determining a strategy for one-to-many persuasion can be encoded using techniques from search-based model engineering, such that evolutionary search can be applied to find a near optimal strategy. This also allows for optimisation of multiple objectives, such as maximising the number of opponents who are persuaded while also minimising the number of arguments shared.

The structure of the underlying argumentation framework — representing the domain knowledge — can also have a significant effect on the time taken to find an effective strategy. Black et al. [2017] consider different structures of frameworks, representing the arguments potentially available to a persuader and its opponent. They show that it is significantly faster to find strategies for bipartite graphs, where the persuader does not need to worry about undermining its own arguments, than it is for graphs that contain cycles, where whether asserting a par-

---

[22][Black et al., 2017] guarantee an optimal *simple* strategy, but such strategies may be outperformed by an optimal *policy*.

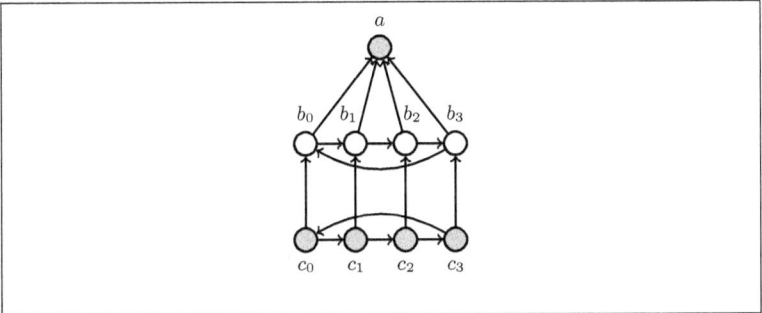

Figure 7: Example of a problem that is particularly challenging for a persuader

ticular argument is helpful or harmful for the persuader depends on the arguments available to the opponent. Consider for example the argumentation framework shown in Figure 7 [Black et al., 2017], where the greyed out nodes are the arguments available to the persuader, the white nodes are the arguments that may be known by the opponent (and so we are assuming that the arguments known to the opponent are some element of $2^{\{b_0,b_1,b_2,b_3\}}$), and the edges represent attacks between arguments. The persuader aims to convince its opponent of the acceptability of argument a. In this case, asserting the argument c_0 could be beneficial to the proponent if the opponent knows the argument b_0 (as c_0 attacks b_0) but if the opponent knows b_1 then asserting c_0 could be detrimental to the proponent's goal (since c_0 attacks c_1, which is the only argument available that attacks b_1). The proponent must therefore take care to consider all the possibilities captured by its opponent model in order to determine whether an argument is likely to be detrimental to achieving its goal.

5.2.3 Benchmarks for evaluating dialogue strategies

The discussion above about about how the structure of the underlying argumentation framework can affect the difficulty of finding effective strategies highlights another challenge for argument dialogues: the lack of benchmark problems for evaluation. The International Competition on Computational Models of Argumentation (ICCMA) provides benchmark problems for the evaluation of abstract argument solvers (see for example [Gaggl et al., 2020]). However, while there are a standard set of decision problems addressed by argument solvers (for example: is a particular argument credulously accepted under the preferred semantics; or return the grounded extension) identifying meaningful benchmarks for argument dialogue systems is more challenging because: these systems are designed with different goals in mind (and even dialogue systems for the same type of dialogue typically employ different protocols); there is no standardised way of representing dialogue systems (see Section 3); there is no agreement on the underlying argumentation formalism used for reasoning; and different assumptions are made about the knowledge available to the strategising agent.

5.3 Using enthymemes in dialogues

Enthymemes are incomplete arguments, where some of the premises and/or the claim of the intended complete argument are omitted. Arguments presented by humans are normally enthymemes [Walton, 1989], and so if we want to support human-agent communication we need to be able to handle enthymemes in dialogues.[23] Humans normally find it easy to understand the intended meaning of an enthymeme. Consider the following example (adapted from [Sperber and Wilson, 1986]):

A: Would you like a coffee?

B: Coffee will keep me awake.

Here, B presents an enthymeme that is missing its claim (since it does not explicitly answer A's question) and also some premises. If A knows that B wants to go to sleep (perhaps B is in their pyjamas ready for bed) then A can infer that B's intended argument was "Coffee will keep me awake, I want to go to sleep, so I would not like a coffee". If, however, A knows that B needs to stay awake (because they have a paper deadline to meet) then A can infer that B's intended argument was "Coffee will keep me awake, I need to stay awake, so I would like a coffee". This illustrates

---

[23]Even if only considering agent-agent communication, we may want to handle enthymemes in order to improve efficiency of the dialogue.

how common knowledge between the participants (specifically, whether B wants to go to sleep or needs to stay awake) can be used to reconstruct the intended complete argument from an enthymeme.

The context can also be important for correctly understanding enthymemes. Consider the following example (also adapted from [Sperber and Wilson, 1986]):

A: Sarita bought The Times.

It is common knowledge[24] that The Times can refer to a copy of the British newspaper The Times, or to the company that publishes The Times, and so there are two possible intended arguments to consider: "Sarita bought The Times, The Times refers to a copy of The Times newspaper, so Sarita bought a copy of The Times newspaper" and "Sarita bought The Times, The Times refers to the company that publishes The Times newspaper, so Sarita bought the company that publishes The Times newspaper". If the statement "Sarita bought the Times" was made during a conversation about how successful a businesswoman Sarita is, one can infer that the most relevant intended argument is the one that concludes that Sarita bought the company that published the The Times. If it was made in response to the statement "I must get a newspaper today", one can infer that the most relevant intended argument is the one claiming that Sarita bought a copy of the newspaper The Times.

Despite their ubiquity in human communication, there are few formal proposals of how enthymemes can be used in computational argumentation dialogues. Key questions to be addressed include:

- How can an agent determine an appropriate enthymeme to put forward, so that the receiver of the enthymeme can accurately reconstruct the intended complete argument?

---

[24]For members of some groups — in New York and much of the rest of the United States most people would take "The Times" to refer to the New York Times, and would call the similarly named British newspaper "The London Times", even though that is not its name. For those people, the reference in the example changes to the New York Times, but is otherwise unchanged since the company that publishes it is also known as "The Times".

- How can an agent receiving an enthymeme accurately identify the intended complete argument?

We have seen in the examples above that common knowledge is important for understanding enthymemes. In [Hunter, 2007] a proposal is made for how the common knowledge can be used both to determine an appropriate enthymeme to send and to identify the intended complete argument. This work, which considers only enthymemes with missing premises (and so cannot handle enthymemes with missing claims), assumes that both the proponent and the receiver of an enthymeme have a model of what they believe is common knowledge between the two. This model assigns a value between 0 and 1 to each element of the domain language, representing the degree to which the agent believes the element can be used as common knowledge, and so anything with a value over a particular threshold can be used as common knowledge. Determining an enthymeme to present is straightforward, one simply removes from the premises of the intended argument anything that one believes can be used as common knowledge. Reconstructing the intended argument involves adding to the enthymeme elements from what is perceived to be the common knowledge in order to build up a complete argument. However, it is not guaranteed that the reconstruction will result in the correct intended argument: the agents' perceptions of the common knowledge may differ, or there may be multiple possible intended arguments to select from.

In the "The Times" example above, where there are multiple possible intended arguments that can be reconstructed from the enthymeme put forward, we see how contextual information about the relevance of arguments can be useful when selecting the appropriate intended argument. Black and Hunter [2012b] extend the proposal discussed above to also support enthymemes with missing claims, taking into account not only the common knowledge but also the relevance of arguments. In addition to modelling what each agent believes can be used as common knowledge, a public agenda is assumed for each agent, which represents a ranking of the agent's information requirements (i.e., propositions such that the agent would like to know if there are reasons to believe they do or do not hold). By harnessing some of the principles of relevance theory [Sperber and Wilson, 2004] (essentially that "relevance of an utterance depends on maximising cognitive effect while minimising cognitive effort" [Black and Hunter, 2012b, p. 56]) Black and Hunter show how one can use both the common knowledge and the agenda to determine enthymemes that are relevant to their recipient, and to overcome some of the ambiguities that arise in selecting the most appropriate intended argument when receiving an enthymeme. Argumentation schemes [Walton and Krabbe, 1996; Macagno et al., 2018] may also be helpful in providing more contextual information that can be used to help determine the appropriate intended argument [Walton and Reed, 2005; Panisson and Bordini, 2017].

Although not the focus of Black and Hunter's work [2012b], they briefly discuss how the information requirements might be obtained. A straightforward case is where an agent asks a specific question, and so explicitly states an information requirement (as in the coffee example above). One might also assume that an agent will typically always be concerned with aspects such as its own welfare and safety, and so will have information requirements to reflect this. It may also be possible to derive information requirements from the context of the dialogue: if an agent is discussing Sarita's success as a businesswoman, one might infer information requirements relating to Sarita's business achievements.

There is also the question of how an agent can develop a view of what can be used as common knowledge. Again, there are some seemingly straightforward cases. If an agent has uttered a proposition, or been present in a dialogue where the proposition was uttered, one may consider that the proposition can be regarded as common knowledge with that agent. However, what if the agent uttered the proposition last week, or last year; can we assume it still remembers, or believes in, the proposition? [Hosseini et al., 2014] considers how an agent x might develop a model of the common knowledge with an agent y taking into account both the direct and indirect information gained about y's knowledge (for example, if y uses some proposition α in a dialogue with x, or some other agent z informs x that y knows α), as well as considering whether there are things that one would typically expect y to know (for example, if y works at a university, one might assume they are familiar with the regulations of that university). [Black and Hunter, 2008] also looks at how an agent x might update its view of the common knowledge with agent y based on their dialogue behaviour. Neither of these works considers the temporal persistence of common knowledge.

So far, our discussion of enthymemes has focused on how an agent can determine an appropriate enthymeme to send such that the receiver is likely to be able to accurately reconstruct the intended argument, and how an agent who receives an enthymeme can identify the argument that was intended; we have not yet said anything about how these enthymemes might actually be used in a dialogue system. The use of enthymemes in dialogues implies the need for locutions that allow agents to recover from misunderstandings: while we hope our processes – for identifying an enthymeme to send and for reconstructing the indented argument – will perform well, misunderstandings will inevitably occur (for example, if there are errors in the common knowledge, or the agents have different perceptions of the relevance of an argument).

There are very few proposals for dialogue systems that support the use of enthymemes. Black and Hunter [2008] present a system for inquiry dialogues that handles enthymemes with missing premises, which includes a **quiz** locution that agents can use to ask for clarification when they cannot reconstruct any complete arguments from a received enthymeme. [Dupin de Saint-Cyr, 2011] and [Xydis et al., 2020] each present a dialogue system that supports enthymemes which may miss premises or their claim, and show that different locutions are needed to handle the misunderstandings that can occur from these different cases. The use of enthymemes in dialogue raises the question of whether, and how, they can be used to give a strategic advantage. When an enthymeme e is moved against an argument a, by omitting some elements of e's intended argument one makes it harder for the receiver of e to identify counter arguments, and to identify whether e is indeed a counter argument for a. Consider the following example from Schopenhauer [1831] (also discussed in [Dupin de Saint-Cyr, 2011]):

> I asserted that the English were supreme in drama. My opponent attempted to give an instance to the contrary, and replied that it was a well-known fact that in music, and consequently in opera, they could do nothing at all. I repelled the attack by reminding him that music was not included in dramatic art, which covered tragedy and comedy alone.

This he knew very well. What he had done was to try to generalise my proposition, so that it would apply to all theatrical representations, and, consequently, to opera and then to music, in order to make certain of defeating me.

Schopenhauer's interpretation here is that his opponent presents an enthymeme in the full knowledge that the enthymeme's intended argument does not in fact counter Schopenhauer's claim, hoping this will go unnoticed by Schopenhauer. As far as we are aware, there are no existing works that explore how enthymemes can be used to give such a strategic advantage in computational argumentation-based dialogues.

6 Conclusions

This chapter has given an introduction to work on argumentation-based dialogue, focusing on what we believe are the key aspects that have been studied so far. We started with an overview of the basic elements of argumentation-based dialogues, the speech acts from which dialogues are formed, the protocols that govern them, the strategies that can be employed within them, and ways that dialogues can be evaluated. We illustrated these ideas with a simple Hamblin-style dialogue game. This was followed by a discussion of approaches to representing dialogues, something that is necessary if we are going to be able to compare different approaches formally, and if we are going to be able to examine the properties of dialogues at a suitably abstract level. We then gave a detailed example of a dialogue system — the value-based deliberation system of [Black and Atkinson, 2011a; Black and Atkinson, 2011b] — before discussing what we believe are the major current challenges in argumentation-based dialogue. These challenges are: multiparty dialogues, dialogue strategy, and handling enthymemes.

There are also a number of other topics that we think are important for achieving the full potential of argumentation-based dialogue systems, but which we have not had the space to discuss here in detail. One thing we have not considered here is where an agent gets its arguments from. For some applications, it may be feasible for a domain expert to manually provide the knowledge that agents can use to construct arguments. However, this is not always practical and in some cases we may need agents that are capable of discovering arguments for themselves, for example using *argument mining* techniques to identify arguments from text on the web (see, for example, [Budzynska and Villata, 2018; Lawrence and Reed, 2020]). We briefly discussed earlier in this chapter the need for benchmark problems for evaluation of argumentation-based dialogue systems, and argument mining techniques may also help us to identify real-world examples for this. Also on the subject of evaluation, if we aim for dialogue systems that can support human-agent interaction then we need to perform experiments with humans in order to ensure, for example, that our formal models of reasoning align with human behaviour (see Chapter 13 of this handbook [Cerutti *et al.*, 2021] for a survey of work in this area) and that the dialogue systems we develop do indeed benefit human users in the ways we envisage. We need also to consider the broader ethical issues associated with developing argumentation-based dialogue systems that can, potentially, affect human users' beliefs or decision making, for example: is it ethical for an agent to withhold some relevant arguments from a human — or even to lie[25] — in order to reach what might be viewed as a better dialogue outcome; is it responsible to develop agents that are capable of persuading a human to change their behaviour; are the dialogue systems we develop likely to disadvantage some subgroups of the population?

As artificial intelligence systems become increasingly ubiquitous, we believe that argumentation-based dialogues have great potential for allowing such systems to engage with each other and with human users in joint reasoning and decision making. An example of the former is [Kodeswaran *et al.*, 2010], and the latter is [Sklar and Azhar, 2015]. In both cases a key advantage is the ease with which such systems can be understood, indeed in [Sklar and Azhar, 2015; Azhar and Sklar, 2017] the system explicitly uses argumentation to explain their reasoning to human users. This ability to support explainable decisions is crucial if we are to be able to trust that artificial intelligence systems are acting in our best interests.

---

[25]The area of mechanism design can support the design of dialogue systems in which there is no incentive to lie, see for example [Rahwan and Larson, 2009].

Acknowledgments

This work was partially supported by EPSRC grants EP/P010105/1, EP/R033722/1, and EP/M01892X/1. The opinions expressed in this chapter are those of the authors and do not necessarily reflect the opinions of the funders.

References

[Amgoud and Parsons, 2001] Leila Amgoud and Simon Parsons. Agent dialogues with conflicting preferences. In *Proceedings of the 8th International Workshop on Agent Theories, Architectures and Languages*, pages 1–15, 2001.

[Amgoud et al., 2000a] Leila Amgoud, Nicolas Maudet, and Simon Parsons. Modelling dialogues using argumentation. In *Proceedings of the 4th International Conference on Multi-Agent Systems*, pages 31–38, 2000.

[Amgoud et al., 2000b] Leila Amgoud, Simon Parsons, and Nicolas Maudet. Arguments, dialogue, and negotiation. In *Proceedings of the 14th European Conference on Artificial Intelligence*, pages 338–342, 2000.

[Atkinson and Bench-Capon, 2007] Katie Atkinson and Trevor J.M. Bench-Capon. Practical reasoning as presumptive argumentation using action based alternating transition systems. *Artificial Intelligence*, 171(10–15):855–874, 2007.

[Atkinson and Bench-Capon, 2021] Katie Atkinson and Trevor Bench-Capon. Value-based argumentation. In Dov Gabbay, Massimiliano Giacomin, Guillermo R. Simari, and Matthias Thimm, editors, *Handbook of Formal Argumentation*, volume 2, chapter 5. College Publications, 2021.

[Atkinson et al., 2013] Katie Atkinson, Priscilla Bench-Capon, and Trevor J.M. Bench-Capon. A strategy for efficient persuasion dialogues. In *Proceedings of the 5th International Conference on Agents and Artificial Intelligence*, pages 332–247, 2013.

[Austin, 1975] John Langshaw Austin. *How to do things with words*. Oxford University Press, 1975.

[Azhar and Sklar, 2017] M Q Azhar and Elizabeth I Sklar. A study measuring the impact of shared decision making in a human-robot team. *International Journal of Robotics Research (IJRR)*, 36:461–482, 2017.

[Bench-Capon, 2002] Trevor J.M. Bench-Capon. Agreeing to differ: Modelling persuasive dialogue between parties without a consensus about values. *Informal Logic*, 22(3):231–245, 2002.

[Black and Atkinson, 2009] Elizabeth Black and Katie Atkinson. Dialogues that account for different perspectives in collaborative argumentation. In *Proceedings of the 8th International Joint Conference on Autonomous Agents and Multi-Agent Systems*, pages 867–874, 2009.

[Black and Atkinson, 2011a] Elizabeth Black and Katie Atkinson. Agreeing what to do. In *Proceedings of the 7th International Workshop on Argumentation in Multi-Agent Systems*, pages 12–30, 2011.

[Black and Atkinson, 2011b] Elizabeth Black and Katie Atkinson. Choosing persuasive arguments for action. In *Proceedings of the 10th International Conference on Autonomous Agents and Multiagent Systems*, pages 849–856, 2011.

[Black and Bentley, 2012] Elizabeth Black and Katie Bentley. An empirical study of a deliberation dialogue system. In *Proceedings of the 1st International Workshop on the Theory and Applications of Formal Argumentation*, pages 132–146, 2012.

[Black and Hunter, 2008] Elizabeth Black and Anthony Hunter. Using enthymemes in an inquiry dialogue system. In *Proceedings of the 7th International Joint Conference on Autonomous Agents and Multi-Agent Systems*, pages 437 – 444, 2008.

[Black and Hunter, 2009] Elizabeth Black and Anthony Hunter. An inquiry dialogue system. *Autonomous Agents and Multi-Agent Systems*, 19(2):173–209, 2009.

[Black and Hunter, 2012a] Elizabeth Black and Anthony Hunter. Executable logic for dialogical argumentation. In *Proceedings of the 20th European Conference on Artificial Intelligence*, pages 15–20, 2012.

[Black and Hunter, 2012b] Elizabeth Black and Anthony Hunter. A relevance-theoretic framework for constructing and deconstructing enthymemes. *Journal of Logic and Computation*, 22(1):55 – 78, 2012.

[Black and Hunter, 2016] Elizabeth Black and Anthony Hunter. Reasons and options for updating an opponent model in persuasion dialogues. In *Proceedings of the 3rd International Workshop on the Theory and Applications of Formal Argumentation*, pages 21–39, 2016.

[Black and Sklar, 2016] Elizabeth Black and Elizabeth I. Sklar. Computational argumentation to support multi-party human-robot interaction: challenges and advantages. In *Proceedings of the Groups in Human-Robot Interaction Workshop: A workshop at the IEEE International Symposium on Robot and Human Interactive Communication*, 2016.

[Black et al., 2014] Elizabeth Black, Amanda J Coles, and Sara Bernardini. Automated planning of simple persuasion dialogues. In *Proceedings of the 15th International Workshop on Computational Logic in Multi-Agent Sys-*

tems, pages 87–104, 2014.
[Black *et al.*, 2017] Elizabeth Black, Amanda J Coles, and Christopher Hampson. Planning for persuasion. In *Proceedings of the 16th Conference on Autonomous Agents and MultiAgent Systems*, pages 933–942, 2017.
[Bonzon and Maudet, 2011] Elise Bonzon and Nicolas Maudet. On the outcomes of multiparty persuasion. In *Proceedings of the 10th International Joint Conference on Autonomous Agents and Multiagent Systems (AAMAS'11)*, pages 47–54, May 2011.
[Budzynska and Villata, 2018] Katarzyna Budzynska and Serena Villata. Processing natural language argumentation. In Pietro Baroni, Dov Gabbay, Massimiliano Giacomin, and Leendert van der Torre, editors, *Handbook of Formal Argumentation, Volume 1*, pages 577 – 627. College Publications, 2018.
[Cerutti *et al.*, 2021] Federico Cerutti, Marcos Cramer, Mathieu Guillaume, Emmanuel Hadoux, Anthony Hunter, and Sylwia Polberg. Empirical cognitive studies about formal argumentation. In Dov Gabbay, Massimiliano Giacomin, Guillermo R. Simari, and Matthias Thimm, editors, *Handbook of Formal Argumentation*, volume 2, chapter 13. College Publications, 2021.
[Dignum and Vreeswijk, 2004] Frank P.M. Dignum and Gerard A.W. Vreeswijk. Towards a testbed for multi-party dialogues. In *Advances in Agent Communication*, volume 2922 of *Lecture Notes in Computer Science*, pages 1955–1955. Springer, 2004.
[Dimopoulos *et al.*, 2019] Yannis Dimopoulos, Jean-Guy Mailly, and Pavlos Moraitis. Argumentation-based negotiation with incomplete opponent profiles. In *Proceedings of the 18th International Conference on Autonomous Agents and Multi-Agent Systems*, pages 1252–1260, 2019.
[Dung, 1995] Phan Minh Dung. On the acceptability of arguments and its fundamental role in nonmonotonic reasoning, logic programming and n-person games. *Artificial Intelligence*, 77:321–357, 1995.
[Dupin de Saint-Cyr, 2011] Florence Dupin de Saint-Cyr. Handling enthymemes in time-limited persuasion dialogs. In *Proceedings of the 5th International Conference on Scalable Uncertainty Management*, pages 149–162, 2011.
[Endriss *et al.*, 2004] Ulle Endriss, Nicolas Maudet, Fariba Sadri, and Francesca Toni. Logic-based agent communication protocols. In F. Dignum, editor, *Advances in Agent Communication*, volume 2922 of *LNAI*, pages 91–107. Springer-Verlag, 2004. Invited contribution.
[Fan and Toni, 2012] Xiuyi Fan and Francesca Toni. Agent strategies for aba-based information-seeking and inquiry dialogues. In *Proceedings of the 20th European Conference on Artificial Intelligence*, pages 324–329, 2012.

[Fernández and Endriss, 2007] Raquel Fernández and Ulle Endriss. Abstract models for dialogue protocols. *Journal of Logic, Language and Information*, 16(2):121–140, 2007.

[Gabbay and Woods, 2001a] Dov M. Gabbay and John Woods. More on non-cooperation in Dialogue Logic. *Logic Journal of the IGPL*, 9(2):321–339, 2001.

[Gabbay and Woods, 2001b] Dov M. Gabbay and John Woods. Non-cooperation in Dialogue Logic. *Synthese*, 127(1-2):161–186, 2001.

[Gaggl et al., 2020] Sarah A. Gaggl, Thomas Linsbichler, Marco Maratea, and Stefan Woltran. Design and results of the second international competition on computational models of argumentation. *Artificial Intelligence*, 279:103193, 2020.

[Gordon, 1994] Thomas F. Gordon. The Pleadings Game: An exercise in computational dialectics. *Artificial Intelligence and Law*, 2:239–292, 1994.

[Governatori et al., 2021] Guido Governatori, Michael J. Maher, and Francesco Olivieri. Strategic argumentation. In Dov Gabbay, Massimiliano Giacomin, Guillermo R. Simari, and Matthias Thimm, editors, *Handbook of Formal Argumentation*, volume 2, chapter 10. College Publications, 2021.

[Hadjinikolis et al., 2013] Christos Hadjinikolis, Yiannis Siantos, Sanjay Modgil, Elizabeth Black, and Peter McBurney. Opponent modelling in persuasion dialogues. In *Proceedings of the 23rd International Joint Conference on Artificial Intelligence*, pages 164–170, 2013.

[Hadoux and Hunter, 2017] Emmanuel Hadoux and Anthony Hunter. Strategic sequences of arguments for persuasion using decision trees. In *Proceedings of the 31st AAAI Conference on Artificial Intelligence*, pages 1128–1134, 2017.

[Hadoux et al., 2015] Emmanuel Hadoux, Aurélie Beynier, Nicolas Maudet, Paul Weng, and Anthony Hunter. Optimization of probabilistic argumentation with Markov decision models. In *Proceedings of the 24th International Joint Conference on Artificial Intelligence*, pages 2004–2010, 2015.

[Hamblin, 1970] Charles L. Hamblin. *Fallacies*. Methuen and Co Ltd, London, UK, 1970.

[Hamblin, 1971] Charles L. Hamblin. Mathematical models of dialogue. *Theoria*, 37:130–155, 1971.

[Hitchcock, 1991] David Hitchcock. Some principles of rational mutual inquiry. In *Proceedings of the 2nd International Conference on Argumentation*, pages 236–243, 1991.

[Hosseini et al., 2014] Seyed Ali Hosseini, Sanjay Modgil, and Odinaldo Rodrigues. Enthymeme construction in dialogues using shared knowledge. In

Proceedings of the 5th International Conference on Computational Models of Argument, pages 325 – 332, 2014.

[Hulstijn, 2000] J. Hulstijn. *Dialogue Models for Inquiry and Transaction*. PhD thesis, Universiteit Twente, Enschede, The Netherlands, 2000.

[Hunter and Polberg, 2017] Anthony Hunter and Sylwia Polberg. Empirical methods for modelling persuadees in dialogical argumentation. In *Proceedings of the 29th IEEE International Conference on Tools with Artificial Intelligence*, pages 382–389, 2017.

[Hunter et al., 2018] Anthony Hunter, Sylwia Polberg, and Nico Potyka. Updating belief in arguments in epistemic graphs. In *Proceedings of the 16th International Conference on Principles of Knowledge Representation and Reasoning*, pages 138–147, 2018.

[Hunter et al., 2021] Anthony Hunter, Sylwia Polberg, Nico Potyka, Tjitze Rienstra, and Matthias Thimm. Probabilistic argumentation: A survey. In Dov Gabbay, Massimiliano Giacomin, Guillermo R. Simari, and Matthias Thimm, editors, *Handbook of Formal Argumentation*, volume 2, chapter 7. College Publications, 2021.

[Hunter, 2007] Anthony Hunter. Real arguments are approximate arguments. In *Proceedings of the 22nd AAAI Conference on Artificial Intelligence*, pages 66–71, 2007.

[Kacpraz et al., 2014] Magdalena Kacpraz, Marcin Dziubinski, and Katarzyna Budzynska. Strategies in dialogues: A game-theoretic approach. In *Proceedings of the 5th International Conference on Computational Models of Argument*, pages 333 – 344, 2014.

[Kakas et al., 2005] Antonis C. Kakas, Nicolas Maudet, and Pavlos Moraitis. Modular representation of agent interaction rules through argumentation. *Autonomous Agents Multi-Agent Systems*, 11(2):189–206, 2005.

[Kodeswaran et al., 2010] Palanivel Kodeswaran, Wenjia Li, Anupam Joshi, Tim Finin, and Filip Perich. Enforcing secure and robust routing with declarative policies. In *Military Communications Conference (MILCOM)*, pages 44–49. IEEE, 2010.

[Kontarinis et al., 2014] Dionysios Kontarinis, Elise Bonzon, Nicolas Maudet, and Pavlos Moraitis. Empirical evaluation of strategies for multiparty argumentative debates. In *15th International Workshop on Computational Logic in Multi-Agent Systems (CLIMA'14)*, pages 105–122, 2014.

[Lawrence and Reed, 2020] John Lawrence and Chris Reed. Argument mining: A survey. *Computational Linguistics*, 45(4):765–818, 2020.

[Loui and Moore, 1997] R. P. Loui and D. M. Moore. Dialogue and deliberation. Report WUCS-97-11, Computer Science and Engineering, Washington

University in St. Louis, 1997.

[Macagno et al., 2018] Fabrizio Macagno, Douglas Walton, and Chris Reed. Argumentation schemes. In Pietro Baroni, Dov Gabbay, Massimiliano Giacomin, and Leendert van der Torre, editors, *Handbook of Formal Argumentation, Volume 1*, pages 519 – 576. College Publications, 2018.

[MacKenzie, 1979a] Jim D. MacKenzie. How to stop talking to tortoises. *Notre Dame Journal of Formal Logic*, 20 (4):705–717, 1979.

[Mackenzie, 1979b] Jim D. Mackenzie. Question-begging in non-cumulative systems. *Journal of Philosophical Logic*, 8:117–113, 1979.

[MacKenzie, 1990] Jim D. MacKenzie. Four dialogue systems. *Studia Logica*, 49 (4):567–583, 1990.

[McBurney and Parsons, 2002] Peter McBurney and Simon Parsons. Games that agents play: A formal framework for dialogues between autonomous agents. *Journal of Logic, Language, and Information*, 11(3):315–334, 2002.

[McBurney et al., 2007] Peter McBurney, David Hitchcock, and Simon Parsons. The eightfold way of deliberation dialogue. *International Journal of Intelligent Systems*, 22(1):95–132, 2007.

[Medellin-Gasque et al., 2013] Rolando Medellin-Gasque, Katie Atkinson, Trevor Bench-Capon, and Peter McBurney. Strategies for question selection in argumentative dialogues about plans. *Argument and Computation*, 4(2):151–179, 2013.

[Mercier and Sperber, 2011] Hugo Mercier and Dan Sperber. Why do humans reason? arguments for an argumentative theory. *Behavioral and brain sciences*, 34(2):57–74, 2011.

[Modgil and Caminada, 2009] Sanjay Modgil and Martin Caminada. Proof theories and algorithms for abstract argumentation frameworks. In *Argumentation in Artificial Intelligence*, pages 105–129. Springer, 2009.

[Modgil and Prakken, 2013] Sanjay Modgil and Henry Prakken. A general account of argumentation with preferences. *Artificial Intelligence*, 195:361–397, 2013.

[Murphy et al., 2016] Josh Murphy, Elizabeth Black, and Michael Luck. A heuristic strategy for persuasion dialogues. In *Proceedings of the 6th International Conference on Computational Models of Argument*, pages 411 – 418, 2016.

[Murphy et al., 2018] Josh Murphy, Alexandru Burdusel, Michael Luck, Steffen Zschaler, and Elizabeth Black. Deriving persuasion strategies using search-based model engineering. In *Proceedings of the 7th International Conference on Computational Models of Argument*, pages 221–232, 2018.

[Oren et al., 2006] Nir Oren, Timothy J. Norman, and Alun Preece. Loose lips sink ships: A heuristic for argumentation. In *Proceedings of the 3rd International Workshop on Argumentation in Multi-Agent Systems*, pages 121 – 134, 2006.

[Panisson and Bordini, 2017] Alison R. Panisson and Rafael H. Bordini. Uttering Only What is Needed: Enthymemes in Multi-Agent Systems. In *Proceedings of the 16th International Conference on Autonomous Agents and Multiagent Systems*, pages 1670–1672, 2017.

[Parsons et al., 2003] Simon Parsons, Michael Wooldridge, and Leila Amgoud. Properties and complexity of formal inter-agent dialogues. *Journal of Logic and Computation*, 13(3):347–376, 2003.

[Parsons et al., 2011] Simon Parsons, Yuqing Tang, Elizabeth I. Sklar, Peter McBurney, and Kai Cai. Argumentation-based reasoning in agents with varying degrees of trust. In *Proceedings of the 10th International Conference on Autonomous Agents and Multi-Agent Systems*, 2011.

[Plotkin, 1981] Gordon D. Plotkin. A structural approach to operational semantics. Technical Report DAIMI FN-19, Computer Science Department, Aarhus University, 1981.

[Prakken and Sartor, 1998] Henry Prakken and Giovanni Sartor. Modelling reasoning with precedents in a formal dialogue game. *Artificial Intelligence and Law*, 6:231–287, 1998.

[Prakken, 2000] Henry Prakken. On dialogue systems with speech acts, arguments, and counterarguments. In *Proceedings of the 7th European Workshop on Logic in Artificial Intelligence*, pages 224–238, 2000.

[Prakken, 2001a] Henry Prakken. Modelling reasoning about evidence in legal procedure. In *Proceedings of the 8th International Conference on Artificial Intelligence and Law*, pages 119–128, 2001.

[Prakken, 2001b] Henry Prakken. Relating protocols for dynamic dispute with logics for defeasible argumentation. *Synthese*, 127:187–219, 2001.

[Prakken, 2005] Henry Prakken. Coherence and flexibility in dialogue games for argumentation. *Journal of Logic and Computation*, 2005.

[Prakken, 2006] Henry Prakken. Formal systems for persuasion dialoge. *Knowledge Engineering Review*, 21(2):163–188, 2006.

[Rach et al., 2018] Nicklas Rach, Wolfgang Minker, and Stefan Ultes. Markov games for persuasive dialogue. In *Proceedings of the 7th International Conference on Computational Models of Argument*, pages 213 – 220, 2018.

[Rach et al., 2019] Niklas Rach, Klaus Weber, Annalena Aicher, Florian Lingenfelser, Elisabeth Andre, and Wolfgang Minker. Emotion recognition based preference modelling in argumentative dialogue systems. In *Proceedings of the 1st International Workshop on Pervasive Computing and Spoken*

Dialogue Systems Technology, pages 838–843, 2019.

[Rahwan and Larson, 2009] Iyad Rahwan and Kate Larson. Argumentation and game theory. In *Argumentation in Artificial Intelligence*, pages 321–339. Springer, 2009.

[Rahwan et al., 2003] Iyad Rahwan, Sarvapalid D. Ramchurn, Nicholas R. Jennings, Peter McBurney, Simon Parsons, and Liz Sonenberg. Argumentation-based negotiation. *Knowledge Engineering Review*, 18(4):343–375, 2003.

[Rahwan et al., 2009] Iyad Rahwan, Philippe Pasquier, Liz Sonenberg, and Frank Dignum. A formal analysis of interest-based negotiation. *Annals of Mathematics and Artificial Intelligence*, 55(3-4):253, 2009.

[Rienstra et al., 2013] Tjitze Rienstra, Matthias Thimm, and Nir Oren. Opponent models with uncertainty for strategic argumentation. In *Proceedings of the 23rd International Joint Conference on Artificial Intelligence International Joint Conference on Artificial Intelligence*, pages 332–338, 2013.

[Rosenfeld and Kraus, 2016a] Ariel Rosenfeld and Sarit Kraus. Providing arguments in discussions on the basis of the prediction of human argumentative behavior. *ACM Trans. Interact. Intell. Syst.*, 6(4):30:1–30:33, 2016.

[Rosenfeld and Kraus, 2016b] Ariel Rosenfeld and Sarit Kraus. Strategical argumentative agent for human persuasion. In *Proceedings of the 22nd European Conference on Artificial Intelligence*, volume 285, pages 320–328. IOS Press, 2016.

[Russell and Norvig, 2020] Stuart Russell and Peter Norvig. *Artificial intelligence: a modern approach*. Pearson Education Limited, 4th edition edition, 2020.

[Sadri et al., 2001] Fariba Sadri, Francesca Toni, and Paolo Torroni. Dialogues for negotiation: Agent varieties and dialogue sequences. In John-Jules Ch. Meyer and Milind Tambe, editors, *Intelligent Agents VIII, 8th International Workshop, ATAL 2001 Seattle, WA, USA, August 1-3, 2001, Revised Papers*, volume 2333 of *Lecture Notes in Computer Science*, pages 405–421. Springer, 2001.

[Schopenhauer, 1831] Arthur Schopenhauer. The art of always being right: 38 ways to win an argument. https://en.wikisource.org/wiki/The_Art_of_Being_Right, 1831. Original title: Die Kunst, Recht zu behalten (Translated by Thomas Saunders in 1896).

[Searle, 1969] John R. Searle. *Speech acts: An essay in the philosophy of language*. Cambridge University Press, 1969.

[Singh, 2000] Munindar P. Singh. A social semantics for agent communication languages. In Frank Dignum and Mark Greaves, editors, *Issues in agent communication*, volume 1916 of *Lecture Notes in Computer Science*, pages 31–45. Springer, Berlin, Heidelberg, 2000.

[Sklar and Azhar, 2015] E. I. Sklar and M. Q. Azhar. Argumentation-based dialogue games for shared control in human-robot systems. *Journal of Human-Robot Interaction*, 4(3):120–148, 2015.

[Sperber and Wilson, 1986] Dan Sperber and Deidre Wilson. *Relevance: Communication and Cognition*. Blackwell Publishing, 1986.

[Sperber and Wilson, 2004] Dan Sperber and Deidre Wilson. Relevance theory. In Laurence R. Horn and Gregory Ward, editors, *The Handbook of Pragmatics*, pages 607 – 632. Blackwell Publishing, 2004.

[Thielscher, 2010] Michael Thielscher. A general game description language for incomplete information games. In *Proceedings of the 24th AAAI Conference on Artificial Intelligence*, pages 994–999, 2010.

[Thimm, 2014] Matthias Thimm. Strategic argumentation in multi-agent systems. *KI*, 28(3):159–168, 2014.

[Torroni, 2002] Paolo Torroni. A study on the termination of negotiation dialogues. In *Proceedings of The 1st International Joint Conference on Autonomous Agents and Multiagent Systems*, pages 1223–1230, 2002.

[Walton and Krabbe, 1996] Douglas N. Walton and Erik C. W. Krabbe. *Commitment in Dialogue. Basic Concepts of Interpersonal Reasoning*. Suny Press, 1996.

[Walton and Reed, 2005] Douglas N. Walton and Chris Reed. Argumentation schemes and enthymemes. *Synthese*, 145:339–370, 2005.

[Walton, 1989] Douglas N. Walton. *Informal Logic: A Handbook for Critical Argumentation*. Cambridge University Press, 1989.

[Wells and Reed, 2012] Simon Wells and Chris Reed. A domain specific language for describing diverse systems of dialogue. *Journal of Applied Logic*, 10(4):309 – 329, 2012.

[Wooldridge, 2009] Michael Wooldridge. *An Introduction to Multiagent Systems*. Wiley, 2nd edition, 2009.

[Xydis *et al.*, 2020] Andreas Xydis, Christopher Hampson, Sanjay Modgil, and Elizabeth Black. Enthymemes in dialogue. In *Proceedings of the 8th International Conference on Computational Models of Argument*, 2020. (In press).

CHAPTER 10

STRATEGIC ARGUMENTATION

GUIDO GOVERNATORI
Data61-CSIRO, Brisbane, Australia
guido.governatori@data61.csiro.au

MICHAEL J. MAHER
Reasoning Research Institute, Canberra, Australia
michael.maher@reasoning.org.au

FRANCESCO OLIVIERI
Griffith University, Brisbane, Australia
francesco.olivieriphd@gmail.com

Abstract

Dialogue games are a dynamic form of argumentation, with multiple parties pooling their arguments with the intention of settling an issue. Such games can have a variety of structures, and may be collaborative or competitive, depending on the motivations of the parties. Strategic argumentation is a class of competitive dialogue games in which two players take turns in contributing their arguments, each attempting to have an issue settled in the way that they would prefer. Thus strategic argumentation games are less about discovering a joint truth than about a player imposing their view on an opponent. They are reflective of legal argumentation.

In the games we study, players have perfect information of the moves players make, but incomplete information on the possible moves (arguments) that other players have available to them. We look both at games using logically structured arguments and games using abstract arguments. We show that playing these games can be computationally hard. We also examine issues of corruption in such games, and discuss approaches to foiling it.

1 Introduction

When two or more parties are engaged in a debate, it is often the case that each party has some information they are not willing to disclose to the other parties. Also, in some cases, the disclosure of some piece of information by one party could prove detrimental for the party, in the sense that the information could be used to prevent the party to reach their aim in the debate, or some of the information disclosed can help the other party to achieve their goal. Accordingly we can provide the following (informal) definition of strategic argumentation.

Definition 1.1. Strategic argumentation *is the problem of determining what arguments (pieces of information) to disclose during a debate in order to achieve the aim a party has in the debate and to prevent the other party from gaining an undesired advantage.*

To illustrate the issue, consider the following argument exchange, first proposed in [Satoh and Takahashi, 2011a]:

Example 1.2. *Let* Pr *and* Op *be the players involved in the following argumentation dialogue (*Pr *and* Op *denote, respectively, the proponent and the opponent):*

Pr_0 : *"You killed the victim."*
Op_1 : *"I did not commit murder! There is no evidence!"*
Pr_1 : *"There is evidence. We found your ID card near the scene."*
Op_2 : *"It is not evidence! I had my ID card stolen!"*
Pr_2 : *"It is you who killed the victim. Only you were near the scene at the time of the murder."*
Op_3 : *"I did not go there. I was at facility A at that time."*
Pr_3 : *"At facility A? Then, it is impossible to have had your ID card stolen since facility A does not allow a person to enter without an ID card."*

We can easily represent arguments of this example with a rule-based *formalism as follows. We have rules R:*

$$r_{\text{Pr}_0}: \Rightarrow murderer(X)$$
$$r'_{\text{Op}_1}: \Rightarrow \neg evidence\_Against(X)$$
$$r''_{\text{Op}_1}: \neg evidence\_Against(X) \Rightarrow \neg murderer(X)$$
$$r_{\text{Pr}_1}: ID(X)\_at\_crime\_scene \Rightarrow evidence\_Against(X)$$
$$r_{\text{Op}_2}: ID(X)\_stolen \Rightarrow \neg evidence\_Against(X)$$
$$r'_{\text{Pr}_2}: \Rightarrow only(X)\_at\_crime\_scene$$
$$r''_{\text{Pr}_2}: only(X)\_at\_crime\_scene \Rightarrow murderer(X)$$
$$r_{\text{Op}_3}: at\_facility\_A(X) \Rightarrow \neg only(X)\_at\_crime\_scene$$
$$r_{\text{Pr}_3}: at\_facility\_A(X) \Rightarrow \neg ID(X)\_stolen$$

and a priority relation $>= \{r_{\text{Op}_2} > r_{\text{Pr}_1}\}$, *where the notation* $r_i : A(r) \Rightarrow C(r)$ *identifies that* r_i *is the name of the rule,* $A(r)$ *is the set of antecedents (possibly empty) while* $C(r)$ *is the conclusion, symbol* $\Rightarrow$ *denotes that the conclusion may be defeated by contrary evidence, as for instance the conflict between* r_{Op_2} *and* r_{Pr_1}, *resolved by* $>$ *(the superiority relation) which allows us to conclude that* $\neg evidence\_Against(X)$ *is the case.*

A feature of this dialogue is that the exchange of arguments reflects an asymmetry of information between the two parties. Each player does not know the other player's knowledge, thus they cannot predict which arguments will be attacked, nor which counterarguments may be employed for attacking their own arguments. In addition, the private information disclosed by a party might eventually be used by the adversary to construct and play justified counterarguments. Thus, Pr3 attacks Op2, but only after Op3 has been given. Thus, the attack Pr3 of the proponent is possible *only when* the opponent discloses some private information through the move Op3 (in this setting, after Op let Pr know that Op was at facility). If we assume that Pr wishes to expose Op's guilt, and Op wishes to hide it, then we can view this dialogue as a game, where a move consists of stating an argument.

This example illustrates a scenario where some of the information disclosed by a party could be detrimental to their aim. This is a common phenomenon in many applications that are suitable to be formally represented by argumentation such as negotiation [Rahwan et al., 2003], brokering [Antoniou et al., 2007], and in the legal domain [Prakken and Sartor, 2015; Governatori et al., 2001]. In a negotiation, the other party could use the information to gain some advantage either on the issue of the negotiation (e.g., price of an item) or on some side effects; in a legal proceeding the opposite party could use the information to win the case. Hence, players in such an argumentation game must be strategic in what arguments they expose, to put themselves in the best position. We refer to such games as *strategic argumentation* games.

Furthermore, in such applications the parties can be represented by agents acting and debating on behalf of their clients, but these agents might not have their client's best interests at heart. This can corrupt the dialogue. For example, suppose the agent for Pr was bribed by Op to omit the claim Pr_2. Then Op_3 would have remained private, and Op's lie would be undiscovered. Similar issues occur whenever we employ an agent, whether human or software.

Technically, games involving privacy are called games of *incomplete information*. As argued in [Governatori et al., 2014], argument games with incomplete information can be modelled by stating that each player has a logical theory, constituting their private knowledge, and which is unknown by the opposite party, and there is an additional theory shared by all parties with the information that is public. A player may build an argument that supports their claim by using some of their private knowledge and the common information; in turn, the other party may construct new arguments by re-using the adversary's disclosed information (along with other pieces of their own private knowledge) in order to defeat the opponent's arguments. In a legal proceeding, we can distinguish between two types of information: the norms in force in the underlying jurisdiction, which are assumed to be known by both parties, and the information, private to each party, on the facts of the case. Accordingly, the legal proceeding can be modelled by three theories, a public one with the common knowledge, encoding the norms of the underlying jurisdiction, plus two private theories: one for each party.

When working with logically structured arguments, the different logical theories are represented by sets of rules (which may include unconditional facts). So, the set R of all rules used to build arguments is partitioned into three (distinct) subsets: a set R_{Com} known by both players, and two subsets R_{Pr} and R_{Op} corresponding, respectively, to Pr's and Op's private knowledge. While the game is evolving, at each turn, a party discloses some of their private arguments and, by doing so, the player reduces their private information ($R_{\mathsf{Pr}}/R_{\mathsf{Op}}$ decreases) with what now becomes part of the new common knowledge base (R_{Com} increases). Consider a setting where $F = \{a, d, f\}$ is the known set of facts (categorical statements), $R_{\mathsf{Com}} = F$ (the facts are common knowledge), and the players have the following sets of rules:

$$R_{\mathsf{Pr}} = \{r_0 : a \Rightarrow b;\ r_1 : d \Rightarrow c;\ r_2 : c \Rightarrow b\}$$
$$R_{\mathsf{Op}} = \{r_3 : c \Rightarrow e;\ r_4 : e, f \Rightarrow \neg b\}.$$

If Pr's intent is to prove b and plays $\{a \Rightarrow b\}$, then Pr wins the game. In fact, Op has no way to prove e and thus r_4 is not active. If, on the other hand, Pr plays $\{d \Rightarrow c,\ c \Rightarrow b\}$ (or even the whole R_{Pr}), this allows Op to succeed. Here, a minimal subset of R_{Pr} is successful. The situation can be reversed for Pr. Replace the sets of private rules with

$$R_{\mathsf{Pr}} = \{a \Rightarrow b;\ d \Rightarrow \neg c\}$$
$$R_{\mathsf{Op}} = \{d, c \Rightarrow \neg b;\ f \Rightarrow c\}.$$

Playing $\{a \Rightarrow b\}$ is now not successful for Pr, while the whole R_{Pr} ensures victory.

Example 1.2 brings out the issues we will address in this chapter: formalizing such dialogues as strategic argumentation games, addressing the difficulty of making a move in a game, and examining the possibility of corruption in such games and means to foil it. We will look at both defeasible logics [Antoniou et al., 2001a] and ASPIC-style structured argumentation [Amgoud et al., 2006; Prakken, 2010a] as languages for expressing arguments. We will also show that the same issues arise if we formulate strategic argumentation in terms of abstract arguments [Dung, 1995]. In looking at corruption, we consider two forms: *espionage* and *collusion*. To counter these possibilities, we examine the use of standards and audit to limit the ability of players to behave corruptly, and the idea of computational resistance to corruption to discourage corruption.

The layout of this chapter is as follows. Section 2 describes the general setting of argumentation and dialogue games. Section 3 provides some technical background on computational complexity, elements of abstract argumentation [Dung, 1995], and a framework for argumentation with logically structured arguments. Section 4 outlines Defeasible Logic and its four main variants. Section 5 presents an instance of the strategic argumentation game with Defeasible Logic as the basis for argumentation, and proves the computational difficulty of playing the game. It extends this result to an instance of structured argumentation under the grounded semantics. Section 6 extends the idea of strategic argumentation further, to abstract argumentation over a variety of semantics. Section 7 investigates how corruption can affect argumentation games, and how it can be countered. Section 8 discusses related work and Section 9 considers possible future directions of this research. Section 10 ends the chapter.

2 Argumentation and Dialogue Games

In this section we briefly describe a general setting of argumentation and dialogue games. In doing so we will not bind concepts such as *argument*, *aim*, *acceptance* or *extension* to a specific meaning, nor specify all details of concepts like argumentation framework. They will be specified more precisely later.

Definition 2.1 (Argumentation framework). *An* argumentation framework AF *is a tuple* $(\mathcal{A}, \mathcal{R})$, *where* $\mathcal{A}$ *is a set of arguments, and* $\mathcal{R}$ *is a collection of relations over* $\mathcal{A}$.

The literature in argumentation theory flourishes with different frameworks describing *what* arguments are, where the two main school of thoughts see them as either *monads* (with no internal structure), or *structured* (made of sub-parts). We will address both schools. For now, we are only interested in saying that there is a function mapping arguments to elements of the language, referred to as *conclusions* (or theses, claims).

Definition 2.2 (Conclusions). *Given an argumentation framework* AF *and a language of expressions* L, *the function* $conc: \mathcal{A} \mapsto 2^L$ *maps each argument to a set of elements of* L. *If* $c_A \in conc(A)$, *then we call* c_A *a* conclusion *of argument* A.

In the monadic view, each argument might have a single, distinct conclusion. In that case, conclusions add nothing to the argumentation framework. In the structured view, an expression might be a conclusion of several arguments, and its negation might also be a conclusion of arguments. Any structured argumentation framework with conclusions can be abstracted to a monadic argumentation framework by simply ignoring its internal structure (and retaining the conclusion function).

For the purposes of this chapter, a semantics maps an argumentation framework to a set of extensions. Each semantics implicitly expresses a criterion for how arguments can coherently be adjudicated together, given an argumentation framework. Each extension in the semantics represents a "reasonable" adjudication, according to that criterion, of the arguments in the argumentation framework. We leave open the details of what an extension is and how it might be represented, but commonly it is a set of arguments or a labelling for arguments (see Section 3.2 for more details of these common representations).

Definition 2.3 (Semantics). *A semantics is a function σ mapping argumentation frameworks to a set of extensions.*

There is a rich array of interactions that are considered dialogues in the argumentation literature [Black et al., 2021] but, as can be seen from the introduction, we have a specific kind of dialogue in mind. We define a *dialogue* as the exchange of arguments between two (or more) parties. We talk of *dialogue games* when we want to analyse the formal properties of the dialogue, using criteria from game theory.

At the beginning of a dialogue game, each agent starts with a private set of arguments but they also share a (possibly empty) set of arguments that are common knowledge[1] to all players. This shared knowledge among the agents will be enriched throughout the game with the arguments played at each turn, as will be clear in the following.

Each player also has an *aim*, the details of which we leave open. Aims might be to have a particular argument accepted in at least one extension, under a particular semantics, or to have the cardinality of

---

[1] By common knowledge we mean, not only that all players have knowledge of the arguments, but also each player knows that the others know, and each knows that the others know that she knows, and so on. [Fagin et al., 2003]

each extension, under a given semantics, be a prime number[2].

Our dialogue games consist of a state and possible changes of state.

Definition 2.4 (Dialogue Game State). *Given a set of agents $Pl_1, \ldots, Pl_n$ (referred to as* players*), a dialogue game state is an argumentation framework $(\mathcal{A}, \mathcal{R})$ where $\mathcal{R}$ contains unary relations $\xi_1, \ldots \xi_n$ on $\mathcal{A}$, one for each player, as well as ξ_{Com} and, possibly, other relations.*

Each unary relation ξ_i defines a subset S_i of $\mathcal{A}$: $S_i = \{a \mid a \in \mathcal{A}, \xi_i(a)\}$. Similarly, $S_{\mathsf{Com}} = \{a \mid a \in \mathcal{A}, \xi_{\mathsf{Com}}(a)\}$. S_{Com} is the set of arguments that are common knowledge to all players, while S_i is the additional set of arguments that player Pl_i knows, but other players don't know she knows (they are *private*).

Thus, a dialogue game state can equally be viewed as a *split argumentation framework* $(\mathcal{A}, S_{\mathsf{Com}}, S_1, \ldots, S_n, \mathcal{R}')$, where $S_{\mathsf{Com}} \cap (\cup_{i=1}^n S_i) = \emptyset$ and $\mathcal{R}'$ is $\mathcal{R} \setminus \{\xi_{\mathsf{Com}}, \xi_1, \ldots, \xi_n\}$.

A dialogue game is a collection of players, each with their own aim, making moves, in turn, to achieve their aim[3].

Definition 2.5 (Dialogue Game). *Given a set of players $Pl_1, \ldots, Pl_n$ and an aim for each player, a* dialogue game *consists of an initial dialogue game state in the form of a split argumentation framework $(\mathcal{A}, S_{\mathsf{Com}}, S_1, \ldots, S_n, \mathcal{R})$, and state transition rules (or* moves*) defined below.*

1. *Players take turns, meaning that only a single player can act at a given turn[4].*

2. *At a given turn k, player Pl_i advances a subset T of its private arguments in order to achieve their aim. If S_{Com}^{k-1} and S_i^{k-1} denote, respectively, the common shared argumentation framework and Pl_i's private argumentation framework at turn $k-1$, then*

---

[2]Admittedly, the latter example is not likely to arise in practice.

[3]Many different types of dialogue have been classified and many protocols have been provided for them; we refer to Chapter 9 of this handbook [Black et al., 2021] for in depth analysis of the various alternatives. In this chapter we restrict ourselves to a minimal and limited view of dialogue games, suitable to define strategic argumentation.

[4]We shall not dwell on the details of how/which players are selected to act at a given turn, as it is outside the scope of this chapter. [Thimm and García, 2010] discusses some other possibilities.

- $S_{\mathsf{Com}}^k = S_{\mathsf{Com}}^{k-1} \cup T$
- $S_i^k = S_i^{k-1} \setminus T$
- $S_j^k = S^{k-1}$ for $j \neq i$

3. The game ends at turn $k+1$, when either: (i) the aim of each player is satisfied, so no player has an incentive to change the state of the game, or (ii) no player with an unsatisfied aim is able to satisfy that aim by making a move.

The state of the dialogue game after turn k is $(\mathcal{A}, S_{\mathsf{Com}}^k, S_1^k, \ldots, S_n^k, \mathcal{R})$. The common argumentation framework then is $CAF^k = (S_{\mathsf{Com}}^k, \mathcal{R})$.

According to the typology of argumentation games in [Thimm and García, 2010], these dialogue games have a dialectical argumentation mechanism and players have no awareness of other players' arguments; agent type is not specified. The games we define below (Definitions 2.6 and 2.7) have an indicator agent type.

If we ignore turn-taking, our dialogue games are *memoryless*: the permitted moves are determined by the current dialogue state, independent of how that state was reached. Other forms of dialogue game may not have this property.

Note that, although the set of common arguments increases monotonically, this game is non-monotonic, meaning that, at any given turn, aims that were satisfied at the previous turn might now be unsatisfied.

Also note that we are considering the relations $\mathcal{R}$ to have a fixed meaning, independent of player's beliefs or perceptions. The *omniscient* argumentation framework corresponding to a dialogue game is $(\mathcal{A}, \mathcal{R})$.

We now formulate a specific type of dialogue games, namely *strategic argumentation dialogues*. In a strategic argumentation dialogue game, we have only *two* players, who take turns in exchanging arguments to accept/reject a topic φ, where $\varphi \in L$. We name one player *Proponent* (Pr), and the other *Opponent* (Op). We shall consider two variants of the strategic argumentation dialogue game: the *symmetric*, and the *asymmetric* strategic argumentation dialogue game. In the symmetric variant, both parties have the burden of proof, that is, the proponent has to establish φ, where the opponent has to establish $\neg \varphi$. (With $\neg \varphi$, we denote the contrary of φ.) In the asymmetric variant, the proponent still has to establish φ, whereas the opponent aims to prevent this.

In the symmetric variant, at one turn, either φ, or $\neg\varphi$, is accepted. If φ is accepted, then it is the opponent's turn; if $\neg\varphi$ is accepted, then is the proponent's turn. At a given turn, the player has two possible courses of action. First, they play a subset of their private argumentation framework (i.e., a non-empty set of arguments). By doing so, they increment the shared argumentation framework with the arguments just played. Second, they pass and admit defeat. This happens when they are not able to change the status of the conclusion. The game ends when a player passes.

Definition 2.6 (Symmetric Strategic Argumentation Dialogue Game). *Consider two players, a proponent* Pr *and an opponent* Op, *an expression* $\varphi \in L$, *a dialogue game state in the form of a split argumentation framework* $(\mathcal{A}, S_{\mathsf{Com}}, S_{\mathsf{Pr}}, S_{\mathsf{Op}}, \mathcal{R})$, *and a conclusion function* conc. *Suppose that there is an argument* $a \in S_{\mathsf{Pr}}$ *such that* $\varphi \in conc(a)$.

Let S^k_{Com}, S^k_{Pr}, *and* S^k_{Op} *denote, respectively, the common knowledge arguments,* Pr*'s private arguments and* Op*'s private arguments after turn* k. *(In particular,* $S^0_{\mathsf{Com}} = S_{\mathsf{Com}}$, $S^0_{\mathsf{Pr}} = S_{\mathsf{Pr}}$, *and* $S^0_{\mathsf{Op}} = S_{\mathsf{Op}}$.*)*

We define a symmetric strategic argumentation dialogue game as a dialogue game where:

1. *The players take turns; if φ is accepted by CAF^0 under semantics σ, then* Op *begins; otherwise* Pr *does so.*

2. *At turn k, if $\neg\varphi$ is accepted in CAF^{k-1}, then it is* Pr*'s turn to play, as follows*

    - Pr *advances a subset of its private arguments $T \subseteq S^{k-1}_{\mathsf{Pr}}$ so that φ is accepted in CAF^k. As a result*
        - $S^k_{\mathsf{Com}} = S^{k-1}_{\mathsf{Com}} \cup T$;
        - $S^k_{\mathsf{Pr}} = S^{k-1}_{\mathsf{Pr}} \setminus T$.
        - $S^k_{\mathsf{Op}} = S^{k-1}_{\mathsf{Op}}$

3. *At turn k, if φ is accepted in CAF^{k-1}, then it is* Op*'s turn to play, as follows*

- Op advances a subset of its private arguments $T \subseteq S_{\mathsf{Op}}^{k-1}$ so that $\neg\varphi$ is accepted in CAF^k. As a result
    - $S_{\mathsf{Com}}^k = S_{\mathsf{Com}}^{k-1} \cup T$;
    - $S_{\mathsf{Pr}}^k = S_{\mathsf{Pr}}^{k-1}$
    - $S_{\mathsf{Op}}^k = S_{\mathsf{Op}}^{k-1} \setminus T$.

4. The game ends at turn $k+1$, when either (i) it is Pr's turn and there is no move for Pr such that CAF^{k+1} accepts φ, in which case Op wins, or (ii) it is Op's turn and there is no move for Op such that CAF^{k+1} accepts $\neg\varphi$, in which case Pr wins.

The only difference in the asymmetric variant with respect to the symmetric one is that, the opponent no longer has the burden of proof: during her turn, Op proposes arguments in order to prevent acceptance of φ, rather than to accept $\neg\varphi$ (see point 3).

Definition 2.7 (Asymmetric Strategic Argumentation Dialogue Game). *Consider two players, a proponent Pr and an opponent Op, an expression $\varphi \in L$, a dialogue game state in the form of a split argumentation framework $(\mathcal{A}, S_{\mathsf{Com}}, S_{\mathsf{Pr}}, S_{\mathsf{Op}}, \mathcal{R})$, and a conclusion function conc.*

Let S_{Com}^k, S_{Pr}^k, and S_{Op}^k denote, respectively, the common knowledge arguments, Pr's private arguments and Op's private arguments after turn k. (In particular, $S_{\mathsf{Com}}^0 = S_{\mathsf{Com}}$, $S_{\mathsf{Pr}}^0 = S_{\mathsf{Pr}}$, and $S_{\mathsf{Op}}^0 = S_{\mathsf{Op}}$.)

We define an asymmetric strategic argumentation dialogue game as a dialogue game where:

1. The players take turns; if φ is accepted by CAF^0 under semantics σ, then Op begins; otherwise Pr does so.

2. At turn k, if φ is not accepted in CAF^{k-1}, then it is Pr's turn to play, as follows

    - Pr advances a subset of its private arguments $T \subseteq S_{\mathsf{Pr}}^{k-1}$ so that φ is accepted in CAF^k. As a result
        - $S_{\mathsf{Com}}^k = S_{\mathsf{Com}}^{k-1} \cup T$;
        - $S_{\mathsf{Pr}}^k = S_{\mathsf{Pr}}^{k-1} \setminus T$;
        - $S_{\mathsf{Op}}^k = S_{\mathsf{Op}}^{k-1}$

3. At turn k, if φ is accepted in CAF^{k-1}, then it is Op's turn to play, as follows

    - Op advances a subset of its private arguments $T \subseteq S_{\mathsf{Op}}^{k-1}$ so that φ is not accepted in CAF^k. As a result
        - $S_{\mathsf{Com}}^k = S_{\mathsf{Com}}^{k-1} \cup T$;
        - $S_{\mathsf{Pr}}^k = S_{\mathsf{Pr}}^{k-1}$
        - $S_{\mathsf{Op}}^k = S_{\mathsf{Op}}^{k-1} \setminus T$.

4. The game ends at turn $k+1$, when either (i) it is Pr's turn and there is no move for Pr such that CAF^{k+1} accepts φ, in which case Op wins, or (ii) it is Op's turn and there is no move for Op such that CAF^{k+1} does not accept φ, in which case Pr wins.

Thus both variants are dialogue games between two players arguing about a conclusion φ on the basis of their common argumentation framework. They leave open the notion of acceptance and the details of the set of relations $\mathcal{R}$, but specify more precisely the aims of the players. From now on, we will use the abbreviations SSA for Symmetric Strategic Argumentation, and AsSA for Asymmetric Strategic Argumentation.

The asymmetric game can be seen in situations where the parties have different proof standards. For example, in a criminal proceeding the prosecution must prove its case "beyond a reasonable doubt", while the defence has only to prevent this. An asymmetric dialogue game was presented in [Eriksson Lundström et al., 2008].

The problems that the players must solve in order to move vary slightly according to the kind of game played (SSA vs. AsSA) and the players (Pr and Op). We formulate them as decision problems as follows:

SSA Problem under Semantics σ

Let $(\mathcal{A}, S_{\mathsf{Com}}^k, S_{\mathsf{Pr}}^k, S_{\mathsf{Op}}^k, \mathcal{R})$ be the split argumentation framework as in Definition 2.6 after turn k, and $\varphi \in L$ be the content of the dispute.

Pr's INSTANCE FOR TURN $k+1$: A split argumentation framework $(\mathcal{A}, S_{\mathsf{Com}}^k, S_{\mathsf{Pr}}^k, S_{\mathsf{Op}}^k, \mathcal{R})$ and an expression $\varphi \in L$.

QUESTION: Does there exist a subset T of S_{Pr}^k such that φ is accepted by CAF^{k+1} under semantics σ?

Op's INSTANCE FOR TURN $k+1$: A split argumentation framework $(\mathcal{A}, S^k_{\mathsf{Com}}, S^k_{\mathsf{Pr}}, S^k_{\mathsf{Op}}, \mathcal{R})$ and an expression $\varphi \in L$.

QUESTION: Does there exist a subset T of S^k_{Op} such that $\neg\varphi$ is accepted by CAF^{k+1} under semantics σ?

Analogously, we can formalise the AsSA Problem.

AsSA Problem under Semantics σ

Let $(\mathcal{A}, S^k_{\mathsf{Com}}, S^k_{\mathsf{Pr}}, S^k_{\mathsf{Op}}, \mathcal{R})$ be the split argumentation framework as in Definition 2.7 after turn k, and $\varphi \in L$ be the content of the dispute.

Pr's INSTANCE FOR TURN $k+1$: A split argumentation framework $(\mathcal{A}, S^k_{\mathsf{Com}}, S^k_{\mathsf{Pr}}, S^k_{\mathsf{Op}}, \mathcal{R})$ and an expression $\varphi \in L$.

QUESTION: Does there exist a subset T of S^k_{Pr} such that φ is accepted by CAF^{k+1} under semantics σ?

Op's INSTANCE FOR TURN $k+1$: A split argumentation framework $(\mathcal{A}, S^k_{\mathsf{Com}}, S^k_{\mathsf{Pr}}, S^k_{\mathsf{Op}}, \mathcal{R})$ and an expression $\varphi \in L$.

QUESTION: Does there exist a subset T of S^k_{Op} such that φ is not accepted by CAF^{k+1} under semantics σ?

In Section 5, we will give an implementation of the strategic argumentation game with *Defeasible Logic* (DL) [Nute, 2001] as the underlying logical framework, and assess the complexity of these problems.

3 Background

In this section we outline the concepts we use involving computational complexity, abstract and structured argumentation. This is not intended to be an introduction to these topics, it is simply a sketch of the concepts, assuming a familiarity with the more common elements. Those with less familiarity with these topics might want to read an introduction first, such as [Johnson, 1990; Dvořák and Dunne, 2017] for computational complexity, [Baroni et al., 2018a; Baroni et al., 2011] for abstract argumentation, and [Prakken, 2010b] for structured argumentation.

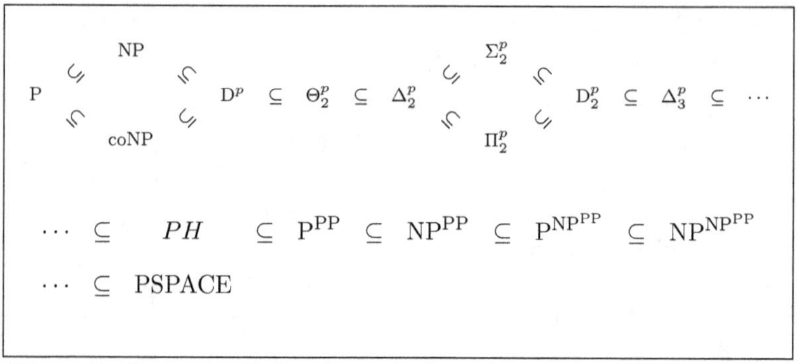

$$P \subseteq \begin{matrix} NP \\ \\ coNP \end{matrix} \subseteq D^p \subseteq \Theta_2^p \subseteq \Delta_2^p \subseteq \begin{matrix} \Sigma_2^p \\ \\ \Pi_2^p \end{matrix} \subseteq D_2^p \subseteq \Delta_3^p \subseteq \cdots$$

$$\cdots \subseteq PH \subseteq P^{PP} \subseteq NP^{PP} \subseteq P^{NP^{PP}} \subseteq NP^{NP^{PP}}$$

$$\cdots \subseteq PSPACE$$

Figure 1: Some complexity classes in the polynomial counting hierarchy, ordered by containment.

3.1 Complexity Classes

When addressing computational complexity we will focus on decision problems, because of their more familiar complexity classes, rather than their functional counterparts, which are more appropriate for many of the computational tasks we will address. We assume familiarity with the polynomial time complexity hierarchy but we will introduce some other complexity classes that we will need, and computational problems that are complete for each class. As is usual, $\mathcal{D}^{\mathcal{C}}$ denotes the class of problems that can be solved with complexity $\mathcal{D}$ if given an oracle for a problem in $\mathcal{C}$.

Within the polynomial hierarchy, a complete problem for Σ_n^p (Π_n^p) is the satisfiability of quantified Boolean formulas (QBF) with quantifiers in the form $\exists\forall\exists\cdots\exists$ (respectively, $\forall\exists\forall\cdots\exists$) with n alternations of quantifiers. PSPACE is the class of decision problems solvable in polynomial space. It contains the entire polynomial hierarchy PH. A complete problem for PSPACE is satisfiability of all quantified Boolean formulas.

D^p is the complexity class of problems that can be expressed as the conjunction of a problem in NP and a problem in coNP. A complete problem for D^p asks, given Boolean formulas ϕ and ψ, is ϕ unsatisfiable and ψ satisfiable? $NP^{D^p} = \Sigma_2^p$. Similarly D_2^p is the conjunction of problems in Σ_2^p and Π_2^p.

Θ_2^p is the class of decision problems solvable by a deterministic polynomial algorithm with $O(\log n)$ calls to an NP oracle. It is equal to $P_{||}^{NP}$, the class of problems solvable by a deterministic polynomial algorithm with non-adaptive calls to an NP oracle. Non-adaptive refers to the restriction that oracle calls cannot depend on the outcome of previous calls. $NP^{\Theta_2^p} = \Sigma_2^p$.

Δ_2^p is equal to P^{NP}. A complete problem for Δ_2^p is, given a Boolean formula ψ, does the lexicographically last satisfying assignment for ψ end with a 1?

PP is, roughly, the class of decision problems that have more accepting paths than rejecting paths. It can be thought of as a decision problem version of the more familiar counting complexity class #P, which addresses absolute counting, while PP addresses relative size of counts. We have $P^{\#P} = P^{PP}$ and $NP^{\#P} = NP^{PP}$. The entire polynomial hierarchy is contained within NP^{PP}. A complete problem for PP, called MAJSAT, is to decide whether a given Boolean formula is satisfied by more than half of the assignments to its variables. This can be expressed via a "counting" quantifier C as satisfying $\mathsf{C}X\ \psi$. Similarly, a complete problem for NP^{PP}, called E-MAJSAT is satisfying formulas $\exists X\mathsf{C}Y\ \psi$. And so on.

The *counting polynomial hierarchy* [Wagner, 1986] extends the polynomial hierarchy by incorporating PP, P^{PP}, NP^{PP}, $coNP^{PP}$, etc. Figure 1 displays containment relations among relevant complexity classes. In addition to the containments displayed, $\Theta_2^p \subseteq PP \subseteq P^{PP}$.

3.2 Abstract Argumentation

Definition 3.1 (Abstract Argumentation Framework). *An abstract argumentation framework is a pair $(\mathcal{A}, \gg)$ where $\mathcal{A}$ is a set of arguments and $\gg$ is a subset of $\mathcal{A} \times \mathcal{A}$, where $(a, b) \in \gg$ denotes that a attacks b.*

An abstract argumentation framework can be represented as a directed graph, where each vertex is an argument, and a directed edge from a to b if a attacks b. An argumentation framework is acyclic if the corresponding directed graph is acyclic.

For the purposes of this chapter, a semantics maps an argumentation framework to a set of extensions, each extension being a set of arguments (the set of arguments accepted in that extension)[5]. When representing the state of an argument in an extension, we will use the labelling approach (see, for example, [Baroni et al., 2018a; Baroni et al., 2011]) in which the argument is labelled either in, out, or undec. That is, an extension E is defined as a function $Lab_E : \mathcal{A} \to \{\text{in}, \text{out}, \text{undec}\}$. Then we can define an extension E as $\{a \in \mathcal{A} \mid Lab_E(a) = \text{in}\}$.

Given an argumentation framework $\mathcal{AF} = (\mathcal{A}, \gg)$, an argument a is said to be *accepted* in an extension E if $Lab_E(a) = \text{in}$, *rejected* in E if $Lab_E(a) = \text{out}$, and *undecided* in E if $Lab_E(a) = \text{undec}$. An extension E is *conflict-free* if no accepted argument is attacked by an accepted argument. An argument a is *defended* by E if every argument that attacks a is attacked by some argument accepted in E. An extension E of $\mathcal{AF}$ is *stable* if it is conflict-free and for every argument $a \in \mathcal{A} \backslash E$ there is an argument in E that attacks a. An extension E of $\mathcal{AF}$ is *complete* if it is conflict-free and, $a \in E$ iff a is defended by E.

The set of complete extensions forms a lower semi-lattice under the containment ordering, and many semantics can be defined directly in terms of this semi-lattice. The least complete extension under the containment ordering exists and is called the *grounded* extension. The *preferred* extensions are the maximal complete extensions under the containment ordering. The *semi-stable* extensions are the complete extensions where the set of arguments labelled with in or out is maximal under the containment ordering. The *ideal* extension is the maximal complete extension contained in all preferred extensions. Similarly, the *eager* extension is the maximal complete extension contained in all semi-stable extensions. These are not necessarily the original definitions of these extensions, but they are equivalent definitions.

We will use $\mathcal{GR}$ to denote the grounded semantics, $\mathcal{ST}$ for the stable semantics, $\mathcal{CO}$ for the complete semantics, $\mathcal{PR}$ for the preferred semantics, $\mathcal{ST}$ for the stable semantics, $\mathcal{SST}$ for the semi-stable semantics, $\mathcal{EA}$ for the eager semantics, and $\mathcal{ID}$ for the ideal semantics.

We say a semantics is *completist* if every argumentation framework is mapped to a set of complete extensions. These semantics will be our main focus. A semantics is *strongly completist* if it is completist and the set of

---

[5]Thus we will not address the gradual and ranking semantics discussed in [Baroni et al., 2018b; Amgoud, 2019].

extensions is determined by the semi-lattice structure of the complete extensions. Among the completist semantics are the grounded, preferred, stable, semi-stable, ideal, eager, and complete semantics. All except the stable semantics are strongly completist. Stable extensions are defined by a property of the individual extension, rather than by a structural property within the semi-lattice of complete extensions, and it turns out there is no equivalent structural definition [Maher, 2016b]. Stable semantics is also exceptional in that some argumentation frameworks have no stable extensions.

Each semantics implicitly expresses a criterion for what arguments can coherently be accepted together, given an argumentation framework. Each extension in the semantics represents a "reasonable" adjudication, according to that criterion, of the arguments in the argumentation framework.

Our restriction to completist semantics is, then, an implicit requirement that reasonable adjudications are conflict-free, defend all the accepted arguments, and accept all the defended arguments.[6] Each of the semantics, except (obviously) the complete semantics, imposes extra requirements, reflecting different emphases: the grounded semantics is highly sceptical, requiring a minimal set of accepted arguments[7]; the preferred semantics requires maximal sets of accepted arguments; the stable semantics requires that no argument is left undecided; the semi-stable semantics requires minimal sets of undecided arguments; the ideal semantics requires accepting only arguments that are accepted in all preferred extensions, and accepting as many of these as possible; the eager semantics requires accepting only arguments that are accepted in all semi-stable extensions, and accepting as many of these as possible.

The grounded, ideal and eager semantics are *unitary*: they contain exactly one extension. Such semantics limit, somewhat, the range of possible strategic aims of players in strategic argumentation, as we will see later.

Structural properties of an argumentation framework can influence the relationship between various semantics, which can make proving the

---

[6] However, we make this restriction in this chapter only for simplicity, and not on the basis that this implicit requirement is justified.

[7] Or, equivalently, accepting only arguments that are accepted in all complete extensions.

computational complexity of some problems easier. An argumentation framework is *well-founded* if there is no infinite sequence of arguments $a_1, a_2, \ldots, a_i, a_{i+1}, \ldots$ such that, for each i, a_{i+1} attacks a_i. Such argumentation frameworks have a single complete extension which must be the grounded extension [Dung, 1995], in which every argument is either accepted or rejected. Every completist semantics for such argumentation frameworks consists of this single extension.

An argument framework is *coherent* if every preferred extension is stable. An argument b *indirectly attacks* an argument a if there is a path of odd length from b to a, and *indirectly defends* a if there is a path of even length from b to a. An argument b is *controversial* wrt a if b indirectly attacks a and indirectly defends a. An argument is *controversial* if it is controversial wrt some argument. An argument framework is *uncontroversial* if there is no controversial argument. An argument framework is *limited controversial* if there is no infinite sequence of arguments $a_1, a_2, \ldots, a_i, \ldots$ such that a_{i-1} is controversial wrt a_i. Dung shows that (Theorem 33 of [Dung, 1995]) every limited controversial argument framework is coherent, and every uncontroversial argument framework is also relatively grounded. An argument framework is *relatively grounded* if intersection of all preferred extensions coincides with the grounded extension.

3.3 Structured Argumentation

Argumentation takes place over a language of expressions, most commonly a language of literals. For definiteness, in this chapter we consider propositional literals.

Definition 3.2 (Language). *The language L of expressions consists of a set of literals. Given a set PROP of propositional atoms, the set of literals is $\text{Lit} = \text{PROP} \cup \{\neg p \mid p \in \text{PROP}\}$. We denote with $\sim p$ the complementary of literal p; if p is a positive literal q, then $\sim p$ is $\neg q$, and if p is a negative literal $\neg q$, then $\sim p$ is q.*

Rules are built out of these expressions. Rules have labels to name them, but these are completely separate from labels used in abstract argumentation.

Definition 3.3 (Rules). *Let* Lab *be a set of rule labels. A rule r with $r \in$ Lab describes the relation between a set of expressions, called the* antecedent *(or* body *or the* premise*) of r and denoted by $A(r)$ (which may be empty) and an expression, called the* consequent*, or* head*, of r and denoted by $C(r)$. Three kind of rules are allowed: strict rules of the form $r : A(r) \to C(r)$, defeasible rules of the form $r : A(r) \Rightarrow C(r)$, and defeaters of the form $r : A(r) \rightsquigarrow C(r)$.*

A strict rule is a rule in the classical sense: whenever the antecedent holds, so does the conclusion. We call a strict rule without antecedent a *fact*, but we often distinguish facts from "true" strict rules that have an antecedent. A defeasible rule is allowed to assert its conclusion unless there is contrary evidence to it. A defeater is a rule that cannot be used to draw any conclusion, but can provide contrary evidence to complementary conclusions. A defeater in this sense [Nute, 1988] can be considered an instance of the general notion of defeater in epistemology: evidence that counts against a belief.

Definition 3.4 (Argumentation Theory). *An argumentation theory D is a structure $(R, >)$, where R is a (finite) set of rules and $> \subseteq R \times R$ is a binary relation on R called the* superiority *relation.*

The relation $>$ describes the relative strength of rules, that is to say, when a single rule may override the conclusion of another rule, and is required to be irreflexive, asymmetric, and acyclic (i.e., its transitive closure is irreflexive). To simplify discussion, we assume no strict rule is inferior to another rule. We use the following abbreviations on R: the set of strict rules in R is denoted by R_s, the set of strict and defeasible rules in R by R_{sd}, the set of defeasible rules by R_d, the set of defeaters by R_{dft}, and $R[q]$ is the set of rules in R whose head is q.

To demonstrate these definitions, we look at a time-honoured example of defeasible reasoning.

Example 3.5. *Consider an argumentation theory consisting of the fol-*

lowing rules

$$
\begin{aligned}
r_1 &: & bird(X) &\Rightarrow fly(X) \\
r_2 &: & penguin(X) &\Rightarrow \neg fly(X) \\
r_3 &: & penguin(X) &\to bird(X) \\
r_4 &: & injured(X) &\rightsquigarrow \neg fly(X) \\
f &: & penguin(tweety) & \\
g &: & bird(freddie) & \\
h &: & injured(freddie) &
\end{aligned}
$$

and a priority relation $r_2 > r_1$.

Here r_1, r_2, r_3, r_4, f are labels and r_3 is (a reference to) a strict rule, while r_1 and r_2 are defeasible rules, r_4 is a defeater, and f, g, h are facts. Thus $R_s = \{r_3, f, g, h\}$ and $R_{sd} = R = \{r_1, r_2, r_3\}$ and $>$ consists of the single tuple (r_2, r_1). The rules express that birds usually fly (r_1), penguins usually don't fly (r_2), that all penguins are birds (r_3), and that an injured animal may not be able to fly (r_4). In addition, we are given the facts that tweety is a penguin, and freddie is an injured bird. Finally, the priority of r_2 over r_1 expresses that when something is both a bird and a penguin (that is, when both rules can fire) it usually cannot fly (that is, only r_2 may fire, it overrules r_1).

By combining the rules in a theory, we can build arguments (we adjust the definition in [Prakken, 2010b] to meet Definition 3.4). In what follows, for a given argument A, **Conc** returns its conclusion, **Sub** returns all its sub-arguments, **Rules** returns all the rules in the argument and, finally, **TopRule** returns the last inference rule in the argument.

Definition 3.6 (Argument). *Let $D = (R, >)$ be an argumentation theory and $\Rightarrow\, \in \{\to, \Rightarrow, \rightsquigarrow\}$. An argument A constructed from D has the form $A_1, \ldots, A_n \Rightarrow_r \psi$, where*

- *A_k is an argument constructed from D, for $1 \leq k \leq n$, and*

- *$r : \mathsf{Conc}(A_1), \ldots, \mathsf{Conc}(A_n) \Rightarrow \psi$ is a rule in R.*

The set of arguments constructed from D is the smallest set of arguments satisfying this condition.

With regard to argument A, the following holds:

$\mathsf{Conc}(A) = \psi$
$\mathsf{Sub}(A) = \mathsf{Sub}(A_1) \cup \cdots \cup \mathsf{Sub}(A_n) \cup \{A\}$
$\mathsf{TopRule}(A) = r : \mathsf{Conc}(A_1), \ldots, \mathsf{Conc}(A_n) \Rightarrow \psi$
$\mathsf{Rules}(A) = \mathsf{Rules}(A_1) \cup \cdots \cup \mathsf{Rules}(A_n) \cup \{\mathsf{TopRule}(A)\}$
$(\mathsf{Rules}(A_1) \cup \cdots \cup \mathsf{Rules}(A_n)) \cap R_{dft} = \emptyset$

If $\mathsf{Rules}(A) \subseteq R_s$ then argument A is strict, otherwise A is defeasible. If $\mathsf{Rules}(A) \cap R_{dft} \neq \emptyset$ then argument A is non-supportive, otherwise it is supportive.

Conflicts between contradictory argument conclusions are resolved on the basis of preferences over arguments using a simple last-link ordering. An argument A is stronger than another argument B (written $A > B$) iff B is defeasible, and either A is strict or $\mathsf{TopRule}(A)$ is stronger than $\mathsf{TopRule}(B)$ ($\mathsf{TopRule}(A) > \mathsf{TopRule}(B)$).

Definition 3.7 (Attacks). *An argument B attacks an argument A iff $\exists A' \in \mathsf{Sub}(A)$ such that $\mathsf{Conc}(B) = \sim\mathsf{Conc}(A')$, and $A' \not> B$.*

We can now define the argumentation framework that is determined by an argumentation theory.

Definition 3.8 (*AF determined by an argumentation theory*). *Let $D = (R, >)$ be an argumentation theory. The argumentation framework determined by D is $(\mathcal{A}, \gg)$, where $\mathcal{A}$ is the set of all arguments constructed from D, and $\gg$ is the attack relation defined above.*

Given this definition of argumentation framework, if D is an argumentation theory, we can abuse notation somewhat and write $\mathcal{GR}(D)$ to denote the grounded extension of the argumentation framework determined by D.

Definition 3.9 (Justified Conclusion). *Given an argumentation theory D, we say a conclusion ψ is justified by D under the grounded semantics iff there exists a supportive argument a in $\mathcal{GR}(D)$ such that $\mathsf{Conc}(a) = \psi$.*

The following example illustrates the notions just introduced.

Example 3.10. *Using the rules from Example 3.5, we have arguments:*

$A_1:\quad \rightarrow_f penguin(tweety)$ \hspace{1em} (*strict argument*)
$A_2:\quad A_1 \rightarrow_{r_3} bird(tweety)$ \hspace{1em} (*strict argument*)
$A_3:\quad A_2 \Rightarrow_{r_1} fly(tweety)$ \hspace{1em} (*defeasible argument*)
$A_4:\quad A_1 \Rightarrow_{r_2} \neg fly(tweety)$ \hspace{1em} (*defeasible argument*)

among others.

If we consider the argument A_3, we have

$$\begin{aligned}
\mathsf{Conc}(A_3) &= fly(tweety) \\
\mathsf{Sub}(A_3) &= \{A_1, A_2, A_3\} \\
\mathsf{TopRule}(A_3) &= r_1 \\
\mathsf{Rules}(A_3) &= \{f, r_1, r_3\}
\end{aligned}$$

A_4 *attacks A_3 because the two arguments have contradictory conclusions and $r_1 \not> r_2$. On the other hand, A_3 does not attack A_4 because $r_2 > r_1$.*

In the argumentation framework determined by this theory there is no argument attacking A_4. Hence A_4 appears in the grounded extension. Since A_4 is a supportive argument, its conclusion $\neg fly(tweety)$ is justified under the grounded semantics.

4 Defeasible Logic

Defeasible Logic (DL) [Nute, 1994] is a rule-based sceptical approach to non-monotonic reasoning. It is based on a logic programming-like language and is a simple, efficient but flexible formalism capable of dealing with many intuitions of non-monotonic reasoning in a natural and meaningful way [Antoniou, 2004].

Defeasible rule languages like defeasible logic have been shown to be useful in representing legal documents and reasoning [Prakken and Sartor, 1998; Antoniou et al., 1999; Reeves et al., 2002; Governatori and Rotolo, 2010; Governatori et al., 2013; Islam and Governatori, 2018; Hashmi et al., 2018]. There are a variety of defeasible logics, which have been argued to represent the different proof standards that apply in legal systems [Governatori, 2011; Governatori and Maher, 2017].

Defeasible logics have much in common with argumentation, but there is only little work substantiating the relationship. [Governatori et al., 2004] characterizes inference in two defeasible logics in terms of argumentation. [Governatori, 2011] maps proof in Carneades [Gordon et al., 2007] at a given proof standard into proof in a defeasible logic. [Lam et al., 2016] showed how to map one instance of $ASPIC^+$ into a defeasible logic. [Maher, 2017b] gave two embeddings of abstract argumentation frameworks $\mathcal{AF}$ into a small subset of defeasible rule languages, implying, in particular, that acceptance in the grounded extension of $\mathcal{AF}$ can be implemented in a wide variety of defeasible logics and other concrete defeasible reasoning formalisms.

In this section we define two defeasible logics, but first we introduce defeasible logic in general.

4.1 Defeasible logic

The language of DL consists of literals and rules. To avoid notational redundancies, we use the same definitions of PROP, Lit, complementary literal, and the same rule types, structure and notation as already introduced in Definition 3.2.

A defeasible theory D is a triple $(F, R, >)$, where $F \subseteq$ Lit is a set of indisputable statements called *facts*, R is a (finite) set of rules, and $> \subseteq R \times R$ is a superiority relation on R as introduced in Definition 3.4.

A *derivation* (or *proof*) is a finite sequence $P = P(1), \ldots, P(n)$ of *tagged literals* of the type $+\Delta q$ (q is definitely provable), $-\Delta q$ (q is definitely refuted), $+d\,q$ (q is defeasibly provable) and $-d\,q$ (q is defeasibly refuted). The proof conditions below define the logical meaning of such tagged literals. Given a proof P, $P(n)$ denotes the n-th element of the sequence, and $P(1..n)$ denotes the first n elements of P. $\pm \Delta$ and $\pm df$ are called *proof tags*. Given $\#$ a proof tag, the notation $D \vdash \pm \# q$ means that there is a proof P in D such that $P(n) = \pm \# q$ for an index n.

In the remainder, we only present the proof conditions for the positive tags: the negative ones are obtained via the principle of *strong negation*. This is closely related to the function that simplifies a formula by moving all negations to an inner most position in the resulting formula, and replaces the positive tags with the respective negative tags, and the other way around [Antoniou et al., 2000a].

The proof conditions for $+\Delta$ describe just forward chaining of strict rules.

$+\Delta$: If $P(n+1) = +\Delta q$ then either
  (1) $q \in F$ or
  (2) $\exists r \in R_s[q]$ s.t. $\forall a \in A(r)$. $+\Delta a \in P(1..n)$.

Literal q is definitely provable if either (1) is a fact, or (2) there is a strict rule for q, whose antecedents have all been definitely proved. Literal q is definitely refuted if (1) is not a fact and (2) every strict rule for q has at least one definitely refuted antecedent. Conceptually, strict derivations are much stronger than defeasible ones: the superiority relation plays no part in them. If we have two strict rules for opposite conclusions whose antecedents are all proven, then the logic will derive both conclusions, which signals an inconsistency within the theory itself.

The conditions to establish a defeasible proof $+d$ have a structure similar to arguments, and are formalised by the following schema.

$+d$: If $P(n+1) = +d\,q$ then either
  (1) $+\Delta q \in P(1..n)$ or
  (2) (2.1) $-\Delta{\sim}q \in P(1..n)$ and
    (2.2) $\exists r \in R_{sd}[q]$ s.t. r is applicable, and
    (2.3) $\forall s \in R[{\sim}q]$. either
      (2.3.1) s is unsupported, or
      (2.3.2) s is defeated.

Intuitively, a rule is applicable if all the literals in the antecedent have previously been proven. Clause (2.3) considers the possible counter-arguments. To derive q, each such counter-argument must be either unsupported, or defeated. A rule is unsupported if it is not possible to give a (valid) justification for at least one of the premises of the rule. The degree of provability of the conclusion we want to obtain determines the meaning of valid justification for a premise. This could vary from a derivation for the premise to a simple chain of rules leading to it. Finally, a rule is defeated if there is an applicable rule stronger than it.

By instantiating the abstract definitions of applicable, supported and defeated, the above structure defines several variants of DL. In particular, we address the distinction between *ambiguity blocking* and *ambiguity propagation*. A literal q is *ambiguous* if (i) there is a chain of reasoning that supports a conclusion q, (ii) one (chain) supporting the complementary conclusion $\sim q$, and (iii) the superiority relation does not resolve this conflict.

Example 4.1. *Consider the defeasible theory $D = (\emptyset, R, \emptyset)$, such that*

$$R = \{r_1: \Rightarrow a, \quad r_2: \Rightarrow b, \quad r_3: \Rightarrow \neg a, \quad r_4: a \Rightarrow \neg b\}.$$

Here a is ambiguous since both r_1 and r_3 are applicable, and there is no superiority between them.

In what follows we shall introduce two variants of DL, the first one supporting ambiguity blocking, and the second one supporting ambiguity propagation. We explain the intuitions behind the two variants by referring to Example 4.1, where a is ambiguous. In a setting where ambiguity is blocked, b is not ambiguous because rule r_2 for b is applicable, whilst r_4 for $\neg b$ is not, since we cannot prove a. On the other hand, in an ambiguity propagating setting, b is ambiguous because a is not disproved, and so the applicability of r_4 is not denied. In this way, the ambiguity is propagated to b.

The ambiguity blocking and ambiguity propagation is a clash in intuitions in non-monotonic reasoning [Touretzky et al., 1987]. However, [Governatori, 2011] argues that the distinction can be used to characterise different proof standards, where ambiguity blocking corresponds to the proof standard of *preponderance of evidence* while ambiguity propagation captures the *beyond reasonable doubt* proof standard. Furthermore, there are scenarios where both intuitions are needed (for different conclusions), and the reasoning for conclusions requiring one of the two proof standard depends on conclusions obtained using the other proof standard. See [Governatori and Maher, 2017] for the details and how to combine the two intuitions.

In the remainder, we shall use ∂ for the proof tag to indicate that a conclusion is defeasibly provable (refutable) under ambiguity blocking, and δ for the corresponding notions under ambiguity propagation.

4.2 Ambiguity Blocking Defeasible Logic

The ambiguity blocking variant of DL was introduced in [Antoniou et al., 2001b] and is captured by the following instantiation of $+d$:

$+\partial$: If $P(n+1) = +\partial q$ then either
  (1) $+\Delta q \in P(1..n)$ or
  (2) (2.1) $-\Delta \sim q \in P(1..n)$ and
      (2.2) $\exists r \in R_{sd}[q]$ s.t. $\forall a \in A(r) + \partial a \in P(1..n)$ and
      (2.3) $\forall s \in R[\sim q]$ either
          (2.3.1) $\exists a \in A(s)$ s.t. $-\partial a \in P(1..n)$ or
          (2.3.2) $\exists t \in R_{sd}[q]$ s.t.
              $\forall a \in A(t) + \partial a \in P(1..n)$ and $t > s$.

To prove $+\partial q$, we have to show that either (1) q is already definitely provable, or (2.2) there is an applicable rule for q and (2.3) for very rule attacking q either (2.3.1) at least one antecedent has been defeasibly refuted, or (2.3.2) the rule is defeated by a (stronger) rule for q.

In other terms, a rule is applicable if all the elements of the body are defeasibly provable. A rule is unsupported if there is an element of the body that is defeasibly refuted. A rule is defeated if it is weaker than an applicable rule. We use $DL(\partial)$ to denote the ambiguity blocking defeasible logic variant.

4.3 Ambiguity Propagating Defeasible Logic

Ambiguity propagation describes a behaviour where ambiguity of a literal is propagated to dependent literals. This is achieved in DL by separating the invalidation of a counterargument from the derivation of tagged literals. To do so, another kind of conclusion, called *support* and denoted by Σ, is introduced [Antoniou et al., 2000b].

$+\Sigma$: If $P(n+1) = +\Sigma q$ then either
  (1) $+\Delta q \in P(1..n)$ or
  (2) (2.1) $-\Delta \sim q \in P(1..n)$ and
      (2.2) $\exists r \in R_{sd}[q]$ s.t.
          (2.2.1) $\forall a \in A(r) + \Sigma a \in P(1..n)$ and
          (2.2.2) $\forall s \in R[\sim q]$ either
              $\exists a \in A(s)$ s.t. $-\delta a \in P(1..n)$, or $s \not> r$.

The condition for $+d$ is thus instantiated as follows:

$+\delta$: If $P(n+1) = +\delta q$ then either
  (1) $+\Delta q \in P(1..n)$ or
  (2) (2.1) $-\Delta \sim q \in P(1..n)$ and
      (2.2) $\exists r \in R_{sd}[q]$ s.t. $\forall a \in A(r) + \delta a \in P(1..n)$ and
      (2.3) $\forall s \in R[\sim q]$ either
        (2.3.1) $\exists a \in A(s)$ s.t. $-\Sigma a \in P(1..n)$ or
        (2.3.2) $\exists t \in R_{sd}[q]$ s.t.
            $\forall a \in A(t) + \delta a \in P(1..n)$ and $t > s$.

The idea is that a conclusion q is supported if (2.1) there is a rule for q such that (2.2.1) all the elements in the antecedent are (at least) supported, and that (2.2.2) all rules for the opposite conclusion have (at least) one premise that has been refuted, or such a rule is not stronger than the rule for q. This means that there is an undefeated argument supporting the conclusion. Then to affirm that a conclusion is provable, we have to provide an argument/rule where all the antecedents are provable, and there is no argument/rule for the opposite that is at least supported. We refer to the ambiguity propagating variant by using DL(δ).

Example 4.1 (Continued). *Consider, again, the theory $D = (\emptyset, R, \emptyset)$, where*

$$R = \{r_1 :\Rightarrow a, \quad r_2 :\Rightarrow b, \quad r_3 :\Rightarrow \neg a, \quad r_4 : a \Rightarrow \neg b\}.$$

By definition of $+\partial$, we obtain the following conclusions from D: $-\partial a$, $-\partial \neg a$, $+\partial b$, $-\partial \neg b$, capturing the ambiguity blocking behaviour of DL(∂). On the other hand, if we compute the consequences of D by using the proof conditions for Σ and δ, we obtain $+\Sigma a$, $+\Sigma \neg a$, $+\Sigma b$, $+\Sigma \neg b$ and thus also $-\delta a$, $-\delta \neg a$, $-\delta b$ and $-\delta \neg b$. In this way, we capture the ambiguity propagation feature of DL(δ).

4.4 Team or Individual Defeat?

The defeasible logics defined above have the property of *team defeat*: the rules for a literal q are compared with the rules for $\sim q$. If each applicable rule for $\sim q$ is inferior to some applicable rule for q, then the rules for q, as a team, overcome the rules for $\sim q$. Thus, q is inferred. In comparison, under *individual defeat* there must be an applicable rule for q that is superior to *all* applicable rules for $\sim q$ in order to overcome the rules for $\sim q$ and infer q. Clearly, any time individual defeat overcomes the rules for $\sim q$, so does team defeat.

To get some intuition about these two forms of defeat we use a variation of an example from [Antoniou et al., 2001b].

Example 4.2. *Consider some rules of thumb about animals and, particularly, mammals. An egg-laying animal is generally not a mammal. Similarly, an animal with webbed feet is generally not a mammal. On the other hand, an animal with fur is generally a mammal. Finally, the monotremes are a subclass of mammal. These rules are represented as defeasible rules below.*

Furthermore, animals with fur and webbed feet are generally mammals, so r_2 should overrule r_4. And monotremes are a class of egg-laying mammals, so r_1 should overrule r_3.

Finally, it happens that a platypus is a furry, egg-laying, web-footed monotreme. Is it a mammal? (That is, is mammal(platypus) a consequence of the defeasible theory below?)

$$r_1 : \quad monotreme(X) \Rightarrow mammal(X)$$
$$r_2 : \quad hasFur(X) \Rightarrow mammal(X)$$
$$r_3 : \quad laysEggs(X) \Rightarrow \neg mammal(X)$$
$$r_4 : \quad webFooted(X) \Rightarrow \neg mammal(X)$$
$$r_1 > r_3$$
$$r_2 > r_4$$
$$monotreme(platypus)$$
$$laysEggs(platypus)$$
$$hasFur(platypus)$$
$$webFooted(platypus)$$

It is obvious that all four rules are applicable to the question of mammal(platypus). Under team defeat, each rule for ¬mammal(platypus) is overcome by some rule for mammal(platypus), so mammal(platypus) *is inferred. However, there is no single rule for mammal(platypus) that overcomes all rules for* mammal(platypus), *so under individual defeat we cannot infer* mammal(platypus) *(nor* ¬mammal(platypus)*).*

Thus, we see that team defeat can be useful in making a justified inference that otherwise would not be made. On the other hand, most expressions of structured argumentation employ individual defeat.

Fortunately, it is easy to adjust the inference conditions for the two logics defined above to obtain individual defeat: we simply replace the sub-conditions (2.3.2) by $r > s$. We denote the individual defeat logics by DL(∂^*) and DL(δ^*). For more discussion of the four variants of defeasible logic discussed here, see [Billington et al., 2010].

Finally, we consider the relationship between these logics. A series of papers [Maher, 2012; Maher, 2013a; Maher, 2013b; Maher, 2014a] investigates the relative expressiveness of variants of Defeasible Logic. In brief, two (defeasible) logics L_1 and L_2 have the same expressiveness iff the two logics simulate each other (where a defeasible logic L_2 simulates a defeasible logic L_1 if there is a polynomial time transformation T that transforms a theory D_1 of L_1 in a theory $D_2 = T(D_1)$ of L_2 such that, for any addition of facts A, all strict and defeasible conclusions of $D_1 \cup A$ are the same as those of $D_2 \cup A$ in L_1). [Maher, 2012; Maher, 2013a] provide polynomial time transformations between each of the four logics defined above.

Theorem 4.3. *[Maher, 2013a] Each of* DL(∂), DL(δ), DL(∂^*), *and* DL(δ^*) *simulates the others.*

5 Strategic Argumentation for Defeasible Logic and Structured Argumentation

We now propose a Defeasible Logic instantiation of the games introduced in Section 2. We shall hence specialise Definitions 2.6 and 2.7 for the instance at hand, and then proceed with the formulation of two problems.

Given a defeasible theory $D = (F, R, >)$, we define the corresponding *split defeasible theory* as $SD = (F_{\text{Com}}, F_{\text{Pr}}, F_{\text{Op}}, R_{\text{Com}}, R_{\text{Pr}}, R_{\text{Op}}, >)$ with $F = F_{\text{Com}} \cup F_{\text{Pr}} \cup F_{\text{Op}}$ and $R = R_{\text{Com}} \cup R_{\text{Pr}} \cup R_{\text{Op}}$. We call the content of dispute discussed by the players the *critical literal*, and note that the arguments brought about by the players will be in the form of defeasible derivations. We assume that each player is informed about the restriction of $>$ to their private rules,

We will have three instances of the definitions of Section 2, owing to the extra expressivity of defeasible logic. Defeasible logic offers the following three ways to express a contrary to $D \vdash +d\,q$: the negation of q can be proved ($D \vdash +d\sim q$); within the logic we can prove that that $+d\,q$ cannot be proved ($D \vdash -d\,q$); and, we cannot prove $+d\,q$ ($D \not\vdash +d\,q$). Thus, if Pr wants to prove q, Op has three possible levels of opposition. The first will lead to a symmetric game, and the third to an asymmetric game. The second falls somewhere in between, and we will call it a *semi-symmetric* game. In the semi-symmetric game Op shoulders a burden of proof, but only to prove that Pr's aim cannot be proved, not to prove the negation of q.

If we consider the asymmetric case corresponds to the Scottish verdict of not proven[8] and the symmetric case corresponds to not guilty, then what is the semi-symmetric case? Technically, in defeasible logic, the distinction between semi-symmetric and asymmetric opposition is caused by a circularity or infinite regress in an argument. Abstractly, it might represent unknowability, or an incapacity of the proceedings/inference rules – inability to decide that l is not provable, even though l, in fact, is not provable (a little bit like Gödel's incompleteness theorem).

The game rules discussed in Section 2 are instantiated as follows. The parties start the game by choosing the critical literal l. Pr has the burden to prove $+d\,l$ by using the remainder of its private rules along with those that currently have been played; Op's final aim is to prove $+d\sim l$ in the symmetric version of the game, to prove $-d\,l$ in the semi-symmetric game, and simply to prevent the proof of $+d\,l$ in the asymmetric game.

Note that, when putting forward an argument, the players: (1) may propose, along with a subset of their private rules, a subset of their private facts to support such rules (see Example 5.2 at the end of this section), and (2) may play an argument whose terminal literal differs

---

[8]Roughly, under this verdict the jury considers the prosecution has not made the case for "guilty", beyond a reasonable doubt, but the defence has not made the case for "innocent". A verdict of *guilty* is given when the jury considers the prosecution has made its case, and *not guilty* when the defence has made its case. See [Barbato, 2005] or the Wikipedia entry for *Not proven*.

from l or $\sim l$ (with the aim to attack/disprove one of the premises of a rule in the proof proving $l/\sim l$).

As the semi-symmetric and asymmetric games differ from the symmetric one only in Op's final aim, to avoid pedantic redundancies we shall provide a single definition for the three games.

Definition 5.1 (SSA (SSSA, AsSA) Game for Defeasible Logic). *Consider two players, a proponent* Pr *and an opponent* Op, *a split defeasible theory* $SD = (F_{\mathsf{Com}}, F_{\mathsf{Pr}}, F_{\mathsf{Op}}, R_{\mathsf{Com}}, R_{\mathsf{Pr}}, R_{\mathsf{Op}}, >)$, *and a critical literal* $l \in L$.

Let F_{Com}^k, R_{Com}^k, F_{Pr}^k, R_{Pr}^k, F_{Op}^k, *and* R_{Op}^k *denote, respectively, the common (knowledge) facts and rules,* Pr*'s private facts and rules, and* Op*'s private facts and rules, after turn* k. *(In particular,* $F_{\mathsf{Com}}^0 = F_{\mathsf{Com}}$, $R_{\mathsf{Com}}^0 = R_{\mathsf{Com}}$, $F_{\mathsf{Pr}}^0 = F_{\mathsf{Pr}}$ $R_{\mathsf{Pr}}^0 = R_{\mathsf{Pr}}$, $F_{\mathsf{Op}}^0 = F_{\mathsf{Op}}$, *and* $R_{\mathsf{Op}}^0 = R_{\mathsf{Op}}$.*) The common defeasible theory at that point is* $D^k = (F_{\mathsf{Com}}^k, R_{\mathsf{Com}}^k, >)$.

We define a symmetric *(resp. semi-symmetric, asymmetric) strategic argumentation game for Defeasible Logic as a dialogue game where:*

1. *The players take turns. If* $D^0 \vdash +d\, l$ *then* Op *begins; otherwise* Pr *does so.*

2. *At turn* k, *if* $D^{k-1} \vdash +d\, \neg l$ *(resp.* $D^{k-1} \vdash -d\, l$ *for the semi-symmetric version,* $D^k \nvdash +d\, l$ *for the asymmetric version), then it is* Pr*'s turn to play, as follows*

   - Pr *advances a subset of its private facts* $\Phi \subseteq F_{\mathsf{Pr}}^{k-1}$ *and rules* $\rho \subseteq R_{\mathsf{Pr}}^{k-1}$ *so that* $D^k \vdash +d\, l$. *As a result*
     - $F_{\mathsf{Com}}^k = F_{\mathsf{Com}}^{k-1} \cup \Phi$ *and* $R_{\mathsf{Com}}^k = R_{\mathsf{Com}}^{k-1} \cup \rho$;
     - $F_{\mathsf{Pr}}^k = F_{\mathsf{Pr}}^{k-1} \setminus \Phi$ *and* $R_{\mathsf{Pr}}^k = R_{\mathsf{Pr}}^{k-1} \setminus \rho$;
     - $R_{\mathsf{Op}}^k = R_{\mathsf{Op}}^{k-1}$.

3. *At turn* k, *if* $D^{k-1} \vdash +d\, l$, *then it is* Op*'s turn to play, as follows*

   - Op *advances a subset of its private* $\Phi \subseteq F_{\mathsf{Op}}^{k-1}$ *and rules* $\rho \subseteq R_{\mathsf{Op}}^{k-1}$ *so that* $D^k \vdash +d\, \neg l$ *(resp.* $D^k \vdash -d\, l$ *for the semi-symmetric version,* $D^k \nvdash +d\, l$ *for the asymmetric version). As a result*

- $F_{\mathsf{Com}}^k = F_{\mathsf{Com}}^{k-1} \cup \Phi$ and $R_{\mathsf{Com}}^k = R_{\mathsf{Com}}^{k-1} \cup \rho$;
- $R_{\mathsf{Pr}}^k = R_{\mathsf{Pr}}^{k-1}$;
- $F_{\mathsf{Op}}^k = F_{\mathsf{Op}}^{k-1} \setminus \Phi$ and $R_{\mathsf{Op}}^k = R_{\mathsf{Op}}^{k-1} \setminus \rho$.

4. *The game ends at turn $k+1$, when either (i) it is Pr's turn and there is no move for Pr such that the common defeasible theory $D^{k+1} \vdash +dl$, in which case Op wins, or (ii) it is Op's turn and there is no move for Op such that the common defeasible theory $D^{k+1} \vdash +d\neg l$ (resp. $D^{k+1} \vdash -dl$ for the semi-symmetric version, $D^k \not\vdash +dl$ for the asymmetric version), in which case Pr wins.*

The corresponding decision problems are as follows.

SSA (SSSA, AsSA) Problem for Defeasible Logic

Let SD^k be a split defeasible theory as in Definition 5.1 after turn k, D^{k+1} be the corresponding common defeasible theory after turn $k+1$, and $l \in L$ be the critical literal.

> Pr's INSTANCE FOR TURN $k+1$: Let F_{Pr}^k and R_{Pr}^k be, respectively, the set of Pr's private facts and rules after turn k, and that the common defeasible theory assume $D^k \vdash +d\neg l$ (resp. $D^k \vdash -dl$ and $D^k \not\vdash +dl$ for the semi-symmetric and asymmetric problems).
>
> QUESTION: Do there exist Φ subset of F_{Pr}^k and ρ subset of R_{Pr}^k such that the common defeasible theory $D^{k+1} \vdash +dl$?
>
> Op's INSTANCE FOR TURN $k+1$: Let F_{Op}^k and R_{Op}^k be, respectively, the set of Op's private facts and rules after turn k, and assume that the common defeasible theory $D^k \vdash +dl$.
>
> QUESTION: Do there exist Φ subset of F_{Op}^k and ρ subset of R_{Op}^k such that the common defeasible theory $D^{k+1} \vdash +d\neg l$ (resp. $D^{k+1} \vdash -dl$ and $D^{k+1} \not\vdash +dl$, for the semi-symmetric and asymmetric problems)?

We explore how these games are played through an example theory that shows how different moves by the players may lead to different result of the game in the symmetric and semi-symmetric/asymmetric variants.

Example 5.2. *Consider* $SD = (F_{\mathsf{Com}}, F_{\mathsf{Pr}}, F_{\mathsf{Op}}, R_{\mathsf{Com}}, R_{\mathsf{Pr}}, R_{\mathsf{Op}}, >)$ *such that*

- $F_{\mathsf{Com}} = \{a\}$ *and* $R_{\mathsf{Com}} = \emptyset$;
- $F_{\mathsf{Pr}} = \{d\}$ *and* $R_{\mathsf{Pr}} = \{r_1 : a \Rightarrow p,\ r_2 : b, d \Rightarrow p\}$;
- $F_{\mathsf{Op}} = \{b, c\}$ *and* $R_{\mathsf{Op}} = \{r_3 : c \Rightarrow \neg p,\ r_4 : b \Rightarrow \neg p\}$; *and*
- $> \{(r_4, r_1), (r_2, r_4)\}$.

The critical literal is p. Pr starts the game and can only advance r_1; the fact that b is not proven makes r_2 unsupported. Consequently, for both variants, $SD^1 \vdash +d\,p$. We now detail the different scenarios for Op wrt the symmetric, semi-symmetric, and asymmetric games.

Symmetric variant. Op considers playing r_3 but realises that is not a legal move. In fact, as r_3 is neither stronger than r_1 nor r_2, by playing it Op would not prove $+d\,\neg p$. By playing r_4, Op must also advance r_4's only premise, b ($SD^2 \vdash +d\,\neg p$ and $SD^2 \vdash +d\,b$). This makes r_2 applicable and allows Pr to play it and win the game.

Semi-symmetric variant. For this variant of the game, Op has the burden to prove $-d\,p$ and plays, again, r_4 ($SD^2 \vdash +d\,\neg p$ and $+d\,\neg p$ implies $-d\,p$). Pr can again play r_2 leading to $SD^3 \vdash +d\,p$, but now if Op plays r_3 (along with c), then $SD^4 \vdash -d\,p$. Pr has no more rules to play and this time Op wins.

Asymmetric variant. This variant of the game unfolds in the same way as the semi-symmetric variant because, for every k, $SD^k \vdash -d\,p$ implies $SD^k \not\vdash +d\,p$.

We can modify the above example to demonstrate the distinction between the semi-symmetric and asymmetric games.

Example 5.3. *Consider the modification of Example 5.2 where r_3 in R_{Op} is replaced by*

$$r_3 : c, \neg p \Rightarrow \neg p$$

Symmetric variant. This variant unfolds in exactly the same way as Example 5.2. Op does not play r_3.

Semi-symmetric variant. For this variant of the game, Pr plays r_1, Op plays r_4, and Pr plays r_2, just as in the symmetric variant. At *this stage* Op *would like to play* r_3 *but, again, this is not a legal move: playing it would not achieve* $SD^4 \vdash -dp$. *Thus* Pr *wins*.

Asymmetric variant. *Again,* Pr *plays* r_1, Op *plays* r_4, *and* Pr *plays* r_2. *However, in this variant* Op *can play* r_3, *because then* $SD^3 \not\vdash +dp$. Pr *has no more moves, so* Op *wins. Alternatively,* Op *could simply play* r_3 *on her first move, to which* Pr *has no response. Thus* Op *wins without exposing* r_4 *and b (and without inducing* Pr *to expose* r_2 *and d)*.

We end this subsection with a brief discussion of fact-based strategic argumentation [Maher, 2014b], a refinement of the strategic argument games where players can only play facts. That is, strategic argument games where $R_{\mathsf{Pr}} = \emptyset$ and $R_{\mathsf{Op}} = \emptyset$. While general strategic argumentation can be a model for legal argumentation in general, this refinement reflects argument about whether regulations have been adhered to. The players are the party subject to the regulations, and the enforcement body for the regulations. R_{Com} represents the regulations, which are fixed. The players can only generate arguments by marshalling facts that support the applicability of clauses in the regulations (i.e. rules) that, in turn, support the player's contentions. This refinement could also be considered a crude partial model for pleadings in civil law (in that it elicits claimed facts from parties), although different in many ways from Gordon's Pleadings Game [Gordon, 1994].

Although this refinement appears to simplify the reasoning required to play the game, in one sense it is no simpler [Maher, 2014b]. Any general strategic argumentation game $SD = (F_{\mathsf{Com}}, F_{\mathsf{Pr}}, F_{\mathsf{Op}}, R_{\mathsf{Com}}, R_{\mathsf{Pr}}, R_{\mathsf{Op}}, >)$ can be reduced to the "simpler" game as follows: for each rule $r_i : \beta \Rightarrow \varphi$ in R_{Pr} we add the rule $r_i : \beta, \alpha(r_i) \Rightarrow \varphi$ to R_{Com} and add the fact $\alpha(r_i)$ to F_{Pr}, where $\alpha(r_i)$ is a new proposition. And similarly for Op. Every move in the resulting game $SD' = (F_{\mathsf{Com}}, F'_{\mathsf{Pr}}, F'_{\mathsf{Op}}, R'_{\mathsf{Com}}, \emptyset, \emptyset, >)$ corresponds to a move of SD, and vice versa.

5.1 Computational Results

We are now ready to show that deciding what arguments to play at a given turn of a dialogue game under Dung's grounded semantics is an NP-complete problem even when the problem of deciding whether a conclusion follows from an argument is computable in polynomial time.

[Governatori et al., 2014] proved that this problem is NP-complete for DL with ambiguity blocking, i.e., DL(∂). We present here an outline of the proof in [Maher, 2014b]. Theorem 5.4 is provided from the viewpoint of Pr. The same result holds for Op.

Theorem 5.4. *The SSA Problem under* DL(∂) *is NP-complete.*

Proof. First, the SSA Problem is polynomially solvable on non-deterministic machines. Consider a dialogue game with sets R^0_{Com}, R^0_{Pr}, R^0_{Op} and the defeasible theory $D^{i-1} = (\emptyset, R^{i-1}_{\text{Com}}, >)$, the theory at turn $i-1$ of a dialogue game. An oracle guesses a set of rules $R^i \subseteq R^{i-1}_{\text{Pr}}$, we compute the consequences of the argumentation theory $D^i = (\emptyset, R^{i-1}_{\text{Com}} \cup R^i, >)$, and we check whether the critical literal is a positive or negative consequence. The computation of consequences can be done in polynomial time [Maher, 2001; Billington et al., 2010].

Second, we reduce 3SAT to the SSA Problem, proving therefore that the problem is NP-hard. Consider a 3SAT formula $\varphi = \bigwedge_{j=1}^{n} C_j$ such that $C_j = \bigvee_{k=1}^{3} x_j^k$. R^i is defined as follows:

1. For each proposition x occurring in φ, R^{i-1}_{Pr} and R^{i-1}_{Op} both contain

$$t_x : \Rightarrow x$$
$$t_{\neg x} : \Rightarrow \neg x.$$

2. For each clause C_j, R^{i-1}_{Com} contains

$$r_j^k : x_j^k \Rightarrow c_j$$

where x_j^k is either a positive literal (x), or a negative literal ($\neg x$).

3. R^{i-1}_{Com} also contains

$$r_{sat} : c_1, \ldots, c_n \Rightarrow sat.$$

For any assignment θ of values to the Boolean variables in φ, let S_θ be the set of x literals that evaluate to true under θ. And for any consistent subset S of x literals, let θ_S be an assignment that evaluates all elements of S to true. We leave it for the reader to verify that if θ satisfies φ then choosing the move S_θ wins for Pr, and if S is a winning move for Pr then S is consistent and θ_S satisfies φ. □

The same result holds for the semi-symmetric and asymmetric games.

Theorem 5.5. *The SSSA and AsSA problems under* $DL(\partial)$ *is NP-complete.*

Proof. The proof is essentially the same as that of Theorem 5.4 except for the case when, at turn i, **Op** must play. In that case, the reduction is identical to the one proposed above, with the only difference that Point 3. now also adds to R^i_{Com} the following rule

$$r_{nsat}: \quad \Rightarrow \neg sat$$

It is trivial to prove that an interpretation satisfies φ iff r_{sat} is applicable iff sat and $\neg sat$ are ambiguous. Thus φ is satisfied iff $-\partial\, sat$ is proved iff $\neg sat$ is not proved. □

While it is possible to define $DL(\partial)$ in terms of an argumentation semantics, the logic corresponding to Dung's grounded semantics is ambiguity-propagating [Governatori et al., 2004; Lam et al., 2016].

The next step is to determine the computational complexity of the problem at hand for the ambiguity propagating variant of DL. The NP-completeness of the strategic argumentation problem under $DL(\delta^*)$ follows immediately from Theorems 4.3, 5.4, and 5.5.

Theorem 5.6. *The SSA, SSSA, and AsSA problems under* $DL(\delta^*)$ *are NP-complete.*

We have the same results for $DL(\partial^*)$ and $DL(\delta)$.

In [Lam et al., 2016], it is shown that the conclusions of an $ASPIC^+$ argumentation theory under grounded semantics are the same as those in $DL(\delta^*)$ (after minor changes to the superiority relation).

Theorem 5.7. *[Lam et al., 2016] Given an $ASPIC^+$ argumentation theory AT, there is a defeasible theory $T(AT)$ such that p is derived under the grounded semantics from AT iff $+\delta^* p$ can be derived from $T(AT)$. Furthermore, all consequences of AT can be computed in time polynomial in the size of AT.*

Thus we can use implementations of $DL(\delta^*)$ to implement $ASPIC^+$ argumentation theories that employ the last-link ordering of arguments and the grounded semantics.

We can solve the strategic argumentation problem by non-deterministically choosing a set R^i of rules and then verifying whether the critical literal p is justified in the argumentation framework determined by D^i, or not. Further, the literals justified by the grounded semantics are computable in polynomial time, as shown above. The strategic argumentation problem is thus in NP.

Now, from Theorems 5.6 and 5.7, we obtain the following result.

Theorem 5.8. *The strategic argumentation problems under the grounded semantics are* NP-*complete.*

6 Strategic Abstract Argumentation

In this section we look beyond the grounded semantics to a wide range of other semantics for abstract argumentation frameworks. After exploring the range of dialogue games that can be played in the context of abstract argumentation, we investigate the possibilities for player aims, and identify the complexity of two computational problems related to playing strategic abstract argumentation games, for selected aims and semantics.

6.1 Strategic Argumentation in the Abstract

We formulate a split argumentation framework in this abstract sense as a tuple
$(\mathcal{A}, \mathcal{A}_{Com}, \mathcal{A}_{Pr}, \mathcal{A}_{Op}, \gg)$ where $\mathcal{A}_{Com}$ is a set of abstract arguments that are common knowledge to the players, $\mathcal{A}_{Pr}$ ($\mathcal{A}_{Op}$) is the set of arguments known to Pr (Op), and $\gg$ is the attack relation over all arguments. Each player knows $\gg$ restricted to the set of arguments the player knows. For example, Pr knows $\gg$ restricted to $(\mathcal{A}_{Com} \cup \mathcal{A}_{Pr}) \times (\mathcal{A}_{Com} \cup \mathcal{A}_{Pr})$. Each player has a *strategic aim* or *desired outcome* (the two terms will be treated as equivalent) that expresses their desired property of the state of the argument framework at the end of the strategic argumentation game.

A strategic abstract argumentation game consists of alternating moves by Pr and Op until one player cannot make a move. In that case the other player wins. Pr starts the game by playing a set of arguments, including a mutually agreed *critical argument* which is the subject of the two players' strategic aims[9]. By "playing a set of arguments" we refer to the transfer of a set of arguments from the player's set of arguments to $\mathcal{A}_{Com}$ such that the revised common argumentation framework $(\mathcal{A}_{Com}, \gg)$ satisfies the player's strategic aim. Thus a move by Pr replaces a split argument framework $(\mathcal{A}_{Com}, \mathcal{A}_{Pr}, \mathcal{A}_{Op}, \gg)$ by a new framework $(\mathcal{A}_{Com} \cup X, \mathcal{A}_{Pr} \backslash X, \mathcal{A}_{Op}, \gg)$, where $X \subseteq \mathcal{A}_{Pr}$ is the set of arguments played by Pr in that move, and the new framework achieves Pr's strategic aim. Similarly, a move by Op transfers arguments from $\mathcal{A}_{Op}$ to $\mathcal{A}_{Com}$. Clearly, if $\mathcal{A}_{Pr}$ or $\mathcal{A}_{Op}$ is finite then the game terminates. We will only consider games where $\mathcal{A}_{Com}$, $\mathcal{A}_{Pr}$ and $\mathcal{A}_{Op}$ are finite.

Thus, a strategic abstract argumentation game is a dialogue game played by two players (Pr and Op). Let $conc$ map arguments to distinct propositions, and let φ be the conclusion of the critical argument. Then the game is an asymmetric strategic argumentation game, as defined in Definition 2.7, where "φ is accepted" is defined as: Pr's aim wrt the critical argument is satisfied.

We assume that the players agree on what is an argument, and whether one argument attacks another. This is implicit in the formulation as a split argumentation framework. But, in theory, there is no reason why the two players should employ the same semantics when they play a strategic argumentation game. For example, Pr might formulate her aim in terms of the preferred semantics, while Op's aim is expressed in terms of the eager semantics. Indeed, it is quite reasonable that different players might perceive the world differently. This is no impediment to the players playing a strategic argumentation game, since the definition of the game only describes moves a player may make, and not the interpretation she puts on the game.

However, there has not been any work on such situations. This is not so surprising when we consider that strategic argumentation is primarily treated as an adversarial game. Real world situations that are modelled by strategic argumentation may need the presence of an adjudicator to enforce any conclusions that result from the game. Such an adjudicator might bring their own perceptions and semantics to the game. Thus, playing in a common semantics could be considered as both players adopting the adjudicator's view of the world.

---

[9]Aims will be discussed in the next subsection.

Similarly, there is no *prima facie* reason why the two players should focus on a single critical argument, rather than have individual, separate foci. The literature has rarely addressed this possibility ([Hadoux et al., 2015] is an exception). However, once we assume that the players agree on a focus, the use of a single critical argument for each player implies no loss of generality. Straightforward constructions can map a disjunction or conjunctions of arguments to a single argument in most semantics[10]. In particular, the arguments supporting the same conclusions can be united in a single argument.

In any case, many of the computational issues discussed in this and the next section depend only on the semantics and the player's aim, and so are still applicable to these less-well-studied forms of strategic argumentation.

Finally, even when addressing the same semantics and critical argument, there is some freedom in the strategic aims of the two players. At one extreme the players might have the same aim and, on the other extreme, have diametrically opposed aims. In between these extremes the players might have different but compatible aims, or have incompatible aims. Aims are discussed in detail in the next subsection. In this chapter we assume that the two players have incompatible aims: it is not possible for both players to achieve their aims simultaneously.

In previous sections we have discussed both symmetric and asymmetric forms of strategic argumentation. In abstract argumentation there is no explicit notion of conclusion and, therefore, no notion of an argument supporting the negation of the conclusion of another. Consequently, symmetric strategic argumentation is not available, in general. We will focus on asymmetric strategic argumentation. That is, whatever Pr's aim is, Op's aim is to prevent it.

In summary, a *strategic abstract argumentation dialogue game* consists of a split abstract argumentation framework, a critical argument, an abstract argumentation semantics, and aims for both Pr and Op. The *play* of the game is a sequence of moves such that each player leaves the game in a state where her strategic aim is satisfied.

---

[10] For example, see Proposition 2 of [Maher, 2016b].

6.2 Players' Aims

The range of strategic aims a player might have is limited under the grounded semantics. But once we consider semantics with multiple extensions a player has a much wider range.

Initially, work on abstract argumentation focussed on credulous and skeptical acceptance. An argument a in argumentation framework AF under semantics σ is *credulously accepted* if it is labelled in in at least one σ-extension. a is *skeptically accepted* if it is labelled in in every σ-extension. These two statuses were inherited from the field of non-monotonic reasoning.

[Wu and Caminada, 2010] extended this work with the notion of justification status. The justification status of an argument a in an argument framework AF is the set of labels a receives in complete extensions. Thus a justification status is a subset of $\{\text{in}, \text{out}, \text{undec}\}$. In general this might lead to $2^3 = 8$ different statuses, but only 6 are possible for the complete semantics [Wu and Caminada, 2010]. Obviously, this approach can be extended to any extension-based semantics [Dvořák, 2011].

[Maher, 2016c; Maher, 2016b] further extended the range of argument statuses to the following, casting these as possible aims of a proponent

1. **Existential:** a is labelled in in at least one σ-extension

2. **Universal:** a is labelled in in all σ-extensions

3. **Unrejected:** a is not labelled out in any σ-extension

4. **Uncontested:** a is labelled in in at least one σ-extension and is not labelled out in any σ-extension

5. **Plurality:** a is labelled in in more σ-extensions than it is labelled out

6. **Majority:** a is labelled in in more σ-extensions than it is not labelled in

7. **Supermajority:** a is labelled in in at least twice as many σ-extensions than it is not labelled in

The last three are called *counting aims*, distinct from the first four which are based on zero/non-zero number of labels, like the justification statuses[11]. In addition, the negation of such conditions and their dual (exchanging the role of in and out), which are plausible aims for the opponent, have also been considered [Maher, 2016b].

But clearly there are many more possibilities. Each of the first four strategic aims can be formulated as a disjunction of justification statuses. So we might consider any disjunction of justification statuses as a potential strategic aim. This would give us $2^8 = 256$ strategic aims. Many of these will be unrealizable under some semantics and/or unrealistic in practice. Under the stable semantics, aims that the argumentation framework has at least one extension or has no extension are also sensible. Further possibilities are aims such as: a is accepted in at least 2 extensions or is universally accepted. There are also many variations possible for the counting aims. For example, [Maher, 2016c] contemplates a weighting on all extensions, with the arguer's aim that the sum of the weights of extensions in which a is labelled in is greater than the sum of weights of the remaining extensions.

Some of the aims seem similar to the ideas behind proof standards that are formalized in [Gordon and Walton, 2009], although those proof standards are formalized in a very different setting. The Existential aim is similar to a *scintilla of evidence*, the Majority and Supermajority correspond to *preponderance of the evidence* and *clear and convincing evidence*, respectively, while the Uncontested aim is like *beyond a reasonable doubt*.[12] The Universal aim corresponds to *beyond a doubt*, in the phrasing of [Farley and Freeman, 1995].

There are some obvious close relationships between these different concepts. a is skeptically accepted iff a has justification status {in} iff a satisfies the Universal aim. Similarly, a is credulously accepted iff a's justification status contains in iff a satisfies the Existential aim. a satisfies the Unrejected aim iff a has justification status {in}, {undec} {in, undec}, or ∅. a satisfies the Uncontested aim iff a has justification

---

[11]A counting utility function was defined in [Thimm and García, 2010], but it counts the number of desired conclusions that appear in all σ-extensions rather than counting the number of σ-extensions in which a conclusion appears.

[12]The Uncontested aim is also similar to the notion of *argumentative inference* in paraconsistent reasoning from maximally consistent sets [Benferhat et al., 1993].

	$\mathcal{GR}$	$\mathcal{ST}$	$\mathcal{CO}$	$\mathcal{PR}$	$\mathcal{SST}$	$\mathcal{EA}$	$\mathcal{ID}$
Existential	in P	NP-c	NP-c	NP-c	Σ_2^p-c	Π_2^p-c	in Θ_2^p
Universal	in P	coNP-c	in P	Π_2^p-c	Π_2^p-c	Π_2^p-c	in Θ_2^p
Unrejected	in P	coNP-c	coNP-c	coNP-c	Π_2^p-c	Σ_2^p-c	in Θ_2^p
Uncontested	in P	coNP-c	D^p-c	D^p-c	D$_2^p$-c	Π_2^p-c	in Θ_2^p
Plurality	in P	PP-c	PP-c	in PP^{NP}	in PP^{NP}	Π_2^p-c	in Θ_2^p
Majority	in P	PP-c	PP-c	in PP^{NP}	in PP^{NP}	Π_2^p-c	in Θ_2^p
Supermajority	in P	PP-c	PP-c	in PP^{NP}	in PP^{NP}	Π_2^p-c	in Θ_2^p

Table 1: Complexity of Aim Verification problem for selected strategic aims and semantics [Maher, 2016b]. For a complexity class $\mathcal{C}$, $\mathcal{C}$-c denotes that the problem is complete for $\mathcal{C}$.

status {in} or {in, undec}. Also, a satisfies the Uncontested aim iff a satisfies the Existential and Unrejected aims.

Furthermore, when a semantics consists of a single extension (in particular, the grounded semantics) credulous and skeptical acceptance are identical, there are only three possible justification statuses for an argument ({in}, {undec}, and {out}), and all but the Unrejected aim, of those listed, are identical. In summary, a unitary semantics greatly simplifies analysis of player aims.

Thus, as we consider a wider range of semantics we must also address a wider range of player aims.

6.3 Computational Problems

We can break down the play of a game into two computational problems: recognising whether (or not) an argumentation framework satisfies a given aim, which is called the Aim Verification problem, and determining what arguments to play in order to leave the game in a state where the given aim is satisfied, the decision form of which is called the Desired Outcome problem. These problems will be different for the different players, because they have different aims.

The problem of verifying that an aim is satisfied by some state of strategic argumentation is a fundamental part of each move in a game.

The Aim Verification Problem

Instance A split argumentation framework $(\mathcal{A}_{Com}, \mathcal{A}_{Pr}, \mathcal{A}_{Op}, \gg)$, an argumentation semantics, a critical argument $a \in \mathcal{A}_{Com}$, and an aim.

Question Is the aim concerning the critical argument satisfied under the given semantics by the argumentation framework $(\mathcal{A}_{Com}, \gg)$?

The complexity of this problem, for a selection of semantics and aims, is presented in Table 1. Given Pr's aim, the complexity of verifying Op's aim is the complement of the complexity of Pr's aim.

These results are derived from existing work on the complexity of credulous and skeptical acceptance in abstract argumentation frameworks for the various semantics (see, for example, [Dunne and Wooldridge, 2009; Woltran, 2014]), and relations between the different aims (Proposition 3 of [Maher, 2016b]). For example, the Uncontested aim is the conjunction of Existential and Unrejected, where the latter is the dual of the negation of Existential. Under the (say) preferred semantics, credulous acceptance is NP-complete. Thus the complexity of Uncontested is a conjunction of NP and coNP, which gives us D^p. Completeness is a straightforward reduction.

For the counting aims, clearly the complexity is in PP^V, where V is the complexity of verifying that a set of arguments forms an extension of the appropriate type[13]. The lower bound for the stable semantics is obtained by reduction from the MAJSAT problem, and the complete semantics is treated by reduction from the stable semantics.

Table 1 only addresses a selected set of strategic aims. When a player has such an aim, their opponent will usually have a quite different aim, one not mentioned in the table. Since we are considering only games where the opponent's aim is the complement of the proponent's aim, the complexity of the Aim Verification problem for Op is the complement of the complexity of the Aim Verification problem for Pr. Thus, for example, under the complete semantics, if Pr has the Existential aim

---

[13]There has been some work done on counting extensions, both on the complexity of counting and identifying tractable cases [Baroni et al., 2010; Fichte et al., 2019]. These works focus on absolute counting, rather than comparing counts (as in the counting aims), so the results are presented in terms of #P rather than PP. Nevertheless, the complexity results are comparable to those for the counting aims in the Aim Verification problem.

then aim verification for Pr is NP-complete, and aim verification for Op is coNP-complete. In general, though, when the opponent's aim is not the complement of the proponent's, the complexity of the two problems is not so directly related.

The Desired Outcome problem [Maher, 2016c] is the problem that a player must solve at each step of a strategic abstract argumentation game. It involves identifying that the player has a legal move, leaving the state of the game in a desired state.

The Desired Outcome Problem for Pr

Instance A split argumentation framework $(\mathcal{A}_{Com}, \mathcal{A}_{Pr}, \mathcal{A}_{Op}, \gg)$ an argumentation semantics, a critical argument $a \in \mathcal{A}_{Com}$, and an aim for Pr.

Question Is there a set $I \subseteq \mathcal{A}_{Pr}$ such that Pr's aim with respect to the critical argument is achieved in the argumentation framework $(\mathcal{A}_{Com} \cup I, \gg)$?

This problem is a generalization of the strategic argumentation problem, as defined in Section 2, which is restricted to accepting the critical argument under the grounded semantics.

It is not difficult to see that the Desired Outcome problem can be solved by a non-deterministic algorithm with an oracle for the Aim Verification problem with Pr's aim. The complexity of this problem, for a selection of semantics and aims, is presented in Table 2.

The complement of this problem decides when Pr does not have a next move. The complexity of this complement problem is clearly the complement of the complexity of the Desired Outcome problem.

We can define the Desired Outcome problem for Op similarly, based on Op's aim. The complexities of the Desired Outcome problems for Pr and Op are not as directly related as is the case for aim verification.

Showing the presence of the Desired Outcome problem in the appropriate complexity class is comparatively straightforward, but showing it is complete in the class requires the construction of argumentation frameworks that extend those used for credulous or skeptical acceptance. An example construction for the Desired Outcome problem with the Universal aim under the stable semantics is shown in Figure 2. In this case

	$\mathcal{GR}$	$\mathcal{ST}$	$\mathcal{CO}$	$\mathcal{PR}$	$\mathcal{SST}$	$\mathcal{EA}$	$\mathcal{ID}$
Existential	NP-c	NP-c	NP-c	NP-c	Σ_2^p-c	Σ_3^p-c	Σ_2^p-c
Universal	NP-c	Σ_2^p-c	NP-c	Σ_2^p-c	Σ_2^p-c	Σ_3^p-c	Σ_2^p-c
Unrejected	NP-c	Σ_2^p-c	Σ_2^p-c	Σ_2^p-c	Σ_2^p-c	Σ_3^p-c	Σ_2^p-c
Uncontested	NP-c	Σ_2^p-c	Σ_2^p-c	Σ_2^p-c	Σ_2^p-c	Σ_3^p-c	Σ_2^p-c
Plurality	NP-c	NP^{PP}-c	NP^{PP}-c	NP^{PP}-c	NP^{PP}-c	Σ_3^p-c	Σ_2^p-c
Majority	NP-c	NP^{PP}-c	NP^{PP}-c	NP^{PP}-c	NP^{PP}-c	Σ_3^p-c	Σ_2^p-c
Supermajority	NP-c	NP^{PP}-c	NP^{PP}-c	NP^{PP}-c	NP^{PP}-c	Σ_3^p-c	Σ_2^p-c

Table 2: Complexity of the Desired Outcome problem for Pr, for selected aims and semantics [Maher, 2016c; Maher, 2016b; Maher, 2016a]. For a complexity class $\mathcal{C}$, $\mathcal{C}$-c denotes that the problem is complete for $\mathcal{C}$.

the problem is Σ_2^p-complete, so we reduce the satisfiability of $\exists X \forall Y\ \psi$ (where ψ is in DNF) formulas to this problem. The diagram has three parts: on the left is the representation of a variable p in X, in the middle is the representation of ψ, and on the right is the representation of a variable q in Y.

In the diagram, the grey nodes are arguments in $\mathcal{A}_{Com}$, and the white nodes (I_p and $I_{\neg p}$) are arguments in $\mathcal{A}_{Pr}$. $\gg$ is described by the directed edges. ($\mathcal{A}_{Op}$ is irrelevant to this problem.) Intuitively, an argument A_s (where s is a literal) accepted in a stable extension corresponds to the literal s being true. The critical argument is A_ψ, and Pr must move so that A_ψ is accepted in all stable extensions. The construction ensures that if Pr plays either both I_p and $I_{\neg p}$ or neither I_p nor $I_{\neg p}$ then either B_p or N_p is accepted and A_ψ is rejected in all stable extensions. Thus, Pr must play only one argument for each p, and this ensures only one of A_p and $A_{\neg p}$ can be accepted. This part of the construction is common to all reductions.

In the diagram, the formula is $\exists p \forall q\ \neg p \vee (p \wedge \neg q)$. It is represented in a slightly roundabout way. The treatment of variables q in Y ensures that both stable extensions containing A_q (i.e. q is true) and stable extensions containing $A_{\neg q}$ (i.e. q is false) are generated. A more formal description of this construction is in the proof of Theorem 7 of [Maher, 2016c].

Given a specific game, we write AV_{Pr} (AV_{Op}) for the Aim Verifica-

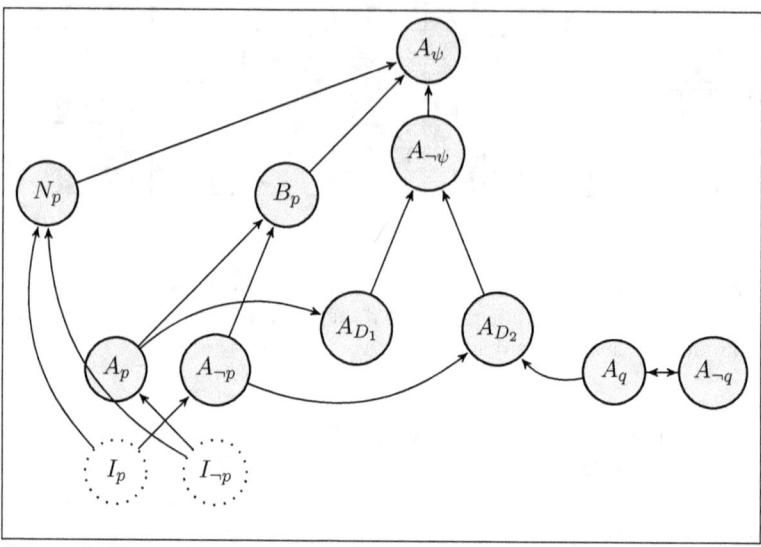

Figure 2: Example construction for the Desired Outcome problem with the Universal aim under the stable semantics

tion problem for Pr's (respectively, Op's) aim. Similarly, DO_{Pr} (DO_{Op}) denotes the Desired Outcome problem for Pr (respectively, Op).

Play begins by Pr playing a set of arguments, including the critical argument, and proceeds by Op and Pr alternately solving their Desired Outcome problem and playing the corresponding set of arguments. Play can extend for, at most, $\min(|\mathcal{A}_{Pr}|, |\mathcal{A}_{Op}|)$ rounds before play terminates, when one player does not make a move. Thus, play for Pr, over the entire game, has a computational cost in $P^{DO_{Pr}}$ while the cost of play for Op is in $P^{DO_{Op}}$ [Maher, 2016b].

The Aim Verification problem is of little interest for the concrete forms of strategic argumentation discussed in Section 5. In those cases the inference problem is polynomial [Maher, 2001; Billington et al., 2010]. Consequently, verifying any of the aims or justification statuses is also polynomial. The Desired Outcome problem corresponds to the SSA, SSSA and AsSA problems in Section 5: they represent the computational cost of making a move, in their respective games. In the case of structured arguments, conceptually the argumentation theory gives rise to an argumentation framework, which can then be interpreted in a chosen semantics. However, this does not mean that the NP-completeness for grounded semantics in Table 2 can be used to prove Theorem 5.8. The difficulty is that there might be greater than polynomially many arguments generated from the argument theory.

7 Corruption in Argumentation

When a game such as strategic argumentation is a model of a real-world situation, we must acknowledge the extra forces and influences that operate upon a player, beyond those of the specific role they have in the game. Often a player is assumed to have no motivations beyond performing their role and conforming to the rules of the game, but this is a rather simplistic view. While games do have rules, we need to consider the possibility that a player breaks the rules, or "cheats".

The context of the game is important in this regard. Organizations have many mechanisms to discourage the risk of corruption of their processes by the individuals performing these processes: managerial oversight, transparency through audit trails, the presence of co-workers, random inspections, etc. Society, as a whole, provides an entire justice system to enforce the rules the society considers important, and to detect and punish violations. When these mechanisms are not available, or are limited, how can we discourage rule-breaking?

[Bartholdi et al., 1989] proposed an answer to this question in the case of vote manipulation: if the computational difficulty of determining what an individual must do to alter the result of an election is too great, a potential vote manipulator may be discouraged from the manipulation, even though he has the opportunity to do it. They called this concept *computational resistance to strategic manipulation*. This insight has spawned a whole subfield of computational social choice [Brandt et al., 2016]. In this section we describe the application of these ideas to strategic argumentation.

Throughout this section, we consider that players are engaged by a client to play the game. A player is expected to adhere to the rules of the game and, in particular, play the game to win for her client. However, while the client is invested in winning the game, the player has various competing incentives. These are the source of the corruption we consider. A player might cheat on behalf of her client, or might sacrifice her client's chances for other incentives. This issue is known in management theory as the *principal-agent problem* or the *agency problem* [Eisenhardt, 1989].

7.1 Corruption and Resistance

Strategic argumentation has relatively few rules, though some are implicit rather than explicitly stated. The players must take turns, but violations of this rule are obvious and, anyway, offer no advantage to the players. A player must make a move if one is available to her. This rule is implicit in the assumption that the player will play her role properly. Such a rule is difficult to enforce without knowledge of the player's arguments. The player's arguments are assumed to be private, but this is also difficult to enforce. We will focus on violations of this privacy[14].

We consider two forms of corruption. The first, *collusion*, arises when one player induces the other to let her win. Such behaviour on its own is straightforward, though illicit, and does not, as such, appear in the game. But it is complicated by the desire of the guilty parties not to be detected. Thus, colluding players must not only ensure the "right" player wins, they must also make sure that an external observer cannot distinguish the collusive play from normal play. If the work to ensure this is computationally more difficult than simply playing the game honestly, then we consider the game to be *resistant to collusion*.

The following example is an instance of collusion.

Example 7.1. *Consider the strategic argument game depicted in Figure 3, where vertices are arguments (grey if they can be played by* Pr, *white for* Op*) and edges are attacks of one argument on another. For concreteness, we assume that we employ the grounded semantics and* Pr*'s strategic aim is that argument* A *is accepted. Normal play would proceed as follows:* Pr *plays* A, Op *plays* B_1 *(thus defeating* A*),* Pr *plays* C *(restoring* A *by defeating* B_1*), and* Op *plays* D *(defeating* C, *and allowing* B_1 *to defeat* A*). Thus, normally,* Pr *loses.*

---

[14]Earlier works that consider privacy include [Cayrol et al., 2002] and [Oren et al., 2006], which have a focus on minimizing the exposure of a player's arguments during play, rather than the loss of privacy by corruption.

10 - STRATEGIC ARGUMENTATION

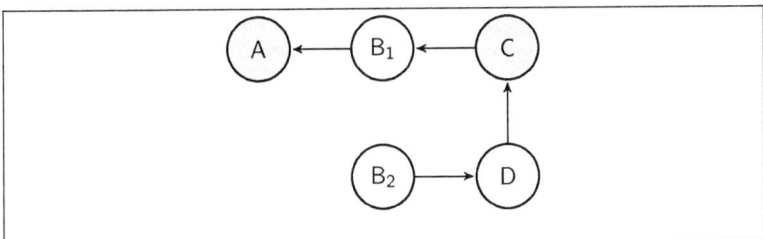

Figure 3: A strategic argumentation game. An argument is grey if it can be played by Pr and white if it can be played by Op.

However, Pr and Op *might collude to ensure* Pr *wins by playing as follows:* Pr *plays* A, Op *plays* B_1 *and* B_2, *and* Pr *plays* C *(restoring* A*).* Pr *now wins because* Op *has no effective move: to play* D *would have no effect because it is defeated by* B_1.

This example also serves to show the difference between collusion and an omniscient argumentation framework $(\mathcal{A}_{Com} \cup \mathcal{A}_{Pr} \cup \mathcal{A}_{Op}, \gg)$. *Under any completist semantics,* A *is accepted in the omniscient argumentation framework, but if* Pr *and* Op *collude to ensure* Op *wins they can do so by following the normal play above.*

The second form of corruption, *espionage*, occurs when, through some means, one player gains knowledge of the other player's arguments. Again, this act is not apparent in the game, but it requires work to develop a strategy, based on that knowledge, to defeat the other player. If this is computationally more difficult than playing the game honestly, then we consider the game to be *resistant to espionage*.

In Example 7.1, the corrupt sequence of moves might also occur if Op committed espionage on Pr in order to ensure Pr wins.

For both forms of resistance, we need to clarify what "computationally more difficult" means. Computational difficulty will be measured in terms of a hierarchy of complexity classes where, although one class might be contained in another, it is often not known that the two classes are distinct. However, if the two classes were equal then part of the (say) polynomial complexity hierarchy would collapse, and this is commonly believed by complexity theorists not to happen. Thus "computationally more difficult" is subject to this commonly-believed assumption. For counting aims we are dealing with the counting polynomial hierarchy, and the corresponding assumption is messier. The topic is less investigated and there are some collapses known within the counting hierarchy. However, those collapses do not affect the containments

$$P^{PP} \subseteq NP^{PP} \subseteq P^{NP^{PP}} \subseteq NP^{NP^{PP}} \subseteq \cdots \subseteq PSPACE$$

The assumption that these containments are strict is the basis of resistance for counting aims.

Inherent in the resistance approach to corruption is the assumption that players will be effectively penalised if their corruption is detected. This assumption relies on issues of governance, lasting identification of the players, and enforcement and scale of penalties, among others. But these issues depend on the context of the game and are beyond the scope of this chapter.

7.2 Computational Problems

We now consider the computational problems that must be solved by players in order to exploit corruption.

Colluders need to to construct an alternating sequence of moves that ends with Pr winning, that is, with Op unable to make a move. This is formalized as follows.

The Winning Sequence Problem for Pr

Instance A split argumentation framework $(\mathcal{A}_{Com}, \mathcal{A}_{Pr}, \mathcal{A}_{Op}, \gg)$ and a desired outcome for Pr.

Question Is there a sequence of moves such that Pr wins?

A similar problem arises when the colluders wish to ensure that Op wins.

The problem for Pr can be solved by nondeterministically generating a sequence of moves, verifying that each move achieves the aim for its player, and verifying that Op has no further move. That is, it can be solved in NP with oracles for AV_{Pr}, AV_{Op} and (the complement of) DO_{Op}. $AV_{Op} = coAV_{Pr}$, since we assume Pr and Op have complementary aims, so the larger of $NP^{AV_{Pr}}$ and $NP^{DO_{Op}}$ is an upper bound for this problem.

In the case of espionage, one player, say Pr, knows her opponent's arguments $\mathcal{A}_{Op}$ and desires a strategy that will ensure Pr wins, no

	$\mathcal{GR}$	$\mathcal{ST}$	$\mathcal{CO}$	$\mathcal{PR}$
Existential	Σ_2^p-c	Σ_3^p-c	Σ_3^p-c	Σ_3^p-c
Universal	Σ_2^p-c	Σ_2^p-c	Σ_2^p-c	Σ_3^p-c
Unrejected	Σ_2^p-c	Σ_2^p-c	Σ_2^p-c	Σ_2^p-c
Uncontested	Σ_2^p-c	Σ_2^p-c	Σ_3^p-c	Σ_3^p-c
Plurality	Σ_2^p-c	$NP^{NP^{PP}}$-c	$NP^{NP^{PP}}$-c	$NP^{NP^{PP}}$-c
Majority	Σ_2^p-c	$NP^{NP^{PP}}$-c	$NP^{NP^{PP}}$-c	$NP^{NP^{PP}}$-c
Supermajority	Σ_2^p-c	$NP^{NP^{PP}}$-c	$NP^{NP^{PP}}$-c	$NP^{NP^{PP}}$-c

	$\mathcal{SST}$	$\mathcal{EA}$	$\mathcal{ID}$
Existential	Σ_4^p-c	Σ_3^p-c	Σ_3^p-c
Universal	Σ_3^p-c	Σ_3^p-c	Σ_3^p-c
Unrejected	Σ_3^p-c	Σ_4^p-c	Σ_3^p-c
Uncontested	Σ_4^p-c	Σ_3^p-c	Σ_3^p-c
Plurality	$NP^{NP^{PP}}$-c	Σ_3^p-c	Σ_3^p-c
Majority	$NP^{NP^{PP}}$-c	Σ_3^p-c	Σ_3^p-c
Supermajority	$NP^{NP^{PP}}$-c	Σ_3^p-c	Σ_3^p-c

Table 3: Complexity of the Winning Sequence problem for Pr for selected aims and semantics [Maher, 2016b].

matter what moves Op makes. A *strategy* for Pr in a split argumentation framework $(\mathcal{A}_{Com}, \mathcal{A}_{Pr}, \mathcal{A}_{Op}, \gg)$ is a function from a set of common arguments to the set of arguments to be played in the next move. A sequence of moves $S_1, T_1, S_2, T_2, \ldots$ resulting in common arguments $\mathcal{A}_{Com}^{Pr,1}, \mathcal{A}_{Com}^{Op,1}, \mathcal{A}_{Com}^{Pr,2}, \mathcal{A}_{Com}^{Op,2}, \ldots$ is *consistent with* a strategy s for Pr if, for every j, $S_{j+1} = s(\mathcal{A}_{Com}^{Op,j}, \mathcal{A}_{Pr})$. A strategy for Pr is *winning* if every valid sequence of moves consistent with the strategy is won by Pr.

The Winning Strategy Problem for Pr

Instance A split argumentation framework $(\mathcal{A}_{Com}, \mathcal{A}_{Pr}, \mathcal{A}_{Op}, \gg)$ and a desired outcome for Pr.

Question Is there a winning strategy for Pr that satisfies the standards?

There is also, of course, the corresponding problem for Op which arises when Op conducts the espionage.

The following result shows that the Winning Strategy problem is PSPACE-complete for all completist semantics and all the aims discussed in this chapter. This is not surprising since, as a result of the espionage, Pr is essentially playing a complete knowledge game and such games are known to be PSPACE-hard, in general.

Theorem 7.2. *[Maher, 2016b] Consider any completist semantics for abstract argumentation, and any of the above aims for* Pr*.*
The Winning Strategy problem is PSPACE-complete.

This theorem applies both to espionage by Pr and espionage by Op. The constructed argumentation framework for this proof is well-founded. Consequently the construction serves for all completist semantics.

7.3 Audit: Standards and Compliance

To investigate collusion, we need to understand what "normal play" looks like and how to recognise it. [Maher, 2017a] proposes that we view this as a matter of audit, with an external body setting standards for play and testing for compliance. In this view there can be multiple standards. We have already seen one standard: that a player must make a move, if she has one (we will call this the *compulsory move* standard). Consequently, colluding players must arrange their play to ensure that the designated loser has no possible moves at the end of the game. Earlier work [Maher, 2016c; Maher, 2016b; Maher, 2016a] implicitly operated under this standard.

However, this standard fails to address obvious collusion, like that in Example 7.1. Thus, additional standards are required. However, a standard can only be justified if it does not interfere with honest play. That is, a player should never face a choice between following the standard and improving her chances of winning. Otherwise, any violation of the standard can be explained away as an attempt to improve those chances.

It is clear that the problem in Example 7.1 stems from Op playing B2. But it is not clear what is an appropriate standard that would prevent this move. Several possibilities suggest themselves:

(1) A player should not play an argument that attacks one of her own (unplayed) arguments, thus causing a self-inflicted injury.

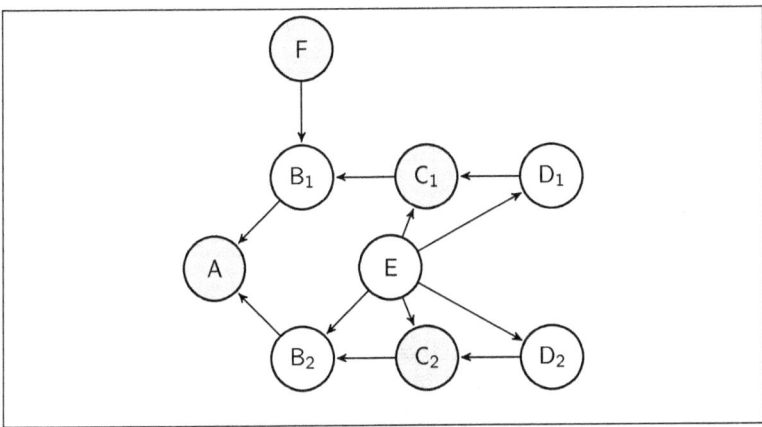

Figure 4: Split argumentation framework demonstrating non-dominance of minimum cardinality moves.

(2) A player should play the smallest number of arguments to achieve her aim[15].

(3) A player should play a subset-minimal set of arguments that achieve her aim.

(1) is clearly too strong to be a standard. If, in Example 7.1 (Figure 3), B_1 also attacked B_2 then following this standard would cause Op to lose immediately. However, when the omniscient argumentation framework is known to a player, [Thimm and García, 2010] prove that this standard (which they call the overcautious selection function) is dominant. Unfortunately, a player cannot be expected to know the omniscient argumentation framework.

(2) is more plausible, but consider the following example from [Maher, 2017a].

Example 7.3. *Consider the strategic argumentation game in Figure 4, and play that proceeds as follows:* Pr *plays* A, Op *plays* B_1 *and* B_2, *and*

---

[15]This is similar to the heuristic of [Oren et al., 2006], though the details of the game are different.

Pr *plays* C_1 *and* C_2. *At this stage* O *must attack both* C_1 *and* C_2, *and she has two alternatives: (1) play* E, *which attacks both* C_1 *and* C_2, *or (2) play both* D_1 *and* D_2, *each attacking one of the* C *arguments. Clearly (1) is the minimum cardinality move. However,* Pr *then responds with* F, *and wins. In (2), the play of* F *is insufficient for* Pr, *since* B_2 *remains undefeated. Hence* Op *wins.*

Thus minimum cardinality is not suitable as a standard, because it can prevent a winning move.

However, [Maher, 2017a] showed that (3) is compatible with normal play: every non-minimal move is dominated by a minimal move[16]. Thus the requirement to play only subset-minimal moves is a suitable standard. It remains open whether there are other standards that could be applied.

In addition, we need to consider how play can be verified as compliant with a standard. This involves issues of which data need to be accessed by the auditor, as well as the computational difficulty of verifying compliance

In terms of accessibility, all that an auditor needs for subset-minimality is an ability to inspect the initial $\mathcal{A}_{Com}$, the sequence of moves, and $\gg$ restricted to the current $\mathcal{A}_{Com}$, all of which can be considered public information. On the other hand, to verify the compulsory move standard requires knowledge of the player's arguments, which is private. Thus an auditor verifying both standards needs access to all aspects of a split argumentation framework. (However, each client might be in a position to audit the compulsory move standard, which would allow the player's arguments to be kept private from the auditor.)

For the auditor, the cost of verifying compliance with the subset-minimality standard involves polynomial many solutions of the Minimal Move problem (see next subsection) for Pr, and the same for Op. In comparison, the compulsory move standard requires a $coDO_L$ check, where L is the loser of the game, to verify that there is no move for L left to play.

For the players, compliance with the subset-minimality standard increases the difficulty of making a move. Not only must they find a move, they must also verify that it is minimal. It also increases the cost to players exploiting collusion: they must arrange the game so that

---

[16]Previous work addressing redundancy or relevance in argumentation includes [Flieger, 2002; Mbarki et al., 2006].

10 - STRATEGIC ARGUMENTATION

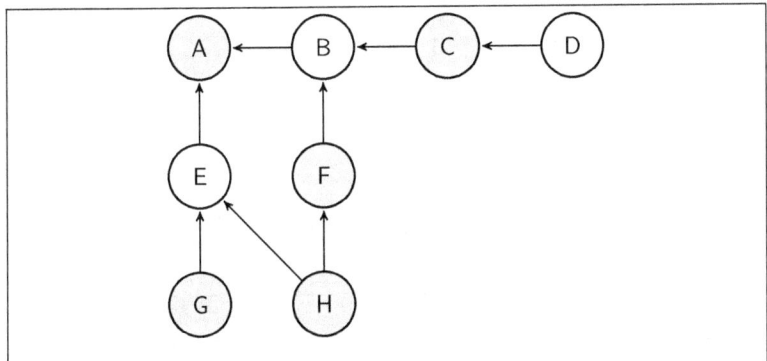

Figure 5: A strategic argumentation game demonstrating weakness of the compulsory move and subset-minimality standards.

their designated player wins, but also ensure that each move is minimal. Furthermore, one easy avenue for exploiting collusion has been eliminated. Consequently, there are games (like Example 7.1) where compliance with both standards ensures that exploitation of collusion cannot be disguised as normal play.

This leads to some questions. Are these two standards sufficient to prevent the disguise of collusion? If not, can we add standards to achieve this goal? Unfortunately, the answer to the first question is no, as the following example shows.

Example 7.4. *Consider the strategic argument game depicted in Figure 5, where arguments in $\mathcal{A}_{Pr}$ are grey and arguments in $\mathcal{A}_{Op}$ are white, and A is the critical argument. If Pr refrains from playing H then Pr will win, since the two arguments attacking A (B and E) can be attacked by Pr's arguments F and G, which cannot be attacked by Op. For example, the sequence of moves:* A, B, F, E, G *results in* Pr *winning.*

On the other hand, the sequence of moves: A, E, H, B, C, D *results in* Op *winning. Thus,* Pr *and* Op *can collude to ensure* Op *wins.*

This example suggests that a variation of (1) above might be needed to detect collusion more thoroughly. Which leads us to the second question: is it possible to impose enough justified standards that no

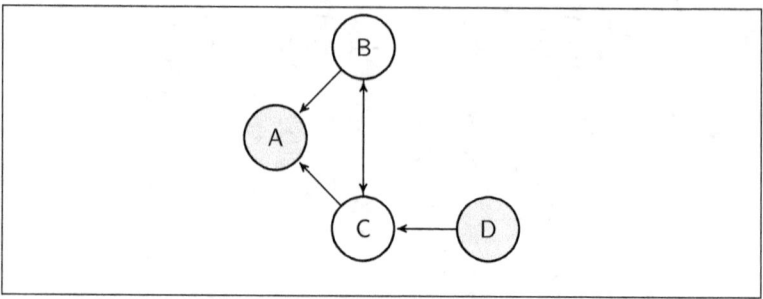

Figure 6: A strategic argumentation game demonstrating that no accumulation of justifiable standards can make all collusion detectable.

collusion can be disguised as compliant play? Again the answer is no.

Consider the argumentation game in Figure 6 under the grounded semantics, where A is the critical argument. After Pr plays A, Op has the choice of playing B or C. Depending on this choice, either Pr or Op will win. If Pr and Op collude they can determine the outcome, but any real restriction imposed by a standard will restrict to one possible outcome, so it cannot be a justified standard. Thus any collusion in this game cannot be detected by imposing justified standards.

Hence, we see that collusion cannot be prevented simply by imposing more and more standards. We must continue to rely on computational difficulty to discourage corruption.

We now take a stab at formalizing these considerations. A *standard* is a restriction on moves a player may make. More precisely, a standard is a function from a player's aim, her private set of arguments ($\mathcal{A}_{\mathsf{Pr}}$ or $\mathcal{A}_{\mathsf{Op}}$), a proposed move (a subset of her private arguments), and the set of arguments $\mathcal{A}_{Com}$, that are common knowledge, to the set $\{permitted, not\_permitted\}$. The standard is *complied with* by a player in the play of a game if each move by the player is permitted by the standard.

A set of standards is *justified* if, for every argumentation game, if for every unpermitted move that achieves the player's aim there is a better (or equal) permitted move that achieves the player's aim. A move m by a player is considered better or equal to another move m' if, for every behaviour of the opposing player, the player can achieve a better or equal outcome of the game by playing m, rather than playing m'. Note that a set of standards might be unjustified even though each standard, individually, is justified. However the combination of the compulsory move and subset-minimality standards is justified.

We say that a strategic argumentation game played under a given finite set of standards has *detectable collusion* if any occurrence of collusion that affects the outcome of the game violates a standard. The set of standards must be finite because an infinite set of standards creates difficulties for compliance verification, both for the players and the auditor. The best that could be done is checks on a random subset of standards. On the face of it, this might be sufficient for the auditor, but if the player has no way to verify her move is compliant with all standards then the auditor cannot reliably infer collusion or incompetence from her failure to comply.

We say that a strategic argumentation game played under a given set of standards is *determined* if all compliant plays of the game lead to the same winner. It appears that collusion is detectable iff the game is determined.

These considerations are similar to the issues in game-theoretic *mechanism design* (see, for example, [Garg et al., 2008]) where the aim of the design is to achieve some social good, such as fairness, honesty, ..., despite the self-interest of the parties involved. Thus there is a strong focus is on a strategy-proof mechanism, where there is no advantage to players in deviating from socially good behaviour. A classic example of mechanism design is two-person cake-cutting, where the mechanism specifies that one player cuts the cake in two, and the other chooses a piece. This mechanism encourages fairness in the division of the cake.

In an argumentation setting, [Rahwan et al., 2009] addresses a version of strategic abstract argumentation (with multiple players) where all players simultaneously play a selection of their arguments, aiming for their focal argument to be accepted. The social good desired is that the arguments accepted after all moves are those that would be accepted if all arguments were available (the omniscient view of the split argumentation framework). That is, roughly, the social good is that arguments are not

	$\mathcal{GR}$	$\mathcal{ST}$	$\mathcal{CO}$	$\mathcal{PR}$	$\mathcal{SST}$	$\mathcal{EA}$	$\mathcal{ID}$
Existential	Res	Res	Res	Res	Res		Res
Universal	Res		Res				Res
Unrejected	Res					Res	Res
Uncontested	Res		Res	Res	Res		Res
Plurality	Res	Res	Res	Res	Res		Res
Majority	Res	Res	Res	Res	Res		Res
Supermajority	Res	Res	Res	Res	Res		Res

Table 4: Resistance to collusion to ensure **Pr** wins, for several aims and semantics [Maher, 2016b]. *Res* denotes that the combination of aim and semantics is computationally resistant to collusion, while a blank denotes that it is not resistant.

hidden[17]. Other work, such as [Schreier et al., 1995; Sakama, 2012], also considers hiding of arguments as unfair or dishonest.

This is a different attitude than in strategic argumentation, which treats argument hiding as an inherent feature of adversarial argumentation. [Rahwan et al., 2009] characterize when their game is strategy-proof, that is, when there is no advantage to players from hiding arguments. It is only in very restrictive circumstances that honesty is the best policy. Their focus is on the game itself. In particular, the self-interest players have derives from their goals within the game. This is in common with most work on mechanism design. In contrast, the work in this section aims at aligning the self-interest of players with their clients, where that self-interest extends *beyond the game itself*. The introduction of standards is an instance of mechanism design, but we have seen that there is no mechanism that allows strategic play and prevents all collusion. Consequently, computational resistance serves as a back-stop, to discourage collusion.

---

[17] An argument a is *hidden* if it is not played, even though a player has it available to play. Sometimes, more specifically, it refers to an argument a that defeats an argument b, but is not played when b is played.

7.4 Resistance to Corruption

Recall that resistance to collusion is based upon the relative computational difficulty of exploiting the corruption, while disguising it, versus the difficulty of playing the game honestly. In other words, we compare the complexity of the Winning Sequence problem with the complexity of normal play as described at the end of subsection 6.3. This comparison is presented in Table 4. While not all combinations of aim and semantics show computational resistance, many do. However, it is notable that three of the aims under the stable semantics do not have resistance to collusion.

This comparison, however, deals only with the initial standard: that a player must play if she has a move. We need to recalculate both the computational cost of normal, honest play and the complexity of the Winning Sequence problem under both standards, in order to determine resistance to collusion when both standards apply. Hence, we need to consider the computational cost of verifying compliance with the subset-minimality standard. The Minimal Move Problem is to verify that a given move is a subset-minimal move.

The Minimal Move Problem for Pr

Instance A split argumentation framework $(\mathcal{A}_{Com}, \mathcal{A}_{Pr}, \mathcal{A}_{Op}, \gg)$, an argumentation semantics, an aim for Pr, and a move $M \subseteq \mathcal{A}_{Pr}$ that achieves the aim for Pr.

Question Is M a minimal set that achieves the aim under the given semantics? That is, is there no subset $N \subset M$ such that Pr's desired outcome is achieved in the argumentation framework $(\mathcal{A}_{Com} \cup N, \gg)$?

It is clear that the complement of this problem can be solved by a non-deterministic algorithm that guesses N and uses an oracle for the Aim Verification problem. Thus the Minimal Move Problem is in coNP^{AV}, where AV is the complexity of the Aim Verification problem. The complexity of the Minimal Move problem for Pr and Op (denoted by MM_{Pr} and MM_{Op}) for selected aims (of Pr) and the grounded and stable semantics is given in Table 5. This is also the work that an auditor must do to verify compliance with the subset-minimality standard. All aims for the grounded semantics lead to the same complexity, so these results have been condensed to a single row.

Honest (i.e. non-corrupt) play under both standards consists of a polynomial number of moves, each involving the search for an effective move, incorporating a verification that the aim is satisfied and the move is minimal. The cost of a single move for Pr is DOM_{Pr}, which is in $NP^{\{AV_{Pr}, MM_{Pr}\}}$ and the total cost of honest play is $P^{DOM_{Pr}}$, and similarly for Op. The total cost of honest play for each player, under the two standards, is shown in Table 5. In some cases the complexity of play has increased as a result of the additional standard, but in other cases it remains the same.

Finally, we must recalculate the cost for collusive play (assuming the players want Pr to win), denoted by WSM. This is the cost of solving the Winning Sequence problem when each player is constrained by the standard to play only subset-minimal moves. The players must search for a sequence of effective minimal moves, and ensure Op has no effective move remaining. Thus WSM is in $NP^{\{AV_{Pr}, MM_{Pr}, AV_{Op}, MM_{Op}, coDO_{Op}\}}$. The complexity of WSM is also given in Table 5. In most cases the additional standard does not change the complexity of solving the Winning Sequence problem.

We can see from the table that, once the subset-minimality standard is incorporated, all aims under the stable semantics are resistant to collusion, an improvement (compare with Table 4).

While the additional standard may increase the cost of playing a strategic argumentation game, it is still not comparable to the cost of solving the Winning Strategy problem. Hence all the completist semantics and all the aims remain resistant to espionage.

Of all the semantics that have been investigated, the naive semantics has an interesting property – it is *corruption-proof*, at least for the non-counting aims [Maher, 2016a]. Under this semantics the extensions are the maximal conflict-free sets. It is corruption-proof because the outcome is determined by the arguments the players have, if they comply with the compulsory move standard. In this sense, the game is strategy-proof. Consequently, if the game has an outcome different from the expected one, we detect corruption/incompetence. But, since every game is determined, this is not a suitable semantics in which to do strategic argumentation.

	MM_{Pr}	MM_{Op}	Hon_{Pr}^{Min}	Hon_{Op}^{Min}	WSM	
Grounded sem.	coNP-c	coNP-c	Δ_2^p-c	Δ_2^p-c	Σ_2^p-c	Res
Stable sem.						
Existential	coNP-c	Π_2^p-c	Δ_2^p-c	Δ_3^p-c	Σ_3^p-c	Res
Universal	Π_2^p-c	coNP-c	Δ_3^p-c	Δ_2^p-c	Σ_3^p-c	Res
Unrejected	Π_2^p-c	coNP-c	Δ_3^p-c	Δ_2^p-c	Σ_3^p-c	Res
Uncontested	Π_2^p-c	coNP-c	Δ_3^p-c	Δ_2^p-c	Σ_3^p-c	Res
Plurality/Majority	coNP^{PP}-c	coNP^{PP}-c	P$^{\text{NP}^{\text{PP}}}$-c	P$^{\text{NP}^{\text{PP}}}$-c	NP$^{\text{NP}^{\text{PP}}}$-c	Res
Supermajority	coNP^{PP}-c	coNP^{PP}-c	P$^{\text{NP}^{\text{PP}}}$-c	P$^{\text{NP}^{\text{PP}}}$-c	NP$^{\text{NP}^{\text{PP}}}$-c	Res

Table 5: Complexity of Minimality problems and normal play with the minimality standard (for Pr and Op), Winning Sequence problems (for Pr), and resistance to collusion (to ensure Pr wins), under the grounded and stable semantics, for selected aims (of Pr) [Maher, 2017a].

7.5 Concrete Argumentation Systems

As we saw in Section 5, the SSA, SSSA, and AsSA problems for DL(∂) and DL(δ) are NP-complete, as are the problems for the ASPIC-like language under the grounded semantics. These correspond to the Desired Outcome problem. It was shown in [Maher, 2014b] that the Winning Strategy problem is PSPACE-complete and the Winning Sequence problem is Σ_2^p-complete for DL(∂); hence, argumentation in DL(∂) is resistant to corruption. These results relied on careful constructions and proofs reliant on the specific logic.

There are many concrete languages, beyond those discussed in Section 5, that can be used to express arguments. There is a wide variety of defeasible logics [Billington *et al.*, 2010; Maier and Nute, 2010; Maier, 2013; Billington, 2019; Maher *et al.*, 2020], languages incorporating inheritance in logic programming [Laenens and Vermeir, 1990; Bossi *et al.*, 1993], other logic programming-based languages [Wan *et* al., 2009; Wan et al., 2015; Kakas et al., 2019; Sakama, 2013], languages inspired by argumentation [Dimopoulos and Kakas, 1995; Verheij, 2003], as well as primitive systems like non-monotonic inheritance networks [Touretzky, 1986]. Unlike systems such as ASPIC [Amgoud et al., 2006; Prakken, 2010b; Wu and Podlaszewski, 2015] and assumption-based argumentation [Bondarenko et al., 1997], these languages are designed independently from – and sometimes prior to – abstract argumentation. Thus the results of this section do not apply directly to such languages, and following the approach of [Maher, 2014b] to establish resistance to corruption would be time-consuming.

However, it was shown in [Maher, 2017b] that many of these concrete languages can encode abstract argumentation frameworks under appropriate semantics. Most of the languages employ the grounded semantics, while DEFLOG [Verheij, 2003], ASPDA [Wan et al., 2015] and a version of NDL [Maier, 2013] employ the stable semantics. Similarly, defeasible logics defined in the framework of [Antoniou et al., 2000a] for a range of logic programming semantics can encode corresponding abstract argumentation frameworks under the corresponding (in the sense of [Caminada et al., 2015]) completist semantics. As a result, the hardness complexity results for these semantics are carried over to the concrete languages. Consequently, many of these languages are shown to be resistant to corruption. See [Maher, 2016c] for details.

8 Related Work

Dialogue games for argumentation describe systems where two opponents argue about the tenability of one or more claims (and thus are in the class of persuasion dialogues [Walton and Krabbe, 1995]). Persuasion dialogues are typically substantive: the participants provide substantive reasons for their claims [Lorenzen and Lorenz, 1978]. As a consequence, the information available during the game evolves, each participant discovering new pieces of information each time the opponent makes new claims.

A structural difference between strategic argumentation and many persuasion dialogues lies within the nature of the reply/counter-argument a player may present: in our setting a participant never asks a *why?*

question to a previous opponent's claim. In fact, the answer to the *why?* question is already provided at the very moment a claim is made: every and each claim is justified/supported by the argument proving it (all the rules in the proof of which the claim is the conclusion). Dialogue systems have been classified based on their structural properties, that is whether a player can make a single or multiple moves in one turn, and whether she is allowed to reply only once or multiple times to the other player's moves. In our game, the turn shifts immediately after a player's move, but this is nonetheless a relaxed constraint given that, during such a move, the player may advance a set of arguments, and not just a single one. Moreover, the player is not obliged to respond to the opponent's last move but she may attack any argument proposed so far (possibly her own if this can prove her claim). It is nonetheless true that our framework is a a sort of *unique/move* protocol (a hybrid version): a player can respond only once before the turn passes to the other player even if, as we have shown, such a response is not limited to a single argument.

We do not allow argument retractions (also known as withdrawals): once an argument is played, it will remain as part of the common rules/knowledge base till the end of the game. But it is clear that such a constraint does not prevent a player attacking one of her previously played arguments. We force a replying move to be structurally *relevant*, that is it must be capable of changing the dialogical status of the critical literal/argument (except for the surrendering move which, instead, gives the victory to the adversary). Even allowing retraction in our framework, the computational complexity does not change: a retraction operation would choose a set of rules/arguments to be discarded; thus there is still a choice to be made. However, retraction would change the nature of the game: in the game of Figure 6, Op would not lose. Furthermore, retraction requires restrictions to ensure games terminate.

On the other hand, within our framework a player is not committed to the arguments she plays. Commitments typically require that moves do not contradict or challenge previous commitments/statements; in our framework, players have commitments only towards the claim at dispute as they may, at any time, advance arguments contradicting their own previous statements.

Our turn-taking is in line with the notion of [Prakken, 2005; Loui, 1998] where "when a player is to move, s/he keeps moving until s/he has changed the status of the initial move his or her way". The sole difference is that we consider the playing of more arguments as a single move, but the essential idea is that even in our framework the player must change the status of the initial claim (the critical literal/argument).

The structure of the arguments defined by our framework is in line with [Prakken, 2005]. The idea of an argumentation theory is that of containing all the arguments that are constructible on the basis of a certain theory or knowledge base.

Our framework is *sound* and *fair* according to definitions given in [Prakken, 2005]. It is sound because if the proponent wins the game, then the current theory actually proves the critical literal. (Symmetrically, if the opponent wins, the theory either fails to prove the critical literal, disproves it, or proves the opposite, depending on the game variant.) The framework also satisfies fairness given that if, at a given turn, the theory proves the critical literal, then proponent is winning the game. (Again, depending on the type of game, we have that if the theory either fails to prove the critical literal, disproves it, or proves the opposite at a given turn, then the opponent is winning the game.)

The conceptual basis of our formalisation that an argument moved at some earlier stage might be a legal counterargument against some later arguments is not a novelty in the literature of the field, and has been adopted in many frameworks [Prakken, 2000; Prakken, 2005].

Our dialogues are coherent (in the sense proposed by [Prakken, 2005, Section 7.1]) since we do not allow players to retract their claims. A participant can play a set of arguments conflicting with some of the moves she has put forward in previous steps, if this helps her in taking advantage of information disclosed by the adversary.

[Devereux and Reed, 2009] describe a rigorous persuasion dialogue game RPD_{GD} obtained by adapting the game RPD_0 of [Walton and Krabbe, 1995], replacing propositional logic as the underlying information carrier with abstract argumentation. It has some features in common with strategic argumentation, including private arguments, alternating moves and strategic play. On the other hand, each move is a single locution, which may be a statement, challenge, or question; the only semantics considered is the grounded semantics; and the roles of Pr and Op are quite different from each other, in comparison to strategic argumentation. [Devereux and Reed, 2009] analyse strategies for their game but it is unclear whether they could be adapted to strategic argumentation.

In game-theoretic terms, a player in a strategic argumentation game has *perfect information* of the structure of the game, the history of the game, and the effects of each move. On the other hand, the players have *incomplete information* of the arguments – and, hence, the possible moves – of adversaries. Most games in the argumentation literature are games of perfect information, while many assume complete information of the adversary, or don't care. For dialogues that are collaborative, seeking to find a joint truth[18], privacy/incomplete information would seem not to matter; for those designed to provide an operational characterization (or proof theory) for specific semantics[19], again it would seem that privacy does not matter. Many works seeking to apply game-theoretic solution concepts, such as Nash equilibria, to argumentation games [Riveret et al., 2008; Procaccia and Rosenschein, 2005; Matt and Toni, 2008; Fan and Toni, 2016] assume players have complete information about an adversary's possible moves, since that is an underlying assumption of Nash equilibria. On the other hand, many argumentation games in the literature are incomplete information games, for example [Roth et al., 2007; Parsons and Sklar, 2005; Rahwan et al., 2009; Satoh and Takahashi, 2011b; Bonzon and Maudet, 2012; Grossi and van der Hoek, 2013].

One way of analysing argumentation games of incomplete information is to frame them as *Bayesian extensive games with observable actions* [Osborne and Rubinstein, 1999, chap. 12]: this is possible because every player observes the move of the other player and uncertainty only derives from an initial move of Chance that distributes private information (rules or arguments) among the players. Hence, Chance selects types for the players by assigning to them possibly different theories from the set of all possible theories constructible from a given language. If this hypothesis is correct, notice that Bayesian extensive games with observable actions allow to simply extend the argumentation models proposed, for example,

---

[18]Such dialogues are known as *inquiry* dialogues [Walton and Krabbe, 1995].

[19]Examples of such work are [Vreeswijk and Prakken, 2000; Amgoud and Cayrol, 2002; Modgil and Caminada, 2009].

in [Riveret et al., 2008; Grossi and van der Hoek, 2013]. Despite this fact, however, complexity results for Bayesian games are far from encouraging (see [Gottlob et al., 2007] for games of strategy). Indeed, it seems that considerations similar to those presented by [Chalkiadakis and Boutilier, 2007] can be applied to argument games: the calculation of the perfect Bayesian equilibrium solution can be tremendously complex due to both the size of the strategy space (as a function of the size of the game tree, and it can be computationally hard to compute it [Dimopoulos et al., 2002]), and the dependence between variables representing strategies and players' beliefs.

Many works, for example [Rienstra et al., 2013; Hadjinikolis et al., 2013] (and see [Thimm, 2014; Black et al., 2021] for more discussion), have addressed the development of a model of the adversary, which can help in developing heuristics for choosing a particular move. Such work does not change the worst-case complexity of making a move, which is NP-hard or worse (see Table 2). Furthermore, even with full knowledge of the adversary, the problem of developing a strategy to beat the adversary is PSPACE-complete (Theorem 7.2).

As mentioned earlier, some work [Rahwan et al., 2009; Sakama, 2012] considers hiding arguments (that is, playing an argument a_1 that you know is defeated by a_2, but keeping a_2 private) to be dishonest or even lying. However, in a game of incomplete knowledge a player does not know which arguments hold in the omniscient argumentation framework, so this attitude seems harsh. In any case, our focus is on strategic arguing, where hiding arguments is acceptable. Those works also address "bullshitting" [Frankfurt, 2005] (the introduction of arguments that the player does not know), which is not acceptable in strategic argumentation. We assign to the adjudicator the responsibility for rejecting such arguments. [Rahwan et al., 2009] shows that, for their single simultaneous move game, honesty is the best policy only in very restrictive circumstances. [Sakama, 2012] identifies some cases in which a player can detect a dishonest adversary, while [Parsons and Sklar, 2005] show that, as the players play more games the probability of a lie being caught by the adversary approaches 1. Apart from these works, which might be considered as addressing corruption isolated to a single player, there seems no discussion of corruption in formal argumentation prior to [Maher, 2014b]. [Schreier et al., 1995] address "argumentational integrity", but this refers to fairness in the performance of general argumentation; they do, however, agree that "pretence of truth" is unfair, and would also consider hidden arguments as "insincere contributions".

A majority of the (persuasion) dialogue and argumentation literature takes the perspective of Dung, which sees arguments as monadic elements. There, arguments are typically abstract: the players know such arguments, can propose one (or a set) of them during a turn of the game, but the players do not know their internal structure. Although for many applications this perspective is admissible and gives good benefits in simplifying the problem, in some cases it results in an oversimplification. Anyway, restricting to abstract argumentation does not reduce the complexity of the problems, in general. We have seen in Section 7.5 that hardness results at the abstract level can be extended to the concrete level. Thus, it seems that the complexity of the problems largely comes from the problems themselves (including semantics and strategic aims) and not from the level of detail of the arguments.

Strategic argumentation can be considered a specific form of collective argumentation [Bodanza et al., 2017] (and judgement aggregation), where different argumentation frameworks contribute to a combined judgement on the arguments. This topic is usually considered in the context of collaboration, but some work considers self-interested agents [Bonzon and Maudet, 2012; Kontarinis et al., 2014]. Strategic argumentation is clearly a framework-wise approach, in the classification of [Bodanza et al., 2017], where argument frameworks are combined, and arguments then evaluated in the result. See Chapter 4 of this handbook [Baumeister et al., 2021] for additional discussion of this topic from a computational social choice perspective.

An approach to argumentation of interest for strategic argumentation is *probabilistic argumentation*. We refer the readers to Chapter 7 of this handbook [Hunter et al., 2021] for an in-depth discussion of this topic. Under the constellations approach to probabilistic argumentation, the key idea is that the existence (or, perhaps, validity) of arguments and attacks is unknown, but there is a probability distribution function describing the likelihood of different possibilities. Such an approach could be a useful refinement for strategic argumentation, allowing the replacement of a complete unknown (the adversary's arguments) with a more detailed model of the adversary. This might provide the basis for a player to choose among different moves.

Within the framework proposed in [Li et al., 2012], probabilities are used to represent the likelihood that arguments and attacks exist. This defines a probability distribution over all possible worlds, where each possible world is an abstract argumentation framework consisting of some subset of the arguments and attacks. Extensions arise, as usual, for a possible world, by applying any of various semantics. In [Li et al., 2012; Fazzinga et al., 2013], the authors tackle the probabilistic counterpart of the problem $VER^\sigma(S)$, that is, the problem $PROB^\sigma_\mathcal{F}(S)$ of computing the probability $Prs^\sigma_\mathcal{F}(S)$ that a set S of arguments is an extension according to a given semantics σ, given a probabilistic argumentation framework $\mathcal{F}$. [Li et al., 2012] suggested that computing the exact value of probability $Prs^\sigma_\mathcal{F}(S)$ requires exponential time, and employed a Monte-Carlo simulation approach to approximate $PROB^\sigma_\mathcal{F}(S)$. However, as far as the admissible and stable semantics are concerned, [Fazzinga et al., 2013]'s results show that the exact value of $Prs^\sigma_\mathcal{F}(S)$ can be determined in polynomial time, without enumerating the possible worlds. Nevertheless, in general the number of extensions is potentially exponential and, for other semantics, the problem is intractable. Consequently, it seems likely that many of the problems arising in strategic probabilistic argumentation will also be difficult.

Finally, there are some works that might appear to be addressing strategic argumentation, but have only weak relevance to the topic. *Strategic manoeuvring* was introduced in [van Eemeren and Houtlosser, 2002] to bridge the gap between dialectical and rhetorical approaches to the study of argumentation [van Eemeren and Houtlosser, 2006]. It refers to "the efforts arguers make in argumentative discourse to reconcile aiming for rhetorical effectiveness with maintaining dialectical standards of reasonableness" [van Eemeren and Houtlosser, 2006]. It was introduced in the context of the pragma-dialectical theory of argumentation [van Eemeren and Grootendorst, 2004; van Eemeren, 2015], which focuses on analysis and evaluation of lingual argumentation. This theory is a much broader view of argumentation than we address here. Nevertheless, there might be links between strategic manoeuvring and strategic argumentation applied to value-based or audience-based argumentation frameworks [Bench-Capon, 2003].

We have already mentioned [Schreier et al., 1995], which addresses ethics of lingual argumentative communication. It proposes standards for lingual argumentation, under the title *argumentational integrity*, and develops a taxonomy of these standards. The standards address rhetoric rather than the relation between arguments, and the notion of integrity does not include corruption (except to the extent already discussed in Section 7.1).

Despite the title, [Dziuda, 2011] analyzes a very different scenario than we do here. In that work, a decision-maker consults an expert, who possibly has an ulterior motive, about deciding between two alternatives. For example, a customer consulting a camera salesman about which camera to buy. The expert has all the arguments (which are informal) for both alternatives, and the decision-maker has none. The game is modelled probabilistically, and the paper performs an equilibrium analysis. Apart from the words "strategic argumentation" and the possibility of a self-interested player, there is no relationship between this work and the work on strategic argumentation presented here.

9 Future Directions

There are multiple avenues for further research in this area.

- The NP-completeness results in Section 5.1 apply to a wide variety of logics whose inference problem can be solved in polynomial time. Other logics, such as those in [Billington, 2011], that have a harder inference problem might result in complexities higher in the polynomial hierarchy. An analysis of such cases could extend the existing results.

- Structured argumentation theories can generate a large number of arguments, possibly infinitely many. This prevents applying the results of Section 7 to structured argumentation directly. For example, we used a different method to prove Theorem 5.8. What is needed is to find a polynomially-sized argumentation framework that is equivalent to the generated argumentation framework for the semantics of interest.

- In this chapter we have focused on a competitive situation, where the two players' aims are inconsistent. However, the basics of strategic argumentation also apply when the player's aims are consistent. In this case, strategic argumentation represents a crude adversarial negotiation. It is worth exploring how concepts from strategic argumentation can be used to analyse such negotiations, both in strategic argumentation games and in other negotiation games.

- Work has focused on two-player games of strategic argumentation. However, there are often more than two stakeholders in an adjudication, and so it would be interesting to see how strategic argumentation can be extended to more players. Among the many issues that would need to be addressed are: the protocol for turn-taking, the criterion for terminating the game, and the possibility of some players cooperating to construct an argument that none of them could construct individually. There is discussion of multi-party dialogues in [Dignum and Vreeswijk, 2003; McBurney and Parsons, 2009; Thimm, 2014]. In general, game play would appear to be more complex because of the potential for shifting alliances between players, and because players might not be compelled to make a move at each opportunity. Corruption might also be more complicated.

- In current work, the players' aims are implicitly assumed to be known and fixed. In some scenarios this might be realistic. However, there are scenarios where the motivations of a player are unclear, and/or may change over the course of argumentation. For example, a defence lawyer might begin with a "not guilty" aim but, if the trial is going badly, change tack to instead aim at a mis-trial. Thus, the extension of strategic argumentation to consider aims as possibly private and flexible/changeable is an interesting one.

- In the treatment of strategic abstract argumentation, the most prominent semantics for Dung's framework have been addressed, but there remain many semantics in the literature for which resistance to corruption is unknown. In addition, the treatment of the subset-minimality standard remains to be done for most semantics.

- The treatment of espionage assumes that full knowledge of an adversary's arguments is obtained. Perhaps the illicit gain of only some knowledge is more realistic. How can this framework be extended to cases where only partial knowledge is obtained? The work of [Dimopoulos et al., 2018] could be a first step in this direction. That paper represents partial knowledge and determines whether a player has the ability to force a desired outcome. However, it will need much expansion, as it only addresses Existential and Universal outcomes, and only for the stable semantics; assumes that the player's control arguments cannot be attacked by partially-known attacks; and does not consider multiple moves.

- Although standards are insufficient to make corruption visible, they can also be useful in guiding heuristic approaches to playing strategic argumentation games. For example, the subset-minimality standard prevents a player needlessly creating an opening for the adversary. ([Oren et al., 2006] employ this as a heuristic in a different dialogue game than the one we have presented.) Thus, it would be helpful to identify more standards, especially those that can be incorporated in heuristics or used to improve a heuristic move.

- The brief discussion of argument retraction in Section 8 deserves expansion. Strategic argumentation with retraction would seem to produce an outcome that is less arbitrary than without retraction, but perhaps the strategic element would be much diminished. Argument retraction would need to be restricted in some way, or an explicit termination rule introduced, otherwise a losing player might be able to prevent termination by repeatedly retracting arguments and then replaying them. Treating such retraction as a disavowal of some or all of the backtracked arguments (i.e. a commitment not to use those arguments in the remainder of the game) might temper the power of retraction and lead to a richer game.

- The notion of resistance to corruption we discussed is based on worst-case complexity, but this is sometimes not reflective of the difficulty of problems that arise in practice. An empirical comparison of the difficulty of solving the problems in practice and a study of approximation algorithms for these problems are needed.

- As observed in subsection 6.1, it can be worthwhile to consider an adjudicator as part of a strategic argumentation game. In this case we might consider whether the adjudicator can be subject to corruption. If the role of the adjudicator is simply to enforce the consequences arrived at by the players then there is nothing *in the game* that allows us to detect corruption.

  However, if we assume that the adjudicator chooses the semantics under which the game will be adjudicated, we have an action by the adjudicator that can be subject to analysis. This leads to quite different games, especially if the adjudicator changes the semantics *during* the playing of the game. While this appears to be rather Kafkaesque, it might be somewhat reflective of some situations where the judiciary can be influenced by other arms of government. The adjudicator then has both the choice of semantics to impose, and the choice of timing of this move. More realistically, [Prakken, 2008] presents a game where the adjudicator plays an active role, based on a detailed model of legal procedure. Perhaps that model is a base on which corruption of adjudication can be investigated.

10 Conclusions

Strategic argumentation is a primarily adversarial approach to dialogue games with incomplete information. It reflects aspects of legal argument. The idea can be applied at a concrete level, as we have demonstrated using defeasible logic rules as the basis for arguments, and at an abstract level, which was demonstrated using Dung's argumentation system.

The key element of strategic argumentation games is each player re-establishing their aim at the end of their turn. The details of the argument framework are not needed at this level of abstraction, only that they can be used to define a notion of acceptance/aim achievement.

Consequently, we have a formulation of strategic argumentation that applies to Dung's notion of argumentation framework [Dung, 1995], but also to bipolar argumentation frameworks [Cayrol and Lagasquie-Schiex, 2005], abstract argumentation frameworks with sets of attacking arguments [Nielsen and Parsons, 2006; Flouris and Bikakis, 2019][20], and preference-based argumentation frameworks [Amgoud and Cayrol, 2002; Baumeister et al., 2015]. If, in the dialogue game $(\mathcal{A}, \mathcal{R})$, we extend $\mathcal{R}$ beyond simply relations on $\mathcal{A}$ then we can have strategic argumentation on constrained argumentation frameworks [Coste-Marquis et al., 2006], weighted argument systems [Dunne et al., 2011], abstract dialectical frameworks [Brewka and Woltran, 2010], and probabilistic argumentation frameworks [Li et al., 2012], and the ideas might well be applicable to other forms of argumentation framework. Similarly, the ideas of strategic argumentation apply to semantics other than Dung-style semantics.

We have also demonstrated how the strategic argumentation framework can be used to address issues of corruption, even when the corrupt behaviour is motivated by rewards extrinsic to the game. We have not much addressed the strategies that a player might employ when playing a strategic argumentation game, although the study of standards in Section 7 provides some guidelines. More information on that topic can be found in Section 5.2 of Chapter 9 in this handbook [Black et al., 2021].

Acknowledgments

We thank the reviewers for their comments, which helped to improve this chapter. Michael Maher has an adjunct position at Griffith University and an honorary position at UNSW.

References

[Amgoud and Cayrol, 2002] Leila Amgoud and Claudette Cayrol. A reasoning model based on the production of acceptable arguments. *Ann. Math. Artif. Intell.*, 34(1-3):197–215, 2002.

---

[20]We have already addressed argumentation with sets of attacking arguments in a non-abstract setting, in Section 5.

[Amgoud et al., 2006] Leila Amgoud, Lianne Bodenstaff, Martin Caminada, Peter McBurney, Simon Parsons, Henry Prakken, Jelle van Veenen, and Gerard Vreeswijk. Final review and report on formal argumentation system. Technical report, 2006.

[Amgoud, 2019] Leila Amgoud. A replication study of semantics in argumentation. In Sarit Kraus, editor, *Proceedings of the Twenty-Eighth International Joint Conference on Artificial Intelligence, IJCAI 2019*, pages 6260–6266. ijcai.org, 2019.

[Antoniou et al., 1999] Grigoris Antoniou, David Billington, and Michael J. Maher. On the analysis of regulations using defeasible rules. In *32nd Annual Hawaii International Conference on System Sciences (HICSS-32)*, 1999.

[Antoniou et al., 2000a] Grigoris Antoniou, David Billington, Guido Governatori, and Michael J. Maher. A flexible framework for defeasible logics. In *AAAI/IAAI*, pages 405–410, 2000.

[Antoniou et al., 2000b] Grigoris Antoniou, David Billington, Guido Governatori, Michael J. Maher, and Andrew Rock. A family of defeasible reasoning logics and its implementation. In *Proc. ECAI-2000*, pages 459–463, 2000.

[Antoniou et al., 2001a] Grigoris Antoniou, David Billington, Guido Governatori, and Michael J. Maher. Representation results for defeasible logic. *ACM Trans. Comput. Log.*, 2(2):255–287, 2001.

[Antoniou et al., 2001b] Grigoris Antoniou, David Billington, Guido Governatori, and Michael J. Maher. Representation results for defeasible logic. *ACM Transactions on Computational Logic*, 2(2):255–286, 2001.

[Antoniou et al., 2007] Grigoris Antoniou, Thomas Skylogiannis, Antonis Bikakis, Martin Doerr, and Nick Bassiliades. Dr-brokering: A semantic brokering system. *Knowledge-Based Systems*, 20(1):61–72, 2007.

[Antoniou, 2004] Grigoris Antoniou. A Discussion of Some Intuitions of Defeasible Reasoning. In George Vouros and Themistoklis Panayiotopoulos, editors, *Methods and Applications of Artificial Intelligence*, volume 3025 of *Lecture Notes in Computer Science*, pages 311–320. Springer Berlin / Heidelberg, 2004.

[Barbato, 2005] Joseph M. Barbato. Scotland's bastard verdict: Intermediacy and the unique three-verdict system. *Indiana International & Comparative Law Review*, 15:543–582, 2005.

[Baroni et al., 2010] Pietro Baroni, Paul E. Dunne, and Massimiliano Giacomin. On extension counting problems in argumentation frameworks. In Pietro Baroni, Federico Cerutti, Massimiliano Giacomin, and Guillermo Ricardo Simari, editors, *Computational Models of Argument: Proceedings of COMMA 2010*, volume 216 of *Frontiers in Artificial Intelligence and Applications*, pages 63–74. IOS Press, 2010.

[Baroni et al., 2011] Pietro Baroni, Martin Caminada, and Massimiliano Giacomin. An introduction to argumentation semantics. *Knowledge Eng. Review*, 26(4):365–410, 2011.

[Baroni et al., 2018a] Pietro Baroni, Martin Caminada, and Massimiliano Giacomin. Abstract argumentation frameworks and their semantics. In Pietro Baroni; Dov Gabbay; Massimiliano Giacomin; Leendert van der Torre, editor, *Handbook on Formal Argumentation*, volume 1, pages 688–767. College Publications, February 2018.

[Baroni et al., 2018b] Pietro Baroni, Antonio Rago, and Francesca Toni. How many properties do we need for gradual argumentation? In *Proceedings of the Thirty-Second AAAI Conference on Artificial Intelligence, (AAAI-18), the 30th innovative Applications of Artificial Intelligence (IAAI-18), and the 8th AAAI Symposium on Educational Advances in Artificial Intelligence (EAAI-18), New Orleans, Louisiana, USA, February 2-7, 2018*, pages 1736–1743, 2018.

[Bartholdi et al., 1989] John J. Bartholdi, Craig A. Tovey, and Michael A. Trick. The computational difficulty of manipulating an election. *Social Choice and Welfare*, 6(3):227–241, 1989.

[Baumeister et al., 2015] Dorothea Baumeister, Daniel Neugebauer, and Jörg Rothe. Verification in attack-incomplete argumentation frameworks. In Toby Walsh, editor, *Algorithmic Decision Theory*, pages 341–358. Springer International Publishing, 2015.

[Baumeister et al., 2021] Dorothea Baumeister, Daniel Neugebauer, and Jörg Rothe. Collective acceptability in abstract argumentation. In Dov Gabbay, Massimiliano Giacomin, Guillermo R. Simari, and Matthias Thimm, editors, *Handbook of Formal Argumentation*, volume 2, chapter 4. College Publications, 2021.

[Bench-Capon, 2003] Trevor J. M. Bench-Capon. Persuasion in practical argument using value-based argumentation frameworks. *J. Log. Comput.*, 13(3):429–448, 2003.

[Benferhat et al., 1993] Salem Benferhat, Didier Dubois, and Henri Prade. Argumentative inference in uncertain and inconsistent knowledge bases. In David Heckerman and E. H. Mamdani, editors, *UAI '93: Proceedings of the Ninth Annual Conference on Uncertainty in Artificial Intelligence*, pages 411–419. Morgan Kaufmann, 1993.

[Billington et al., 2010] David Billington, Grigoris Antoniou, Guido Governatori, and Michael J. Maher. An inclusion theorem for defeasible logics. *ACM Trans. Comput. Log.*, 12(1):6, 2010.

[Billington, 2011] David Billington. A defeasible logic for clauses. In *AI 2011: Advances in Artificial Intelligence - 24th Australasian Joint Conference,*

Proceedings, pages 472–480, 2011.

[Billington, 2019] David Billington. *Factual and Plausible Reasoning*, volume 81 of *Studies in Logic*. College Publications, 2019.

[Black et al., 2021] Elizabeth Black, Nicolas Maudet, and Simon Parsons. Argumentation-based dialogue. In Dov Gabbay, Massimiliano Giacomin, Guillermo R. Simari, and Matthias Thimm, editors, *Handbook of Formal Argumentation*, volume 2, chapter 9. College Publications, 2021.

[Bodanza et al., 2017] Gustavo Adrian Bodanza, Fernando Tohmé, and Marcelo Auday. Collective argumentation: A survey of aggregation issues around argumentation frameworks. *Argument & Computation*, 8(1):1–34, 2017.

[Bondarenko et al., 1997] Andrei Bondarenko, Phan Minh Dung, Robert A. Kowalski, and Francesca Toni. An abstract, argumentation-theoretic approach to default reasoning. *Artif. Intell.*, 93:63–101, 1997.

[Bonzon and Maudet, 2012] Elise Bonzon and Nicolas Maudet. On the outcomes of multiparty persuasion. In *Argumentation in Multi-Agent Systems - 8th International Workshop, ArgMAS 2011*, volume 7543 of *Lecture Notes in Computer Science*, pages 86–101. Springer, 2012.

[Bossi et al., 1993] Annalisa Bossi, Michele Bugliesi, Maurizio Gabbrielli, Giorgio Levi, and Maria Chiara Meo. Differential logic programming. In *Conference Record of the Twentieth Annual ACM SIGPLAN-SIGACT Symposium on Principles of Programming Languages*, pages 359–370, 1993.

[Brandt et al., 2016] Felix Brandt, Vincent Conitzer, Ulle Endriss, Jérôme Lang, and Ariel D. Procaccia, editors. *Handbook of Computational Social Choice*. Cambridge University Press, 2016.

[Brewka and Woltran, 2010] Gerhard Brewka and Stefan Woltran. Abstract dialectical frameworks. In *Principles of Knowledge Representation and Reasoning: Proceedings of the Twelfth International Conference, KR 2010, Toronto, Ontario, Canada, May 9-13, 2010*, 2010.

[Caminada et al., 2015] Martin Caminada, Samy Sá, João Alcântara, and Wolfgang Dvořák. On the equivalence between logic programming semantics and argumentation semantics. *Int. J. Approx. Reasoning*, 58:87–111, 2015.

[Cayrol and Lagasquie-Schiex, 2005] Claudette Cayrol and Marie-Christine Lagasquie-Schiex. On the acceptability of arguments in bipolar argumentation frameworks. In *Symbolic and Quantitative Approaches to Reasoning with Uncertainty, 8th European Conference, ECSQARU 2005, Proceedings*, pages 378–389, 2005.

[Cayrol et al., 2002] Claudette Cayrol, Sylvie Doutre, Marie-Christine Lagasquie-Schiex, and Jérôme Mengin. "Minimal defence": a refinement of the preferred semantics for argumentation frameworks. In *9th International*

Workshop on Non-Monotonic Reasoning (NMR 2002), Proceedings, pages 408–415, 2002.

[Chalkiadakis and Boutilier, 2007] Georgios Chalkiadakis and Craig Boutilier. Coalitional bargaining with agent type uncertainty. In Manuela M. Veloso, editor, *IJCAI*, pages 1227–1232, 2007.

[Coste-Marquis et al., 2006] Sylvie Coste-Marquis, Caroline Devred, and Pierre Marquis. Constrained argumentation frameworks. In *Proceedings, Tenth International Conference on Principles of Knowledge Representation and Reasoning*, pages 112–122, 2006.

[Devereux and Reed, 2009] Joseph Devereux and Chris Reed. Strategic argumentation in rigorous persuasion dialogue. In *Argumentation in Multi-Agent Systems, 6th International Workshop, ArgMAS 2009*, pages 94–113, 2009.

[Dignum and Vreeswijk, 2003] Frank Dignum and Gerard Vreeswijk. Towards a testbed for multi-party dialogues. In Frank Dignum, editor, *Advances in Agent Communication, International Workshop on Agent Communication Languages, ACL 2003, Melbourne, Australia, July 14, 2003*, volume 2922 of *Lecture Notes in Computer Science*, pages 212–230. Springer, 2003.

[Dimopoulos and Kakas, 1995] Yannis Dimopoulos and Antonis C. Kakas. Logic programming without negation as failure. In *Proceedings of the 1995 International Symposium on Logic Programming*, pages 369–384, 1995.

[Dimopoulos et al., 2002] Yannis Dimopoulos, Bernhard Nebel, and Francesca Toni. On the computational complexity of assumption-based argumentation for default reasoning. *Artif. Intell.*, 141(1/2):57–78, 2002.

[Dimopoulos et al., 2018] Yannis Dimopoulos, Jean-Guy Mailly, and Pavlos Moraitis. Control argumentation frameworks. In Sheila A. McIlraith and Kilian Q. Weinberger, editors, *Proceedings of the Thirty-Second AAAI Conference on Artificial Intelligence, (AAAI-18)*, pages 4678–4685. AAAI Press, 2018.

[Dung, 1995] P.M. Dung. On the acceptability of arguments and its fundamental role in nonmonotonic reasoning, logic programming and n-person games. *Artificial Intelligence*, 77(2):321–358, 1995.

[Dunne and Wooldridge, 2009] Paul E. Dunne and Michael Wooldridge. Complexity of abstract argumentation. In I. Rahwan and G.R. Simari, editors, *Argumentation in Artificial Intelligence*, pages 85–104. Springer, 2009.

[Dunne et al., 2011] Paul E. Dunne, Anthony Hunter, Peter McBurney, Simon Parsons, and Michael J. Wooldridge. Weighted argument systems: Basic definitions, algorithms, and complexity results. *Artif. Intell.*, 175(2):457–486, 2011.

[Dvořák and Dunne, 2017] Wolfgang Dvořák and Paul E. Dunne. Computational problems in formal argumentation and their complexity. *IfCoLog Journal of Logics and their Applications*, 4(8), 2017.

[Dvořák, 2011] Wolfgang Dvořák. On the complexity of computing the justification status of an argument. In Sanjay Modgil, Nir Oren, and Francesca Toni, editors, *Theorie and Applications of Formal Argumentation - First International Workshop, TAFA*, volume 7132 of *Lecture Notes in Computer Science*, pages 32–49. Springer, 2011.

[Dziuda, 2011] Wioletta Dziuda. Strategic argumentation. *J. Econ. Theory*, 146(4):1362–1397, 2011.

[Eisenhardt, 1989] Kathleen M. Eisenhardt. Agency theory: An assessment and review. *The Academy of Management Review*, 14(1):57–74, 1989.

[Eriksson Lundström et al., 2008] Jenny Eriksson Lundström, Guido Governatori, Subhasis Thakur, and Vineet Padmanabhan. An asymmetric protocol for argumentation games in defeasible logic. In Aditya Ghose and Guido Governatori, editors, *10 Pacific Rim International Workshop on Multi-Agents*, volume 5044 of *LNAI*, pages 219–231, Heidelberg, 2008. Springer.

[Fagin et al., 2003] Ronald Fagin, Joseph Y. Halpern, Yoram Moses, and Moshe Y. Vardi. *Common Knowledge Revisited*. MIT Press, 2003.

[Fan and Toni, 2016] Xiuyi Fan and Francesca Toni. On the interplay between games, argumentation and dialogues. In *Proceedings of the 2016 International Conference on Autonomous Agents & Multiagent Systems, Singapore, May 9-13, 2016*, pages 260–268, 2016.

[Farley and Freeman, 1995] Arthur M. Farley and Kathleen Freeman. Burden of proof in legal argumentation. In *Proceedings of the Fifth International Conference on Artificial Intelligence and Law, ICAIL '95*, pages 156–164, 1995.

[Fazzinga et al., 2013] Bettina Fazzinga, Sergio Flesca, and Francesco Parisi. On the complexity of probabilistic abstract argumentation. In *IJCAI 2013, Proceedings of the 23rd International Joint Conference on Artificial Intelligence, Beijing, China, August 3-9, 2013*, 2013.

[Fichte et al., 2019] Johannes Klaus Fichte, Markus Hecher, and Arne Meier. Counting complexity for reasoning in abstract argumentation. In *The Thirty-Third AAAI Conference on Artificial Intelligence, AAAI*, pages 2827–2834. AAAI Press, 2019.

[Flieger, 2002] Johannes C. Flieger. Relevance and minimality in systems of defeasible argumentation. Technical report, Imperial College, London, 2002.

[Flouris and Bikakis, 2019] Giorgos Flouris and Antonis Bikakis. A comprehensive study of argumentation frameworks with sets of attacking arguments. *Int. J. Approx. Reason.*, 109:55–86, 2019.

[Frankfurt, 2005] Harry G. Frankfurt. *On Bullshit*. Princeton University Press, 2005.

[Garg et al., 2008] Dinesh Garg, Yadati Narahari, and Sujit Gujar. Foundations of mechanism design: A tutorial. *Sadhana*, 33(2), 2008.

[Gordon and Walton, 2009] Thomas F. Gordon and Douglas Walton. Proof burdens and standards. In I. Rahwan and G.R. Simari, editors, *Argumentation in Artificial Intelligence*, pages 239–260. Springer, 2009.

[Gordon et al., 2007] Thomas F. Gordon, Henry Prakken, and Douglas Walton. The Carneades model of argument and burden of proof. *Artif. Intell.*, 171(10-15):875–896, 2007.

[Gordon, 1994] Thomas F. Gordon. The pleadings game: An exercise in computational dialectics. *Artif. Intell. Law*, 2:239–292, December 1994.

[Gottlob et al., 2007] Georg Gottlob, Gianluigi Greco, and Toni Mancini. Complexity of pure equilibria in bayesian games. In Manuela M. Veloso, editor, *IJCAI*, pages 1294–1299, 2007.

[Governatori and Maher, 2017] Guido Governatori and Michael J. Maher. Annotated defeasible logic. *Theory Pract. Log. Program.*, 17(5-6):819–836, 2017.

[Governatori and Rotolo, 2010] Guido Governatori and Antonino Rotolo. Changing legal systems: legal abrogations and annulments in defeasible logic. *Logic Journal of IGPL*, 18(1):157–194, 2010.

[Governatori et al., 2001] Guido Governatori, Marlon Dumas, Arthur H.M. ter Hofstede, and Phillipa Oaks. A formal approach to protocols and strategies for (legal) negotiation. In Henry Prakken, editor, *Proceedings of the 8th International Conference on Artificial Intelligence and Law*, pages 168–177. IAAIL, ACM Press, 2001.

[Governatori et al., 2004] Guido Governatori, Michael J. Maher, Grigoris Antoniou, and David Billington. Argumentation semantics for defeasible logics. *Journal of Logic and Computation*, 14(5):675–702, 2004.

[Governatori et al., 2013] Guido Governatori, Francesco Olivieri, Antonino Rotolo, and Simone Scannapieco. Computing strong and weak permissions in defeasible logic. *Journal of Philosophical Logic*, 42(6):799–829, 2013.

[Governatori et al., 2014] Guido Governatori, Francesco Olivieri, Simone Scannapieco, Antonino Rotolo, and Matteo Cristani. Strategic argumentation is NP-complete. In *Proc. ECAI 2014*. IOS Press, 2014.

[Governatori, 2011] Guido Governatori. On the relationship between Carneades and Defeasible Logic. In *The 13th International Conference on Artificial Intelligence and Law, Proceedings of the Conference*, pages 31–40, 2011.

[Grossi and van der Hoek, 2013] Davide Grossi and Wiebe van der Hoek. Audience-based uncertainty in abstract argument games. In *IJCAI'13*, pages

143–149. AAAI Press, 2013.

[Hadjinikolis et al., 2013] Christos Hadjinikolis, Yiannis Siantos, Sanjay Modgil, Elizabeth Black, and Peter McBurney. Opponent modelling in persuasion dialogues. In *IJCAI 2013, Proceedings of the 23rd International Joint Conference on Artificial Intelligence*, 2013.

[Hadoux et al., 2015] Emmanuel Hadoux, Aurélie Beynier, Nicolas Maudet, Paul Weng, and Anthony Hunter. Optimization of probabilistic argumentation with markov decision models. In Qiang Yang and Michael J. Wooldridge, editors, *Proceedings of the Twenty-Fourth International Joint Conference on Artificial Intelligence, IJCAI 2015, Buenos Aires, Argentina, July 25-31, 2015*, pages 2004–2010. AAAI Press, 2015.

[Hashmi et al., 2018] Mustafa Hashmi, Guido Governatori, Ho-Pun Lam, and Moe Thandar Wynn. Are we done with business process compliance: state of the art and challenges ahead. *Knowl. Inf. Syst.*, 57(1):79–133, 2018.

[Hunter et al., 2021] Anthony Hunter, Sylwia Polberg, Nico Potyka, Tjitze Rienstra, and Matthias Thimm. Probabilistic argumentation: A survey. In Dov Gabbay, Massimiliano Giacomin, Guillermo R. Simari, and Matthias Thimm, editors, *Handbook of Formal Argumentation*, volume 2, chapter 7. College Publications, 2021.

[Islam and Governatori, 2018] Mohammad Badiul Islam and Guido Governatori. RuleRS: a rule-based architecture for decision support systems. *Artif. Intell. Law*, 26(4):315–344, 2018.

[Johnson, 1990] David S. Johnson. A catalog of complexity classes. In *Handbook of Theoretical Computer Science, Volume A: Algorithms and Complexity*, pages 67–161. Elsevier, 1990.

[Kakas et al., 2019] Antonis C. Kakas, Pavlos Moraitis, and Nikolaos I. Spanoudakis. *GORGIAS*: Applying argumentation. *Argument & Computation*, 10(1):55–81, 2019.

[Kontarinis et al., 2014] Dionysios Kontarinis, Elise Bonzon, Nicolas Maudet, and Pavlos Moraitis. On the use of target sets for move selection in multi-agent debates. In *ECAI 2014 - 21st European Conference on Artificial Intelligence*, pages 1047–1048, 2014.

[Laenens and Vermeir, 1990] Els Laenens and Dirk Vermeir. A fixpoint semantics for ordered logic. *J. Log. Comput.*, 1(2):159–185, 1990.

[Lam et al., 2016] Ho-Pun Lam, Guido Governatori, and Régis Riveret. On aspic$^+$ and defeasible logic. In *Computational Models of Argument - Proceedings of COMMA 2016*, pages 359–370, 2016.

[Li et al., 2012] Hengfei Li, Nir Oren, and Timothy J. Norman. Probabilistic argumentation frameworks. In Sanjay Modgil, Nir Oren, and Francesca

Toni, editors, *Theorie and Applications of Formal Argumentation - First International Workshop, TAFA 2011. Revised Selected Papers*, volume 7132 of *Lecture Notes in Computer Science*, pages 1–16. Springer, 2012.

[Lorenzen and Lorenz, 1978] Paul Lorenzen and Kuno Lorenz. *Dialogische Logik*. Darmstadt, 1978.

[Loui, 1998] Ronald Prescott Loui. Process and policy: Resource-bounded nondemonstrative reasoning. *Computational Intelligence*, 14(1):1–38, 1998.

[Maher et al., 2020] Michael J. Maher, Ilias Tachmazidis, Grigoris Antoniou, Stephen Wade, and Long Cheng. Rethinking defeasible reasoning: A scalable approach. *Theory and Practice of Logic Programming*, 20(4):552–586, 2020.

[Maher, 2001] Michael J. Maher. Propositional defeasible logic has linear complexity. *Theory and Practice of Logic Programming*, 1(6):691–711, 2001.

[Maher, 2012] Michael J. Maher. Relative expressiveness of defeasible logics. *Theory and Practice of Logic Programming*, 12(4-5):793–810, 2012.

[Maher, 2013a] Michael J. Maher. Relative expressiveness of defeasible logics II. *Theory and Practice of Logic Programming*, 13:579–592, 2013.

[Maher, 2013b] Michael J. Maher. Relative Expressiveness of Well-Founded Defeasible Logics. In Stephen Cranefield and Abhaya Nayak, editors, *AI 2013: Advances in Artificial Intelligence*, volume 8272 of *Lecture Notes in Computer Science*, pages 338–349. Springer International Publishing, 2013.

[Maher, 2014a] Michael J. Maher. Comparing defeasible logics. In Torsten Schaub, Gerhard Friedrich, and Barry O'Sullivan, editors, *ECAI 2014 - 21st European Conference on Artificial Intelligence*, volume 263 of *Frontiers in Artificial Intelligence and Applications*, pages 585–590. IOS Press, 2014.

[Maher, 2014b] Michael J. Maher. Complexity of exploiting privacy violations in strategic argumentation. In Duc Nghia Pham and Seong-Bae Park, editors, *PRICAI 2014: Trends in Artificial Intelligence - 13th Pacific Rim International Conference on Artificial Intelligence*, volume 8862 of *Lecture Notes in Computer Science*, pages 523–535. Springer, 2014.

[Maher, 2016a] Michael J. Maher. Corrupt strategic argumentation: The ideal and the naive. In *AI 2016: Advances in Artificial Intelligence*, pages 17–28, 2016.

[Maher, 2016b] Michael J. Maher. Resistance to corruption of general strategic argumentation. In *Proc. Int. Conf. Principles and Practice of Multi-Agent Systems*, pages 61–75, 2016.

[Maher, 2016c] Michael J. Maher. Resistance to corruption of strategic argumentation. In *AAAI Conference on Artificial Intelligence*, pages 1030–1036, 2016.

[Maher, 2017a] Michael J. Maher. Corruption and audit in strategic argumentation. Technical report, Reasoning Research Institute, 2017.

[Maher, 2017b] Michael J. Maher. Relating concrete defeasible reasoning formalisms and abstract argumentation. *Fundam. Inform.*, 155(3):233–260, 2017.

[Maier and Nute, 2010] Frederick Maier and Donald Nute. Well-founded semantics for defeasible logic. *Synthese*, 176(2):243–274, 2010.

[Maier, 2013] Frederick Maier. Interdefinability of defeasible logic and logic programming under the well-founded semantics. *Theory and Practice of Logic Programming*, 13:107–142, 2013.

[Matt and Toni, 2008] P. Matt and F. Toni. A game-theoretic measure of argument strength for abstract argumentation. In *JELIA 2008*, volume 5293 of *LNCS*, pages 285–297. Springer, 2008.

[Mbarki et al., 2006] Mohamed Mbarki, Jamal Bentahar, and Bernard Moulin. Specification and complexity of strategic-based reasoning using argumentation. In *Argumentation in Multi-Agent Systems, Third International Workshop, ArgMAS 2006, Hakodate, Japan, May 8, 2006, Revised Selected and Invited Papers*, pages 142–160, 2006.

[McBurney and Parsons, 2009] Peter McBurney and Simon Parsons. Dialogue games for agent argumentation. In Guillermo Ricardo Simari and Iyad Rahwan, editors, *Argumentation in Artificial Intelligence*, pages 261–280. Springer, 2009.

[Modgil and Caminada, 2009] Sanjay Modgil and Martin Caminada. Proof theories and algorithms for abstract argumentation frameworks. In *Argumentation in Artificial Intelligence*, pages 105–129. Springer, 2009.

[Nielsen and Parsons, 2006] Søren Holbech Nielsen and Simon Parsons. A generalization of Dung's abstract framework for argumentation: Arguing with sets of attacking arguments. In *Argumentation in Multi-Agent Systems, Third International Workshop, ArgMAS 2006*, pages 54–73, 2006.

[Nute, 1988] Donald Nute. Defeasible reasoning: A philosophical analysis in Prolog. In James H. Fetzer, editor, *Aspects of Artificial Intelligence*, volume 1 of *Studies in Cognitive Systems*. Springer, 1988.

[Nute, 1994] Donald Nute. Defeasible logic. In Dov M. Gabbay, Chris J. Hogger, and J. Allen Robinson, editors, *Handbook of Logic in Artificial Intelligence and Logic Programming*, volume 3. Oxford University Press, 1994.

[Nute, 2001] Donald Nute. Defeasible logic: Theory, Implementation and Applications. In *Proceedings of the 14th International Conference on Applications of Prolog (INAP 2001)*, pages 151–169. Springer, Berlin, 2001.

[Oren et al., 2006] Nir Oren, Timothy J. Norman, and Alun D. Preece. Loose lips sink ships: a heuristic for argumentation. In *Proceedings of the Third*

International Workshop on Argumentation in Multi-Agent Systems (ArgMAS 2006), pages 121–134, 2006.

[Osborne and Rubinstein, 1999] M. J. Osborne and A. Rubinstein. *A Course in Game Theory*. MIT Press, 1999.

[Parsons and Sklar, 2005] Simon Parsons and Elizabeth Sklar. How agents alter their beliefs after an argumentation-based dialogue. In *Argumentation in Multi-Agent Systems, Second International Workshop, ArgMAS 2005*, pages 297–312, 2005.

[Prakken and Sartor, 1998] Henry Prakken and Giovanni Sartor. Modelling reasoning with precedents in a formal dialogue game. *Artif. Intell. Law*, 6(2-4):231–287, 1998.

[Prakken and Sartor, 2015] Henry Prakken and Giovanni Sartor. Law and logic: A review from an argumentation perspective. *Artif. Intell.*, 227:214–245, 2015.

[Prakken, 2000] Henry Prakken. Relating protocols for dynamic dispute with logics for defeasible argumentation. *Synthese*, 127:2001, 2000.

[Prakken, 2005] Henry Prakken. Coherence and flexibility in dialogue games for argumentation. *J. Log. Comput.*, 15(6):1009–1040, 2005.

[Prakken, 2008] Henry Prakken. A formal model of adjudication dialogues. *Artif. Intell. Law*, 16(3):305–328, 2008.

[Prakken, 2010a] Henry Prakken. An abstract framework for argumentation with structured arguments. *Argument and Computation*, 1:93–124, 2010.

[Prakken, 2010b] Henry Prakken. An abstract framework for argumentation with structured arguments. *Argument and Computation*, 1(2):93–124, 2010.

[Procaccia and Rosenschein, 2005] Ariel D. Procaccia and Jeffrey S. Rosenschein. Extensive-form argumentation games. In Marie-Pierre Gleizes, Gal A. Kaminka, Ann Nowé, Sascha Ossowski, Karl Tuyls, and Katja Verbeeck, editors, *EUMAS 2005 - Proceedings of the Third European Workshop on Multi-Agent Systems*, pages 312–322. Koninklijke Vlaamse Academie van Belie voor Wetenschappen en Kunsten, 2005.

[Rahwan et al., 2003] Iyad Rahwan, Sarvapali D. Ramchurn, Nicholas R. Jennings, Peter Mcburney, Simon Parsons, and Liz Sonenberg. Argumentation-based negotiation. *The Knowledge Engineering Review*, 18(4):343–375, 2003.

[Rahwan et al., 2009] Iyad Rahwan, Kate Larson, and Fernando A. Tohmé. A characterisation of strategy-proofness for grounded argumentation semantics. In *IJCAI*, pages 251–256, 2009.

[Reeves et al., 2002] Daniel M. Reeves, Michael P. Wellman, and Benjamin N. Grosof. Automated negotiation from declarative contract descriptions. *Computational Intelligence*, 18(4):482–500, 2002.

[Rienstra et al., 2013] Tjitze Rienstra, Matthias Thimm, and Nir Oren. Opponent models with uncertainty for strategic argumentation. In *IJCAI 2013, Proceedings of the 23rd International Joint Conference on Artificial Intelligence*, 2013.

[Riveret et al., 2008] Régis Riveret, Henry Prakken, Antonino Rotolo, and Giovanni Sartor. Heuristics in argumentation: A game theory investigation. In *COMMA 2008*, pages 324–335. IOS Press, 2008.

[Roth et al., 2007] Bram Roth, Régis Riveret, Antonino Rotolo, and Guido Governatori. Strategic argumentation: a game theoretical investigation. In *The Eleventh International Conference on Artificial Intelligence and Law, Proceedings of the Conference*, pages 81–90, 2007.

[Sakama, 2012] Chiaki Sakama. Dishonest arguments in debate games. In *Computational Models of Argument - Proceedings of COMMA 2012, Vienna, Austria, September 10-12, 2012*, pages 177–184, 2012.

[Sakama, 2013] Chiaki Sakama. Debate games in logic programming. In *Declarative Programming and Knowledge Management - Declarative Programming Days, KDPD 2013*, pages 185–201, 2013.

[Satoh and Takahashi, 2011a] K. Satoh and K. Takahashi. A semantics of argumentation under incomplete information. In *Proceedings of Jurisn 2011*, 2011.

[Satoh and Takahashi, 2011b] Ken Satoh and Kazuko Takahashi. A semantics of argumentation under incomplete information. In *JURISIN*, pages 86–97, 2011.

[Schreier et al., 1995] Margrit Schreier, Norbert Groeben, and Ursula Christmann. "That's Not Fair!" argumentational integrity as an ethics of argumentative communication. *Argumentation*, 9:267–289, 1995.

[Thimm and García, 2010] Matthias Thimm and Alejandro Javier García. On strategic argument selection in structured argumentation systems. In *Argumentation in Multi-Agent Systems - 7th International Workshop, ArgMAS 2010*, pages 286–305, 2010.

[Thimm, 2014] Matthias Thimm. Strategic argumentation in multi-agent systems. *Künstliche Intell.*, 28(3):159–168, 2014.

[Touretzky et al., 1987] David S. Touretzky, John F. Horty, and Richmond H. Thomason. A Clash of Intuitions: The Current State of Nonmonotonic Multiple Inheritance Systems. In *Proceedings of the 10th international joint conference on Artificial intelligence (IJCAI'87)*, pages 476–482, San Francisco, CA, USA, 1987. Morgan Kaufmann Publishers Inc.

[Touretzky, 1986] David S. Touretzky. *The Mathematics of Inheritance Systems*. Morgan Kaufmann, 1986.

[van Eemeren and Grootendorst, 2004] Frans H. van Eemeren and R. Grootendorst, editors. *A Systematic Theory of Argumentation. The Pragma-Dialectical Approach.* Cambridge University Press, 2004.

[van Eemeren and Houtlosser, 2002] Frans H. van Eemeren and Peter Houtlosser. Strategic maneuvering: Maintaining a delicate balance. In Frans H. van Eemeren and Peter Houtlosser, editors, *Dialectic and Rhetoric: The Warp and Woof of Argumentation Analysis.* Kluwer Academic Publishers, 2002.

[van Eemeren and Houtlosser, 2006] Frans H. van Eemeren and Peter Houtlosser. Strategic maneuvering: A synthetic recapitulation. *Argumentation*, 20:381–392, 2006.

[van Eemeren, 2015] Frans H. van Eemeren, editor. *Reasonableness and Effectiveness in Argumentative Discourse, Fifty Contributions to the Development of Pragma-Dialectics*, volume 27 of *Argumentation Library.* Springer, 2015.

[Verheij, 2003] Bart Verheij. DefLog: on the logical interpretation of prima facie justified assumptions. *J. Log. Comput.*, 13(3):319–346, 2003.

[Vreeswijk and Prakken, 2000] Gerard Vreeswijk and Henry Prakken. Credulous and sceptical argument games for preferred semantics. In *Logics in Artificial Intelligence, European Workshop, JELIA 2000*, volume 1919 of *Lecture Notes in Computer Science*, pages 239–253. Springer, 2000.

[Wagner, 1986] Klaus W. Wagner. The complexity of combinatorial problems with succinct input representation. *Acta Inf.*, 23(3):325–356, 1986.

[Walton and Krabbe, 1995] Douglas Walton and Erik C. W. Krabbe. *Commitment in Dialogue: Basic Concepts of Interpersonal Reasoning.* State University of New York Press, 1995.

[Wan et al., 2009] Hui Wan, Benjamin N. Grosof, Michael Kifer, Paul Fodor, and Senlin Liang. Logic programming with defaults and argumentation theories. In *ICLP*, pages 432–448, 2009.

[Wan et al., 2015] Hui Wan, Michael Kifer, and Benjamin N. Grosof. Defeasibility in answer set programs with defaults and argumentation rules. *Semantic Web*, 6(1):81–98, 2015.

[Woltran, 2014] Stefan Woltran. Abstract argumentation – all problems solved? presentation at ECAI-14, 2014. https://www.dbai.tuwien.ac.at/staff/woltran/ecai2014.pdf.

[Wu and Caminada, 2010] Yining Wu and Martin Caminada. A labelling-based justification status of arguments. *Studies in Logic*, 3(4):12–29, 2010.

[Wu and Podlaszewski, 2015] Yining Wu and Mikolaj Podlaszewski. Implementing crash-resistance and non-interference in logic-based argumentation. *J. Log. Comput.*, 25(2):303–333, 2015.

CHAPTER 11

ON THE INCREMENTAL COMPUTATION OF SEMANTICS IN DYNAMIC ARGUMENTATION

GIANVINCENZO ALFANO
Department of Informatics, Modeling, Electronics and System Engineering, University of Calabria, Italy
g.alfano@dimes.unical.it

SERGIO GRECO
Department of Informatics, Modeling, Electronics and System Engineering, University of Calabria, Italy
greco@dimes.unical.it

FRANCESCO PARISI
Department of Informatics, Modeling, Electronics and System Engineering, University of Calabria, Italy
fparisi@dimes.unical.it

GERARDO I. SIMARI
*Department of Computer Science and Engineering, Universidad Nacional del Sur (UNS) & Institute for Computer Science and Engineering (ICIC UNS-CONICET),
Bahia Blanca, Argentina*
gis@cs.uns.edu.ar

GUILLERMO R. SIMARI
*Department of Computer Science and Engineering, Universidad Nacional del Sur (UNS) & Institute for Computer Science and Engineering (ICIC UNS-CONICET),
Bahia Blanca, Argentina*
grsimari@gmail.com

ALFANO, GRECO, PARISI, SIMARI, SIMARI

Abstract

Argumentation frameworks model dynamic situations where arguments and their relationships (*e.g.,* attacks) frequently change over time. As a consequence, the sets of conclusions (*e.g.,* extensions of abstract argumentation frameworks, or warranted literals for structured argumentation frameworks) often need to be computed again after performing an update. However, as most of the argumentation semantics proposed so far suffer from high computational complexity, computing the set of conclusions from scratch is costly in general. In this work, we address the problems of efficiently recomputing extensions of dynamic abstract argumentation frameworks and warranted literals in dynamic defeasible knowledge bases. In particular, we first present an incremental algorithmic solution whose main idea is that of using an initial extension and the update to identify a (potentially small) portion of an abstract argumentation framework, which is sufficient to compute an extension of the updated framework. We discuss the incremental technique for the computation of extensions in different abstract argumentation frameworks and then explore how similar ideas carry over to the computation of warranted literals in defeasible logic programming.

1 Introduction

Computational Argumentation is an established research field in the area of Knowledge Representation and Reasoning (KR) [Bench-Capon and Dunne, 2007; Rahwan and Simari, 2009; Baroni *et al.*, 2011a; Modgil and Prakken, 2011; Eiter *et al.*, 2015], which is central in Artificial Intelligence (AI). An (abstract) argumentation framework [Dung, 1995] is a simple, yet powerful formalism for modeling disputes between agents. The formal meaning of an argumentation framework is given in terms of argumentation semantics, which intuitively tell us the sets of arguments (referred to as *extensions*) that can be accepted together to support a point of view in a discussion. For an abstract argumentation framework, an argument is an abstract entity whose role is entirely determined by its relationships with other arguments. In contrast, DeLP [Garcia and Simari, 2004] is a well-known argumentation formalism where arguments have an explicit structure as they derive from a knowledge base (DeLP program) consisting of facts and strict and defeasible rules. By considering the structure of arguments, *i.e.*, their inner construction, it becomes possible to analyze reasons for and against a conclusion closely, and the *warrant status* of a claim in the context of a knowledge base represents the main output of a dialectical process.

Although the ideas underlying abstract and structured argumentation frameworks are intuitive and straightforward, most of the argumentation semantics proposed so far suffer from high computational complexity [Dunne and Wooldridge, 2009; Dunne, 2009; Dvorak and Woltran, 2010; Kröll *et al.*, 2017; Cecchi *et al.*, 2006]. Most research in the domain of formal argumentation (both in the abstract and structured settings) have focused on *static* frameworks (*i.e.*, frameworks whose structure does not change over time), whereas argumentation frameworks are frequently used for modeling dynamic systems [Baumann, 2011; Falappa *et al.*, 2011; Liao *et al.*, 2011; Baroni *et al.*, 2014b; Charwat *et al.*, 2015; Bistarelli *et al.*, 2018a; Bistarelli *et al.*, 2018b]; since, as a matter of fact, the argumentation process is inherently dynamic, this is not surprising. For instance, consider how many times we change our minds after learning something new about a situation that is the focus of our reasoning. There is evidence of that in social network threads [Kökciyan *et al.*, 2016], where users frequently post new arguments against or supporting other posts, often made by the same users that change their minds. Surprisingly, the definition of evaluation algorithms and the analysis of the computational complexity taking into account such dynamic aspects have been mostly neglected, whereas, in these situations, incremental computation techniques could significantly improve performance. In many cases, especially when few updates at a time are performed, the changes made to a framework can result in small changes to the set of its conclusions—extensions of abstract argumentation frameworks; warranted literals for structured argumentation—and recomputing the whole semantics from scratch can be avoided.

The following is a summary of the contributions of this work:

- By focusing on the most popular argumentation semantics for abstract frameworks, *i.e.*, *complete*, *preferred*, *stable*, *ideal*, and *grounded*, we present a general approach for incrementally solving the following computational task: given an argumentation framework AF, an extension for AF under semantics σ, and an update u, obtain an extension of the updated argumentation framework $u(AF)$ under σ. In other words, we explore the possibility of incrementally solving the task σ-SE of the International Competition on Computational Models of Argumentation (ICCMA) [Thimm and Villata, 2017]: given an argumentation framework, obtain some σ-extension. The technique consists of the following main steps: (i) identification of the *influenced set*, which intuitively consists of the set of arguments whose acceptance status may change after performing an update; (ii) identification of a (possibly) smaller argumentation framework, called *reduced* argumentation framework, based on the influenced set and additional information provided by the initial extension; (iii) using *any* non-incremental algorithm to compute an extension of the reduced argumentation framework; and (iv) obtaining the final extension by merging a portion of the initial extension with the one computed for the reduced argumentation framework.

- We show that the main idea behind the above-described incremental approach can be adapted to *extended* abstract argumentation frameworks, *i.e.*, bipolar argumentation frameworks allowing the presence of attacks and supports, as well as argumentation frameworks with second-order interactions (*e.g.*, attacks towards attacks). This is achieved by leveraging meta-argumentation approaches, which provide ways to transform a more general abstract framework into a Dung framework.

- Intending to minimize wasted effort in the computation of the warrant status of literals of a DeLP program after performing an update, we summarize the necessary elements to develop the updating techniques in DeLP's structured argumentation. Particularly, we focus only on literals that are potentially affected by a given update (namely, *influenced* and *core* literals), and avoids the computation of the status of *inferable* and *preserved* literals.

Organization. As a prelude, we first briefly recall basic notions of abstract argumentation frameworks [Dung, 1995] and then introduce updates in Section 2. The incremental technique for recomputing an extension of an updated abstract argumentation framework under different semantics is presented in Section 3. The main idea behind the above-described incremental approach is then adapted to cope with extended argumentation frameworks in Section 4. Next, in Section 5, we discuss the critical aspects of the technique dealing with structured argumentation in an easy-to-read manner. Related work is discussed in Section 6, and conclusions and directions for future work are drawn in Section 7.

2 Abstract Argumentation Frameworks and Updates

We assume the existence of a set $Args$ of *arguments*. An *(abstract) argumentation framework* [Dung, 1995] is a pair $\langle Ar, att \rangle$, where $Ar \subseteq Args$ is a finite set of *arguments*, and $att \subseteq Ar \times Ar$ is a binary relation over Ar whose elements are called *attacks*. Thus, an argumentation framework can be viewed as a directed graph where nodes correspond to arguments and edges correspond to attacks.

Example 2.1 (Running example for abstract argumentation). Let $AF_0 = \langle Ar_0, att_0 \rangle$ be an argumentation framework, where $Ar_0 = \{a, b, c, d, e, f, g, h\}$ and $att_0 = \{(a,b), (b,a), (b,c), (c,c), (d,a), (d,e), (e,d), (b,e), (f,e), (g,d), (g,h), (h,e), (h,f)\}$. The argumentation framework AF_0 is shown in Figure 1.

Given an argumentation framework $\langle Ar, att \rangle$ and arguments $a, b \in Ar$, we say that a *attacks* b iff $(a,b) \in att$, and that a set $S \subseteq Ar$ attacks b iff there is $a \in S$ attacking b. We use $S^+ = \{b \mid \exists a \in S : (a,b) \in att\}$ to denote the set of all arguments that are attacked by S.

Moreover, we say that $S \subseteq Ar$ *defends* a iff $\forall b \in Ar$ such that b attacks a, there is $c \in S$ such that c attacks b. A set $S \subseteq Ar$ of arguments is said to be:

- *conflict-free* if there are no $a, b \in S$ such that a *attacks* b;

- *admissible* if it is conflict-free and it defends all its arguments.

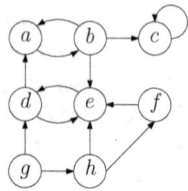

Figure 1: AF_0 of Example 2.1.

An argumentation semantics specifies the criteria for identifying a set of arguments, called *extension*, that can be considered "reasonable" together. A *complete extension* ($\mathcal{CO}$) is an admissible set that contains all the arguments that it defends. A complete extension S is said to be:

- *preferred* ($\mathcal{PR}$) iff it is maximal (w.r.t. $\subseteq$);
- *stable* ($\mathcal{ST}$) iff it attacks every argument in $A \setminus S$;
- *grounded* ($\mathcal{GR}$) iff it is minimal (w.r.t. $\subseteq$).
- *ideal* ($\mathcal{ID}$) iff it is contained in every preferred extension and it is maximal (w.r.t. $\subseteq$).

Given an argumentation framework AF and a semantics $\sigma \in \{\mathcal{CO}, \mathcal{PR}, \mathcal{ST}, \mathcal{GR}, \mathcal{ID}\}$, we use $\mathcal{E}_\sigma(AF)$ to denote the set of σ-extensions for AF, i.e., the set of extensions for AF according to the given semantics σ.

All the above-mentioned semantics except the stable admit at least one extension, and the grounded and ideal admits exactly one extension [Dung, 1995; Dung et al., 2007; Caminada, 2006]. Grounded and ideal semantics are called *deterministic* or *unique status* as $|\mathcal{E}_{\mathcal{GR}}(AF)| = |\mathcal{E}_{\mathcal{ID}}(AF)| = 1$, whereas the other above recalled semantics are called *nondeterministic* or *multiple status*. For any AF AF, it holds that $\mathcal{E}_{\mathcal{ST}}(AF) \subseteq \mathcal{E}_{\mathcal{PR}}(AF) \subseteq \mathcal{E}_{\mathcal{CO}}(AF)$, $\mathcal{E}_{\mathcal{GR}}(AF) \subseteq \mathcal{E}_{\mathcal{CO}}(AF)$, and $\mathcal{E}_{\mathcal{ID}}(AF) \subseteq \mathcal{E}_{\mathcal{CO}}(AF)$.

Example 2.2. *The set of admissible sets for the argumentation framework AF_0 shown in Figure 1 is* $\{\emptyset, \{b\}, \{g\}, \{a,g\}, \{b,g\}, \{f,g\},$

σ	$\mathcal{E}_\sigma(AF_0)$	$\mathcal{E}_\sigma(AF)$
$\mathcal{CO}$	$\{\{f,g\},\{a,f,g\},\{b,f,g\}\}$	$\{\{g\},\{a,g\},\{b,f,g\}\}$
$\mathcal{PR}$	$\{\{a,f,g\},\{b,f,g\}\}$	$\{\{a,g\},\{b,f,g\}\}$
$\mathcal{ST}$	$\{\{b,f,g\}\}$	$\{\{b,f,g\}\}$
$\mathcal{ID}$	$\{\{f,g\}\}$	$\{\{g\}\}$
$\mathcal{GR}$	$\{\{f,g\}\}$	$\{\{g\}\}$

Table 1: Sets of extensions for AF_0 and $AF = +(c,f)(AF_0)$.

$\{a,g,f\}$, $\{b,g,f\}\}$, and the set $\mathcal{E}_\sigma(AF_0)$ of extensions, with $\sigma \in \{\mathcal{CO}, \mathcal{PR}, \mathcal{ST}, \mathcal{ID}, \mathcal{GR}\}$ is as reported in the second column of Table 1.

2.1 Labelling and Status of Arguments

The argumentation semantics can be also defined in terms of *labelling* [Baroni et al., 2011a]. A labelling for an argumentation framework $\langle Ar, att \rangle$ is a total function $\mathcal{L}ab : Ar \to \{\text{in}, \text{out}, \text{undec}\}$ assigning to each argument a label. $\mathcal{L}(a) = \text{in}$ means that argument a is accepted, $\mathcal{L}(a) = \text{out}$ means that a is rejected, while $\mathcal{L}(a) = \text{undec}$ means that a is undecided.

Let $in(\mathcal{L}) = \{a \mid a \in Ar \land \mathcal{L}(a) = \text{in}\}$, $out(\mathcal{L}) = \{a \mid a \in Ar \land \mathcal{L}(a) = \text{out}\}$, and $un(\mathcal{L}) = \{a \mid a \in Ar \land \mathcal{L}(a) = \text{undec}\}$. In the following, we also use the triple $\langle in(\mathcal{L}), out(\mathcal{L}), un(\mathcal{L}) \rangle$ to represent the labelling $\mathcal{L}$.

Given an argumentation framework $AF = \langle Ar, att \rangle$, a labelling $\mathcal{L}$ for AF is said to be *admissible (or legal)* if $\forall a \in in(\mathcal{L}) \cup out(\mathcal{L})$ it holds that

(i) $\mathcal{L}(a) = \text{out}$ iff $\exists b \in Ar$ such that $(b,a) \in att$ and $\mathcal{L}(b) = \text{in}$; and

(ii) $\mathcal{L}(a) = \text{in}$ iff $\mathcal{L}(b) = \text{out}$ for all $b \in Ar$ such that $(b,a) \in att$.

Moreover, $\mathcal{L}$ is a *complete* labelling iff conditions (i) and (ii) hold for all $a \in Ar$.

Between complete extensions and complete labellings there is a bijective mapping defined as follows: for each extension E there is a unique labelling $\mathcal{L} = \langle E, E^+, Ar \setminus (E \cup E^+) \rangle$ and for each labelling $\mathcal{L}$ there is a

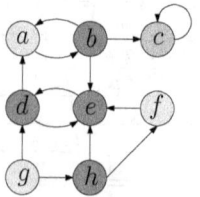 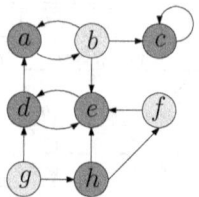

Figure 2: Labelling $\mathcal{L}$ corresponding to the preferred extensions $E_{\mathcal{PR}} \in \mathcal{E}_{AF_0}(\mathcal{PR}) = \langle \{a,f,g\}, \{b,d,e,h\}, \{c\} \rangle$ (left-hand side) and $E'_{\mathcal{PR}} \in \mathcal{E}_{AF_0}(\mathcal{PR}) = \langle \{b,f,g\}, \{a,d,e,h\}, \{c\} \rangle$ (right-hand side). A green (resp., red, orange) node x is such that $\mathcal{L}(x) =$ in (resp., out, undec).

unique extension $in(\mathcal{L})$. We say that $\mathcal{L}$ is the labelling *corresponding* to E.

Example 2.3. *Continuing from Example 2.2, $\langle \{a,f,g\}, \{b,d,e,h\}, \{c\} \rangle$ is the labelling corresponding to the preferred extension $E_{\mathcal{PR}} = \{a,f,g\}$, as shown in Figure 2.*

In the following, we say that the *status of an argument* a w.r.t. a labelling $\mathcal{L}$ (or its corresponding extension $in(\mathcal{L})$) is in (resp., out, undec) iff $\mathcal{L}(a) =$ in (resp., $\mathcal{L}(a) =$ out, $\mathcal{L}(a) =$ undec). We will avoid to mention explicitly the labelling (or the extension) whenever it is understood.

2.2 Updating a Dung Argumentation Framework

An argumentation framework typically models a temporary situation, and new arguments and attacks can be added or retracted to take into account new available knowledge.

Performing an update on an argumentation framework AF_0 means modifying it into an argumentation framework AF by adding or removing arguments or attacks. We use $+(a,b)$, with $a,b \in Ar_0$ and $(a,b) \notin att_0$, (resp. $-(a,b)$, with $(a,b) \in att_0$) to denote the addition (resp. deletion) of an attack (a,b), and $u(AF_0)$ to denote the application of update $u = \pm(a,b)$ to AF_0 (where $\pm$ means either $+$ or $-$). Applying an update u to an argumentation framework implies that its

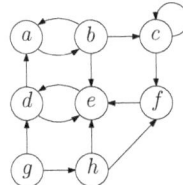

Figure 3: $AF = +(c,f)(AF_0)$

semantics (set of extensions or labellings). Table 1 reports the sets of extensions for the argumentation framework AF_0 of Figure 1 and for $AF = +(c,f)(AF_0)$ of Figure 3 which is obtained from AF_0 by performing the update $+(c,f)$.

Concerning the addition (resp. deletion) of a set of isolated arguments, it is easy to see that if AF is obtained from AF_0 through the addition (resp. deletion) of a set S of isolated argument, then, let E_0 be an extension for AF_0, $E = E_0 \cup S$ (resp. $E = E_0 \setminus S$) is an extension for AF that can be trivially computed. Of course, if arguments in S are not isolated, for addition we can first add isolated arguments and then add attacks involving these arguments, while for deletion we can first delete all attacks involving arguments in S. Thus we do not consider these kinds of update in the following.

3 Incremental Computation of Extensions in Dynamic Argumentation Frameworks

We tackle the problem of incrementally computing extensions of dynamic argumentation frameworks: given an initial extension and an update (or a set of updates), we devise a technique for computing an extension of the updated argumentation framework under five well-known semantics (*i.e.*, *complete, preferred, stable, grounded*, and *ideal*).

The idea, initially proposed in [Greco and Parisi, 2016a; Greco and Parisi, 2016b] and then developed in [Alfano et al., 2017a], is that of identifying a reduced (updated) argumentation framework sufficient to compute an extension of the whole argumentation framework and use state-of-the-art algorithms to recompute an extension of the reduced argumentation framework only.

For the sake of presentation, we first present the technique for semantics $\sigma \in \{\mathcal{CO}, \mathcal{PR}, \mathcal{ST}, \mathcal{GR}\}$, and then show how to deal with the ideal semantics in Section 3.3, since the definition of the reduced argumentation framework for the ideal semantics is different from that for the other semantics.

We first give some sufficient conditions ensuring that a given σ-extension for an argumentation framework AF continues to be a σ-extension for the updated argumentation framework $u(AF)$. Then, we introduce the *influenced set* that intuitively consists of the set of arguments whose status may change after performing an update.

Updates Preserving a Given Initial Extension

Given an update $\pm(a,b)$ and an initial extension E_0 corresponding to $\mathcal{L}_0$, for each pair of initial statuses $\mathcal{L}_0(a)$ and $\mathcal{L}_0(b)$ of the arguments involved in the update, Tables 2 and 3 tell us the semantics for which E_0 is still an extension after the update, as stated in the following proposition.

Proposition 3.1 (Irrelevant Updates [Alfano et al., 2017b]). *Let AF_0 be an argumentation framework, σ a semantics, $E_0 \in \mathcal{E}_\sigma(AF_0)$ an extension of AF_0 under semantics σ, $\mathcal{L}_0$ the labelling corresponding to E_0, and u an update. If σ is in the cell $\langle \mathcal{L}_0(a), \mathcal{L}_0(b) \rangle$ of Table 2 and $u = +(a,b)$ (resp., of Table 3 and $u = -(a,b)$), then $E_0 \in \mathcal{E}_\sigma(u(AF_0))$.*

The results in Tables 2 and 3 concerning the grounded semantics follow from those in [Boella et al., 2009a; Boella et al., 2009b], where the principles according to which the grounded extension does not change when attacks are added or removed have been studied.

In the following, given an argumentation framework AF_0 and a σ-extension E_0 for it, we say that an update u is *irrelevant* w.r.t. E_0 and σ iff the conditions of Proposition 3.1 hold. Otherwise, u is said to be *relevant*.

Example 3.2. *Consider AF_0 of Figure 1 and its sets of extensions listed in the second column of Table 1. $E_0 = \{b, f, g\}$ is an extension according to semantics $\sigma \in \{\mathcal{CO}, \mathcal{PR}, \mathcal{ST}\}$. Thus, $\mathcal{L}_0(c) = $ out and $\mathcal{L}_0(f) = $ in, and using Proposition 3.1 it follows that for update $u = +(c,f)$ E_0 is*

update $+(a,b)$		$\mathcal{L}_0(b)$		
		in	undec	out
$\mathcal{L}_0(a)$	in			$\mathcal{CO}, \mathcal{PR}, \mathcal{ST}, \mathcal{GR}$
	undec		$\mathcal{CO}, \mathcal{GR}$	$\mathcal{CO}, \mathcal{PR}, \mathcal{GR}$
	out	$\mathcal{CO}, \mathcal{PR}, \mathcal{ST}$	$\mathcal{CO}, \mathcal{GR}$	$\mathcal{CO}, \mathcal{PR}, \mathcal{ST}, \mathcal{GR}$

Table 2: Cases for which $E_0 \in \mathcal{E}_\sigma(u(AF_0))$ for $u = +(a,b)$.

update $-(a,b)$		$\mathcal{L}_0(b)$		
		in	undec	out
$\mathcal{L}_0(a)$	in	NA	NA	
	undec	NA		$\mathcal{CO}, \mathcal{PR}, \mathcal{GR}$
	out	$\mathcal{CO}, \mathcal{PR}, \mathcal{ST}, \mathcal{GR}$	$\mathcal{CO}, \mathcal{PR}, \mathcal{GR}$	$\mathcal{CO}, \mathcal{PR}, \mathcal{ST}, \mathcal{GR}$

Table 3: Cases for which $E_0 \in \mathcal{E}_\sigma(u(AF_0))$ for $u = -(a,b)$.

still an extension of $u(AF_0)$ (see the last column of Table 1). Thus $+(c,f)$ is irrelevant w.r.t. E_0 and σ.

In contrast, $+(c,f)$ is relevant w.r.t. $E_0 = \{a, f, g\}$ and any semantics (in this case $\mathcal{L}_0(c) = $ undec and $\mathcal{L}_0(f) = $ in, and no semantics is listed in the cell $\langle$undec, in$\rangle$ of Table 2).

It is important to note that Tables 2 and 3 are not meant to be exhaustive, as more conditions can be found for which a σ-extension is preserved after an update. For instance, for the grounded semantics, the initial extension is preserved also if $\mathcal{L}_0(a) = $ out and $\mathcal{L}_0(b) = $ in and argument a of updated $+(a,b)$ is not reachable from b. Here we provided a simple set of conditions that can be easily checked by just looking at the initial labelling $\mathcal{L}_0$. The technique for the incremental computation can be trivially extended by considering a more general set of such conditions.

Influenced Set

Given an argumentation framework, an update, and an initial σ-extension of the considered framework, the influenced set consists of the arguments whose acceptance status (according to the semantics σ) may change after performing the update. For irrelevant updates, the influenced set will be empty, as in this case, the initial extension can be immediately returned as an extension of the updated argumentation framework. If none of the conditions of Proposition 3.1 hold (*i.e.*, the update is relevant), then the influenced set may turn out to be not empty. In such case, the influenced set will be used to delineate a portion of the argumentation framework, called *reduced* argumentation framework, that we will use to recompute (a portion of) an extension for the updated argumentation framework.

Given an argumentation framework $AF = \langle Ar, att \rangle$ and an argument $b \in Ar$, we use $Reach_{AF}(b)$ to denote the set of arguments that are reachable from b in the graph AF.

Definition 3.3 (Influenced Set [Alfano et al., 2017b]). Let $AF = \langle Ar, att \rangle$ be an argumentation framework, $u = \pm(a,b)$, E an extension of AF under semantics $\sigma \in \{\mathcal{CO}, \mathcal{PR}, \mathcal{ST}, \mathcal{ID}, \mathcal{GR}\}$, and let

- $INF_0(u, AF, E) = \begin{cases} \emptyset & \text{if } u \text{ is irrelevant w.r.t. } E \text{ and } \sigma \text{ or } \exists (z,b) \in att \\ & \text{s.t. } z \neq a \wedge z \in E \ \wedge \ z \notin Reach_{AF}(b); \\ \{b\} & \text{otherwise}; \end{cases}$

- $INF_{i+1}(u, AF, E) = INF_i(u, AF, E) \ \cup \ \{y \mid \exists (x,y) \in att \text{ s.t. } x \in INF_i(u, AF, E) \wedge \nexists(z,y) \in att \text{ s.t. } z \in E \wedge z \notin Reach_{AF}(b)\}$.

The *influenced set* of u w.r.t. AF and E is $INF(u, AF, E) = INF_n(u, AF, E)$ such that $INF_n(u, AF, E) = INF_{n+1}(u, AF, E)$.

Thus, the set of arguments that are influenced by an update of b's status are those that can be reached from b without using any intermediate argument y whose status is known to be **out** because it is determined by an argument $z \in E$ that is not reachable from (and thus not influenced by) b.

Example 3.4. *Consider the argumentation framework $AF_0 = \langle Ar_0, att_0 \rangle$ of Figure 1 and the update $u = +(c, f)$. We have that $Reach_{AF_0}(f) = Ar_0 \setminus \{g, h\}$. The influenced set depends on the initial extension chosen. For the (preferred) extension $\{b, f, g\}$ of Example 3.2, we have that the influenced set is empty as u is irrelevant. In contrast, for the (preferred) extension $E_0 = \{a, f, g\}$, the influenced set is $INF(u, AF_0, E_0) = \{f, e\}$. Indeed, $d \notin INF(u, AF_0, E_0)$ since it is attacked by $g \in E_0$ which*

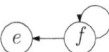

Figure 4: $RAF(+(c,f), AF_0, \{a, f, g\})$

is not reachable from f. Thus the arguments that can be reached from d do not belong to $INF(u, AF_0, E_0)$. If we consider the initial grounded extension $\{f, g\}$, then $\{f, e\}$ turns out once again to be the influenced set.

Reduced Argumentation Framework

Given the influenced set, we define a subgraph, called *reduced* argumentation framework, that will be used to compute the status of the influenced arguments, thus providing an extension that will be combined with that of initial argumentation framework to obtain an extension of the updated argumentation framework, for every semantics $\sigma \in \{\mathcal{CO}, \mathcal{PR}, \mathcal{ST}, \mathcal{GR}\}$.

For any argumentation framework $AF = \langle Ar, att \rangle$ and set $S \subseteq Ar$ of arguments, we denote with $AF{\downarrow}_S = \langle S, att \cap (S \times S) \rangle$ the subgraph of AF induced by arguments in S. Moreover, given two argumentation frameworks $AF_1 = \langle Ar_1, att_1 \rangle$ and $AF_2 = \langle Ar_2, att_2 \rangle$, we denote as $AF_1 \sqcup AF_2 = \langle Ar_1 \cup Ar_2, att_1 \cup att_2 \rangle$ the *union* of the two argumentation frameworks.

Definition 3.5 (Reduced Argumentation Framework [Alfano et al., 2017b]). *Let $AF_0 = \langle Ar_0, att_0 \rangle$ be an argumentation framework, $E_0 \in \mathcal{E}_\sigma(AF_0)$ an extension for AF_0 under a semantics $\sigma \in \{\mathcal{CO}, \mathcal{PR}, \mathcal{ST}, \mathcal{GR}\}$, and $u = \pm(a,b)$ an update. Let $AF = \langle Ar, att \rangle$ be the argumentation framework updated using u. The reduced argumentation framework for AF_0 w.r.t. E_0 and u (denoted as $RAF(u, AF_0, E_0)$) is as follows.*

- *$RAF(u, AF_0, E_0)$ is empty if $INF(u, AF_0, E_0)$ is empty.*

- *$RAF(u, AF_0, E_0) = AF{\downarrow}_{INF(u, AF_0, E_0)} \sqcup AF_1 \sqcup AF_2$ where:*

  (i) AF_1 is the union of the frameworks $\langle \{a,b\}, \{(a,b)\} \rangle$ s.t. $(a,b) \in att$, $a \notin INF(u, AF_0, E_0)$, $a \in E_0$, and $b \in INF(u, AF_0, E_0)$;

(ii) AF_2 is the union of the frameworks $\langle \{c\}, \{(c,c)\} \rangle$ s.t. there is $(e,c) \in att$, $e \notin INF(u, AF_0, E_0)$, $e \notin (E_0 \cup E_0^+)$, and $c \in INF(u, AF_0, E_0)$.

Hence, the argumentation framework $RAF(u, AF_0, E_0)$ contains, in addition to the subgraph of $u(AF_0)$ induced by $INF(u, AF_0, E_0)$, additional nodes and edges containing needed information on the "external context", *i.e.*, information about the status of arguments which are attacking some argument in $INF(u, AF_0, E_0)$. Specifically, if there is in $u(AF_0)$ an edge from an uninfluenced node a whose status in **in** to an influenced node b, then we add the edge (a, b) so that, as a does not have incoming edges in $RAF(u, AF_0, E_0)$, its status is confirmed to be **in**. Moreover, if there is in $u(AF_0)$ an edge from an uninfluenced node e to an influenced node c such that e is **undec**, we add edge (c, c) to $RAF(u, AF_0, E_0)$ so that the status of c cannot be **in**. Using fake arguments/attacks to represent external contexts has been exploited in [Baroni et al., 2014a] where decomposability properties of argumentation semantics are investigated.

Example 3.6. *For our running example, if $E_0 = \{a, f, g\}$ and $u = +(c, f)$, the reduced argumentation framework $RAF(+(c, f), AF_0, E_0)$ consists of the subgraph induced by $INF(u, AF_0, E_0) = \{f, e\}$ plus the edge (f, f) as there is the attack (c, f) in the updated argumentation framework from a non influenced argument c labelled as* **undec** *toward the influenced argument f. Hence, $RAF(+(c, f), AF_0, E_0) = \langle \{e, f\}, \{(f, f), (f, e)\} \rangle$ as shown in Figure 4.*

The following theorem states that, for every semantics $\sigma \in \{\mathcal{CO}, \mathcal{PR}, \mathcal{ST}, \mathcal{GR}\}$, an extension for the updated argumentation framework can be obtained by the union of an extension of the reduced argumentation framework and the projection of the initial extension on the uninfluenced part.

Theorem 3.7 ([Alfano et al., 2017b]). *Let AF_0 be an argumentation framework, $AF = u(AF_0)$ be the argumentation framework resulting from performing update $u = \pm(a, b)$ on AF_0, and $E_0 \in \mathcal{E}_\sigma(AF_0)$ be an extension for AF_0 under a semantics $\sigma \in \{\mathcal{CO}, \mathcal{PR}, \mathcal{ST}, \mathcal{GR}\}$. If $\mathcal{E}_\sigma(RAF(u, AF_0, E_0))$ is not empty, then there is an extension $E \in$*

$\mathcal{E}_\sigma(AF)$ for the updated argumentation framework AF such that $E = (E_0 \setminus INF(u, AF_0, E_0)) \cup E_d$, where E_d is a σ-extension for reduced argumentation framework $RAF(u, AF_0, E_0)$.

Example 3.8. *Continuing with our example, for the preferred semantics, let $E_0 = \{a, f, g\}$ and $u = +(c, f)$, we have that $INF(u, AF_0, E_0) = \{f, e\}$, and $RAF(+(c, f), AF_0, E_0) = \langle \{e, f\}, \{(f, f), (f, e)\} \rangle$. Thus, using the theorem, there is an extension E of the updated argumentation framework such that $E = (\{a, f, g\} \setminus \{f, e\}) \cup E_d$ where $E_d = \emptyset$ is a preferred extension of the reduced argumentation framework. In fact, $E = \{a, g\} \in \mathcal{E}_{\mathcal{PR}}(u(AF_0))$.*

It is worth noting that the set of extensions of an argumentation framework can be empty only for the stable semantics. Thus, in the case that this happens for the reduced argumentation framework (*i.e.*, $\mathcal{E}_\sigma(RAF(u, AF_0, E_0)) = \emptyset$), the theorem does not give a method to determine an extension of the updated argumentation framework, as shown in the following example.

Example 3.9. *Consider the two stable extensions $\{a, c\}$ and $\{a, d, e\}$ for AF_0 and the update $u = +(d, d)$. Depending on the initial extension, the influenced set is either $INF(u, AF, \{a, c\}) = \emptyset$ (as u is irrelevant w.r.t. $\{a, c\}$ and $\mathcal{ST}$) or $INF(u, AF, \{a, d, e\})$
$= \{d\}$. Thus, starting from the extension $\{a, c\}$ we directly know $\{a, c\}$ is a stable extension of the updated argumentation framework. However, starting from $\{a, d, e\}$, the reduced argumentation framework will be $RAF(u, AF_0, \{a, d, e\}) = \langle \{d\}, \{(d, d)\} \rangle$, which has no stable extension. In this case, the theorem does not provide a stable extension of the updated argumentation framework, thought a stable extension exists: that obtained by starting from the initial extension $\{a, c\}$.*

Note that, if we consider the preferred semantics, for which the starting extensions are again $\{a, c\}$ and $\{a, d, e\}$, a preferred extension of the updated argumentation framework can be obtained no matter what starting extension is chosen. In particular, as the preferred extension for reduced argumentation framework $\langle \{d\}, \{(d, d)\} \rangle$ is the empty set, it follows that $(\{a, d, e\} \setminus \{d\}) \cup \emptyset = \{a, e\}$ is a preferred extension of the updated argumentation framework.

Algorithm 1 Incr-Alg(AF_0, u, σ, E_0, Solver$_\sigma$) [Alfano et al., 2017b]
Input: $AF_0 = \langle Ar_0, att_0 \rangle$,
 update $u = \pm(a, b)$,
 semantics $\sigma \in \{\mathcal{CO}, \mathcal{PR}, \mathcal{ST}, \mathcal{GR}\}$,
 extension $E_0 \in \mathcal{E}_\sigma(AF_0)$,
 function Solver$_\sigma(AF)$ returning a σ-extension of AF if it exists, $\bot$ otherwise;
Output: A σ-extension $E \in \mathcal{E}_\sigma(u(AF_0))$ if it exists, $\bot$ otherwise;
1: $S = INF(u, AF_0, E_0)$;
2: **if** $(S = \emptyset)$ **then**
3:     **return** E_0;
4: **end if**
5: $AF_d = RAF(u, AF_0, E_0)$;
6: Let $E_d =$ Solver$_\sigma(AF_d)$;
7: **if** $(E_d \neq \bot)$ **then**
8:     **return** $E = (E_0 \setminus S) \cup E_d$;
9: **else**
10:     **return** Solver$_\sigma(u(AF_0))$;
11: **end if**

3.1 Incremental Algorithm

Algorithm 1 computes an extension of an updated argumentation framework [Alfano et al., 2017b]. Besides taking as input an initial argumentation framework AF_0, an update u, a semantics $\sigma \in \{\mathcal{CO}, \mathcal{PR}, \mathcal{ST}, \mathcal{GR}\}$, and an extension $E_0 \in \mathcal{E}_\sigma(AF_0)$, it also takes as input a function that computes a σ-extension for an argumentation framework, if any. In particular, function Solver$_\sigma(AF)$ will be used to compute an extension of the reduced argumentation framework, which will be then combined with the portion of the initial extension that does not change in order to obtain an extension for the updated argumentation framework (as stated in Theorem 3.7).

More in detail, Algorithm 1 works as follows. First, the influenced set of AF_0 w.r.t. update u and the given initial extension E_0 is computed (Line 1). If it is empty, then E_0 will be still an extension of the updated argumentation framework under the given semantics σ, and thus it is

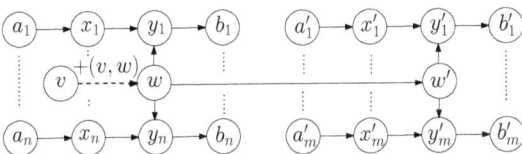

Figure 5: Simulating multiple updates by a single one.

returned (Line 3). Otherwise, the reduced argumentation framework AF_d is computed at Line 5, and function Solver_σ is invoked to compute a σ-extension of AF_d, if any. If $\sigma \in \{\mathcal{CO}, \mathcal{PR}, \mathcal{GR}\}$, then AF_d will have an extension E_d, which is combined with $E_0 \setminus S$ at Line 8 to get an extension for the updated argumentation framework. For the stable semantics, if $\mathcal{E}_{\mathcal{ST}}(RAF(u, AF_0, E_0))$ is not empty, then the algorithm proceeds as for the other semantics (Line 8). Otherwise, function Solver_σ is invoked to compute a stable extension of the whole updated argumentation framework $u(AF_0)$, if any.

The soundness and completeness of the algorithm follows from the result of Theorem 3.7 and the soundness and completeness of function Solver_S used.

Theorem 3.10 (Soundness and Completeness [Alfano et al., 2017b]). *Let AF_0 be an argumentation framework, $u = \pm(a, b)$, and $E_0 \in \mathcal{E}_\sigma(AF_0)$ an extension for AF_0 under $\sigma \in \{\mathcal{CO}, \mathcal{PR}, \mathcal{ST}, \mathcal{GR}\}$. If Solver_σ is sound and complete then Algorithm 1 computes $E \in \mathcal{E}_\sigma(u(AF_0))$ if $\mathcal{E}_\sigma(u(AF_0))$ is not empty, otherwise it returns $\perp$.*

3.2 Applying Multiple Updates Simultaneously

The approach described in the previous section extends to the case of multiple updates, *i.e.*, set of updates performed simultaneously. In fact, performing a set of updates $U = \{+(a_1, b_1), \ldots, +(a_n, b_n), -(a'_1, b'_1), \ldots, -(a'_m, b'_m)\}$ on AF_0 can be reduced to performing a single update $+(v, w)$ on an argumentation framework $AF_{E_0}^U$ whose definition depends on both the set of updates U and the initial σ-extension E_0, as explained in what follows.

Given a set U of updates for an argumentation framework AF_0, and a σ-extension E_0 for AF_0, we use U^* to denote the subset of U consisting of the relevant updates (that is, the updates in U for which the conditions of Proposition 3.1 do not hold).

The argumentation framework $AF_{E_0}^U$ for applying a set U^* of relevant updates is obtained from AF_0 by (i) adding arguments x_i, y_i and the chain of attacks between a_i and b_i as shown in Figure 5, for each update $+(a_i, b_i) \in U^*$; (ii) replacing each attack (a'_j, b'_j) in AF_0 with the chain of attacks between a'_j and b'_j as shown in Figure 5, for each update $-(a_j, b_j) \in U^*$; and (iii) adding the new arguments v, w, w' and the attacks involving them as shown in Figure 5. The following definition considers a general set of updates which includes also irrelevant updates.

Definition 3.11 (AF for applying a set of updates [Greco and Parisi, 2016b]).
Let $AF_0 = \langle Ar_0, att_0 \rangle$ be an argumentation framework, and E_0 a σ-extension for AF_0. Let

- $att^+ = \{(a_1, b_1), \ldots, (a_n, b_n)\} \subseteq (Ar_0 \times Ar_0) \setminus att_0$, and
- $att^- = \{(a'_1, b'_1), \ldots, (a'_m, b'_m)\} \subseteq att_0$

such that $att^+ \cap att^- = \emptyset$ be two sets of attacks.

Let $U = \{+(a_i, b_i) \mid (a_i, b_i) \in att^+\} \cup \{-(a_j, b_j) \mid (a_j, b_j) \in att^-\}$ be a set of updates, and $U^* \subseteq U$ be the set of relevant updates w.r.t. E_0 and σ. Then, $AF_{E_0}^U = \langle Ar^U, att^U \rangle$ denotes the argumentation framework obtained from AF_0 as follows:

- $Ar^U = Ar_0 \cup \{x_i, y_i \mid +(a_i, b_i) \in U^*\} \cup \{x'_j, y'_j \mid -(a_j, b_j) \in U^*\} \cup \{v, w, w'\}$, where all $x_i, y_i, x'_j, y'_j, w, w'$, and v are new arguments not occurring in Ar_0, and

- $att^U = (att_0 \setminus att^-) \cup \{(a_i, b_i) \mid +(a_i, b_i) \in (U \setminus U^*)\} \cup$
  $\{(a_i, x_i), (x_i, y_i), (y_i, b_i) \mid +(a_i, b_i) \in U^*\} \cup$
  $\{(a_j, x'_j), (x'_j, y'_j), (y'_j, b_j) \mid -(a_j, b_j) \in U^*\} \cup$
  $\{(w, y_i) \mid +(a_i, b_i) \in U^*\} \cup$
  $\{(w', y'_j) \mid -(a_j, b_j) \in U^*\} \cup \{(w, w')\}$.

It is worth noting that, in the definition above, each argument x_i, y_i, x'_i, and y'_i is assumed to be unique and non-identical for every attack (a_i, b_i).

The following theorem states that every extension of the argumentation framework AF obtained by performing on AF_0 all the updates in U corresponds to an extension of $+(v,w)(AF_{E_0}^U)$, where $+(v,w)$ is a single attack update.

Theorem 3.12 ([Greco and Parisi, 2016b; Alfano et al., 2017b]). *Let $AF_0 = \langle Ar_0, att_0 \rangle$ be an argumentation framework, E_0 a σ-extension for AF_0, and U a set of updates. Let AF be the argumentation framework obtained from AF_0 by performing all updates in U on it. Then, for any semantics $\sigma \in \{\mathcal{CO}, \mathcal{PR}, \mathcal{ST}, \mathcal{GR}\}$ $E \in \mathcal{E}_\sigma(AF)$ iff there is $E^U \in \mathcal{E}_\sigma(+(v,w)(AF_{E_0}^U))$ such that $E^U \cap Ar_0 = E$.*

3.3 Dealing with the Ideal Semantics

Algorithm 1 can be extended to deal with the ideal semantics. The only difference is that we need a new definition of reduced argumentation framework since, as illustrated in the following example, that of Definition 3.5 does not work for the ideal semantics.

Example 3.13. *Consider the argumentation framework $AF_0 = \langle \{a,b,c,d\}, \{(a,b), (b,a), (c,d), (d,c), (a,c), (b,c)\} \rangle$ and the update $u = -(b,c)$. The ideal extension of AF_0 is $E_0 = \{d\}$ (i.e., arguments a and b are both labeled as* undec*). The influenced set is $INF(u, AF_0, E) = \{c,d\}$. However, the RAF obtained by applying Definition 3.5 is $\langle \{c,d\}, \{(c,c), (c,d), (d,c)\} \rangle$, its ideal extension is $\{d\}$, and applying the result of Theorem 3.7 we would obtain that $\{d\}$ is still the ideal extension for $u(AF_0)$. But this is not correct, as the ideal extension for $u(AF_0)$ is the empty set.*

Before defining the reduced framework for the ideal semantics, we define the paths providing the information on the "context" outside the influenced set $INF(u, AF, E)$ that needs to be added to determine the new status of the arguments influenced by update u w.r.t. the ideal extension E.

Given an argumentation framework $AF = \langle Ar, att \rangle$ with ideal extension E and a set $S \subseteq Ar$, we use $Node(AF, S, E)$ (resp. $Edge(AF, S, E)$) to denote a set of arguments $x_1, \ldots, x_n$ (resp. attacks $(x_1, x_2), \ldots, (x_{n-1}, x_n)$) in AF such that there is a path $x_1 \ldots x_n$ in AF with $x_n \in S$,

$x_1, \ldots, x_{n\text{-}1} \notin S$ and $x_1, \ldots, x_{n\text{-}1} \notin E \cup E^+$ (i.e., $x_1, \ldots, x_{n\text{-}1}$ are **undec**). Essentially, if S is the influenced set of an update, to determine the status of nodes in S we must also consider all nodes and attacks occurring in paths (of any length) ending in S whose nodes outside S are all labeled as **undec**. The motivation to also consider the paths ending in S is that some of the undecided arguments occurring in these paths could be labelled **in** or **out** in some preferred labelling and, therefore, together they could determine a change in the status of nodes in S.

Definition 3.14. (Reduced Argumentation Framework for Ideal Semantics [Greco and Parisi, 2016b; Alfano et al., 2019a])
Let $AF_0 = \langle Ar_0, att_0 \rangle$ be an argumentation framework, E_0 be the ideal extension for AF_0, and $u = \pm(a,b)$ an update. Let $AF = \langle Ar, att \rangle$ be the argumentation framework updated by using u. The reduced argumentation framework for AF_0 w.r.t. E_0 and u (denoted as $RAF_{\mathcal{ID}}(u, AF_0, E_0)$) is as follows.

- $RAF_{\mathcal{ID}}(u, AF_0, E_0)$ is empty if $INF(u, AF_0, E_0)$ is empty.

- $RAF_{\mathcal{ID}}(u, AF_0, E_0) = AF{\downarrow}_{INF(u,AF_0,E)} \sqcup AF_1 \sqcup AF_2$ where:

    (i) AF_1 is the union of the frameworks $\langle \{a,b\}, \{(a,b)\} \rangle$ such that $(a,b) \in att$ and $a \notin INF(u, AF_0, E_0)$, $a \in E_0$, and $b \in INF(u, AF_0, E_0)$;

    (ii) AF_2 is the union of the frameworks $\langle Node(AF, INF(u, AF_0, E_0), E_0)$ and $Edge(AF, INF(u, AF_0, E_0), E_0) \rangle$.

Example 3.15. For the argumentation framework AF_0 of the running example (see Figures 1 and 3), where the initial ideal extension is $E_0 = \{f, g\}$ and $u = +(c, f)$, the reduced argumentation framework $RAF_{\mathcal{ID}}(+(c, f), AF_0, E_0)$ is defined as the subgraph induced by $INF(u, AF_0, E_0) = \{f, e\}$ plus the sub-graph consisting of the paths of undecided arguments ending in the influenced set, that is, $AF_2 = \langle \{a, b, c\}, \{(a,b), (b,a), (b,c), (c,c), (c,f)\} \rangle$. Hence, we have $RAF_{\mathcal{ID}}(+(c, f), AF_0, E_0) = \langle \{a, b, c, e, f\}, \{(a,b), (b,a), (b,c), (c,c), (c,f), (f,e)\} \rangle$. The ideal extension of the reduced framework is the empty set.

It can be shown that the result of Theorem 3.7 also holds for the case of the ideal semantics [Alfano et al., 2019a]. By applying that result, we obtain that the (updated) ideal extension for the updated argumentation framework of Example 3.15 is $(\{f,g\} \setminus \{f,e\}) \cup \emptyset = \{g\}$ (see Table 1).

Example 3.16. *Consider again the argumentation framework AF_0 and the update u of Example 3.13, where the ideal extension of AF_0 is $E_0 = \{d\}$ and $INF(u, AF_0, E) = \{c, d\}$.*

Thus, $RAF_{\mathcal{ID}}(u, AF_0, E_0) = AF\!\downarrow_{INF(u,AF_0,E_0)} \sqcup AF_1 \sqcup AF_2$ where:

- $AF\!\downarrow_{INF(u,AF_0,E)} = \langle \{c,d\}, \{(c,d), (d,c)\} \rangle$,

- $AF_1 = \langle \emptyset, \emptyset \rangle$ *and*

- $AF_2 = \langle \{a,b,c\}, \{(a,b), (b,a), (a,c)\} \rangle$.

So $RAF_{\mathcal{ID}}(u, AF_0, E_0) = \langle \{a,b,c,d\}, \{(a,b), (b,a), (c,d), (d,c), (a,c)\} \rangle$, and its ideal extension is $\emptyset$. Thus, using the result of Theorem 3.7, we obtain that the ideal extension for the updated argumentation framework $u(AF_0)$ is the empty set.

Finally, Algorithm 1 can be used to compute the updated ideal extension of a given argumentation framework by using $AF_d = RAF_{\mathcal{ID}}(u, AF_0, E_0)$ at Line 5 and an external solver that computes the ideal extension of the reduced argumentation framework.

In the next two sections, we will deal with other possible ways to apply the incremental algorithm in other approaches to formal (computational) argumentation. First, Section 4 deals with bipolarity and extended argumentation frameworks, while Section 5 centers on Defeasible Logic Programming (DeLP) as a structured argumentation formalism.

4 Bipolarity and Second-Order Attacks

Dung's framework has been extended along several dimensions; for instance, see [Baroni *et al.*, 2009; Modgil, 2009; Villata *et al.*, 2012]. The proposed incremental approach can be applied to different kinds of abstract argumentation frameworks that extend Dung's model. The main idea is that of using meta-argumentation approaches, which provide ways to transform a more general abstract framework into a Dung framework, and apply the incremental technique on the meta argumentation framework [Alfano et al., 2017a; Alfano et al., 2018a; Alfano et al., 2018b].

Bipolarity in argumentation is discussed in [Amgoud et al., 2004], where a survey of the use of bipolarity is given, as well as a formal definition of bipolar argumentation frameworks, which extend Dung's concept of argumentation framework by also including the relation of support between arguments. The notion of support has been found to be useful in many application domains, including decision making [Amgoud et al., 2005]. Several interpretations of the notion of support have been proposed in the literature [Amgoud et al., 2004; Cayrol and Lagasquie-Schiex, 2005; Cayrol and Lagasquie-Schiex, 2009; Cayrol and Lagasquie-Schiex, 2010; Boella et al., 2010; Villata et al., 2012] (see [Cohen et al., 2014] for a comprehensive survey). In this work, we focus on *deductive* support [Boella et al., 2010; Villata et al., 2012] which is intended to capture the following intuition: if argument a supports argument b then the acceptance of a implies the acceptance of b, and thus the non-acceptance of b implies the non-acceptance of a. However, the approach presented in this section can be adapted to work also with *necessary* support [Nouioua and Risch, 2011; Nouioua and Risch, 2010; Baroni et al., 2011b] due to the duality between these two kinds of interpretations of the support relation [Cohen et al., 2014]. The acceptability of arguments in the presence of a support relation was first investigated in [Cayrol and Lagasquie-Schiex, 2005]. Later on, to handle bipolarity in argumentation, [Cayrol and Lagasquie-Schiex, 2009; Cayrol and Lagasquie-Schiex, 2010] proposed an approach based on using the concept of *coalition* of arguments, where sets of arguments are considered as a group that plays the role of an argument and defeats then occur between coalitions. However, when considering a deductive interpretation of support [Boella et al., 2010; Villata et al., 2012], coalitions may lead to counter-intuitive results [Cohen et al., 2014]; nevertheless, they are useful in contexts where support is interpreted differently.

Furthermore, other abstract argumentation frameworks have been considered, such as Extended Argumentation Frameworks, which extend bipolar argumentation frameworks by modelling (apart from attacks/-supports between arguments) also attacks towards an attack or a support (called second-order attacks). Thanks to a meta argumentation approach, an extended argumentation framework can be converted into an abstract argumentation framework by using additional meta-arguments as well as attacks between them to model supports and second-order attacks.

In the following, we discuss how to extend the incremental technique to deal with extended argumentation frameworks. An *Extended Argumentation Framework* [Boella et al., 2010] is a quadruple $\langle Ar, att, sup, s\text{-}att \rangle$, where where (i) $Ar \subseteq \mathcal{A}rgs$, (ii) $att \subseteq Ar \times Ar$, (iii) $sup \subseteq Ar \times Ar$ is a binary relation over Ar whose elements are called *supports*, (iv) $att \cap sup = \emptyset$, and (v) $s\text{-}att$ is a binary relation over $Ar \times (att \cup sup)$ whose elements are called *second-order attacks*.

In the following, a second-order attack from an argument a to an attack (b, c) will be denoted as $(a \twoheadrightarrow (b \rightarrow c))$, while an attack from an argument a to a support (b, c) will be denoted as $(a \twoheadrightarrow (b \Rightarrow c))$. Thus, a Dung argumentation framework is an extended argumentation framework of the form $\langle Ar, att, \emptyset, \emptyset \rangle$, while a bipolar argumentation framework is extended argumentation framework of the form $\langle Ar, att, sup, \emptyset \rangle$.

Example 4.1. Consider the extended argumentation framework $EF_0 = \langle Ar_0, att_0, sup_0, s\text{-}att_0 \rangle$ where:

- $Ar_0 = \{a, b, c, d, e, f\}$ is the set of arguments;

- $att_0 = \{(a, c), (c, b), (b, d), (d, e), (e, d), (e, e), (e, f)\}$ is the set of attacks;

- $sup_0 = \{(a, b)\}$ is the set of supports; and

- $s\text{-}att_0 = \{(a, (b, d))\}$ is the set of second-order attacks.

The corresponding graph is shown in Figure 6, where second-order attacks are drawn using double-headed arrows.

The semantics of an extended argumentation framework can be given by means of the following meta argumentation framework.

Definition 4.2 (Meta Argumentation Framework [Boella et al., 2010]). *The meta argumentation framework for* $EF = \langle Ar, att, sup, s\text{-}att \rangle$ *is* $MF = \langle Ar^m, att^m \rangle$ *where:*

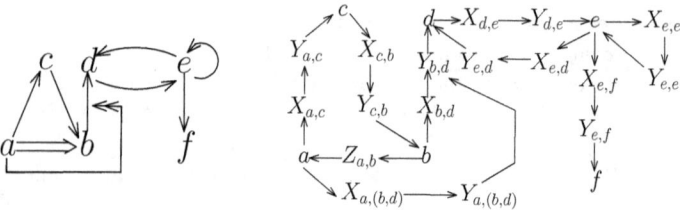

Figure 7: Meta framework for EF_0 of Example 4.1.

- $Ar^m = Ar \cup \{X_{a,b}, Y_{a,b} \mid (a,b) \in att\} \cup \{Z_{a,b} \mid (a,b) \in sup\} \cup \{X_{a,(b,c)}, Y_{a,(b,c)} \mid (a,(b,c)) \in s\text{-}att, (b,c) \in att\}$

- $att^m = \{(a, X_{a,b}), (X_{a,b}, Y_{a,b}), (Y_{a,b}, b) \mid (a,b) \in att\} \cup$
  $\{(b, Z_{a,b}), (Z_{a,b}, a) \mid (a,b) \in sup\} \cup$
  $\{(a, X_{a,(b,c)}), (X_{a,(b,c)}, Y_{a,(b,c)}), (Y_{a,(b,c)}, Y_{b,c}) \mid (a,(b,c)) \in s\text{-}att, (b,c) \in att\} \cup$
  $\{(a, X_{a,(b,c)}), (X_{a,(b,c)}, Y_{a,(b,c)}), (Y_{a,(b,c)}, Z_{b,c}) \mid (a,(b,c)) \in s\text{-}att, (b,c) \in sup\}$

The meaning of meta-arguments $X_{a,b}$, $Y_{a,b}$ and $Z_{a,b}$ is as follows. $X_{a,b}$ represents the fact that the corresponding attack (a,b) is "negligible" in the extended argumentation framework—it belongs to an extension of the meta argumentation framework iff a does not belong to an extension of the extended argumentation framework. On the other hand, $Y_{a,b}$ represents the fact that (a,b) is "significant" in the extended argumentation framework, and it belongs to an extension of the meta argumentation framework iff argument b does not. Finally, meta-argument $Z_{a,b}$ represents a support relation between a and b: it does not belong to an extension for the meta argumentation framework iff the supported argument b is accepted in the deductive model of support.

Moreover, a second order attack of the form $(a \twoheadrightarrow (b \to c))$ is encoded as an attack towards the meta-argument $Y_{b,c}$ (that represents the fact that (b,c) is "significant"), while an attack of the form $(a \twoheadrightarrow (b \Rightarrow c))$ is encoded as an attack toward the meta-argument $Z_{b,c}$. The meta argumentation framework for the extended argumentation framework of Example 4.1 is shown in Figure 7.

Extensions for an extended argumentation framework EF are obtained from extensions for its meta argumentation framework MF: E is an σ-extension for EF iff $E^m \in \mathcal{E}_\sigma(MF)$ and $E = E^m \cap Ar$, where Ar is the set of arguments of EF. Using this relationship, the notion of labelling can be extended to extended argumentation frameworks as well. As done in [Boella et al., 2010], in the following we will focus on the preferred and stable semantics. However, the technique can be also applied to grounded, ideal, and complete semantics by means of meta argumentation approach.

Example 4.3. *For the meta argumentation framework MF of Figure 7, we have the following preferred extensions (which are also stable extensions): (i) $\{a, b, d, f, Y_{a,c}, X_{c,b}, Y_{d,e}, Y_{a,(b,d)}, X_{e,e}, X_{e,d}, X_{e,f},\}$, which corresponds to the extension $\{a,b,d,f\}$ of the extended argumentation framework of Example 4.1, and (ii) $\{c, d, f, X_{a,c}, Y_{c,b}, Z_{a,b}, X_{b,d}, Y_{d,e} X_{e,e}, X_{e,d}, X_{e,f}, X_{a,(b,d)}\}$, which corresponds to the extension $\{c,d,f\}$ of the extended argumentation framework.*

Updates over Extended Argumentation Frameworks For extended argumentation frameworks we also consider updates consisting of additions and deletions of support relations and second-order attacks, in addition to the attack updates considered for Dung's frameworks. Specifically, the addition (resp., deletion) of a support relation from an argument a to an argument b will be denoted as $+(a \Rightarrow b)$ (resp. $-(a \Rightarrow b)$). Analogously, the addition (resp., deletion) of a second-order attack from an argument a to an attack (b, c) will be denoted as $+(a \twoheadrightarrow (b \rightarrow c))$ (resp., $-(a \twoheadrightarrow (b \rightarrow c))$). Finally, if (b, c) is a support, then the update will be denoted as $+(a \twoheadrightarrow (b \Rightarrow c))$ (resp., $-(a \twoheadrightarrow (b \Rightarrow c))$). We use $u(EF_0)$ to denote the extended argumentation framework resulting from the application of update u to an initial extended framework EF_0.

We introduce the compact argumentation framework for performing an update on extended argumentation frameworks—it will be used in a variant of Algorithm 1 for the incremental computation. The definition builds on (the compact version of) that proposed in [Boella et al., 2010] and considers additional meta-arguments and attacks that will allow us to simulate addition updates to be performed on the extended argumentation framework by means of single updates performed on the corresponding (compact) meta argumentation framework.

Definition 4.4 (Compact Argumentation Framework [Alfano et al., 2018b]). *Let $EF = \langle Ar, att, sup, s\text{-}att \rangle$ be an extended argumentation framework, and u an update of one of the following forms:*

- $u = \pm(e \to f)$
- $u = \pm(e \Rightarrow f)$
- $u = \pm(e \twoheadrightarrow (g \to h))$
- $u = \pm(e \twoheadrightarrow (g \Rightarrow h))$.

The compact argumentation framework for EF w.r.t. u is $CF(EF, u) = \langle Ar^m, att^m \rangle$ where:

- $Ar^m = A \cup \{Z_{a,b} \mid (a,b) \in sup\} \cup$
  $\{X_{c,d}, Y_{c,d} \mid (e,(c,d)) \in s\text{-}att, (c,d) \in att\} \cup$
  $\{Z_{e,f} \mid u = +(e \Rightarrow f)\} \cup$
  $\{X_{g,h}, Y_{g,h} \mid u = +(e \twoheadrightarrow (g \to h))\}$

- $att^m = att \setminus \{(g,h) \mid u = +(e \twoheadrightarrow (g \to h))\} \cup$
  $\{(g, X_{g,h}), (X_{g,h}, Y_{g,h}), (Y_{g,h}, h) \mid u = +(e \twoheadrightarrow (g \to h))\} \cup$
  $\{(b, Z_{a,b}), (Z_{a,b}, a) \mid (a,b) \in sup\} \cup$
  $\{(e, Z_{a,b}) \mid (e,(a,b)) \in s\text{-}att, (a,b) \in sup\} \cup$
  $\{(c, X_{c,d}), (X_{c,d}, Y_{c,d}), (Y_{c,d}, d), (e, Y_{c,d}) \mid$
  $(e,(c,d)) \in s\text{-}att, (c,d) \in att\} \cup$
  $\{(f, Z_{e,f}) \mid u = +(e \Rightarrow f)\}.$

Besides the meta-arguments $Z_{a,b}$ of Definition 4.2, and the attacks involving those arguments, the above meta argumentation framework contains meta-arguments $X_{c,d}, Y_{c,d}$ for encoding second order attacks in $s\text{-}att$ toward attacks $(c,d) \in att$. In fact, an attack $e \twoheadrightarrow (a \Rightarrow b)$ in $s\text{-}att$ toward a support is encoded as an attack from e toward $Z_{a,b}$ in the meta argumentation framework, while $e \twoheadrightarrow (c \to d)$ in $s\text{-}att$ is encoded as an attack from e toward $Y_{c,d}$ in the meta argumentation framework (which contains also the attacks $(c, X_{c,d}), (X_{c,d}, Y_{c,d}), (Y_{c,d}, d)$). Moreover, meta-arguments $Z_{e,f}$ and $X_{g,h}, Y_{g,h}$, are added to the meta argumentation framework for encoding, respectively, the addition of a second order attack toward a support $(e, f) \in sup$ or toward an attack $(g, h) \in att$. In the latter case, meta-arguments $X_{g,h}$ and $Y_{g,h}$ along with the set of

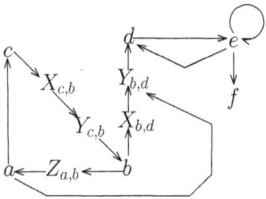

Figure 8: Compact argumentation framework for the extended argumentation framework EF_0 of Figure 6 w.r.t. the update $u = +(d \twoheadrightarrow (c \to b))$.

attacks $\{(g, X_{g,h}), (X_{g,h}, Y_{g,h}), (Y_{g,h}, h)\}$ are used to simulate the attack $g \to h$ which is attacked by e in the extended argumentation framework. This enables the definition of simple attack updates to simulate second-order attack updates.

Example 4.5. *The compact argumentation framework for the EAF EF_0 of Figure 6 w.r.t. the update $u = +(d \twoheadrightarrow (c \to b))$ is shown in Figure 8. Herein, the attacks involving the meta-arguments $X_{b,d}$ and $Y_{b,d}$ allow us to simulate the second order attack $a \twoheadrightarrow (b \to d)$. Moreover, the attacks involving the meta-arguments $X_{c,b}$ and $Y_{c,b}$ are added to enable the simulation of the second-order update u by a single attack update on the meta argumentation framework.*

We now define updates on the meta argumentation framework.

Definition 4.6 (Updates for the meta argumentation framework [Alfano et al., 2018b]). Let $EF = \langle Ar, att, sup, s\text{-}att \rangle$ be an extended argumentation framework, and u an update for EF. The corresponding update u^m for the compact argumentation framework $CF(EF, u)$ is as follows:

$u^m =$	$u^m =$
$+(Z_{e,f} \to e)$ if $u = +(e \Rightarrow f)$	$-(Z_{e,f} \to e)$ if $u = -(e \Rightarrow f))$
$+(c \to d)$ if $u = +(c \to d)$	$-(c \to d))$ if $u = -(c \to d))$
$+(e \to Y_{g,h})$ if $u = +(e \twoheadrightarrow (g \to h))$	$-(e \to Y_{g,h})$ if $u = -(e \twoheadrightarrow (g \to h))$
$+(e \to Z_{a,b})$ if $u = +(e \twoheadrightarrow (a \Rightarrow b))$	$-(e \to Z_{a,b})$ if $u = -(e \twoheadrightarrow (a \Rightarrow b))$

For instance, continuing with Example 4.5, given the extended argumentation framework EF_0 of Figure 6 and the update $u = +(d \twoheadrightarrow (c \to$

b)), we have that update u^m for the compact argumentation framework $CF(EF_0, u)$ shown in Figure 8 is $u^m = +(d \to Y_{c,b})$.

Finally, given an initial extension for an extended argumentation framework and an update, we define the initial labelling for the corresponding compact argumentation framework as follows.

Definition 4.7 (Corresponding initial labelling [Alfano et al., 2018b]). *Given an extended argumentation framework $EF_0 = \langle Ar, att, sup, s\text{-}att \rangle$ and a initial labelling $\mathcal{L}_0$, the corresponding initial labelling $\mathcal{L}_0^m$ for the compact argumentation framework $CF(EF_0, u) = \langle Ar^m, att^m \rangle$ is as follows:*

$\forall a \in Ar \cap Ar^m :$	$\mathcal{L}_0^m(a) = \mathcal{L}_0(a);$	
$\forall X_{a,b} \in Ar^m :$	$\mathcal{L}_0^m(X_{a,b}) = \text{in}$	$\text{if } \mathcal{L}_0(a) = \text{out}$
	$\mathcal{L}_0^m(X_{a,b}) = \text{out}$	$\text{if } \mathcal{L}_0(a) = \text{in}$
	$\mathcal{L}_0^m(X_{a,b}) = \text{undec}$	$\text{if } \mathcal{L}_0(a) = \text{undec}$
$\forall Y_{a,b} \in Ar^m :$	$\mathcal{L}_0^m(Y_{a,b}) = \text{in}$	$\text{if (i) } \mathcal{L}_0^m(X_{a,b}) = \text{out and}$ $\text{(ii) } \forall c \in Ar \text{ s.t. } (c, (a, b)) \in s\text{-}att,$ $\mathcal{L}_0(c) = \text{out}$
	$\mathcal{L}_0^m(Y_{a,b}) = \text{out}$	$\text{if (i) } \mathcal{L}_0^m(X_{a,b}) = \text{in or}$ $\text{(ii) } \exists c \in Ar \mid (c, (a, b)) \in s\text{-}att$ $\text{and } \mathcal{L}_0(c) = \text{in}$
	$\mathcal{L}_0^m(Y_{a,b}) = \text{undec},$	otherwise.

$\forall Z_{a,b} \in Ar^m :$	$\mathcal{L}_0^m(Z_{a,b}) = \text{in}$	$\text{if (i) } \mathcal{L}_0(b) = \text{out and}$ $\text{(ii) } \forall c \in Ar \text{ s.t. } (c, (a, b)) \in s\text{-}att,$ $\mathcal{L}_0(c) = \text{out}$
	$\mathcal{L}_0^m(Z_{a,b}) = \text{out}$	$\text{if (i) } \mathcal{L}_0(b) = \text{in or}$ $\text{(ii) } \exists c \in Ar \mid (c, (a, b)) \in s\text{-}att$ $\text{and } \mathcal{L}_0(c) = \text{in}$
	$\mathcal{L}_0^m(Z_{a,b}) = \text{undec},$	otherwise.

For instance, given the initial preferred extension $E_0 = \{a, b, d, f\}$ of the extended argumentation framework EF_0 of Example 4.1, the initial labelling for the compact argumentation framework $CF(EF_0, +(d \to Y_{c,b}))$ of Figure 8 is such that $\mathcal{L}_0^m(a) = \mathcal{L}_0(a) = \text{in}$, $\mathcal{L}_0^m(c) = \mathcal{L}_0(c) = \text{out}$, $\mathcal{L}_0^m(X_{c,b}) = \text{in}$, and $\mathcal{L}_0^m(Y_{c,b}) = \text{out}$. Also, we have that $\mathcal{L}_0^m(b) =$

$\mathcal{L}_0(b) = \text{in}$, $\mathcal{L}_0^m(X_{b,d}) = \text{out}$, $\mathcal{L}_0^m(Y_{b,d}) = \text{out}$ since $\mathcal{L}_0^m(a) = \mathcal{L}_0(a) = \text{in}$, and $\mathcal{L}_0^m(d) = \mathcal{L}_0(d) = \text{in}$.

Algorithm 2 Incr-EAF(EF_0, u, E_0, σ, Solver$_\sigma$)

Input: Extended argumentation framework $EF_0 = \langle Ar_0, att_0, sup_0, satt_0 \rangle$,
update u over EF_0,
an initial σ-extension E_0 for EF_0,
semantics $\sigma \in \{\mathcal{PR}, \mathcal{ST}\}$,
function Solver$_\sigma(AF)$ that returns an σ-extension of AF if it exists, and $\bot$ otherwise;
Output: An σ-extension E for $u(EF_0)$ if it exists, $\bot$ otherwise;
1: **if** $checkProp(EF_0, u, E_0, \sigma)$ **then**
2:     return E_0;
3: **end if**
4: Let $CF_0 = CF(EF_0, u)$ be the compact argumentation framework for EF_0 w.r.t. u (cf. Definition 4.4);
5: Let u^m be the update for CF_0 corresponding to u (cf. Definition 4.6);
6: Let E_0^m be the initial σ-extension for CF_0 corresponding to E_0;
7: Let $E^m = $ Incr-Alg(CF_0, u^m, σ, E_0^m, Solver$_\sigma$);
8: **if** ($E^m \neq \bot$) **then**
9:     return $E = (E^m \cap Ar_0)$;
10: **else**
11:     return $\bot$;
12: **end if**

Incremental Algorithm for Extended Argumentation Frameworks We are now ready to present the algorithm for extending the incremental technique to the case of extended argumentation frameworks. Given an extended argumentation framework EF_0, a semantics $\sigma \in \{\mathcal{PR}, \mathcal{ST}\}$, an extension $E_0 \in \mathcal{E}_\sigma(EF_0)$, and an update u of the form $u = \pm(a \Rightarrow b)$, $u = \pm(a \to b)$, $u = \pm(e \twoheadrightarrow (c \Rightarrow d))$, or $u = \pm(e \twoheadrightarrow (c \to d))$, Algorithm 2 computes an extension E of the updated extended argumentation framework $u(EF_0)$, if it exists [Alfano et al., 2018b]. The algorithm works as follows. It first checks if the initial extension E_0 is still an extension of the updated extended argumentation framework at Line 1, where $checkProp(EF_0, u, E_0, \sigma)$ is a function returning *true* iff the update is irrelevant—the interested reader can find the conditions under which an update for an extended argumentation framework is irrelevant in [Alfano et al., 2018b]. If this is the case, it immediately returns the initial extension. Otherwise, it computes the compact argumentation framework CF_0 (Line 4), the update u^m for CF_0 (Line 5), and the initial σ-extension E_0^m for CF_0 (Line 6). Next, it invokes function Incr-Alg (*i.e.*, Algorithm 1). Incr-Alg takes as input the parameters CF_0, u^m, σ, E_0^m, and Solver$_\sigma$, where Solver$_\sigma$ is an external solver that can compute an σ-extension for the input argumentation framework. Finally, the extension of the updated extended argumentation framework (if any) is obtained by projecting out the extension E^m returned by Incr-Alg over the set of arguments Ar_0 of the initial extended argumentation framework (Line 9).

From a computational point of view, in the worst case (that is, when every argument is influenced, and thus the RAF collapses to be the updated framework), Algorithm 1 and Algorithm 2 have the same computational complexity as the corresponding task in the static setting under the considered AF semantics. It is worth noting that the overhead of computing the influenced set and the RAF is polynomial in the input framework's size.

The use of the incremental techniques discussed in this section and the previous one become significant in practice. In fact, in [Alfano et al., 2017b] it is shown that Algorithm 1 outperforms state-of-the-art solvers that compute the extensions from scratch for single updates by two orders of magnitude on average, and it remains faster than the competitors even when recomputing an extension after performing updates simultaneously. Moreover, [Alfano et al., 2018b] reports on an experimental analysis showing that Algorithm 2 also outperforms by two orders of magnitude the computation from scratch on EAFs, where solvers from scratch taking as input the (compact) Dung argumentation frameworks resulting from the transformation of the candidate EAF (cf. Definition 4.4) are used. Finally, the experimental results concerning the use of both Algorithm 1 and Algorithm 2 also revealed that the improvements of using incremental techniques become larger as the computation from scratch becomes more challenging.

5 Incremental Computation in Defeasible Logic Programming

In [Besnard et al., 2014], four frameworks that consider the structure of arguments were presented. Two of them, ASPIC$^+$ [Modgil and Prakken, 2014] and ABA [Toni, 2014], build the set of all possible arguments from the knowledge base and then rely on using one of the possible Dung semantics to decide on the acceptance of arguments. The other two—Logic-Based Deductive Argumentation [Besnard and Hunter, 2014] and DeLP [Garcia and Simari, 2014]—only build the arguments involved in answering the query. These last two frameworks exhibit several differences [Besnard et al., 2014]—among them is the base logic used as a knowledge representation language: [Besnard and Hunter, 2014] relies on propositional logic, requiring a theorem prover to solve queries; on the other hand, DeLP [Garcia and Simari, 2014] adopts an extension of logic programming, which is a computational framework per se. To better understand the differences among the frameworks mentioned above, we refer the interested reader to [Garcia et al., 2020a], where a variant of DeLP using the grounded semantics is also discussed.

A fundamental distinction between DeLP and the other three frameworks, which significantly affects a query's resolution, rests on how attacks between arguments are described. DeLP considers two forms of defeat: *proper* and *blocking*; the former is akin to Dung's attack [Dung, 1995], whereas the latter presents a different behavior since the two arguments that are part of the blocking defeat relation, attacker and attackee, are defeated (hence the use of the term *blocking defeater*). Of course, this could be modeled in Dung's graphs as a mutual attack, but the DeLP mechanism forbids, in a properly formed dialogue, the use of two blocking defeaters successively because the introduction of another blocking defeater is unnecessary since the first two are already defeated. Moreover, to find the answers required by the query, other considerations of dialogical nature are taken into account, strengthening the reasoning process by forbidding common dialogical fallacies; these characteristics have been reflected in the development of a game-based semantics [Viglizzo et al., 2009].

In this section, we focus on the incremental computation in the context of structured argumentation. Particularly, we discuss an incremental technique [Alfano et al., 2018d; Alfano et al., 2018c] for *Defeasible Logic Programming (DeLP)* [Garcia and Simari, 2004; Garcia and Simari, 2014] which shares the same underlying ideas and the goal of avoiding wasted effort as in the (incremental) technique previously discussed for AFs. Given that our primary focus is on the changes in the structure of the arguments used to answer a query, we have considered the DeLP language; however, the ideas here developed can inspire similar techniques for other structured argumentation frameworks such as ABA and ASPIC$^+$. Next, we will summarize the necessary elements to develop the updating techniques in DeLP's structured argumentation; see [Alfano et al., 2018c] for an extended presentation.

5.1 Defeasible Logic Programming and Updates

A DeLP program $\mathcal{P} = (\Pi, \Delta)$ consists of sets Π and Δ of *strict* and *defeasible* rules defined using elements of a set *Lit* of literals, that are ground atoms obtained from a set **At** of atoms. $Lit_\mathcal{P}$ denotes the set of literals occurring in a rule of $\mathcal{P}$, and the symbol "$\sim$" represents strong negation; for any literal $\alpha \in Lit$ the formula $\sim\sim\alpha$ is considered equivalent to α and can be used for denoting it. Particularly, given the literals $\alpha_0, \alpha_1, \ldots, \alpha_n$, a strict rule $\alpha_0 \leftarrow \alpha_1, \ldots, \alpha_n$ (with $n \geq 0$) represents non-defeasible information, while *defeasible* rules $\alpha_0 \prec \alpha_1, \ldots, \alpha_n$ (with $n > 0$) represent tentative information, *i.e.*, information that can be used if nothing can be posed against it. Given a strict or defeasible rule r, we use $head(r)$ to denote α_0, and $body(r)$ to denote the set of literals $\{\alpha_1, \ldots, \alpha_n\}$. Strict rules with empty body will also be called *facts*.[1]

As an example of DeLP program, let us consider $\mathcal{P}_1 = (\Pi_1, \Delta_1)$, where:

$\Pi_1 = \{x, y, z, (w \leftarrow y)\}$ is the set of strict rules (and facts), and

$\Delta_1 = \{(a \prec w), (a \prec z), (\sim a \prec z), (b \prec a), (b \prec z), (c \prec b, x),$

---

[1] With a little abuse of notation, in the following we will denote a fact $(\alpha \leftarrow)$ simply by α.

$(\sim c \mathrel{-\!\!\prec} b), (d \mathrel{-\!\!\prec} \sim c)\}$ is the set of defeasible rules.

Given a DeLP program $\mathcal{P} = (\Pi, \Delta)$ and a literal $\alpha \in \mathit{Lit}_\mathcal{P}$, a *(defeasible) derivation* for α w.r.t. $\mathcal{P}$ is a finite sequence $\alpha_1, \alpha_2, \ldots, \alpha_n = \alpha$ of literals such that (i) each literal α_i is in the sequence because there exists a (strict or defeasible) rule $r \in \mathcal{P}$ with head α_i and body $\alpha_{i_1}, \alpha_{i_2}, \ldots, \alpha_{i_k}$ such that $i_j < i$ for all $j \in [1, k]$, and (ii) there do not exist two literals α_i and α_j such that $\alpha_j = \sim\!\alpha_i$. A derivation is said to be a *strict derivation* if only strict rules are used.

A program $\mathcal{P}$ is *contradictory* if and only if there exist defeasible derivations for at least two complementary literals α and $\sim\!\alpha$ from $\mathcal{P}$. We assume that Π (the strict part of $\mathcal{P}$) is not contradictory. However, complementary literals can be derived from $\mathcal{P}$ when defeasible rules are used in the derivation. Two literals α and β are said to be *contradictory* if (i) neither $\Pi \cup \{\alpha\}$ nor $\Pi \cup \{\beta\}$ strictly derive a pair of complementary literals, whereas (ii) $\Pi \cup \{\alpha, \beta\}$ does. Pairs of complementary literals are clearly contradictory; a set of literals is said to be contradictory if it contains two contradictory literals.

Considering the program $\mathcal{P}_1$, the literal c can be derived using the following sets of rules and facts: $\{(c \mathrel{-\!\!\prec} x, b), (x), (b \mathrel{-\!\!\prec} a), (a \mathrel{-\!\!\prec} w), (w \leftarrow y), (y)\}$; the derivation (y, w, a, b, x, c) describes how rules can be applied to derive c. However, the set of rules $\Pi_1 \cup \Delta_1$ is contradictory since also $\sim c$ can be derived using the rules: $\{(\sim c \mathrel{-\!\!\prec} b), (b \mathrel{-\!\!\prec} a), (a \mathrel{-\!\!\prec} w), (w \leftarrow y), (y)\}$. The non-contradictory set of literals that can be derived from Π_1 is $\{x, y, w, z\}$.

DeLP incorporates a defeasible argumentation formalism for the treatment of contradictory knowledge, allowing the identification of conflicting pieces of knowledge, and a *dialectical process* is used for deciding which information prevails as warranted. This process involves the construction and evaluation of arguments that either support or interfere with a user-issued query. An *argument* $\mathcal{A}$ for a literal α is a couple $\langle \mathcal{A}, \alpha \rangle$ where $\mathcal{A}$ is a set of defeasible rules representing a derivation that is (i) supported by facts, (ii) non-contradictory, and (iii) $\subseteq$-minimal (*i.e.*, there is no proper subset of $\mathcal{A}$ satisfying both (i) and (ii)). As an example, $\langle \mathcal{A}_1, c \rangle = \langle \{(c \mathrel{-\!\!\prec} x, b), (b \mathrel{-\!\!\prec} a), (a \mathrel{-\!\!\prec} w), (w \leftarrow y)\}, c \rangle$ and $\langle \mathcal{A}_2, \sim\!a \rangle = \langle \{(\sim\!a \mathrel{-\!\!\prec} z)\}, \sim\!a \rangle$ are two arguments that can be obtained from the program $\mathcal{P}_1$. An argument $\langle \mathcal{A}, \alpha \rangle$ is said to be a *sub-argument* of $\langle \mathcal{A}', \alpha' \rangle$ if $\mathcal{A} \subseteq \mathcal{A}'$.

The main task of DeLP is establishing *warranted* literals. A literal α is said to be warranted if there exists an undefeated argument $\langle \mathcal{A}, \alpha \rangle$. To determine if an argument $\langle \mathcal{A}, \alpha \rangle$ is undefeated, *defeaters* for $\langle \mathcal{A}, \alpha \rangle$ are considered, and since reinstatement could happen when all of $\mathcal{A}$'s possible defeaters are defeated, the process continues considering defeaters for $\mathcal{A}$'s defeaters, and so on. To define defeaters, to decide when an attack is successful, we require a comparison criterion $\succ$ over arguments, which is irreflexive and asymmetric. As the comparison criterion is a modular part of the argumentation inference engine, we will abstract away from this criterion and simply assume the existence of a comparison criterion $\succ$ between arguments: $\langle \mathcal{A}, \alpha \rangle \succ \langle \mathcal{B}, \beta \rangle$ meaning that argument $\langle \mathcal{A}, \alpha \rangle$ is preferred to $\langle \mathcal{B}, \beta \rangle$. Intuitively, an argument $\langle \mathcal{A}, \alpha \rangle$ *attacks* an argument $\langle \mathcal{B}, \beta \rangle$ when there is a sub-argument $\langle \mathcal{C}, \gamma \rangle$ of $\langle \mathcal{B}, \beta \rangle$, such that α and γ are contradictory. When the attacker satisfies that $\langle \mathcal{C}, \gamma \rangle$ is not preferred to $\langle \mathcal{A}, \alpha \rangle$ (*i.e.,* $\langle \mathcal{C}, \gamma \rangle \not\succ \langle \mathcal{A}, \alpha \rangle$), the attacker is called a *defeater*. A defeater $\langle \mathcal{A}, \alpha \rangle$ for $\langle \mathcal{B}, \beta \rangle$ will be referred to as a *proper defeater* if $\langle \mathcal{A}, \alpha \rangle \succ \langle \mathcal{C}, \gamma \rangle$; otherwise, it will be called a *blocking defeater*.

An other part of the dialectical process is the construction of the so called *dialectical tree*, which is used to decide the warrant status of a literal. A dialectical tree contains all the possible acceptable argumentation lines (namely, sequences of defeating arguments) that can be constructed from the given argument that sits on the root of that tree as paths from the root to the leaves. (see [Chesñevar and Simari, 2007] for a discussion). More in detail, a dialectical tree for an argument $\langle \mathcal{A}, \alpha \rangle$ is a tree-like structure where nodes are arguments and the root node is $\langle \mathcal{A}, \alpha \rangle$. Each root-to-leaf path in the tree is an *acceptable argumentation line*, which is a finite sequence of arguments that satisfy the following four constraints: (*i*) every argument of the sequence defeats its predecessor; (*ii*) the arguments in odd (resp., even) positions of the sequence does not contradict the strict part of the program; (*iii*) two blocking defeaters cannot appear one immediately after the other in the sequence; and (*iv*) arguments cannot appear twice in the sequence (also when appearing as sub-arguments).

Therefore, it is interesting to note that a dialectical tree for an argument represents the exhaustive dialectical analysis for that argument. Each dialectical tree is then marked to obtain the status of the literal α in the argument at its root through a bottom-up marking procedure, consisting in i) marking all leaves of the tree as UNDEFEATED; then, ii) every non-leaf node is marked as DEFEATED if and only if at least one of its children is marked as UNDEFEATED, otherwise it is marked as UNDEFEATED. Thus, if there exists a marked dialectical tree whose root contains an argument for α, which is marked as UNDEFEATED, we will say that α is *warranted*[2]. Considering the program $\mathcal{P}_1$, only x, y, z, w, and b are warranted.

Given a DeLP program $\mathcal{P}$, we define a total function $\mathcal{S}_\mathcal{P} : Lit \to \{$in, out, undec$\}$ assigning a *status* to each literal w.r.t. $\mathcal{P}$ as follows: $\mathcal{S}_\mathcal{P}(\alpha) =$ in if α is warranted; $\mathcal{S}_\mathcal{P}(\alpha) =$ out if $\mathcal{S}_\mathcal{P}(\sim\alpha) =$ in; $\mathcal{S}_\mathcal{P}(\alpha) =$ undec if neither $\mathcal{S}_\mathcal{P}(\alpha) =$ in nor $\mathcal{S}_\mathcal{P}(\alpha) =$ out. For literals not occurring in the program we also say that their status is unknown.

Updates. An *update* for a DeLP program $\mathcal{P} = \langle \Pi, \Delta \rangle$ modifies $\mathcal{P}$ into a new program $\mathcal{P}' = \langle \Pi', \Delta' \rangle$ by adding or removing a strict or a defeasible rule r. In particular, we allow the removal of any rule r of $\mathcal{P}$ through an update, and consider the addition of a rule r such that $body(r) \subseteq Lit_\mathcal{P}$ and $head(r) \subseteq Lit$, thus allowing also the addition of a rule whose head is a literal not belonging to $Lit_\mathcal{P}$. Given a DeLP program $\mathcal{P}$ and a strict or defeasible rule r, we use $u = +r$ (resp., $u = -r$) to denote a rule addition (resp., deletion) update to be performed on $\mathcal{P}$, obtaining the DeLP-program $u(\mathcal{P})$ resulting from the application of update u to $\mathcal{P}$. In the following, we assume that any update u is feasible, meaning that i) we only remove (resp. add) strict or defeasible rules appearing (resp., not appearing) in the given program $\mathcal{P}$, and ii) guaranteeing that the strict part of the updated program $u(\mathcal{P})$ will not be contradictory.

---

[2]The system available at the following link allows us to compare the abstract semantics with that of DeLP: https://hosting.cs.uns.edu.ar/~daqap/client/index.html; see [Leiva et al., 2019] for a description.

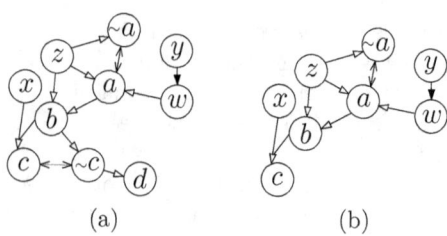

Figure 9: (a) $G(\mathcal{P}_1)$ for program $\mathcal{P}_1$; (b) $G(\mathcal{P}_1')$ for program $\mathcal{P}_1' = -(\sim c \prec b)(\mathcal{P}_1)$.

5.2 Incremental Computation of Warranted Literals

We first introduce the concept of labeled directed hypergraph associated with a DeLP program, which is central to our incremental approach.

Given a program $\mathcal{P}$, the corresponding labelled hypergraph $G(\mathcal{P}) = \langle N, H \rangle$ consists of a set N of nodes and a set H of labelled hyper-edges (Src, t, l), where Src is a possibly empty set called the *source set*, t is called the *target node*, and $l \in \{\texttt{def}, \texttt{str}, \texttt{cfl}\}$ is a label associated to the hyper-edge. Literals for which there exists a strict derivation in Π are immediately added to the set N of nodes of $G(\mathcal{P})$. Then, for each (strict or defeasible) rule whose body is in N, the head is added to N, and a (**str** or **def**) labelled hyper-edge corresponding to the (strict or defeasible) rule is added to the set H of hyper-edges. Finally, there is a pair of (**cfl**) labelled hyper-edges for each pair of complementary literals appearing as nodes in the hypergraph.

The hypergraph $G(\mathcal{P}_1)$ for the DeLP program $\mathcal{P}_1$ is shown in Figure 9(a) where ↔ (resp. ⇠ and ◂−) denotes hyper-edges labeled as **cfl** (resp. **def** and **str**).

We say that there is a path from a literal β to a literal α, if either (i) there exists a hyper-edge whose source set contains β and whose target is α, or (ii) there exists a literal γ and also there exist paths from β to γ and from γ to α. Moreover, we say that a node y is *reachable* from a set X of nodes if there exists a path from some x in X to y.

Given some update u, we denote with $G(u, \mathcal{P})$ the labeled hypergraph $G(u^+(\mathcal{P}))$ or $G(u^-(\mathcal{P}))$, depending on whether u consists of an insertion or deletion, respectively. The reason of this difference is that, to determine the set of literals whose status may change by deleting a rule r, we need to consider the hypergraph also containing the hyperedge derived from r.

Given a DeLP-program $\mathcal{P}$ and an update u, our incremental approach for recomputing the status of the literals after performing u consists of the following steps.

- Firstly, it is checked whether the update u is *irrelevant*, that is all literals in Lit are preserved. In such a case the initial status $\mathcal{S}_\mathcal{P}$ is returned.

- If u is not irrelevant, we need to:

  (i) compute the set of literals that are "influenced" by the update;

  (ii) among the influenced literals determine the subset of literals whose status may change after performing the update (called *core literals*). The status of uninfluenced literals does not change after the update.

  (iii) compute the updated status of the core literals; and

  (iv) determine the updated status of the *inferable* literals, *i.e.*, the literals whose status can be immediately determined from the status of the core literals.

The identification of relevant and irrelevant updates, as well influenced, preserved, core, and inferable literals is discussed below. In [Alfano et al., 2018d; Alfano et al., 2018c], it is shown that, in practice, the algorithm resulting from applying the above-mentioned steps turns out to be much more efficient than recomputing everything from scratch.

Irrelevant updates. Sufficient conditions guaranteeing that the status of each literal in the updated program is the same as that of the initial program are investigated in [Alfano et al., 2018c]. In these cases we say that the update u is *irrelevant*. One of these conditions holds whenever we add (resp. remove) a defeasible rule whose head's status is **in** (resp. **out**) w.r.t. the initial program. However, this does not hold for updates concerning strict rules. In these cases, we need to makes use of the hypergraph associated with a DeLP program, as well as the status of the literals *related* to an update.

A literal is said to be related to a given update $u = \pm r$ and program $\mathcal{P}$ if it can be reached from $head(r)$ in the labelled hypergraph $G(u, \mathcal{P})$ by navigating forward each rules and backward strict rules only, until no new related literals can be found. We call *deductive closure* of facts and strict rules of a program $\mathcal{P}$ the set of literals that are facts in $\mathcal{P}$ or can be derived from the strict part Π of $\mathcal{P}$. Given this, an update $u = \pm r$ is irrelevant if either (i) $head(r)$ does not belong to $G(u, \mathcal{P})$; or (ii) either $head(r)$ or $\sim head(r)$ appears in the deductive closure of facts and strict rules of both programs $\mathcal{P}$ and $u(\mathcal{P})$; or (iii) at least one literal in the body of r is either **out** or not related to u. Recomputing the status of the updated program's literals can be avoided if an irrelevant update is performed.

Relevant updates and influenced set. We now consider the computation of the status of literals for updates which have not been identified as irrelevant. An update is *relevant* whenever it causes the status of at least one literal to change. That is, even if for relevant updates the status of some literals may not change, and therefore for those literals, their status does not need to be recomputed when the update is performed. To avoid wasted effort, we determine the subset of literals whose status needs to be recomputed after an update. Towards this end, we discuss the concept of *influenced set*, which consists of the set of literals that are related to a given update u and program $\mathcal{P}$ but using only labeled hyper-edges whose corresponding rules are such that (i) the head (or its complement) is not in the deductive closures of both $\mathcal{P}$ and $u(\mathcal{P})$, and (ii) the body does not contain a literal that is not related to u and $\mathcal{P}$ and such that its status is **out**—intuitively, the other hyper-edges can be ignored as they correspond to rules whose head does not change status. For instance, for the program $\mathcal{P}_1$ and update $u = -(\sim c \prec b)$, the influenced set consists only of the literals b, c, $\sim c$, and d.

The notion of influenced set for DeLP programs is conceptually similar to the influenced set of Definition 3.3 for abstract argumentation frameworks (where arguments have no internal structure). Although

11 - On the Incremental Computation of Semantics

the aim is analogous, here we deal with incremental computation of the status of *structured* arguments, and consider a notion of influenced set w.r.t. an update for a DeLP program and its status that we then apply to (hyper)graphs representing DeLP programs, from which structured arguments are derived. A significant difference is that here we have both strict and defeasible rules meaning that to determine a portion of the hypergraph that contains nodes corresponding to literals whose status may change, we need to navigate strict edges both forward and backward. As an example, consider the DeLP program $\mathcal{P}_\chi = \langle \Pi_\chi, \Delta_\chi \rangle$ where $\Pi_\chi = \{f_1, f_2, a \leftarrow b, \sim a \leftarrow c\}$ and $\Delta_\chi = \{b \prec f_1\}$, and let $u = +(c \prec f_2)$ be an update yielding the updated DeLP program $\mathcal{P}'_\chi$. The influenced set is $\{c, \sim a, a, b\}$. Observe that b is included in the influenced set by navigating backward via the (hyper)edge corresponding to the strict rule $a \leftarrow b$, while the other literals are reached by forward reachability. Note that including b is important as its status changes (it is **undec** w.r.t. $\mathcal{P}'_\chi$, it was **in** w.r.t. $\mathcal{P}_\chi$).

Preserved, core, and inferable literals. Using the influenced set we can identify the preserved literals, *i.e.*, the literals whose status does not change after performing a relevant update. This set consists of the literals (*i.e.*, nodes) of the updated hypergraph that are not influenced. The status of a literal for which there is no argument in the (updated) program may depend only on the status of its complementary literal—we call such literals *inferable* and use them to define what we call *core literals*. The core literals for an update u are those in $Lit_{\mathcal{P}'}$ that are influenced but are not inferable, where $\mathcal{P}'$ is the updated program. The status of an inferred literal w.r.t. the updated program can be either **out** or **undec**, and if it is **out** it is entailed by the status of a core or preserved literal that is **in**. Finally, the status of the literals not in $Lit_{\mathcal{P}'}$ can be readily determined to be **undec**.

Considering the program $\mathcal{P}_1$ and update $u = -(\sim c \prec b)$, $\sim c$ and d are the only inferable literals, while b and c are core literals. The (updated) status of the inferable literal $\sim c$ is **out** as it is entailed by the (updated) status of its complementary literal, which is **in**; the status of the inferable literal d remains **undec**.

Efficiency. The incremental technique discussed in this section has been the subject of analysis in [Alfano *et al.*, 2018c; Alfano *et al.*, 2018d], which report on a set of experiments comparing the incremental approach with full recomputation from scratch (that is, the direct computation of the status of all the literals in an updated DeLP program using the DeLP-Solver). It turned out that the incremental approach significantly outperforms computation from scratch. Specifically, the incremental algorithm takes only a few seconds for DeLP programs, while the approach from scratch takes almost 2 minutes.

6 Related Work

Overviews of key concepts in argumentation theory and formal models of argumentation in the field of Artificial Intelligence are presented in [Bench-Capon and Dunne, 2007; Besnard and Hunter, 2008; Rahwan and Simari, 2009; Atkinson *et al.*, 2017]. Further discussion regarding uses of computational argumentation as an Agreement Technology can be found in [Modgil *et al.*, 2013].

A comprehensive introduction to the semantics of static abstract argumentation frameworks can be found in [Baroni *et al.*, 2011a]. Although the idea underlying abstract argumentation frameworks is intuitive and straightforward, most of the semantics proposed so far suffer from a high computational complexity [Dunne and Wooldridge, 2009; Dunne, 2009; Dvorak *et al.*, 2010; Dvorak and Woltran, 2010; Fazzinga *et al.*, 2013; Fazzinga *et al.*, 2015a; Fazzinga *et al.*, 2016; Fazzinga *et al.*, 2015b]. Complexity bounds and evaluation algorithms for abstract argumentation frameworks have been intensely studied in the literature, but most of this research focused on static frameworks, whereas, in practice, argumentation frameworks are dynamic systems [Capobianco *et al.*, 2005; Falappa *et al.*, 2011; Baumann, 2011; Liao *et al.*, 2011; Baroni *et al.*, 2014b; Charwat *et al.*, 2015]. In fact, in general, an AF represents a temporary situation, and new arguments and attacks can be added/retracted to model new available knowledge. For instance, for disputes among users of online social networks [Kökciyan *et al.*, 2016], arguments/attacks are continuously added or removed by users to express their point of view in response to the last move made by the adversaries (often disclosing as few arguments/attacks as possible).

There have been several significant efforts aimed at coping with the dynamic aspects of abstract argumentation. In [Boella et al., 2009a; Boella et al., 2009b], the authors have investigated the principles according to which a grounded extension of a Dung abstract argumentation framework does not change when the set of arguments/attacks is changed. Meanwhile, in [Bisquert et al., 2013] a synthesis is presented concerning the characterization of changes based on the work presented in [Cayrol et al., 2008; Cayrol et al., 2010; Bisquert et al., 2011; Bisquert et al., 2012] where the evolution of the set of extensions after performing a change operation is studied; here, a change operation can be about adding or removing one interaction or adding or removing one argument and a set of interactions.

Dynamic argumentation has been applied to the decision-making mechanisms of an autonomous agent by [Amgoud and Vesic, 2012], where it is studied how the acceptability of arguments evolves when a new argument is added to the decision system. Other relevant works on dynamic aspects of Dung's argumentation frameworks include the following. [Baumann, 2011] has proposed an approach exploiting the concept of the splitting of logic programs to deal with dynamic argumentation. The technique considers weak expansions of the initial argumentation framework, where added arguments never attack previous ones. [Baumann and Brewka, 2010] have investigated whether and how it is possible to modify a given argumentation framework so that a desired set of arguments becomes an extension, whereas [Oikarinen and Woltran, 2011] have studied equivalence between two argumentation frameworks when further information (another argumentation framework) is added to both argumentation frameworks. [Baumann, 2012] has focused on expansions where new arguments and attacks may be added, but the attacks among the old arguments remain unchanged, while [Baumann, 2014] have characterized update and deletion equivalence, where adding/deleting arguments/attacks is allowed (deletions were not considered by [Oikarinen and Woltran, 2011; Baumann, 2012]).

Several approaches for dividing argumentation frameworks into subgraphs have been explored in the context of dynamic argumentation frameworks. The division-based method, proposed in [Liao et al., 2011] and then refined in [Baroni et al., 2014b], divides the updated framework into two parts: *affected* and *unaffected*, where only the status of affected arguments is recomputed after updates. Using the results introduced in [Liao et al., 2011], the work presented in [Liao and Huang, 2013] investigated the efficient evaluation of the justification status of a subset of arguments in an argumentation framework (instead of the whole set of arguments), and proposed an approach based on answer-set programming for *local* computation. In [Liao, 2013], an argumentation framework is decomposed into a set of strongly connected components, yielding sub-argumentation frameworks located in layers, which are then used for incrementally computing the semantics of the given argumentation framework by proceeding layer by layer. Then, [Xu and Cayrol, 2015] introduced a matrix representation of argumentation frameworks and proposed a matrix reduction that, when applied to dynamic argumentation frameworks, resembles the division-based method in [Liao et al., 2011].

Changes in bipolar argumentation frameworks have been studied in the work [Cayrol and Lagasquie-Schiex, 2015], where it is shown how the addition of one argument together with one support that involves that argument (and without introducing any attack) impacts the extensions of the updated bipolar argumentation framework. However, these works do not address the incremental computation in dynamic bipolar argumentation frameworks, nor in extended argumentation frameworks modeling attacks towards the attack relation [Modgil, 2007; Baroni et al., 2009] and defeasible support [Boella et al., 2010].

There have been fewer attempts to consider the dynamics of the defeasible argumentation in the field of structured argumentation [Besnard et al., 2014]. As in the abstract argumentation case, there have been some works following the belief revision approach. In [Falappa et al., 2002], the issue of modifying strict rules to become defeasible was analyzed in the context of revisions effected over a knowledge base, while in [Moguillansky et al., 2013] the authors thoroughly explored the different cases that may occur when a DeLP program is modified by adding, deleting, or changing its elements. Neither of these works explored the implementation issues related to the problems studied here. Regarding implementations of approaches focusing on improving the tractability of determining the status of pieces of knowledge, in [Capobianco et al., 2005; Capobianco and Simari, 2009], the authors consider several alternatives to avoid recomputing warrants. In [Deagustini et al., 2013], the authors focus on challenges arising in the development of recommender systems, addressing them via the design of novel architectures that improve the computation of answers. Finally, [Gottifredi et al., 2013] makes use of heuristics designed to improve efficiency, and [Testerink et al., 2019] deals with the computational complexity of performing recalculations in a structured argumentation setting by relying on an approximation algorithm.

We believe that the set of ideas proposed in this work may be a forerunner of similar techniques for the optimization of other structured argumentation frameworks such as, for example, ABA and ASPIC$^+$. Regarding ABA, the construction of deductions is very similar to that of arguments for DeLP, although the way arguments attack each other is different. Therefore, similarly, the ABA framework could be represented using hypergraphs (where assumptions may be modeled as defeasible facts) to identify irrelevant updates and restrict the hypergraph to compute the semantics of updated programs efficiently. The similarities between DeLP and ASPIC$^+$ are even more substantial: both have a distinction between strict and defeasible inference rules, and both use comparison criteria to resolve attacks into defeats; however, while ASPIC$^+$ evaluates arguments with the standard AF semantics, DeLP has a special-purpose definition of argument evaluation [Garcia et al., 2020b]. Therefore, the ideas developed here can be of inspiration to optimize the incremental computation of the semantics of ASPIC$^+$ programs.

7 Conclusions and Future Work

We have reviewed techniques for the incremental and efficient computation in dynamic abstract argumentation and defeasible knowledge bases. In the case of abstract argumentation, we have presented a technique enabling any non-incremental algorithm to be used as an incremental one for computing some extension of dynamic argumentation frameworks.

The algorithm identifies a tighter portion of the updated argumentation framework to be examined for recomputing the semantics. The incremental algorithm proposed for Dung's frameworks enables a technique for the incremental computation of extensions of dynamic frameworks incorporating supports and second-order attacks (that we called extended argumentation frameworks). Recently, in [Alfano et al., 2020b], we have investigated incremental techniques for the ASAF framework [Gottifredi et al., 2018], where attacks and support relations of any order are considered. For the case of structured argumentation, we have discussed an algorithm able to incrementally solve the problem determining the warrant status of literals in a DeLP program which is updated by adding or deleting strict or defeasible rules. The experimental analysis performed in [Alfano et al., 2017b; Alfano et al., 2018b; Alfano et al., 2018d; Alfano et al., 2018c] showed that, in practice, the incremental approach, for both the cases of abstract and structured frameworks, turns out to be much more efficient than recomputing everything from scratch.

The notions behind the use of an incremental approach can be extended further, as done in [Alfano et al., 2019a; Alfano and Greco, 2021], where an incremental technique was recently proposed aimed at determining whether a given argument is skeptically preferred accepted in dynamic argumentation frameworks by exploiting the concept of influenced and reduced argumentation frameworks presented here in Section 3. Future work will be devoted to extending our technique to cope with other argumentation frameworks [Alfano et al., 2020d; Alfano et al., 2020e] and other computational problems [Alfano et al., 2020a; Alfano et al., 2019b; Alfano et al., 2021]. It would be interesting to deal with different interpretations of the support relation, *e.g.*, that one in [Cayrol and Lagasquie-Schiex, 2009; Cayrol and Lagasquie-Schiex, 2010] where a meta argumentation approach is also adopted to deal with bipolarity. We plan to address the problem of incrementally enumerating *all* the extensions of an abstract argumentation framework. Following [Alfano et al., 2019a; Alfano et al., 2020c], devising an incremental computation approach for the skeptical/credulous acceptance in dynamic argumentation frameworks, and its extensions (*e.g.*, bipolar argumentation frameworks and ASAFs), is another intriguing direction for future work. Finally, we believe the basic ideas presented for the case of structured argumentation could carry over to other frameworks, *e.g.*, ASPIC$^+$ or ABA; this is another research direction we are planning to pursue in future work.

Acknowledgments

The authors are grateful to the participants of the workshop *Current Trends in Formal Argumentation* (held at the University Centre of Bertinoro from November 3rd to 6th, 2019), and in particular to Matthias Thimm, for discussions on the limitations of the technique previously presented in [Alfano et al., 2018c] that improved the characterization of the irrelevant updates, and of the set of influenced literals, which are on the basis of the incremental approach for DeLP. Moreover, the authors wish to thank the reviewers for their comments and suggestions that helped us in improving the paper. Finally, this work was partly supported in Argentina by Universidad Nacional del Sur (UNS) under grant PGI 24/ZN34, Consejo Nacional de Investigaciones Cientificas y Técnicas (CONICET), and Agencia Nacional de Promoción Cientifica y Tecnológica under grant PICT-2018-0475.

References

[Alfano and Greco, 2021] G. Alfano and S. Greco. Incremental skeptical preferred acceptance in dynamic argumentation frameworks. *IEEE Intelligent Systems*, 2021.

[Alfano et al., 2017a] Gianvincenzo Alfano, Sergio Greco, and Francesco Parisi. Computing stable and preferred extensions of dynamic bipolar argumentation frameworks. In *Proc. of Workshop on Advances In Argumentation In Artificial Intelligence co-located with XVI International Conference of the Italian Association for Artificial Intelligence (AI*IA)*, pages 28–42, 2017.

[Alfano et al., 2017b] Gianvincenzo Alfano, Sergio Greco, and Francesco Parisi. Efficient computation of extensions for dynamic abstract argumentation frameworks: An incremental approach. In *Proc. of International Joint Conference on Artificial Intelligence (IJCAI)*, pages 49–55, 2017.

[Alfano et al., 2018a] Gianvincenzo Alfano, Sergio Greco, and Francesco Parisi. Computing extensions of dynamic abstract argumentation frameworks with second-order attacks. In *IDEAS*, pages 183–192, 2018.

[Alfano et al., 2018b] Gianvincenzo Alfano, Sergio Greco, and Francesco Parisi. A meta-argumentation approach for the efficient computation of stable and preferred extensions in dynamic bipolar argumentation frameworks. *Intelligenza Artificiale*, 12(2):193–211, 2018.

[Alfano et al., 2018c] Gianvincenzo Alfano, Sergio Greco, Francesco Parisi, Gerardo I. Simari, and Guillermo R. Simari. An incremental approach to structured argumentation over dynamic knowledge bases. In *Principles of Knowledge Representation and Reasoning: Proceedings of the Sixteenth International Conference, KR*, pages 78–87, 2018.

[Alfano et al., 2018d] Gianvincenzo Alfano, Sergio Greco, Francesco Parisi, Gerardo I. Simari, and Guillermo R. Simari. Incremental computation of warranted arguments in dynamic defeasible argumentation: The rule addition case. In *Proc. of the Symposium on Applied Computing (SAC)*, pages 911–917, 2018.

[Alfano et al., 2019a] Gianvincenzo Alfano, Sergio Greco, and Francesco Parisi. An efficient algorithm for skeptical preferred acceptance in dynamic argumentation frameworks. In *Proceedings of the Twenty-Eighth International Joint Conference on Artificial Intelligence, IJCAI 2019, Macao, China, August 10-16, 2019*, pages 18–24, 2019.

[Alfano et al., 2019b] Gianvincenzo Alfano, Sergio Greco, and Francesco Parisi. On scaling the enumeration of the preferred extensions of abstract argumentation frameworks. In *Proceedings of the 34th ACM/SIGAPP Symposium on Applied Computing, SAC 2019, Limassol, Cyprus, April 8-12, 2019*, pages 1147–1153. ACM, 2019.

[Alfano et al., 2020a] Gianvincenzo Alfano, Marco Calautti, Sergio Greco, Francesco Parisi, and Irina Trubitsyna. Explainable acceptance in probabilistic abstract argumentation: Complexity and approximation. In *Proceedings of the 17th International Conference on Principles of Knowledge Representation and Reasoning, KR 2020, Rhodes, Greece, September 12-18, 2020*, pages 33–43, 2020.

[Alfano et al., 2020b] Gianvincenzo Alfano, Andrea Cohen, Sebastian Gottifredi, Sergio Greco, Francesco Parisi, and Guillermo Ricardo Simari. Dynamics in abstract argumentation frameworks with recursive attack and support relations. In *Proc. of European Conference on Artificial Intelligence (ECAI)*, pages 577–584, 2020.

[Alfano et al., 2020c] Gianvincenzo Alfano, Sergio Greco, and Francesco Parisi. Computing skeptical preferred acceptance in dynamic argumentation frameworks with recursive attack and support relations. In *Proceedings of COMMA*, volume 326 of *Frontiers in Artificial Intelligence and Applications*, pages 67–78. IOS Press, 2020.

[Alfano et al., 2020d] Gianvincenzo Alfano, Sergio Greco, Francesco Parisi, and Irina Trubitsyna. On the semantics of abstract argumentation frameworks: A logic programming approach. *Theory Pract. Log. Program.*, 20(5):703–718, 2020.

[Alfano et al., 2020e] Gianvincenzo Alfano, Sergio Greco, Francesco Parisi, and Irina Trubitsyna. On the semantics of recursive bipolar afs and partial stable models. In *Proceedings of the Workshop on Advances In Argumentation In Artificial Intelligence 2020 co-located with the 19th International Conference of the Italian Association for Artificial Intelligence (AIxIA 2020), Online, November 25-26, 2020*, volume 2777 of *CEUR Workshop Proceedings*, pages 16–30. CEUR-WS.org, 2020.

[Alfano et al., 2021] Gianvincenzo Alfano, Sergio Greco, Francesco Parisi, and Irina Trubitsyna. Argumentation frameworks with strong and weak constraints: Semantics and complexity. In *Proc. of AAAI*, page (to appear), 2021.

[Amgoud and Vesic, 2012] Leila Amgoud and Srdjan Vesic. Revising option status in argument-based decision systems. *Journal of Logic and Computation*, 22(5):1019–1058, 2012.

[Amgoud et al., 2004] Leila Amgoud, Claudette Cayrol, and Marie-Christine Lagasquie-Schiex. On the bipolarity in argumentation frameworks. In *Proc. of International Workshop on Non-Monotonic Reasoning (NMR)*, pages 1–9, 2004.

[Amgoud et al., 2005] Leila Amgoud, Jean-François Bonnefon, and Henri Prade. An argumentation-based approach to multiple criteria decision. In *Proc. of European Conference on Symbolic and Quantitative Approaches to Reasoning and Uncertainty (ECSQARU)*, pages 269–280, 2005.

[Atkinson et al., 2017] Katie Atkinson, Pietro Baroni, Massimiliano Giacomin, Anthony Hunter, Henry Prakken, Chris Reed, Guillermo R. Simari, Matthias Thimm, and Serena Villata. Towards artificial argumentation. *AI Magazine*, 38(3):25–36, 2017.

[Baroni et al., 2009] Pietro Baroni, Federico Cerutti, Massimiliano Giacomin, and Giovanni Guida. Encompassing attacks to attacks in abstract argumentation frameworks. In *Proc. of European Conference on Symbolic and Quantitative Approaches to Reasoning with Uncertainty (ECSQARU)*, pages 83–94, 2009.

[Baroni et al., 2011a] Pietro Baroni, Martin Caminada, and Massimiliano Giacomin. An introduction to argumentation semantics. *The Knowledge Engineering Review*, 26(4):365–410, 2011.

[Baroni et al., 2011b] Pietro Baroni, Federico Cerutti, Massimiliano Giacomin, and Giovanni Guida. AFRA: Argumentation Framework with Recursive

Attacks. *IJAR*, 52(1):19–37, 2011.

[Baroni et al., 2014a] Pietro Baroni, Guido Boella, Federico Cerutti, Massimiliano Giacomin, Leendert W. N. van der Torre, and Serena Villata. On the input/output behavior of argumentation frameworks. *Artificial Intelligence*, 217:144–197, 2014.

[Baroni et al., 2014b] Pietro Baroni, Massimiliano Giacomin, and Beishui Liao. On topology-related properties of abstract argumentation semantics. A correction and extension to dynamics of argumentation systems: A division-based method. *Artificial Intelligence*, 212:104–115, 2014.

[Baumann and Brewka, 2010] Ringo Baumann and Gerhard Brewka. Expanding argumentation frameworks: Enforcing and monotonicity results. In *Proc. of Third International Conference on Computational Models of Argument (COMMA)*, pages 75–86, 2010.

[Baumann, 2011] Ringo Baumann. Splitting an argumentation framework. In *Proc. of International Conference on Logic Programming and Non-Monotonic Reasoning (LPNMR)*, pages 40–53, 2011.

[Baumann, 2012] Ringo Baumann. Normal and strong expansion equivalence for argumentation frameworks. *Artificial Intelligence*, 193:18–44, 2012.

[Baumann, 2014] Ringo Baumann. Context-free and context-sensitive kernels: Update and deletion equivalence in abstract argumentation. In *Proc. of European Conference on Artificial Intelligence (ECAI)*, pages 63–68, 2014.

[Bench-Capon and Dunne, 2007] Trevor J. M. Bench-Capon and Paul E. Dunne. Argumentation in artificial intelligence. *Artificial Intelligence*, 171(10 - 15):619 – 641, 2007.

[Besnard and Hunter, 2008] Philippe Besnard and Anthony Hunter. *Elements of Argumentation*. MIT Press, 2008.

[Besnard and Hunter, 2014] Philippe Besnard and Anthony Hunter. Constructing argument graphs with deductive arguments: A tutorial. *Argument & Computation*, 5(1):5–30, 2014.

[Besnard et al., 2014] Philippe Besnard, Alejandro J. Garcia, Anthony Hunter, Sanjay Modgil, Henry Prakken, Guillermo R. Simari, and Francesca Toni. Introduction to structured argumentation. *Argument & Computation – Special Issue: Tutorials on Structured Argumentation*, 5(1):1–4, 2014.

[Bisquert et al., 2011] Pierre Bisquert, Claudette Cayrol, Florence Dupin de Saint-Cyr, and Marie-Christine Lagasquie-Schiex. Change in argumentation systems: Exploring the interest of removing an argument. In *Proceedings of the 5th International Conference on Scalable Uncertainty Management (SUM)*, pages 275–288, 2011.

[Bisquert et al., 2012] Pierre Bisquert, Claudette Cayrol, Florence Dupin de Saint-Cyr, and Marie-Christine Lagasquie-Schiex. Changement dans un système d'argumentation : suppression d'un argument. *Rev. d'Intelligence Artif.*, 26(3):225–253, 2012.

[Bisquert et al., 2013] Pierre Bisquert, Claudette Cayrol, Florence Dupin de Saint-Cyr, and Marie-Christine Lagasquie-Schiex. Characterizing change in abstract argumentation systems. In *Trends in Belief Revision and Argumentation Dynamics*, volume 48, pages 75–102. 2013.

[Bistarelli et al., 2018a] Stefano Bistarelli, Francesco Faloci, Francesco Santini, and Carlo Taticchi. Studying dynamics in argumentation with Rob. In *COMMA*, pages 451–452, 2018.

[Bistarelli et al., 2018b] Stefano Bistarelli, Lars Kotthoff, Francesco Santini, and Carlo Taticchi. Containerisation and dynamic frameworks in iccma'19. In *Proceedings of the Second International Workshop on Systems and Algorithms for Formal Argumentation (SAFA 2018) co-located with the 7th International Conference on Computational Models of Argument (COMMA 2018), Warsaw, Poland, September 11, 2018.*, pages 4–9, 2018.

[Boella et al., 2009a] Guido Boella, Souhila Kaci, and Leendert W. N. van der Torre. Dynamics in argumentation with single extensions: Abstraction principles and the grounded extension. In *Proc. of European Conference on Symbolic and Quantitative Approaches to Reasoning and Uncertainty (ECSQARU)*, pages 107–118, 2009.

[Boella et al., 2009b] Guido Boella, Souhila Kaci, and Leendert W. N. van der Torre. Dynamics in argumentation with single extensions: Attack refinement and the grounded extension. In *Proc. of Sixth Int. Workshop on Argumentation in Multi-Agent Systems (ArgMAS)*, pages 150–159, 2009.

[Boella et al., 2010] Guido Boella, Dov M. Gabbay, Leendert W. N. van der Torre, and Serena Villata. Support in abstract argumentation. In *Computational Models of Argument: Proceedings of COMMA 2010, Desenzano del Garda, Italy, September 8-10, 2010.*, pages 111–122, 2010.

[Caminada, 2006] Martin Caminada. Semi-stable semantics. In *Proc. of 1st International Conference on Computational Models of Argument (COMMA)*, pages 121–130, 2006.

[Capobianco and Simari, 2009] Marcela Capobianco and Guillermo R. Simari. A proposal for making argumentation computationally capable of handling large repositories of uncertain data. In *Proc. of Scalable Uncertainty Management*, pages 95–110, 2009.

[Capobianco et al., 2005] Marcela Capobianco, Carlos I. Chesñevar, and Guillermo R. Simari. Argumentation and the dynamics of warranted beliefs in changing environments. *Autonomous Agents and Multi-Agent Systems*, 11(2):127–151, 2005.

[Cayrol and Lagasquie-Schiex, 2005] Claudette Cayrol and Marie-Christine Lagasquie-Schiex. On the acceptability of arguments in bipolar argumentation frameworks. In *Proc. of European Conference on Symbolic and Quantitative Approaches to Reasoning and Uncertainty (ECSQARU)*, pages 378–389, 2005.

[Cayrol and Lagasquie-Schiex, 2009] Claudette Cayrol and Marie-Christine Lagasquie-Schiex. Bipolar abstract argumentation systems. In *Argumentation in Artificial Intelligence*, pages 65–84. 2009.

[Cayrol and Lagasquie-Schiex, 2010] Claudette Cayrol and Marie-Christine Lagasquie-Schiex. Coalitions of arguments: A tool for handling bipolar argumentation frameworks. *International Journal of Intelligent System*, 25(1):83–109, 2010.

[Cayrol and Lagasquie-Schiex, 2015] Claudette Cayrol and Marie-Christine Lagasquie-Schiex. Change in abstract bipolar argumentation systems. In *Proc. of International Conference on Scalable Uncertainty Management (SUM)*, pages 314–329, 2015.

[Cayrol et al., 2008] Claudette Cayrol, Florence Dupin de Saint-Cyr, and Marie-Christine Lagasquie-Schiex. Revision of an argumentation system. In *Proc. of International Conference on Principles of Knowledge Representation and Reasoning (KR)*, pages 124–134, 2008.

[Cayrol et al., 2010] Claudette Cayrol, Florence Dupin de Saint-Cyr, and Marie-Christine Lagasquie-Schiex. Change in abstract argumentation frameworks: Adding an argument. *Journal of Artificial Intelligence Research*, 38:49–84, 2010.

[Cecchi et al., 2006] Laura A. Cecchi, Pablo R. Fillottrani, and Guillermo R. Simari. On the complexity of DeLP through game semantics. In *Proc. of the 11th International Workshop on Nonmonotonic Reasoning, Lake District, GBP*, pages 386–394, 2006.

[Charwat et al., 2015] Günther Charwat, Wolfgang Dvorák, Sarah A. Gaggl, Johannes P. Wallner, and Stefan Woltran. Methods for solving reasoning problems in abstract argumentation - A Survey. *Artificial Intelligence*, 220:28–63, 2015.

[Chesñevar and Simari, 2007] Carlos I. Chesñevar and Guillermo R. Simari. Modelling inference in argumentation through labelled deduction: Formalization and logical properties. *Logica Universalis*, 1(1):93–124, 2007.

[Cohen et al., 2014] Andrea Cohen, Sebastian Gottifredi, Alejandro J. Garcia, and Guillermo R. Simari. A survey of different approaches to support in argumentation systems. *The Knowledge Engineering Review*, 29(5):513–550, 2014.

[Deagustini et al., 2013] Cristhian A. D. Deagustini, Santiago E. Fulladoza Dalibón, Sebastian Gottifredi, Marcelo A. Falappa, Carlos I. Chesñevar, and Guillermo R. Simari. Relational databases as a massive information source for defeasible argumentation. *Knowledge-Based Systems*, 51:93–109, 2013.

[Dung et al., 2007] Phan Minh Dung, Paolo Mancarella, and Francesca Toni. Computing ideal sceptical argumentation. *Artificial Intelligence*, 171(10-15):642–674, 2007.

[Dung, 1995] Phan Minh Dung. On the acceptability of arguments and its fundamental role in nonmonotonic reasoning, logic programming and n-person games. *Artificial Intelligence*, 77(2):321–358, 1995.

[Dunne and Wooldridge, 2009] Paul E. Dunne and Michael Wooldridge. Complexity of abstract argumentation. In *Argumentation in Artificial Intelligence*, pages 85–104. 2009.

[Dunne, 2009] Paul E. Dunne. The computational complexity of ideal semantics. *Artificial Intelligence*, 173(18):1559–1591, 2009.

[Dvorak and Woltran, 2010] Wolfgang Dvorak and Stefan Woltran. Complexity of semi-stable and stage semantics in argumentation frameworks. *Information Processing Letters*, 110(11):425–430, 2010.

[Dvorak et al., 2010] Wolfgang Dvorak, Reinhard Pichler, and Stefan Woltran. Towards fixed-parameter tractable algorithms for argumentation. In *Proc. of European Conference on Symbolic and Quantitative Approaches to Reasoning and Uncertainty (ECSQARU)*, 2010.

[Eiter et al., 2015] Thomas Eiter, Hannes Strass, Miroslaw Truszczynski, and Stefan Woltran, editors. *Advances in Knowledge Representation, Logic Programming, and Abstract Argumentation*, volume 9060. Springer, 2015.

[Falappa et al., 2002] Marcelo A. Falappa, Gabriele Kern-Isberner, and Guillermo R. Simari. Explanations, belief revision and defeasible reasoning. *Artif. Intell.*, 141(1/2):1–28, 2002.

[Falappa et al., 2011] Marcelo A. Falappa, Alejandro J. Garcia, Gabriele Kern-Isberner, and Guillermo R. Simari. On the evolving relation between belief revision and argumentation. *The Knowledge Engineering Review*, 26(1):35–43, 2011.

[Fazzinga et al., 2013] Bettina Fazzinga, Sergio Flesca, and Francesco Parisi. Efficiently estimating the probability of extensions in abstract argumentation. In *Proc. of International Conference on Scalable Uncertainty Management (SUM)*, pages 106–119, 2013.

[Fazzinga et al., 2015a] Bettina Fazzinga, Sergio Flesca, and Francesco Parisi. On the complexity of probabilistic abstract argumentation frameworks. *ACM Transactions on Computational Logic*, 16(3):22, 2015.

[Fazzinga et al., 2015b] Bettina Fazzinga, Sergio Flesca, Francesco Parisi, and Adriana Pietramala. PARTY: A mobile system for efficiently assessing the probability of extensions in a debate. In *Proc. of International Conference on Database and Expert Systems Applications (DEXA)*, pages 220–235, 2015.

[Fazzinga et al., 2016] Bettina Fazzinga, Sergio Flesca, and Francesco Parisi. On efficiently estimating the probability of extensions in abstract argumentation frameworks. *International Journal of Approximate Reasoning*, 69:106–132, 2016.

[Garcia and Simari, 2004] Alejandro J. Garcia and Guillermo R. Simari. Defeasible logic programming: An argumentative approach. *Theory and Practice of Logic Programming (TPLP)*, 4(1-2):95–138, 2004.

[Garcia and Simari, 2014] Alejandro J. Garcia and Guillermo R. Simari. Defeasible logic programming: DeLP-servers, contextual queries, and explanations for answers. *Argument & Computation*, 5(1):63–88, 2014.

[Garcia et al., 2020a] Alejandro J. Garcia, Henry Prakken, and Guillermo R. Simari. A comparative study of some central notions of ASPIC$^+$ and DeLP. *Theory and Practice of Logic Programming*, 20(3):358–390, 2020.

[Garcia et al., 2020b] Alejandro Javier Garcia, Henry Prakken, and Guillermo Ricardo Simari. A comparative study of some central notions of ASPIC+ and delp. *Theory Pract. Log. Program.*, 20(3):358–390, 2020.

[Gottifredi et al., 2013] Sebastian Gottifredi, Nicolas D. Rotstein, Alejandro J. Garcia, and Guillermo R. Simari. Using argument strength for building dialectical bonsai. *Annals of Mathematics and Artificial Intelligence*, 69(1):103–129, 2013.

[Gottifredi et al., 2018] Sebastian Gottifredi, Andrea Cohen, Alejandro J. Garcia, and Guillermo R. Simari. Characterizing acceptability semantics of argumentation frameworks with recursive attack and support relations. *Artif. Intell.*, 262:336–368, 2018.

[Greco and Parisi, 2016a] Sergio Greco and Francesco Parisi. Efficient computation of deterministic extensions for dynamic abstract argumentation frameworks. In *Proc. of European Conference on Artificial Intelligence (ECAI)*, pages 1668–1669, 2016.

[Greco and Parisi, 2016b] Sergio Greco and Francesco Parisi. Incremental computation of deterministic extensions for dynamic argumentation frameworks. In *Proc. of European Conference On Logics In Artificial Intelligence (JELIA)*, pages 288–304, 2016.

[Kökciyan et al., 2016] Nadin Kökciyan, Nefise Yaglikci, and Pinar Yolum. Argumentation for resolving privacy disputes in online social networks: (extended abstract). In *Proc. of International Conference on Autonomous*

11 - ON THE INCREMENTAL COMPUTATION OF SEMANTICS

Agents and Multiagent Sytems (AAMAS), pages 1361–1362, 2016.

[Kröll et al., 2017] Markus Kröll, Reinhard Pichler, and Stefan Woltran. On the complexity of enumerating the extensions of abstract argumentation frameworks. In *Proc. of IJCAI*, pages 1145–1152, 2017.

[Leiva et al., 2019] Mario A. Leiva, Gerardo I. Simari, Sebastian Gottifredi, Alejandro J. Garcia, and Guillermo R. Simari. DAQAP: defeasible argumentation query answering platform. In *Proceedings of the 13th International Conference on Flexible Query Answering Systems (FQAS)*, pages 126–138, 2019.

[Liao and Huang, 2013] Beishui Liao and Huaxin Huang. Partial semantics of argumentation: basic properties and empirical results. *Journal of Logic and Computation*, 23(3):541–562, 2013.

[Liao et al., 2011] Beishui Liao, Li Jin, and Robert C. Koons. Dynamics of argumentation systems: A division-based method. *Artificial Intelligence*, 175(11):1790–1814, 2011.

[Liao, 2013] Beishui Liao. Toward incremental computation of argumentation semantics: A decomposition-based approach. *Annals of Mathematics and Artificial Intelligence*, 67(3-4):319–358, 2013.

[Modgil and Prakken, 2011] Sanjay Modgil and Henry Prakken. Revisiting preferences and argumentation. In *Proc. of International Joint Conference on Artificial Intelligence (IJCAI)*, pages 1021–1026, 2011.

[Modgil and Prakken, 2014] Sanjay Modgil and Henry Prakken. The ASPIC$^+$ framework for structured argumentation: A tutorial. *Argument & Computation*, 5(1):31–62, 2014.

[Modgil et al., 2013] Sanjay Modgil, Francesca Toni, Floris Bex, Ivan Bratko, Carlos I. Chesnevar, Wolfgang Dvorak, Marcelo A. Falappa, Xiuyi Fan, Sarah Alice Gaggl, Alejandro J. Garcia, Maria P. Gonzalez, Thomas F. Gordon, Joao Leite, Martin Mouzina, Chris Reed, Guillermo R. Simari, Stefan Szeider, Paolo Torroni, and Stefan Woltran. *Agreement Technologies*, volume 8 of *Law, Governance and Technology*, chapter 21: The Added Value of Argumentation: Examples and Challenges, pages 357–404. Springer, 2013.

[Modgil, 2007] Sanjay Modgil. An abstract theory of argumentation that accommodates defeasible reasoning about preferences. In *Proc. of European Conference on Symbolic and Quantitative Approaches to Reasoning with Uncertainty (ECSQARU)*, pages 648–659, 2007.

[Modgil, 2009] Sanjay Modgil. Reasoning about preferences in argumentation frameworks. *Artificial Intelligence*, 173(9-10):901–934, 2009.

[Moguillansky et al., 2013] Martin O. Moguillansky, Nicolas D. Rotstein, Marcelo A. Falappa, Alejandro J. Garcia, and Guillermo R. Simari. Dynamics of knowledge in DeLP through argument theory change. *TPLP*, 13(6):893–957, 2013.

[Nouioua and Risch, 2010] Farid Nouioua and Vincent Risch. Bipolar argumentation frameworks with specialized supports. In *Proc. of ICTAI*, pages 215–218, 2010.

[Nouioua and Risch, 2011] Farid Nouioua and Vincent Risch. Argumentation frameworks with necessities. In *Proc. of SUM*, pages 163–176, 2011.

[Oikarinen and Woltran, 2011] Emilia Oikarinen and Stefan Woltran. Characterizing strong equivalence for argumentation frameworks. *Artificial Intelligence*, 175(14-15):1985–2009, 2011.

[Rahwan and Simari, 2009] Iyad Rahwan and Guillermo R. Simari. *Argumentation in Artificial Intelligence*. Springer Publishing Company, Incorporated, 1st edition, 2009.

[Testerink *et al.*, 2019] Bas Testerink, Daphne Odekerken, and Floris Bex. A method for efficient argument-based inquiry. In *Proceedings of the 13th International Conference on Flexible Query Answering Systems (FQAS)*, pages 114–125, 2019.

[Thimm and Villata, 2017] Matthias Thimm and Serena Villata. The first international competition on computational models of argumentation: Results and analysis. *Artificial Intelligence*, 252:267–294, 2017.

[Toni, 2014] Francesca Toni. A tutorial on assumption-based argumentation. *Argument & Computation*, 5(1):89–117, 2014.

[Viglizzo *et al.*, 2009] Ignacio D. Viglizzo, Fernando A. Tohmé, and Guillermo R. Simari. *Annals of Mathematics and Artificial Intelligence*, 57(2):181–204, 2009.

[Villata *et al.*, 2012] Serena Villata, Guido Boella, Dov M. Gabbay, and Leendert W. N. van der Torre. Modelling defeasible and prioritized support in bipolar argumentation. *Annals of Mathematics and Artificial Intelligence*, 66(1-4):163–197, 2012.

[Xu and Cayrol, 2015] Yuming Xu and Claudette Cayrol. The matrix approach for abstract argumentation frameworks. In *Proc. of International Workshop on Theory and Applications of Formal Argumentation (TAFA)*, pages 243–259, 2015.

Part III

Meta Investigations

Chapter 12

Logic-Based Approaches to Formal Argumentation

Ofer Arieli
School of Computer Science, The Academic College of Tel-Aviv, Israel
oarieli@mta.ac.il

AnneMarie Borg
Department of Information and Computing Sciences, Utrecht University, The Netherlands
a.borg@uu.nl

Jesse Heyninck
Department of Computer Science, TU Dortmund, Germany
jesse.heyninck@tu-dortmund.de

Christian Strasser
Institute of Philosophy II, Ruhr University Bochum, Germany
christian.strasser@rub.de

Abstract

We study the logical foundations of Dung-style argumentation frameworks. Logic-based methods in the context of argumentation theory are described from two perspectives: (a) a survey of logic-based instantiations of argumentation frameworks, their properties and relations, and (b) a review of logical methods for the study of argumentation dynamics. In this chapter we restrict ourselves to Tarskian logics, based on (propositional) languages and corresponding (constructive) semantics or syntactic rule-based systems.

1 Motivation, Introduction and Scope

The purpose of this chapter is to study the *logical* foundations of formal argumentation and highlight its role in the modeling of defeasible reasoning. For this, we assume the availability of an underlying *logic* (that is, a pair of a formal propositional language and a corresponding (reflexive, monotonic, and transitive) consequence relation), upon which argumentation-based formalisms are defined. We then study logic-based approaches to formal argumentation from two perspectives. One perspective is concerned with instantiations of argumentation frameworks by logic-based formalisms. The need to instantiate Dung's abstract argumentation frameworks [Dung, 1995] by deductive (or, more generally, structured) approaches is well acknowledged in the literature (see, e.g., [Caminada and Wu, 2011; Prakken, 2018; Prakken and Winter, 2018] for some papers on the subject), and is primarily motivated by giving logical justifications to the notions of arguments and counter-arguments. Moreover, several fundamental mathematical and philosophical notions that cannot be studied in an abstract context (or at least not natural to this context), can be investigated in a logic-based setting. This includes, for example, properties such as consistency, maximal consistency [Rescher and Manor, 1970], deductive closure [Caminada and Amgoud, 2007], logical omniscience, and so forth, as well as inference principles that can be related to general patterns of non-monotonic and paraconsistent reasoning, and which are better suited to a deductive (logic-based) setting.

The second perspective taken in this chapter is related to the use of logic-based machinery to describe (that is, represent and reason with) argumentation-based dynamics. Indeed, the availability of an underlying 'core' logic triggers a wide variety of methods for formally expressing argumentation-related processes. For instance, since modal logics allow to qualify statements, alethic arguments (about necessity and possibility), epistemic ones (about knowledge and belief) [Hintikka, 2005; Ditmarsch *et al.*, 2015], and deontic phrases (about obligations and permissios) [von Wright, 1951; Gabbay *et al.*, 2013; Straßer and Arieli, 2019] can be expressed, giving rise to different applications in linguistics, security and game theory (see e.g., [Blackburn *et al.*, 2006] and [Ditmarsch et al., 2015]). Also, the presence of an underlying logic allows for incorporation of proof-theoretical methods [Arieli and Straßer, 2019] and related structural methodologies [Grossi, 2013] to reason with argumentation frameworks and characterize their properties (see also [Gabbay and Gabbay, 2015; Gabbay and Gabbay, 2016]).

This chapter is divided into two parts according to the two perspectives described above. The first part of the chapter, given in Section 2, is focused on the first perspective, namely: a study of logic-based approaches to formal argumentation. The formalisms that are investigated in this part are those that are based on some underlying (core) *logic* (in the traditional sense of this notion, described in Definition 1 and Remark 1). This means, in particular, that not only the arguments in these formalisms have a particular structure (as opposed to abstract argumentation frameworks [Dung, 1995; Baroni et al., 2018], where an abstraction is made of the structure of arguments), but also that their validity can be logically justified. It follows that not all the formalisms under the umbrella of structured argumentation will be considered in this chapter, but only those that are based on specific core logics.

To study the logical instantiations of formal Dung-style argumentation, we first recall, in Section 2.2, three central approaches that correspond to this line of research: logic-based deductive methods [Besnard and Hunter, 2001; Arieli and Straßer, 2015; Besnard and Hunter, 2018], assumption-based argumentation [Bondarenko et al., 1997; Toni, 2014; Čyras et al., 2018] and ASPIC [Prakken, 2010; Modgil and Prakken, 2014; Modgil and Prakken, 2018]. Then, in Section 2.3, we consider the main properties of each approach, in particular: its relation to reasoning with maximal consistency, the rationality postulates that it satisfies, and the inference principles validated by the induced entailment relations. Finally, in Section 2.4, we study relations among these approaches, as well as their relations to other defeasible reasoning methods.

The second part of this chapter describes logic-based methods for representing and reasoning with argumentation dynamics. In this chapter, by 'dynamics' we mean *processes* in the context of a fixed argumentative framework.[1] Basic notions and concepts such as conflict-

---

[1] A similar terminology is sometimes used in the context of revising argumentation

ing arguments, defending arguments, and Dung-style extensions are expressed by logical formulas, and corresponding reasoning processes, based on proof-theoretical methods, are described. The representations are divided between those that are based on propositional languages or their extensions by quantifications (Section 3.1), and those that incorporate modal operators (Section 3.2). The reasoning machinery described in this chapter (Section 3.3) is again one that takes into account the logical relationships among the arguments (although it can be easily adjusted to abstract entities). It can be seen as an extension of Gentzen-type proof calculi [Gentzen, 1934], in which the dynamics of arguments are taken into consideration, and so the proofs are dynamic, in the sense that a derived argument can be retracted in light of more-recently derived counter-arguments [Arieli and Straßer, 2016; Arieli and Straßer, 2019].

We conclude the chapter with some final remarks (Section 4) and proofs of unpublished results (in the appendix). The general structure of this chapter is sketched in Figure 1.

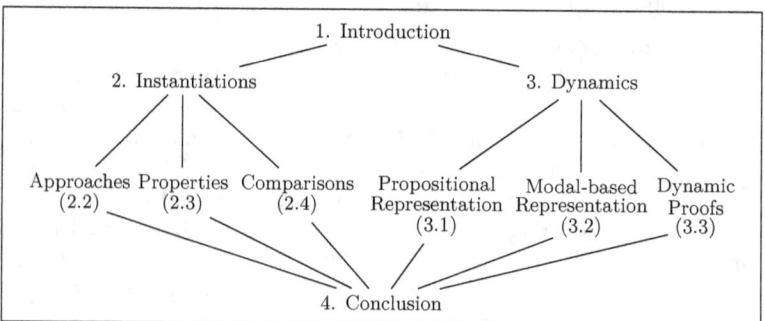

Figure 1: Schematic structure of the chapter

We note, finally, that due to the broad scope of this chapter, some parts of it may be viewed as "second-order" surveys, pointing to other reviews on specific sub-topics of this chapter. In some other parts we

---

frameworks, see also Chapters 8 and 11 in this handbook [Baumann et al., 2021; Alfano et al., 2021].

give more detailed descriptions on specific formalisms. We do so mainly for illustrating our points, but this should not be taken as a preference of one method over the others.

2 Logical Instantiations

The first part of this chapter is devoted to logic-based instantiations of formal argumentation. We describe different approaches to logical argumentation (Section 2.2), consider some of their properties (Section 2.3), and review the (known) relations among them (Section 2.4). First, we recall some common notions and notations.

2.1 Preliminaries

In what follows we shall assume that the underlying language $\mathcal{L}$ is propositional. Sets of formulas are denoted by $\mathcal{S}, \mathcal{T}$, finite sets of formulas are denoted by $\Gamma, \Delta, \Pi, \Theta$, formulas are denoted by $\phi, \psi, \delta, \gamma$, and atomic formulas are denoted by p, q, r, all of which can be primed or indexed. The set of all the atomic formulas of $\mathcal{L}$ is denoted $\mathsf{Atoms}(\mathcal{L})$, and the set of the (well-formed) formulas of $\mathcal{L}$ is denoted $\mathsf{WFF}(\mathcal{L})$.

All the approaches to formal argumentation considered in this chapter assume an underlying logic that forms the basis for specifying arguments and counter-arguments. The next definition is thus at the heart of our study.

Definition 1 (logic). *A (propositional) logic is a pair* $\mathfrak{L} = \langle \mathcal{L}, \vdash \rangle$*, where* $\mathcal{L}$ *is a propositional language, and* $\vdash$ *is a (Tarskian, [Tarski, 1941]) consequence relation for a language* $\mathcal{L}$*, that is: a binary relation between sets of formulas and formulas in* $\mathcal{L}$*, satisfying the following conditions:*

- *Reflexivity: if* $\psi \in \mathcal{S}$ *then* $\mathcal{S} \vdash \psi$.

- *Monotonicity: if* $\mathcal{S} \vdash \psi$ *and* $\mathcal{S} \subseteq \mathcal{S}'$ *then* $\mathcal{S}' \vdash \psi$.

- *Transitivity: if* $\mathcal{S} \vdash \psi$ *and* $\mathcal{S}', \psi \vdash \phi$ *then* $\mathcal{S}, \mathcal{S}' \vdash \phi$.[2]

---

[2] As usual, we use the notation $\mathcal{S}, \mathcal{S}'$ on the left-hand side of the entailment symbol to denote $\mathcal{S} \cup \mathcal{S}'$. In case of singletons we shall usually omit the parenthesis and abbreviate $\mathcal{S} \cup \{\psi\}$ by $\mathcal{S}, \psi$.

In what follows we also assume that a consequence relation satisfies some further standard conditions:

- *Structurality: for every $\mathcal{L}$-substitution θ,[3] if $\mathcal{S} \vdash \psi$ then $\theta(\mathcal{S}) \vdash \theta(\psi)$.*

- *Non-Triviality: $p \nvdash q$ for every two distinct atomic formulas p and q.*

- *Finitariness: if $\mathcal{S} \vdash \psi$, there is a finite set $\Gamma \subseteq \mathcal{S}$ such that $\Gamma \vdash \psi$.*

Structurality means closure under substitutions of formulas. Non-triviality is convenient for excluding trivial logics, and finitariness is often essential for practical reasoning, such as being able to form arguments (based on a finite number of assumptions) for entailments with a possibly infinite number of premises.

To some extent, Definition 1 determines the instantiations covered in Section 2.2 (and the scope of the whole chapter in general): not only that the arguments should have a specific structure (unlike, e.g., arguments in abstract argumentation frameworks that are of a purely abstract nature), but they should be based on (i.e., justified by) some underlying logic as well (see also Definitions 4 and 5).[4] As indicated in Definition 1, in the sequel we shall consider (arbitrary) propositional logics, although most of the formalisms can be easily extended to more generic logics (including first-ordered ones), since the relevant ideas and approaches can be represented at this level.

In what follows we shall assume that the language $\mathcal{L}$ contains at least the following connectives and constant:

a $\vdash$-*negation* $\neg$, satisfying: $p \nvdash \neg p$ and $\neg p \nvdash p$ (for every atomic p),

---

[3]That is, θ is a finite set of pairs $\{(p_1, \psi_1), \ldots (p_n, \psi_n)\}$, where for every $1 \leq i \leq n$, p_i is an atom and ψ_i is an $\mathcal{L}$-formula, such that for every $\mathcal{L}$-formula ϕ, the $\mathcal{L}$-formula $\theta(\phi)$ is obtained from ϕ by replacing in it each occurrence of p_i by ψ_i ($i = 1\ldots, n$). We denote $\theta(\mathcal{S}) = \{\theta(\phi) \mid \phi \in \mathcal{S}\}$.

[4]Note that this means that some approaches to structured argumentation whose underlying formalisms do not meet the conditions of Definition 1 are not covered in Section 2.2, such as defeasible logic programming [García and Simari, 2004] and instances of ASPIC$^+$ where neither strict nor defeasible rules are based on a logic in the sense of Definition 1.

a $\vdash$-*conjunction* $\wedge$, satisfying: $\mathcal{S} \vdash \psi \wedge \phi$ iff $\mathcal{S} \vdash \psi$ and $\mathcal{S} \vdash \phi$,

a $\vdash$-*disjunction* $\vee$, satisfying: $\mathcal{S}, \phi \vee \psi \vdash \sigma$ iff $\mathcal{S}, \phi \vdash \sigma$ and $\mathcal{S}, \psi \vdash \sigma$,

a $\vdash$-*implication* $\supset$, satisfying: $\mathcal{S}, \phi \vdash \psi$ iff $\mathcal{S} \vdash \phi \supset \psi$,

a $\vdash$-*falsity* F, satisfying: $\mathsf{F} \vdash \psi$ for every formula ψ.[5]

In what follows, we shall abbreviate $(\phi \supset \psi) \wedge (\psi \supset \phi)$ by $\phi \leftrightarrow \psi$. For a set of formulas $\mathcal{S}$ we denote $\neg \mathcal{S} = \{\neg \psi \mid \psi \in \mathcal{S}\}$, and for a finite set of formulas Γ we denote by $\bigwedge \Gamma$ (respectively, by $\bigvee \Gamma$) the conjunction (respectively, the disjunction) of all the formulas in Γ. The powerset of $\mathcal{L}$ is denoted by $\wp(\mathcal{L})$. Now,

- We say that an $\mathcal{L}$-formula ψ is a $\vdash$-*theorem*, if $\emptyset \vdash \psi$.

- The $\vdash$-*transitive closure* of a set $\mathcal{S}$ of $\mathcal{L}$-formulas is defined by $Cn_\vdash(\mathcal{S}) = \{\psi \mid \mathcal{S} \vdash \psi\}$.

- We shall say that a set $\mathcal{S}$ is $\vdash$-*consistent* if $\mathcal{S} \not\vdash \mathsf{F}$. In particular, if $\mathcal{S}$ is *not* $\vdash$-consistent (i.e, if it is $\vdash$-inconsistent), it is trivialized with respect to $\vdash$ in the sense that $Cn_\vdash(\mathcal{S})$ consists of every formula in $\mathcal{L}$. Note, in particular, that if $\mathcal{S}$ is $\vdash$-inconsistent, then $\mathcal{S} \vdash \neg \bigwedge \Gamma$ for $\Gamma \subseteq \mathcal{S}$.

When $\vdash$ is clear from the context we will sometimes omit it from the notations above (and say that a formula is a theorem, a set of formulas is consistent, and write $Cn(\mathcal{S})$ for the $\vdash$-transitive closure $\mathcal{S}$).

Remark 2. *To all of the instantiations considered here there are extensions in which the language contains also non-logical components such as priorities among the arguments. As we concentrate on purely logical approaches, these extensions will not be covered in this chapter.*

Definition 3 (explosive/contrapositive logic). *A logic $\mathfrak{L} = \langle \mathcal{L}, \vdash \rangle$ is* explosive, *if for every $\mathcal{L}$-formula ψ the set $\{\psi, \neg \psi\}$ is $\vdash$-inconsistent.*[6] *We say that $\mathfrak{L}$ is* contrapositive, *if (a) $\vdash \neg \mathsf{F}$ and (b) for every nonempty Γ and ψ it holds that $\Gamma \vdash \neg \psi$ iff for every $\phi \in \Gamma$ we have: $\Gamma \setminus \{\phi\}, \psi \vdash \neg \phi$.*

---

[5]In particular, F is not a standard atomic formula, since $\mathsf{F} \vdash \neg \mathsf{F}$.

[6]That is, $\psi, \neg \psi \vdash \mathsf{F}$. Thus, in explosive logics every formula follows from complementary assumptions.

2.2 Central Approaches to Logical Argumentation

In this section we review some central approaches to logical argumentation. Further details about these approaches, related approaches, and relevant references can be found in [Prakken and Vreeswijk, 2002; Besnard et al., 2014; Besnard and Hunter, 2018; Prakken, 2018].

2.2.1 Logic-Based Methods

A. Arguments. Some of the first works on logic-based formal argumentation used classical logic (CL) as the underlying base logic to generate arguments. This indeed is the most common approach in the study and implementation of such argumentation frameworks. To avoid trivial reasoning in such cases, the set of assumptions of an argument (the so-called argument's *support*) is assumed to be consistent and frequently also minimal, in the sense that no proper subset of the argument's support entails the argument's conclusion (see [Besnard and Hunter, 2001; Besnard and Hunter, 2009; Gorogiannis and Hunter, 2011; Besnard and Hunter, 2014; Besnard and Hunter, 2018]). This leads to the following definition:

Definition 4 (classical argument). *A classical argument is a pair* $A = \langle \Gamma, \psi \rangle$, *where* Γ *is a finite set of formulas in the language of* $\{\neg, \vee, \wedge, \supset, \mathsf{F}\}$ *(with their usual bivalent interpretations), such that: (1)* $\Gamma \vdash_{\mathsf{CL}} \psi$ *(namely:* ψ *follows, according to classical logic, from* Γ*), (2)* Γ *is* $\vdash_{\mathsf{CL}}$-*consistent, and (3) for no* $\Gamma' \subsetneq \Gamma$ *it holds that* $\Gamma' \vdash_{\mathsf{CL}} \psi$.

A more general view of arguments (which will be taken here) allows to base arguments on arbitrary logics, and relaxes the two assumptions (consistency and minimality) on their supports (see, e.g, [Arieli and Straßer, 2015; Besnard and Hunter, 2018]):[7]

Definition 5 (argument). *Given a logic* $\mathfrak{L} = \langle \mathcal{L}, \vdash \rangle$, *an* $\mathfrak{L}$-*argument (an argument, for short) is a pair* $A = \langle \Gamma, \psi \rangle$, *where* Γ *is a finite set of* $\mathcal{L}$-*formulas and* ψ *is an* $\mathcal{L}$-*formula, such that* $\Gamma \vdash \psi$. *We denote the set of all* $\mathfrak{L}$-*arguments by* $\mathsf{Arg}_{\mathfrak{L}}$.

---

[7]See, e.g., [Arieli and Straßer, 2020] for a comparison of Definitions 4 and 5.

In what follows, we shall usually denote arguments by A, B, C, etc., possibly primed or indexed. Now:

- Given an argument $A = \langle \Gamma, \psi \rangle$, we shall call Γ the *support set* (or the *premise set*) of A, and ψ the *conclusion* (or the *claim*) of A, denoting them by $\mathsf{Sup}(A)$ and $\mathsf{Conc}(A)$, respectively. For a set S of arguments, we denote: $\mathsf{Sup}(\mathsf{S}) = \bigcup_{A \in \mathsf{S}} \mathsf{Sup}(A)$ and $\mathsf{Conc}(\mathsf{S}) = \{\mathsf{Conc}(A) \mid A \in \mathsf{S}\}$.

- The set of the $\mathfrak{L}$-arguments whose supports are subsets of $\mathcal{S}$ is denoted by $\mathsf{Arg}_{\mathfrak{L}}(\mathcal{S})$. That is: $\mathsf{Arg}_{\mathfrak{L}}(\mathcal{S}) = \{A \in \mathsf{Arg}_{\mathfrak{L}} \mid \mathsf{Sup}(A) \subseteq \mathcal{S}\}$.

- Given an argument $A \in \mathsf{Arg}_{\mathfrak{L}}$, its set of *sub-arguments* is denoted by $\mathsf{Sub}(A)$. That is: $\mathsf{Sub}(A) = \{B \in \mathsf{Arg}_{\mathfrak{L}} \mid \mathsf{Sup}(B) \subseteq \mathsf{Sup}(A)\}$.

Remark 6. *An alternative notation for an argument $\langle \Gamma, \psi \rangle$ is $\Gamma \Rightarrow \psi$ (where $\Rightarrow$ is a new symbol, not appearing in the language of Γ and ψ). The latter resembles the way sequents are denoted in the context of proof theory [Gentzen, 1934]. This notation is frequently used in sequent-based argumentation (see, e.g., [Arieli and Straßer, 2015; Arieli and Straßer, 2019]) to emphasize the fact that the only requirement on Γ and ψ to form an argument is that the latter follows, according to the base logic, from the former.*

B. Attacks. Disagreements between arguments are often described in terms of counter-arguments. It is often said that a counter-argument *attacks* the argument that it challenges.[8] Attacks between arguments are usually described in terms of *attack rules* (with respect to the underlying logic). Table 1 lists some of them. Other attack rules between classical arguments are described e.g. in [Gorogiannis and Hunter, 2011] and [Besnard and Hunter, 2018, Section 5.2]. For a variety of attacks in terms of sequents we refer to [Arieli and Straßer, 2015]. Attack rules incorporating modalities are introduced in [Straßer and Arieli, 2019].

---

[8]Sometimes, mainly when priorities among arguments are introduced, or in the context of specific types of attacks, the term "defeat" is used for "successful attacks".

Rule Name	Acronym	Attacking Argument	Attacked Argument	Attack Conditions
Defeat	Def	$\langle \Gamma_1, \psi_1 \rangle$	$\langle \Gamma_2, \psi_2 \rangle$	$\vdash \psi_1 \supset \neg \bigwedge \Gamma_2$
Direct Defeat	DirDef	$\langle \Gamma_1, \psi_1 \rangle$	$\langle \{\gamma_2\} \cup \Gamma_2', \psi_2 \rangle$	$\vdash \psi_1 \supset \neg \gamma_2$
Undercut	Ucut	$\langle \Gamma_1, \psi_1 \rangle$	$\langle \Gamma_2' \cup \Gamma_2'', \psi_2 \rangle$	$\vdash \psi_1 \leftrightarrow \neg \bigwedge \Gamma_2'$
Canonical Undercut	CanUcut	$\langle \Gamma_1, \psi_1 \rangle$	$\langle \Gamma_2, \psi_2 \rangle$	$\vdash \psi_1 \leftrightarrow \neg \bigwedge \Gamma_2$
Direct Undercut	DirUcut	$\langle \Gamma_1, \psi_1 \rangle$	$\langle \{\gamma_2\} \cup \Gamma_2', \psi_2 \rangle$	$\vdash \psi_1 \leftrightarrow \neg \gamma_2$
Consistency Undercut	ConUcut	$\langle \emptyset, \neg \bigwedge \Gamma_2' \rangle$	$\langle \Gamma_2' \cup \Gamma_2'', \psi_2 \rangle$	
Rebuttal	Reb	$\langle \Gamma_1, \psi_1 \rangle$	$\langle \Gamma_2, \psi_2 \rangle$	$\vdash \psi_1 \leftrightarrow \neg \psi_2$
Defeating Rebuttal	DefReb	$\langle \Gamma_1, \psi_1 \rangle$	$\langle \Gamma_2, \psi_2 \rangle$	$\vdash \psi_1 \supset \neg \psi_2$
Big Argument Attack	BigArgAt	$\langle \Gamma_1, \psi_1 \rangle$	$\langle \{\gamma_2\} \cup \Gamma_2', \psi_2 \rangle$	$\vdash \bigwedge \Gamma_1 \supset \neg \gamma_2$

Table 1: Some attack rules. The support sets of the attacked arguments are assumed to be nonempty (to avoid attacks on theorems).

Rules like those specified in Table 1 form attack schemes that are applied to particular arguments according to the underlying logic. For instance, when classical logic is the underlying formalism, the attacks of $\langle p, p \rangle$ on $\langle \neg p, \neg p \rangle$ and of $\langle \neg p, \neg p \rangle$ on $\langle p \wedge q, p \rangle$[9] are obtained by applications of the Defeat rule (or other rules in the table). When an attack rule $\mathcal{R}$ is applied we shall sometimes say that its attacking argument $\mathcal{R}$-attacks the attacked argument.

Remark 7. *Clearly, the rules in Table 1 are related. The relations among some of the rules for classical arguments are considered in [Gorogiannis and Hunter, 2011] and [Besnard and Hunter, 2018, Section 5.2]. Figure 2 shows that for any base logic as defined in Definition 1 these relations (together with other relations for ConUcut and BigArgAt) hold also for the more general definition of argument (Definition 5). In this figure, an arrow from $\mathcal{R}_1$ to $\mathcal{R}_2$ means that $\mathcal{R}_1 \subseteq \mathcal{R}_2$.*

---

[9] Here and in what follows we omit the set signs when the support of the arguments are singletons.

12 - LOGIC-BASED APPROACHES TO FORMAL ARGUMENTATION

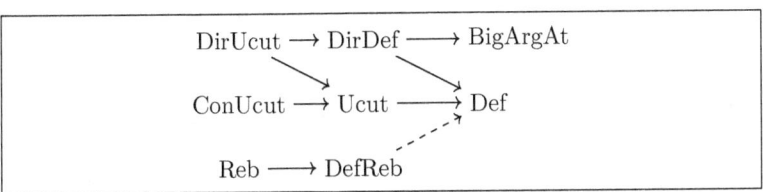

Figure 2: Relations between attack relations from Table 1 (for any base logic). The dashed arrow concerns contrapositive base logics.

C. Argumentation Frameworks. A logical argumentation formalism may be represented as an argumentation framework in the style of Dung [Dung, 1995]. This is defined next.

Definition 8 (logical argumentation framework). *Let $\mathfrak{L} = \langle \mathcal{L}, \vdash \rangle$ be a logic and $\mathcal{A}$ a set of attack rules with respect to $\mathfrak{L}$. Let also $\mathcal{S}$ be a set of $\mathcal{L}$-formulas. The (logical) argumentation framework for $\mathcal{S}$, induced by $\mathfrak{L}$ and $\mathcal{A}$, is the pair $\mathcal{AF}_{\mathfrak{L},\mathcal{A}}(\mathcal{S}) = \langle \mathrm{Arg}_{\mathfrak{L}}(\mathcal{S}), \mathit{Attack}(\mathcal{A}) \rangle$, where $\mathrm{Arg}_{\mathfrak{L}}(\mathcal{S})$ is the set of the $\mathfrak{L}$-arguments whose supports are subsets of $\mathcal{S}$, and $\mathit{Attack}(\mathcal{A})$ is a relation on $\mathrm{Arg}_{\mathfrak{L}}(\mathcal{S}) \times \mathrm{Arg}_{\mathfrak{L}}(\mathcal{S})$, defined by $(A_1, A_2) \in \mathit{Attack}(\mathcal{A})$ iff there is some $\mathcal{R} \in \mathcal{A}$ such that A_1 $\mathcal{R}$-attacks A_2.*

Argumentation frameworks that are induced by classical logic (and some attack rules), and whose arguments are classical (Definition 4), are called *classical (logical) argumentation frameworks*.

In what follows, somewhat abusing the notations, we shall sometimes identify the relation $\mathit{Attack}(\mathcal{A})$ with $\mathcal{A}$. To simplify the notations, we shall also frequently omit the subscripts $\mathfrak{L}$ and $\mathcal{A}$ in $\mathcal{AF}_{\mathfrak{L},\mathcal{A}}(\mathcal{S})$, and just write $\mathcal{AF}(\mathcal{S})$.

Example 9. Let $\mathcal{AF}_{\mathsf{CL}}(\mathcal{S}) = \langle \mathrm{Arg}_{\mathsf{CL}}(\mathcal{S}), \mathit{Attack}(\mathcal{A}) \rangle$ be a logical argumentation framework for the set $\mathcal{S} = \{p, q, \neg p \vee \neg q, r\}$, based on classical logic (CL), and in which $\mathit{Attack}(\mathcal{A})$ is obtained from the attack rules in $\mathcal{A}$, where $\{\mathrm{ConUcut}\} \subseteq \mathcal{A} \subseteq \{\mathrm{DirDef}, \mathrm{DirUcut}, \mathrm{ConUcut}\}$. The following arguments are in $\mathsf{Arg}_{\mathsf{CL}}(\mathcal{S})$:

$A_1 = \langle r, r \rangle$
$A_2 = \langle p, p \rangle$
$A_3 = \langle q, q \rangle$
$A_4 = \langle \neg p \vee \neg q, \neg p \vee \neg q \rangle$
$A_5 = \langle p, \neg((\neg p \vee \neg q) \wedge q) \rangle$
$A_6 = \langle q, \neg((\neg p \vee \neg q) \wedge p) \rangle$
$A_7 = \langle \{p, q\}, p \wedge q \rangle$
$A_8 = \langle \{\neg p \vee \neg q, q\}, \neg p \rangle$
$A_9 = \langle \{\neg p \vee \neg q, p\}, \neg q \rangle$
$A_\top = \langle \emptyset, \neg(p \wedge q \wedge (\neg p \vee \neg q)) \rangle$
$A_\bot = \langle \{p, q, \neg p \vee \neg q\}, \neg r \rangle$

Figure 3 shows a graphical representation of part of the logical argumentation framework with direct defeat and consistency undercut as the attack rules. Here, nodes represent arguments, and directed edges represent attacks (the direction of an edge represents the direction of the attack that it represents).

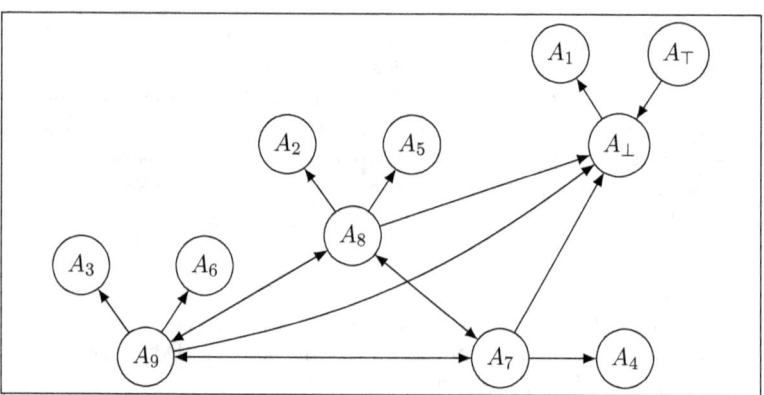

Figure 3: Part of the framework from Example 9.

D. Dung's Semantics. Given an argumentation framework, a key issue in its understanding is the question what combinations of arguments (called *extensions*) can collectively be accepted from this framework. According to Dung [Dung, 1995], this is determined as follows:

Definition 10 (extension-based semantics). *Let* $\mathcal{AF}(\mathcal{S}) = \langle \mathsf{Arg}_{\mathcal{L}}(\mathcal{S}),$ *Attack*$(\mathcal{A}) \rangle$ *be a logical argumentation framework, and let* $\mathcal{E} \cup \{A\} \subseteq$

$\mathsf{Arg}_{\mathfrak{L}}(\mathcal{S})$. Below, maximality and minimality are taken with respect to the subset relation.

- We say that $\mathcal{E}$ attacks an argument A, if there is an argument $B \in \mathcal{E}$ that attacks A (that is, $(B, A) \in \mathsf{Attack}(\mathcal{A})$). The set of arguments in $\mathsf{Arg}_{\mathfrak{L}}(\mathcal{S})$ that are attacked by $\mathcal{E}$ (called the range of $\mathcal{E}$) is denoted $\mathcal{E}^+$.

- We say that $\mathcal{E}$ defends A, if $\mathcal{E}$ attacks every argument in $\mathsf{Arg}_{\mathfrak{L}}(\mathcal{S})$ that attacks A.

- The set $\mathcal{E}$ is called conflict-free with respect to $\mathcal{AF}(\mathcal{S})$, if it does not attack any of its elements (i.e., $\mathcal{E}^+ \cap \mathcal{E} = \emptyset$). A set that is maximally conflict-free with respect to $\mathcal{AF}(\mathcal{S})$ is called a naive extension of $\mathcal{AF}(\mathcal{S})$.

- An admissible extension of $\mathcal{AF}(\mathcal{S})$ is a subset of $\mathsf{Arg}_{\mathfrak{L}}(\mathcal{S})$ that is conflict-free with respect to $\mathcal{AF}(\mathcal{S})$ and defends all of its elements. A complete extension of $\mathcal{AF}(\mathcal{S})$ is an admissible extension of $\mathcal{AF}(\mathcal{S})$ that contains all the arguments that it defends.

- The minimal complete extension of $\mathcal{AF}(\mathcal{S})$ is called the grounded extension of $\mathcal{AF}(\mathcal{S})$ and a maximal complete extension of $\mathcal{AF}(\mathcal{S})$ is called a preferred extension of $\mathcal{AF}(\mathcal{S})$. A complete extension $\mathcal{E}$ of $\mathcal{AF}(\mathcal{S})$ is called a stable extension of $\mathcal{AF}(\mathcal{S})$ if $\mathcal{E} \cup \mathcal{E}^+ = \mathsf{Arg}_{\mathfrak{L}}(\mathcal{S})$.

- We write $\mathsf{Naive}(\mathcal{AF}(\mathcal{S}))$ [respectively: $\mathsf{Adm}(\mathcal{AF}(\mathcal{S}))$, $\mathsf{Cmp}(\mathcal{AF}(\mathcal{S}))$, $\mathsf{Prf}(\mathcal{AF}(\mathcal{S}))$, $\mathsf{Stb}(\mathcal{AF}(\mathcal{S}))$] the set of all the naive [respectively: admissible, complete, preferred, stable] extensions of $\mathcal{AF}(\mathcal{S})$ and $\mathsf{Grd}(\mathcal{AF}(\mathcal{S}))$ for the unique grounded extension of $\mathcal{AF}(\mathcal{S})$.

Remark 11. In [Dung, 1995], preferred extensions are defined as the maximally admissible sets and stable extensions are the conflict-free extensions whose range consists of all the arguments not in the extension. It is well known that these definitions are equivalent to the ones in Definition 10. Furthermore, stable extensions are preferred (but not necessarily vice-versa), and as is shown in [Dung, 1995, Theorem 25], the grounded extension of an argumentation framework is unique. For more properties of the extensions defined above, further references, and other types of extensions, see, e.g., [Baroni and Giacomin, 2009; Baroni et al., 2011; Baroni et al., 2018].

Skeptical and credulous approaches for making inferences from the above-mentioned extensions are defined as follows:

Definition 12 (extension-based entailments). *Let* $\mathcal{AF}(\mathcal{S}) = \langle \mathrm{Arg}_{\mathcal{L}}(\mathcal{S}),$ $\mathrm{Attack}(\mathcal{A}) \rangle$ *be a logical argumentation framework, and let* $\mathsf{Sem} \in \{\mathsf{Naive},$ $\mathsf{Cmp}, \mathsf{Grd}, \mathsf{Prf}, \mathsf{Stb}\}$. *We denote:*

- $\mathcal{S} \mid\!\sim^{\mathcal{L},\mathcal{A}}_{\mathsf{Grd}} \psi$ *if there is an argument* $\langle \Gamma, \psi \rangle \in \mathsf{Grd}(\mathcal{AF}_{\mathcal{L},\mathcal{A}}(\mathcal{S}))$,[10] [11]

- $\mathcal{S} \mid\!\sim^{\mathcal{L},\mathcal{A}}_{\cup\mathsf{Sem}} \psi$ *if there is an argument* $\langle \Gamma, \psi \rangle \in \bigcup \mathsf{Sem}(\mathcal{AF}_{\mathcal{L},\mathcal{A}}(\mathcal{S}))$,

- $\mathcal{S} \mid\!\sim^{\mathcal{L},\mathcal{A}}_{\cap\mathsf{Sem}} \psi$ *if there is an argument* $\langle \Gamma, \psi \rangle \in \bigcap \mathsf{Sem}(\mathcal{AF}_{\mathcal{L},\mathcal{A}}(\mathcal{S}))$,

- $\mathcal{S} \mid\!\sim^{\mathcal{L},\mathcal{A}}_{\Cap\mathsf{Sem}} \psi$ *if for every* $\mathcal{E} \in \mathsf{Sem}(\mathcal{AF}_{\mathcal{L},\mathcal{A}}(\mathcal{S}))$ *there is an argument* $\langle \Gamma, \psi \rangle \in \mathcal{E}$.

Example 13. *Consider again the argumentation framework* $\mathcal{AF}_{\mathsf{CL}}(\mathcal{S})$ *from Example 9, where* $\mathcal{S} = \{r, p, q, \neg p \lor \neg q\}$. *In the notations of that example (see also Figure 3), the grounded extension of* $\mathcal{AF}_{\mathsf{CL}}(\mathcal{S})$ *is* $\mathrm{Arg}_{\mathsf{CL}}(\{A_\top, A_1\})$, *and the naive/preferred/stable extensions on* $\mathcal{AF}_{\mathsf{CL}}(\mathcal{S})$ *are* $\mathrm{Arg}_{\mathsf{CL}}(\mathcal{E}_i)$ $(i \in \{1,2,3\})$, *where:*

- $\mathcal{E}_1 = \{A_\top, A_1, A_2, A_3, A_5, A_6, A_7\}$,
- $\mathcal{E}_2 = \{A_\top, A_1, A_3, A_4, A_6, A_8\}$,
- $\mathcal{E}_3 = \{A_\top, A_1, A_2, A_4, A_5, A_9\}$.

It follows that for every entailment $\mid\!\sim$ *considered in Definition 12 we have that* $\mathcal{S} \mid\!\sim r$. *The other formulas in* $\mathcal{S}$ *can only be credulously inferred: for every* $\psi \in \mathcal{S} - \{r\}$ *and* $\mathsf{Sem} \in \{\mathsf{Naive}, \mathsf{Prf}, \mathsf{Stb}\}$ *we have that* $\mathcal{S} \mid\!\sim_{\cup\mathsf{Sem}} \psi$, *but* $\mathcal{S} \not\mid\!\sim_{\cap\mathsf{Sem}} \psi$, $\mathcal{S} \not\mid\!\sim_{\Cap\mathsf{Sem}} \psi$, *and* $\mathcal{S} \not\mid\!\sim_{\mathsf{Grd}} \psi$. *Note, moreover,*

---

[10]We make a distinction between the grounded semantics and the other types of semantics, since unlike the other types, the grounded extension is unique (recall Remark 11).

[11]Recall that by the definition of $\mathsf{Grd}(\mathcal{AF}_{\mathcal{L},\mathcal{A}}(\mathcal{S}))$ it holds that $\Gamma \subseteq \mathcal{S}$. The same note holds for the other items in this definition.

that for instance $\mathcal{S} \mathrel{|\!\!\sim}_{\cap\mathsf{Sem}} p \vee q$ (but $\mathcal{S} \mathrel{|\!\!\not\sim}_{\cap\mathsf{Sem}} p \vee q$), since at least one of p or q (but not both) follows from each preferred/stable extension, from which $p \vee q$ is inferred.

The next example, taken from [Straßer and Arieli, 2019], demonstrates the usefulness of incorporating modalities for having logic-based argumentative approaches to normative reasoning.

Example 14. *Consider the following example by Horty [Horty, 1994]:*

When a meal is served (m), one should not eat with fingers (f). However, if the meal is asparagus (a), one should eat with fingers.

This scenario may be represented by the deontic logic SDL *(standard deontic logic, i.e., the normal modal logic* KD*), where the modal operator* O *intuitively represents obligations. In this setting, the statements above may be expressed, respectively, by the formulas $m \supset \mathsf{O}\neg f$ and $(m \wedge a) \supset \mathsf{O}f$. Now, in case that asparagus is indeed served $(m \wedge a)$ one expects to derive the (unconditional) obligation to eat with fingers $(\mathsf{O}f)$ rather than not to eat with fingers $(\mathsf{O}\neg f)$.*

This is a paradigmatic case of specificity: *a more specific obligation cancels (or overrides) a less specific obligation. An attack rule that reflects this intuition may be expressed as follows:*

Specificity Undercut (SpecUcut):
$\langle \Gamma \cup \{\phi \supset \mathsf{O}\psi\}, \neg(\phi' \supset \mathsf{O}\psi') \rangle$ attacks $\langle \Gamma' \cup \{\phi' \supset \mathsf{O}\psi'\}, \sigma \rangle$ *if the following conditions are met: (i) $\Gamma \vdash \phi$, (ii) $\phi \vdash \phi'$, and (iii) $\psi \vdash \neg\psi'$.*

Condition (i) expresses that the conditional $\phi \supset \mathsf{O}\psi$ is 'triggered' in view of Γ, Condition (ii) expresses that ϕ is logically at least as strong as ϕ' (i.e., the former is more specific than the latter), and Condition (iii) indicates that the conditionals have conflicting conclusions (after filtering the modalities).

We thus consider an argumentation framework that is based on the following set:

$$\mathcal{S} = \{m,\ a,\ m \supset \mathsf{O}\neg f,\ (m \wedge a) \supset \mathsf{O}f\}.$$

Some arguments in $\mathsf{Args}_{\mathsf{SDL}}(\mathcal{S})$ are listed in Figure 4 (right). Figure 4 (left) shows an attack diagram where the sole attack rule is SpecUcut.

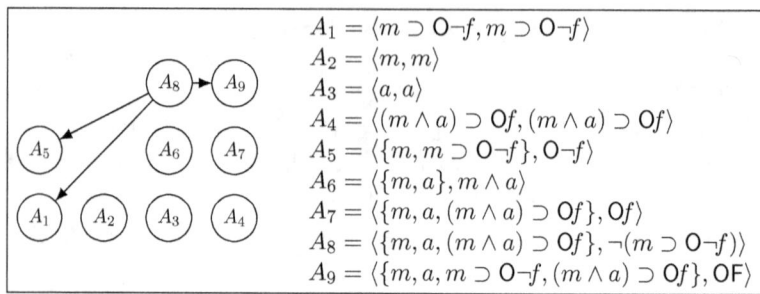

Figure 4: (Part of) the normative argumentation framework of Example 14.

It follows that we have the following expected deductions for every entailment $\mid\sim$ in Definition 12:

- $\mathcal{S} \not\mid\sim \mathsf{O}\neg f$. Indeed, one cannot derive $\mathsf{O}\neg f$, since the application of Modus Ponens to $m \supset \mathsf{O}\neg f$ (depicted by argument A_5) gets attacked by A_8.

- $\mathcal{S} \mid\sim \mathsf{O}f$. Indeed, A_7 is not attacked by an argument in $\mathsf{Arg}_{\mathsf{SDL}}(\mathcal{S})$, thus it is part of every grounded, preferred, and stable extension of the underlying normative argumentation framework, and so its descendant follows from $\mathcal{S}$. (Note that A_7 is attacked by SDL-derivable arguments, but none of them is in $\mathsf{Arg}_{\mathsf{SDL}}(\mathcal{S})$).

We refer to [Straßer and Arieli, 2019] for further examples of well-known puzzles, treated by SDL-based argumentation frameworks.

Remark 15. *Clearly, whenever a framework $\mathcal{AF}_{\mathfrak{L},\mathcal{A}}(\mathcal{S})$ has Sem-extensions, it holds that if $\mathcal{S} \mid\sim^{\mathfrak{L},\mathcal{A}}_{\mathsf{nSem}} \psi$ then $\mathcal{S} \mid\sim^{\mathfrak{L},\mathcal{A}}_{\mathsf{mSem}} \psi$. Also, if $\mathcal{S} \mid\sim^{\mathfrak{L},\mathcal{A}}_{\mathsf{mSem}} \psi$ then $\mathcal{S} \mid\sim^{\mathfrak{L},\mathcal{A}}_{\mathsf{USem}} \psi$ (thus both types of skeptical reasoning entail credulous reasoning). The converses, however, do not hold. Example 13 shows that for $\mathsf{Sem} \in \{\mathsf{Prf}, \mathsf{Stb}\}$, $\mid\sim^{\mathfrak{L},\mathcal{A}}_{\mathsf{USem}} \not\subseteq \mid\sim^{\mathfrak{L},\mathcal{A}}_{\mathsf{nSem}}$, and $\mid\sim^{\mathfrak{L},\mathcal{A}}_{\mathsf{USem}} \not\subseteq \mid\sim^{\mathfrak{L},\mathcal{A}}_{\mathsf{mSem}}$, and $\mid\sim^{\mathfrak{L},\mathcal{A}}_{\mathsf{mSem}} \not\subseteq \mid\sim^{\mathfrak{L},\mathcal{A}}_{\mathsf{nSem}}$. To see another example for the latter, consider the logical argumentation framework $\mathcal{AF}_{\mathfrak{L},\mathcal{A}}(\mathcal{S}')$, where $\mathcal{S}' = \{p \wedge q, p \wedge \neg q\}$, $\mathfrak{L} = \mathsf{CL}$, and $\mathcal{A} = \{\mathsf{Ucut}\}$. Then $\mathcal{S}' \mid\sim^{\mathfrak{L},\mathcal{A}}_{\mathsf{mSem}} p$ but $\mathcal{S}' \not\mid\sim^{\mathfrak{L},\mathcal{A}}_{\mathsf{nSem}} p$ (because $\bigcap \mathsf{Sem}(\mathcal{AF}_{\mathfrak{L},\mathcal{A}}(\mathcal{S}'))$ consists only of tautological arguments, i.e., those with empty support sets).*

Proposition 16. *Let $\mathcal{AF}(\mathcal{S})$ be a logical argumentation framework for a finite $\mathcal{S}$, based on a contrapositive logic $\mathfrak{L}$ and the set $\mathcal{A} = \{DirUcut, ConUcut\}$. Then:*

1. $\mathcal{S} \hspace{0.1em}\vert\!\sim_{\mathsf{Grd}}^{\mathfrak{L},\mathcal{A}} \psi$ *iff* $\mathcal{S} \hspace{0.1em}\vert\!\sim_{\cap\mathsf{Prf}}^{\mathfrak{L},\mathcal{A}} \psi$ *iff* $\mathcal{S} \hspace{0.1em}\vert\!\sim_{\cap\mathsf{Stb}}^{\mathfrak{L},\mathcal{A}} \psi$.

2. $\mathcal{S} \hspace{0.1em}\vert\!\sim_{\mathsf{UPrf}}^{\mathfrak{L},\mathcal{A}} \psi$ *iff* $\mathcal{S} \hspace{0.1em}\vert\!\sim_{\mathsf{UStb}}^{\mathfrak{L},\mathcal{A}} \psi$.

3. $\mathcal{S} \hspace{0.1em}\vert\!\sim_{\cap\mathsf{Prf}}^{\mathfrak{L},\mathcal{A}} \psi$ *iff* $\mathcal{S} \hspace{0.1em}\vert\!\sim_{\cap\mathsf{Stb}}^{\mathfrak{L},\mathcal{A}} \psi$.

The above proposition is shown in [Arieli et al., 2019], and some variations of it are proved in [Arieli et al., 2018]. As mentioned there, the assumptions on the logic and the attack rules are essential for the proposition to hold.

2.2.2 The ASPIC System

ASPIC$^+$ [Prakken, 2010; Modgil and Prakken, 2013] is another well-known approach to structured argumentation, based on some underlying logic. It contains (at least) two types of premises: axioms (which cannot be questioned) and ordinary premises (which can be questioned/attacked). Also, there are two types of rules: strict and defeasible. The latter, unlike strict rules, allow for exceptions. A wide variety of research has been done on ASPIC$^+$, both from a theoretical perspective (e.g., rationality postulates were introduced in [Caminada and Amgoud, 2007] for ASPIC, an earlier version of ASPIC$^+$, and the use of preferences has been investigated in [Modgil and Prakken, 2013]) and from an application perspective (See [Modgil and Prakken, 2018, Section 6] for an overview). We refer to [Modgil and Prakken, 2014; Modgil and Prakken, 2018] for extensive surveys on ASPIC$^+$ and related approaches. Unless otherwise stated, the definitions in this section are taken from [Modgil and Prakken, 2018] (the chapter on ASPIC$^+$ in the first volume of the handbook).

Remark 17. *As noted in Remark 2, we only discuss purely logical instances of logical argumentation frameworks. For ASPIC$^+$ this means that we do not take into account any ordering over the defeasible elements.*

Definition 18 (ASPIC-based argumentation system). *An argumentation system is a tuple* $AS = \langle \mathcal{L}, ^-, \mathcal{R}, n \rangle$, *where:*

- $\mathcal{L}$ *is a propositional language,*
- $^-$ *is a contrariness function from* $\mathcal{L}$ *to* $2^{\mathcal{L}} \setminus \emptyset$, [12]
- $\mathcal{R} = \langle \mathcal{R}_s, \mathcal{R}_d \rangle$ *consists of strict* $(\mathcal{R}_s)$ *and defeasible* $(\mathcal{R}_d)$ *inference rules of the form* $\phi_1, \ldots, \phi_n \to \phi$ *and* $\phi_1, \ldots, \phi_n \Rightarrow \phi$ *respectively, such that* $\mathcal{R}_s \cap \mathcal{R}_d = \emptyset$,
- $n : \mathcal{R}_d \to \mathsf{WFF}(\mathcal{L})$ *is a (possibly partial) function assigning names to defeasible rules.*

The contrariness function allows to specify conflicts between elements of the language. Strict rules are deductive in the sense that the truth of their premises $\phi_1, \ldots, \phi_n$ necessarily implies the truth of their antecedent ϕ. Unlike strict rules, a defeasible rule warrants the truth of its conclusion only provisionally: its application can be retracted in case counter-arguments are encountered. A naming function associates a name $n(r)$ with some of the defeasible rules in $\mathcal{R}_d$. This will facilitate the formulation of the attack form *undercut* (see below).

Definition 19 (ASPIC theory). *A knowledge-base in an argumentation system* $AS = \langle \mathcal{L}, ^-, \mathcal{R}, n \rangle$ *is a pair* $\mathcal{K} = \langle \mathcal{K}_n, \mathcal{K}_p \rangle$ *of* $\mathcal{L}$-*formulas that consists of two disjoint sets:* $\mathcal{K}_n$ *(the axioms) and* $\mathcal{K}_p$ *(the ordinary premises). An ASPIC argumentation theory is a pair* $AT = \langle AS, \mathcal{K} \rangle$, *where AS is an argumentation system and* $\mathcal{K}$ *is a knowledge-base in AS.*

Arguments in ASPIC$^+$ differ from arguments in logic-based argumentation frameworks. These are *inference trees* that are constructed from the rules of the argumentation system and the formulas in the knowledge base:

Definition 20 (ASPIC argument). *An ASPIC-argument A on the basis of an ASPIC-theory AT is of one of the following forms:*

---

[12]In many publications, a distinction is made between *contraries* and *contradictories*. This distinction mainly plays a role when preferences over defeasible rules are taken into account and therefore is left out of this survey.

1. ϕ, if $\phi \in \mathcal{K}_n \cup \mathcal{K}_p$. In this case we denote:
   Prem$(A) = \{\phi\}$;
   Conc$(A) = \phi$;
   Sub$(A) = \{\phi\}$;
   Rules$(A) =$ DefRules$(A) =$ TopRules$(A) = \emptyset$.

2. $A_1, \ldots, A_n \to \psi$, if $A_1, \ldots, A_n$ are ASPIC-arguments such that there exists a strict rule of the form Conc$(A_1), \ldots,$ Conc$(A_n) \to \psi$ in $\mathcal{R}_s$. In this case we denote:
   Prem$(A) =$ Prem$(A_1) \cup \ldots \cup$ Prem(A_n);
   Conc$(A) = \psi$;
   Sub$(A) =$ Sub$(A_1) \cup \ldots \cup$ Sub$(A_n) \cup \{A\}$;
   Rules$(A) =$ Rules$(A_1) \cup \ldots \cup$ Rules$(A_n) \cup \{$Conc$(A_1), \ldots,$ Conc$(A_n) \to \psi\}$;
   TopRules$(A) = \bigcup_{B \in \mathsf{Sub}(A)}$ TopRules(B);
   DefRules$(A) = \{r \in \mathcal{R}_d \mid r \in$ Rules$(A)\}$.

3. $A_1, \ldots, A_n \Rightarrow \psi$, if $A_1, \ldots, A_n$ are ASPIC-arguments such that there exists a defeasible rule of the form Conc$(A_1), \ldots,$ Conc$(A_n) \Rightarrow \psi$ in $\mathcal{R}_d$. In this case we denote:
   Prem$(A) =$ Prem$(A_1) \cup \ldots \cup$ Prem(A_n);
   Conc$(A) = \psi$;
   Sub$(A) =$ Sub$(A_1) \cup \ldots \cup$ Sub$(A_n) \cup \{A\}$;
   Rules$(A) =$ Rules$(A_1) \cup \ldots \cup$ Rules$(A_n) \cup \{$Conc$(A_1), \ldots,$ Conc$(A_n) \Rightarrow \psi\}$;
   TopRules$(A) = \{$Conc$(A_1), \ldots,$ Conc$(A_n) \Rightarrow \psi\}$;
   DefRules$(A) = \{r \in \mathcal{R}_d \mid r \in$ Rules$(A)\}$.

We denote the set of arguments that can be constructed from an argumentation theory $AT = \langle AS, \mathcal{K} \rangle$ by Arg(AT).

Example 21. Let $AS = \langle \mathcal{L}, ^-, \mathcal{R}, n \rangle$ be an argumentation system, where $\mathcal{L}$ is a standard propositional language with Atoms$(\mathcal{L}) = \{p, q, r, n(r_1)\}$, $\overline{\phi} = \{\psi \mid \psi \equiv \neg \phi\}$ for any $\mathcal{L}$-formula ϕ, the rules in $\mathcal{R}_s$ coincide with those of classical logic in the sense that $\phi_1, \ldots, \phi_n \to \phi \in \mathcal{R}_s$ iff $\{\phi_1, \ldots, \phi_n\} \vdash_{\mathsf{CL}} \phi$ for $\mathcal{L}$-formulas $\phi_1, \ldots, \phi_n, \phi$, and

$$\mathcal{R}_d = \{r_1 : p \Rightarrow \neg q; \; r_2 : q \Rightarrow \neg n(r_1)\}, \quad \mathcal{K}_p = \{p, q, r\}, \quad \mathcal{K}_n = \emptyset$$

Among others, the following ASPIC-arguments can be constructed:

$A_1 : r$ $\qquad A_4 : A_2 \Rightarrow \neg q$ $\qquad A_7 : A_2, A_4 \to p \wedge \neg q$
$A_2 : p$ $\qquad A_5 : A_3 \Rightarrow \neg n(r_1)$ $\qquad A_8 : A_3, A_4 \to \neg r$
$A_3 : q$ $\qquad A_6 : A_2, A_3 \to p \wedge q$ $\qquad A_9 : A_3, A_4 \to \neg p$

In ASPIC⁺ arguments can be attacked on their defeasible rules (undercut), on conclusions of sub-arguments whose top-rule is defeasible (rebuttal) and on their ordinary premises (undermine attack):

Definition 22 (ASPIC-attack). *An ASPIC-argument A attacks an ASPIC-argument B iff A undercuts, rebuts or undermines B, where:*

- *A undercuts B (on B') iff*

$$\mathsf{Conc}(A) \in \overline{n(\mathsf{Conc}(B_1),\ldots,\mathsf{Conc}(B_n) \Rightarrow \phi)}$$

for some $B' \in \mathsf{Sub}(B)$ of the form $B_1,\ldots,B_n \Rightarrow \phi$;

- *A rebuts B (on B') iff $\mathsf{Conc}(A) \in \overline{\phi}$ for some $B' \in \mathsf{Sub}(B)$ of the form $B''_1,\ldots,B''_n \Rightarrow \phi$.*

- *A undermines B (on B') iff $\mathsf{Conc}(A) \in \overline{\phi}$ for some $B' = \phi$, for some $\phi \in \mathsf{Prem}(B) \cap \mathcal{K}_p$.*

Remark 23. *Note that attacks in ASPIC⁺ always target defeasible elements of the attacked argument: undercuts attack a defeasible rule (for this the naming function was instrumental), rebuts always attack in the head of a defeasible rule, and undermining always targets defeasible premises. Also note the difference in terminology to logic-based argumentation: the* undercut *attack in the context of ASPIC⁺ is quite different from the undercut attack for logic-based argumentation (see Table 1). The latter resembles more* undermining*-attacks in the context of ASPIC⁺.*

Now, Dung-style argumentation frameworks are defined in ASPIC⁺ as follows:

Definition 24 (ASPIC argumentation framework). *Let $AT = \langle AS, \mathcal{K} \rangle$ be an ASPIC argumentation theory. An (ASPIC) argumentation framework, defined by AT, is a pair $\mathcal{AF}(AT) = \langle \mathsf{Arg}(AT), \mathsf{Attack} \rangle$, where:*

- Arg(AT) *is the set of ASPIC-arguments constructed from AT, as in Definition 20; and*

- $(X, Y) \in$ Attack *iff X attacks Y, as in Definition 22.*[13]

Example 25 (Example 21 continued). *In the argumentation theory from Example 21, we have that:*

- A_5 *undercuts* A_4, A_7, A_8 *and* A_9 *(all of them on A_4),*

- A_4 *undermines* A_3, A_5, A_6, A_8 *and* A_9 *(all on A_3),*

- A_3 *rebuts* A_4, A_7, A_8 *and* A_9 *(all on A_4).*

There are more attacks between $A_1, \ldots, A_9$ besides the ones listed here: the full attack relation between these arguments is shown in Figure 5.

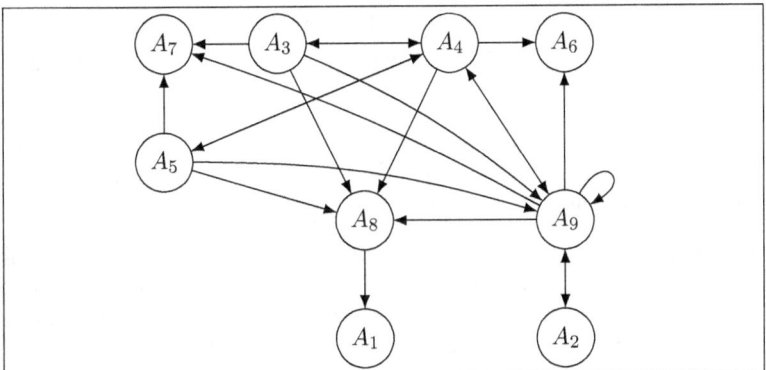

Figure 5: Part of the framework from Example 25.

Dung-style semantics, as defined in Definition 10, can now be applied to the frameworks defined above as well. For example, given $\mathcal{AF}(\text{AT}) = \langle \text{Arg}(\text{AT}), \text{Attack} \rangle$, $\mathcal{E} \subseteq \text{Arg}(\text{AT})$ is an admissible extension of $\mathcal{AF}(\text{AT})$ if it is conflict-free with respect to $\mathcal{AF}(\text{AT})$ and defends all of its elements. Similarly, $\mathcal{E}$ is a complete extension of $\mathcal{AF}(\text{AT})$ if it

---

[13]Note that, unlike logic-based argumentation, where frameworks may differ in their attack rules, in ASPIC systems always *all* the possible attack rules are applied.

is an admissible extension of $\mathcal{AF}(\mathrm{AT})$ that contains all the arguments it defends. Like before, we will denote by $\mathsf{Sem}(\mathcal{AF}(\mathrm{AT}))$ all the Sem-extensions of $\mathcal{AF}(\mathrm{AT})$, for $\mathsf{Sem} \in \{\mathsf{Naive}, \mathsf{Adm}, \mathsf{Cmp}, \mathsf{Grd}, \mathsf{Prf}, \mathsf{Stb}\}$.

The next definition is a counterpart, for the ASPIC$^+$ system, of Definition 12:

Definition 26 (ASPIC extension-based entailments). *Let $\mathcal{AF}(AT) = \langle \mathrm{Arg}(AT), \mathit{Attack} \rangle$ be an argumentation framework for some argumentation theory AT and let $\mathsf{Sem} \in \{\mathsf{Grd}, \mathsf{Cmp}, \mathsf{Prf}, \mathsf{Stb}, \mathsf{Naive}\}$. Then:*

- *$AT \mathrel{\|\!\sim}_{\cup\mathsf{Sem}} \psi$ if there is an argument $A \in \bigcup \mathsf{Sem}(\mathcal{AF}(AT))$ with $\mathsf{Conc}(A) = \psi$. In this case it is said that ψ is credulously justified;*

- *$AT \mathrel{\|\!\sim}_{\cap\mathsf{Sem}} \psi$ if there is an argument $A \in \bigcap \mathsf{Sem}(\mathcal{AF}(AT))$ with $\mathsf{Conc}(A) = \psi$. In this case it is said that ψ is skeptically justified;*

- *$AT \mathrel{\|\!\sim}_{\cap\!\!\!\cap\mathsf{Sem}} \psi$ if for every $\mathcal{E} \in \mathsf{Sem}(\mathcal{AF}(AT))$ there is an argument $A \in \mathcal{E}$ with $\mathsf{Conc}(A) = \psi$. In this case it is said that ψ is weakly skeptically justified.*

As any Dung-style argumentation framework has a single grounded extension, the entailments $\mathrel{\|\!\sim}_{\cap\mathsf{Grd}}$, $\mathrel{\|\!\sim}_{\cup\mathsf{Grd}}$ and $\mathrel{\|\!\sim}_{\cap\!\!\!\cap\mathsf{Grd}}$ coincide, we will therefore sometimes omit the initial symbol from the subscript.

Remark 27. *Unlike standard consequence relations (Definition 1) and the extension-based entailments for the logic-based approach (Definition 12), which are relations between sets of formulas and formulas, the entailments above are relations between argumentation theories and formulas. This will not cause any confusion in what follows.*

Example 28 (Example 25 continued). *In the argumentation framework from Example 25 shown in Figure 5, for the ASPIC argumentation theory AT from Example 21, we have that $\mathsf{Grd}(\mathcal{AF}(AT)) = \emptyset$.[14] It is easy to see that there are two preferred extensions for this framework: one contains (among others) the arguments A_1, A_2, A_4 and A_7 and the other contains (among others) A_1, A_2, A_3, A_5 and A_6. Therefore, the following conclusions can be derived for $\mathsf{Sem} = \mathsf{Prf}$:*

---

[14]Recall that we identify $\mathsf{Grd}(\mathcal{AF}(\mathrm{AT}))$ with its single set.

- $AT \hspace{0.5mm}\mid\hspace{-1mm}\sim_{\cap Prf} \phi$ iff $\phi \in Cn(r \wedge p)$, since A_1 and A_2 occur in each preferred extension;

- $AT \hspace{0.5mm}\mid\hspace{-1mm}\sim_{\cap Prf} \neg q \vee (\neg n(r_1) \wedge q)$ since A_4 occurs in one preferred extension and A_5 and A_3 in the other preferred extension;

- $AT \hspace{0.5mm}\mid\hspace{-1mm}\sim_{\cup Prf} \phi$ for $\phi \in \{p, \neg q, q\}$ (among others), since each of the arguments besides A_8 and A_9 from Example 21 is part of at least one preferred extension.

Remark 29. *A similar result as that of Proposition 16 in the previous section is not available for ASPIC systems, since in the presence of odd attack cycles some preferred extensions may not attack all arguments in their complement (and therefore might not be stable). This can also lead to settings in which no stable extension exist. This is demonstrated in the next example.*

Example 30. *As in our previous example, let $\mathcal{R}_s$ be instantiated by classical logic. Let also $\overline{\phi} = \{\neg \phi\}$ for every formula ϕ, $\mathcal{K} = \langle \emptyset, \emptyset \rangle$, and let $\mathcal{R}_d$ consist of the following three rules: $r_1 : \Rightarrow \neg n(r_2)$, $r_2 : \Rightarrow \neg n(r_3)$, $r_3 : \Rightarrow \neg n(r_1)$. Note that, for instance, the arguments*

$$A_1 : \Rightarrow \neg n(r_2), \quad A_2 : \Rightarrow \neg n(r_3), \quad A_3 : \Rightarrow \neg n(r_1)$$

are involved in an odd attack cycle (of length 3). As a consequence, neither of the three arguments can be part of an admissible extension. Thus, the only preferred extension will consist of all strict arguments (which conclude classical theorems). Clearly, this extension will not be able to attack the three arguments above, and thus it is not stable.

We note, nevertheless, that there are instances of ASPIC$^+$ for which a similar result to that of Proposition 16 is available. This is especially the case when ASPIC$^+$ is instantiated by a contrapositive strict rule base, when the contrariness operator is defined by the negation of the language and no undercutting arguments can be generated from the knowledge base. See further discussions in Sections 2.3.1 and 2.4.

2.2.3 Assumption-Based Argumentation

Assumption-based argumentation (ABA, [Bondarenko et al., 1997]) is another prominent formalism for logical argumentation. It was introduced in the 1990s as a computational framework to capture and generalize default and defeasible reasoning, inspired by Dung's semantics for abstract argumentation and by logic programming with its dialectical interpretation of the acceptability of negation-as-failure assumptions based on "no-evidence-to-the-contrary". In this section we recall the basic definitions that are related to this approach. For extensive surveys on ABA and related approaches, we refer to [Dung et al., 2009; Toni, 2014; Čyras et al., 2017; Čyras et al., 2018]. ABA-based implementations are surveyed in [Cerutti et al., 2018, Section 3.2].

Definition 31 (assumption-based framework). *An* assumption-based framework *(in short: ABF) is a tuple* $\mathcal{ABF} = \langle \mathcal{L}, \mathcal{R}, \mathcal{A}, \sim \rangle$ *where:*

- $\mathcal{L}$ *is a (propositional) language,*

- $\mathcal{R}$ *is a set of* strict rules, *whose elements are of the form* $\psi_1, \ldots, \psi_n \to \psi$, *where* ψ, ψ_i ($1 \leq i \leq n$) *are* $\mathcal{L}$*-formulas,*

- $\mathcal{A}$ *is a nonempty set of* $\mathcal{L}$*-formulas, called the* defeasible (or candidate) assumptions, *and*

- $\sim\, : \mathcal{A} \to \wp(\mathcal{L})$ *is a* contrariness operator, *assigning a finite set of* $\mathcal{L}$*-formulas to every defeasible assumption in* $\mathcal{L}$.[15]

Somewhat like the rules in ASPIC, rules in ABFs can be chained to form *deductions*. Given a set $\mathcal{S} \subseteq \mathcal{A}$ of defeasible assumptions, an $\mathcal{S}$-based deduction may be viewed as a proof, i.e., a sequence of $\mathcal{L}$-formulas, where each element of the sequence is either a formula in $\mathcal{S}$ or is obtained from previous elements in the sequence by an application of a rule $\mathcal{R}$, just like an application of Modus Ponens.

Definition 32 ($\vdash_\mathcal{R}$). *Let* $\mathcal{R}$ *be a set of inference rules over* $\mathcal{L}$. *We write* $\mathcal{S} \vdash_\mathcal{R} \psi$ *if there is an* $\mathcal{S}$*-deduction, based on the rules in* $\mathcal{R}$, *that*

---

[15] Note that the contrariness operator is not a connective of $\mathcal{L}$, as it is restricted only to the candidate assumptions.

culminates in ψ, i.e., there is a sequence $\phi_1,\ldots,\phi_n$ of $\mathcal{L}$-formulas such that $\phi_n = \psi$ and for each $1 \leq i \leq n$, $\phi_i \in \mathcal{S}$ or there are $\phi_{i_1},\ldots,\phi_{i_m}$ for which $i_1,\ldots,i_m < i$ and $\phi_{i_1},\ldots,\phi_{i_m} \to \phi_i \in \mathcal{R}$.

For instance, if $p \to q \in \mathcal{R}$, then $p \vdash_\mathcal{R} q$.

As in logic-based argumentation and ASPIC, (defeasible) assertions in an ABF may be attacked in the presence of counter (defeasible) information. This is described in the next definition.

Definition 33 (attacks in ABFs). *Let $\mathcal{ABF} = \langle \text{Atoms}(\mathcal{L}), \mathcal{R}, \mathcal{A}, \sim \rangle$ be an assumption-based framework, and let $\mathcal{S}, \mathcal{T} \subseteq \mathcal{A}$, $\psi \in \mathcal{A}$. We say that $\mathcal{S}$ attacks ψ if there are $\mathcal{S}' \subseteq \mathcal{S}$ and $\phi \in {\sim}\psi$ such that $\mathcal{S}' \vdash_\mathcal{R} \phi$. Accordingly, $\mathcal{S}$ attacks $\mathcal{T}$ if $\mathcal{S}$ attacks some $\psi \in \mathcal{T}$.*

Remark 34. *In contrast to most of the logical argumentation frameworks defined in the preceding sections (as well as other approaches to structured argumentation, such as DeLP [García and Simari, 2004]), in which attacks are defined between individual arguments, in ABA systems attacks are defined between sets of assumptions. This may be viewed as a higher level of abstraction, operating on equivalence classes that consist of arguments generated from the same assumptions.*

Using the above notion of attack, Dung-style semantics is defined on ABFs just as in Definition 10. The only difference is that an extension $\mathcal{E}$ in an ABF is required to be *closed* with respect to the rules in $\mathcal{R}$, namely: $\mathcal{E} = Cn_{\vdash_\mathcal{R}}(\mathcal{E}) \cap \mathcal{A}$. Thus, for instance, for $\mathcal{S} \subseteq \mathcal{A}$ we say that

- $\mathcal{S}$ is *conflict-free* (with respect to $\mathcal{ABF}$) iff $\mathcal{S}$ does not attack itself.

- $\mathcal{S}$ *defends* (with respect to $\mathcal{ABF}$) a set $\mathcal{S}' \subseteq \mathcal{A}$ iff for every closed set $\mathcal{S}^\star$ that attacks $\mathcal{S}'$, $\mathcal{S}$ attacks $\mathcal{S}^\star$.

- $\mathcal{S}$ is *admissible* (with respect to $\mathcal{ABF}$) iff it is closed, conflict-free, and defends itself. An admissible set is called *complete*, if it does not defend any of its proper supersets.

- $\mathcal{S}$ is *stable* (with respect to $\mathcal{ABF}$) iff it is closed, conflict-free and attacks every $\phi \in \mathcal{A} \setminus \mathcal{S}$.

In ABA it is usual to refer also to the intersection of all the complete extensions of an ABF, which is called the *well-founded* extension of that ABF.

Like before, we denote by Naive($\mathcal{ABF}$) [respectively: Adm($\mathcal{ABF}$), Cmp($\mathcal{ABF}$), Grd($\mathcal{ABF}$), Prf($\mathcal{ABF}$), Stb($\mathcal{ABF}$), WF($\mathcal{ABF}$)] the set of all the naive [respectively: admissible, complete, grounded, preferred, stable, well-founded] extensions of $\mathcal{ABF}$.[16]

If every set of assumptions $\mathcal{S} \subseteq \mathcal{A}$ is $\vdash_\mathcal{R}$-closed, the ABF is called *flat*. In [Bondarenko et al., 1997] it is shown that most of the relations between the Dung extensions considered in Remark 11 carry on to flat ABFs (see also [Čyras et al., 2018, Theorems 2.12 and 2.14], and [Heyninck and Straßer, 2021a] for prioritized settings). For non-flat ABFs, however, some of these relations cease to hold. For instance, there may be non-flat ABFs without complete extensions (cf. Item 2 of Proposition 38).

The following form of ABFs is considered in [Heyninck and Arieli, 2018; Heyninck and Arieli, 2019b; Heyninck and Arieli, 2020b]:

Definition 35 (simple contrapositive ABFs). *A contrapositive assumption-based framework is a tuple* $\mathcal{ABF} = \langle \mathfrak{L}, \Gamma, \Delta, \sim \rangle$ *where:*

- $\mathfrak{L} = \langle \mathcal{L}, \vdash \rangle$ *is an explosive and contrapositive logic,*[17]

- Γ *(the* strict assumptions*) and* Δ *(the* candidate/defeasible assumptions*) are distinct (countable) sets of* $\mathcal{L}$-*formulas, where the former is assumed to be* $\vdash$-*consistent and the latter is assumed to be nonempty,*

- $\sim : \Delta \to \wp(\mathcal{L})$ *is a* contrariness operator, *assigning a finite set of* $\mathcal{L}$-*formulas to every defeasible assumption in* Δ, *such that for every* $\vdash$-*consistent* $\psi \in \Delta$ *it holds that* $\psi \not\vdash \bigwedge \sim \psi$ *and* $\bigwedge \sim \psi \not\vdash \psi$.

---

[16] Note that, as observed in [Heyninck and Arieli, 2020b], the grounded extension of an ABF may not be unique, thus (unlike the previous cases) this time Grd($\mathcal{ABF}$) is not an extension but a set of extensions.

[17] Classical logic CL, intuitionistic logic, the central logic in the family of constructive logics, and standard modal logics are all explosive and contrapositive logics.

A contrapositive ABF is called simple, if its language $\mathcal{L}$ contains a negation $\neg$, and for every $\psi \in \mathcal{A}$, $\sim\psi = \{\neg\psi\}$.

Given a simple contrapositive assumption-based framework $\mathcal{ABF} = \langle \mathcal{L}, \Gamma, \Delta, \sim \rangle$, the notion of attack and Dung-style semantics are defined as before, with the obvious adjustments using the consequence relation $\vdash$ of the base logic instead of the entailment $\vdash_{\mathcal{R}}$. For instance,

- $\mathcal{S} \subseteq \Delta$ attacks $\psi \in \Delta$ iff $\Gamma, \mathcal{S} \vdash \phi$ for some $\phi \in \sim\psi$. Accordingly, $\mathcal{S}$ attacks $\mathcal{T}$ if $\mathcal{S}$ attacks some $\psi \in \mathcal{T}$,

- $\mathcal{S} \subseteq \Delta$ is closed in $\mathcal{ABF}$ if $\mathcal{S} = \Delta \cap Cn_{\vdash}(\Gamma \cup \mathcal{S})$.

The other semantic notions remain exactly as before.

Given a (simple, contrapositive) assumption-based framework $\mathcal{ABF}$ and Sem $\in$ {Naive, WF, Grd, Prf, Stb}, we denote:

Definition 36 (ABA extension-based entailments).

- $\mathcal{ABF} \mid\!\sim_{\cup\mathsf{Sem}} \psi$ iff $\Gamma, \mathcal{E} \vdash \psi$ for some $\mathcal{E} \in \mathsf{Sem}(\mathbf{ABF})$.

- $\mathcal{ABF} \mid\!\sim_{\cap\mathsf{Sem}} \psi$ iff $\Gamma, \bigcap \mathsf{Sem}(\mathbf{ABF}) \vdash \psi$.

- $\mathcal{ABF} \mid\!\sim_{\cap\!\!\!\cap\mathsf{Sem}} \psi$ iff $\Gamma, \mathcal{E} \vdash \psi$ for every $\mathcal{E} \in \mathsf{Sem}(\mathbf{ABF})$.

The entailment relations in Definition 36 are again different from those in Definitions 1 and 12, as they are defined on ABFs and formulas (cf. Remark 27). Like before, this will not cause any confusion in the sequel.

Example 37. Let $\mathcal{L} = \mathsf{CL}$, $\Gamma = \emptyset$, $\Delta = \{p, \neg p, q\}$, and $\sim\psi = \{\neg\psi\}$ for every formula ψ. A corresponding attack diagram is shown in Figure 6.[18]

Here, $\mathsf{Naive}(\mathcal{ABF}) = \mathsf{Prf}(\mathcal{ABF}) = \mathsf{Stb}(\mathcal{ABF}) = \{\{p,q\}, \{\neg p, q\}\}$, and therefore $\mathcal{ABF} \mid\!\sim_{\circ\mathsf{Sem}} q$ for every $\circ \in \{\cup, \cap, \cap\!\!\!\cap\}$ and Sem $\in$ {Naive, Prf, Stb}.

Some interesting properties of simple contrapositive ABFs are given next (see [Heyninck and Arieli, 2018; Heyninck and Arieli, 2019b; Heyninck and Arieli, 2020b]).

---

[18] For reasons that will become apparent in the sequel (see Remark 41), we include in the diagram only *closed sets*. Thus, the set $\{p, \neg p\}$ is omitted from the diagram.

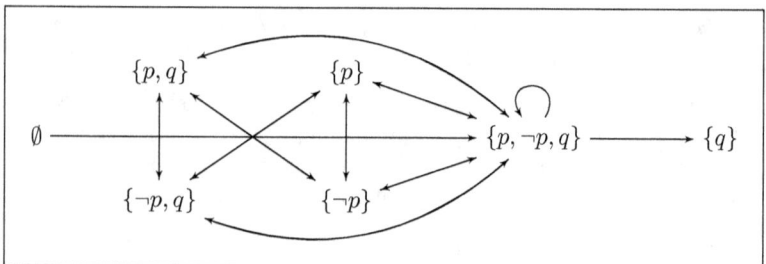

Figure 6: An attack diagram for Example 37

Proposition 38. Let $\mathcal{ABF} = \langle \mathfrak{L}, \Gamma, \Delta, \neg \rangle$ be a simple contrapositive ABF. Then:

1. $\mathsf{Naive}(\mathcal{ABF}) = \mathsf{Prf}(\mathcal{ABF}) = \mathsf{Stb}(\mathcal{ABF})$.

2. If $\mathsf{F} \in \Delta$ then $\mathsf{Grd}(\mathcal{ABF}) = \mathsf{WF}(\mathcal{ABF})$.

The next example shows that the condition in Item 2 of the last proposition is indeed necessary:

Example 39. Let $\mathfrak{L}$ be an explosive logic, $\Delta = \{p, \neg p, q\}$ and $\Gamma = \{s, s \supset q\}$. Note that the emptyset is not admissible, since it is not closed (indeed, $\Gamma \vdash q$). Also, $\{q\}$ is not admissible since $p, \neg p, q \vdash \neg q$.[19] The two minimal complete extensions here are $\{p, q\}$ and $\{\neg p, q\}$, thus there is no unique grounded extension in this case.

Corollary 40. Let $\mathcal{ABF}$ be a simple contrapositive ABF, and let $\circ \in \{\Cap, \cup, \cap\}$. For every ψ we have that: $\mathcal{ABF} \hspace{0.1em}\mid\hspace{-0.5em}\sim_{\circ\mathsf{Naive}} \psi$ iff $\mathcal{ABF} \hspace{0.1em}\mid\hspace{-0.5em}\sim_{\circ\mathsf{Prf}} \psi$ iff $\mathcal{ABF} \hspace{0.1em}\mid\hspace{-0.5em}\sim_{\circ\mathsf{Stb}} \psi$. Moreover, if $\mathsf{F} \in \Delta$ then $\mathcal{ABF} \hspace{0.1em}\mid\hspace{-0.5em}\sim_{\circ\mathsf{Grd}} \psi$ iff $\mathcal{ABF} \hspace{0.1em}\mid\hspace{-0.5em}\sim_{\circ\mathsf{WF}} \psi$.

Remark 41. *Interestingly, as shown in [Heyninck and Arieli, 2018], the closure requirement is redundant in the definition of extensions of simple contrapositive ABFs. Thus, for instance, if $\mathcal{E} \subseteq \Delta$ is conflict-free and attacks every $\psi \in \Delta \setminus \mathcal{E}$ then it is closed (so closure is assured in the definition of stable extensions), a maximally conflict-free subset of Δ is*

---

[19]Note that q is also attacked by $\{p, \neg p\}$ and does not counterattack it. However, $\{p, \neg p\}$ is not closed, and for admissibility checking it is enough to consider only closed sets (see also Remark 41).

closed (thus closure is guaranteed in the definition of naive extensions), and so forth. For grounded and well-founded semantics, the closure requirement is redundant only if $\mathsf{F} \in \Delta$.

Remark 42. *In [Heyninck and Straßer, 2021a] other classes of ABFs are studied. It is shown there that also for so-called* well-behaved *ABFs, the preferred and stable extension coincide. Well-behaved ABFs are flat ABFs that satisfy a slightly weaker notion of contraposition than the one above, and a property called* sanity *that says that if* $\sim\phi$ *follows from a set of assumptions* Δ *then it follows from* $\Delta \setminus \{\phi\}$ *(which is also satisfied by contrapositive ABFs). Otherwise, no restrictions on the underlying language are imposed.*[20]

2.3 Properties of the Frameworks and Their Entailments

In order to evaluate and compare the various approaches to logical argumentation, different properties and postulates have been introduced in the literature. In this section we consider the three logical argumentation methods of Section 2.2 in light of these criteria. We do so from three perspectives:

- relations to reasoning with maximal consistency, following [Rescher and Manor, 1970] (Section 2.3.1),

- rationality postulates for argumentative reasoning, following [Caminada and Amgoud, 2007] (Section 2.3.2), and

- inference principles for non-monotonic reasoning, following [Kraus et al., 1990] (Section 2.3.3).

In what follows we review the main results in the literature concerning the above-mentioned issues. We recall that it is not the purpose of this survey to resolve open questions or particular cases that were not addressed so far,[21] thus we do not pretend to have an exhaustive coverage of the subject.

---

[20]For technical details we refer to the paper whose main focus is to study and compare systems of prioritized ABFs.

[21]The only exception are the (yet unpublished) results in the appendix of the chapter, which appear in a paper that is currently under review.

2.3.1 Relations to Reasoning with Maximal Consistency

Reasoning with maximally consistent subsets (MCS), see [Rescher and Manor, 1970], is a well-known approach to handle inconsistencies within non-monotonic reasoning. The idea is to derive conclusions from inconsistent knowledge-bases, by considering the maximally consistent subsets of these knowledge bases. This idea has been applied in a variety of research directions within artificial intelligence, e.g.: knowledge-based integration systems [Baral et al., 1991], consistency operators for belief revision [Konieczny and Pérez, 2002] and computational linguistics [Malouf, 2007].

The relation between reasoning with maximally consistent subsets and formal argumentation has been studied extensively since this possibility was raised in [Cayrol, 1995]. In what follows we survey some of the main results relating MCS-based reasoning and the logic-based methods of the previous section. For a more extensive overview of the subject we refer to [Arieli et al., 2018; Arieli et al., 2019].

Reasoning with maximally consistent subsets of the premises is based on the following definition:

Definition 43 ($\mathsf{MCS}_{\mathfrak{L}}(\mathcal{S}), \mathsf{MCS}_{\mathfrak{L}}^{\mathcal{S}'}(\mathcal{S})$). Let $\mathfrak{L} = \langle \mathcal{L}, \vdash \rangle$ be a logic and let $\mathcal{S}', \mathcal{S}$ be sets of $\mathcal{L}$-formulas (intuitively, $\mathcal{S}'$ are the strict assumptions and $\mathcal{S}$ are the defeasible ones).

- $\mathsf{MCS}_{\mathfrak{L}}(\mathcal{S})$ is the set of the maximally $\vdash$-consistent subsets of $\mathcal{S}$. More specifically, $\mathsf{MCS}_{\mathfrak{L}}(\mathcal{S}) = \{\mathcal{T} \subseteq \mathcal{S} \mid \mathcal{T}$ is $\vdash$-consistent and for every $\mathcal{T}'$ such that $\mathcal{T} \subsetneq \mathcal{T}' \subseteq \mathcal{S}$, $\mathcal{T}'$ is $\vdash$-inconsistent$\}$.

- $\mathsf{MCS}_{\mathfrak{L}}^{\mathcal{S}'}(\mathcal{S})$ is the set of the maximally $\vdash$-consistent subsets of $\mathcal{S}$, given $\mathcal{S}'$. More specifically, $\mathsf{MCS}_{\mathfrak{L}}^{\mathcal{S}'}(\mathcal{S}) = \{\mathcal{T} \subseteq \mathcal{S} \mid \mathcal{T} \cup \mathcal{S}'$ is $\vdash$-consistent and for every $\mathcal{T}'$ such that $\mathcal{T} \subsetneq \mathcal{T}' \subseteq \mathcal{S}$, $\mathcal{T}' \cup \mathcal{S}'$ is $\vdash$-inconsistent$\}$.

The second item in the definition above, which defines maximally consistent subsets w.r.t. a set of strict assumptions, is known from [Makinson, 2003] as *default assumptions*. Some of the corresponding entailment relations are defined in [Makinson, 2003] as well, which is similar to those in Definitions 12, 26 and 36:

Definition 44 (MCS-based entailments). Let $\mathfrak{L} = \langle \mathcal{L}, \vdash \rangle$ be a logic and let $\mathcal{S}', \mathcal{S}$ be sets of $\mathcal{L}$-formulas. We denote:

- $\mathcal{S}', \mathcal{S} \mathrel{\vert\!\sim}^{\mathfrak{L}}_{\cap\mathsf{mcs}} \psi$ iff $\psi \in Cn_{\mathfrak{L}}(\mathcal{S}' \cup \bigcap \mathsf{MCS}^{\mathcal{S}'}_{\mathfrak{L}}(\mathcal{S}))$;
- $\mathcal{S}', \mathcal{S} \mathrel{\vert\!\sim}^{\mathfrak{L}}_{\cap\!\!\cap\mathsf{mcs}} \psi$ iff $\psi \in \bigcap_{\mathcal{T} \in \mathsf{MCS}^{\mathcal{S}'}_{\mathfrak{L}}(\mathcal{S})} Cn_{\mathfrak{L}}(\mathcal{S}' \cup \mathcal{T})$;
- $\mathcal{S}', \mathcal{S} \mathrel{\vert\!\sim}^{\mathfrak{L}}_{\cup\mathsf{mcs}} \psi$ iff $\psi \in \bigcup_{\mathcal{T} \in \mathsf{MCS}^{\mathcal{S}'}_{\mathfrak{L}}(\mathcal{S})} Cn_{\mathfrak{L}}(\mathcal{S}' \cup \mathcal{T})$.

In the definition above, $\mathcal{S}'$ is the set of the strict assumptions, and $\mathcal{S}$ is the set of defeasible assumptions. When $\mathcal{S}' = \emptyset$ we shall just omit it. In this case we have that:

- $\mathcal{S} \mathrel{\vert\!\sim}^{\mathfrak{L}}_{\cap\mathsf{mcs}} \psi$ iff $\psi \in Cn_{\mathfrak{L}}(\bigcap \mathsf{MCS}_{\mathfrak{L}}(\mathcal{S}))$;
- $\mathcal{S} \mathrel{\vert\!\sim}^{\mathfrak{L}}_{\cap\!\!\cap\mathsf{mcs}} \psi$ iff $\psi \in \bigcap_{\mathcal{T} \in \mathsf{MCS}_{\mathfrak{L}}(\mathcal{S})} Cn_{\mathfrak{L}}(\mathcal{T})$;
- $\mathcal{S} \mathrel{\vert\!\sim}^{\mathfrak{L}}_{\cup\mathsf{mcs}} \psi$ iff $\psi \in \bigcup_{\mathcal{T} \in \mathsf{MCS}_{\mathfrak{L}}(\mathcal{S})} Cn_{\mathfrak{L}}(\mathcal{T})$.

Example 45. Suppose that the base logic is classical logic (i.e., $\mathfrak{L} = \mathsf{CL}$).

- Let $\mathcal{S} = \{p, \neg p, q\}$. Then $\bigcap \mathsf{MCS}_{\mathsf{CL}}(\mathcal{S}) = \{q\}$, thus $\mathcal{S} \mathrel{\vert\!\sim}^{\mathsf{CL}}_{\cap\mathsf{mcs}} q$ but $\mathcal{S} \mathrel{\not\vert\!\sim}^{\mathsf{CL}}_{\cap\mathsf{mcs}} p$ and $\mathcal{S} \mathrel{\not\vert\!\sim}^{\mathsf{CL}}_{\cap\mathsf{mcs}} \neg p$.
- Let $\mathcal{S} = \{p \wedge q, \neg p \wedge q\}$. Then $\bigcap \mathsf{MCS}_{\mathsf{CL}}(\mathcal{S}) = \emptyset$, thus $\mathcal{S} \mathrel{\vert\!\sim}^{\mathsf{CL}}_{\cap\mathsf{mcs}} \psi$ only if ψ is a classical theorem. On the other hand, $\mathcal{S} \mathrel{\vert\!\sim}^{\mathsf{CL}}_{\cap\!\!\cap\mathsf{mcs}} q$ (and still $\mathcal{S} \mathrel{\not\vert\!\sim}^{\mathsf{CL}}_{\cap\!\!\cap\mathsf{mcs}} p$ and $\mathcal{S} \mathrel{\not\vert\!\sim}^{\mathsf{CL}}_{\cap\!\!\cap\mathsf{mcs}} \neg p$).
- It is easy to verify that for any $\mathcal{S}$, if $\mathcal{S} \mathrel{\vert\!\sim}^{\mathfrak{L}}_{\cap\mathsf{mcs}} \psi$ then $\mathcal{S} \mathrel{\vert\!\sim}^{\mathfrak{L}}_{\cap\!\!\cap\mathsf{mcs}} \psi$. As the previous item shows, the converse does not hold.

The next result relates MCS-based entailments and entailments that are induced by argumentation frameworks that are based on classical logic:

Proposition 46. ([Arieli et al., 2018, Propositions 4.3]),[Borg et al., 2021, Theorem 5][22] Let $\mathcal{AF}_{\mathfrak{L},\mathcal{A}}(\mathcal{S})$ be a logic-based argumentation framework, where $\mathfrak{L}$ is classical logic and $\emptyset \subset \mathcal{A} \subseteq \{\mathsf{Ucut}, \mathsf{Def}\}$. Then:

---

[22]The results in [Borg et al., 2021] are phrased in the more general context of hypersequent-based argumentation. Since standard sequent calculi are special instances of hypersequent calculi, the results are applicable also to sequent-based argumentation.

- $\mathcal{S} \hspace{0.1em}\mid\hspace{-0.5em}\sim_{\mathsf{Grd}}^{\mathcal{L},\mathcal{A}} \psi$ iff $\mathcal{S} \hspace{0.1em}\mid\hspace{-0.5em}\sim_{\cap\mathsf{Prf}}^{\mathcal{L},\mathcal{A}} \psi$ iff $\mathcal{S} \hspace{0.1em}\mid\hspace{-0.5em}\sim_{\cap\mathsf{Stb}}^{\mathcal{L},\mathcal{A}} \psi$ iff $\mathcal{S} \hspace{0.1em}\mid\hspace{-0.5em}\sim_{\cap\mathsf{mcs}}^{\mathcal{L}} \psi$.
- $\mathcal{S} \hspace{0.1em}\mid\hspace{-0.5em}\sim_{\mathsf{UPrf}}^{\mathcal{L},\mathcal{A}} \psi$ iff $\mathcal{S} \hspace{0.1em}\mid\hspace{-0.5em}\sim_{\mathsf{UStb}}^{\mathcal{L},\mathcal{A}} \psi$ iff $\mathcal{S} \hspace{0.1em}\mid\hspace{-0.5em}\sim_{\mathsf{Umcs}}^{\mathcal{L}} \psi$.

If $\mathcal{A} = \{\mathsf{DirUcut}\}$, we have that:

- $\mathcal{S} \hspace{0.1em}\mid\hspace{-0.5em}\sim_{\cap\mathsf{Prf}}^{\mathcal{L},\mathcal{A}} \psi$ iff $\mathcal{S} \hspace{0.1em}\mid\hspace{-0.5em}\sim_{\cap\mathsf{Stb}}^{\mathcal{L},\mathcal{A}} \psi$ iff $\mathcal{S} \hspace{0.1em}\mid\hspace{-0.5em}\sim_{\cap\mathsf{mcs}}^{\mathcal{L}} \psi$.

Example 47. *By the last proposition, the correspondence between the examples in Remark 15 and those of Example 45 is not coincidental.*

We refer to [Arieli et al., 2018] for many other results concerning the relations between reasoning with maximal consistency and logic-based argumentation (or, more precisely, sequent-based argumentation, a specific form of logic-based argumentation – see Remark 6).

The relation between ABA and maximally consistent subsets has been studied, e.g., in [Borg, 2020; Heyninck and Arieli, 2018; Heyninck and Arieli, 2020b; Heyninck and Straßer, 2021a]. In particular, a similar result as the one above is shown for simple contrapositive assumption-based frameworks (recall Definition 35).

Proposition 48. *([Heyninck and Arieli, 2018, Theorems 1 and 3] and [Borg, 2020, Theorem 3]) Let $\mathcal{ABF} = \langle \mathfrak{L}, \Gamma, \Delta, \sim \rangle$ be a simple contrapositive assumption-based framework. Then:*

- $\mathcal{ABF} \hspace{0.1em}\mid\hspace{-0.5em}\sim_{\cap\mathsf{Prf}}^{\mathcal{L},\mathcal{A}} \psi$ iff $\mathcal{ABF} \hspace{0.1em}\mid\hspace{-0.5em}\sim_{\cap\mathsf{Stb}}^{\mathcal{L},\mathcal{A}} \psi$ iff $\Gamma, \Delta \hspace{0.1em}\mid\hspace{-0.5em}\sim_{\cap\mathsf{mcs}}^{\mathcal{L}} \psi$.
- $\mathcal{ABF} \hspace{0.1em}\mid\hspace{-0.5em}\sim_{\mathsf{UPrf}}^{\mathcal{L},\mathcal{A}} \psi$ iff $\mathcal{ABF} \hspace{0.1em}\mid\hspace{-0.5em}\sim_{\mathsf{UStb}}^{\mathcal{L},\mathcal{A}} \psi$ iff $\Gamma, \Delta \hspace{0.1em}\mid\hspace{-0.5em}\sim_{\mathsf{Umcs}}^{\mathcal{L}} \psi$.
- *If* $\mathsf{F} \in \Delta$ *then* $\mathcal{ABF} \hspace{0.1em}\mid\hspace{-0.5em}\sim_{\mathsf{Grd}}^{\mathcal{L},\mathcal{A}} \psi$ iff $\Gamma, \Delta \hspace{0.1em}\mid\hspace{-0.5em}\sim_{\cap\mathsf{mcs}}^{\mathcal{L}} \psi$.

If $\mathfrak{L}$ is contrapositive then:

- $\mathcal{ABF} \hspace{0.1em}\mid\hspace{-0.5em}\sim_{\cap\mathsf{Prf}}^{\mathcal{L},\mathcal{A}} \psi$ iff $\mathcal{ABF} \hspace{0.1em}\mid\hspace{-0.5em}\sim_{\cap\mathsf{Stb}}^{\mathcal{L},\mathcal{A}} \psi$ iff $\Gamma, \Delta \hspace{0.1em}\mid\hspace{-0.5em}\sim_{\cap\mathsf{mcs}}^{\mathcal{L}} \psi$.

Remark 49. *A result similar to the one of Proposition 48 is obtained in [Heyninck and Straßer, 2021a] for what is called there* well-behaved *assumption-based frameworks, which among other things requires closure of the underlying inference rules under contraposition. It is shown that for well-behaved assumption-based frameworks,*

$$\mathsf{MCS}_\mathfrak{L}(\mathcal{ABF}) = \mathsf{Prf}(\mathcal{ABF}) = \mathsf{Stb}(\mathcal{ABF}).$$

By including priorities, the results are further generalized to cover preferred subtheories [Brewka, 1989].

Example 50. *Recall Example 37 with the assumption-based framework for $\mathfrak{L} = \mathsf{CL}$, $\Gamma = \emptyset$, $\Delta = \{p, \neg p, q\}$ and $\sim\!\psi = \{\neg\psi\}$ for every formula ψ. Since $\mathsf{Naive}(\mathcal{ABF}) = \mathsf{Prf}(\mathcal{ABF}) = \mathsf{Stb}(\mathcal{ABF}) = \{\{p,q\}, \{\neg p, q\}\}$, we have $\mathcal{ABF} \mathrel{\vdash\mkern-10mu\sim}_{\mathsf{osem}} q$ for $\circ \in \{\Cap, \cup, \cap\}$ and $\mathsf{Sem} \in \{\mathsf{Naive}, \mathsf{Prf}, \mathsf{Stb}\}$. In view of Proposition 48 and Remark 49 it is not surprising that $\mathsf{MCS}_{\mathsf{CL}}(\mathcal{S}) = \{\{p, q\}, \{\neg p, q\}\}$.*

We turn now to MCS-based reasoning and ASPIC systems. In [Modgil and Prakken, 2013, §5.3.2] it is shown that Brewka's *preferred subtheories* [Brewka, 1989] are an instance of ASPIC$^+$. Since no preference ordering is considered in this chapter, preferred subtheories correspond to maximally consistent subsets. The following proposition states this result in terms of sets of formulas.

Proposition 51. *([Modgil and Prakken, 2013, Th. 34]) Let $\mathcal{AF}(AT) = \langle \mathsf{Arg}(AT), \mathsf{Attack} \rangle$ be an ASPIC-argumentation framework for some ASPIC-argumentation theory AT, based on a propositional language $\mathcal{L}$, a set $\mathcal{S}$ of $\mathcal{L}$-formulas, and where the rules are all strict. Suppose further that $\Gamma \to \gamma \in \mathcal{R}$ iff γ follows according to classical logic from Γ. Let $\mathsf{Arg}(\Delta) \subseteq \mathsf{Arg}(AT)$ be the arguments constructed from premises in Δ. Then:*

- *If Δ is a maximally consistent subset of $\mathcal{S}$, then $\mathsf{Arg}(\Delta)$ is a stable extension of $\mathcal{AF}(AT)$.*

- *If $\mathcal{E}$ is a stable extension of $\mathcal{AF}(AT)$, then $\bigcup_{A \in \mathcal{E}} \mathsf{Prem}(A)$ is a maximally consistent subset of $\mathcal{S}$.*

Example 52. *To illustrate the last result consider the ASPIC argumentation system $AS = \langle \mathcal{L}, ^-, \mathcal{R}, n \rangle$, where $\mathcal{L}$ is a propositional language with $\mathsf{Atoms}(\mathcal{L}) = \{p, q\}$, the rules in $\mathcal{R}_s$ coincide with those of classical logic as in Example 21, $\mathcal{K}_p = \{p, \neg p, q\}$, $\mathcal{K}_n = \emptyset$, and $\overline{\phi} = \{\neg\phi\}$ for any $\mathcal{L}$-formula ϕ. Among others, the following ASPIC-arguments can be constructed:*

$$A_1 : p \qquad A_2 : \neg p \qquad A_3 : q \qquad A_4 : A_1, A_2 \to \neg q$$

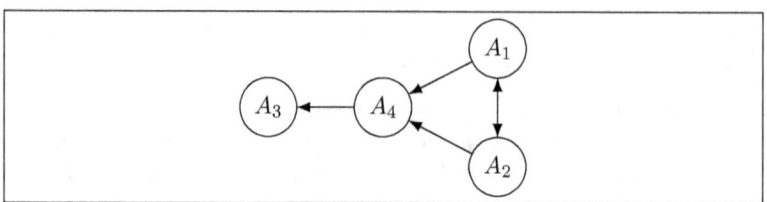

Figure 7: Part of the framework from Example 52.

The corresponding attack diagram is given in Figure 7.

$\mathcal{AF}(AT)$ *has two stable extensions, one containing among others A_1 and A_3 and the second containing among others A_2 and A_3. As expected in view of Proposition 51, we see that these correspond to the two maximally consistent subsets of $\{p, \neg p, q\}$, namely: $\{p, q\}$ and $\{\neg p, q\}$.*

Remark 53. *It is interesting to note that unlike some other frameworks (cf., e.g., Propositions 46 and 48), the grounded extension in the ASPIC framework of Example 52 does* not *contain the free formula q. This is since the inconsistent argument A_4 causes interferent behavior for the grounded semantics (see Section 2.3.2.B for more details).*

While the result in Proposition 51 above is about ASPIC-frameworks with only strict rules, one may also consider maximal consistent sets of formulas in the context of defeasible rules. In [Heyninck and Straßer, 2021b], maximal consistent sets of defeasible rules are defined as follows:

Definition 54 (MCS(AT)). *Let* AT $= \langle \text{AS}, \mathcal{K} \rangle$ *be an ASPIC argumentation theory, where* $\mathcal{K} = \langle \mathcal{K}_n, \mathcal{K}_p \rangle$, AS $= \langle \mathcal{L}, \overline{}, \mathcal{R}, n \rangle$, *and* $\mathcal{R} = \mathcal{R}_d \cup \mathcal{R}_s$. *We define:*

- $\mathcal{R}_d^{\mathcal{K}} = \mathcal{R}_d \cup \{\Rightarrow \phi \mid \phi \in \mathcal{K}_p\}$.

- *A set of defeasible rules* $\mathcal{D} \subseteq \mathcal{R}_d^{\mathcal{K}}$ *is AT-inconsistent iff there are $\mathcal{L}$-formulas ϕ and $\psi \in \overline{\phi}$, for which* $\mathcal{K}_n \vdash_{\mathcal{R}_s \cup \mathcal{D}} \psi$ *and* $\mathcal{K}_n \vdash_{\mathcal{R}_s \cup \mathcal{D}} \phi$. *Otherwise, $\mathcal{D}$ is AT-consistent.*[23]

- *A rule* $r = \psi_1, \ldots, \psi_n \Rightarrow \phi \in \mathcal{R}_d^{\mathcal{K}}$ *is triggered by some* $\mathcal{D} \subseteq \mathcal{R}_d^{\mathcal{K}}$ *if* $\mathcal{K}_n \vdash_{\mathcal{R}_s \cup \mathcal{D}} \psi_i$ *for each* $1 \leq i \leq n$.

---

[23]Maximally consistent sets of defeasible rules also play a role in constrained input/output logics, see [Makinson and Van Der Torre, 2001]

- $\hat{\wp}(\mathcal{R}_d^{\mathcal{K}})$ is the set of all $\mathcal{D} \subseteq \mathcal{R}_d^{\mathcal{K}}$ such that every $r \in \mathcal{D}$ is triggered by $\mathcal{D}$.

- MCS(AT) is the set of all $\subseteq$-maximal consistent $\mathcal{D} \in \hat{\wp}(\mathcal{R}_d^{\mathcal{K}})$.

Example 55. *Let* AT $= \langle \text{AS}, \mathcal{K} \rangle$ *be an ASPIC argumentation theory, where* AS $= \langle \mathcal{L}, \overline{}, \mathcal{R}, n \rangle$, $\mathcal{R}_d = \{r_1 : \top \Rightarrow p,\ r_2 : p \Rightarrow q,\ r_3 : \top \Rightarrow \neg q\}$, $\mathcal{R}_s$ *is induced by classical logic, and* $\mathcal{K} = \emptyset$. *Then,*

- $\hat{\wp}(\mathcal{R}_d^{\mathcal{K}}) = \{\{r_1\}, \{r_1, r_2\}, \{r_1, r_2, r_3\}, \{r_1, r_3\}, \{r_3\}\}$, *and*
- MCS(AT) $= \{\{r_1, r_2\}, \{r_1, r_3\}\}$.

Note that $\{r_2, r_3\} \notin \hat{\wp}(\mathcal{R}_d^{\mathcal{K}})$ *since* r_2 *is not triggered by this set. Furthermore,* $\{r_1, r_2, r_3\} \in \hat{\wp}(\mathcal{R}_d^{\mathcal{K}}) \setminus$ MCS(AT) *since the set is inconsistent.*

For the next result we need also the following definition:

Definition 56 (contrapositive ASPIC theory, Arg($\mathcal{D}$)). *Let* AT $= \langle \text{AS}, \mathcal{K} \rangle$ *be an ASPIC argumentation theory as in the previous definition. Then:*

- AT *is contrapositive if it satisfies*

    S1 *If* $\Delta, \psi \vdash_{\mathcal{R}_s} \phi'$ *for some* $\phi' \in \overline{\phi}$ *then* $\Delta, \phi \vdash_{\mathcal{R}_s} \psi'$ *for some* $\psi' \in \overline{\psi}$; *and*

    S2 *If* $\Delta \vdash_{\mathcal{R}_s} \phi'$ *for some* $\phi' \in \overline{\phi}$ *then* $\Delta \setminus \{\phi\} \vdash_{\mathcal{R}_s} \phi'$.

- *For* $\mathcal{D} \in \hat{\wp}(\mathcal{R}_d^{\mathcal{K}})$, *we define:* Arg($\mathcal{D}$) $= \{A \in \text{Arg(AT)} \mid \text{DefRules}(A) \subseteq \mathcal{D} \cap \mathcal{R}_d\}$.

We get the following representation theorem for ASPIC$^+$ frameworks without undercut attacks:

Proposition 57. ([Heyninck and Straßer, 2021b, Theorem 6]) *For any contrapositive ASPIC argumentation theory* AT *without undercut attacks, it holds that:*

$$\text{Prf}(\mathcal{AF}(\text{AT})) = \text{Stb}(\mathcal{AF}(\text{AT})) = \{\text{Arg}(\mathcal{D}) \mid \mathcal{D} \in \text{MCS(AT)}\}.$$

Example 58 (Example 55 continued). *In Example 55 we have the two stable resp. preferred extensions* Arg($\{r_1, r_2\}$) *and* Arg($\{r_1, r_3\}$).

Maximal consistency is also related to properties of extensions and of argumentation semantics, as will be shown in the next section. Here we only comment on one such property, which is directly related to the maximally consistent subsets of the premises.

Remark 59. *Consider the following property, investigated in [Amgoud and Besnard, 2010; Vesic, 2013]:*

$$\mathsf{MCS}_{\mathsf{CL}}(\mathcal{S}) = \{\mathit{Sup}(\mathcal{E}) \mid \mathcal{E} \in \mathit{Sem}(\mathcal{AF}(\mathcal{S}))\}.$$

It is shown that in classical argumentation frameworks (i.e., those that consist of classical arguments in the sense of Definition 4), the equation above is met for both the stable (i.e, when Sem = Stb*) and preferred (*Sem = Prf*) semantics, and when the attack relation is either DirDef, DirUcut, or BigArgAt, while for the other attacks (Def, Ucut, Reb, DefReb) the above property ceases to hold.*

Other properties of the attack relations, as well as properties of the extensions and of the induced entailments will be considered in the next sections.

2.3.2 Rationality Postulates for Argumentative Reasoning

Since the introduction of the rationality postulates for ASPIC in [Caminada and Amgoud, 2007], they have become a standard to assess approaches to structured argumentation. The postulates state that the conclusions of a framework should be closed under its strict rules (in approaches without a distinction between strict and defeasible rules, this simply means closure under the rules of the system), that the set of conclusions should be consistent, and that the set of formulas that is the result of the closure of the conclusions should be consistent as well. Another property states that an extension should also contain all the sub-arguments of its arguments. These postulates may formally be defined as follows:

Definition 60 (rationality postulates for extensions). *Let* $\mathcal{AF} = \langle \mathsf{Arg}, \mathsf{Attack} \rangle$ *be an argumentation framework,* $\mathfrak{L} = \langle \mathcal{L}, \vdash \rangle$ *a logic,* Sem *a semantics for it and* $\mathcal{E} \in \mathsf{Sem}(\mathcal{AF})$. *Then* $\mathcal{AF}$ *satisfies:*

- sub-argument closure, *iff for all $A \in \mathcal{E}$, $\mathsf{Sub}(A) \subseteq \mathcal{E}$;*
- closure, *iff $Cn_{\mathcal{L}}(\mathsf{Conc}(\mathcal{E})) = Conc(\mathcal{E})$;*
- direct consistency, *iff $\mathsf{Conc}(\mathcal{E})$ is $\vdash$-consistent; and*
- indirect consistency, *iff $Cn_{\mathcal{L}}(\mathsf{Conc}(\mathcal{E}))$ is $\vdash$-consistent.*

In [Caminada and Amgoud, 2007] it was shown that, if an argumentation framework $\mathcal{AF}$ satisfies indirect consistency, it satisfies direct consistency as well and if $\mathcal{AF}$ satisfies closure and direct consistency, it also satisfies indirect consistency.

Following [Caminada and Amgoud, 2007], many related rationality postulates were introduced in the literature, some of them will be discussed in what follows. While the postulates in [Caminada and Amgoud, 2007] are mainly concerned with the properties of the extensions of a framework (under certain semantics), there are other postulates that are related to the inferences relations induced by the frameworks. For instance, the *non-interference* and *crash-resistance* postulates, introduced in [Caminada *et al.*, 2011], guarantee that the entailment relation of argumentation frameworks do not collapse in view of inconsistent information. Next, we formalize these postulates.

For the next definitions, we say that two sets $\mathcal{S}_1, \mathcal{S}_2$ of $\mathcal{L}$-formulas are *syntactically disjoint* iff $\mathsf{Atoms}(\mathcal{S}_1) \cap \mathsf{Atoms}(\mathcal{S}_2) = \emptyset$.[24] This will be denoted by $\mathcal{S}_1 \mid \mathcal{S}_2$.

Definition 61 (rationality postulates for inferences). *Let $\mathord{\sim} \subseteq \wp(\mathcal{L}) \times \mathcal{L}$.*

- *We say that $\mathord{\sim}$ satisfies* non-interference, *iff for every two sets $\mathcal{S}_1, \mathcal{S}_2$ of $\mathcal{L}$-formulas, and every $\mathcal{L}$-formula ϕ such that $\mathcal{S}_1 \cup \{\phi\} \mid \mathcal{S}_2$, it holds that $\mathcal{S}_1 \mathrel{\sim} \phi$ iff $\mathcal{S}_1, \mathcal{S}_2 \mathrel{\sim} \phi$.*

- *We say that $\mathord{\sim}$ satisfies* crash-resistance *iff there is no $\mathord{\sim}$-contaminating set $\mathcal{S}$ of $\mathcal{L}$-formulas, where a set $\mathcal{S}$ such that $\mathsf{Atoms}(\mathcal{S}) \subsetneq \mathsf{Atoms}(\mathcal{L})$, is called* contaminating *(w.r.t. $\mathord{\sim}$), if for every $\mathcal{S}'$ such that $\mathcal{S} \mid \mathcal{S}'$ and for every $\mathcal{L}$-formula ϕ, it holds that $\mathcal{S} \mathrel{\sim} \phi$ iff $\mathcal{S}, \mathcal{S}' \mathrel{\sim} \phi$.*

---

[24] Recall that $\mathsf{Atoms}(\mathcal{S})$ denotes the set of atoms occurring in the formulas of $\mathcal{S}$.

Remark 62. *In [Caminada et al., 2011] it is shown that crash-resistance follows from non-Interference under some very weak criteria on the monotonic base logic.*

Note, for instance, that the consequence relation $\vdash_{\mathsf{CL}}$ of classical logic does not satisfy either of the properties of Definition 61. Indeed, where $\mathcal{S}_2$ is inconsistent, non-interference is violated, and any inconsistent set is $\vdash_{\mathsf{CL}}$-contaminating. We refer to [Caminada et al., 2011] for more discussion on non-interference and crash-resistance.

Since rationality postulates are an important indicator of the usefulness of an argumentation system, extensive research has been conducted on the properties a system should satisfy in order for the rationality postulates to be satisfied. In the remainder of this section we will discuss the results of this research for the three approaches to logical argumentation frameworks discussed earlier.

A. Rationality postulates for logic-based methods

There are many studies on the properties of logic-based frameworks, including those in [Gorogiannis and Hunter, 2011; Amgoud and Besnard, 2013; Amgoud, 2014; Borg and Straßer, 2018; Arieli et al., 2020; Borg et al., 2021]. Below, we survey the main results, starting with the postulates that are concerned with the properties of the attack rules and then those that are related to the properties of extensions and extension-based inferences.

Studies on requirements on the attack relation of a classical argumentation framework to fulfill rationality postulates are presented in [Amgoud and Besnard, 2010; Vesic, 2013]. The conditions considered in those work are presented next.

Definition 63 (attack relation properties). *Let $\mathcal{AF}(\mathcal{S}) = \langle \mathsf{Arg}(\mathcal{S}), \mathsf{Attack} \rangle$ be a classical argumentation framework. Then Attack is called:*

- *conflict-dependent, iff for each $(A, B) \in \mathsf{Attack}$, $\mathsf{Sup}(A) \cup \mathsf{Sup}(B) \vdash \mathsf{F}$;*

- *conflict-sensitive, iff for each $A, B \in \mathsf{Arg}(\mathcal{S})$, if $\mathsf{Sup}(A) \cup \mathsf{Sup}(B) \vdash \mathsf{F}$ then $(A, B) \in \mathsf{Attack}$;*

- valid, *iff for each $\mathcal{E} \subseteq \mathsf{Arg}(\mathcal{S})$, if $\mathcal{E}$ is conflict-free, then $\mathsf{Sup}(\mathcal{E})$ is consistent;*

- conflict-complete, *iff for every minimally inconsistent set $\mathcal{T} \subseteq \mathcal{S}$, for every $\mathcal{T}_1, \mathcal{T}_2 \subseteq \mathcal{T}$ such that $\mathcal{T}_1 \neq \emptyset, \mathcal{T}_2 \neq \emptyset$ and $\mathcal{T}_1 \cup \mathcal{T}_2 = \mathcal{T}$ and for every $A \in \mathsf{Arg}(\mathcal{S})$ with $\mathsf{Sup}(A) = \mathcal{T}_1$ there is an argument $B \in \mathsf{Arg}(\mathcal{S})$ with $\mathsf{Sup}(B) = \mathcal{T}_2$ such that $(B, A) \in \mathit{Attack}$;*

- symmetric, *iff when $(A, B) \in \mathit{Attack}$ also $(B, A) \in \mathit{Attack}$.*

We refer to [Amgoud and Besnard, 2010; Vesic, 2013] for a discussion on these properties and the relations among them. Table 2 summarizes which of the properties above are satisfied by the attack rules from Table 1.[25]

Attack rule	conflict-dependent	conflict-sensitive	valid	conflict-complete	symmetric
Def	✓	×	×	✓	×
DirDef	✓	×	×	×	×
Ucut	✓	×	×	✓	×
DirUcut	✓	×	×	×	×
ConUcut	✓	×	×	×	×
Reb	✓	×	×	×	✓
DefReb	✓	×	×	×	✓
Reb ∪ DirUcut	✓	×	×	×	×
BigArgAt	✓	×	×	×	×

Table 2: The satisfiability of the properties from Definition 63 for attack rules in Table 1.

---

[25] Note that, in this context, Reb ∪ DirUcut is the only union of attack rules considered in the literature.

Another study on the properties of attack relations in logic-based argumentation frameworks is given in [Gorogiannis and Hunter, 2011]. Again, the study refers to classical argumentation framework, that is: the arguments meet the restrictions in Definition 4. An overview over various necessary and sufficient conditions on the attack relations considered in [Gorogiannis and Hunter, 2011] is given in Table 3.

Necessary conditions on attacks		
If $(A, B) \in$ Attack, then	$\{\mathsf{Conc}(A)\} \cup \mathsf{Sup}(B) \vdash \mathsf{F}.$	(D1)
	there is a $\phi \in \mathsf{Sup}(B)$ s.t. $\mathsf{Conc}(A) \vdash \neg\phi.$	(D1')
	$\mathsf{Conc}(A) \vdash \neg\mathsf{Conc}(B).$	(D1'')
	$\neg\mathsf{Conc}(A) \vdash \bigwedge \mathsf{Sup}(B),$	(D5)
	there is a $\phi \in \mathsf{Sup}(B)$ s.t. $\neg\mathsf{Conc}(A) \vdash \phi.$	(D5')
	$\neg\mathsf{Conc}(A) \vdash \mathsf{Conc}(B),$	(D5'')
	there is a $\Gamma \subseteq \mathsf{Sup}(B)$ s.t. $\vdash \neg\mathsf{Conc}(A) \equiv \bigwedge \Gamma.$	(D5''')
Sufficient conditions on attacks		
$(C, B) \in$ Attack if $(A, B) \in$ Attack and	$\vdash \mathsf{Conc}(A) \equiv \mathsf{Conc}(C)$	(D2)
	$\mathsf{Conc}(C) \vdash \mathsf{Conc}(A)$	(D2')
$(A, C) \in$ Attack, if $(A, B) \in$ Attacks and	$\vdash \mathsf{Sup}(B) = \mathsf{Sup}(C)$	(D3)
	$\mathsf{Sup}(B) \subseteq \mathsf{Sup}(C)$	(D3')
There is a C such that $\mathsf{Conc}(A) \vdash \mathsf{Conc}(C)$ and $(C, B) \in$ Attack, if	$\{\mathsf{Conc}(A)\} \cup \mathsf{Sup}(B) \vdash \mathsf{F}$	(D6)
	there is a $\phi \in \mathsf{Sup}(B)$ s.t. $\mathsf{Conc}(A) \vdash \neg\phi$	(D6')
	$\mathsf{Conc}(A) \vdash \neg\mathsf{Conc}(B)$	(D6'')
$(A, B) \in$ Attack if	there is a $\Gamma \subseteq \mathsf{Sup}(B)$ s.t. $\vdash \mathsf{Conc}(A) \equiv \neg \bigwedge \Gamma$	(D6''')
Sufficient and necessary conditions on attacks		
$(A, B) \in$ Attack iff $(A', B') \in$ Attack, if	$\vdash A \equiv A'$ and $\vdash B \equiv B'$	(D0)

Table 3: Conditions on the attack relations in [Gorogiannis and Hunter, 2011].

Proposition 64. ([Gorogiannis and Hunter, 2011, Prop. 6 and 10])
With $\mathcal{AF}(\mathcal{S}) = \langle \mathsf{Arg}(\mathcal{S}), \mathit{Attack} \rangle$ being a classical argumentation framework:

- *Table 4, summarizes which of the postulates from Table 3 hold for the attack rules from Table 1.*

- *Table 5 summarizes by which of the postulates from Table 3 the different attack relations are characterized.*

	Def	DirDef	Ucut	DirUcut	CanUcut	Reb	DefReb
D0	✓	✓	✓	✓	✓	✓	✓
D1	✓	✓	✓	✓	✓	✓	✓
D2	✓	✓	✓	✓	✓	✓	✓
D2′	✓	✓	×	×	×	×	✓
D3	✓	✓	✓	✓	✓	×	×
D3′	✓	✓	✓	✓	×	×	×

Table 4: Overview of the constraints on the attack relation (Table 3) that are satisfied by the rules from Table 1 (Based on [Gorogiannis and Hunter, 2011, Table 1 and Proposition 6])

Remark 65. *The interplay between logical principles about argumentation, on the one hand, and inference principles as studied in proof theory, on the other hand, is also studied in [Corsi and Fermüller, 2017]. In that paper a series of logical principles of attack relations in argumentation frameworks is stated, and their collection leads to a characterization of classical logical consequence relations that only involves argumentation frameworks. We refer to [Corsi and Fermüller, 2017] and [Corsi and Fermüller, 2019] for further details.*

We turn now to postulates concerning the extensions of logic-based argumentation frameworks. Definition 66 lists a series of rationality postulates studied in, e.g., [Caminada and Amgoud, 2007; Gorogiannis and

	D1, D6	**D1′, D6′**	**D1″, D6″**	**D6‴**
D2′	Def	DirDef	DefReb	-
D2	CanUcut (D5)	DirUcut (D5′)	Reb (D5″)	-
-	-	-	-	Ucut (D5‴)

Table 5: Overview of the attack relation postulates from Table 3 that characterize the attack rules from Table 1. An attack rule is characterized by the conjunction of the attack relation postulates from the appropriate row, column and (where applicable) the cell. For example, the attack rule is direct undercut iff the attack relation postulates D1′, D2, D5′ and D6′ are satisfied (Based on [Gorogiannis and Hunter, 2011, Table 2 and Proposition 10]).

Hunter, 2011; Amgoud and Besnard, 2013; Amgoud, 2014; Amgoud and Besnard, 2010; Arieli et al., 2020].[26]

Definition 66 (extension-based postulates). *Let $\mathcal{AF}(\mathcal{S}) = \langle \mathsf{Arg}(\mathcal{S}), \mathsf{Attack} \rangle$ be an argumentation framework for $\mathcal{S}$, based on a logic $\mathfrak{L} = \langle \mathcal{L}, \vdash \rangle$, and let $\mathsf{Free}_\mathfrak{L}(\mathcal{S}) = \bigcap \mathsf{MCS}_\mathfrak{L}(\mathcal{S})$.[27] The following postulates are defined with respect to the Sem-extensions of $\mathcal{AF}(\mathcal{S})$.*

Postulates on Individual Extensions, where $\mathcal{E} \in \mathsf{Sem}(\mathcal{AF}(\mathcal{S}))$:

- Support consistency: $\bigcup_{A \in \mathcal{E}} \mathsf{Sup}(A) \not\vdash \mathsf{F}$;
- Consistency: $\bigcup_{A \in \mathcal{E}} \mathsf{Conc}(A) \not\vdash \mathsf{F}$;
- Closure under support: *if $\mathsf{Sup}(A) \subseteq \mathsf{Sup}(\mathcal{E})$ then $A \in \mathcal{E}$;*
- Exhaustiveness: *if $\mathsf{Sup}(A) \cup \{\mathsf{Conc}(A)\} \subseteq \mathsf{Conc}(\mathcal{E})$, then $A \in \mathcal{E}$;*
- Strong exhaustiveness: *if $\mathsf{Sup}(A) \subseteq \mathsf{Conc}(\mathcal{E})$, then $A \in \mathcal{E}$;*
- Support inclusion: $\mathsf{Sup}(\mathcal{E}) \subseteq \mathsf{Conc}(\mathcal{E})$;

---

[26] We use naming conventions from [Amgoud, 2014; Arieli et al., 2020].

[27] When the underlying logic is clear from the context, we shall just write $\mathsf{Free}(\mathcal{S})$.

- Limited [strong] exhaustiveness: *[strong] exhaustiveness restricted to extensions $\mathcal{E}$ with $\bigcup \mathsf{Sup}(\mathcal{E}) \neq \emptyset$.*

Semantic-Wide Postulates:

- Core support consistency: $\bigcup_{A \in \bigcap \mathsf{Sem}(\mathcal{AF}(\mathcal{S}))} \mathsf{Sup}(A) \nvdash \mathsf{F}$;
- Core conclusion consistency: $\bigcap_{\mathcal{E} \in \mathsf{Sem}(\mathcal{AF}(\mathcal{S}))} \mathsf{Conc}(\mathcal{E}) \nvdash \mathsf{F}$;
- Core consistency: $\bigcup_{A \in \bigcap \mathsf{Sem}(\mathcal{AF}(\mathcal{S}))} \mathsf{Conc}(A) \nvdash \mathsf{F}$;
- Core closure:

$$\bigcap_{\mathcal{E} \in \mathsf{Sem}(\mathcal{AF}(\mathcal{S}))} \mathsf{Conc}(\mathcal{E}) = Cn_{\mathfrak{L}} \left(\bigcap_{\mathcal{E} \in \mathsf{Sem}(\mathcal{AF}(\mathcal{S}))} \mathsf{Conc}(\mathcal{E}) \right);$$

- Non-triviality: *there is an $\mathcal{S}$ for which* $\mathsf{Arg}(\mathcal{S}) \setminus \mathsf{Arg}(\mathsf{Free}(\mathcal{S})) \neq \emptyset$ *and* $\mathsf{Arg}(\mathcal{S}) \neq \bigcup \mathsf{Sem}(\mathcal{AF}(\mathcal{S}))$;
- Free precedence: $\mathsf{Arg}(\mathsf{Free}(\mathcal{S})) \subseteq \bigcap \mathsf{Sem}(\mathcal{AF}(\mathcal{S}))$;
- Maximal consistency: $\mathsf{Sem}(\mathcal{AF}(\mathcal{S})) = \{\mathsf{Arg}(\mathcal{T}) \mid \mathcal{T} \in MCS_{\mathfrak{L}}(\mathcal{S})\}$;
- Stability: $\mathsf{Stb}(\mathcal{AF}(\mathcal{S})) \neq \emptyset$;
- Strong stability: $\mathsf{Stb}(\mathcal{AF}(\mathcal{S})) = \mathsf{Prf}(\mathcal{AF}(\mathcal{S}))$.

We start with the results in [Gorogiannis and Hunter, 2011]:

Proposition 67. *Let $\mathcal{AF}(\mathcal{S}) = \langle \mathsf{Arg}(\mathcal{S}), Attack \rangle$ be a classical argumentation framework. Table 6 summarizes which of the (semantic-wide) postulates from Definition 66 are satisfied in $\mathcal{AF}(\mathcal{S})$ with respect to a semantic Sem and the conditions in Table 3.*

Another investigation of the rationality postulates in Definition 66 for logic-based argumentation appears in [Amgoud, 2014] and [Amgoud and Besnard, 2013]. Again, it is assumed that the supports of the arguments are consistent and minimal with respect to the subset relation. The core logic may be any explosive propositional logic, and the attack relations are divided according to the properties they have, which are specified in Definition 63 and in the following definition (see also [Amgoud, 2014, Definition 12]):

Postulate	Semantics	1,2,6	1',2,6'	1',2,6''	1,2,6'''
Free precedence	Sem_1	✓	✓	✓	✓
Non-triviality	Sem_2	✗	✗	✗	✗
Non-triviality	Grd	✓	✓	✓	✓
Core support consistency	Sem_1	✓	✓	✗	✓
$\mathsf{Grd}(\mathcal{AF}(\mathcal{S})) = \mathsf{Free}(\mathsf{Arg}(\mathcal{S}))$	Grd	✓	✓	✗	✓
Consistency	Grd	✓	✓	✗	✓
Consistency	Sem_1	✗	+D3' ✓	✗	✗

Table 6: Overview results of the (semantic-wide) postulates from Definition 66 that are satisfied by argumentation frameworks with semantics Sem (where $\mathsf{Sem}_1 \in \{\mathsf{Grd}, \mathsf{Cmp}, \mathsf{Prf}, \mathsf{Stb}\}$ and $\mathsf{Sem}_2 \in \{\mathsf{Cmp}, \mathsf{Prf}, \mathsf{Stb}\}$) and attacks satisfying the conditions in Table 3 (In the table, +D3' denotes that the attack postulate D3' is also required, in addition the postulates D1', D2 and D6').

Definition 68 (postulates R_1 and R_2 for attack rules). *Let $\mathcal{R}$ be an attack relation. The following conditions are verified with respect to every set $\mathcal{S}$ of $\mathcal{L}$-formulas:*[28]

R_1 *for every $A, B, C \in \mathsf{Arg}(\mathcal{S})$ such that $\mathsf{Sup}(A) \subseteq \mathsf{Sup}(B)$, it holds that if $(A, C) \in \mathcal{R}$ then $(B, C) \in \mathcal{R}$;*

R_2 *for every $A, B, C \in \mathsf{Arg}(\mathcal{S})$ such that $\mathsf{Sup}(A) \subseteq \mathsf{Sup}(B)$, it holds that if $(C, A) \in \mathcal{R}$ then $(C, B) \in \mathcal{R}$.*[29]

Proposition 69. *Let $\mathcal{AF}(\mathcal{S})$ be an argumentation framework, for some explosive propositional logic $\mathfrak{L} = \langle \mathcal{L}, \vdash \rangle$ and where the arguments are $\vdash$-consistent and $\subseteq$-minimal. Table 7 summarizes the results from [Amgoud and Besnard, 2013; Amgoud, 2014]. In particular, it shows which postulates are satisfied under the conditions in the left-most column.*[30]

---

[28] As usual, we exchange between the rule name and the corresponding relation.

[29] Note that R_2 corresponds to $D3'$ in Table 3.

[30] Note that the results in Table 7 refer also to the ideal (Idl) and the semi-stable (SStb) semantics. We refer to [Amgoud and Besnard, 2013; Amgoud, 2014], as well as to [Baroni and Giacomin, 2009; Baroni et al., 2011; Baroni et al., 2018] for their definitions.

In [Arieli et al., 2020] and its extension in [Borg, 2019, Chapter 4] many of the postulates from Definitions 61 and 66 are investigated for sequent-based argumentation [Arieli and Straßer, 2015]. In particular, the arguments may be of the general form of Definition 5 (no constraints are posed on their supports). Also, the base logic is any logic satisfying the standard rules in Table 8. Therefore, the characterizations in [Arieli et al., 2020] hold not only for classical logic, but also for many other logics, including intuitionistic logic and several modal logics.

Three classes of argumentation frameworks are studied:

- $\mathcal{AF}^{\mathsf{sub}}_{\mathfrak{L},\mathcal{A}}(\mathcal{S})$: frameworks based on Defeat and/or Undercut, therefore it holds that $\mathcal{A} \cap \{\mathsf{Def}, \mathsf{Ucut}\} \neq \emptyset$;

- $\mathcal{AF}^{\mathsf{dir}}_{\mathfrak{L},\mathcal{A}}(\mathcal{S})$: frameworks based on some and only direct attack rules, that is: $\emptyset \neq \mathcal{A} \subseteq \{\mathsf{DirDef}, \mathsf{DirUcut}\}$;

- $\mathcal{AF}^{\mathsf{con}}_{\mathfrak{L},\mathcal{A}}(\mathcal{S})$: frameworks that, in addition to only direct attack rules, are based on Consistency Undercut, i.e., $\{\mathsf{ConUcut}\} \subsetneq \mathcal{A} \subseteq \{\mathsf{ConUcut}, \mathsf{DirDef}, \mathsf{DirUcut}\}$.

Proposition 70. ([Arieli et al., 2020, Theorem 1]) *Let $\mathfrak{L} = \langle \mathcal{L}, \vdash \rangle$ be a logic in which the rules of Table 8 are satisfied. Table 9 lists which rationality postulates are satisfied by the three classes of frameworks defined above, and with respect to which semantics* Sem $\in \{$Grd, Cmp, Prf, Stb$\}$.

Remark 71. *The columns of $\mathcal{AF}^{\mathsf{dir}}_{\mathfrak{L},\mathcal{A}}(\mathcal{S})$ and $\mathcal{AF}^{\mathsf{con}}_{\mathfrak{L},\mathcal{A}}(\mathcal{S})$ in Table 9 show that all the postulates are compatible (that is, they can be satisfied together).*

In [Borg and Straßer, 2018], relevance in structured argumentation is studied. In particular it is investigated, under which conditions the entailment relation induced by a framework of structured argumentation is robust under the addition of irrelevant information, i.e., information that can already be derived from it (semantic irrelevance) or information that is syntactically unrelated to the already available information (syntactic irrelevance). Rather than taking one of the main approaches to structured argumentation, a simple argumentation setting is introduced, into which the other approaches can be translated. The main results on syntactic relevance are based on the notion of *pre-relevance*,

	P1	P2	P3	P4	P5	P6	P7
$\mathsf{Sem}(\mathcal{AF}(\mathcal{S})) = \emptyset$				✓		✓	
$\mathsf{Sem}(\mathcal{AF}(\mathcal{S})) = \emptyset + \mathsf{Cn}_{\mathfrak{L}}(\emptyset) \neq \emptyset$	×	×		✓		✓	
$\mathsf{Sem}(\mathcal{AF}(\mathcal{S})) = \emptyset + \mathsf{Free}(\mathcal{S}) \neq \emptyset$				✓		✓	×
$\mathsf{Sem}(\mathcal{AF}(\mathcal{S})) \neq \emptyset + \mathcal{E} = \mathsf{Arg}(\mathsf{Supp}(\mathcal{E}))$			✓				
$\mathsf{Cn}_{\mathfrak{L}} \neq \emptyset + \mathsf{Sem} = \mathsf{Adm}$	×						
Closure	✓	✓					
Consistency				✓		✓	
Support consistency				✓	✓	✓	
Support consistency Conflict-dependent			Naive	✓	✓	✓	
Support cons. + Confl.-dep. + $\mathsf{Stb}(\mathcal{AF}(\mathcal{S})) \neq \emptyset$			Stb	✓	✓	✓	
Consistency + Sub arg. closure			✓	✓	✓	✓	
Consistency + $\mathcal{E} = \mathsf{Arg}(\mathsf{Supp}(\mathcal{E}))$				✓	✓	✓	
Conflict dependent							Sem_2
Conflict dependent + Sensitive			Sem_1				Sem_2
Conflict dependent + Symmetric + $|C| > 2$				×			Sem_2
Exhaustive + $\mathcal{E} = \mathsf{Arg}(\mathsf{Supp}(\mathcal{E}))$	✓	✓		✓	✓	✓	
$R_1 + R_2$			Sem_1				
R_2			Stb				

Table 7: Overview of the results from [Amgoud and Besnard, 2013; Amgoud, 2014], under the assumptions stated in Proposition 69. Legend of the postulates: P1 = closure, P2 = core closure, P3 = sub-argument closure, P4 = consistency, P5 = support consistency, P6 = core conclusion consistency, P7 = free precedence. Also, $\mathsf{Sem}_1 \in \{\mathsf{Grd}, \mathsf{Cmp}, \mathsf{Prf}, \mathsf{Idl}, \mathsf{Stb}, \mathsf{SStb}\}$ and $\mathsf{Sem}_2 \in \{\mathsf{Grd}, \mathsf{Prf}, \mathsf{Idl}, \mathsf{SStb}\}$. The condition $|C| > 2$ denotes that there is a minimal conflict of three or more formulas. Only the results from [Amgoud and Besnard, 2013; Amgoud, 2014] are shown: ✓ indicates that the postulate is satisfied for all considered semantics, Sem indicates that the postulate is satisfied for the particular semantics, × indicates that the postulate is not satisfied and an empty box indicates that the result is unknown, under the given conditions.

Rule Name	Rule's Conditions	Rule's Conclusion
Reflexivity		$\langle \psi, \psi \rangle$
Monotonicity	$\langle \Gamma, \Delta \rangle$	$\langle \Gamma \cup \Gamma', \Delta \cup \Delta' \rangle$
Transitivity	$\langle \Gamma_1, \Delta_1 \cup \{\psi\} \rangle$, $\langle \Gamma_2 \cup \{\psi\}, \Delta_2 \rangle$	$\langle \Gamma_1 \cup \Gamma_2, \Delta_1 \cup \Delta_2 \rangle$
Left-$\wedge$	$\langle \Gamma \cup \{\psi\} \cup \{\phi\}, \Delta \rangle$	$\langle \Gamma \cup \{\psi \wedge \phi\}, \Delta \rangle$
Right-$\wedge$	$\langle \Gamma, \Delta \cup \{\psi\} \rangle$, $\langle \Gamma, \Delta \cup \{\phi\} \rangle$	$\langle \Gamma, \Delta \cup \{\psi \wedge \phi\} \rangle$
Left-$\neg$	$\langle \Gamma, \Delta \cup \{\psi\} \rangle$	$\langle \Gamma \cup \{\neg\psi\}, \Delta \rangle$
Right-$\neg$	$\langle \Gamma \cup \{\psi\}, \Delta \rangle$	$\langle \Gamma, \Delta \cup \{\neg\psi\} \rangle$

Table 8: Rules for the base logics in Proposition 70.

which is related to *basic relevance* known from relevance logics [Avron, 2014]. Intuitively, a consequence relation satisfies pre-relevance, if the derived conclusion can be derived from a relevant (w.r.t. shared atoms) subset of the antecedents.

Definition 72 (pre-relevance). *A consequence relation $\vdash \;\subseteq \wp(\mathcal{L}) \times \mathcal{L}$ satisfies pre-relevance, if for each disjoint sets $\mathcal{S}_1 \cup \{\phi\} \mid \mathcal{S}_2$, if $\mathcal{S}_1, \mathcal{S}_2 \vdash \phi$ then there is some $\mathcal{S}'_1 \subseteq \mathcal{S}_1$ such that $\mathcal{S}'_1 \vdash \phi$.*

Example 73. *We list some entailment relations that satisfy pre-relevance:*

- *the consequence relation of the (semi-)relevance logic RM ([Avron, 2016, Proposition 6.5]),*

- *the entailment $\vdash_{\mathsf{CL}}^{\top}$ that is the restriction of $\vdash_{\mathsf{CL}}$ to pairs (Γ, ϕ), for which it holds that $\nvdash_{\mathsf{CL}} \neg \bigwedge \Gamma$, and*

- *the entailment $\vdash_{\cup_{\mathsf{mcs}}}^{\mathsf{CL}}$ (Definition 44).*[31]

---

[31] In [Wu and Podlaszewski, 2014] $\vdash_{\mathsf{CL}}^{\top}$ is used to obtain a crash-resistant version of ASPIC, and, similarly, in [Grooters and Prakken, 2016] the authors make use of $\vdash_{\cup_{\mathsf{mcs}}}^{\mathsf{CL}}$ also for ASPIC.

Postulate	$\mathcal{AF}^{dir}_{\mathfrak{L},\mathcal{A}}(\mathcal{S})$	$\mathcal{AF}^{con}_{\mathfrak{L},\mathcal{A}}(\mathcal{S})$	$\mathcal{AF}^{sub}_{\mathfrak{L},\mathcal{A}}(\mathcal{S})$
Closure	✓	✓	×
Closure under support	✓	✓	×
Sub-argument closure	✓	✓	✓
Support inclusion	✓	✓	✓
Consistency	✓	✓	×
Support consistency	✓	✓	×
Maximal consistency	Prf, Stb	Prf, Stb	×
Exhaustiveness	Prf, Stb	✓	×
Limited exhaustiveness	✓	✓	×
Strong exhaustiveness	Prf, Stb	✓	×
Limited strong exhaustiveness	✓	✓	×
Free precedence	Prf, Stb	✓	✓
Limited free precedence	✓	✓	✓
Stability	✓	✓	✓
Strong stability	✓	✓	✓
Non-interference	Prf, Stb	✓	✓
Crash-resistance	Prf, Stb	✓	✓

Table 9: Postulates satisfaction (Proposition 70, originally presented in [Arieli et al., 2020]) for Sem ∈ {Grd, Cmp, Prf, Stb}. Cells with ✓ indicate no conditions for the postulate, otherwise specific semantics with respect to which the postulate holds are indicated. Cells with × mean that the postulate does not hold. In case of non-interference and crash-resistance the base logic is assumed to be uniform.[32]

The following proposition follows from [Borg and Straßer, 2018, Th. 1].

---

[32]A logic $\mathfrak{L} = \langle \mathcal{L}, \vdash \rangle$ is called *uniform* [Łos and Suszko, 1958; Urquhart, 2001], if for every two sets $\mathcal{S}_1, \mathcal{S}_2$ of $\mathcal{L}$-formulas and an $\mathcal{L}$-formula ψ it holds that $\mathcal{S}_1 \vdash \psi$ iff $\mathcal{S}_1, \mathcal{S}_2 \vdash \psi$ and $\mathcal{S}_2$ is a $\vdash$-consistent set such that $\mathsf{Atoms}(\mathcal{S}_2) \cap \mathsf{Atoms}(\mathcal{S}_1 \cup \{\psi\}) = \emptyset$.

Proposition 74. *Let $\vdash$ be a pre-relevant consequence relation over the language $\mathcal{L}$, $\mathcal{S}$ be a set of $\mathcal{L}$-sentences, $\mathsf{Arg}_\vdash(\mathcal{S}) = \{\langle \Gamma, \phi \rangle \mid \Gamma \vdash \phi\}$, Attack is induced by direct attack rules (such as DirDef and/or DirUcut) and let $\mathcal{AF}(\mathcal{S}) = \langle \mathsf{Arg}_\vdash(\mathcal{S}), Attack \rangle$ be the corresponding argumentation framework. Then the relation $\mid\!\!\sim_{\star\mathsf{Sem}}$ satisfies non-interference for $\star \in \{\cap, \cap\!\!\!\!\!\cap, \cup\}$ and $\mathsf{Sem} \in \{\mathsf{Grd}, \mathsf{Cmp}, \mathsf{Prf}\}$.*

Remark 75. *Like the examples in items 2 and 3 of Example 73, consequence relations $\vdash$ considered in Proposition 74 need not be induced by a logic in the technical sense of Definition 1. In fact, as is demonstrated in [Borg and Straßer, 2018], structured argumentation frameworks such as ASPIC and ABA can be translated into the $\vdash$-based argumentation frameworks of Proposition 74.*

B. Rationality postulates for ASPIC$^+$

Discussions on rationality postulates for ASPIC$^+$ can be found, among others, in [Modgil and Prakken, 2013; Modgil and Prakken, 2018; Caminada, 2018b]. For the completeness of the presentation we recall here some of the main results. For this, we need two notions, introduced in [Modgil and Prakken, 2018] and [Dung and Thang, 2014], respectively.

Definition 76 (well-formed argumentation framework). *An ASPIC argumentation framework defined by an ASPIC argumentation theory $AT = \langle AS, \mathcal{K} \rangle$, where $AS = \langle \mathcal{L}, ^-, \mathcal{R}, n \rangle$ and $\mathcal{K} = \mathcal{K}_n \cup \mathcal{K}_p$, is called well-formed, if whenever ϕ is a contrary of ψ (i.e., $\phi \in \overline{\psi}$ while $\psi \notin \overline{\phi}$), then $\psi \notin \mathcal{K}_n$ and ψ is not the consequent of a strict rule.*

Definition 77 (self-contradiction axiom; closure under transposition). *An ASPIC argumentation framework $\mathcal{AF}(AT) = \langle \mathsf{Arg}(AT), Attack \rangle$, defined by an ASPIC argumentation theory $AT = \langle AS, \mathcal{K} \rangle$, where $AS = \langle \mathcal{L}, ^-, \mathcal{R}, n \rangle$ and $\mathcal{K} = \mathcal{K}_n \cup \mathcal{K}_p$ satisfies:*

- *the self-contradiction axiom, if for each minimally inconsistent set $\mathcal{S}$ of $\mathcal{L}$-formulas it holds that $\{\neg \phi \mid \phi \in \mathcal{S}\} \subseteq Cn_{\mathcal{R}_s}(\mathcal{S})$;*[33]

---

[33] A set $\mathcal{S}$ of $\mathcal{L}$-formulas is *minimally inconsistent* if there is some formula ϕ such that $\phi \in \mathsf{Cn}_{\mathcal{R}_S}(\mathcal{S})$ and $\overline{\phi} \in \mathsf{Cn}_{\mathcal{R}_S}(\mathcal{S})$, and for each $\mathcal{S}' \subsetneq \mathcal{S}$ no such ϕ exists.

- closure under transposition, *if for each $\phi_1, \ldots, \phi_n \to \phi \in \mathcal{R}_s$, for each $i \in \{1, \ldots, n\}$, $\phi_1, \ldots, \phi_{i-1}, \neg\phi, \phi_{i+1}, \ldots, \phi_n \to \neg\phi_i \in \mathcal{R}_s$ as well.*

Proposition 78. ([Dung and Thang, 2014],[Modgil and Prakken, 2018]) *Let $\mathcal{AF}(AT) = \langle \text{Arg}(AT), \text{Attack} \rangle$ be an argumentation framework and let $\mathcal{E} \in \text{Cmp}(\mathcal{AF}(AT))$. Table 10 lists the rationality postulates that are satisfied under the different conditions of Definitions 76 and 77.*

Postulate	−	Well-formed framework	Self-contradiction axiom	Closure under transposition
Sub-argument closure	✓	✓	✓	✓
Closure	✗	✓	✓	✓
Direct consistency	✗	✓	✓	✓
Indirect consistency	✗	✓	✓	✓

Table 10: Overview of the postulates that are satisfied by ASPIC⁺ argumentation frameworks given some condition on the set of strict rules and a contrary relation. The column titled − denotes that there are no requirements placed on the framework.

Remark 79. *The results in [Modgil and Prakken, 2018] are given for prioritized frameworks (i.e., with a preference relation defined on the arguments of $\mathcal{AF}(AT)$). However, since the non-prioritized setting is a special case of the prioritized setting, the results still apply here.*

The satisfaction of the non-interference and crash-resistance postulates for ASPIC⁺ are not so straightforward. For example, when the strict rules are based on classical logic, explosion might still occur. See [Caminada, 2018b] for an extensive discussion on non-interference and crash-resistance for ASPIC⁺. One of the challenges when trying to resolve these issues is that the postulates from [Caminada and Amgoud, 2007] should still be satisfied by the resulting framework.

Several variants of ASPIC⁺ have been proposed in the literature, some of them satisfy non-interference and crash-resistance. An overview of some of these systems, the settings in which they have been studied and the postulates that they satisfy, can be found in Table 11.[34]

System	Priorities	Incons. arg. filtered	Direct consistency	Closure	Crash resistance
ASPIC$^+$	Yes	No	✓	✓	×
ASPIC Lite	No	Yes	Cmp	Cmp	Cmp
ASPIC Lite	Yes	Yes	Cmp	×	Cmp
ASPIC*	Yes	No	✓	✓†	✓†
ASPIC$^-$	Yes	No	✓	✓	×
ASPIC$^-$	No	Yes	✓	✓	✓
ASPIC$^-$	Yes	Yes	✓	×	✓
ASPIC$^\ominus$	Yes	No	Grd	Grd	Grd

Table 11: Overview of the different variants to ASPIC$^+$ and the conditions under which some of the postulates are satisfied. "Yes" means that the results also hold when taking into account priorities over the defeasible rules, whereas "no" means that when priorities are taken into account, counter-examples to the results exist. In columns 4–6, ✓ denotes that the postulate is satisfied, × denotes that the postulate is not satisfied, and Cmp [resp. Grd] denotes that the postulate is studied and satisfied for complete [resp. grounded] semantics Finally, ✓† denotes that a weaker variant of the postulate is satisfied.

Remark 80. *Below are some further explanations and notes that are related to the results in Table 11.*

- *The variant* ASPIC Lite, *introduced in [Wu and Podlaszewski, 2014], is obtained by filtering all inconsistent arguments out of the argumentation framework. An argument A is inconsistent if $\{\mathsf{Conc}(B) \mid B \in \mathsf{Sub}(A)\}$ is inconsistent. It is then shown that*

---

[34] As for ASPIC$^+$ with filtering out inconsistent arguments: no results are known, even though ASPIC Lite is its subsystem.

non-interference and crash-resistance are satisfied for complete semantics, while the postulates from [Caminada and Amgoud, 2007] are still satisfied as well. For the proof it is necessary that at least one extension exists. Among others, that is why other semantics are not discussed in that particular paper. Moreover, it is shown that the results do not hold when preferences are introduced.

- A weaker version of crash-resistance, called non-triviality, is discussed in [Grooters and Prakken, 2016]. This variant, called ASPIC*, restricts the application of strict rules. In particular, chaining of strict rules and applying strict rules to inconsistent sets of antecedents is not allowed.

- ASPIC⁻ [Caminada et al., 2014] is a variant of ASPIC⁺ that uses the attack form of unrestricted rebut. Its violation of non-interference is shown in [Heyninck and Straßer, 2017]. Closure is also violated if inconsistent arguments are filtered out, in the presence of priorities.

- Another variant of ASPIC⁺ with unrestricted rebut, called ASPIC$^\ominus$, is studied in [Heyninck and Straßer, 2017] and [Heyninck and Straßer, 2019]. In ASPIC$^\ominus$, the notion of unrestricted rebut is generalized such that an argument can attack another argument if its conclusion claims that a subset of the commitments of the attacked argument are not tenable together. It is shown that the resulting framework ASPIC$^\ominus$, where the priority relation is a preorder using the so-called weakest link principle, satisfies the rationality postulates from both [Caminada and Amgoud, 2007] and [Caminada et al., 2011] under grounded semantics.

C. Rationality postulates for ABA

Recall from Section 2.2.3 that an extension is a set of assumptions (more precisely, $\mathcal{E} \subseteq \mathcal{A}$ for every extension $\mathcal{E}$ of an assumption-based framework $\mathcal{ABF} = \langle \mathcal{L}, \mathcal{R}, \mathcal{A}, \sim \rangle$) that is also closed with respect to the rules in $\mathcal{R}$ (i.e., $\mathcal{E} = Cn_{\vdash_\mathcal{R}}(\mathcal{E})$). From this it follows immediately that the closure postulate from Definition 60 is satisfied. Thus, from the rationality postulates in [Caminada and Amgoud, 2007], it remains to show consistency. In the context of flat assumption-based argumentation frameworks, this postulate can be defined as follows [Čyras and Toni, 2016a; Heyninck and Straßer, 2021a]:

> **Consistency:** for all extensions $\mathcal{E}$, it holds that there are no $\phi, \psi \in \mathcal{E}$ such that $\phi \in \sim\!\psi$.[35]

In the non-prioritized setting, as discussed in this chapter, it follows immediately that extensions for any of the considered semantics are consistent, since otherwise these would not be conflict-free (recall the definition of attack in assumption-based frameworks, Definition 33). However, as shown in e.g., [Čyras and Toni, 2016a; Heyninck and Straßer, 2021a], whether an assumption-based framework satisfies consistency in a prioritized setting depends on the definition of the preference ordering and the notion of conflict-freeness. A discussion of this is beyond the scope of this chapter.[36]

The rationality postulates for inferences (recall Definition 61) have been studied for simple contrapositive assumption-based frameworks (recall Definition 35) in [Heyninck and Arieli, 2020b]. Note that, since the entailment relation for assumption-based frameworks is defined for frameworks and not for sets of formulas (as in the case of the discussed logic-based approaches), the notion of syntactically disjoint sets of formulas has to be lifted to assumption-based frameworks. Two assumption-based frameworks $\mathcal{ABF}_1 = \langle \mathfrak{L}, \Gamma_1, \Delta_1, \sim_1 \rangle$ and $\mathcal{ABF}_2 = \langle \mathfrak{L}, \Gamma_2, \Delta_2, \sim_2 \rangle$ are *syntactically disjoint* if $(\Gamma_1 \cup \Delta_1) \mid (\Gamma_2 \cup \Delta_2)$. Besides this new notion of syntactically disjointness, the definitions of non-interference and crash-resistance remain the same as for logic-based argumentation and the ASPIC-family.

---

[35]Since [Čyras and Toni, 2016a; Heyninck and Straßer, 2021a] restrict their attention to flat assumption-based argumentation frameworks, this notion of consistency is equivalent to the following formulation, which bears closer similarities to *indirect consistency*: for all extensions $\mathcal{E}$, it holds that there are no $\phi, \psi \in \mathcal{L}$ s.t. $\mathcal{E} \vdash_S \phi$ and $\mathcal{E} \vdash_S \psi$ and $\phi \in \sim\!\psi$.

[36]In contexts where besides the contrariness relation there are other negations (e.g., when translating extended logic programs into ABA), various notions of consistency may have to be considered (see e.g., [Wakaki, 2017]).

Proposition 81. *([Heyninck and Arieli, 2020b, Theorems 7 and 8]) Let $\mathcal{ABF} = \langle \mathfrak{L}, \Gamma, \Delta, \sim \rangle$ be a simple contrapositive assumption-based framework. Table 12 lists under what conditions the corresponding entailment relations satisfy non-interference for Sem $\in$ {Naive, Prf, Stb, Grd, WF}.*

Entailment	−	$F \in Ab$
$\mathrel{\mid\!\sim}_{\cap\text{Sem}}$	Sem $\in$ {Naive, Prf, Stb}	Sem $\in$ {Naive, Prf, Stb}
$\mathrel{\mid\!\sim}_{\cup\text{Sem}}$	Sem $\in$ {Naive, Prf, Stb}	Sem $\in$ {Naive, Prf, Stb}
$\mathrel{\mid\!\sim}_{\cap\text{Sem}}$	×	Sem $\in$ {Grd, WF}

Table 12: Results from [Heyninck and Arieli, 2020b] concerning the conditions and semantics under which simple-contrapositive assumption-based frameworks satisfy non-interference. × denotes that non-interference is not satisfied for any Sem $\in$ {Naive, Prf, Stb, Grd, WF}.

In [Borg and Straßer, 2018] it is shown that

- ABA frameworks with domain-specific rules and whose contrariness relation do not introduce syntactic discontinuities, i.e., for all formulas ϕ we have that $\text{Atoms}(\sim \phi) \subseteq \text{Atoms}(\phi)$, satisfy non-interference, and

- ABA frameworks whose inference rules $\mathcal{R}$ are induced by logics $\mathfrak{L} = \langle \mathcal{L}, \vdash \rangle$ for which $\vdash$ is pre-relevant (see Definition 72), i.e., $\phi_1, \ldots, \phi_n \to \psi \in \mathcal{R}$ iff $\phi_1, \ldots, \phi_n \vdash \psi$, satisfy non-interference.

2.3.3 Inference Principles for Non-Monotonic Reasoning

Next, we examine the argumentation-based entailment relations from Definitions 12, 26 and 36, relative to general patterns for non-monotonic reasoning, originally studied in [Shoham, 1988], [Gabbay, 1985], [Kraus et al., 1990; Lehmann and Magidor, 1992], and [Makinson, 1994]. These works study how to adjust the set of conclusions (which may be reduced, not necessarily increased) upon a growth in the set of assumptions. In our case, since the assumptions are divided to strict premises and defeasible premises, it will be useful to distinguish between the two ways of increasing the set of premises: we shall use the operator ⊎ for the addition of strict premises and ⋓ for the addition of defeasible premises. Accordingly, we define:

Definition 82 (premise addition). *Let $S = \langle S_s, S_d \rangle$ be a pair of sets of formulas in a language $\mathcal{L}$.*[37] *We denote:*

- $S \cupplus \phi = \langle S_s, S_d \rangle \cupplus \phi = \langle S_s, S_d \cup \{\phi\} \rangle$,
- $S \uplus \phi = \langle S_s, S_d \rangle \uplus \phi = \langle S_s \cup \{\phi\}, S_d \rangle$.

Note that logic-based argumentation is considered here only with respect to defeasible assumptions, therefore ⊎ will not be used in that context, and the meaning of ⋓ in case the logic-based argumentation is simply the union, ∪. For the other formalisms, addition of premises is defined as follows:

Definition 83 (premise addition in ASPIC). *Let $AT = \langle \langle \mathcal{L}, ^-, \mathcal{R}, n \rangle, \langle \mathcal{K}_n, \mathcal{K}_p \rangle \rangle$ be an ASPIC argumentation theory, and let ϕ be an $\mathcal{L}$-formula. We define:*

- $AT \cupplus \phi = \langle \langle \mathcal{L}, ^-, \mathcal{R}, n \rangle, \mathcal{K} \cupplus \phi \rangle$ *(where $\phi \notin \mathcal{K}_n$),*
- $AT \uplus \phi = \langle \langle \mathcal{L}, ^-, \mathcal{R}, n \rangle, \mathcal{K} \uplus \phi \rangle$ *(where $\phi \notin \mathcal{K}_p$).*

Definition 84 (premise addition in ABA). *Let $\mathcal{ABF} = \langle \mathcal{L}, \mathcal{R}, \mathcal{A}, \sim \rangle$ be an assumption-based argumentation framework, and let ϕ be an $\mathcal{L}$-formula. We define:*

- $\mathcal{ABF} \cupplus \phi = \langle \mathcal{L}, \mathcal{R} \setminus \{\Theta \to \phi \mid \Theta \subset \mathsf{WFF}(\mathcal{L})\}, \mathcal{A} \cup \{\phi\}, \sim \rangle$,[38]
- $\mathcal{ABF} \uplus \phi = \langle \mathcal{L}, \mathcal{R} \cup \{\to \phi\}, \mathcal{A} \setminus \{\phi\}, \sim \rangle$.[39]

---

[37] The subscripts 's' and 'd' indicate that, intuitively, the first component consists of the strict premises and the second component is the set of defeasible premises.

[38] Removing $\Theta \to \phi$ from Γ ensures that $\mathcal{ABF} \cupplus \phi$ is flat if so is $\mathcal{ABF}$, and is proposed in [Cyras and Toni, 2016b]. Furthermore, we let $\sim\phi = \emptyset$ and $\sim\psi$ is defined as in the original $\mathcal{ABF}$ for any $\psi \in \mathcal{A}$.

[39] $\sim\psi$ is defined as in the original $\mathcal{ABF}$ for any $\psi \in \mathcal{A} \setminus \{\phi\}$.

Let $\mathcal{ABF} = \langle \mathfrak{L}, \Gamma, \Delta, \sim \rangle$ be a (simple) contrapositive assumption-based argumentation framework, and let ϕ be an $\mathcal{L}$-formula. We define:

- $\mathcal{ABF} \uplus\!\!\!\!\cup\, \phi = \langle \mathfrak{L}, \Gamma, \Delta \cup \{\phi\}, \sim \rangle$,

- $\mathcal{ABF} \uplus \phi = \langle \mathfrak{L}, \Gamma \cup \{\phi\}, \Delta, \sim \rangle$.[40]

Using the operators $\uplus$ and $\uplus\!\!\!\!\cup$ we can now consider known postulates for non-monotonic reasoning, adjusted to the two types of information updates. To make the presentation more compact we will define the properties for ASPIC, ABA, MCS-based reasoning and logic-based argumentation in one definition. For this we call a *knowledge base* one of the following:

⋄ an ASPIC argumentation theory $\text{AT} = \langle\langle \mathcal{L}, \overline{}, \mathcal{R}, n \rangle, \langle \mathcal{K}_n, \mathcal{K}_p \rangle\rangle$,

⋄ an assumption-based framework $\mathcal{ABF}$,

⋄ a set of $\mathcal{L}$-formulas for logic-based argumentation with a logic $\mathfrak{L} = \langle \mathcal{L}, \vdash \rangle$, or

⋄ a pair of $\mathcal{L}$-formulas $\langle \mathcal{S}', \mathcal{S} \rangle$ in MCS-based reasoning and a logic $\mathfrak{L} = \langle \mathcal{L}, \vdash \rangle$.

In the context of a fixed language $\mathcal{L}$ resp. a fixed logic $\mathfrak{L} = \langle \mathcal{L}, \vdash \rangle$ resp. a fixed set of strict rules $\mathcal{R}_s$, it will also be useful to consider *empty knowledge bases*, written $\text{KB}_\emptyset$ and denoting, the argumentation theory $\text{AT} = \langle\langle \mathcal{L}, \overline{}, \emptyset, n \rangle, \langle \emptyset, \emptyset \rangle\rangle$ in the context of ASPIC, resp. the assumption-based framework $\langle \mathcal{L}, \mathcal{R}_s, \emptyset, \emptyset \rangle$ in the context of assumption-based argumentation, resp. the pair of empty sets of $\mathcal{L}$-formulas $\langle \emptyset, \emptyset \rangle$ in the context of MCS-based reasoning, resp. the empty set of $\mathcal{L}$-formulas in the context of logic-based argumentation.

Definition 85 (properties for non-monotonic reasoning). Let $\mathfrak{L} = \langle \mathcal{L}, \vdash \rangle$ be a propositional logic, KB a knowledge base, ϕ, ψ, σ $\mathcal{L}$-formulas, and $\sqcup \in \{\uplus\!\!\!\!\cup, \uplus\}$. For an entailment relation $\mathrel{\vert\!\sim} \,\subseteq \wp(\mathcal{L}) \times \mathcal{L}$ we define:

$\sqcup$-Cautious Reflexivity ($\sqcup$-**CREF**): $\text{KB}_\emptyset \sqcup \phi \mathrel{\vert\!\sim} \phi$ where ϕ is $\vdash$-consistent.

$\sqcup$-Reflexivity ($\sqcup$-**REF**): $\text{KB}_\emptyset \sqcup \phi \mathrel{\vert\!\sim} \phi$.

---

[40]Since in the context of simple contrapositive assumption-based frameworks is is not necessary to restrict attention to flat assumption-based frameworks, ϕ is not removed from Δ.

Right Weakening *(**RW**)*: If $\text{KB} \mathrel{|\!\sim} \phi$ and $\phi \vdash \psi$ then $\text{KB} \mathrel{|\!\sim} \psi$.

⊔-Cautious Monotonicity *(**⊔-CM**)*: If $\text{KB} \mathrel{|\!\sim} \phi$ and $\text{KB} \mathrel{|\!\sim} \psi$ then $\text{KB} \sqcup \{\phi\} \mathrel{|\!\sim} \psi$.

⊔-Cautious Cut *(**⊔-CC**)*: If $\text{KB} \mathrel{|\!\sim} \psi$ and $\text{KB} \sqcup \{\psi\} \mathrel{|\!\sim} \phi$ then $\text{KB} \mathrel{|\!\sim} \phi$.

⊔-Left Logical Equivalence *(**⊔-LLE**)*: If $\vdash \phi \equiv \psi$ and $\text{KB} \sqcup \phi \mathrel{|\!\sim} \sigma$ then $\text{KB} \sqcup \psi \mathrel{|\!\sim} \sigma$.

⊔-OR *(**⊔-OR**)*: If $\text{KB} \sqcup \phi \mathrel{|\!\sim} \delta$ and $\text{KB} \sqcup \psi \mathrel{|\!\sim} \delta$ then $\text{KB} \sqcup \{\phi \vee \psi\} \mathrel{|\!\sim} \delta$.

⊔-Rational Monotonicity *(**⊔-RM**)*: If $\text{KB} \mathrel{|\!\sim} \psi$ and $\text{KB} \mathrel{|\!\not\sim} \neg\phi$ then $\text{KB} \sqcup \phi \mathrel{|\!\sim} \psi$.[41]

Remark 86. *We refer to [Kraus et al., 1990; Lehmann and Magidor, 1992] for a detailed discussion on* **CM**, **RW**, **LLE**, **OR**, *and* **RM** *and to [Gabbay, 1985] for a discussion on* **CC**. *All of these properties are well-known and have been extensively examined in different contexts and for different purposes involving inference in a non-monotonic way.*

Some interesting variations of these properties have been considered in the literature but have, to the best ouf our knowledge, not been studied for argumentative consequence relations. For example, an interesting weaker variant of cautious monotony is known as very cautious monotony **(VCM)** *[Hawthorne and Makinson, 2007] or conjunctive cautious monotony [Bochman, 2013] and is defined as follows: if* $\Gamma \mathrel{|\!\sim} \phi \wedge \psi$ *then* $\Gamma \sqcup \phi \mathrel{|\!\sim} \psi$. *This variant has not been studied yet in structured argumentation.*

Another variation is semi-monotonicity **(SM)** *[Antoniou, 1998], stating that when adding defeasible information, every extension (according to a given semantics) of the original framework is a subset of some extension of the supplemented framework. For more variants of the properties discussed here, we refer the reader to [Bochman, 2013; Eichhorn, 2018] in which many more variants are discussed and studied.*

The properties in Definition 85 are often gathered for defining systems for non-monotonic inference.

---

[41]In ASPIC this has to be rephrased in terms of the contrariness relation instead of negation: If $\text{KB} \mathrel{|\!\sim} \psi$ and $\text{KB} \mathrel{|\!\not\sim} \phi'$ for all $\phi' \in \overline{\phi}$, then $\text{KB} \sqcup \phi \mathrel{|\!\sim} \psi$.

Definition 87 (systems for non-monotonic inference). *Let* $\sqcup \in \{\talloblong, \uplus\}$. *We say that an entailment* $\vdash$ *is:*

- $\sqcup$-*cumulative, if it satisfies* $\sqcup$-**REF**, **RW**, $\sqcup$-**LLE**, $\sqcup$-**CM** *and* $\sqcup$-**CC**.

- $\sqcup$-*cautiously cumulative, if it satisfies* $\sqcup$-**CREF**, **RW**, $\sqcup$-**LLE**, $\sqcup$-**CM** *and* $\sqcup$-**CC**.

- $\sqcup$-*(cautiously) preferential, if it is* $\sqcup$-*(cautiously) cumulative and satisfies* $\sqcup$-**OR**.

- $\sqcup$-*(cautiously) rational, if it is* $\sqcup$-*(cautiously) preferential and satisfies* $\sqcup$-**RM**.

Table 13 classifies the argumentation-based entailment relations according to Definition 87.[42]

The positive results presented in Table 13 follow from the representational results in Propositions 46, 48 and 51, using the next two propositions:

Proposition 88. *Let* $\mathfrak{L} = \langle \mathcal{L}, \vdash \rangle$ *be a propositional logic. The entailments* $\vdash^{\mathfrak{L}}_{\cap\mathsf{mcs}}$ *and* $\vdash^{\mathfrak{L}}_{\Cap\mathsf{mcs}}$ *are* $\talloblong$-*cautiously cumulative and* $\uplus$-*cumulative.*

Proposition 89. *Let* $\mathfrak{L} = \langle \mathcal{L}, \vdash \rangle$ *be a propositional logic and let* $\sqcup \in \{\talloblong, \uplus\}$. *The entailment* $\vdash^{\mathfrak{L}}_{\Cap\mathsf{mcs}}$ *is* $\sqcup$-*preferential.*

Proofs of the last two propositions are given in Appendix A.

Remark 90. *Some of the results in Table 13 have been shown before. For instance, in [Benferhat et al., 1993] it is shown that* $\vdash^{\mathsf{CL}}_{\Cap\mathsf{mcs}^{\emptyset}_{\mathfrak{L}}}$ *is* $\talloblong$-*preferential, the results for simple contrapositive ABFs are shown in [Heyninck and Arieli, 2018], and the results concerning the* $\talloblong$-*cautious cumulativity and the non* $\talloblong$-*cautious preferentiality of* $\vdash^{\mathfrak{L},\mathcal{A}_{\mathsf{UD}}}_{\cap\mathsf{gps}}$ *follow from [Arieli and Straßer, 2019, Proposition 16 and Note 10].*

---

[42]Since the credulous entailment is often monotonic (see [Benferhat *et al.*, 1997] for MCS-based reasoning and [Borg *et al.*, 2021, Proposition 8] for argumentation-based reasoning), the results in Table 13 refer to skeptical entailments.

System.	MCS reasoning $\mid\!\sim^{\mathcal{L}}_{\cap mcs}$	$\mid\!\sim^{\mathcal{L}}_{\cap mcs}$	logic-based arg. $\mid\!\sim^{\mathcal{L},\mathcal{A}_{UD}}_{\cap gps}$	$\mid\!\sim^{CL,DirUcut}_{\cap ps}$	simple contrap. $\mid\!\sim_{\cap ps}$	$\mid\!\sim_{\cap gps}$	ABA $\mid\!\sim_{Grd}$ (†)	ASPIC $\mid\!\sim_{\cap Stb}$ (‡)
⊔-ccum.	Yes	Yes	Yes	Yes	Yes	Yes	Yes	Yes
⊎-cum.	Yes	Yes	–	–	Yes	Yes	Yes	Yes
⊔-cpref.	No	Yes	No	Yes	Yes	No	No	Yes
⊎-pref.	No	Yes	–	–	Yes	No	No	Yes
⊔-crat.	No	No	No	No	No	No	No	No
⊎-rat.	No	No	–	–	No	No	No	No

Table 13: Overview over the properties of non-monotonic inference. In the table, "(c)cum." means "(cautiously) cumulative", "(c)pref." means "(cautiously) preferential", and "(c)rat." means "(cautiously) rational". We let: $\emptyset \subset \mathcal{A}_{UD} \subseteq \{\text{Ucut}, \text{Def}\}$, $\text{gps} \in \{\text{Grd}, \text{Prf}, \text{Stb}\}$, and $\text{ps} \in \{\text{Prf}, \text{Stb}\}$, Also, (†) means that $F \in \Delta$, (‡) means "without defeasible rules", and "–" means that the property is not applicable in the context of the given entailment.

Counter-examples for ⊔-**OR** which justify the negative results in Table 13 are easy to find. We give some examples for MCS-based reasoning, which in view of the cited representational results immediately generalize for the listed argumentation systems in Table 13.

Example 91 (Counter-Example, ⊎-OR, $\mid\!\sim_{\cap mcs}$). *Suppose that the underlying logic $\mathcal{L}$ is classical logic, and let $\mathcal{S} = \{\neg p \wedge r, \neg q \wedge r\}$. In this case we have:*

- $\langle \{p\}, \mathcal{S} \rangle \mid\!\sim^{\mathcal{L}}_{\cap mcs} r$, *since* $\mathsf{MCS}^{\{p\}}(\mathcal{S}) = \{\{\neg q \wedge r\}\}$,

- $\langle \{q\}, \mathcal{S} \rangle \mid\!\sim^{\mathcal{L}}_{\cap mcs} r$, *since* $\mathsf{MCS}^{\{q\}}(\mathcal{S}) = \{\{\neg p \wedge r\}\}$, *while*

- $\langle \{p \vee q\}, \mathcal{S} \rangle \not\mid\!\sim^{\mathcal{L}}_{\cap mcs} r$, *since* $\mathsf{MCS}^{\{p \vee q\}}(\mathcal{S}) = \{\{\neg p \wedge r\}, \{\neg q \wedge r\}\}$.

Example 92 (Counter-example, ⊔-OR, $\mid\!\sim_{\cap mcs}$). *Suppose again that the underlying logic $\mathcal{L}$ is classical logic, and let $\mathcal{S} = \{\neg p, \neg q, \neg p \supset r, \neg q \supset r\}$. Then we have:*

- $\langle \emptyset, \mathcal{S} \cup \{p\}\rangle \mathrel{\mid\!\sim}^{\mathfrak{L}}_{\cap\mathsf{mcs}} r$,
  since $\mathsf{MCS}^\emptyset(\mathcal{S} \cup \{p\}) = \{\{p, \neg q, \neg p \supset r, \neg q \supset r\}, \{\neg p, \neg q, \neg p \supset r, \neg q \supset r\}\}$ and thus $\bigcap \mathsf{MCS}^\emptyset(\mathcal{S} \cup \{p\}) = \{\neg q, \neg p \supset r, \neg q \supset r\}$,

- $\langle \emptyset, \mathcal{S} \cup \{q\}\rangle \mathrel{\mid\!\sim}^{\mathfrak{L}}_{\cap\mathsf{mcs}} r$,
  since $\mathsf{MCS}^\emptyset(\mathcal{S} \cup \{q\}) = \{\{\neg p, q, \neg p \supset r, \neg q \supset r\}, \{\neg p, \neg q, \neg p \supset r, \neg q \supset r\}\}$ and thus $\bigcap \mathsf{MCS}^\emptyset(\mathcal{S} \cup \{q\}) = \{\neg p, \neg p \supset r, \neg q \supset r\}$, while

- $\langle \emptyset, \mathcal{S} \cup \{p \vee q\}\rangle \mathrel{\not\mid\!\sim}^{\mathfrak{L}}_{\cap\mathsf{mcs}} r$,
  since $\mathsf{MCS}^\emptyset(\mathcal{S} \cup \{p \vee q\}) = \{\{p \vee q, \neg p, \neg p \supset r, \neg q \supset r\}, \{p \vee q, \neg q, \neg p \supset r, \neg q \supset r\}, \{\neg p, \neg q, \neg p \supset r, \neg q \supset r\}\}$ and thus

$$\bigcap \mathsf{MCS}^\emptyset(\mathcal{S} \cup \{p \vee q\}) = \{\neg p \supset r, \neg q \supset r\}.$$

Example 93 (Counter-example, ⊔-**RM**, $\mathrel{\mid\!\sim}_{\cap\mathsf{mcs}}$). *Let $\mathfrak{L}$ be classical logic and $\mathcal{S} = \{r, p \wedge q \wedge \neg r, (p \wedge r) \supset \neg q, \neg p \wedge q\}$. We have $\mathsf{MCS}^\emptyset(\mathcal{S}) = \{\{r, (p \wedge r) \supset \neg q, \neg p \wedge q\}, \{p \wedge q \wedge \neg r, (p \wedge r) \supset \neg q\}\}$. One of the two elements of $\mathsf{MCS}^\emptyset(\mathcal{S})$ does not imply $\neg p$, while both of them imply q. Thus, $\langle \emptyset, \mathcal{S}\rangle \mathrel{\mid\!\sim}_{\cap\mathsf{mcs}} q$ and $\langle \emptyset, \mathcal{S}\rangle \mathrel{\not\mid\!\sim}_{\cap\mathsf{mcs}} \neg p$.*

Now, consider $\langle \emptyset, \mathcal{S} \cup \{p\}\rangle$ and $\langle \{p\}, \mathcal{S}\rangle$. We have:

- $\mathsf{MCS}^\emptyset(\mathcal{S} \cup \{p\}) = \{\{r, (p \wedge r) \supset \neg q, \neg p \wedge q\}, \{p \wedge q \wedge \neg r, (p \wedge r) \supset \neg q, p\}, \{r, p, (p \wedge r) \supset \neg q\}\}$ *and*

- $\mathsf{MCS}^{\{p\}}(\mathcal{S}) = \{\{p \wedge q \wedge \neg r, (p \wedge r) \supset \neg q\}, \{r, (p \wedge r) \supset \neg q\}\}$.

*As a consequence, $\langle \emptyset, \mathcal{S} \cup \{p\}\rangle \mathrel{\not\mid\!\sim}_{\cap\mathsf{mcs}} q$ and $\langle \{p\}, \mathcal{S}\rangle \mathrel{\not\mid\!\sim}_{\cap\mathsf{mcs}} q$. Thus, neither ⊎-**RM** nor ⊍-**RM** holds in this case.*

Not so many results on inferential properties are known for fragments of ASPIC$^+$ and ABA that are beyond those that coincide with reasoning with maximally consistent subsets. To the best of our knowledge, for ABA frameworks, inferential behavior for these fragments has only been studied in [Heyninck and Straßer, 2021a], where the following results are shown:

Remark 94. *For flat ABFs that are not necessarily simple contrapositive but whose strict rule set is contrapositive (see Remark 49), [Heyninck and Straßer, 2021a] show the following additional results:*

- $\mathrel{\mid\!\sim}_{\cap\mathsf{Grd}}$ *satisfies* ⊎-**CM** *and* ⊎-**CC**

- $\mathrel{\mid\!\sim}_{\cap\mathsf{Prf}}$ *satisfies* ⊎-**CC**

- *if* $\mathcal{ABF}$ *is well-behaved (recall Remark 49), then* $\mathrel{\mid\!\sim}_{\cap\mathsf{sem}}$ *satisfies* ⊎-**CM** *for* sem $\in \{\mathsf{Prf}, \mathsf{Stb}\}$.[43]

Another study of inferential behavior of assumption-based argumentation is given in [Čyras and Toni, 2015] (in [Cyras and Toni, 2016b] it is extended to ABA$^+$), where yet another set of postulates is studied. For example, cautious cut and cautious monotony are defined in [Čyras and Toni, 2015] as follows:

Definition 95. *Given* $\mathcal{ABF} = \langle \mathcal{L}, \mathcal{R}, \mathcal{A}, \sim \rangle$, *for an arbitrary extension* $\mathcal{E} \in \mathsf{Sem}(\mathcal{ABF})$ $\mathcal{L}$-*formula* $\phi \notin \mathcal{A}$, *and* $\sqcup \in \{\mathbb{U}, \uplus\}$, *we define:*

⊔-**SCC:** *If* $\phi \in Cn_{\vdash_\mathcal{R}}(\mathcal{E})$, *then for every* $\mathcal{E}' \in \mathsf{Sem}(\mathcal{ABF} \sqcup \phi)$, $Cn_{\vdash_\mathcal{R}}(\mathcal{E}) \subseteq Cn_{\vdash_\mathcal{R}}(\mathcal{E}')$.

⊔-**WCC:** *If* $\phi \in Cn_{\vdash_\mathcal{R}}(\mathcal{E})$, *then for some* $\mathcal{E}' \in \mathsf{Sem}(\mathcal{ABF} \sqcup \phi))$, $Cn_{\vdash_\mathcal{R}}(\mathcal{E}) \subseteq Cn_{\vdash_\mathcal{R}}(\mathcal{E}')$.

⊔-**SCM:** *If* $\phi \in Cn_{\vdash_\mathcal{R}}(\mathcal{E})$, *then for every* $\mathcal{E}' \in \mathsf{Sem}(\mathcal{ABF} \sqcup \phi))$, $Cn_{\vdash_\mathcal{R}}(\mathcal{E}) \supseteq Cn_{\vdash_\mathcal{R}}(\mathcal{E}')$.

⊔-**WCM:** *If* $\phi \in Cn_{\vdash_\mathcal{R}}(\mathcal{E})$, *then for some* $\mathcal{E}' \in \mathsf{Sem}(\mathcal{ABF} \sqcup \phi))$, $Cn_{\vdash_\mathcal{R}}(\mathcal{E}) \supseteq Cn_{\vdash_\mathcal{R}}(\mathcal{E}')$.

It can be shown that, for each $\sqcup \in \{\mathbb{U}, \uplus\}$, ⊔-**CC** and ⊔-**CM**, defined for $\mathrel{\mid\!\sim}_{\cup\mathsf{Sem}}$, imply, respectively, ⊔-**WCC** and ⊔-**WCM** (and, obviously, ⊔-**SCC** and ⊔-**SCM** also respectively imply the two latter rules).

The following proposition and examples are shown in [Čyras and Toni, 2015]:

Proposition 96. *For each* $\sqcup \in \{\mathbb{U}, \uplus\}$,

- *grounded semantics satisfies* ⊔-**SCC** *and* ⊔-**SCM**,

---

[43] The satisfaction of the postulates for $\mathrel{\mid\!\sim}_{\cap\mathsf{sem}}$ and $\mathrel{\mid\!\sim}_{\cup\mathsf{sem}}$-entailments are not studied in [Heyninck and Straßer, 2021a], and neither is satisfaction of properties such as ⊎-**REF**, ⊎-**LLE**, **RW** or ⊎-**OR**. The same holds for any of the $\mathbb{U}$-properties.

- *preferred and stable semantics satisfy* ⊔-**WCC** *and* ⊔-**WCM**.

Here are counter-examples to ⊎-**SCC** and ⊎-**SCM** for preferred and stable semantics:

Example 97. Let $\mathcal{ABF} = \langle \{p, q, r, p', q', r', s\}, \mathcal{R}, \mathcal{A}, \sim \rangle$ with

$\mathcal{A} = \{p, q, r\}$,

$\mathcal{R} = \{p \to q';\ r \to p';\ q \to p';\ q \to s;\ s \to r'\}$, and

$\sim x = \{x'\}$ for any $x \in \mathcal{A}$.

A fragment of the attack diagram of this ABF is given in Figure 8a. Here $\{q\}$ is the unique preferred and stable extension and $\{q\} \vdash_{\mathcal{R}} s$. Consider now $\mathcal{ABF} \uplus \{s\}$ (see Figure 8b for a fragment of the attack diagram). Now there are two preferred (and stable) extensions: $\{q\}$ and $\{p\}$. Since $Cn_{\mathcal{R}}(\{p\}) \not\subseteq Cn_{\mathcal{R}}(\{q\})$, it follows that ⊎-**SCM** is violated. Likewise, since $Cn_{\mathcal{R}}(\{p\}) \not\supseteq Cn_{\mathcal{R}}(\{q\})$, it follows that ⊎-**SCC** is violated.

Notice that this example is also a counter-example to ⊎-**CM** for $\vdash_{\cap \mathsf{Sem}}$ with $\mathsf{Sem} \in \{\mathsf{Prf}, \mathsf{Stb}\}$, as $\mathcal{ABF} \vdash_{\cap \mathsf{Sem}} s$ and $\mathcal{ABF} \vdash_{\cap \mathsf{Sem}} q$, yet $\mathcal{ABF} \uplus \{s\} \not\vdash_{\cap \mathsf{Sem}} q$.

Here are counter-examples to ⊍-**SCC** and ⊍-**SCM** for the preferred semantics:

Example 98. Let $\mathcal{ABF}$ be as in Example 97. Observe that:

$\mathcal{ABF} \uplus \{s\} = \langle \{p, q, r, s, p', q', r', s'\}, \mathcal{R}, \mathcal{A}, \sim \rangle$, with

$\mathcal{A}' = \{p, q, r, s\}$,

$\mathcal{R}' = \{p \to q';\ r \to p';\ q \to p';\ s \to r'\}$, and

$\sim x = \{x'\}$ for any $x \in \mathcal{A}$.

A fragment of the attack diagram of this ABF is given in Figure 8c.

The framework $\mathcal{ABF} \uplus \{s\}$ has two preferred (and stable) extensions, namely $\{q, s\}$ and $\{p, s\}$. In this case ⊍-**SCM** is violated, since $Cn_{\mathcal{R}}(\{q\}) \not\subseteq Cn_{\mathcal{R}}(\{p, s\})$. Likewise, ⊍-**SCC** is violated, since $Cn_{\mathcal{R}}(\{q\}) \not\supseteq Cn_{\mathcal{R}}(\{p, s\})$.

As in Example 97, this example can also be seen to be a counter-example to ⊍-**CM**.

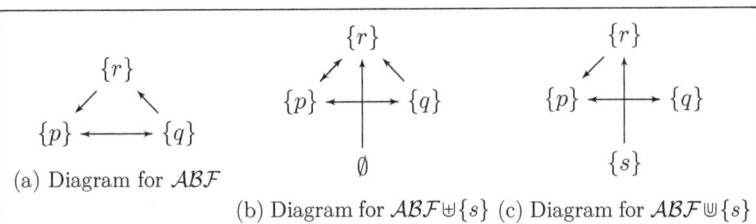

Figure 8: Attack diagrams for Examples 97 and 98. To avoid clutter only attacks from minimal sets are included.

In [Li et al., 2018], inference properties are studied for ASPIC⁺. However, right weakening, left logical equivalence and reflexivity are defined there in a different way. In more detail, [Li et al., 2018] study the following alternative versions of these rules:

Definition 99 (alternative inference properties). *Given an ASPIC argumentation theory $AT = \langle\langle \mathcal{L}, ^-, \mathcal{R}, n\rangle, (\mathcal{K}_n, \mathcal{K}_p)\rangle$, $\mathcal{L}$-formulas ϕ, ψ, an operator $\sqcup \in \{\uplus, \cup\!\!\!\cup\}$ and an entailment relation $\mid\!\sim$ as in Definition 26, we say that $\mid\!\sim$ satisfies:*

REF^d *if $\phi \in \mathcal{K}_p$ then $AT \mid\!\sim \phi$*

REF^s *if $\phi \in \mathcal{K}_n$ then $AT \mid\!\sim \phi$*

RW^d *if $AT \mid\!\sim \phi$ and $\phi \Rightarrow \psi \in \mathcal{R}_d$ then $AT \mid\!\sim \psi$*

RW^s *if $AT \mid\!\sim \phi$ and $\phi \to \psi \in \mathcal{R}_s$ then $AT \mid\!\sim \psi$*

⊔-LLE^d *if $\phi \Rightarrow \psi \in \mathcal{R}_d$, $\psi \Rightarrow \phi \in \mathcal{R}_d$ and $AT \sqcup \phi \mid\!\sim \sigma$ then $AT \sqcup \psi \mid\!\sim \sigma$*

⊔-LLE^s *if $\phi \to \psi \in \mathcal{R}_s$, $\psi \to \phi \in \mathcal{R}_s$ and $AT \sqcup \phi \mid\!\sim \sigma$ then $AT \sqcup \psi \mid\!\sim \sigma$*

Notice that **RW** implies **RW^s** and ⊔-**LLE** implies ⊔-**LLE^s** (for any $\sqcup \in \{\uplus, \cup\!\!\!\cup\}$), **REF^s** implies ⊎-**REF** (but not vice versa) and **REF^d** implies ⋓-**REF** (but not vice versa).

The main positive results of [Li et al., 2018] are the following:

Proposition 100.

- $\mathrel|\!\sim_{\cap\mathsf{Grd}}$ *satisfies* **REF**^{s}, **RW**^{s}, **LLE**^{s}, $\cup$-**CM** *and* $\cup$-**CC**.

- $\mathrel|\!\sim_{\cap\mathsf{Prf}}$ *satisfies* **REF**^{s}, **RW**^{s}, **LLE**^{s} *and* $\cup$-**CC**.

- $\mathrel|\!\sim_{\cap\mathsf{Prf}}$ *and* $\mathrel|\!\sim_{\cup\mathsf{Prf}}$ *satisfy* **REF**^{s}, **RW**^{s} *and* **LLE**^{s}.

We conclude this section by making some observations on both the significance of satisfaction or violations of the properties discussed in this section and the current state of the art. On one hand, there is a long tradition in non-monotonic logic which claims or assumes the properties for cumulative inference relations to "constitute a basic set of principles that any reasonable account of defaults must obey" [Geffner and Pearl, 1992]. As such, the satisfaction of such properties can be seen as a minimal condition on any formalization of non-monotonic reasoning. However, the generality of this claim has been put into doubt by, e.g. Bochman [Bochman, 2005; Bochman, 2006; Bochman, 2013], who posits a distinction between *explanatory* and *preferential* reasoning, where only for the latter cumulativity is feasible. Furthermore, some of the properties considered in this section are not outside of controversy, such as rational monotony (cf., for instance, [Stalnaker, 1994]). In sum, we submit that the feasibility of the postulates for non-monotonic reasoning depends on the precise context of application. Once this is decided, the results in this section offer some indications of which formalisms are appropriate for specific needs.

Finally, it is evident from this survey that the formalizations of the properties differ greatly in different works, making it difficult to compare results and transfer them between systems. Therefore, we think that it is an important direction for future work to study the relations between the different formulations of the properties studied in this section, and – more generally – to express some other criteria for relating and comparing the different approaches to logic-based argumentations, as well as their relations to other forms of non-monotonic reasoning. Some steps in this direction are reviewed in the next section.

2.4 Comparative Study

In this section we review some results concerning the inter-relations among the three logic-based approaches to formal argumentation considered in Section 2.2, as well as some of their connections to related methods to defeasible reasoning.

2.4.1 Relations among the Logic-Based Approaches

From the descriptions of logic-based argumentation, assumption-based argumentation and ASPIC$^+$ given above, the similarities of the frameworks are clear: they all use the same *pipeline-methodology* where an argumentation framework is constructed from the following components:

- a core (base) logic that determines the underlying language and the consequence relation for the arguments,

- attack rules relating arguments with counterarguments,

- a knowledge-base, encoding the set of the 'global' assumptions of the framework,

- an argumentation semantics, according to which sets of jointly acceptable arguments and their respective accepted conclusions are determined.

However, the formalisms outlined in Section 2.2 clearly differ in the specific ways formal substance is given to this general methodology. Table 14 gives an overview of the specific instantiations of the main argumentative concepts by logic-based argumentation (LBA), assumption-based argumentation (ABA) and ASPIC$^+$.

Concept	LBA	ABA	ASPIC$^+$
Knowledge-base	$\mathcal{S}$ and $\mathfrak{L}$	$\langle \mathcal{L}, \mathcal{R}_s, \mathcal{K}_p, \sim \rangle$	$\langle \mathcal{L}, ^-, \mathcal{R}, n \rangle$, $\langle \mathcal{K}_n, \mathcal{K}_p \rangle$
Arguments	support-conclusion pairs	sets of assumptions	proof trees
Attacks	various	direct defeat	undermining, rebut, undercut

Table 14: Argumentative concepts and their instantiations in logic-based frameworks

An important question that arises in such a comparison is concerned with the impact of the different choices on the resulting inference relation. Such a question can be partly answered by considering the exact relationship between the formalisms under consideration. This can be done in several ways, for instance by

1. comparing the inference relations associated with the respective formalisms,

2. investigating translations between the different formalisms, and

3. comparing the relative expressivity of the different formalisms.

Several works, including [Prakken, 2010; Arieli *et al.*, 2018; Heyninck and Arieli, 2018; Borg, 2020; Heyninck and Arieli, 2020b; Heyninck and Straßer, 2021a], have concluded that logic-based argumentation, assumption-based argumentation and ASPIC$^+$ agree on what we could call a *core fragment*, namely when the underlying (strict) base logic is classical logic (or even any contrapositive Tarskian logic), and the defeasible assumptions are some propositional formulas. Indeed, it follows from Propositions 16, 46 and 48 that all three frameworks give rise to the same inference relation for the above-mentioned fragment and that this core fragment coincides with MCS-based reasoning.

When moving away from this core fragment, the formalisms start to behave in fundamentally different ways. First, it should be noted that logic-based argumentation as represented here, is restricted to (usually contrapositive) Tarskian logics, where the knowledge-base consists of defeasible propositional formulas.[44] In contrast, ABA and ASPIC$^+$, do allow to use not only defeasible, but also strict assumptions. Moreover, ASPIC$^+$ allows to reason with defeasible rules in addition to defeasible premises, i.e., with ASPIC$^+$ one can make inferences from knowledge bases that ABA cannot handle.

As we will describe below, there are ways to express defeasible rules with the help of defeasible premises and strict rules, but it seems equally

---

[44]We note that this restriction can be lifted by adding strict assumptions and applying the attack rules only on the defeasible arguments. See [Borg, 2020] for the details. Here we follow the main line of research so far that combines logic-based framework with defeasible information only.

interesting to compare the inferential behavior of ABA and ASPIC⁺ for knowledge bases whose only defeasible elements are premises. In [Prakken, 2010, Corollary 8.10] it is shown that given a *flat* assumption-based framework $\mathcal{ABF} = \langle \mathsf{Atoms}(\mathcal{L}), \mathcal{R}, \mathcal{A}, \sim \rangle$ (i.e, when for no $\Theta \cup \{\theta\} \subseteq \mathcal{A}$, $\Theta \vdash_\mathcal{R} \theta$.), the ASPIC-based argumentation framework $\mathrm{AT}_{\mathcal{ABF}} = \langle \langle \mathsf{Atoms}(\mathcal{L}), \bar{}, \langle \mathcal{R}, \emptyset \rangle, n \rangle, \langle \emptyset, \mathcal{K} \rangle \rangle$ gives rise to the same inferences.

Proposition 101. *Let $\mathcal{ABF} = \langle \mathsf{Atoms}(\mathcal{L}), \mathcal{R}, \mathcal{A}, \sim \rangle$ be a flat assumption-based framework. Consider the ASPIC-based argumentation framework $\mathrm{AT}_{\mathcal{ABF}} = \langle \langle \mathsf{Atoms}(\mathcal{L}), \bar{}, \langle \mathcal{R}, \emptyset \rangle, n \rangle, \langle \emptyset, \mathcal{K} \rangle \rangle$ for arbitrary n*[45] *and where $\bar{}$ is defined by $\bar{\phi} = \sim\phi$ for any $\phi \in \mathcal{A}$ and $\bar{\phi} = \emptyset$ otherwise. Then for any $\dagger \in \{\cup, \cap, \mathbin{\text{⋒}}\}$ and $\mathsf{Sem} \in \{\mathsf{Grd}, \mathsf{Prf}, \mathsf{Cmp}, \mathsf{Stb}\}$, $\mathcal{ABF} \mathrel{\vert\!\sim}_{\dagger\mathsf{Sem}} \psi$ iff $\mathrm{AT}_{\mathcal{ABF}} \mathrel{\vert\!\sim}_{\dagger\mathsf{Sem}} \psi$.*

It follows that for knowledge-bases with a flat rule-base and any semantics subsumed by complete semantics ABA and ASPIC⁺ provide the same inferences. However, for non-flat knowledge-bases, this correspondence breaks down, as demonstrated by the next example.

Example 102. *Let $\mathsf{Atoms}(\mathcal{L}) = \{p, q\}$, $\mathcal{R} = \{p \to q\}$, and $\mathcal{ABF} = \langle \{p, q\}, \mathcal{R}, \{p, q\}, \sim \rangle$ where $\sim p = \emptyset$ and $\sim q = \{q\}$. For this ABF, the unique preferred extension is $\emptyset$. Indeed, $\{p\}$ is not admissible since it is not closed (since $\{p\} \vdash_\mathcal{R} q$) and any set containing q is not admissible (since q attacks itself). If we move to ASPIC⁺ we have the argumentation theory $\mathrm{AT}_{\mathcal{ABF}} = \langle \langle \{p, q\}, \bar{}, \langle \mathcal{R}, \emptyset \rangle, n \rangle, \langle \emptyset, \{p, q\} \rangle \rangle$, and the arguments $A = \langle p \rangle$, $B = \langle q \rangle$, $C = A \to q$.*

There is an attack from B to itself and from C to B. Notice furthermore that C is unattacked (Recall here that no rebuttals are possible in the heads of strict rules, which is why C does not rebut itself). This means that $\{A, C\}$ is the unique stable and preferred extensions.

It is perhaps interesting to note that $\{A, C\}$ presents a violation of the rationality postulate of consistency from [Caminada, 2018a] (see Section 2.3.2, and in particular definition 60). It is an open question if there are any differences in inferential behavior between ASPIC⁺ and non-flat ABA for knowledge-bases whose extensions satisfy all the rationality postulates.

---

[45] Note that n can be safely ignored since the set of defeasible rules $\mathcal{R}_d$ is empty.

Translation methods. Given both the conceptual differences (as displayed in Table 14) and the diverging inferential behavior of LBA, ABA and ASPIC$^+$, the correspondences described above have been supplemented by *translations* among the formalisms. Particular attention has been paid to translations from ASPIC$^+$ into ABA. Conceptually, this corresponds to asking if one can model defeasible rules as defeasible premises. Such a question has been answered positively in [Dung and Thang, 2014] and [Heyninck and Straßer, 2016], sharing the same underlying idea: given an ASPIC-based argumentation framework $\langle \mathcal{L}, ^-, \mathcal{R}_s \cup \mathcal{R}_d, n \rangle$, the underlying language $\mathcal{L}$ is extended to $\mathcal{L}'$ as to contain a name $N(r)$ for every $r \in \mathcal{R}_d$. This name is then added as a defeasible assumption in the ABF.[46] The strict rule-base is then supplemented with rules that ensure that the names of the defeasible rules are handled adequately in the argumentative inference process. In particular, for every rule $r = \phi_1, \ldots, \phi_n \Rightarrow \psi \in \mathcal{R}_d$, the following rules are added (resulting in R($\mathcal{R}_d$)):[47]

- $N(r), \phi_1, \ldots, \phi_n \to \psi$, which ensures that ψ is (defeasibly) derivable from $\{\phi_1, \ldots, \phi_n\}$;

- $\overline{\psi} \to \overline{N(r)}$ which enables an attack on $N(r)$ if the contrary of the consequent of r is derivable (thus mirroring rebuttal);

- $\overline{n(r)} \to \overline{N(r)}$, which enables an attack on $N(r)$ if $\overline{n(r)}$ is derivable (thus mirroring undercut).

In [Heyninck and Straßer, 2016] it is shown that this translation is adequate for *flat argumentation theories* for admissible, preferred and stable semantics. In [Dung and Thang, 2014], it is shown that their translation is adequate for any semantics subsumed by complete semantics. In the following, given a flat[48] argumentation theory $AT = \langle \langle \mathcal{L}, ^-, \mathcal{R}_s \cup \mathcal{R}_d, n \rangle, \langle \mathcal{K}_n, \mathcal{K}_n \rangle \rangle$, let

$\mathcal{ABF}(AT) = \langle \mathcal{L}, \mathcal{R}_s \cup \text{R}(\mathcal{R}_d) \cup \{\to \phi \mid \phi \in \mathcal{K}_n\}, \mathcal{K}_p \cup \{N(r) \mid r \in \mathcal{R}_d\}, \sim \rangle$

---

[46] In [Dung and Thang, 2014] the language is also extended with an atom not ψ for every $\phi_1, \ldots, \phi_n \Rightarrow \psi$ such that in the translated ABF, not ψ is a defeasible assumption similar to negation as failure.

[47] For simplicity, we denote by $\overline{\phi}$ any $\phi' \in \overline{\phi}$.

[48] An argumentation theory $AT = \langle \langle \mathcal{L}, ^-, \mathcal{R}_s \cup \mathcal{R}_d, n \rangle, \langle \mathcal{K}_n, \mathcal{K}_p \rangle \rangle$ is *flat* if there is no $A \in \text{Arg}(AT)$ such that $\text{Conc}(A) \in \mathcal{K}_p \setminus \text{Prem}(A)$.

We now recall the adequacy result from [Heyninck and Straßer, 2016]

Proposition 103. *Given a flat argumentation theory AT, $\dagger \in \{\cap, \cup, \cupplus\}$, and $\mathsf{Sem} \in \{\mathsf{Prf}, \mathsf{Stb}\}$: $AT \mid\!\sim_{\dagger\mathsf{Sem}} \phi$ iff $\mathcal{ABF}(AT) \mid\!\sim_{\dagger\mathsf{Sem}} \phi$.*

No adequate translation is known for non-flat argumentation theories.

Expressivity, Complexity and Representation of Arguments.
A third way to compare the logic-based approaches to formal argumentation considered in this chapter is by studying their expressiveness. In other words, one may compare the answers to the question: "what kind of problems can be solved by this formalism" [Strass, 2014]. In terms of feasibility, this often boils down to questions of computational complexity. In that respect, we note that while the complexity of ABA has been studied in [Dimopoulos *et al.*, 2002], for LBA and ASPIC$^+$ similar complexity results are missing. As noted in [Modgil and Prakken, 2018], the complexity of these formalisms is indeed an important open question.

Another point of difference between the formalisms is related to how exactly arguments are represented. In ASPIC$^+$ and logic-based argumentation, arguments are formed for specific conclusions. In ABA, on the other hand, nodes of an argumentation graph are made up of sets of assumptions, without a specific conclusion. In this sense, ABA can be said to operate on the level of equivalence classes of arguments with the same support. For this reason, given a finite set of defeasible assumptions, ABA will give rise to an argumentation graph bounded by the size of the power set of the set of defeasible assumptions. Logic-based argumentation and ASPIC$^+$, on the other hand, might still generate an infinite argumentation graph since the underlying base logic might generate an infinite set of conclusions for every set of defeasible assumptions. On the other hand, this also means that in ASPIC$^+$ and logic-based argumentation, all the possible conclusions are present in the argumentation graph, whereas in ABA these conclusions have to still be derived. Altogether, we can summarize this difference as follows: ABA represents arguments in a more compact way, which has both positive aspects (e.g. boundedness of the argumentation graph) and negative aspects (e.g. some information might not be readily present in the argumentation graph). In [Amgoud et al., 2011], a procedure is developed to compute a finite *core* of a logic-based argumentation system, which returns all the results of the original system. Similarly, in [Arieli and Straßer, 2019] congruence relations (and their corresponding structures) are discussed for argumentation frameworks in the context of sequent-based argumentation, e.g., based on equivalent support sets of arguments. For $ASPIC^+$, the problem of having infinite number of arguments out of a finite set of assumptions is avoided in [D'Agostino and Modgil, 2018; D'Agostino and Modgil, 2020] in the context of dialectical argumentation frameworks and depth-bounded logics. This approach involves preferences among arguments and is concentrated on classical logic as the base logic of the framework.

2.4.2 Connections to Other Approaches

Next, we discuss relations between the logic-based argumentation formalisms presented in this chapter and other formalisms for defeasible reasoning. Clearly, it is not possible to formally and fully define here all the related formalisms, thus in what follows we just give some general description of each related formalism, together with some references for further reading. This means also that we will not be able to express the relations between the formalisms in detail, but instead we shall provide the general underlying ideas and references to papers where the relations are fully described.

It was arguably one of the goals of Dung in [Dung, 1995] to show that the way conflicts are handled in abstract argumentation theory correspond to the way conflicts are handled in many different kinds of formalisms for defeasible reasoning. In [Dung, 1995], Dung showed that this is the case by proving *representation results* for several formalisms for defeasible reasoning. He showed how to construct argumentation graphs for several such formalisms in a way that is both intuitive and gives rise to an adequate representation when applying the abstract argumentation semantics to the resulting argumentation graph.

Since then, various additional argumentative characterizations of formalisms for defeasible reasoning have been proposed. We have already mentioned in Section 2.3.1 argumentative characterizations of reasoning with maximal consistent subsets [Rescher and Manor, 1970] by logic-based argumentation, assumption-based argumentation and ASPIC$^+$. In the rest of this section we use these formalisms for argumentative characterizations of *adaptive logics* [Batens, 2007; Straßer, 2014], *default assumptions* [Makinson, 2003], *logic programming* [Apt, 1990], *default logic* [Reiter, 1980] and *autoepistemic logic* [Moore, 1985]. An illustration of these relations in given in Figure 9 at the end of this section.

A. Adaptive Logics Adaptive logics offer a general framework for defeasible reasoning. A plethora of forms of defeasible reasoning has been explicated in the adaptive logic framework. Some examples are: the modeling of abduction (e. g., [Meheus and Batens, 2006; Gauderis and Van De Putte, 2012]), inductive generalization (e. g., [Batens and Haesaert, 2003; Batens, 2006]), default reasoning (e. g., [Straßer, 2012]), reasoning from incompatible obligations (e. g., [Beirlaen and Straßer, 2014; Van De Putte and Straßer, 2013]), causal discovery (e. g., [Van Dyck, 2004]), diagnostic reasoning (e. g., [Weber and Provijn, 1999]), reasoning with vague predicates (e. g., [Van Kerckhove and Vanackere, 2003]), etc.

Adaptive logics are equipped with a dynamic proof theory extending a Tarskian core logic with a set of retractable inferences which are associated with defeasible assumptions. More specifically, these assumptions are sets of formulas of a predefined 'abnormal' form that are assumed to be false in the given inference. When an assumption turns out to be dubious in view of a premise set, the inference associated with it gets retracted.

Semantically, adaptive logics are based on preferential semantics that are adequate relative to the dynamic proof theory. Given a Tarskian core logic $\mathfrak{L}$, not all the $\mathfrak{L}$-models of the premises are considered when determining the consequences, but only a sub-class is "selected", namely those models which are "sufficiently normal". Different types of adaptive logics follow different *strategies* that offer specifications of what it means to be sufficiently normal. For instance, in adaptive logics that follows the minimal abnormality strategy, those models are selected for which there are no models that verify less abnormal formulas.

As shown in [Heyninck and Straßer, 2016], there is a straightforward translation of the framework of adaptive logics into ABA: given an adaptive logic $\mathbf{AL} = \langle \mathfrak{L}, \Omega \rangle$, where $\mathfrak{L} = \langle \mathcal{L}, \vdash \rangle$ is a Tarskian logic and $\Omega \subseteq \mathcal{L}$ is a set of abnormalities, and a set of premises Γ, the corresponding ABF is defined as $\mathcal{ABF}_{\mathbf{AL}} = \langle \mathfrak{L}, \Gamma, \{\neg \phi \mid \phi \in \Omega\}, \sim \rangle$, where $\sim \neg \phi = \phi$. It is shown that for preferred, naive and stable semantics, this translation is adequate to represent different types of adaptive strategies.

B. Logic Programming Logic programming (LP) is one of the most popular approaches to knowledge representation and has been widely studied, implemented and applied [Apt, 1990]. (Propositional) logic programs are set of rules of the form:

$$\phi_1 \vee \ldots \vee \phi_n \leftarrow \psi_1, \ldots, \psi_m, \sim \psi_{m+1}, \ldots \sim \psi_{m+l}$$

where ϕ_i, ψ_j are formulas for any $1 \leq i \leq n$ and $1 \leq j \leq m + l$. The left-hand side of the implication is call the rule's *head* and the right-hand side of the implication is the rule's *body*. Now,

- If in all the rules of the program, every ϕ_i ($1 \leq i \leq n$) and ψ_j ($1 \leq j \leq m+1$) is atomic, the program is called a *disjunctive logic program*, and

- If, in addition, $n \leq 1$ for every rule in the program, the program is called *normal*.

There are many ways of giving semantics to logics programs. One of the better-known one is based on the notion of a *reduct*, which is a set of rules that is calculated on the basis of a set of atoms. For example,

> the *Gelfond-Lifschitz reduct* [Gelfond and Lifschitz, 1991] $\frac{\mathcal{P}}{\Delta}$, of a normal logic program $\mathcal{P}$ with respect to a set of atoms Δ, is constructed as follows: $\phi \leftarrow \psi_1, \ldots, \psi_m \in \frac{\mathcal{P}}{\Delta}$ iff $\phi \leftarrow \psi_1, \ldots, \psi_m, \sim \psi_{m+1}, \ldots \sim \psi_{m+l} \in \mathcal{P}$ and $\psi_i \notin \Delta$ for any $m < i \leq m + l$.

Based on such a reduct, the semantics of logic programming then describe ways to select sets of atoms which count as models. For example,

> the *stable model semantics* says that a set of atoms is a stable model if it is the minimal model of its own Gelfond-Lifschitz reduct.[49]

---

[49]That is, Δ is a stable model of $\mathcal{P}$ if for every $p \leftarrow q_1, \ldots q_n \in \frac{\mathcal{P}}{\Delta}$, either $p \in \Delta$ or $q_i \notin \Delta$ for some $1 \leq i \leq n$, and there is no $\Delta' \subsetneq \Delta$ with the same property.

The translation of logic programming into assumption-based argumentation has been the subject of several publications (e.g., [Schulz and Toni, 2015; Dung et al., 2016; Caminada and Schulz, 2017; Heyninck and Arieli, 2019a]). The basic idea underlying all of these publications is the same: the set of assumptions is made up of negated atoms, and the contrary of a negated atom is the positive atom. The (strict) rules consist of the rules of the logic programs. Thus, a set of negated atoms will attack a negated atom if the logic program and the attacking set allows to derive the positive version of the attacked negated atom. Therefore, the underlying idea is to assume the 'absence' of any atom A appearing in the logic program (the defeasible assumptions), unless, on the basis of attacks derived by the programs rules, some set of assumptions indicates that A holds.

The correspondence results in Table 15 where proven in [Caminada and Schulz, 2017] for normal logic programs.

ABA Extension	LP Model
complete	stable (3-valued)
grounded	well-founded
preferred	regular
stable	stable (2-valued)
ideal	ideal

Table 15: Correspondence between model of normal logic programs and extensions of ABA frameworks

Remark 104. *It is interesting to note that L-stable models (i.e. 3-valued stable models that are maximal w.r.t. atoms assigned a definite truth value) do not correspond to semi-stable sets of assumptions (see [Caminada and Schulz, 2017, Example 13]), although both of these semantics are based on the same idea of maximizing the assignment of determinate truth values.*

The results above were extended in [Heyninck and Arieli, 2019a] to disjunctive logic programming under stable model semantics. Furthermore, argumentative characterizations of the so-called *well-justified*

[Shen et al., 2014] and *well-founded* [Wang et al., 2012] semantics of *general* or *first-order* logic programs (i.e., logic programs where any first-order formula can occur in the head or the body of a rule) are provided in [Dung et al., 2016]. These generalizations are based on the same idea as [Caminada and Schulz, 2017]: the assumptions consist of negated atoms and attacks occur when the attacking set allows to derive the positive version of the attacked (negated) atom. What changes, however, is the derivability relation used to determine if attacks occur. For example, in [Heyninck and Arieli, 2019a] in addition to allowing for *modus ponens* on the rules of the program as in [Caminada and Schulz, 2017], one has also to allow for *reasoning by cases* and *resolution* in the derivations. Likewise, in [Dung et al., 2016] both *modus ponens* on the rules of the program and any deduction valid in first-order logic are allowed. In [Wakaki, 2017] *extended logic programs* [Gelfond and Lifschitz, 1991] under three- and two-valued stable semantics are translated into assumption-based argumentation. These translations have been used to obtain explanations of (non-)derivability of literals in [Schulz and Toni, 2016] and explaining and characterizing inconsistencies of logic programs [Schulz et al., 2015].

C. Default Logics Reiter's default logic [Reiter, 1980] has also been translated in assumption-based argumentation. Again, here we just we recall the basics of default logic in an informal way. Defaults are objects of the form
$$\frac{\phi \ : \ M\psi_1, \ldots, M\psi_n}{\psi}.$$
Here, $\phi, \psi_1, \ldots, \psi_n, \psi$ are formulas in the language, and the intuitive meaning of this expression is the following:

> if ϕ holds, and none of $\neg\psi_1, \ldots, \neg\psi_n$ is provable, then normally one may suppose that ψ also holds.

An *extension* of a set of defaults Δ is a set of formulas Θ, such that Θ is a fixed point under the operator ∇_Δ, i.e., $\nabla_\Delta(\Theta) = \Theta$, where the operator ∇_Δ is defined as follows: given a set Θ, $\nabla_\Delta(\Theta)$ is the smallest set such that:

1. for every $\frac{\phi:M\psi_1,...,M\psi_n}{\psi} \in \Delta$, if $\phi \in \Theta$ and $\neg\psi_i \notin \Theta$ for every $1 \leqslant i \leqslant n$, then $\psi \in \nabla_\Delta(\Theta)$, and

2. $\nabla_\Delta(\Theta) = Cn(\nabla_\Delta(\Theta))$,

The translation into ABA proposed in [Bondarenko et al., 1997] works as follows: the language is that of classical logic extended with $M\phi$ for any $\phi \in \mathcal{L}$. The assumptions are $M\phi$ for any $\phi \in \mathcal{L}$, i.e., we assume (defeasibly) that for any formula $\phi \in \mathcal{L}$, its negation is not provable. The rules are generated by taking the default rules together with a set of rules that captures (classical) first-order logic. Finally, the contrary of $M\phi$ is defined as $\neg\phi$ (recall that $M\phi$ is interpreted as $\neg\phi$ not being provable): a positive proof of $\neg\phi$ gives us a counter-argument to the assumption $M\phi$.

In [Bondarenko et al., 1997] it is shown that under this translation, stable extensions in ABA correspond to Reiter's default extensions. An interesting open question is whether similar results hold for other semantics for default logic, such as those from [Brewka and Gottlob, 1997; Antonelli, 1999; Denecker et al., 2003].

D. Autoepistemic Logics Moore's autoepistemic logic [Moore, 1985] is another well-established formalisms for defeasible reasoning. It involves theories consisting of formulas in a doxastic language, which is typically the closure $\mathcal{L}^\mathbf{L}$ of a propositional language $\mathcal{L}$ under a belief operator $\mathbf{L}$. The intuitive meaning of $\mathbf{L}\phi$ is that 'ϕ is believed'. Thus, autoepistemic logic is a formal logic for the representation and reasoning of knowledge about knowledge, and theories containing formulas of the form $\mathbf{L}\phi$ are viewed are representing "knowledge of a perfect, rational, introspective agent" [Moore, 1985; Konolige, 1988; Bogaerts, 2015]. An autoepistemic theory $\Delta \subseteq \mathcal{L}^\mathbf{L}$ represents both positive and negative introspection of a logically perfect agent, i.e., $\phi \in \Delta$ iff $\mathbf{L}\phi \in \Delta$ and $\phi \notin \Delta$ iff $\neg\mathbf{L}\phi \in \Delta$. Autoepistemic logic has been shown to have connections to many other formalisms for defeasible reasoning, such as several variants of default and priority logic [Janhunen, 1998], and several classes of logic programming [Marek and Truszczyński, 1993].

A translation of autoepistemic logics to ABA frameworks is provided in [Bondarenko et al., 1997]. According to this translation, the set of assumptions is made up of the assumption of both negative and positive knowledge: $Ab = \{\mathbf{L}\phi, \neg\mathbf{L}\phi \mid \phi \in \mathcal{L}\}$. Thus, both negative and positive knowledge are assumed equally plausible. However, there are asymmetric treatments when it comes to the definition of contraries: the contrary of positive knowledge $\mathbf{L}\phi$ is the negative knowledge (or absence of knowledge) of $\neg\mathbf{L}\phi$ (i.e., $\overline{\mathbf{L}\phi} = \neg\mathbf{L}\phi$). The contrary of absence of knowledge of a formula, on the other hand, is the formula itself, that is: $\overline{\neg\mathbf{L}\phi} = \phi$. The rule-base is a set of rules capturing first-order logic, but formulated over the modal language $\mathcal{L}^{\mathbf{L}}$. It is interesting to note, however, that within the rule-base, no rules for the modal operator are defined. Under this translation, the strict premises consist of the autoepistemic theory Δ. [Bondarenko et al., 1997] shows that stable extensions of the translation in ABA correspond to the so-called *consistent stable expansions* [Moore, 1985] of the translated autoepistemic theory. For other semantics, no correspondences are known.

Figure 9 provides a schematic description of the relations among the formalisms described in this section.

Besides the translations discussed above, we mention the following additional translations which are beyond the scope of this paper:

- In [Borg, 2020] a generalization of sequent-based argumentation, called *assumptive sequent-based argumentation*, is shown to capture assumption-based argumentation, adaptive logics and default assumptions.

- We note that in [Caminada and Schulz, 2017] it is also shown that assumption-based argumentation can be translated in logic-programming.

- In [Caminada et al., 2015] translations from normal logic programming to abstract argumentation and vice-versa have been presented which are adequate for most (but not all) argumentation semantics.

- In [Heyninck and Arieli, 2020a] it is shown that *approximation fixpoint theory* [Denecker et al., 2000], a general approach to the study of non-monotonic reasoning, can be translated into assumption-based argumentation. This allows for the straightforward

12 - Logic-Based Approaches to Formal Argumentation

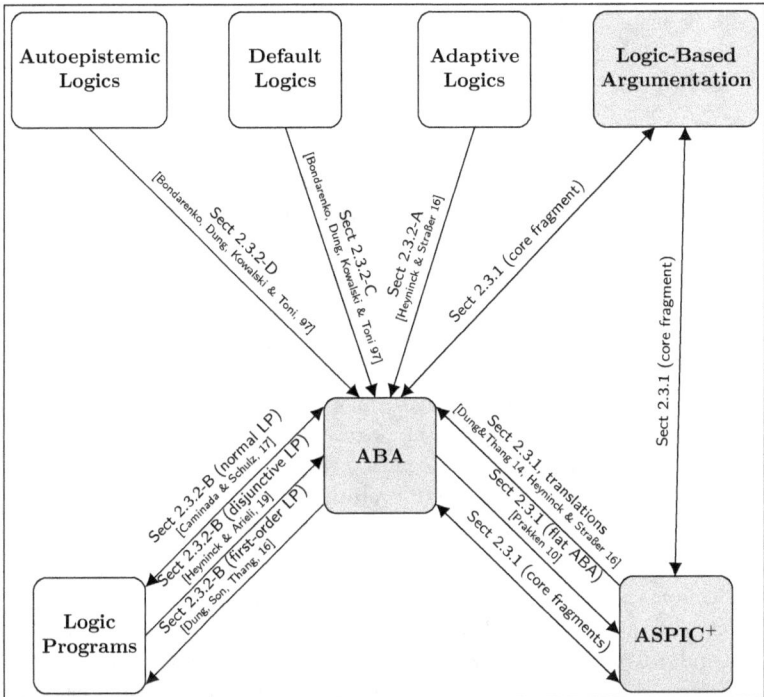

Figure 9: Argumentative representations of formalisms for modeling defeasible reasoning, presented in Sectrion 2.4

translation of many semantic variations on logic programming, default logic and auto-epistemic logic into assumption-based argumentation.

- Relationships (and further references) of ASPIC$^+$ to defeasible logic programming [García and Simari, 2004], classical logical argumentation frameworks (see the paragraph below Definition 8) and prioritized formalisms, such as Brewka's preferred subtheories [Brewka, 1989] and prioritized default logic [Brewka, 1994], are described in [Modgil and Prakken, 2014; Modgil and Prakken, 2018].

- Translations of *abstract dialectical frameworks* [Brewka *et al.*, 2017] into logic programming respectively system Z [Geffner and Pearl, 1992] are shown in [Strass, 2013] respectively [Heyninck *et al.*, 2020].

3 Logical Methods for Studying Argumentation Dynamics

There are a variety of methods for studying the dynamics of argumentation systems.[50] This includes, among others, dialectic games (see [Modgil and Caminada, 2009]), discussions [Caminada, 2018a], and, to some extent, even machine learning algorithms [Budzynska and Villata, 2018]. Other approaches involve formal (logic) programming methods, such as reductions to answer set programs (ASP), defeasible logic programs (DeLP) and constraint satisfaction problems (CSP) (see, e.g., [Cerutti *et al.*, 2017] for a description of these methods and further references).

The common ground of the methods that are described in this section (following the scope of this chapter) is that all of them assume the availability of an underlying Tarskian logic and apply related formal methods (e.g., satisfiability of formulas in the underlying language or proof procedures that allow to make inferences by derivation sequences). In the first two subsections (3.1 and 3.2) we survey several logic-basic representation methods that are adequate for expressing the selection of arguments in view of argumentation semantics and epistemic notions such as beliefs and their justifications in an argumentative setting. In the last subsection (3.3) we consider proof-theoretic methods that are adequate for structured argumentation.

3.1 Representation Methods Based on [Quantified] Propositional Languages

As indicated in, e.g., [Besnard and Doutre, 2004] and [Egly and Woltran, 2006], given a finite argumentation framework, computing its admissible

---

[50] Recall that 'dynamics' means here processes of a (fixed) argumentative framework and not its revision.

sets or its complete extensions can be done by a straightforward encoding, in propositional classical logic, of the requirements in the fourth item of Definition 10. Indeed, given an abstract argumentation framework $\mathcal{AF}$, one may associate a propositional atom with every argument in $\mathcal{AF}$ (in what follows, to ease the notations, we shall use the same symbol for an argument and its propositional variable), and accordingly construct the following formula:

$$\mathsf{ADM}(\mathcal{AF}) = \bigwedge_{p \in \mathsf{Arg}} \left((p \supset \bigwedge_{(q,p) \in \mathit{Attack}} \neg q) \land (p \supset \bigwedge_{(q,p) \in \mathit{Attack}} (\bigvee_{(r,q) \in \mathit{Attack}} r)) \right).^{51}$$

Clearly, the arguments of an admissible set of $\mathcal{AF}$ correspond to the atoms that are verified (i.e., those that are assigned the truth value 'true') by a model of $\mathsf{ADM}(\mathcal{AF})$ and every model of $\mathsf{ADM}(\mathcal{AF})$ is associated with an admissible set of $\mathcal{AF}$, the elements of which correspond to the verified atoms of the model. Similar considerations hold for the following formula, representing the complete extensions of $\mathcal{AF}$:

$$\mathsf{CMP}(\mathcal{AF}) = \bigwedge_{p \in \mathsf{Arg}} \left((p \supset \bigwedge_{(q,p) \in \mathit{Attack}} \neg q) \land (p \leftrightarrow \bigwedge_{(q,p) \in \mathit{Attack}} (\bigvee_{(r,q) \in \mathit{Attack}} r)) \right).$$

Another, more informative way, of representing admissible and/or complete extensions, is to turn to signed formulas (and so to an underlying three-valued semantics). By this, it is possible not only to identify the arguments in the extensions (those that are verified by the models of the formulas), but also identify the arguments that are attacked by the extensions (those that are falsified by the models of the formulas). Briefly, the idea is to associate every argument in the framework with a *pair* $\langle p^+, p^- \rangle$ of ("signed") atoms, the truth values of which describe the status of the associated argument: accepted (p^+ is verified, p^- is falsified), rejected (p^+ is falsified, p^- is verified), and undecided (both p^+ and p^-

---

[51] Recall that $\bigwedge \emptyset = \mathsf{T}$ (truth) and $\bigvee \emptyset = \mathsf{F}$ (falsity).

are falsified).[52] [53] [54]

Now, consider the following formula :

$$\mathsf{CMP}^{\pm}(\mathcal{AF}) = \bigwedge_{\langle p^+, p^-\rangle \in \mathsf{Arg}} \left\{ \begin{array}{l} \left((p^+ \wedge \neg p^-) \supset \bigwedge_{(\langle q^+, q^-\rangle, \langle p^+, p^-\rangle) \in \mathit{Attack}}(\neg q^+ \wedge q^-)\right), \quad (1) \\ \left((\neg p^+ \wedge p^-) \supset \bigvee_{(\langle q^+, q^-\rangle, \langle p^+, p^-\rangle) \in \mathit{Attack}}(q^+ \wedge \neg q^-)\right), \quad (2) \\ \left((\neg p^+ \wedge \neg p^-) \supset \right. \\ \quad \left(\neg (\bigwedge_{(\langle q^+, q^-\rangle, \langle p^+, p^-\rangle) \in \mathit{Attack}}(\neg q^+ \wedge q^-)) \wedge \right. \\ \quad \left.\left. \neg (\bigvee_{(\langle q^+, q^-\rangle, \langle p^+, p^-\rangle) \in \mathit{Attack}}(q^+ \wedge \neg q^-)) \right)\right), \quad (3) \\ \neg (p^+ \wedge p^-) \quad\quad\quad\quad\quad\quad\quad\quad\quad\quad\quad\quad\quad\quad (4) \end{array} \right\}.$$

- the subformula denoted by (1) states that any argument that attacks an accepted argument must be rejected,

- the subformula denoted by (2) states that any rejected argument must be attacked by at least one accepted argument,

- the subformula denoted by (3) states that for undecided arguments the previous conditions do not hold,[55] and

- the subformula denoted by (4) states that an argument may be either accepted, rejected, or undecided (i.e., a fourth state depicted by $p^+ \wedge p^-$ is excluded).

The next proposition (proved in [Arieli and Caminada, 2013]) shows the one-to-one correspondence between the models of $\mathsf{CMP}^{\pm}(\mathcal{AF})$ and the complete extensions of $\mathcal{AF}$.

---

[52] The superscripts $+$ and $-$ have several meaning in different contexts, as A^+ (respectively, A^-) denotes the set of arguments that are attacked by (respectively, that attack) A. This notational overloading will not cause any confusion in what follows. Signed formulas were used in the context of inconsistency-tolerant reasoning in [Besnard and Schaub, 1998].

[53] Again, we freely switch between an argument and the pair of atomic formulas that is associated with it, so a pair $\langle p^+, p^-\rangle$ of (signed) atoms also stands for an argument in the framework.

[54] For a representation in terms of four-valued semantics, where both p^+ and p^- may be verified, we refer to [Arieli, 2016].

[55] These three subformulas state conditions that correspond to Caminada's *complete labeling* (see [Baroni et al., 2018]). See also Remark 106.

Proposition 105. *Let $\mathcal{AF} = \langle \mathsf{Arg}, \mathit{Attack} \rangle$ be an argumentation framework. Then:*

- *For every complete extension $\mathcal{E} \in \mathsf{Cmp}(\mathcal{AF})$ there is a model $\mathcal{M}$ of $\mathsf{CMP}^{\pm}(\mathcal{AF})$ such that*
  - $\mathsf{In}(\mathcal{M}) = \{\langle p^+, p^- \rangle \mid \mathcal{M}(p^+) = t,\ \mathcal{M}(p^-) = f\} = \mathcal{E}$,
  - $\mathsf{Out}(\mathcal{M}) = \{\langle p^+, p^- \rangle \mid \mathcal{M}(p^+) = f,\ \mathcal{M}(p^-) = t\} = \mathcal{E}^+$,
  - $\mathsf{Undec}(\mathcal{M}) = \{\langle p^+, p^- \rangle \mid \mathcal{M}(p^+) = f,\ \mathcal{M}(p^-) = f\} = \mathsf{Arg} \setminus (\mathcal{E} \cup \mathcal{E}^+)$.

- *For every model $\mathcal{M}$ of $\mathsf{CMP}^{\pm}(\mathcal{AF})$ there is a complete extension $\mathcal{E} \in \mathsf{Cmp}(\mathcal{AF})$ such that*
  - $\mathcal{E} = \mathsf{In}(\mathcal{M}) = \{\langle p^+, p^- \rangle \mid \mathcal{M}(p^+) = t,\ \mathcal{M}(p^-) = f\}$
  - $\mathcal{E}^+ = \mathsf{Out}(\mathcal{M}) = \{\langle p^+, p^- \rangle \mid \mathcal{M}(p^+) = f,\ \mathcal{M}(p^-) = t\}$,
  - $\mathsf{Arg} \setminus (\mathcal{E} \cup \mathcal{E}^+) = \mathsf{Undec}(\mathcal{M}) = \{\langle p^+, p^- \rangle \mid \mathcal{M}(p^+) = f,\ \mathcal{M}(p^-) = f\}$.

Remark 106. *The notations in the first bullet of Proposition 105 are not accidental, as they correspond to the three types of assignments (in, out, undec) of the complete labeling of $\mathcal{AF}$.*[56] *Moreover, as shown in [Arieli and Caminada, 2013], all the results in this section carry on to labeling semantics.*

As an immediate consequence of the last proposition we get a representation of the stable extension of $\mathcal{AF}$. Indeed, as a stable extension is a set $\mathcal{E} \subseteq \mathsf{Arg}$ such that $\mathsf{Arg} = \mathcal{E} \cup \mathcal{E}^+$, by the last proposition we just have to add a requirement that $\mathsf{Undec}(\mathcal{M}) = \emptyset$ for every model $\mathcal{M}$ of a theory. This can be easily done by adding the following 'excluded middle' condition:

$$\mathsf{EM}^{\pm}(\mathcal{AF}) = \bigwedge_{\langle p^+, p^- \rangle \in \mathsf{Arg}} (p^+ \vee p^-)$$

Corollary 107. *Let $\mathcal{AF} = \langle \mathsf{Arg}, \mathit{Attack} \rangle$ be an argumentation framework. Then:*

---

[56]Labeling semantics for argumentation frameworks is described, e.g., in [Baroni et al., 2018].

- *For every $\mathcal{E} \in \mathsf{Stb}(\mathcal{AF})$ there is a model $\mathcal{M}$ of $\mathsf{CMP}^{\pm}(\mathcal{AF}) \cup \{\mathsf{EM}^{\pm}(\mathcal{AF})\}$ such that $\mathsf{In}(\mathcal{M}) = \mathcal{E}$ and $\mathsf{Out}(\mathcal{M}) = \mathcal{E}^{+}$.*

- *For every model $\mathcal{M}$ of $\mathsf{CMP}^{\pm}(\mathcal{AF}) \cup \{\mathsf{EM}^{\pm}(\mathcal{AF})\}$ there is a stable extension $\mathcal{E} \in \mathsf{Stb}(\mathcal{AF})$ such that $\mathcal{E} = \mathsf{In}(\mathcal{M})$ and $\mathcal{E}^{+} = \mathsf{Out}(\mathcal{M})$.*

When it comes to other types of extensions like grounded or preferred extensions, propositional formulas in classical logic are not sufficient for the representation, since the definitions of such extensions involve qualitative or comparative considerations. One way of dealing with this is to incorporate quantifiers in the language. As is shown in [Egly and Woltran, 2006; Arieli and Caminada, 2013; Diller et al., 2015; Arieli, 2016], for this purpose first-order languages are not necessary, and it is sufficient to remain in the propositional level, by using *quantified Boolean formulas*. For this, we extend the underlying language with universal and existential quantifiers $\forall, \exists$ over propositional variables.

Intuitively, the meaning of a quantified Boolean formula (QBF) of the form $\exists p \forall q \psi$ is that there exists a truth assignment of p such that for every truth assignment of q, ψ is true. Clearly, every QBF is associated with a logically equivalent propositional formula, thus ultimately we are still at the propositional level. This may be formally defined as follows:

Definition 108 (QBF-related notions). *Consider a QBF Ψ.*

- *An occurrence of an atom p in Ψ is called* free *if it is not in the scope of a quantifier $\mathsf{Q}p$, for $\mathsf{Q} \in \{\forall, \exists\}$.*

- *We denote by $\Psi[\phi_1/p_1, \ldots, \phi_n/p_n]$ the uniform substitution of each free occurrence of a variable (atom) p_i in Ψ by a formula ϕ_i, for $i = 1, \ldots, n$, and denote by T and F the propositional constants for truth and falsity (respectively).*[57]

- *Valuations over QBFs are, as usual, functions that assign truth values to the propositional variables (the atomic formulas) in the QBFs, and are extended to complex formulas as follows:*

---

[57]That is, for every valuation ν it holds that $\nu(\mathsf{T}) = t$ and $\nu(\mathsf{F}) = f$.

$\nu(\neg\psi) = \neg\nu(\psi),$
$\nu(\psi \circ \phi) = \nu(\psi) \circ \nu(\phi) \text{ for } \circ \in \{\wedge, \vee, \supset\},$
$\nu(\forall p\, \psi) = \nu(\psi[\mathsf{T}/p]) \wedge \nu(\psi[\mathsf{F}/p]),$
$\nu(\exists p\, \psi) = \nu(\psi[\mathsf{T}/p]) \vee \nu(\psi[\mathsf{F}/p]).$

Preferred extensions of an argumentation framework $\mathcal{AF}$ with n arguments that correspond to the n pairs $\{\langle p_1^+, p_1^-\rangle, \ldots, \langle p_n^+, p_n^-\rangle\}$ may now be represented by the following QBF:

$$\begin{aligned}\mathsf{PRF}^\pm(\mathcal{AF}) = {}& \mathsf{CMP}^\pm(\mathcal{AF})(p_1^+, p_1^-, \ldots, p_n^+, p_n^-) \wedge \\ & \forall q_1^+, q_1^-, \ldots, q_n^+, q_n^- \Big(\mathsf{CMP}^\pm(\mathcal{AF})(q_1^+, q_1^-, \ldots, q_n^+, q_n^-) \supset \\ & \mathsf{INC}_\subseteq^\pm(p_1^+, p_1^-, \ldots, p_n^+, p_n^-, q_1^+, q_1^-, \ldots, q_n^+, q_n^-)\Big).\end{aligned}$$

Here, $\mathsf{CMP}^\pm(\mathcal{AF})(p_1^+, p_1^-, \ldots, p_n^+, p_n^-)$ is the formula $\mathsf{CMP}^\pm(\mathcal{AF})$ considered previously, but with the free variables $p_1^+, p_1^-, \ldots p_n^+, p_n^-$, and

$$\begin{aligned}&\mathsf{INC}_\subseteq^\pm(p_1^+, p_1^-, \ldots p_n^+, p_n^-, q_1^+, q_1^-, \ldots q_n^+, q_n^-) = \\ & \bigwedge_i\Big((p_i^+ \wedge \neg p_i^-) \supset (q_i^+ \wedge \neg q_i^-)\Big) \supset \bigwedge_i\Big((q_i^+ \wedge \neg q_i^-) \supset (p_i^+ \wedge \neg p_i^-)\Big).\end{aligned}$$

Intuitively, a model $\mathcal{M}$ of $\mathsf{PRF}^\pm(\mathcal{AF})$ should satisfy two requirements: the condition in the first line of the formula (i.e., $\mathsf{CMP}^\pm(\mathcal{AF})$) assures that the pairs $\langle p^+, p^-\rangle$ that are verified by $\mathcal{M}$ correspond to a complete extension of $\mathcal{AF}$. The condition on the second and the third line ($\mathsf{CMP}^\pm(\mathcal{AF}) \supset \mathsf{INC}_\subseteq^\pm(\mathcal{AF})$) assures that this set of pairs is not strictly $\subset$-included in another set that forms a complete extension of $\mathcal{AF}$. We thus have:

Proposition 109. ([Arieli and Caminada, 2013]) *Let* $\mathcal{AF} = \langle \mathsf{Arg}, \mathsf{Attack}\rangle$ *be an argumentation framework. Then:*

- *For every preferred extension* $\mathcal{E} \in \mathsf{Prf}(\mathcal{AF})$ *there is a model* $\mathcal{M}$ *of* $\mathsf{PRF}^\pm(\mathcal{AF})$ *such that* $\mathsf{In}(\mathcal{M}) = \mathcal{E}$, $\mathsf{Out}(\mathcal{M}) = \mathcal{E}^+$, *and* $\mathsf{Undec}(\mathcal{M}) = \mathsf{Arg} \setminus (\mathcal{E} \cup \mathcal{E}^+)$.

- *For every model* $\mathcal{M}$ *of* $\mathsf{PRF}^\pm(\mathcal{AF})$ *there is a preferred extension* $\mathcal{E} \in \mathsf{Prf}(\mathcal{AF})$ *such that* $\mathcal{E} = \mathsf{In}(\mathcal{M})$, $\mathcal{E}^+ = \mathsf{Out}(\mathcal{M})$, *and* $\mathsf{Arg} \setminus (\mathcal{E} \cup \mathcal{E}^+) = \mathsf{Undec}(\mathcal{M})$.

In a similar way it is possible to represent the grounded semantics as well as other types of comparative Dung-type extensions, such as semi-stable semantics, eager semantic, ideal semantics, and so forth (see [Arieli and Caminada, 2013]). In [Diller et al., 2015] similar QBF-based representations are used for representing extensions of abstract dialectical frameworks [Brewka et al., 2017], and in [Arieli, 2016] they are used for representing conflict-tolerant semantics. It follows that off-the-shelf SAT-solvers and/or QBF-solvers may be used for computing argumentation-based entailments by Dung semantics.

Another approach based on propositional logic is taken in [Straßer and Šešelja, 2010]. Again, arguments are represented by propositional letters in a finite set **Atoms**. The language of propositional logic is enriched with a connective $\twoheadrightarrow$ characterized by the axiom scheme $(\phi \wedge (\phi \twoheadrightarrow \psi)) \supset \neg \psi$ to express argumentative attack. The fact that an argument ψ (in **Atoms**) is defeated is then expressed by:

$$\mathsf{def}\,\psi = \bigvee_{\phi \in \mathsf{Atoms}} (\phi \wedge (\phi \twoheadrightarrow \psi)).$$

In order to express admissible semantics, i.e., the idea that the selected arguments have to defend themselves from all attacks, the following axiom is used:

$$(\phi \wedge (\psi \twoheadrightarrow \phi)) \supset \mathsf{def}\,\psi.$$

The logic $\mathfrak{L}_\mathsf{A} = \langle \mathcal{L}^{\twoheadrightarrow}_{\mathsf{Atoms}}, \vdash_\mathsf{A} \rangle$ is axiomatized by classical propositional logic enriched with the three discussed axiom schemes. In order to characterize complete extensions, $\mathfrak{L}_\mathsf{A}$ is enriched with

$$\bigwedge_{\phi \in \mathsf{Atoms}} ((\phi \twoheadrightarrow \psi) \supset \mathsf{def}\,\phi) \supset \psi$$

resulting in $\mathfrak{L}_\mathsf{C} = \langle \mathcal{L}^{\twoheadrightarrow}_{\mathsf{Atoms}}, \vdash_\mathsf{C} \rangle$, expressing that if an argument is defended then it is selected.[58]

---

[58]The presentation of the logics in [Straßer and Šešelja, 2010] is slightly simplified in that the original systems also capture argumentative changes, that is, a dynamic proof theory is presented that allows for the addition of new arguments and new argumentative attacks "on-the-fly". For a similar approach see our discussion in Section 3.3.

Similar to the approach in QBL, in order to characterize grounded and preferred semantics, more formal machinery needs to be employed. Instead of quantifiers, in [Straßer and Šešelja, 2010] the preferential semantics of adaptive logics is used (recall Section 2.4.2-A). That means, for the grounded [preferred] semantics those $\mathfrak{L}_C$-interpretations are selected in which the least [most] atoms are true. As shown in [Van De Putte, 2013], the selection semantics underlying adaptive logics can also be expressed in terms of maximal consistent subsets.

Given our previous discussion of MCS-based reasoning, we therefore state the following corollary from [Straßer and Šešelja, 2010, Theorem 1]: Given a logic $\mathfrak{L} = \langle \mathcal{L}, \vdash \rangle$ and sets $\mathcal{T}$ and $\mathcal{T}'$ of $\mathcal{L}$-sentences, let in the following proposition $\mathsf{MC}_\mathfrak{L}^\mathcal{T}(\mathcal{T}')$ be the set of all maximally $\vdash$-consistent sets $\mathcal{S}$ of $\mathcal{L}$-sentences for which: (a) $\mathcal{T} \subseteq \mathcal{S}$, and (b) there is no $\vdash$-consistent set $\mathcal{S}'$ of $\mathcal{L}$-sentences for which both $(\mathcal{S} \cap \mathcal{T}') \subsetneq (\mathcal{S}' \cap \mathcal{T}')$ and $\mathcal{T} \subseteq \mathcal{S}'$.

Proposition 110. *Let $\mathcal{AF} = \langle \mathsf{Args}, \mathsf{Attack} \rangle$ be an abstract argumentation framework based on a finite set of arguments. Consider the language $\mathcal{L}_{\mathsf{Args}}^{\vec{}}$ and let $\Gamma = \{\phi \twoheadrightarrow \psi \mid (\phi, \psi) \in \mathsf{Attacks}\} \cup \{\neg(\phi \twoheadrightarrow \psi) \mid (\phi, \psi) \notin \mathsf{Attacks}\}$. We have:*

- $\mathsf{Adm}(\mathcal{AF}) = \{\mathsf{Atoms}(\mathcal{S}) \mid \mathcal{S} \in \mathsf{MCS}_{\mathfrak{L}_A}^\Gamma(\mathcal{L}_{\mathsf{Args}}^{\vec{}})\}$
  (In other words, $\mathcal{T} \in \mathsf{Adm}(\mathcal{AF})$ iff there is a maximally $\mathfrak{L}_A$-consistent set of sentences $\mathcal{S}$ for which $\Gamma \subseteq \mathcal{S}$ and $\mathcal{T} = \mathsf{Atoms}(\mathcal{S})$),

- $\mathsf{Cmp}(\mathcal{AF}) = \{\mathsf{Atoms}(\mathcal{S}) \mid \mathcal{S} \in \mathsf{MCS}_{\mathfrak{L}_C}^\Gamma(\mathcal{L}_{\mathsf{Args}}^{\vec{}})\}$,

- $\mathsf{Grd}(\mathcal{AF}) = \mathsf{Atoms}(\mathcal{S})$ *where* $\{\mathcal{S}\} = \mathsf{MC}_{\mathfrak{L}_C}^\Gamma(\{\neg\phi \mid \phi \in \mathsf{Atoms}\})$,

- $\mathsf{Prf}(\mathcal{AF}) = \{\mathsf{Atoms}(\mathcal{S}) \mid \mathcal{S} \in \mathsf{MC}_{\mathfrak{L}_A}^\Gamma(\mathsf{Atoms})\} = \{\mathsf{Atoms}(\mathcal{S}) \mid \mathcal{S} \in \mathsf{MC}_{\mathfrak{L}_C}^\Gamma(\mathsf{Atoms})\}$,

- $\mathsf{SStb}(\mathcal{AF}) = \{\mathsf{Atoms}(\mathcal{S}) \mid \mathcal{S} \in \mathsf{MC}_{\mathfrak{L}_C}^\Gamma(\{\phi \vee \mathsf{def}\phi \mid \phi \in \mathsf{Atoms}\})\}$.[59]

We note, finally, that the presentation in this section is by no means exhaustive, but rather meant to illustrate the way logical propositional

---

[59]$\mathsf{SStb}(\mathcal{AF})$ is the set of the *semi-stable extensions* of $\mathcal{AF}$, that is: the complete extensions $\mathcal{E}$ such that $\mathcal{E} \cup \mathcal{E}^+$ is maximal among all the complete extensions of $\mathcal{AF}$.

formulas may be used for encoding the dynamics of argumentation-based reasoning. Among other approaches that are based on a Tarskian logic we recall the ones in [Gabbay and Gabbay, 2016] and [Fandinno and del Cerro, 2018] based on intuitionistic logic, in [Dyrkolbotn, 2014] based on Łukasiewicz logic, in [Dvořák et al., 2012] based on monadic second order logic, in [Gabbay, 2011] and [Gabbay and Gabbay, 2015] based on classical logic, and in [de Saint-Cyr et al., 2016] based on first-order logic with finite domains. We refer to [Besnard et al., 2020] for a recent comprehensive survey on the subject (see in particular Sections 4–8 therein, which are relevant to the material in this chapter), where also a variety of implementations are described (summarized in [Besnard et al., 2020, Table 4]).

3.2 Representation Methods Based on Modal Languages

In this section we consider several systems for reasoning about argumentation in a modal logical context. We distinguish two major purposes these systems serve:

1. The first goal, which is shared among all the presented systems and discussed in Section 3.2.1, is to express underlying notions of abstract argumentation, such as attacks and semantic selections, in the object language via modal operators.

2. The second goal, discussed in Section 3.2.2, is to integrate central notions underlying argumentative reasoning with those expressing argumentation dynamics in Item 1, for instance, propositional attitudes such as belief and endorsement, and justification. In this way, the presented logics offer a comprehensive logical model of (meta)argumentation and its dynamics.

We start with the basic settings of [Boella et al., 2005; Caminada and Gabbay, 2009; Grossi, 2010; Villata et al., 2012; Grossi, 2013], which are concerned with meta-argumentative reasoning, and then move on to some frameworks that include epistemic considerations [Grossi and van der Hoek, 2014; Shi et al., 2018].

3.2.1 Argumentation Logics

Grossi in [Grossi, 2010; Grossi, 2013] defines *argumentation models* to reason about argumentative situations. An argumentation model $\mathcal{M}$ based on an argumentation framework $\mathcal{AF} = \langle \mathsf{Args}, \rightarrow \rangle$[60] is a tuple $\langle \mathsf{Args}, \leftarrow, v \rangle$, where $\leftarrow$ is the inverted version of $\rightarrow$ (that is, $A \leftarrow B$ iff $B \rightarrow A$). The pair $\langle \mathsf{Args}, \leftarrow \rangle$ constitutes a Kripkean possible world frame where arguments provide the points connected by the accessibility relation $\leftarrow$. As usual, the assignment v associates each propositional atom with a set of points (arguments) in which they hold.

In the following, we enrich the propositional language by two unary modalities. Thus, formulas in the language are defined by the following BNF:[61]

$$\phi \; := \; \mathsf{Atoms} \; | \; \neg \phi \; | \; \phi \wedge \phi \; | \; \Box_a \phi \; | \; \Box_u \phi \; | \; \mathsf{F}$$

where Atoms is a set of propositional atoms of the language. The diamond-versions of the given modal operators are defined as usual: $\Diamond_a =_{df} \neg \Box_a \neg$ and $\Diamond_u =_{df} \neg \Box_u \neg$. Other propositional connectives, such as implication $\supset$, disjunction $\vee$, and the propositional constant T for truth are defined as usual in classical propositional logic.

Validity for atoms and propositional connectives is defined in the usual way. Similarly, the modal operators $\Box_a$ and $\Box_u$ function like a usual necessitation and universal necessitation operator. For a model $\mathcal{M} = \langle \mathsf{Args}, \leftarrow, v \rangle$ and an argument $A \in \mathsf{Args}$, we define:

- $\mathcal{M}, A \models \Box_a \phi$ iff for all $B \in \mathsf{Args}$ for which $A \leftarrow B$ we have $\mathcal{M}, B \models \phi$. Since worlds are identified with arguments, this definition is understood as follows: all attackers B of the argument A have the property ϕ.

- $\mathcal{M}, A \models \Box_u \phi$ iff for all $B \in \mathsf{Args}$, $\mathcal{M}, B \models \phi$. In words: all the arguments $B \in \mathsf{Args}$ have the property ϕ.[62]

---

[60]To keep the original notations, we use in this section the arrow sign for designating the attack relation.

[61]We use the $\Box$-notation in our language since we will later on generalize this logic to a product logic where the argumentation-related modalities will provide the vertical axis.

[62]Thus, if $\mathcal{M}, A_0 \models \Box_u \phi$ for some A_0 then $\mathcal{M}, A \models \Box_u \phi$ for every $A \in \mathsf{Args}$.

- $\mathcal{M} \models \phi$ iff for all $A \in \mathsf{Args}$ it holds that $\mathcal{M}, A \models \phi$. The set of all formulas ϕ for which $\mathcal{M} \models \phi$ is denoted by $[\![\phi]\!]_{\mathcal{M}}$ (the subscript is removed when the context disambiguates).

In sum, since there are no frame conditions, we are dealing with models of the modal logic K enriched with universal modality.

Example 111. *Consider the argumentation framework and the assignment v presented in Figure 10.*

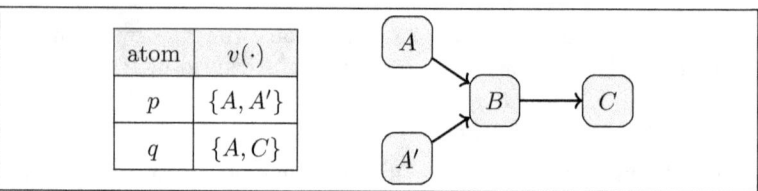

Figure 10: Left: the assignment of Example 111; Right: the argumentation framework of Example 111

In this case, we have:

- *$\mathcal{M}, A \models \Box_a \mathsf{F}$ and $\mathcal{M}, A' \models \Box_a \mathsf{F}$, expressing that A and A' have no attackers.*

- *$\mathcal{M}, B \models \Diamond_a \Box_a \mathsf{F}$, expressing that there is an attacker against which B cannot be defended (since this attacker has no attackers).*

- *$\mathcal{M}, C \models \Box_a \Diamond_a \mathsf{T}$ and $\mathcal{M}, C \models \Box_a \Diamond_a p$, expressing that for all attackers of C there is a defender (either A or A')*

More generally, we have for any $x \in \mathsf{Args}$:

- *$\mathcal{M}, x \models \Box_u((p \vee q) \supset \Box_a \Diamond_a (p \vee q))$, expressing that the set $\{A, A', C\}$ (consisting of the worlds in which $p \vee q$ holds) attacks all its attackers.*

As the following proposition shows, the induced logic is expressive enough to characterize several standard semantics.

Proposition 112. ([Grossi, 2010, p. 411]) *Let* $\mathcal{AF} = \langle \mathsf{Args}, \rightarrow \rangle$ *and* $\mathcal{M} = \langle \mathsf{Args}, \leftarrow, v \rangle$. *For* $[\![\phi]\!]_{\mathcal{M}} = \mathcal{E} \subseteq \mathsf{Args}$, *it holds that:*

$$\mathcal{M} \models \mathsf{sem}(\phi) \text{ iff } \mathcal{E} \in \mathsf{Sem}(\mathcal{AF}),$$

where the correspondence between the formula sem *and the semantics* Sem *is the following:*

Sem	$\mathsf{sem}(\phi)$
Adm	$\boxdot_u(\phi \supset (\boxdot_a \neg \phi \wedge \boxdot_a \Diamond_a \phi))$
Cmp	$\boxdot_u((\phi \supset \boxdot_a \neg \phi) \wedge (\phi \leftrightarrow \boxdot_a \Diamond_a \phi))$
Stb	$\boxdot_u(\phi \leftrightarrow \boxdot_a \neg \phi)$

Example 113. *In Example 111 we have, for instance, that:*

- $\mathcal{M} \models \mathsf{adm}(p)$, *since* $\{A, A'\}$ *is admissible, while*

- $\mathcal{M} \models \neg\mathsf{cmp}(p)$ *and* $\mathcal{M} \models \neg\mathsf{stb}(p)$, *since* $\{A, A'\}$ *is neither complete nor stable, and*

- $\mathcal{M} \models \mathsf{cmp}(p \vee q)$ *and* $\mathcal{M} \models \mathsf{stb}(p \vee q)$, *since* $\{A, A', C\}$ *is complete and stable.*

The logic, however, lacks the resources to express argumentation semantics that are based on minimality or maximality assumptions, such as grounded and preferred semantics. We recall (see [Dung, 1995]) that the grounded extension is characterized by the least fixed point of the function

$$\mathsf{defended}: \wp(\mathsf{Args}) \to \wp(\mathsf{Args}),$$

which maps a set $\mathcal{S}$ of arguments to the set of all arguments in Args that are defended by $\mathcal{S}$. Now, recall from our example that $\boxdot_a \Diamond_a$ expresses argumentative defense in the logic, i.e., $\mathcal{M}, A \models \boxdot_a \Diamond_a \phi$ iff $[\![\phi]\!]_{\mathcal{M}}$ defends A. We thus need to characterize the formula ψ for which $[\![\psi]\!]_{\mathcal{M}}$ is minimal such that $[\![\psi]\!]_{\mathcal{M}} = [\![\boxdot_a \Diamond_a \psi]\!]_{\mathcal{M}}$. For this purpose one can enrich the argumentation logic by a fixpoint μ-operator (see [Bradfield and Stirling, 2001] for an introduction to modal μ-calculi), defined as follows: [63]

$$\mathcal{M}, A \models \mu p.\phi(p) \text{ iff } A \in \bigcap \{\mathcal{S} \in \wp(\mathsf{Args}) \mid [\![\phi]\!]_{\mathcal{M}[p:=\mathcal{S}]} \subseteq \mathcal{S}\},$$

where $\mathcal{M}[p := \mathcal{S}] = \langle \mathsf{Args}, \leftarrow, v'\rangle$, $v'_{\mathsf{Atoms}\setminus\{p\}} = v_{\mathsf{Atoms}\setminus\{p\}}$, and $v'(p) = \mathcal{S}$.[64]

In [Grossi, 2013] Grossi tackles preferred and semi-stable semantics[65] by means of a second-order formalization:

$$\mathcal{M}, A \models \exists p.\phi(p) \text{ iff there is an } \mathcal{S} \subseteq \mathsf{Args} \text{ such that } \mathcal{M}_{[p:=\mathcal{S}]}, A \models \phi(p).$$

The following proposition is shown in [Grossi, 2010] for the grounded semantics and in [Grossi, 2013] for the preferred and semi-stable semantics:[66]

Proposition 114. *Denote by $\phi \sqsubseteq_{\mathsf{u}} \psi$ the formula $\Box_{\mathsf{u}}(\phi \supset \psi)$ and denote by $\phi \sqsubset_{\mathsf{u}} \psi$ the formula $(\phi \sqsubseteq_{\mathsf{u}} \psi) \wedge \neg(\psi \sqsubseteq_{\mathsf{u}} \phi)$. Let ϕ be a formula such that $[\![\phi]\!]_{\mathcal{M}} = \mathcal{E} \subseteq \mathsf{Args}$. It holds that:*

$$\mathcal{M} \models \mathsf{sem}(\phi) \text{ iff } \mathcal{E} \in \mathsf{Sem}(\mathcal{AF}),$$

where the correspondence between the formula sem and the semantics Sem is the following:

Sem	$\mathsf{sem}(\phi)$
Grd	$\mathsf{cmpl}(\phi) \wedge \forall q.(\mathsf{cmpl}(q) \supset \phi \sqsubseteq_{\mathsf{u}} q)$
Prf	$\mathsf{cmpl}(\phi) \wedge \neg \exists q.(\mathsf{cmpl}(q) \wedge \phi \sqsubset_{\mathsf{u}} q))$
SStb	$\mathsf{cmpl}(\phi) \wedge \neg \exists q.((\phi \vee \Diamond_{\mathsf{a}}\phi) \sqsubset_{\mathsf{u}} (q \vee \Diamond_{\mathsf{a}} q)))$

---

[63] All systems introduced in this section have an adequate axiomatization (see e.g. [Grossi, 2010]), which we omit for space reasons.

[64] If $\mathcal{A}$ is a set of atoms and v is a valuation, $v_{\mathcal{A}}$ denotes the restriction of v to the atoms in $\mathcal{A}$.

[65] Recall Footnote 59.

[66] See below for the treatment of preferred extensions in [Shi et al., 2018] in terms of a fixpoint μ-operator.

In [Caminada and Gabbay, 2009], Caminada and Gabbay also use argumentation models, but proceed differently when characterizing argumentation semantics. Let p_i, p_o and p_u be three atoms which are intended to represent the three argument labels in, out, and undec. We can now elegantly express the characteristic requirements of complete labelings:[67]

1. $\mathcal{M}, A \models (\Box_a \mathsf{F} \vee \Box_a p_o) \supset p_i$ expresses that if A is not attacked ($\Box_a \mathsf{F}$) or all attackers of A are out ($\Box_a p_o$), then A is in;

2. $\mathcal{M}, A \models \Diamond_a p_i \supset p_o$ expresses that if A is attacked by an argument that is in, then A is out;

3. $\mathcal{M}, A \models \Box_a(p_o \vee p_u) \wedge \Diamond_a p_u \supset p_u$ expresses that if A has only attackers that are out or undec and at least one attacker is undec, then A is undec as well;

4. $\mathcal{M}, A \models (p_i \vee p_o \vee p_u) \wedge \neg(p_i \wedge p_o) \wedge \neg(p_i \wedge p_u) \wedge \neg(p_o \wedge p_u)$ expresses that A has exactly one label.

By restricting argumentation models to those that satisfy Items 1–4 (at every argument A), we can, for instance, characterize the grounded extension as follows, where again $\mathcal{AF} = \langle \mathsf{Args}, \rightarrow \rangle$: If for every model $\mathcal{M}$ in the restricted class based on the frame $\langle \mathsf{Args}, \leftarrow \rangle$ we have $\mathcal{M}, B \models p_i$ then $B \in \mathsf{Grd}(\mathcal{AF})$, and vice versa. Other semantics are represented in [Caminada and Gabbay, 2009] by techniques from circumscription logic.

A different approach is taken in [Boella et al., 2005] and [Villata et al., 2012]. The starting point there is again an argumentation framework $\mathcal{AF} = \langle \mathsf{Args}, \rightarrow \rangle$, but instead of treating arguments as possible worlds in a Kripkean frame as in the previous approaches, the set of worlds is now given by $\wp(\mathsf{Args})$. Again, the accessibility relation encodes argumentative attacks.

Denote by $\rightarrow^\wp$ the following lifting of $\rightarrow$ to $\wp(\mathsf{Args}) \times \wp(\mathsf{Args})$: we write $\mathcal{S} \rightarrow^\wp \mathcal{S}'$ iff there is an $A \in \mathcal{S}$ and a $B \in \mathcal{S}'$ such that $A \rightarrow B$. Let also $\rightarrow^\wp_C = (\wp(\mathsf{Args}) \times \wp(\mathsf{Args})) \setminus \rightarrow^\wp$ be the complement of $\rightarrow^\wp$. Figure 11 shows a simple example.

---

[67] Recall Remark 106. See [Caminada, 2006] and [Baroni et al., 2018] for a characterization of argumentation semantics in terms of labelings.

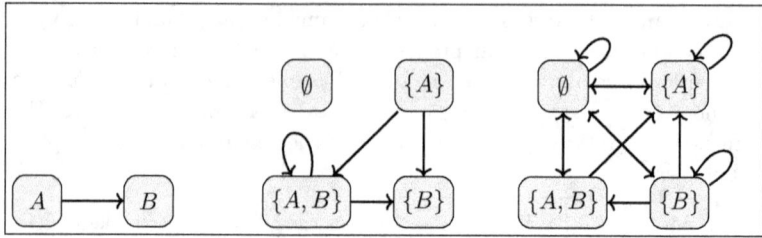

Figure 11: Left: The attack diagram for $\mathcal{AF} = \langle \{A, B\}, \rightarrow \rangle$, where $\rightarrow = \{(a, b)\}$; Middle: Graph for $\rightarrow^\wp$; Right: Graph for $\rightarrow_C^\wp$.

The formal language is similar to the ones given above, except that now the propositional atoms corresponds directly to the abstract arguments:

$$\phi ::= \mathsf{Args} \mid \neg\phi \mid \phi \wedge \phi \mid \Box_u \phi \mid \Box_a \phi$$

The truth conditions of propositional connectives are as usual. We define:

- $\mathcal{M}, \mathcal{S} \models A$ iff $A \in \mathcal{S}$. This expresses that a is a member of the currently considered set of arguments;

- $\mathcal{M}, \mathcal{S} \models \Box_a \phi$ iff for all $\mathcal{S}'$ for which $\mathcal{S} \rightarrow_C^\wp \mathcal{S}'$, it holds that $\mathcal{M}, \mathcal{S}' \models \phi$. This expresses that ϕ holds for all sets of arguments $\mathcal{S}'$ not attacked by $\mathcal{S}$.

- $\mathcal{M}, \mathcal{S} \models \Box_u \phi$ iff for all $\mathcal{S} \in \wp(\mathsf{Args})$ it holds that $\mathcal{M}, \mathcal{S} \models \phi$. This expresses that all sets of arguments have the property ϕ.

Just like the previous formalisms, at its core also this logic is K enriched with a universal modality. The logic allows us to express core concepts of abstract argumentation such as attack and defense:

- $\mathcal{M} \models \Box_u(A \supset \Box_a \neg B)$ expresses that A attacks B,

- $\mathcal{M} \models \Box_u(\bigwedge \mathcal{S} \supset \Box_a \neg \bigwedge \mathcal{S}')$ expresses that some argument $A \in \mathcal{S}$ attacks some argument $A' \in \mathcal{S}'$,

- $\mathcal{M} \models \Box_{\mathsf{u}} \bigwedge_{\mathcal{S}' \in \wp(\mathsf{Args})} (\Box_{\mathsf{u}}(\bigwedge \mathcal{S}' \supset \Box_{\mathsf{a}} \neg A) \supset \Box_{\mathsf{u}}(\bigwedge \mathcal{S} \supset \Box_{\mathsf{a}} \neg \bigwedge \mathcal{S}'))$
  expresses that the set of arguments $\mathcal{S}$ defends the argument A.[68]

In a series of articles Gabbay and various co-authors investigate logical characterizations of argumentation frameworks. In [Gabbay and Gabbay, 2015] and [Gabbay and Gabbay, 2016] the basic idea is similar to the systems presented above: arguments are represented by propositional atoms, and the fact that an argument A attacks argument B is represented by the formula $A \supset \sim B$, in which $\supset$ is an implication and $\sim$ is a negation of the underlying logic. Different core logics are considered:

- In [Gabbay and Gabbay, 2016] the underlying logic is the intuitionistic logic G_3, whose Kripkean models consist of two linearly ordered worlds (also known as Here-and-There logic [Pearce, 2006]).

- In [Gabbay and Gabbay, 2015] the underlying logic is classical and $\sim$ is a strong negation N, for which $\sim p \supset \neg p$ but not necessarily vice versa (where $\neg$ is the classical negation).[69] N can be used to express different argument label/statuses: a holds if a is **in**, $\mathsf{N}a$ holds if a is **out** and $\neg a \wedge \neg \mathsf{N}a$ holds if a is **undec**.

Remark 115. *The negation N in the second item also has an elegant modal characterization in the logic* CNN *[Gabbay and Gabbay, 2015].*

---

[68]To express this, the set Args is supposed to be finite (otherwise a second-order approach is needed). In order to express properties of specific semantics the authors enhance their modal logic by unary non-normal modal operators. We refer to [Villata et al., 2012] for further details.

[69]An earlier characterization of Dung-style argumentation in classical logic has been presented in [Gabbay, 2011] for stable semantics (as well as for complete semantics in a 3-valued setting). The only logical connective in the presented system is the "Peirce-Quine-Dung dagger" $\Downarrow$, a generalization of the Peirce-Quine dagger or of NOR: $\Downarrow \Delta$ is true iff $\bigvee \Delta$ is false. The attack relation corresponds in this representation to the direct subformula relation (which is generalized to equivalence classes in order to deal with attack cycles): note that if $\Downarrow \Delta$ is true all members of Δ are false and, vice versa, if some member of Δ is true, $\Downarrow \Delta$ is false. In this context Gabbay also develops a "geometric concept of proof" which concerns inference rules (such as geometrical modus ponens) that operate on patterns of a given attack diagram and which are adequate to a given proof procedure in the Peirce-Quine-Dung-Dagger logic. Similar to the modal systems discussed here, the logic in [Gabbay, 2011] offers several generalizations, such as quantifiers, higher-order attacks, etc.

Like G_3, there are two worlds in the underlying pointed Kripkean models, just now for each world the other world is the only accessible one. The modal truth conditions for N are then spelled out by: $N\phi$ holds in one world iff $\neg\phi$ holds in the other. Similarly to intuistionistic possible worlds models (including those of G_3), models of CNN are constrained by a "monotony" requirement on $\models$: if p holds at the actual world, it necessarily holds at the other world as well. However, if p holds at the non-actual world, it need not hold at the actual world, although the actual world is accessible.

The translations of a given argumentation framework into the language of G_3 (see Equation (1)) or of CNN (see Equation (2)) are also similar for both systems, where for each $x \in \mathsf{Args}$, $x^- = \{y \in \mathsf{Args} \mid y \to x\}$ and the formula n in Equation (1), introduced to identify the actual world, can be defined by $\bigwedge_{x \in \mathsf{Args}}(x \vee \neg x)$:[70]

$$\bigwedge_{x \in \mathsf{Args}} \left(\begin{array}{c} \overbrace{(x \supset (\mathsf{n} \vee \bigwedge_{y \in x^-} \neg y))}^{\text{if in, all attackers out}} \wedge \overbrace{(\bigwedge_{y \in x^-} \neg y \supset (\mathsf{n} \vee x))}^{\text{if all attackers out, then in}} \\ \wedge \underbrace{(\neg x \supset (\mathsf{n} \vee \bigvee_{y \in x^-} y))}_{\text{if out, some attackers in}} \wedge \underbrace{(\bigvee_{y \in x^-} y \supset (\mathsf{n} \vee \neg x))}_{\text{if some attackers in, then out}} \end{array} \right) \quad (1)$$

$$\bigwedge_{x \in \mathsf{Args}} \left(\begin{array}{c} \overbrace{(\bigwedge_{y \in x^-} \mathsf{N}y \leftrightarrow x)}^{x \text{ in iff all attackers out}} \wedge \overbrace{((\bigwedge_{y \in x^-} \neg y \wedge \bigvee_{y \in x^-} \neg \mathsf{N}y) \supset (\neg x \wedge \neg \mathsf{N}x))}^{\text{if all attackers not in and some und, then und}} \\ \wedge \underbrace{(\bigwedge_{x \in y^-} x \supset \mathsf{N}y)}_{x \text{ attacks } y} \end{array} \right) \quad (2)$$

In both systems (i.e., the Kripkean semantics for G_3 and in CNN), we can, for each atom, identify one of the truth-assignment patterns in (the

---

[70] Clearly, like previous encodings, the translations presuppose a finite set of arguments.

12 - LOGIC-BASED APPROACHES TO FORMAL ARGUMENTATION

left part of) Table 16 relative to the two worlds in a given model. These patterns correspond to argument labels as indicated in the same table. This means that the models of the translated argumentation frameworks are one-to-one related to the complete labelings of the framework. As a consequence, the entailed atoms characterize the grounded extension. Stable semantics can be characterized by demanding excluded middle $p \vee \sim p$ (where again in the case of G_3 $\sim$ is intuitionistic negation and in the case of CNN it is strong negation).

	G_3 / CNN			LN1			
	in	out	undec	in	out	undec	
w_1	1	0	0	1	0	1	w_1
w_2	1	0	1	1	0	0	w_2
				1	0	1	w_3

Table 16: Overview: truth-value assignment pattern and argument labelings. Note that in G_3 and CNN two worlds are used, while in LN1 there are three worlds.

We illustrate this by means of the argumentation framework in Figure 12.

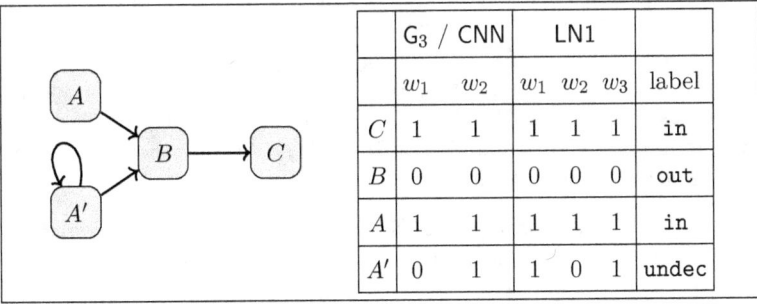

	G_3 / CNN		LN1			
	w_1	w_2	w_1	w_2	w_3	label
C	1	1	1	1	1	in
B	0	0	0	0	0	out
A	1	1	1	1	1	in
A'	0	1	1	0	1	undec

Figure 12: Example for the characterizations of the given AF on the left in the logics G_3, CNN and LN1

A related approach is introduced in [Gabbay, 2009] and [Caminada and Gabbay, 2009], where argumentation frameworks are characterize in terms of provability logic[71] and argumentation labelings are modeled in terms of fixed points of modal formulas. The underlying logic **LN1** is given by **K4**, enhanced with:

- Löb's axiom ($\Diamond \phi \supset \Diamond(\phi \wedge \Box \neg \phi)$),

- an axiom of linearity (($\Diamond \phi \wedge \Diamond \psi) \supset (\Diamond(\phi \wedge \psi) \vee \Diamond(\phi \wedge \Diamond \psi) \vee \Diamond(\psi \wedge \Diamond \phi))$), and

- some axioms characterizing the behavior of atoms: ($p \supset \Box(\neg p \supset \Box p)$, $\Box(\Box \bot \vee p) \leftrightarrow \Box p$ and $\Box(\Box \bot \vee \neg p) \leftrightarrow \Box \neg p$).

Pointed **LN1** models are such that the accessibility relation $<$ forms finite linear chains starting with the actual world. Additionally, it is required that if all non-endpoints of $<$ agree on the assignment of an atom, then the endpoint takes over the same assignment.

Let $G\phi = \phi \wedge \Box \phi$. Argumentation frameworks are translated into the language of **LF1** as follows:

$$G\left(\Box\bot \vee \bigwedge_{\substack{x \in \text{Args} \\ x^- \neq \emptyset}} \left(x \leftrightarrow \bigwedge_{y \in x^-} \Diamond \neg y\right)\right) \wedge \bigwedge_{\substack{x \in \text{Args} \\ x^- = \emptyset}} Gx \qquad (3)$$

In [Gabbay, 2009] it is shown that there is a one-to-one correspondence between LP1-models of the formula in Equation (3), whose states form chains of length 3, and complete labelings of the given argumentation framework. As was the case for G_3 and **CNN**, we can again uniquely associate argument labels with valuation patterns at the given possible worlds (see the right-hand side of Table 16). We show how this plays out in our example in Figure 12.

Remark 116. *The logics* G_3, **CNN** *and* **LN1** *can readily express higher-order and joint attacks, as well as argument quantifiers. We refer to the original papers for more details.*

---

[71] A similar approach was used in [Gabbay, 1990] for cyclic logic programs.

3.2.2 Belief, Informativeness and Awareness

One of the advantages of using modal argumentation logics is the possibility to integrate epistemic modalities. In this section we demonstrate this.

Grossi and van der Hoek [Grossi and van der Hoek, 2014] propose a modal product logic (see [Gabbay and Shehtman, 1998]) in which the argumentation logic from [Grossi, 2010; Grossi, 2013] (see our discussion in the previous section) provides one ingredient and a **KD45** epistemic logic provides another. The latter have frames of the form $\langle \mathcal{S}, \mathcal{P} \rangle$, where $\mathcal{S}$ is a set of (epistemic) states and $\mathcal{P} \subseteq \mathcal{S}$ is a non-empty subset of $\mathcal{S}$, namely those that a given agent considers possible. A frame of the product logic is then the product of an epistemic frame $\langle \mathcal{S}, \mathcal{P} \rangle$ and an argumentation frame $\langle \mathcal{A}, \leftarrow \rangle$. The domain of a model $\mathcal{M}$ of the product logic is the Cartesian product between epistemic states and arguments ($\mathcal{S} \times \mathsf{Args}$) and its assignment function v associates propositional atoms with sets of state-argument pairs in its domain. One can picture the workings of such a product logic in terms of a chess-board with epistemic states providing the x-axis and arguments providing the y-axis (see Example 117 below for a concrete illustration). The epistemic modality, $\boxminus_\mathsf{b}$, and its universal cousin, $\boxminus_\mathsf{u}$, move along the x-axis while keeping arguments fixed. The argumentative modality $\boxminus_\mathsf{a}$ and $\boxminus_\mathsf{u}$, move along the y-axis while keeping states fixed:

- $\mathcal{M}, (s, A) \models \Box_\mathsf{a} \phi$ iff for all $B \in \mathsf{Args}$ such that $A \leftarrow B$, we have: $\mathcal{M}, (s, B) \models \phi$

- $\mathcal{M}, (s, A) \models \Box_\mathsf{u} \phi$ iff for all $B \in \mathsf{Args}$, we have: $\mathcal{M}, (s, B) \models \phi$

- $\mathcal{M}, (s, A) \models \boxminus_\mathsf{b} \phi$ iff for all $s' \in \mathcal{P}$, we have: $\mathcal{M}, (s', A) \models \phi$.

- $\mathcal{M}, (s, A) \models \boxminus_\mathsf{u} \phi$ iff for all $s' \in \mathcal{S}$, we have: $\mathcal{M}, (s', A) \models \phi$.

Grossi and van der Hoek also introduce a designated symbol/atom σ to signify that an argument A supports an epistemic state s in case $\mathcal{M}, (s, A) \models \sigma$.

To illustrate these definitions, we take a look at an example.

Example 117. *Consider the following argumentative scenario (inspired by [Modgil, 2009] and [Grossi, 2010]):*

Default (*C*) *It was sunny yesterday, so it will be sunny today.*

Pete (*B*) *Currently there are thick clouds, it is going to rain and storm.*

CNN (*A*) *The weather report of the CNN reports sunny but windy weather.*

FOX (*A′*) *The weather report of FOX news reports sunny and calm weather.*

We use the atoms w for it "being windy", s for it "being sunny", and CNN, FOX, and Pete are atoms that indicate sources of information.

We consider the epistemic states $\mathcal{S} = \{s_1, s_2, s_3\}$ where the possible epistemic states of our agent are $\mathcal{P} = \{s_1, s_2\}$. Figure 13 illustrates the situation. On the y-axis we find our four arguments where the arrows between them illustrate the inverted(!) attack relation. On the x-axis we find the epistemic state, where the possible epistemic states in $\mathcal{P}$ are highlighted.

- *Highlighted in boxes along the x axis are properties of arguments that are robust under changes of the epistemic state. For instance,*
  - $\mathcal{M}, (s_i, A) \models$ CNN *for all* $1 \leq i \leq 3$, *which indicates that argument A is based on evidence from CNN.*
  - *Similarly, argument A' is based on evidence from FOX, etc.*

- *Highlighted in boxes along the y-axis are properties of epistemic states that are robust under changes of the considered argument. For instance,*
  - $\mathcal{M}, (s_1, x) \models s \wedge \neg w$ *for all* $x \in \{A, A', B, C\}$, *which expresses that according to state s_1 we have calm and sunny weather.*

- *The symbol σ indicates which arguments support which epistemic states. For instance,*
  - $\mathcal{M}, (s_2, A') \models \sigma$ *meaning that argument A' supports state s_2.*

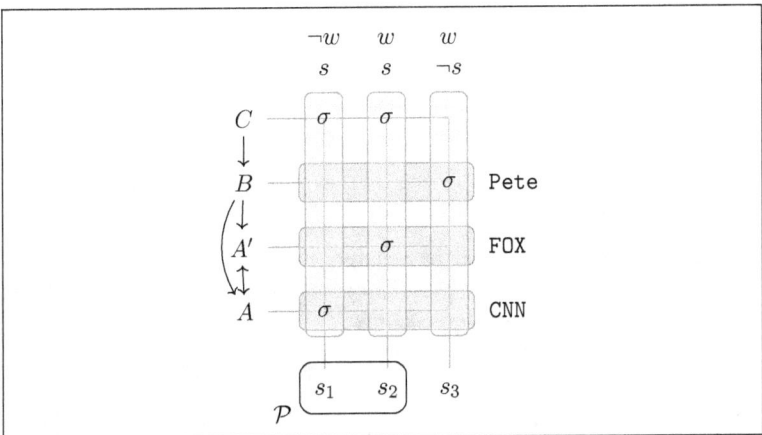

Figure 13: Model $\mathcal{M}$ in for Example 117. The vertical [horizontal] boxes represent properties of states [arguments] that are robust under changes of the considered arguments [states].

In the given system we can express properties that concern information states that involve both beliefs and argumentative properties, such as:

- $\mathcal{M} \models (\neg s \wedge \sigma) \supset \boxdot_{\mathsf{a}}(\mathsf{CNN} \wedge \mathsf{FOX})$ meaning that if an argument supports "not sunny" then all attackers of it rely on CNN or FOX.

- $\mathcal{M} \models \boxminus_{\mathsf{b}}(s \wedge ((w \wedge \sigma) \supset (\mathsf{FOX} \vee \Diamond_{\mathsf{a}} \mathsf{Pete})))$ meaning that our agent believes s and that if an argument supports windy weather then it relies on FOX or it is attacked by an argument that relies on Pete.

Grossi and van der Hoek enrich this framework further by an *endorsement operator* $\boxdot_{\mathsf{e}}$ that works similar to $\boxminus_{\mathsf{b}}$ except that it operates on the y-axis and therefore concerns arguments rather than epistemic states: instead of fixing a set of possible belief states we now fix a set of endorsed arguments $\mathcal{E} \subseteq \mathsf{Args}$ and define:

- $\mathcal{M}, (s, A) \models \boxdot_{\mathsf{e}} \phi$ iff for all $a \in \mathcal{E}$, $\mathcal{M}, (s, a) \models \phi$.

This way it is possible to formally characterize several types of argumentation-based beliefs:

- $\mathsf{SB}\phi = \boxminus_b(\boxdot_u\phi \wedge \Diamond_u\sigma)$ expressing an (argumentatively) supported belief in ϕ,

- $\mathsf{EB}\phi = \boxminus_b(\boxdot_u\phi \wedge \Diamond_e\sigma)$ expressing an endorsed supported belief in ϕ, and

- $\mathsf{JB}(\phi,\psi) = \boxminus_b(\boxdot_u\phi \wedge \Diamond_e(\sigma \wedge \boxminus_u\psi))$ expressing a belief in ϕ, justified by a belief in ψ.[72]

Example 118. *Suppose that in Example 117 we have six agents, Anne, Bill, Chris, Dan, Eli, and Fay that endorse different arguments and have different beliefs. We have, for instance:*

	Anne	Bill	Chris	Dan	Eli	Fay
Endorsed arguments	$\{A',C\}$	$\{C\}$	$\{A\}$	$\{B\}$	$\{A',C\}$	$\{B\}$
Possible belief states	$\{s_2\}$	$\{s_1,s_2\}$	$\{s_1\}$	$\{s_1,s_2\}$	$\{s_3\}$	$\{s_3\}$
$\mathsf{SB}s$	Yes	Yes	Yes	Yes	No	No
$\mathsf{EB}s$	Yes	Yes	Yes	No	No	No
$\mathsf{JB}(s,\mathsf{FOX})$	Yes	No	No	No	No	No
$\mathsf{JB}(s,\mathsf{CNN})$	No	No	Yes	No	No	No
$\mathsf{JB}(\neg s,\mathsf{Pete})$	No	No	No	No	No	Yes

While in the framework of Grossi and van der Hoek belief and argumentative considerations are treated by independent modalities, in [Shi et al., 2018] beliefs are dependent on the underlying argumentative structure. For this they consider *argumentation-support models* which are defined as product modal logics similar to the models discussed above. Let us highlight some differences. First, the language in [Shi et al., 2018] does not allow for arbitrary nesting of modalities. The underlying grammar is defined as follows:

$$\alpha := \top \mid p \mid \neg\alpha \mid \alpha \wedge \alpha \mid \boxminus_u\alpha \mid \boxdot_u\beta$$
$$\beta := \top \mid \Box_a\alpha \mid \neg\beta \mid \beta \wedge \beta \mid \boxdot_a^\alpha\beta \mid \mathsf{Gfp}^\alpha$$

---

[72]In this definition also a universal belief modality is used, which is defined as usual.

While α-formulas express facts about possible worlds, β-formulas describe arguments. To explain the meaning of the different modal operators, let us take a look at the semantics.

For this we take a closer look at the argumentation-support models introduced. An argumentation-support model is given by $\langle \mathcal{S}, \mathsf{Args}, \{\leftarrow_X \mid X \subseteq \mathcal{S}\}, v_\mathsf{s}, v_\mathsf{a}\rangle$, where $\mathcal{S}$ is a (non-empty) set of (factual) states, Args is a set of arguments, for each $X \subseteq \mathcal{S}$, $\leftarrow_X$ is a contextualized (inverted) attack relation, and v_s [respectively, v_a] associates propositional atoms [respectively, arguments] with [non-empty] sets of states.[73] [74] Just like in [Grossi and van der Hoek, 2014], formulas are evaluated at state-argument pairs. For all classical connectives this works as expected (e.g., $\mathcal{M}, (s, A) \models p$ iff $s \in v_s(p)$, and, $\mathcal{M}, (s, A) \models \phi_1 \wedge \phi_2$ iff $\mathcal{M}, (s, A) \models \phi_1$ and $\mathcal{M}, (s, A) \models \phi_2$, etc.). Let us therefore take a look at the modal operators.

First, we notice that the attack modality $\square_\mathsf{a}^\alpha$ is contextualized to formulas α expressing claims that are disputed in the respective attacks.

- $\mathcal{M}, (s, A) \models \square_\mathsf{a}^\phi \psi$ iff for all B with $A \leftarrow_{[\![\phi]\!]_\mathcal{M}} B$, it holds $\mathcal{M}, (s, B) \models \psi$ (where $[\![\phi]\!]_\mathcal{M} = \{s' \in \mathcal{S} \mid \mathcal{M}, (s, C) \models \phi$ for any $C \in \mathsf{Args}\}$). In words: all attackers B of the argument A in a dispute about the claim ϕ satisfy ψ (where, just like in the product logics of [Grossi and van der Hoek, 2014] discussed above, we keep the given state fixed).

The authors consider several constraints on this relation:

1. $A \leftarrow_X B$ iff $A \leftarrow_{W \setminus X} B$. Clearly, if the attack concerns the question whether X is the case, it will equally concern the question whether $W \setminus X$ is the case.

2. If $A \leftarrow_X B$ then

    (a) $v_\mathsf{a}(A) \subseteq X$ or $v_\mathsf{a}(A) \subseteq W \setminus X$, and

    (b) $v_\mathsf{a}(A) \subseteq X$ implies $v_\mathsf{a}(B) \subseteq W \setminus X$.

---

[73] Note the difference of this approach to the models of [Grossi and van der Hoek, 2014], in which there is only one assignment function $v : \mathsf{Atoms} \to \wp(\mathcal{S} \times \mathsf{Args})$.

[74] In [Shi et al., 2017] and in a similar setting the same authors propose a topological semantics to model evidence supporting arguments.

The attacked argument will either support X or $W \setminus X$ and the attacking argument should have an opposite stance.

3. If $A \leftarrow_X B$ and $v_a(A) \subseteq Y \subset X$, then $A \leftarrow_Y B$. If B attacks A concerning the claim X and A supports the stronger claim Y, then B also attacks A on the stronger claim.

The universal vertical and horizontal modalities $\boxplus_u$ and $\boxminus_u$ are analogous to those in [Grossi and van der Hoek, 2014] discussed above. For the $\square_a$ modality we have:

- $\mathcal{M}, (s, A) \models \square_a \alpha$ iff $v_a(A) \subseteq \llbracket \alpha \rrbracket_\mathcal{M}$, meaning that the considered argument A supports the claim α.

Also, Shi et al. enhance the logic with a μ-operator Gfp^α (similar to [Grossi, 2010], see the discussion in the previous section) to express membership in admissible extensions:[75]

- $\mathcal{M}, (s, A) \models \mathsf{Gfp}^\phi$ iff A is in an admissible set of arguments in the argumentation framework $\langle \mathsf{Args}, \rightarrow_{\llbracket \phi \rrbracket_\mathcal{M}} \rangle$.

An agent believes in α in case there is an admissible argument for α and there is no admissible argument for $\neg \alpha$. This can be expressed by putting

$$B\alpha := \Diamond_u(\underbrace{\square_a \alpha}_{\text{it supports } \alpha} \wedge \underbrace{\mathsf{Gfp}^\alpha}_{\text{it is admissible}}) \wedge \neg \Diamond_u(\underbrace{\square_a \neg \alpha}_{\text{it supports } \neg \alpha} \wedge \underbrace{\mathsf{Gfp}^{\neg \alpha}}_{\text{it is admissible}}).$$

with overbraces "there is an argument s.t. ..." and "there is no argument s.t. ..."

Example 119. *Consider again the scenario in Example 117. Given a set of states $\mathcal{S} = \{s_1, s_2, s_3\}$ we let our assignments be as in Table 17. We then get, for instance, where $1 \leq i \leq 3$,*

- $\mathcal{M}, (s_i, x) \models \mathsf{Gfp}^s \wedge \square_a s$ *for* $x \in \{A, A', C\}$*, while* $\mathcal{M}, (s_i, B) \not\models \mathsf{Gfp}^s$ *and* $\mathcal{M}, (s_i, B) \not\models \square_a s$

- $\mathcal{M}, (s_i, A') \models \mathsf{Gfp}^{s \wedge w} \wedge \square_a(s \wedge w)$ *and* $\mathcal{M}, (s_i, A) \models \mathsf{Gfp}^{\neg(s \wedge w)} \wedge \square_a \neg(s \wedge w)$

---

[75]Gfp^α is the greatest postfix point of $\boxplus_a^\alpha \Diamond_a^\alpha$. See [Shi et al., 2018] for an axiomatization. Note also that the discussion in [Shi et al., 2018] is restricted to uncontroversial argumentation frameworks (see also [Dung, 1995] for a definition).

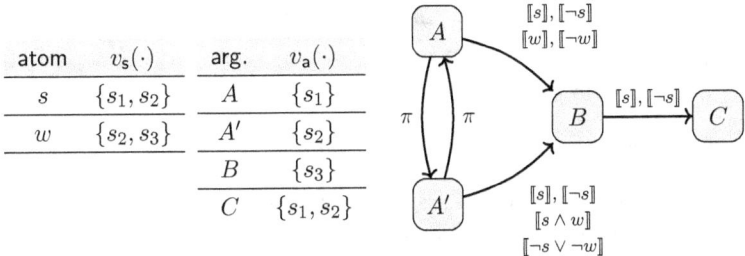

Table 17: Left and Middle: Assignments for Example 119; Right: The attack-diagrams for the contextualized attack relations. Arrows exist for each of the listed labels (e.g., $B \to_{[\![s]\!]} C$ and $B \to_{[\![\neg s]\!]} C$), where π is a placeholder for $[\![s \wedge w]\!], [\![\neg s \vee \neg w]\!], [\![w]\!]$ and $[\![\neg w]\!]$.

- $\mathcal{M} \models \mathsf{B}s$ while $\mathcal{M} \not\models \mathsf{B}(s \wedge w)$.

The systems presented above have the merit of allowing for argumentation-based approaches to belief and justification, which allow for new and interesting insights. E.g., for all of Grossi's and van der Hoek's belief types (SB, EB and JB) negative introspection fails for beliefs that are not supported by arguments, but succeeds otherwise. That is (where $\mathsf{XB} \in \{\mathsf{SB}, \mathsf{EB}\}$), while:

$\not\models \neg\mathsf{XB}\phi \supset \mathsf{XB}\neg\mathsf{XB}\phi$, and

$\not\models \neg\mathsf{JB}(\phi, \psi) \supset \mathsf{JB}(\neg\mathsf{JB}(\phi, \psi), \psi)$

we have (see [Grossi and van der Hoek, 2014, Proposition 6])

$\models (\neg\mathsf{XB}\phi \wedge \boxminus_\mathsf{b} \Diamond_\mathsf{e} \sigma) \supset \mathsf{XB}\neg\mathsf{XB}\phi$, and

$\models (\neg\mathsf{JB}(\phi, \psi) \wedge \boxminus_\mathsf{b} \Diamond_\mathsf{e} (\sigma \wedge \boxminus_\mathsf{u}\psi)) \supset \mathsf{JB}(\neg\mathsf{JB}(\phi, \psi), \psi)$

Similarly, in Shi et al.'s system the aggregation of beliefs fails, i.e., $\not\models (\mathsf{B}\alpha \wedge \mathsf{B}\alpha') \supset \mathsf{B}(\alpha \wedge \alpha')$, which may give rise to applications to paradoxes, respectively difficult scenarios, such as the lottery or the preface paradox.

3.3 Reasoning with Dynamic Derivations

Although the satisfiability methods described in the previous sections are logic-based, from a pure logical perspective they have some drawbacks:

- In many of the described formalisms, the encoding of the arguments are by propositional variables, thus arguments are treated as abstract entities. As such, these methods are more adequate to abstract argumentation [Baroni et al., 2018] than to structured argumentation. Put differently, if these methods are applied to argumentation frameworks such as the ones considered in Section 2, the construction of the frameworks and the reasoning methods are distinguished: first the arguments and the attacks among them are produced, and only then the satisfiability-based methods can be applied on them.

- Even more serious is the fact that many of these methods are applicable only to *finite* argumentation frameworks, as for the encoding of the formulas a finite set of arguments is assumed. As such, these methods are suitable only for some logical instantiations (assumption-based frameworks, for instance), but not for all of them (e.g., logic-based argumentation frameworks which are infinite since so are the transitive closures of sets of assertions).

In this section we describe an alternative method to reasoning with logic-based argumentation, which overcomes the two shortcomings of the other approach described above: it is applicable to infinite frameworks and is affected by the logical content of the arguments and the attack rules.

Let $\mathcal{AF}_{\mathfrak{L},\mathcal{A}}(\mathcal{S}) = \langle \mathsf{Arg}_{\mathfrak{L}}(\mathcal{S}), \mathit{Attack}(\mathcal{A}) \rangle$ be a logical argumentation framework (Definition 8) and let $\mathcal{P}$ be a sound and complete proof system for $\mathfrak{L}$.[76] The idea is to use (inference) rules in $\mathcal{P}$ for deriving new

---

[76]$\mathcal{P}$ may be a Hilbert-type proof system, a Gentzen-type sequent calculus, a natural deduction system, a semantic tableaux system, or any other proof method that is based on finite sequences (or trees) of finite syntactical expressions which are based on the underlying language (see e.g. Section 1.3 of [Avron et al., 2018] for a general definition of such proof systems). Here we concentrate on sequent calculi, since a sequent is in fact a multiple-conclusion argument. For the other kinds of proof systems some simple modifications of the definitions in what follows are needed.

arguments from already derived ones, and to use (attack) rules in $\mathcal{A}$ for excluding derived arguments, when opposing arguments are also derived. This gives rise to the notion of *dynamic proofs* (or dynamic derivations), which are intended for explicating the actual non-monotonic flavor of reasoning processes in a logical argumentation framework. The main idea behind these formalisms is that, unlike 'standard' proof methods, an argument can be challenged (and possibly withdrawn) by a counter-argument, and so a certain argument may be considered as not accepted at a certain stage of the proof, even if it were considered accepted in an earlier stage of the proof. It is only when an argument is 'finally derived' (in the sense that will be explained later on) that it can be safely concluded by the dynamic proof. In the rest of this section we elaborate on this idea (full details and formal definitions can be found in [Arieli and Straßer, 2019]).

A proof system in our case is determined by a *proof setting* $\mathfrak{S} = \langle \mathfrak{L}, \mathcal{P}, \mathcal{A} \rangle$ consisting of a logic $\mathfrak{L}$, a corresponding sound and complete proof calculus $\mathcal{P}$ for producing $\mathfrak{L}$-arguments, and a set $\mathcal{A}$ of attack rules for eliminating (undefended) attacked arguments. An argument $\langle \mathcal{S}, \psi \rangle$ that is eliminated (i.e., is attacked by an application of a rule in $\mathcal{A}$) will be denoted in what follows by $\cancel{\langle \mathcal{S}, \psi \rangle}$.

Definition 120 (proof tuple). *A (proof) tuple is a triple $T = \langle i, A, \mathsf{J} \rangle$, where i (the tuple's index) is a natural number, $A \in \{\langle \Gamma, \Delta \rangle, \cancel{\langle \Gamma, \Delta \rangle}\}$ (the tuple's argument) is a (possibly attacked) multiple-conclusion argument,[77, 78] and J (the tuple's justification) is a string, consisting of a rule name followed by a sequence of numbers.[79] In the sequel we shall sometimes identify a proof tuple with its argument.*

Definition 121 (simple derivation). *Let $\mathfrak{S} = \langle \mathfrak{L}, \mathcal{P}, \mathcal{A} \rangle$ be a proof setting. A simple $\mathfrak{S}$-derivation based on a set $\mathcal{S}$ of formulas in $\mathcal{L}$, is a finite*

---

[77]Thus Δ, the conclusion of A, is a finite set of formulas and not just a formula. (In classical logic, Δ may be replaced by its disjunction $\bigvee \Delta$.) When Δ is a singleton we shall omit the parentheses and identify A with a standard argument in the sense of Definition 5.

[78]When the underlying calculus is Hilbert-type or based on a natural deduction system, A may be just a formula (corresponding to the rule conclusions is those proof systems) rather than an argument.

[79]This string indicates what rule has to be applied, and on what tuples, in order to derive T.

sequence $\mathcal{D}_{\mathfrak{S}}(\mathcal{S}) = \langle T_1, \ldots T_m \rangle$ of proof tuples, where each $T_i \in \mathcal{D}$ is of either of the following forms:

- $T_i = \langle i, A, \mathsf{J} \rangle$, where $\mathsf{J} = \text{``}\mathcal{R}\ i_1, \ldots, i_n\text{''}$ and A is obtained by applying the inference rule $\mathcal{R} \in \mathcal{P}$ on the arguments of the tuples $T_{i_1}, \ldots T_{i_k}$ ($i_1, \ldots, i_n < i$).

- $T_i = \langle i, A, \mathsf{J} \rangle$, where $\mathsf{J} = \text{``}\mathcal{R}\ i_1, \ldots, i_n\text{''}$ and A is obtained by applying the elimination rule $\mathcal{R} \in \mathcal{A}$ on the arguments of the tuples $T_{i_1}, \ldots T_{i_k}$ ($i_1, \ldots, i_n < i$). In this case both the attacked argument A and the attacking argument A_{i_1} should be elements of $\mathsf{Arg}_{\mathfrak{L}}(\mathcal{S})$.[80]

Tuples of the first form are called introducing tuples and those of the second form are called eliminating tuples.

Example 122. Let $\mathcal{P}$ be Gentzen's proof system LK for classical logic. Table 18 presents this system in terms of (multiple-conclusion) arguments.

Consider now the set of assumptions $\mathcal{S} = \{\neg p, p, q\}$ (see also Example 37). Figure 14 presents a simple derivation with respect to LK and Ucut as the sole attack rule. To simplify the reading, in this and other derivations in the rest of the paper we shall sometimes use abbreviations or omit some details, e.g. the tuple signs in proof steps.

Note that in this derivation Tuple 8 represents a Ucut-attack of the argument in Tuple 7 on the argument in Tuple 1 (where the former serves also as the justification of the attack), and Tuple 11 represents a Ucut-attack of the argument in Tuple 1 on the argument in Tuple 7, justified by the arguments in Tuples 9 and 10. Thus, Tuples 8 and 11 are eliminating while the other tuples are introducing.

Not all the attacks in a simple derivation should be successful, since if the attacking argument is itself attacked by another argument (i.e., it appears in an eliminating tuple) the attack may not be validated. The iterative process in Figure 15 checks this, and evaluates each tuple's argument: **Elim** is the status of an eliminated argument whose attacker is not already eliminated, **Attack** means that the argument attacks an argument whose status is **Elim**, and **Accept** is the status of a derived

Rule Name	Acronym	Rule's conditions	Rule's conclusion
Axiom			$\langle \psi, \psi \rangle$
Weakening		$\langle \mathcal{S}, \mathcal{T} \rangle$	$\langle \mathcal{S} \cup \mathcal{S}', \mathcal{T} \cup \mathcal{T}' \rangle$
Cut		$\langle \mathcal{S}_1, \mathcal{T}_1 \cup \{\psi\} \rangle$, $\langle \mathcal{S}_2 \cup \{\psi\}, \mathcal{T}_2 \rangle$	$\langle \mathcal{S}_1 \cup \mathcal{S}_2, \mathcal{T}_1 \cup \mathcal{T}_2 \rangle$
Left-$\wedge$	[$\wedge$L]	$\langle \mathcal{S} \cup \{\psi\} \cup \{\phi\}, \mathcal{T} \rangle$	$\langle \mathcal{S} \cup \{\psi \wedge \phi\}, \mathcal{T} \rangle$
Right-$\wedge$	[$\wedge$R]	$\langle \mathcal{S}, \mathcal{T} \cup \{\psi\} \rangle$, $\langle \mathcal{S}, \mathcal{T} \cup \{\phi\} \rangle$	$\langle \mathcal{S}, \mathcal{T} \cup \{\psi \wedge \phi\} \rangle$
Left-$\vee$	[$\vee$L]	$\langle \mathcal{S} \cup \{\psi\}, \mathcal{T} \rangle$, $\langle \mathcal{S} \cup \{\phi\}, \mathcal{T} \rangle$	$\langle \mathcal{S} \cup \{\psi \vee \phi\}, \mathcal{T} \rangle$
Right-$\vee$	[$\vee$R]	$\langle \mathcal{S}, \mathcal{T} \cup \{\psi\} \cup \{\phi\} \rangle$	$\langle \mathcal{S}, \mathcal{T} \cup \{\psi \vee \phi\} \rangle$
Left-$\supset$	[$\supset$L]	$\langle \mathcal{S}, \mathcal{T} \cup \{\psi\} \rangle$, $\langle \mathcal{S} \cup \{\phi\}, \mathcal{T} \rangle$	$\langle \mathcal{S} \cup \{\psi \supset \phi\}, \mathcal{T} \rangle$
Right-$\supset$	[$\supset$R]	$\langle \mathcal{S} \cup \{\psi\}, \mathcal{T} \cup \{\phi\} \rangle$	$\langle \mathcal{S}, \mathcal{T} \cup \{\psi \supset \phi\} \rangle$
Left-$\neg$	[$\neg$L]	$\langle \mathcal{S}, \mathcal{T} \cup \{\psi\} \rangle$	$\langle \mathcal{S} \cup \{\neg\psi\}, \mathcal{T} \rangle$
Right-$\neg$	[$\neg$R]	$\langle \mathcal{S} \cup \{\psi\}, \mathcal{T} \rangle$	$\langle \mathcal{S}, \mathcal{T} \cup \{\neg\psi\} \rangle$

Table 18: Arguments construction rules according to LK.

argument whose status is not Elim.

Definition 123 ((strongly) coherent derivation). *A simple derivation $\mathcal{D}$ is coherent, if there is no argument that eliminates another argument and that is eliminated itself. Formally:* Attack$(\mathcal{D}) \cap$ Elim$(\mathcal{D}) = \emptyset$. *We say that $\mathcal{D}$ is strongly coherent, if*

$$\mathsf{Sup}(\mathsf{Attack}(\mathcal{D})) = \bigcup_{A \in \mathsf{Attack}(\mathcal{D})} \mathsf{Sup}(A)$$

is consistent.[81]

Example 124 (Example 122 continued). *Consider the simple derivation $\mathcal{D}$ of Example 122.*

---

[80]This prevents situations in which, e.g., $\langle \neg p, \neg p \rangle$ Ucut-attacks $\langle p, p \rangle$, although $\mathcal{S} = \{p\}$.

[81]As shown in [Arieli *et al.*, 2018], in the proof setting $\mathfrak{S} = \langle \mathsf{CL}, LK, \{\mathsf{Ucut}\} \rangle$, strong coherence implies coherence (but not vice-versa).

1.	$\langle p, p \rangle$	Axiom
2.	$\langle \emptyset, \{p, \neg p\} \rangle$	Right-$\neg$, 1
3.	$\langle \emptyset, p \vee \neg p \rangle$	Right-$\vee$, 2
4.	$\langle p \vee \neg p, \neg(p \wedge \neg p) \rangle$	...
5.	$\langle \neg(p \wedge \neg p), p \vee \neg p \rangle$	...
6.	$\langle q, q \rangle$	Axiom
7.	$\langle \neg p, \neg p \rangle$	Axiom
8.	~~$\langle p, p \rangle$~~	Ucut, 7, 7, 7, 1
9.	$\langle p, \neg\neg p \rangle$	...
10.	$\langle \neg\neg p, p \rangle$	...
11.	~~$\langle \neg p, \neg p \rangle$~~	Ucut, 1, 9, 10, 7

Figure 14: A derivation with respect to LK and Ucut, based on $\mathcal{S} = \{\neg p, p, q\}$

- When considering only the simple derivation consisting of lines 1–8 we have that $\langle q, q \rangle, \langle \neg p, \neg p \rangle \in$ Accept, Attack $= \{\langle \neg p, \neg p \rangle\}$ and Elim $= \{\langle p, p \rangle\}$.

- When considering the simple derivation consisting of lines 1–11 we have that $\langle q, q \rangle$, $\langle p, p \rangle \in$ Accept, Attack $= \{\langle p, p \rangle\}$ and Elim $= \{\langle \neg p, \neg p \rangle\}$. Note that when the algorithm in Figure 15 reaches line 8, $\langle p, p \rangle$ is not added to Elim since its attacking argument $\langle \neg p, \neg p \rangle$ is already in Elim at that point.[82]

In particular, in each step the derivation that is obtained is both coherent and strongly coherent.

Now we can define what dynamic derivations are.

---

[82]This is so, since the evaluation process progresses backwards, from the last tuple to the first one, so $\langle \neg p, \neg p \rangle$ is already eliminated in the first evaluation step, following line 11.

```
Input: a simple derivation 𝒟.
let Attack := Elim := Derived := ∅;
while (𝒟 is not empty) do {
    if the last element in 𝒟 introduces an argument A, then
        add A to the set Derived;
    if the last element in 𝒟 is an attack of A₁ ∉ Elim on A₂, then
        add A₁ to Attack and A₂ to Elim;
    remove the last element from 𝒟 }
let Accept := Derived − Elim;
Output: Attack, Elim, Accept.
```

Figure 15: Evaluation of a simple derivation.

Definition 125 (dynamic derivation). *Let $\mathfrak{S} = \langle \mathfrak{L}, \mathcal{P}, \mathcal{A} \rangle$ be a proof setting. A dynamic $\mathfrak{S}$-derivation based on a set $\mathcal{S}$ of formulas in $\mathcal{L}$, is an $\mathcal{S}$-based simple $\mathfrak{S}$-derivation $\mathcal{D}_\mathfrak{S}(\mathcal{S})$ which is of one of the following forms:*

a) $\mathcal{D}_\mathfrak{S}(\mathcal{S}) = \langle T \rangle$, where $T = \langle 1, A, \mathsf{J} \rangle$ is a proof tuple.

b) $\mathcal{D}_\mathfrak{S}(\mathcal{S})$ is obtained by adding to a dynamic derivation a sequence of introducing tuples whose arguments are not in $\mathsf{Elim}(\mathcal{D}_\mathfrak{S}(\mathcal{S}))$.

c) $\mathcal{D}_\mathfrak{S}(\mathcal{S})$ is obtained by adding to a dynamic derivation a sequence of eliminating tuples where the attacking arguments are in $\mathsf{Arg}_\mathfrak{L}(\mathcal{S})$ and are not attacked by arguments in the set $\mathsf{Accept}(\mathcal{D}_\mathfrak{S}(\mathcal{S})) \cap \mathsf{Arg}_\mathfrak{L}(\mathcal{S})$. The attacks must be based on arguments that are proved in $\mathcal{D}_\mathfrak{S}(\mathcal{S})$.[83]

One may think of a dynamic derivation as a proof that progresses over derivation steps. At each step the current derivation is extended

---

[83]This condition assures that the attacks are 'sound': the attacking arguments are not counter-attacked by an accepted $\mathcal{S}$-based argument.

by a 'block' of introduced arguments or eliminated arguments. As a result, the statuses of the arguments in the derivation are updated. In particular, a derived argument may be eliminated in light of new derived arguments, but also the other way around is possible: an eliminated argument may be 'restored' if its attacking argument is counter-attacked. It follows that previously accepted data may not be accepted anymore (and vice-versa) until and unless new derived information revises the state of affairs.

Example 126 (Examples 122 and 124, continued). *The simple derivation of Example 122 is also a dynamic derivation. Example 124 demonstrates the dynamic nature of this derivation. For instance, although the argument $\langle \neg p, \neg p \rangle$ is derived in Step 7 of the derivation, it is eliminated in Step 11 of the derivation as a consequence of an Undercut attack, initiated by $\langle p, p \rangle$.*

The next definition, of the outcomes of a dynamic derivation, indicates when it is 'safe' to conclude that a derived argument must hold under any circumstances.

Definition 127 (final derivability). *Let $\mathfrak{S} = \langle \mathcal{L}, \mathcal{P}, \mathcal{A} \rangle$ be a proof setting and let $\mathcal{S}$ be a set of $\mathcal{L}$-formulas.*

- *A formula ψ is finally derived in a coherent dynamic $\mathfrak{S}$-derivation $\mathcal{D}_\mathfrak{S}(\mathcal{S})$, if for some $\Gamma \subseteq \mathcal{S}$ the argument $A = \langle \Gamma, \psi \rangle$ is in $\mathsf{Arg}_\mathcal{L}(\mathcal{S}) \cap \mathsf{Accept}(\mathcal{D}_\mathfrak{S}(\mathcal{S}))$, and for every coherent dynamic derivation $\mathcal{D}'_\mathfrak{S}(\mathcal{S})$ extending $\mathcal{D}_\mathfrak{S}(\mathcal{S})$ (i.e., any dynamic derivation whose prefix is $\mathcal{D}_\mathfrak{S}(\mathcal{S})$), still $A \in \mathsf{Accept}(\mathcal{D}'_\mathfrak{S}(\mathcal{S}))$.*

- *A formula ψ is sparsely finally derived in a strongly coherent dynamic $\mathfrak{S}$-derivation $\mathcal{D}_\mathfrak{S}(\mathcal{S})$, if for some $\Gamma \subseteq \mathcal{S}$ the argument $A = \langle \Gamma, \psi \rangle$ is in $\mathsf{Arg}_\mathcal{L}(\mathcal{S}) \cap \mathsf{Accept}(\mathcal{D}_\mathfrak{S}(\mathcal{S}))$, and for every strongly coherent dynamic derivation $\mathcal{D}'_\mathfrak{S}(\mathcal{S})$ that extends $\mathcal{D}_\mathfrak{S}(\mathcal{S})$ there is some $\Gamma' \subseteq \mathcal{S}$ such that the argument $A' = \langle \Gamma', \psi \rangle$ is in $\mathsf{Arg}_\mathcal{L}(\mathcal{S}) \cap \mathsf{Accept}(\mathcal{D}'_\mathfrak{S}(\mathcal{S}))$.*

Thus, final derivability means that an argument is derived and accepted in a valid dynamic derivation and remains in this status in every extension of the derivation. Sparse final derivability is a weaker notion, meaning that if an argument A is derived and accepted in a valid dynamic derivation, in every extension of that derivation the conclusion of A is a conclusion of a derived and accepted argument.

Definition 128 ($\vdash_\cap^\mathfrak{S}$, $\vdash_\text{\tiny{⋒}}^\mathfrak{S}$). *Let* $\mathfrak{S} = \langle \mathfrak{L}, \mathfrak{C}, \mathcal{A} \rangle$ *be a proof setting,* $\mathcal{S}$ *a set of* $\mathcal{L}$*-formulas, and* ψ *an* $\mathcal{L}$*-formula.*

- $\mathcal{S} \vdash_\cap^\mathfrak{S} \psi$ *iff there is a* $\mathfrak{S}$*-derivation based on* $\mathcal{S}$*, in which* ψ *is finally derived.*

- $\mathcal{S} \vdash_\text{\tiny{⋒}}^\mathfrak{S} \psi$ *iff there is a* $\mathfrak{S}$*-derivation based on* $\mathcal{S}$*, in which* ψ *is sparsely finally derived.*

Example 129.

a) q *is finally derived (and so also sparsely finally derived) in the derivation of Figure 14 where* $\mathfrak{S} = \langle \mathsf{CL}, LK, \{Ucut\} \rangle$ *and* $\mathcal{S} = \{p, \neg p, q\}$. *Indeed, the only arguments in* $\mathsf{Arg}_\mathsf{CL}(\mathcal{S})$ *that can potentially Ucut-attack* $\langle q, q \rangle$ *are of the form* $\langle \{p, \neg p\}, \psi \rangle$ *or* $\langle \{p, \neg p, q\}, \psi \rangle$*, where* ψ *is logically equivalent to* $\neg q$*. However, such arguments are counter-attacked by the argument* $\langle \emptyset, p \vee \neg p \rangle$*, obtained in Tuple 3 of the derivation. It follows, by the conditions in Item (c) of Definition 125, that no eliminating tuple in which* $\langle q, q \rangle$ *is attacked can be derived in any extension of the derivation above, thus* q *is finally derived in this derivation.*

We have, then, that $\{p, \neg p, q\} \vdash_\star^\mathfrak{S} q$*, while* $\{p, \neg p, q\} \not\vdash_\star^\mathfrak{S} p$ *and* $\{p, \neg p, q\} \not\vdash_\star^\mathfrak{S} \neg p$*, for any* $\star \in \{\cap, ⋒\}$.

b) *To see the need for sparse final derivability, let again* $\mathfrak{S} = \langle \mathsf{CL}, LK, \{Ucut\} \rangle$ *and consider the set* $\mathcal{S}' = \{p \wedge q, \neg p \wedge q\}$*. Note that both* $A_1 = \langle p \wedge q, q \rangle$ *and* $A_2 = \langle \neg p \wedge q, q \rangle$ *are LK-derivable in this case, but neither of them is finally derivable, since any* $\mathfrak{S}$*-derivation that includes them can be extended with derivations of* $A_3 = \langle \neg p \wedge q, \neg(p \wedge q) \rangle$ *and* $A_4 = \langle p \wedge q, \neg(\neg p \wedge q) \rangle$ *that respectively Ucut-attack* A_1 *and* A_2*. Note, however, that these attacks cannot be applied* simultaneously*, since the attackers* A_3 *and* A_4 *counter-attack each other. It follows that in each extension of the derivation either* A_1 *or* A_2 *is accepted, and so* q *is sparsely finally derived from* $\mathcal{S}'$.

We have, then, that $\{p \wedge q, \neg p \wedge q\} \mid\!\sim_{\cap}^{\mathfrak{S}} q$ (and it is easy to verify that $\{p \wedge q, \neg p \wedge q\} \not\mid\!\sim_{\cap}^{\mathfrak{S}} p$ and $\{p \wedge q, \neg p \wedge q\} \not\mid\!\sim_{\cap}^{\mathfrak{S}} \neg p$).

The next proposition, proven in [Arieli et al., 2018], provides some soundness and completeness results for entailments by dynamic proofs (Definition 128) and entailments induced by Dung-semantics (Definition 12), and relates both of these entailments to reasoning with maximal consistency (Definition 44).

Proposition 130. *Let* $\mathfrak{S} = \langle \mathsf{CL}, LK, \{Ucut\} \rangle$ *be a proof setting. Then for every finite set* $\mathcal{S}$ *of formulas and formula* ψ, *it holds that:*

- $\mathcal{S} \mid\!\sim_{\cap}^{\mathfrak{S}} \psi$ *iff* $\mathcal{S} \mid\!\sim_{\cap\mathsf{mcs}}^{\mathsf{CL}} \psi$ *iff* $\mathcal{S} \mid\!\sim_{\mathsf{Grd}}^{\mathsf{CL},\{Ucut\}} \psi$ *iff* $\mathcal{S} \mid\!\sim_{\cap\mathsf{Prf}}^{\mathsf{CL},\{Ucut\}} \psi$ *iff* $\mathcal{S} \mid\!\sim_{\cap\mathsf{Stb}}^{\mathsf{CL},\{Ucut\}} \psi$.

- $\mathcal{S} \mid\!\sim_{\Cap}^{\mathfrak{S}} \psi$ *iff* $\mathcal{S} \mid\!\sim_{\Cap\mathsf{mcs}}^{\mathsf{CL}} \psi$ *iff* $\mathcal{S} \mid\!\sim_{\Cap\mathsf{Prf}}^{\mathsf{CL},\{Ucut\}} \psi$ *iff* $\mathcal{S} \mid\!\sim_{\Cap\mathsf{Stb}}^{\mathsf{CL},\{Ucut\}} \psi$.

We refer to [Arieli et al., 2018] for further related results, where e.g. the base logic is not necessarily classical logic and the attack is not necessarily Undercut.

Example 131. *The first item of Example 129 demonstrates the first two items of the last proposition for* $\mathcal{S} = \{p, \neg p, q\}$ *(Examples 122 and 126), as* $\bigcap \mathsf{MCS}_{\mathsf{CL}}(\mathcal{S}) = \{q\}$. *The second item of Example 129 exemplifies the second item of Proposition 130, where* $\mathcal{S}' = \{p \wedge q, \neg p \wedge q\}$ *is the set of assertions.*

Some other approaches for reasoning with logic-based (structured) argumentation frameworks are the following:[84]

- For logic-based methods whose arguments are classical (Definition 4), already the construction of arguments poses serious computational challenges, since the finding of a *minimal* subset of a set of formulas that implies the consequent is in the second level of the polynomial hierarchy [Eiter and Gottlob, 1995]. Algorithms for constructing classical arguments and counter-arguments can be found e.g. in [Efstathiou and Hunter, 2011].

---

[84] As indicated before, description of algorithms for reasoning with argumentation frameworks which are not logic-based, including those for abstract argumentation frameworks, are not in the scope of the current chapter. For the latter, see e.g. the surveys in [Modgil and Caminada, 2009] and [Cerutti et al., 2017].

- Common computational machineries of logic-based argumentation frameworks are based on dispute trees and dispute derivations [Dung et al., 2006; Dung et al., 2007], both of which can be represented as games between proponent and opponent players. For some illustrations and an overview of their use in ABA frameworks, see [Dung et al., 2009, Section 5] and [Čyras et al., 2018, Section 5].

- Illustrations of reasoning with ASPIC$^+$ can be found, e.g., in [Modgil and Prakken, 2014, Section 4.5]; Inference engines for APSIC$^+$ are surveyed (with relevant further references) in [Modgil and Prakken, 2018, Section 6].

In [Straßer and Šešelja, 2010] a similar dynamic proof theory to the one discussed above has been presented, but for abstract argumentation instead of structured argumentation. It allows for the addition of new arguments and new argumentative attacks in an ongoing open-ended proof of an adaptive logic. The finally derivable propositional atoms are those that are in the intersection of a given semantics. The latter are characterized in terms of different adaptive proof strategies.

4 Concluding Remarks

Formal argumentation theory is by now a well-established and still extensively growing research area, even when restricted to its applications in Artificial Intelligence. There is no wonder, then, that it has many branches with different disciplines, some of them went as far as pure graph-theoretical approaches, treating argumentation frameworks as directed graphs, and so viewing their nodes (that is, the arguments) as totally abstract entities. In this chapter, we have taken to some extent the opposite approach, arguing that a meaningful and solid argumentation-based system must have a *logic* behind it, which takes a primary role not only in the construction of argumentation frameworks, but is also essential for the specification of their dynamics and deductive methods of reasoning. In Sections 2 and 3 we demonstrated, respectively, the fundamental role that logic may (and should) have in relation to these two aspects of formal argumentation systems. Indeed, the common ground of all the approaches surveyed in this chapter is that they are logically developed methodologies towards formal argumentation systems. We believe that this is crucial for justifying the outcomes of such systems in a logical and rational way.

Acknowledgments

We thank Marcos Cramer and the reviewers of this chapter for their helpful comments. Ofer Arieli is supported by the Israel Science Foundation (Grant No. 550/19). Jesse Heyninck is supported by the German National Science Foundation under the DFG-project CAR (Conditional Argumentative Reasoning) KE-1413/11-1.

References

[Alfano et al., 2021] Gianvincenzo Alfano, Sergio Greco, Francesco Parisi, Gerardo I. Simari, and Guillermo R. Simari. On the incremental computation of semantics in dynamic argumentation. In Dov Gabbay, Massimiliano Giacomin, Guillermo R. Simari, and Matthias Thimm, editors, *Handbook of Formal Argumentation*, volume 2, chapter 11. College Publications, 2021.

[Amgoud and Besnard, 2010] Leila Amgoud and Philippe Besnard. A formal analysis of logic-based argumentation systems. In Amol Deshpande and Anthony Hunter, editors, *Proceedings of the 4th International Conference Scalable Uncertainty Management (SUM'10)*, volume 6379 of *Lecture Notes in Computer Science*, pages 42–55. Springer, 2010.

[Amgoud and Besnard, 2013] Leila Amgoud and Philippe Besnard. Logical limits of abstract argumentation frameworks. *Journal of Applied Non-Classical Logics*, 23(3):229–267, 2013.

[Amgoud et al., 2011] Leila Amgoud, Philippe Besnard, and Srdjan Vesic. Identifying the core of logic-based argumentation systems. In *Proceedings of the IEEE 23rd International Conference on Tools with Artificial Intelligence (ICTAI'11)*, pages 633–636. IEEE Computer Society, 2011.

[Amgoud, 2014] Leila Amgoud. Postulates for logic-based argumentation systems. *International Journal of Approximate Reasoning*, 55(9):2028–2048, 2014.

[Antonelli, 1999] G. Aldo Antonelli. A directly cautious theory of defeasible consequence for default logic via the notion of general extension. *Artificial Intelligence*, 109(1–2):71–109, 1999.

[Antoniou, 1998] Grigoris Antoniou. Studying properties of classes of default logics. *Journal of Experimental & Theoretical Artificial Intelligence*, 10(4):495–505, 1998.

[Apt, 1990] Krzysztof R. Apt. Logic programming. In Jan van Leeuwen, editor, *Handbook of Theoretical Computer Science, Volume B: Formal Models and Semantics*, pages 493–574. Elsevier and MIT Press, 1990.

[Arieli and Caminada, 2013] Ofer Arieli and Martin Caminada. A QBF-based formalization of abstract argumentation semantics. *Journal of Applied Logic*, 11(2):229–252, 2013.

[Arieli and Straßer, 2015] Ofer Arieli and Christian Straßer. Sequent-based logical argumentation. *Argument & Computation*, 6(1):73–99, 2015.

[Arieli and Straßer, 2016] Ofer Arieli and Christian Straßer. Deductive argumentation by enhanced sequent calculi and dynamic derivations. In Mario Benevides and René Thiemann, editors, *Proceedings of the 10th Workshop on Logical and Semantic Frameworks with Applications, (LSFA'15)*, volume 323 of *Electronic Notes in Theoretical Computer Science*, pages 21–37, 2016.

[Arieli and Straßer, 2019] Ofer Arieli and Christian Straßer. Logical argumentation by dynamic proof systems. *Theoretical Computer Science*, 781:63–91, 2019.

[Arieli and Straßer, 2020] Ofer Arieli and Christian Straßer. On minimality and consistency tolerance in logical argumentation frameworks. In Henry Prakken, Stefano Bistarelli, Francesco Santini, and Carlo Taticchi, editors, *Proceedings of the 8th International Conference on Computational Models of Argument (COMMA'20)*, volume 326 of *Frontiers in Artificial Intelligence and Applications*, pages 91–102. IOS Press, 2020.

[Arieli et al., 2018] Ofer Arieli, AnneMarie Borg, and Christian Straßer. Reasoning with maximal consistency by argumentative approaches. *Journal of Logic and Computation*, 28(7):1523–1563, 2018.

[Arieli et al., 2019] Ofer Arieli, AnneMarie Borg, and Jesse Heyninck. A review of the relations between logical argumentation and reasoning with maximal consistency. *Annals of Mathematics and Artificial Intelligence*, 87(3):187–226, 2019.

[Arieli et al., 2020] Ofer Arieli, AnneMarie Borg, and Christian Straßer. Tuning logical argumentation frameworks: A postulate-derived approach. In Roman Barták and Eric Bell, editors, *Proceedings of the 33rd International Florida Artificial Intelligence Research Society Conference (FLAIRS'20)*, pages 557–562. AAAI Press, 2020.

[Arieli, 2016] Ofer Arieli. On the acceptance of loops in argumentation frameworks. *Journal of Logics and Computation*, 26(4):1203–1234, 2016.

[Avron et al., 2018] Arnon Avron, Ofer Arieli, and Anna Zamansky. *Theory of Effective Propositional Paraconsistent Logics*, volume 75 of *Studies in Logic*. College Publications, 2018.

[Avron, 2014] Arnon Avron. What is relevance logic? *Annals of Pure and Applied Logic*, 165(1):26–48, 2014.

[Avron, 2016] Arnon Avron. RM and its nice properties. In Katalin Bimbó, editor, *J. Michael Dunn on Information Based Logics*, volume 8 of *Outstanding Contributions to Logic*, pages 15–43. Springer, 2016.

[Baral et al., 1991] Chitta Baral, Sarit Kraus, and Jack Minker. Combining multiple knowledge bases. *IEEE Transactions on Knowledge and Data Engineering*, 3(2):208–220, 1991.

[Baroni and Giacomin, 2009] Pietro Baroni and Massimiliano Giacomin. Semantics of abstract argument systems. In Guillermo Simari and Iyad Rahwan, editors, *Argumentation in Artificial Intelligence*, pages 25–44. Springer, 2009.

[Baroni et al., 2011] Pietro Baroni, Martin Caminada, and Massimiliano Giacomin. An introduction to argumentation semantics. *The Knowledge Engineering Review*, 26(4):365–410, 2011.

[Baroni et al., 2018] Pietro Baroni, Martin Caminada, and Massimiliano Giacomin. Abstract argumentation frameworks and their semantics. In Pietro Baroni, Dov Gabbay, Massimiliano Giacomin, and Leon van der Torre, editors, *Handbook of Formal Argumentation*, pages 159–236. College Publications, 2018.

[Batens and Haesaert, 2003] Diderik Batens and Lieven Haesaert. On classical adaptive logics of induction. *Logique et Analyse*, 44(173–175):255–290, 2003.

[Batens, 2006] Diderik Batens. On a logic of induction. *L&PS – Logic & Philosophy of Science*, IV(1):3–32, 2006.

[Batens, 2007] Diderik Batens. A universal logic approach to adaptive logics. *Logica Universalis*, 1(1):221–242, 2007.

[Baumann et al., 2021] Ringo Baumann, Sylvie Doutre, Jean-Guy Mailly, and Johannes P. Wallner. Enforcement in formal argumentation. In Dov Gabbay, Massimiliano Giacomin, Guillermo R. Simari, and Matthias Thimm, editors, *Handbook of Formal Argumentation*, volume 2, chapter 8. College Publications, 2021.

[Beirlaen and Straßer, 2014] Mathieu Beirlaen and Christian Straßer. Non-monotonic reasoning with normative conflicts in multi-agent deontic logic. *Journal of Logic and Computation*, 24(6):1179–1207, 2014.

[Benferhat et al., 1993] Salem Benferhat, Claudette Cayrol, Didier Dubois, Jérôme Lang, and Henri Prade. Inconsistency management and prioritized syntax-based entailment. In Ruzena Bajcsy, editor, *Proceedings of the 13th International Joint Conference on Artificial Intelligence (IJCAI'93)*, pages 640–647. Morgan Kaufmann, 1993.

[Benferhat et al., 1997] Salem Benferhat, Didier Dubois, and Henri Prade. Some syntactic approaches to the handling of inconsistent knowledge bases: A comparative study part 1: The flat case. *Studia Logica*, 58(1):17–45, 1997.

[Besnard and Doutre, 2004] Philippe Besnard and Sylvie Doutre. Checking the acceptability of a set of arguments. In James P. Delgrande and Torsten Schaub, editors, *Proceedings of the 10th International Workshop on Non-Monotonic Reasoning (NMR'04)*, pages 59–64, 2004.

[Besnard and Hunter, 2001] Philippe Besnard and Anthony Hunter. A logic-based theory of deductive arguments. *Artificial Intelligence*, 128(1–2):203–235, 2001.

[Besnard and Hunter, 2009] Philippe Besnard and Anthony Hunter. Argumentation based on classical logic. In Guillermo Simari and Iyad Rahwan, editors, *Argumentation in Artificial Intelligence*, pages 133–152. Springer, 2009.

[Besnard and Hunter, 2014] Philippe Besnard and Anthony Hunter. Constructing argument graphs with deductive arguments: a tutorial. *Argument & Computation*, 5(1):5–30, 2014.

[Besnard and Hunter, 2018] Philippe Besnard and Anthony Hunter. A review of argumentation based on deductive arguments. In Pietro Baroni, Dov Gabay, Massimiliano Giacomin, and Leon van der Torre, editors, *Handbook of Formal Argumentation*, pages 437–484. College Publications, 2018.

[Besnard and Schaub, 1998] Philippe Besnard and Torsten Schaub. Signed systems for paraconsistent reasoning. *Journal of Automated Reasoning*, 20:191–213, 1998.

[Besnard et al., 2014] Philippe Besnard, Alejandro García, Antony Hunter, Sanjay Modgil, Henry Prakken, Guillermo Simari, and Francesca Toni. Introduction to structured argumentation. *Argument & Computation*, 5(1):1–4, 2014.

[Besnard et al., 2020] Philippe Besnard, Claudette Cayrol, and Marie-Christine Lagasquie-Schiex. Logical theories and abstract argumentation: A survey of existing works. *Journal of Argument and Computation*, 11(1-2):41–102, 2020.

[Blackburn et al., 2006] Patrick Blackburn, Johan van Benthem, and Frank Wolter. *Handbook of Modal Logic*. Studies in Logic and Practical Reasoning. Elsevier Science, 2006.

[Bochman, 2005] Alexander Bochman. *Explanatory Nonmonotonic Reasoning*, volume 4 of *Advances in Logic*. World Scientific, 2005.

[Bochman, 2006] Alexander Bochman. Two paradigms of nonmonotonic reasoning. In *Proceedings of the International Symposium on Artificial Intelligence and Mathematics (ISAIM'06)*, 2006.

[Bochman, 2013] Alexander Bochman. *A logical theory of nonmonotonic inference and belief change*. Springer Science & Business Media, 2013.

[Boella et al., 2005] Guido Boella, Joris Hulstijn, and Leendert Van Der Torre. A logic of abstract argumentation. In Simon Parsons, Nicolas Maudet, Pavlos Moraitis, and Iyad Rahwan, editors, *Proceedings of the 2nd International Workshop on Argumentation in Multi-Agent Systems (ArgMAS'05)*, volume 4049 of *Lecture Notes in Computer Science*, pages 29–41. Springer, 2005.

[Bogaerts, 2015] Bart Bogaerts. *Groundedness in logics with a fixpoint semantics*. PhD thesis, Informatics Section, Department of Computer Science, Katholieke Universiteit Leuven, 2015.

[Bondarenko et al., 1997] Andrei Bondarenko, Phan Minh Dung, Robert Kowalski, and Francesca Toni. An abstract, argumentation-theoretic approach to default reasoning. *Artificial Intelligence*, 93(1):63–101, 1997.

[Borg and Straßer, 2018] AnneMarie Borg and Christian Straßer. Relevance in structured argumentation. In Jérôme Lang, editor, *Proceedings of the 27th International Joint Conference on Artificial Intelligence (IJCAI'18)*, pages 1753–1759. ijcai.org, 2018.

[Borg et al., 2021] AnneMarie Borg, Christian Straßer, and Ofer Arieli. A generalized proof-theoretic approach to structured argumentation by hypersequent calculi. *Studia Logica*, 109(1):167–238, 2021.

[Borg, 2019] AnneMarie Borg. *Modeling Defeasible Reasoning from an Argumentative Angle, With Emphasis on Sequent-Based Argumentation*. PhD thesis, Ruhr Universität Bochum, 2019.

[Borg, 2020] AnneMarie Borg. Assumptive sequent-based argumentation. *Journal of Applied Logics-IfCoLog Journal of Logics and their Applications*, 7(3):227–294, 2020.

[Bradfield and Stirling, 2001] Julian C. Bradfield and Colin Stirling. Modal logics and mu-calculi: An introduction. In Jan A. Bergstra, Alban Ponse, and Scott A. Smolka, editors, *Handbook of Process Algebra*, pages 293–330. North-Holland / Elsevier, 2001.

[Brewka and Gottlob, 1997] Gerhard Brewka and Georg Gottlob. Well-founded semantics for default logic. *Fundamenta Informaticae*, 31(3, 4):221–236, 1997.

[Brewka et al., 2017] Gerhard Brewka, Stefan Ellmauthaler, Hannes Strass, Johannes Peter Wallner, and Stefan Woltran. Abstract dialectical frameworks. an overview. *Journal of Applied Logics-IfCoLog Journal of Logics and their Applications*, 4(8):2263–2317, 2017.

[Brewka, 1989] Gerhard Brewka. Preferred subtheories: An extended logical framework for default reasoning. In Natesa Sridharan, editor, *Proceedings of the 11th International Joint Conference on Artificial Intelligence (IJCAI'89)*, pages 1043–1048. Morgan Kaufmann, 1989.

[Brewka, 1994] Gerhard Brewka. Adding priorities and specificity to default logic. In Craig MacNish, David Pearce, and Luís Moniz Pereira, editors, *Proceedings of the European Workshop on Logics in Artificial Intelligence, European Workshop, (JELIA'94)*, volume 838 of *Lecture Notes in Computer Science*, pages 247–260. Springer, 1994.

[Budzynska and Villata, 2018] Katabzyna Budzynska and Serena Villata. Processing natural language argumentation. In Pietro Baroni, Dov Gabbay, Massimiliano Giacomin, and Leon van der Torre, editors, *Handbook of Formal Argumentation*, pages 577–627. College Publications, 2018.

[Caminada and Amgoud, 2007] Martin Caminada and Leila Amgoud. On the evaluation of argumentation formalisms. *Artificial Intelligence*, 171(5):286–310, 2007.

[Caminada and Gabbay, 2009] Martin Caminada and Dov Gabbay. A logical account of formal argumentation. *Studia Logica*, 93(2):109–145, 2009.

[Caminada and Schulz, 2017] Martin Caminada and Claudia Schulz. On the equivalence between assumption-based argumentation and logic programming. *Journal of Artificial Intelligence Research*, 60:779–825, 2017.

[Caminada and Wu, 2011] Martin Caminada and Yining Wu. On the limitations of abstract argumentation. In *Proceedings of the 23rd Benelux Conference on Artificial Intelligence (BNAIC'11)*, 2011.

[Caminada et al., 2011] Martin Caminada, Walter Carnielli, and Paul Dunne. Semi-stable semantics. *Journal of Logic and Computation*, 22(5):1207–1254, 2011.

[Caminada et al., 2014] Martin Caminada, Sanjay Modgil, and Nir Oren. Preferences and unrestricted rebut. In Simon Parsons, Nir Oren, Chris Reed, and Federico Cerutti, editors, *Proceedings of the 5th International Conference on Computation Models of Argument (COMMA'14)*, Frontiers in Artificial Intelligence and Applications 266, pages 209–220. IOS Press, 2014.

[Caminada et al., 2015] Martin Caminada, Samy Sá, João Alcântara, and Wolfgang Dvořák. On the equivalence between logic programming semantics and argumentation semantics. *International Journal of Approximate Reasoning*, 58:87–111, 2015.

[Caminada, 2006] Martin Caminada. On the issue of reinstatement in argumentation. In Michael Fisher, Wiebe van der Hoek, Boris Konev, and Alexei

Lisitsa, editors, *Proceedings of the 10th European Conference on Logics in Artificial Intelligence (JELIA'06)*, volume 4160 of *Lecture Notes in Computer Science*, pages 111–123. Springer, 2006.

[Caminada, 2018a] Martin Caminada. Argumentation semantics as formal discussion. In Pietro Baroni, Dov Gabay, Massimiliano Giacomin, and Leon van der Torre, editors, *Handbook of Formal Argumentation*, pages 487–518. College Publications, 2018.

[Caminada, 2018b] Martin Caminada. Rationality postulates: Applying argumentation theory for non-monotonic reasoning. In Pietro Baroni, Dov Gabay, Massimiliano Giacomin, and Leon van der Torre, editors, *Handbook of Formal Argumentation*, pages 771–795. College Publications, 2018.

[Cayrol, 1995] Claudette Cayrol. On the relation between argumentation and non-monotonic coherence-based entailment. In *Proceedings of the 14th International Joint Conference on Artificial Intelligence (IJCAI'95)*, pages 1443–1448. Morgan Kaufmann, 1995.

[Cerutti et al., 2017] Federico Cerutti, Sarah Alice Gaggl, Matthias Thimm, and Johannes Peter Wallner. Foundations of implementations for formal argumentation. *Journal of Applied Logics-IfCoLog Journal of Logics and their Applications*, 4(8):2623–2706, 2017.

[Cerutti et al., 2018] Federico Cerutti, Sarah Alice Gaggl, Matthias Thimm, and Johannes Peter Wallner. Foundations of implementations for formal argumentation. In Pietro Baroni, Dov Gabbay, Massimiliano Giacomin, and Leon van der Torre, editors, *Handbook of Formal Argumentation*, pages 689–767. College Publications, 2018.

[Corsi and Fermüller, 2017] Esther Anna Corsi and Christian G. Fermüller. Logical argumentation principles, sequents, and nondeterministic matrices. In *Proceedings of the 6th International Workshop on Logic, Rationality, and Interaction (LORI'17)*, volume 10455 of *Lecture Notes in Computer Science*, pages 422–437. Springer, 2017.

[Corsi and Fermüller, 2019] Esther Anna Corsi and Christian G. Fermüller. Connecting fuzzy logic and argumentation frames via logical attack principles. *Soft Computing*, 23(7):2255–2270, 2019.

[Čyras and Toni, 2015] Kristijonas Čyras and Francesca Toni. Non-monotonic inference properties for assumption-based argumentation. In Elizabeth Black, Sanjay Modgil, and Nir Oren, editors, *Proceedings of the 3rd International Workshop on Theory and Applications of Formal Argument (TAFA'15)*, volume 9524 of *Lecture Notes in Computer Science*, pages 92–111. Springer, 2015.

[Čyras and Toni, 2016a] Kristijonas Čyras and Francesca Toni. ABA+: assumption-based argumentation with preferences. In Chitta Baral, James

Delgrande, and Frank Wolter, editors, *Proceedings of the 15th International Conference on Principles of Knowledge Representation and Reasoning (KR'16)*, pages 553–556. AAAI Press, 2016.
[Cyras and Toni, 2016b] Kristijonas Cyras and Francesca Toni. Properties of ABA+ for non-monotonic reasoning. *arXiv preprint arXiv:1603.08714*, 2016.
[Čyras et al., 2017] Kristijonas Čyras, Xiuyi Fan, Claudia Schulz, and Francesca Toni. Assumption-based argumentation: Disputes, explanations, preferences. *Journal of Applied Logics-IfCoLog Journal of Logics and their Applications*, 4(8):2407–2455, 2017.
[Čyras et al., 2018] Kristijonas Čyras, Xiuyi Fan, Claudia Schulz, and Francesca Toni. Assumption-based argumentation: Disputes, explanations, preferences. In Pietro Baroni, Dov Gabbay, Massimiliano Giacomin, and Leon van der Torre, editors, *Handbook of Formal Argumentation*, pages 2407–2456. College Publications, 2018.
[D'Agostino and Modgil, 2018] Marcello D'Agostino and Sanjay Modgil. Classical logic, argumentation and dialectic. *Artificial Intelligence*, 262:15–51, 2018.
[D'Agostino and Modgil, 2020] Marcello D'Agostino and Sanjay Modgil. A fully rational account of structured argumentation under resource bounds. In Christian Bessiere, editor, *Proceedings of the 29th International Joint Conference on Artificial Intelligence (IJCAI'20)*, pages 1841–1847. ijcai.org, 2020.
[de Saint-Cyr et al., 2016] Florence Dupin de Saint-Cyr, Pierre Bisquert, Claudette Cayrol, and Marie-Christine Lagasquie-Schiex. Argumentation update in YALLA (yet another logic language for argumentation). *International Journal of Approximate Reasoning*, 75:57–92, 2016.
[Denecker et al., 2000] Marc Denecker, Victor Marek, and Mirosław Truszczyński. Approximations, stable operators, well-founded fixpoints and applications in nonmonotonic reasoning. In *Logic-based Artificial Intelligence*, volume 597 of *The Springer International Series in Engineering and Computer Science*, pages 127–144. Springer, 2000.
[Denecker et al., 2003] Marc Denecker, Victor W Marek, and Mirosław Truszczyński. Uniform semantic treatment of default and autoepistemic logics. *Artificial Intelligence*, 143(1):79–122, 2003.
[Diller et al., 2015] Martin Diller, Johannes Peter Wallner, and Stefan Woltran. Reasoning in abstract dialectical frameworks using quantified boolean formulas. *Argument & Computation*, 6(2):149–177, 2015.
[Dimopoulos et al., 2002] Yannis Dimopoulos, Bernhard Nebel, and Francesca Toni. On the computational complexity of assumption-based argumentation for default reasoning. *Artificial Intelligence*, 141(1-2):57–78, 2002.

[Ditmarsch et al., 2015] Hans van Ditmarsch, Joseph Halpern, Wiebe van der Hoek, and Barteld Kooi. *Handbook of Epistemic Logic*. College Publications, 2015.

[Dung and Thang, 2014] Phan Minh Dung and Phan Minh Thang. Closure and consistency in logic-associated argumentation. *Journal of Artificial Inteligence Research*, 49:79–109, 2014.

[Dung et al., 2006] Phan Minh Dung, Robert Kowalski, and Francesca Toni. Dialectic proof procedures for assumption-based, admissible argumentation. *Artificial Intelligence*, 170(2):114–159, 2006.

[Dung et al., 2007] Phan Minh Dung, Paolo Mancarella, and Francesca Toni. Computing ideal sceptical argumentation. *Artificial Intelligence*, 171(10-15):642–674, 2007.

[Dung et al., 2009] Phan Minh Dung, Robert Kowalski, and Francesca Toni. Assumption-based argumentation. In Guillermo Simari and Iyad Rahwan, editors, *Argumentation in Artificial Intelligence*, pages 199–218. Springer, 2009.

[Dung et al., 2016] Phan Minh Dung, Tran Cao Son, and Phan Minh Thang. Argumentation-based semantics for logic programs with first-order formulae. In Matteo Baldoni, Amit Chopra, Tran Cao Son, Katsutoshi Hirayama, and Paolo Torroni, editors, *Proceedings of the 19th International Conference on Principles and Practice of Multi-Agent Systems (PRIMA'16)*, volume 9862 of *Lecture Notes in Computer Science*, pages 43–60. Springer, 2016.

[Dung, 1995] Phan Minh Dung. On the acceptability of arguments and its fundamental role in nonmonotonic reasoning, logic programming and n-person games. *Artificial Intelligence*, 77(2):321–357, 1995.

[Dvořák et al., 2012] Wolfgang Dvořák, Stefan Szeider, and Stefan Woltran. Abstract argumentation via monadic second order logic. In Eyke Hüllermeier, Sebastian Link, Thomas Fober, and Bernhard Seeger, editors, *Proceedings of the 6th International Conference on Scalable Uncertainty Management (SUM'12)*, volume 7520 of *Lecture Notes in Computer Science*, pages 85–98. Springer, 2012.

[Dyrkolbotn, 2014] Sjur Dyrkolbotn. On a formal connection between truth, argumentation and belief. In Margot Colinet, Sophia Katrenko, and Rasmus K. Rendsvig, editors, *Pristine Perspectives on Logic, Language, and Computation - (ESSLLI '12) and (ESSLLI'13) Student Sessions. Selected Papers*, volume 8607 of *Lecture Notes in Computer Science*, pages 69–90. Springer, 2014.

[Efstathiou and Hunter, 2011] Vasiliki Efstathiou and Anthony Hunter. Algorithms for generating arguments and counterarguments in propositional logic. *International Journal of Approximate Reasoning*, 52(6):672–704, 2011.

[Egly and Woltran, 2006] Uwe Egly and Stefan Woltran. Reasoning in argumentation frameworks using quantified boolean formulas. In Paul E. Dunne and Trevor J. M. Bench-Capon, editors, *Proceedings of the 1st International Conference on Computational Models of Argument (COMMA'06)*, volume 144 of *Frontiers in Artificial Intelligence and Applications*, pages 133–144. IOS Press, 2006.

[Eichhorn, 2018] Christian Eichhorn. *Rational Reasoning with Finite Conditional Knowledge Bases*. Springer, 2018.

[Eiter and Gottlob, 1995] Thomas Eiter and Georg Gottlob. The complexity of logic-based abduction. *Journal of the ACM*, 42(1):3–42, 1995.

[Fandinno and del Cerro, 2018] Jorge Fandinno and Luis Fariñas del Cerro. Constructive logic covers argumentation and logic programming. In Michael Thielscher, Francesca Toni, and Frank Wolter, editors, *Proceedings of the 16th International Conference on Principles of Knowledge Representation and Reasoning (KR'18)*, pages 128–137. AAAI Press, 2018.

[Gabbay and Gabbay, 2015] Dov Gabbay and Murdoch Gabbay. The attack as strong negation, Part I. *Logic Journal of the IGPL*, 23(6):881–941, 2015.

[Gabbay and Gabbay, 2016] Dov Gabbay and Murdoch Gabbay. The attack as intuitionistic negation. *Logic Journal of the IGPL*, 24(5):807–837, 2016.

[Gabbay and Shehtman, 1998] Dov Gabbay and Valentin Shehtman. Products of modal logics, Part 1. *Logic Journal of the IGPL*, 6(1):73–146, 1998.

[Gabbay et al., 2013] Dov Gabbay, John Horty, and Xavier Parent. *Handbook of Deontic Logic and Normative Systems*. College Publications, 2013.

[Gabbay, 1985] Dov Gabbay. Theoretical foundations for non-monotonic reasoning in expert systems. In Krzysztof R. Apt, editor, *Proceedings of the Conference on Logics and models of concurrent systems*, volume 13 of *NATO ASI Series*, pages 439–457. Springer, 1985.

[Gabbay, 1990] Dov Gabbay. Modal provability foundations for negation by failure. In Peter Schroeder-Heister, editor, *Proceedings of the International Workshop on Extensions of Logic Programming*, volume 475 of *Lecture Notes in Computer Science*, pages 179–222. Springer, 1990.

[Gabbay, 2009] Dov Gabbay. Modal provability foundations for argumentation networks. *Studia Logica*, 93(2–3):181–198, 2009.

[Gabbay, 2011] Dov Gabbay. Dung's argumentation is essentially equivalent to classical propositional logic with the Peirce–Quine dagger. *Logica Universalis*, 5(2):255–318, 2011.

[García and Simari, 2004] Alejandro García and Guillermo Simari. Defeasible logic programming: an argumentative approach. *Theory and Practice of Logic Programming*, 4(1–2):95–138, 2004.

[Gauderis and Van De Putte, 2012] Tjerk Gauderis and Frederik Van De Putte. Abduction of generalizations. *Theoria. An International Journal for Theory, History and Foundations of Science*, 27(3):345–363, 2012.

[Geffner and Pearl, 1992] Hector Geffner and Judea Pearl. Conditional entailment: Bridging two approaches to default reasoning. *Artificial Intelligence*, 53(2-3):209–244, 1992.

[Gelfond and Lifschitz, 1991] Michael Gelfond and Vladimir Lifschitz. Classical negation in logic programs and disjunctive databases. *New generation computing*, 9(3-4):365–385, 1991.

[Gentzen, 1934] Gerhard Gentzen. Untersuchungen über das logische Schließen I, II. *Mathematische Zeitschrift*, 39:176–210, 405–431, 1934.

[Gorogiannis and Hunter, 2011] Nikos Gorogiannis and Anthony Hunter. Instantiating abstract argumentation with classical logic arguments: Postulates and properties. *Artificial Intelligence*, 175(9–10):1479–1497, 2011.

[Grooters and Prakken, 2016] Diana Grooters and Henry Prakken. Two aspects of relevance in structured argumentation: Minimality and paraconsistency. *Journal of Artificial Inteligence Research*, 56:197–245, 2016.

[Grossi and van der Hoek, 2014] Davide Grossi and Wiebe van der Hoek. Justified beliefs by justified arguments. In Chitta Baral, Giuseppe De Giacomo, and Thomas Eiter, editors, *Proceedings of the 14th International Conference on Principles of Knowledge Representation and Reasoning (KR'14)*. AAAI Press, 2014.

[Grossi, 2010] Davide Grossi. On the logic of argumentation theory. In Wiebe van der Hoek, Gal Kaminka, Yves Lespérance, Michael Luck, and Sandip Sen, editors, *Proceedings of the 9th International Conference on Autonomous Agents and Multiagent Systems (AAMAS'10)*, pages 409–416. IFAAMAS, 2010.

[Grossi, 2013] Davide Grossi. Abstract argument games via modal logic. *Synthese*, 190:5–29, 2013.

[Hawthorne and Makinson, 2007] James Hawthorne and David Makinson. The quantitative/qualitative watershed for rules of uncertain inference. *Studia Logica*, 86(2):247–297, 2007.

[Heyninck and Arieli, 2018] Jesse Heyninck and Ofer Arieli. On the semantics of simple contrapositive assumption-based argumentation frameworks. In Sanjay Modgil, Katarzyna Budzynska, and John Lawrence, editors, *Proceedings of the 7th International Conference on Computation Models of Argument (COMMA'18)*, volume 305 of *Frontiers in Artificial Intelligence and Applications*, pages 9–20. IOS Press, 2018.

[Heyninck and Arieli, 2019a] Jesse Heyninck and Ofer Arieli. An argumentative characterization of disjunctive logic programming. In Paulo Moura Oliveira, Paulo Novais, and Luís Paulo Reis, editors, *Proceedings of the 19the Conference on Progress in Artificial Intelligenc (EPIA'19)*, volume 11805 of *Lecture Notes in Computer Science*, pages 526–538. Springer, 2019.

[Heyninck and Arieli, 2019b] Jesse Heyninck and Ofer Arieli. Simple contrapositive assumption-based frameworks. In Marcello Balduccini, Yuliya Lierler, and Stefan Woltran, editors, *Proceedings of the 15th International Conference on Logic Programming and Nonmonotonic Reasoning (LPNMR'19)*, volume 11481 of *Lecture Notes in Computer Science*, pages 75–88. Springer, 2019.

[Heyninck and Arieli, 2020a] Jesse Heyninck and Ofer Arieli. Argumentative reflections of approximation fixpoint theory. *Computational Models of Argument: Proceedings of COMMA 2020*, 326:215, 2020.

[Heyninck and Arieli, 2020b] Jesse Heyninck and Ofer Arieli. Simple contrapositive assumption-based frameworks. *Journal of Approximate Reasoning*, 121:103–124, 2020.

[Heyninck and Straßer, 2016] Jesse Heyninck and Christian Straßer. Relations between assumption-based approaches in nonmonotonic logic and formal argumentation. In *Proceedings of the 16th International Workshop on Non-Monotonic Reasoning (NMR'16)*, 2016.

[Heyninck and Straßer, 2017] Jesse Heyninck and Christian Straßer. Revisiting unrestricted rebut and preferences in structured argumentation. In Carles Sierra, editor, *Proceedings of the 26th International Joint Conference on Artificial Intelligence, (IJCAI'17)*, pages 1088–1092. ijcai.org, 2017.

[Heyninck and Straßer, 2019] Jesse Heyninck and Christian Straßer. A fully rational argumentation system for preordered defeasible rules. In Edith Elkind, Manuela Veloso, Noa Agmon, and Matthew Taylor, editors, *Proceedings of the 18th International Conference on Autonomous Agents and MultiAgent Systems (AAMAS'19)*, pages 1704–1712. International Foundation for Autonomous Agents and Multiagent Systems, 2019.

[Heyninck and Straßer, 2021a] Jesse Heyninck and Christian Straßer. A comparative study of assumption-based approaches to reasoning with priorities. *Journal of Applied Logic – IfCoLog Journal of Logics and their Applications*, 8(3):737–808, 2021.

[Heyninck and Straßer, 2021b] Jesse Heyninck and Christian Straßer. Rationality and maximal consistent sets for a fragment of ASPIC$^+$ without undercut. *Argument and Computation*, 12(1):3–47, 2021.

[Heyninck et al., 2020] Jesse Heyninck, Gabriele Kern-Isberner, and Matthias Thimm. On the correspondence between abstract dialectical frameworks and nonmonotonic conditional logics. In *Proceedings of the 33rd International*

Florida Artificial Intelligence Research Society Conference (FLAIRS'20), 2020.

[Hintikka, 2005] Jaakko Hintikka. *Knowledge and Belief: An Introduction to the Logic of the Two Notions*. Texts in philosophy. King's College Publications, 2005. Reprint.

[Horty, 1994] John Horty. Moral dilemmas and nonmonotonic logic. *Journal of Philosophical Logic*, 23:35–65, 1994.

[Janhunen, 1998] Tomi Janhunen. On the intertranslatability of autoepistemic, default and priority logics, and parallel circumscription. In Jürgen Dix, Luis Fariñas del Cerro, and Ulrich Furbach, editors, *Proceedings of the European Workshop on Logics in Artificial Intelligence (JELIA'98)*, volume 1489 of *Lecture Notes in Computer Science*, pages 216–232. Springer, 1998.

[Konieczny and Pérez, 2002] Sébastien Konieczny and Ramón Pino Pérez. Merging information under constraints: a logical framework. *Journal of Logic and Computation*, 12(5):773–808, 2002.

[Konolige, 1988] Kurt Konolige. On the relation between default and autoepistemic logic. 35(3):343–382, 1988.

[Kraus *et al.*, 1990] Sarit Kraus, Daniel Lehmann, and Menachem Magidor. Nonmonotonic reasoning, preferential models and cumulative logics. *Artificial Intelligence*, 44(1):167 – 207, 1990.

[Lehmann and Magidor, 1992] Daniel Lehmann and Menachem Magidor. What does a conditional knowledge base entail? *Artificial Intelligence*, 55:1–60, 1992.

[Li *et al.*, 2018] Zimi Li, Nir Oren, and Simon Parsons. On the links between argumentation-based reasoning and nonmonotonic reasoning. In Elizabeth Black, Sanjay Modgil, and Nir Oren, editors, *Proceedings of the 4th International Workshop on Theory and Applications of Formal Argument (TAFA'17)*, volume 10757 of *Lecture Notes in Computer Science*, pages 67–85. Springer, 2018.

[Łos and Suzsko, 1958] Jerzy Łos and Roman Suzsko. Remarks on sentential logics. *Indagationes Mathematicae*, 20:177–183, 1958.

[Makinson and Van Der Torre, 2001] David Makinson and Leendert Van Der Torre. Constraints for Input/Output logics. *Journal of Philosophical Logic*, 30:155–185, 2001.

[Makinson, 1994] David Makinson. General patterns in nonmonotonic reasoning. In D. Gabbay, C. Hogger, and J. Robinson, editors, *Handbook of Logic in Artificial Intelligence and Logic Programming*, volume 3, pages 35–110. Oxford Science Publications, 1994.

[Makinson, 2003] David Makinson. Bridges between classical and nonmonotonic logic. *Logic Journal of the IGPL*, 11(1):69–96, 2003.

[Malouf, 2007] Rob Malouf. Maximal consistent subsets. *Computational Linguistics*, 33(2):153–160, 2007.

[Marek and Truszczyński, 1993] Wiktor Marek and Mirosław Truszczyński. Reflexive autoepistemic logic and logic programming. In Luís Moniz Pereira and Anil Nerode, editors, *Proceedings of the 2nd International Workshop on Logic Programming and Nonmonotonic Reasoning (LPNMR'93)*, pages 115–131. MIT Press, 1993.

[Meheus and Batens, 2006] Joke Meheus and Diderik Batens. A formal logic for abductive reasoning. *Logic Journal of IGPL*, 14(2):221–236, 2006.

[Modgil and Caminada, 2009] Sanjay Modgil and Martin Caminada. Proof theories and algorithms for abstract argumentation frameworks. In Guillermo Simari and Iyad Rahwan, editors, *Argumentation in Artificial Intelligence*, pages 105–132. Springer, 2009.

[Modgil and Prakken, 2013] Sanjay Modgil and Henry Prakken. A general account of argumentation with preferences. *Artificial Intelligence*, 195:361–397, 2013.

[Modgil and Prakken, 2014] Sanjay Modgil and Henry Prakken. The ASPIC+ framework for structured argumentation: a tutorial. *Argument & Computation*, 5(1):31–62, 2014.

[Modgil and Prakken, 2018] Sanjay Modgil and Henry Prakken. Abstract rule-based argumentation. In Pietro Baroni, Dov Gabay, Massimiliano Giacomin, and Leon van der Torre, editors, *Handbook of Formal Argumentation*, pages 287–364. College Publications, 2018.

[Modgil, 2009] Sanjay Modgil. Reasoning about preference in argumentation frameworks. *Artificial Intelligence*, 173(9-10):901–934, 2009.

[Moore, 1985] Robert Moore. Semantic considerations on nonmonotonic logic. *Artificial Intelligence*, 25(1):75–94, 1985.

[Pearce, 2006] David Pearce. Equilibrium logic. *Annals of Mathematics and Artificial Intelligence*, 47(1–2):3–41, 2006.

[Prakken and Vreeswijk, 2002] Henry Prakken and Gerard Vreeswijk. Logics for defeasible argumentation. In Dov Gabbay and Franz Guenthner, editors, *Handbook of Philosophical Logic*, pages 219–318. Springer, 2002.

[Prakken and Winter, 2018] Henry Prakken and Michiel de Winter. Abstraction in argumentation: Necessary but dangerous. In Sanjay Modgil, Katarzyna Budzynska, and John Lawrence, editors, *Proceedings of the 7th International Conference on Computation Models of Argument (COMMA'18)*, volume 305 of *Frontiers in Artificial Intelligence and Applications*, pages 85–96. IOS Press, 2018.

[Prakken, 2010] Henry Prakken. An abstract framework for argumentation with structured arguments. *Argument & Computation*, 1(2):93–124, 2010.

[Prakken, 2018] Henry Prakken. Historical overview of formal argumentation. In Pietro Baroni, Dov Gabay, Massimiliano Giacomin, and Leon van der Torre, editors, *Handbook of Formal Argumentation*, pages 75–143. College Publications, 2018.

[Reiter, 1980] Raymond Reiter. A logic for default reasoning. *Artificial Intelligence*, 13(1-2):81–132, 1980.

[Rescher and Manor, 1970] Nicholas Rescher and Ruth Manor. On inference from inconsistent premises. *Theory and Decision*, 1:179–217, 1970.

[Schulz and Toni, 2015] Claudia Schulz and Francesca Toni. Logic programming in assumption-based argumentation revisited-semantics and graphical representation. In Blai Bonet and Sven Koenig, editors, *Proceedings of the 29th AAAI Conference on Artificial Intelligence (AAAI'15)*, pages 1569–1575. AAAI Press, 2015.

[Schulz and Toni, 2016] Claudia Schulz and Francesca Toni. Justifying answer sets using argumentation. *Theory and Practice of Logic Programming*, 16(1):59–110, 2016.

[Schulz et al., 2015] Claudia Schulz, Ken Satoh, and Francesca Toni. Characterising and explaining inconsistency in logic programs. In Francesco Calimeri, Giovambattista Ianni, and Miroslaw Truszczynski, editors, *Proceedings of the 13th International Conference on Logic Programming and Nonmonotonic Reasoning (LPNMR'15)*, volume 9345 of *Lecture Notes in Computer Science*, pages 467–479. Springer, 2015.

[Shen et al., 2014] Yi-Dong Shen, Kewen Wang, Thomas Eiter, Michael Fink, Christoph Redl, Thomas Krennwallner, and Jun Deng. FLP answer set semantics without circular justifications for general logic programs. *Artificial Intelligence*, 213:1–41, 2014.

[Shi et al., 2017] Chenwei Shi, Sonja Smets, and Fernando R. Velázquez-Quesada. Argument-based belief in topological structures. In Jérôme Lang, editor, *Proceedings of the 16th Conference on Theoretical Aspects of Rationality and Knowledge (TARK'17)*, volume 251 of *EPTCS*, pages 489–503, 2017.

[Shi et al., 2018] Chenwei Shi, Sonja Smets, and Fernando R. Velázquez-Quesada. Beliefs supported by binary arguments. *Journal of Applied Non-Classical Logics*, 28(2–3):165–188, 2018.

[Shoham, 1988] Yoav Shoham. *Reasoning About Change: Time and Causation from the Standpoint of Artificial Intelligence*. MIT Press, 1988.

[Stalnaker, 1994] Robert Stalnaker. What is a nonmonotonic consequence relation? *Fundamenta Informaticae*, 21(1, 2):7–21, 1994.

[Strass, 2013] Hannes Strass. Approximating operators and semantics for abstract dialectical frameworks. *Artificial Intelligence*, 205:39–70, 2013.

[Strass, 2014] Hannes Strass. On the relative expressiveness of argumentation frameworks, normal logic programs and abstract dialectical frameworks. *arXiv preprint arXiv:1405.0805*, 2014.

[Straßer and Arieli, 2019] Christian Straßer and Ofer Arieli. Normative reasoning by sequent-based argumentation. *Journal of Logic and Computation*, 29(3):387–415, 2019.

[Straßer and Šešelja, 2010] Christian Straßer and Dunja Šešelja. Towards the proof-theoretic unification of Dung's argumentation framework: an adaptive logic approach. *Journal of Logic and Computation*, 21(2):133–156, 2010.

[Straßer, 2012] Christian Straßer. Adaptively applying modus ponens in conditional logics of normality. *Journal of Applied Non-Classical Logics*, 22(1–2):125–148, 2012.

[Straßer, 2014] Christian Straßer. *Adaptive Logic and Defeasible Reasoning. Applications in Argumentation, Normative Reasoning and Default Reasoning*. Springer, 2014.

[Tarski, 1941] Alfred Tarski. *Introduction to Logic*. Oxford University Press, 1941.

[Toni, 2014] Francesca Toni. A tutorial on assumption-based argumentation. *Argument & Computation*, 5(1):89–117, 2014.

[Urquhart, 2001] Alasdair Urquhart. Many-valued logic. In Dov Gabbay and Franz Guenthner, editors, *Handbook of Philosophical Logic*, volume II, pages 249–295. Kluwer, 2001. Second edition.

[Van De Putte and Straßer, 2013] Frederik Van De Putte and Christian Straßer. A logic for prioritized normative reasoning. *Journal of Logic and Computation*, 23(3):563–583, 2013.

[Van De Putte, 2013] Frederik Van De Putte. Default assumptions and selection functions: A generic framework for non-monotonic logics. In Félix Castro, Alexander Gelbukh, and Miguel González, editors, *Proceedings of the 12th Mexican International Conference on Advances in Artificial Intelligence and Its Applications (MICAI'13)*, volume 8265 of *Lecture Notes in Computer Science*. Springer, 2013.

[Van Dyck, 2004] Maarten Van Dyck. Causal discovery using adaptive logics. Towards a more realistic heuristics for human causal learning. *Logique et Analyse*, 185–188:5–32, 2004.

[Van Kerckhove and Vanackere, 2003] Bart Van Kerckhove and Guido Vanackere. Vagueness-adaptive logic: A pragmatical approach to sorites paradoxes. *Studia Logica*, 75(3):383–411, 2003.

[Vesic, 2013] Srdjan Vesic. Identifying the class of maxi-consistent operators in argumentation. *Journal of Artificial Intelligence Research*, 47:71–93, 2013.

[Villata et al., 2012] Serena Villata, Guido Boella, Dov M Gabbay, Leendert van der Torre, and Joris Hulstijn. A logic of argumentation for specification and verification of abstract argumentation frameworks. *Annals of Mathematics and Artificial Intelligence*, 66(1-4):199–230, 2012.

[von Wright, 1951] Georg von Wright. Deontic logic. *Mind*, 60(237):1–15, 1951.

[Wakaki, 2017] Toshiko Wakaki. Assumption-based argumentation equipped with preferences and its application to decision making, practical reasoning, and epistemic reasoning. *Computational Intelligence*, 33(4):706–736, 2017.

[Wang et al., 2012] Yisong Wang, Fangzhen Lin, Mingyi Zhang, and Jia-Huai You. A well-founded semantics for basic logic programs with arbitrary abstract constraint atoms. In Jörg Hoffmann and Bart Selman, editors, *Proceedings of the 26th AAAI Conference on Artificial Intelligence (AAAI'2012)*, pages 835–841. AAAI Press, 2012.

[Weber and Provijn, 1999] Erik Weber and Dagmar Provijn. A formal analysis of diagnosis and diagnostic reasoning. *Logique et Analyse*, 42(165–166):161–180, 1999.

[Wu and Podlaszewski, 2014] Yining Wu and Mikołaj Podlaszewski. Implementing crash-resistance and non-interference in logic-based argumentation. *Journal of Logic and Computation*, 25(2):303–333, 2014.

A Proofs

Below we provide proofs to propositions that appear in the chapter and to the best of our knowledge have not been fully proven yet in the literature.

Proposition 88. *Let* $\mathfrak{L} = \langle \mathcal{L}, \vdash \rangle$ *be a propositional logic. The entailments* $\mathrel{\vphantom{\sim}\smash{\mid\hspace{-0.55em}\sim}}^{\mathfrak{L}}_{\cap\mathsf{mcs}}$ *and* $\mathrel{\vphantom{\sim}\smash{\mid\hspace{-0.55em}\sim}}^{\mathfrak{L}}_{\cap\mathsf{mcs}}$ *are* ⊎-*cautiously cumulative and* ⊎-*cumulative.*

Proof. The properties ⊔-**(C)REF**, **RW**, ⊔-**LLE** follow directly from Definition 44. Note that for ⊎, full reflexivity does not hold since for an ⊢-inconsistent formula ϕ, $\mathsf{MCS}^{\emptyset}(\{\phi\}) = \{\emptyset\}$. The properties ⊔-**CC** and ⊔-**CM** follow for $\mathrel{\vphantom{\sim}\smash{\mid\hspace{-0.55em}\sim}}^{\mathfrak{L}}_{\cap\mathsf{mcs}}$ and $\mathrel{\vphantom{\sim}\smash{\mid\hspace{-0.55em}\sim}}^{\mathfrak{L}}_{\cap\mathsf{mcs}}$ by Lemma 132 and Corollary 133. We paradigmatically show the case for $\mathrel{\vphantom{\sim}\smash{\mid\hspace{-0.55em}\sim}}^{\mathfrak{L}}_{\cap\mathsf{mcs}}$ and ⊔ = ⊎: Suppose that $\mathcal{S}', \mathcal{S} \mathrel{\vphantom{\sim}\smash{\mid\hspace{-0.55em}\sim}}^{\mathfrak{L}}_{\cap\mathsf{mcs}} \psi$. Then the following equivalences hold: $\mathcal{S}', \mathcal{S} \mathrel{\vphantom{\sim}\smash{\mid\hspace{-0.55em}\sim}}^{\mathfrak{L}}_{\cap\mathsf{mcs}} \phi$, iff

$\bigcap \mathsf{MCS}_{\mathfrak{L}}^{\mathcal{S}'}(\mathcal{S}) \vdash \phi$, iff (by Corollary 133 and since $\bigcap \mathsf{MCS}_{\mathfrak{L}}^{\mathcal{S}'}(\mathcal{S}) \vdash \psi$ by the supposition) $\bigcap \mathsf{MCS}_{\mathfrak{L}}^{\mathcal{S}'}(\mathcal{S} \cup \{\psi\}) \vdash \phi$, iff $\mathcal{S}', \mathcal{S} \cup \{\psi\} \hspace{0.2em}\big|\!\!\!\sim_{\bigcap\mathsf{mcs}} \phi$. □

Lemma 132. *If* $\langle \mathcal{S}', \mathcal{S}\rangle \hspace{0.2em}\big|\!\!\!\sim_{\bigcap\mathsf{mcs}} \psi$. *Then:*
1. $\mathsf{MCS}_{\mathfrak{L}}^{\mathcal{S}'}(\mathcal{S} \cup \{\psi\}) = \{\mathcal{T} \cup \{\psi\} \mid \mathcal{T} \in \mathsf{MCS}_{\mathfrak{L}}^{\mathcal{S}'}(\mathcal{S})\}$, *and*
2. $\mathsf{MCS}_{\mathfrak{L}}^{\mathcal{S}'}(\mathcal{S}) = \mathsf{MCS}_{\mathfrak{L}}^{\mathcal{S}' \cup \{\psi\}}(\mathcal{S})$.

Proof. Item 1, $\subseteq$: Suppose that $\mathcal{T} \in \mathsf{MCS}_{\mathfrak{L}}^{\mathcal{S}'}(\mathcal{S} \cup \{\psi\})$. Thus, $\mathcal{T} \cap \mathcal{S}$ is a $\vdash$-consistent subset of $\mathcal{S}$, given $\mathcal{S}'$. Assume that there is a $\mathcal{T}' \in \mathsf{MCS}_{\mathfrak{L}}^{\mathcal{S}'}(\mathcal{S})$ such that $\mathcal{T} \cap \mathcal{S} \subsetneq \mathcal{T}'$. By the supposition, $\mathcal{T}' \vdash \psi$. Thus, $\mathcal{T}' \cup \{\psi\}$ is a $\vdash$-consistent subset of $\mathcal{S} \cup \{\psi\}$, given $\mathcal{S}'$. But since $\mathcal{T} \subsetneq \mathcal{T}' \cup \{\psi\}$, this is a contradiction to the $\subseteq$-maximal consistency of $\mathcal{T}$. Thus, $\mathcal{T} \cap \mathcal{S} \in \mathsf{MCS}_{\mathfrak{L}}^{\mathcal{S}'}(\mathcal{S})$. By the assumption again, $\mathcal{T} \vdash \psi$, and so $\mathcal{T} = (\mathcal{T} \cap \mathcal{S}) \cup \{\psi\}$ is an element of the set in the right-hand side of the equation of Item 1.

Item 1, $\supseteq$: Suppose that $\mathcal{T} \in \mathsf{MCS}_{\mathfrak{L}}^{\mathcal{S}'}(\mathcal{S})$. Thus, $\mathcal{T}$ is a $\vdash$-consistent subset of $\mathcal{S}$, given $\mathcal{S}'$. Since $\langle \mathcal{S}', \mathcal{S}\rangle \hspace{0.2em}\big|\!\!\!\sim_{\bigcap\mathsf{mcs}} \psi$, we have that $\mathcal{T}, \mathcal{S}' \vdash \psi$ and so $\mathcal{T} \cup \{\psi\}$ is a $\vdash$-consistent subset of $\mathcal{S} \cup \{\psi\}$, given $\mathcal{S}'$. Assume for a contradiction that there is a proper superset $\mathcal{T}' \supsetneq (\mathcal{T} \cup \{\psi\})$ such that $\mathcal{T}' \in \mathsf{MCS}_{\mathfrak{L}}^{\mathcal{S}'}(\mathcal{S} \cup \{\psi\})$. Then, $\mathcal{T} \subsetneq (\mathcal{T}' \cap \mathcal{S})$ and $\mathcal{T}' \cap \mathcal{S}$ is a $\vdash$-consistent subset of $\mathcal{S}$ given $\mathcal{S}'$, which contradicts the $\subseteq$-maximal consistency of $\mathcal{T}$.

Item 2, $\supseteq$: Suppose that $\mathcal{T} \in \mathsf{MCS}^{\mathcal{S}' \cup \{\psi\}}(\mathcal{S})$. Thus, $\mathcal{T}$ is a $\vdash$-consistent subset of $\mathcal{S}$ given $\mathcal{S}' \cup \{\psi\}$, and so also given $\mathcal{S}'$. Assume that there is a set $\mathcal{T}' \in \mathsf{MCS}_{\mathfrak{L}}^{\mathcal{S}'}(\mathcal{S})$ such that $\mathcal{T} \subsetneq \mathcal{T}'$. Thus, $\mathcal{T}'$ is $\vdash$-inconsistent with ψ (given $\mathcal{S}'$) since otherwise $\mathcal{T}'$ is $\vdash$-consistent with $\mathcal{S}$ given $\mathcal{S}' \cup \{\psi\}$ in contrast to $\mathcal{T} \in \mathsf{MCS}_{\mathfrak{L}}^{\mathcal{S}' \cup \{\psi\}}(\mathcal{S})$. Thus, $\mathcal{T}', \mathcal{S}', \psi \vdash \mathsf{F}$. By the main supposition also $\mathcal{T}', \mathcal{S}' \vdash \psi$. Thus, by transitivity, $\mathcal{T}', \mathcal{S}' \vdash \mathsf{F}$ which is a contradiction to the choice of $\mathcal{T}'$. Thus, $\mathcal{T} \in \mathsf{MCS}^{\mathcal{S}'}(\mathcal{S})$.

Item 2, $\subseteq$: The proof is similar to that of the previous item. Briefly, suppose that $\mathcal{T} \in \mathsf{MCS}_{\mathfrak{L}}^{\mathcal{S}'}(\mathcal{S})$. Since $\langle \mathcal{S}', \mathcal{S}\rangle \hspace{0.2em}\big|\!\!\!\sim_{\bigcap\mathsf{mcs}} \psi$, necessarily $\mathcal{T}$ is a $\vdash$-consistent subset of $\mathcal{S}$, given $\mathcal{S}' \cup \{\psi\}$, and trivially then $\mathcal{T} \in \mathsf{MCS}_{\mathfrak{L}}^{\mathcal{S}' \cup \{\psi\}}(\mathcal{S})$. □

The following corollary follows immediately in view of the fact that $\hspace{0.2em}\big|\!\!\!\sim_{\bigcap\mathsf{mcs}}^{\mathfrak{L}}$ is contained in $\hspace{0.2em}\big|\!\!\!\sim_{\bigcap\mathsf{mcs}}^{\mathfrak{L}}$.

Corollary 133. *If* $\langle \mathcal{S}', \mathcal{S}\rangle \hspace{0.2em}\big|\!\!\!\sim_{\bigcap\mathsf{mcs}} \psi$ *then Items 1 and 2 of Lemma 132 hold.*

Proposition 89. Let $\mathfrak{L} = \langle \mathcal{L}, \vdash \rangle$ be a propositional logic and let $\sqcup \in \{\talloblong, \uplus\}$. The entailment $\mid\!\sim^{\mathfrak{L}}_{\cap\mathsf{mcs}}$ is $\sqcup$-preferential.

Proof. The proposition follows by Proposition 88 and Lemma 134. □

Lemma 134. $\mid\!\sim^{\mathfrak{L}}_{\cap\mathsf{mcs}}$ satisfies $\sqcup$-*OR*.

Proof. We first consider the case $\sqcup = \talloblong$. Suppose that $\langle \mathcal{S}', \mathcal{S} \cup \{\phi_1\}\rangle \mid\!\sim^{\mathfrak{L}}_{\cap\mathsf{mcs}} \psi$ and $\langle \mathcal{S}', \mathcal{S} \cup \{\phi_2\}\rangle \mid\!\sim^{\mathfrak{L}}_{\cap\mathsf{mcs}} \psi$. Let $\mathcal{T} \in \mathsf{MCS}^{\mathcal{S}'}_{\mathfrak{L}}(\mathcal{S} \cup \{\phi_1 \vee \phi_2\})$ and $\mathcal{T}' = \mathcal{T} \cap \mathcal{S}$. If $\mathcal{T}'$ is $\vdash$-inconsistent with $\phi_1 \vee \phi_2$, then $\mathcal{T}' \in \mathsf{MCS}^{\mathcal{S}'}_{\mathfrak{L}}(\mathcal{S} \cup \{\phi_1\}) \cap \mathsf{MCS}^{\mathcal{S}'}_{\mathfrak{L}}(\mathcal{S} \cup \{\phi_2\})$ and $\mathcal{T} = \mathcal{T}'$. By the supposition $\mathcal{T}', \mathcal{S}' \vdash \psi$ and so $\mathcal{T}, \mathcal{S}' \vdash \psi$.

If $\mathcal{T}'$ is $\vdash$-consistent with both ϕ_1 and ϕ_2, then $\mathcal{T}' \cup \{\phi_1\} \in \mathsf{MCS}^{\mathcal{S}'}_{\mathfrak{L}}(\mathcal{S} \cup \{\phi_1\})$, $\mathcal{T}' \cup \{\phi_2\} \in \mathsf{MCS}^{\mathcal{S}'}_{\mathfrak{L}}(\mathcal{S} \cup \{\phi_2\})$, and $\mathcal{T} = \mathcal{T}' \cup \{\phi_1 \vee \phi_2\}$. By the supposition $\mathcal{T}', \phi_1, \mathcal{S}' \vdash \psi$ and $\mathcal{T}', \phi_2, \mathcal{S}' \vdash \psi$. Hence, $\mathcal{T}', \phi_1 \vee \phi_2, \mathcal{S}' \vdash \psi$ and so $\mathcal{T}, \mathcal{S}' \vdash \psi$.

If $\mathcal{T}'$ is $\vdash$-consistent with ϕ_1 but is not $\vdash$-consistent with ϕ_2, then $\mathcal{T}' \cup \{\phi_1\} \in \mathsf{MCS}^{\mathcal{S}'}_{\mathfrak{L}}(\mathcal{S} \cup \{\phi_1\})$, $\mathcal{T} = \mathcal{T}' \cup \{\phi_1 \vee \phi_2\}$, and $\mathcal{S}', \mathcal{T}', \phi_2 \vdash \mathsf{F}$. Thus $\mathcal{S}', \mathcal{T}', \phi_2 \vdash \psi$. By the supposition also $\mathcal{T}', \phi_1, \mathcal{S}' \vdash \psi$ and thus $\mathcal{T}', \phi_1 \vee \phi_2, \mathcal{S}' \vdash \psi$. Hence, $\mathcal{T}, \mathcal{S}' \vdash \psi$.

The case that $\mathcal{T}'$ is $\vdash$-consistent with ϕ_2 but $\vdash$-inconsistent with ϕ_1 is analogous.

Since our case distinction is exhaustive and in every case that $\mathcal{T}, \mathcal{S}' \vdash \psi$, we have $\langle \mathcal{S}', \mathcal{S} \cup \{\phi_1 \vee \phi_2\}\rangle \mid\!\sim^{\mathfrak{L}}_{\cap\mathsf{mcs}} \psi$.

We now consider the case $\sqcup = \uplus$. Suppose that $\langle \mathcal{S}' \cup \{\phi_1\}, \mathcal{S}\rangle \mid\!\sim_{\cap\mathsf{mcs}} \psi$ and also $\langle \mathcal{S}' \cup \{\phi_2\}, \mathcal{S}\rangle \mid\!\sim_{\cap\mathsf{mcs}} \psi$. Let $\mathcal{T} \in \mathsf{MCS}^{\mathcal{S}' \cup \{\phi_1 \vee \phi_2\}}_{\mathfrak{L}}(\mathcal{S})$. Thus, $\mathcal{T}$ is $\vdash$-consistent with $\phi_1 \vee \phi_2$ in the context of $\mathcal{S}'$. Then, $\mathcal{T}$ is $\vdash$-consistent with ϕ_1 or with ϕ_2. Without loss of generality suppose the former. Hence, $\mathcal{T} \in \mathsf{MCS}^{\mathcal{S}' \cup \{\phi_1\}}_{\mathfrak{L}}(\mathcal{S})$. By the supposition, $\mathcal{T}, \mathcal{S}', \phi_1 \vdash \psi$. If $\mathcal{T}$ is $\vdash$-consistent with ϕ_2 in the context of $\mathcal{S}'$, also $\mathcal{T} \in \mathsf{MCS}^{\mathcal{S}' \cup \{\phi_2\}}_{\mathfrak{L}}(\mathcal{S})$, and so $\mathcal{T}, \mathcal{S}', \phi_2 \vdash \psi$. Otherwise, $\mathcal{T}, \mathcal{S}', \phi_2 \vdash \mathsf{F}$ and thus $\mathcal{T}, \mathcal{S}', \phi_2 \vdash \psi$. In any case, since $\vee$ is a disjunction with respect to $\vdash$, it holds that $\mathcal{T}, \mathcal{S}', \phi_1 \vee \phi_2 \vdash \psi$. Thus, $\langle \mathcal{S}' \cup \{\phi_1 \vee \phi_2\}, \mathcal{S}\rangle \mid\!\sim^{\mathfrak{L}}_{\cap\mathsf{mcs}} \psi$. □

CHAPTER 13

EMPIRICAL COGNITIVE STUDIES ABOUT FORMAL ARGUMENTATION

FEDERICO CERUTTI
Department of Information Engineering, University of Brescia, Italy
federico.cerutti@unibs.it

MARCOS CRAMER
Faculty of Computer Science, TU Dresden, Germany
marcos.cramer@tu-dresden.de

MATHIEU GUILLAUME
Centre for Research in Cognition and Neurosciences, Université Libre de Bruxelles, Belgium
maguilla@ulb.ac.be

EMMANUEL HADOUX
Department of Computer Science, University College London, UK
e.hadoux@ucl.ac.uk

ANTHONY HUNTER
Department of Computer Science, University College London, UK
anthony.hunter@ucl.ac.uk

SYLWIA POLBERG
School of Computer Science and Informatics, Cardiff University, UK
polbergs@cardiff.ac.uk

Abstract

The evaluation of the adequacy of approaches to formal argumentation is often done through instantiations with other established formalisms, such as logic programming or non-monotonic logic. Furthermore, new developments are frequently motivated with examples of use cases that call for the additional features. While such evaluation approaches might be useful and technically sound, they often fail to show to what degree and under what circumstances they reflect human reasoning. In order to address this challenge, in recent years multiple empirical cognitive studies have been conducted to test the relationship between human behaviour and the formal models of abstract and structured argumentation. In this chapter we describe, compare and discuss these studies, taking into account their different methodological approaches. Furthermore we discuss their relevance and potential benefits for formal argumentation, and we review various open questions that are left for future research in this area.

1 Introduction

In the previous chapters of this handbook [Gabbay et al., 2021], formal argumentation has been introduced as a logical machinery for handling defeasible reasoning, citing examples that are palatable to human readers. However, this opens the question of whether formal argumentation should be the *prescriptive* model of defeasible reasoning, or a *descriptive* one. Each of these perspectives has its merits. Prescriptive formal argumentation is valuable in regulated settings, such as legal cases, political debates, or diagnosis support in medicine. However, there are various other settings that are much less regulated, or when only one of the involved parties has to adhere to certain guidelines or protocols. Common examples include patient-doctor interactions, such as persuading the patient to stop smoking or to finish a course of antibiotics. These call for more descriptive models, particularly when misusing prescriptive ones for such scenarios can be inefficient or even harmful [Nguyen and Masthoff, 2008]. Furthermore, formal argumentation could serve as a bridge between prescriptive and descriptive approaches to reasoning. For example, it has the potential to explain descriptively how prescriptive judgements on reasoning come to be accepted. And in the case of scenarios like patient-doctor interactions, formal argumentation has the potential to produce recommendations (conceivable as *prescriptions* of a novel kind) as to which forms of arguments are known to be persuasive.

In order to ensure that the models of formal argumentation are suited for such applications to human reasoning and to bridging the gap between prescriptive and descriptive approaches, the empirically founded descriptive methodology needs to be applied to these models. This chapter describes advances that have been made in recent years in this research thread. Based on multiple studies comparing various argumentation formalisms to actual human reasoning, this chapter will describe how these formalisms perform – according to a variety of metrics – when compared to how lay people use and evaluate arguments. The ultimate research question here, first proposed in [Rahwan et al., 2010], is to quantify the descriptive quality of formal argumentation mechanisms compared to human argumentation, as well as compared to human reasoning, which can be viewed as a special case of monological argumentation (see, for example, [Mercier and Sperber, 2011]).

We start our investigation by considering probably the most famous case of formal argumentation machinery, namely Dung's argumentation framework, originally proposed in [Dung, 1995], as well as its qualities in describing human reasoning in Section 2. Since its debut, Dung's framework has been extended in various ways in order to incorporate more facets of human reasoning, some with a particular aim of taking a step back from the prescriptive approach and offer a more descriptive perspective. Thus, we also look closer at these initiatives.

Since defeasible reasoning is, ultimately, about handling uncertainty in the world, it was only a matter of time that probabilistic extensions to Dung's argumentation framework were proposed [Li et al., 2011; Hunter, 2012; Thimm, 2012; Hunter, 2013; Hunter and Thimm, 2017]. Section 3.1 shows the results of empirical studies using them as descriptive accounts of of human reasoning.

Dung's framework owns a part of its popularity to its simplicity as it considers arguments as atomic entities, and permits only the attack relation between them. An important subarea of argumentation expands on this by considering additional kinds of interactions that can happen between arguments. The most popular in this regard are works that distinguish between attacks and supports [Cayrol and Lagasquie-Schiex, 2013]: in Section 3.2 we review studies investigating the descriptive quality of bipolar argumentation frameworks.

Dung's argumentation frameworks consider arguments as atomic entities. However, this is not always adequate, because the arguments that humans produce can have an internal structure. The connection between the internal structure of arguments and the attack relation assumed in Dung's argumentation framework is studied by the formalisms of structured argumentation. In Section 4 we report studies that the connection of these formalisms of structured argumentation, such as ASPIC [Prakken, 2010], to human reasoning.

The descriptive approach has been applied to multiple formalisms of human reasoning, argumentation and persuasion, not just to formalisms that lie strictly within the bounds of formal argumentation as defined by the first and the present volume of the Handbook of Formal Argumentation. While this work is outside the scope of this chapter, there are some relevant connections to the scope of this chapter. Therefore we have included an extensive discussion of related work in Section 5, namely on cognitive biases in logical reasoning tasks (Section 5.1), non-monotonic reasoning (Section 5.2), persuasion (Section 5.3), emotions (Section 5.4), argumentation schemes (Section 5.5), Bayesian argumentation (Section 5.6), and argumentation-based judgment aggregation (Section 5.7).

In Section 6, we conclude the chapter with a brief discussion of the commonalities of the studies considered in this chapter, which lie mostly in the shared methodological approach that has the potential to be developed further and become more relevant to research in formal argumentation in the future.

2 Human Reasoning and Dung Frameworks

In his seminal paper, Dung [1995] introduced abstract argumentation as a method for giving a unified account of multiple approaches in non-monotonic logic as well as of some problems from other areas. This method was explicitly motivated by the reference to how humans evaluate the acceptability of an argument based on the evaluation of all potential counterarguments. At this point, the only connection between abstract argumentation and actual human argumentation was this motivational link. But as computational argumentation established itself as a subfield of AI, more and more researchers began to apply the methodology of abstract argumentation in a way that presupposed the existence of some viable connections to actual human reasoning, e.g. for developing tools that support humans in the organization, evaluation or production of arguments, see [Cerutti *et al.*, 2018] for a survey.

This development naturally gave rise to the research question – first made explicit by Rahwan *et al.* [2010] – whether the approach of abstract argumentation has any definite connections to actual human argumentation or reasoning that could be measured through cognitive empirical studies. In the meantime multiple studies have approached this research question. Some of these works have researched the connections between actual human reasoning and the notion of *argumentation framework* as introduced by Dung [1995]. Others have explored the connections between human reasoning and some of the various extensions of Dung's original notion of argumentation frameworks. In the current section, we focus on studies of the first kind by giving an overview over their findings and over what still needs to be explored in the future. The works of the second kind will be considered in the next section.

2.1 Preliminaries About Dung Frameworks

In this section we define certain background notions from abstract argumentation theory as introduced by Dung [Dung, 1995] and as explained in its current state-of-the-art form by Baroni *et al.* [Baroni *et al.*, 2018].

Definition 2.1. *A Dung framework, also called* argumentation framework *of* AF*, is a finite directed graph* $AF = \langle Ar, att \rangle$ *in which the set* Ar *of vertices is considered to represent arguments and the set* att *of edges is considered to represent the attack relation between arguments, i.e. the relation between a counterargument and the argument that it attacks.*

Given an argumentation framework, we want to choose the sets of arguments for which it is rational and coherent to accept them together. A set of arguments that may be accepted together is called an *extension*. Multiple *argumentation semantics* have been defined in the literature, i.e. multiple different ways of defining extensions given an argumentation framework. Before we consider specific argumentation semantics, we first give a formal definition of the notion of *argumentation semantics*:

Definition 2.2. *An* argumentation semantics *is a function σ that maps any AF $AF = \langle Ar, att \rangle$ to a set $\sigma(AF)$ of subsets of Ar. The elements of $\sigma(AF)$ are called σ-extensions of AF.*

Remark 2.3. *We usually define an argumentation semantics σ by specifying criteria which a subset of Ar has to satisfy in order to be a σ-extension of AF.*

In this chapter we consider the *complete, grounded, preferred, semi-stable, stable, stage, CF2, stage2* and *SCF2 semantics*. The first five are based on the notion of *admissibility* and are therefore called *admissibility-based semantics*. The last five always choose extensions that are *naive extensions*, i.e. maximal conflict-free sets of arguments, which is why they are called *naive-based semantics*. Note that the stable semantics is the only semantics that belongs to both categories (at the price of not providing any extension at all in some scenarios). Apart from these nine semantics, we also define *naive extensions* and *SCOOC-naive extensions*, as we need them for our definition of CF2 and SCF2 semantics respectively.

Of the nine semantics defined in this section, SCF2 is the one that has been most recently introduced in the literature [Cramer and van der Torre, 2019], and is thus the only one that is not covered in [Baroni et al., 2018]. Since it is the least well-known of the semantics considered in this section, we provide some intuitions about it: SCF2 semantics is based on the principle of *Strong Completeness Outside Odd Cycles*, abbreviated *SCOOC*. Informally, the SCOOC principle says that if an argument a and its attackers are not in an odd cycle, then an extension not containing any of a's attackers must contain a. The principle is based on the idea that it is generally desirable that an argument that is not attacked by any argument in a given extension should itself be in that extension. While it is possible to ensure this generally desirable property in AFs without odd cycles, this is not the case for AFs involving an odd cycle. The idea behind the SCOOC principle is to still satisfy this property as much as possible, i.e. whenever the argument under consideration and its attackers are not in an odd cycle. The SCF2 semantics is defined in a similar way as the already well-known CF2 semantics, with the difference being that the SCOOC principle is ensured to be satisfied in each strongly connected component; Cramer and van der Torre [2019] have shown that this way the SCOOC principle turns out to be also satisfied globally.

Definition 2.4. *An att-path is a sequence* $\langle a_0, \ldots, a_n \rangle$ *of arguments where*
$(a_i, a_{i+1}) \in att$ *for* $0 \leq i < n$ *and where* $a_j \neq a_k$ *for* $0 \leq j < k \leq n$ *with either* $j \neq 0$ *or* $k \neq n$. *An* odd att-cycle *is an att-path* $\langle a_0, \ldots, a_n \rangle$ *where* $a_0 = a_n$ *and* n *is odd.*

Definition 2.5. *Let* $AF = \langle Ar, att \rangle$ *be an AF, and let* $S \subseteq Ar$. *We write* $AF|_S$ *for the restricted AF* $\langle S, att \cap (S \times S) \rangle$. *The set* S *is called* conflict-free *iff there are no arguments* $b, c \in S$ *such that* b *attacks* c *(i.e. such that* $(b, c) \in att$). *Argument* $a \in Ar$ *is* defended *by* S *iff for every* $b \in Ar$ *such that* b *attacks* a *there exists* $c \in S$ *such that* c *attacks* b. *We define* $S^+ = \{a \in Ar \mid S \text{ attacks } a\}$ *and* $S^- = \{a \in Ar \mid a \text{ attacks some } b \in S\}$. *We define* S *to be* strongly complete outside odd cycles *iff for every argument* $a \in Ar$, *if no argument in* $\{a\} \cup \{a\}^-$ *is in an odd att-cycle and* $S \cap \{a\}^- = \emptyset$, *then* $a \in S$.

- S *is a* complete extension *of AF iff it is conflict-free, it defends all its arguments and it contains all the arguments it defends.*

- S *is a* stable extension *of AF iff it is conflict-free and it attacks all the arguments of* $Ar \setminus S$.

- S *is the* grounded extension *of AF iff it is a subset-minimal complete extension of AF.*

- S *is a* preferred extension *of AF iff it is a subset-maximal complete extension of AF.*

- S *is a* semi-stable extension *of AF iff it is a complete extension and there exists no complete extension* S_1 *such that* $S \cup S^+ \subset S_1 \cup S_1^+$.

- S *is a* stage extension *of AF iff* S *is a conflict-free set and there exists no conflict-free set* S_1 *such that* $S \cup S^+ \subset S_1 \cup S_1^+$.

- *S is a* naive extension *of AF iff S is a subset-maximal conflict-free set.*

- *S is a SCOOC-naive extension iff S is subset-maximal among the conflict-free subsets of Ar that are strongly complete outside odd cycles.*

The idea behind CF2, stage2 and SCF2 semantics is that we partition the AF into *strongly connected components* and recursively evaluate it, component by component, using a procedure called the *simplified SCC-recursive scheme*. For defining this scheme, we first need some auxiliary notions:

Definition 2.6. *Let $AF = \langle Ar, att \rangle$ be an AF, and let $a, b \in Ar$. We define $a \sim b$ iff either $a = b$ or there is an att-path from a to b and there is an att-path from b to a. The equivalence classes under the equivalence relation $\sim$ are called* strongly connected components *(SCCs) of AF. We denote the set of SCCs of AF by* SCCS_{AF}. *Given $S \subseteq Ar$, we define $D_F(S) := \{b \in Ar \mid \exists a \in S : (a, b) \in att \wedge a \not\sim b\}$.*

Definition 2.7. *Let σ be an argumentation semantics. The argumentation semantics $scc(\sigma)$ is defined as follows. Let $AF = \langle Ar, att \rangle$ be an AF, and let $S \subseteq Ar$. Then S is an $scc(\sigma)$-extension of AF iff either*

- $|\text{SCCS}_{AF}| = 1$ *and S is a σ-extension of AF, or*
- $|\text{SCCS}_{AF}| > 1$ *and for each $C \in \text{SCCS}_{AF}$, $S \cap C$ is an $scc(\sigma)$-extension of $AF|_{C \backslash D_A F(S)}$.*

Definition 2.8. *We define* CF2, stage2 *and* SCF2 *semantics as follows:*

- CF2 semantics *is defined to be scc(naive).*

- stage2 *semantics is defined to be scc(stage).*

- *Given an AF $AF = \langle Ar, att \rangle$, a set $S \subseteq Ar$ is called a SCF2 extension of AF iff S is a scc(SCOOC-naive)-extension of $AF|_{Ar'}$, where $Ar' := \{a \in Ar \mid (a, a) \notin att\}$.*

Most argumentation semantics allow for the possibility of multiple extensions, so that the status of an argument depends on the choice of extension. For some purposes a single status for each argument is needed. One way to do this is through the notion of a *justification status* as defined in [Wu and Caminada, 2010] and [Baroni *et al.*, 2018], whose terminology we follow where possible. Here we focus on the *strongly accepted*, *strongly rejected* and *weakly undecided* justification statuses:

Definition 2.9. *Let $AF = \langle Ar, att \rangle$ be an AF, let σ be an argumentation semantics such that AF has at least one σ-extension, and let $a \in A$ be an argument. We say that a is strongly accepted with respect to σ iff for every σ-extension E of F, $a \in E$. We say that a is strongly rejected with respect to σ iff for every σ-extension E of F, some $b \in E$ attacks a. We say that a is weakly undecided iff it is neither strongly accepted nor strongly rejected.*

2.2 Empirical Cognitive Studies About Dung Frameworks

The argumentation semantics that have been proposed in the literature share some features while they differ in other respects. One feature that all major argumentation semantics have in common is the way in which they treat *simple reinstatement*, namely the fact that they all give the justification status *strongly accepted* to argument a in the AF depicted in Figure 1. One feature on which various semantics differ is the way they treat *floating reinstatement* and *3-cycle reinstatement*, i.e. the justification status that they give to argument d in the AF depicted in Figure 2 and to argument h in the AF depicted in Figure 3: Argument d is weakly undecided with respect to the grounded semantics and the complete semantics, but is strongly accepted with respect to the other seven semantics defined above. Argument h is weakly undecided with respect to the grounded, complete, preferred and semi-stable semantics, but is strongly accepted with respect to the CF2, stage, stage2 and SCF2 semantics. (In stable semantics, the AF depicted in Figure 3 has no extension, so that the notion of a justification status cannot be meaningfully applied.)

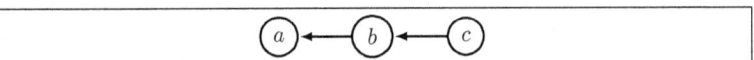

Figure 1: Simple reinstatement of argument a.

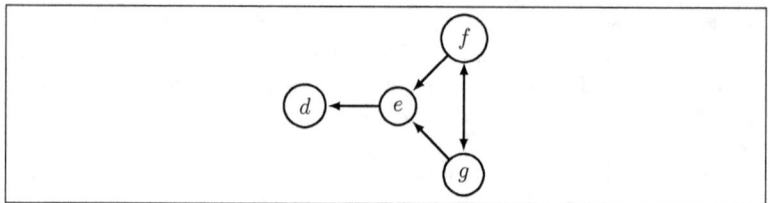

Figure 2: Floating reinstatement of argument d.

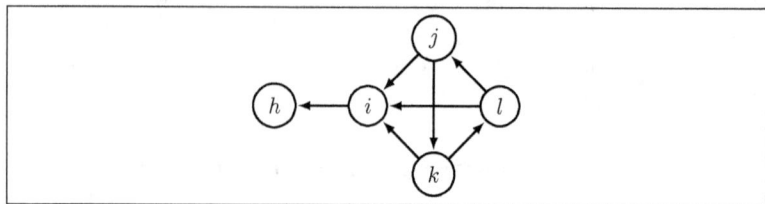

Figure 3: 3-cycle reinstatement of argument h.

This observation gives rise to two research questions with respect to the connection between actual human reasoning and abstract argumentation:

1. Do the features that all major argumentation semantics have in common (e.g. simple reinstatement) correspond to some cognitively real feature of human reasoning?

2. Which argumentation semantics can best predict human evaluation of arguments?

The first two studies that approached these research questions were performed by Rahwan et al. [2010]. The specific goal of their studies was to test how humans evaluate simple reinstatement and floating reinstatement. Their two studies involved 20 and 47 participants that were randomly approached in offices and public spaces in Dubai. Participants were shown between one and four natural language arguments and asked to assess the conclusion of the highlighted argument, using a seven-point Likert scale anchored at *certainly false* and *certainly true*. The natural language arguments were designed to correspond to the arguments in the simple reinstatement or floating reinstatement AFs depicted in Figures 1 and 2 respectively. Some participants were only shown a part of those arguments. The highlighted argument that participants had to judge always corresponded to argument a or d in these two AFs.

Here is an example of an argument set used in Rahwan *et al.*'s studies as a natural language analogue of the simple reinstatement AF depicted in Figure 1:

(a) The battery of Alex's car is not working. Therefore, Alex's car will halt.

(b) The battery of Alex's car has just been changed today. Therefore, the battery of Alex's car is working.

(c) The garage was closed today. Therefore, the battery of Alex's car has not been changed today.

The results of their study suggest that confidence in the conclusion of an argument is highest in the case of an unattacked argument, lowest in the case of an argument attacked by an unattacked argument, and takes a medium value in the case of a reinstated argument, i.e. an argument that is attacked by an attacked argument. The study could find no difference between the confidence in the conclusion of a simply reinstated argument as compared to the confidence in the conclusion of a floating-reinstated argument.

Concerning the two research questions stated above, these results can be interpreted as follows:

1. The first research question gets a partially positive response. The simple reinstatement of an argument increased the confidence in the conclusion of an argument compared to the case when the argument was attacked by an unattacked argument, as is suggested by all major argumentation semantics. But this response is only partially positive, because the confidence in that conclusion did not rise back to the level of an unattacked argument, something that cannot be explained with any of the major extension-based argumentation semantics.

2. Since all standard semantics agree on the evaluation of simple reinstatement, only the results on floating reinstatement could distinguish between different semantics. Therefore only limited claims could be made concerning the second research question about which semantics best predicts human judgments. The fact that floating reinstatement is treated in the same way as simple reinstatement suggests that grounded and complete semantics are worse at predicting human behaviour than the other seven argumentation semantics defined above.

Being the first study to investigate the cognitive plausibility of the formalisms from argumentation theory, it laid the foundations for further work in this area. Cramer and Guillaume [Cramer and Guillaume, 2018b; Cramer and Guillaume, 2019] performed two further studies that expanded Rahwan *et al.*'s. One of the stated aims of these studies was to overcome some limitations of Rahwan *et al.*'s methodology.

For example, Rahwan *et al.* [2010] did not empirically test their assumption that the natural language argument sets that they designed actually correspond to the intended AFs. This limitation is especially pressing in light of the fact that the attacks that they intended to be unidirectional were based on conflicts between the conclusion of the attacking argument and the premise of the attacked argument, without any indication of a preference. In the frameworks of structured argumentation from the ASPIC family [Modgil and Prakken, 2018; Caminada *et al.*, 2014], such underminings without preferences always give rise to bidirectional attacks. An empirical study that we discuss in Section 4.4 has confirmed that humans are more likely to interpret such underminings without preferences as bidirectional attacks than as unidirectional attacks [Cramer and Guillaume, 2018a].

To overcome this limitation of Rahwan *et al.*'s study, Cramer and Guillaume first performed a study about the directionality of attacks in natural language argumentation [Cramer and Guillaume, 2018a]. Since this study compares human reasoning to predictions of structured argumentation frameworks like ASPIC+ [Modgil and Prakken, 2018], it is discussed in detail in Section 4.4 rather than here. For the current purpose, it is enough to explain that this study introduced the notion of an *attack type* between natural language arguments and discovered that some attack types are systematically interpreted as unidirectional attacks, while others are mostly interpreted as bidirectional attacks, and a third class of attack types is interpreted as a unidirectional attack by some participants and as a bidirectional attack by others participants. For their two subsequent studies on the connection between human reasoning and abstract argumentation, Cramer and Guillaume used the attack types of the first two kinds, as they had been confirmed to have a certain stable interpretation concerning the directionality of the attack relation between them.

For their first study, Cramer and Guillaume [2018b] used the findings of their prestudy on the directionality of attacks in natural language argumentation [Cramer and Guillaume, 2018a] to design the sets of three to five natural language arguments that correspond to the simple reinstatement AF, the floating reinstatement AF or the 3-cycle reinstatement AF. Including the 3-cycle reinstatement AF allowed them to distinguish some semantics that Rahwan *et al.*'s study could not distinguish. The study involved arguments based on three different thematic contexts: arguments based on news reports, arguments based on scientific publications, and arguments based on the precision of a calculation tool. As an example, here is the argument set of the scientific context corresponding to the floating reinstatement AF depicted in Figure 2:

(d) Specimen A consists only of amylase. The 1972 Encyclopaedia of Biochemistry states that amylase is an enzyme. So specimen A consists of an enzyme.

(e) A peer-reviewed research article by Smith et al. from 2006 presented new findings that amylase is not an enzyme. Therefore no specimen consisting only of amylase consists of an enzyme.

(f) A study that the Biology Laboratory of Harvard University has published in 2011 corrects mistakes made in the study by Smith et al. and concludes that amylase is a biologically active enzyme.

(g) A study that the Biochemistry Laboratory of Oxford University has published in 2011 corrects mistakes made in the study by Smith et al. and concludes that amylase is a biologically inactive enzyme.

The study was conducted with 130 undergraduate students from the University of Luxembourg. Participants were first asked to draw the attack relation between the given arguments and then to assess the acceptability of each argument by indicating either that they *accept* the argument, that they *reject* it, or that they consider it *undecided*. The limitation to three possible responses instead of a seven-point Likert scale as in Rahwan *et al.*'s study was justified by the fact that this allows for a direct comparison of human responses with the three justification statuses of arguments that we defined at the end of Section 2.1.

For both tasks, a group discussion methodology was applied to stimulate more rational thinking: Participants first responded to the task individually, next they collaboratively discussed their responses with their peers, and finally they provided an updated individual response.

The results of this study suggest that human judgements about simple reinstatement are in line with the predictions of all major semantics, thus providing a positive response to the first research question for the case of simple reinstatement. Concerning the second research question, the study suggests that CF2, stage, stage2 and SFC2 semantics predict human behaviour best, as they were the only semantics that could predict the majority responses for all arguments in all three AFs considered in this study. Preferred and semi-stable semantics fail to predict human responses in the case of 3-cycle resinstatement, while grounded and complete semantics fail to predict human responses for both floating reinstatement and 3-cycle reinstatement. (Stable semantics is disregarded here, as it does not make a meaningful prediction for 3-cycle reinstatement.)

In a second study involving 61 undergraduate students, Cramer and Guillaume [2019] modified their methodology in order to be able to study human assessments of twelve different AFs of three to eight arguments each. For this purpose they designed a fictional scenario in which arbitrary argumentation frameworks could be constructed in a uniform way. This allowed them to include enough different and sufficiently complex AFs to distinguish between all major argumentation semantics.

The arguments were set in the following fictional context: participants were located on an imaginary island, faced with conflicting information coming from various islanders, and they had to evaluate the arguments provided in order to hopefully find the location(s) of the buried treasure(s). All the attacks between the arguments were based on information that a certain islander is not trustworthy. As an example, here is the argument set corresponding to the floating reinstatement AF depicted in Figure 2:

(d) Islander Olivia says that there is a treasure buried near the eastern tip of the island. So we should dig up the sand near the eastern tip of the island.

(e) Islander Neil says that islander Olivia is not trustworthy and that there is a treasure buried between the two oak trees. So we should not trust what Olivia says, and we should dig up the sand between the two oak trees.

(f) Islander Lisa says that islander Mila and islander Neil are not trustworthy and that there is a treasure buried on the peak of the mountain. So we should not trust what Mila and Neil say, and we should dig up the sand on the peak of the mountain.

(g) Islander Mila says that islander Lisa and islander Neil are not trustworthy and that there is a treasure buried next to the old wall. So we should not trust what Lisa and Neil say, and we should dig up the sand next to the old wall.

In this study, the notion of an attack relation was explained in advance to participants and the intended attack relation was shown to them together with the natural language arguments, in order to ensure that participants do not overlook attacks in the case of the more complex argumentation frameworks. As in the first study, the participants assessed arguments in a three-valued way (*accept, reject* or *undecided*).

The results of this second study suggest that SCF2, CF2 and grounded semantics are significantly better at predicting human judgements than preferred, semi-stable, stage and stage2 semantics. The differences between SCF2, CF2 and grounded were not significant in this study. (Again, stable semantics is disregarded here, because there were multiple AFs with no stable extension.)

Note the apparent mismatch between the results of the three studies with respect to the grounded semantics. While the grounded semantics was not a good predictor of human reasoning in the first two studies, it was among the three best predictors in the third study. Cramer and Guillaume [2019] explain this apparent mismatch by pointing out that their second study used more complex argumentation frameworks, which made the reasoning task cognitively more challenging and therefore led to more participants making use of the simplifying strategy of choosing *undecided* whenever there is some reason for doubt.[1]

2.3 Outlook on Human Reasoning and Dung Frameworks

Considering the results of the three studies in this area together, the two research questions introduced above can be partially answered as follows:

1. *Do the features that all major argumentation semantics have in common (e.g. simple reinstatement) correspond to some cognitively real feature of human reasoning?*

   This research question has only been addressed for the case of simple reinstatement, not for other features that the major argumentation semantics have in common. The existing studies suggest that the way simple reinstatement is treated by all the major argumentation semantics does indeed correspond to some cognitively

---

[1] One possible way in which this explanation hypothesized by Cramer and Guillaume could be empirically tested is by designing a study in which participants have four instead of three possible responses for each argument:

1. "There are convincing reasons for accepting the argument."
2. "There are convincing reasons for rejecting the argument."
3. "There are both reasons for accepting the argument as well as reasons for rejecting it, and the information provided is not enough to decide which of these reasons should be preferred."
4. "I don't know which of these three responses is most rational."

This way the ambiguity of the *undecided* response in Cramer and Guillaume's studies is removed, because the participants have to decide whether they are making an informed judgment about the undecidedness of the argument or whether they are unsure what the correct answer is. If Cramer and Guillaume's hypothesized explanation is correct, complex argumentation frameworks should give rise to more frequent "don't know" responses, whereas informed undecidedness (response 3) should be better predicted by SCF2 or CF2 semantics than by grounded semantics.

real feature of human reasoning. However, the fact that a simply reinstated argument is treated in the same way as an unattacked argument might be an oversimplification.

2. *Which argumentation semantics can best predict human evaluation of arguments?*

   Taken together, the results suggest that SCF2 and CF2 semantics are the best predictors of human evaluation of arguments. For complex argumentation frameworks, the grounded semantics is also a good predictor.

Given that the responses to these two research questions are based only on three studies, all of which have some limitations in their methodology, they should be considered preliminary answers that might have to be updated by future studies.

We see two main avenues for future research in this area: On the one hand, we could attempt to design empirical studies in such a way that their findings can be compared to the set of extensions provided by each semantics rather than just to the justification status of each argument. One way this could be done is by showing participants a set of arguments some of which are highlighted, and to ask the participants whether it would be rational to accept the highlighted arguments together while not accepting any other argument from the set. Another avenue for future research is to broaden our perspective on the first research question by considering further features that all major argumentation semantics share, e.g. that the number of attackers of an argument has no impact on the status of that argument, as long as all attackers are of the same kind (e.g. all of them are unattacked, or all of them are attacked in an equivalent way).

3 Human Reasoning and Extended Frameworks of Formal Argumentation

3.1 Empirical studies about probabilistic argumentation

Argumentation is subject to various kinds of uncertainty that can arise due to imperfections of the agents involved in a given situation, incompleteness of the available information, the types of arguments we have at hand, and much more. This can lead to, for instance, doubts concerning the structure of the graph, acceptance of arguments, or how these change when we use argumentation in dialogues and dynamic settings. One of the prominent approaches for handling such lack of confidence is probabilistic argumentation, which often provides the means of quantifying the level of uncertainty we are dealing with. The two most prominent approaches within this area are the constellations approach and the epistemic approach [Hunter, 2013], discussed in more detail in Chapter 7 of this handbook [Hunter et al., 2021]:

Constellations approach It is based on a probability distribution over the subgraphs of the argument graph ([Hunter, 2012] which extends [Dung and Thang, 2010] and [Li et al., 2011]), and can be used to represent the uncertainty over the structure of the graph (i.e. whether a particular argument or attack appears in the argument graph under consideration).

Epistemic approach It is based on a probability distribution over the subsets of the arguments [Thimm, 2012; Hunter, 2013; Hunter and Thimm, 2017]. It can be used to represent the uncertainty over which arguments are believed to be accepted. The epistemic approach can be constrained (using axioms or postulates) to be consistent with Dung's semantics (see Section 2.1), but it can also be used as a potentially valuable alternative to Dung's dialectical semantics [Thimm, 2012; Hunter, 2013].

A further approach is based on labellings for arguments using *in*, *out*, and *undecided*, from [Caminada and Gabbay, 2009], augmented with *off* for denoting that the argument does not occur in the graph [Riveret and Governatori, 2016]. A probability distribution over labellings can be used to give a form of probabilistic argumentation that overlaps with the constellations and epistemic approaches.

Hence, there are some interesting proposals for bringing probability theory into argumentation. However, empirical verification of probabilistic argumentation is an open research question and in this section we will discuss the relevant work in that area.

3.1.1 Flu Vaccine Study

In order to investigate the real-world plausibility of the constellations and epistemic approaches, Polberg and Hunter [Polberg and Hunter, 2018] undertook an exploratory study involving two dialogues concerning flu shots (one of which is presented in Table 1). The dialogues were created using statements found on the NHS and CDC websites (information sections as well as FAQ) and anti-vaccine forums, so based on information that was prepared for or widely available to the public. 40 responses were gathered per dialogue[2] using crowd-sourcing techniques. This means that the participants were members of the general public, not argumentation specialists. The dialogues proceeded in steps and after each step, the participants were given the following tasks:

Agreement The participants were asked to state how much they agree or disagree with a given statement. They were allowed to choose one of the seven options (*Strongly Agree, Agree, Somewhat Agree, Neither Agree nor Disagree, Somewhat Disagree, Disagree, Strongly Disagree*) or select the answer *Don't Know*.

Explanation The participants were then asked to explain the chosen level of agreement for every statement, especially any reasons for disagreement that had not been mentioned in the dialogue.

Relation The participants were asked to state how they viewed the relation between the statements. For every listed pair, they could say whether one statement was *A good reason against, A somewhat good reason against, Somewhat related, but can't say how, A somewhat good reason for, A good reason for* the other statement or select the answer *N/A* (i.e. that the statements were unrelated).

Awareness The participants were asked which of the presented statements they had been familiar with prior to the experiment. This task was given only after the last step of the dialogue.

The answers provided by the participants were then used to analyze the argument graphs generated from the responses, whether the

---
[2]We note that the original number of responses was much greater, but a portion of participants has been disqualified due to failed language and attention checks used in the experiment

Steps	Person		Statement
1 to 5	P1	A	Hospital staff members do not need to receive flu shots.
1 to 5	P2	B	Hospital staff members are exposed to the flu virus a lot. Therefore, it would be good for them to receive flu shots in order to stay healthy.
2 to 5	P1	C	The virus is only airborne and it is sufficient to wear a mask in order to protect yourself. Therefore, a vaccination is not necessary.
3 to 5	P2	D	The flu virus is not just airborne, it can be transmitted through touch as well. Hence, a mask is insufficient to protect yourself against the virus.
4 to 5	P1	E	The flu vaccine causes flu in order to gain immunity. Making people sick, who otherwise might have stayed healthy, is unreasonable.
5	P2	F	The flu vaccine does not cause flu. It only has some side effects, such as headaches, that can be mistaken for flu symptoms.

Table 1: A five-step dialogue between persons P1 and P2. This exchange starts with P1 claiming that hospital staff do not need to receive flu shots, to which P2 objects. The two counterarguments of P1 are then defeated by P2. The table presents at which steps a given statement was visible, who uttered it and what was its content.

agreement or disagreement with the statements adhered to epistemic postulates, effect of the agreement with an argument on the relations carried out by it and vice versa, and changes in opinions on arguments occurring during the dialogue.

In what follows we discuss how the findings affect the constellation and epistemic approaches.

3.1.2 Reflections on the Constellation Approach

The dialogues in the flu vaccine study proceeded in steps, by which we understand that at every step one or more arguments were added to the existing ones. At every such stage, the users stated how they viewed relations between the visible arguments. The *Relation* task answers were used to construct the graphs declared by the participants which were then compared between each other as well as to the intended graph for each dialogue stage, i.e. the graph depicting the minimal set of relations the authors have considered reasonable for a given set of arguments. Examples of such graphs can be seen in Figure 4. The similarities and disparities between all these graphs allow us to draw some conclusions for the constellation approach.

The results of the study show that in general, the most common graph (i.e. the graph that was declared by the greatest number of participants) of a given dialogue stage contained the intended graph for that stage. However, the most common graphs were not the only graph produced by the participants and there was still a significant portion of people that were of different opinions. As visible in Table 2, it was seen that people may interpret the statements and the relations between them differently and without adhering to the intended relations. Furthermore, as was in various cases made apparent by the answers provided in the *Explanation task*, their personal knowledge can affect their perception and evaluation of the dialogue.

There is therefore some uncertainty as to how people view relationships between arguments, and these different views can affect how they behave during a given argument exchange. The very purpose of the constellation approach is to be able to model such uncertainties concerning the topology of the argument graphs. Hence, the data from the study supports the use of the constellation approach to probabilistic argumentation for modelling the argument graphs representing the views of dialogue participants.

3.1.3 Reflections on the Epistemic Approach

The answers provided by the participants in the *Agreement* task were used to create belief distributions corresponding to these answers (e.g.

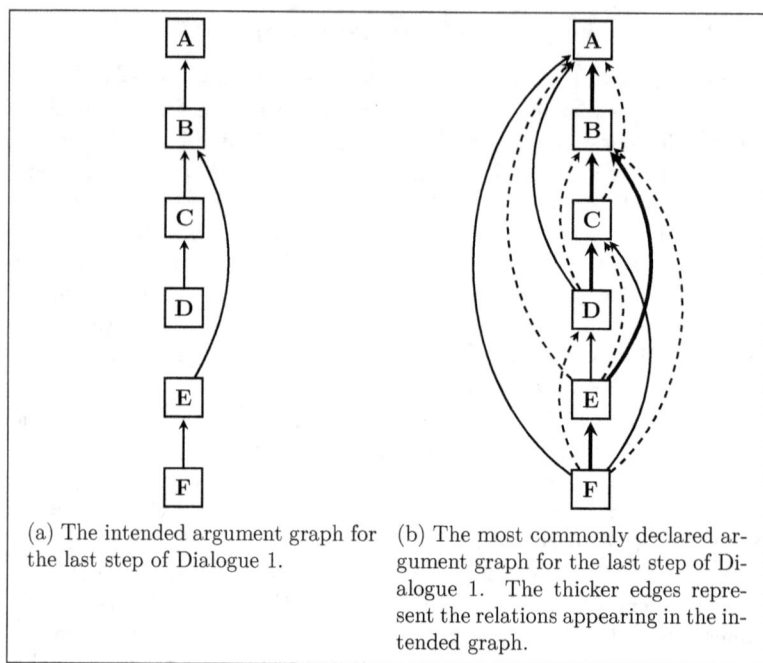

(a) The intended argument graph for the last step of Dialogue 1.

(b) The most commonly declared argument graph for the last step of Dialogue 1. The thicker edges represent the relations appearing in the intended graph.

Figure 4: The intended and the common graphs for the last step of Dialogue 1 from Table 1. Solid edges stand for attack and dashed for support.

Strong agreement response was mapped to belief of 1, representing complete belief). These were later evaluated in terms of the adherence to the epistemic postulates and of the number of different degrees of belief the participants used.

Let us start with the epistemic postulates (see also Chapter 7 of this handbook [Hunter *et al.*, 2021]). While classical semantics tend to represent a number of properties at the same time, a single postulate tends to focus on a single aspect. They therefore allow a more detailed view on the participant behaviour and can allow us to analyze the cases in which classic semantics may fail to explain it, thus providing more feedback to argumentation-based systems than classical semantics do.

Relation	Attacking	Supporting	Dependent	N/A
(B, A)	60.50	31.50	8	0
(C, A)	29.38	63.75	5.63	1.25
(C, B)	68.75	20	7.50	3.75
(D, A)	53.33	30.83	15	0.83
(D, B)	10	84.17	5.83	0
(D, C)	66.67	25	8.33	0
(E, A)	30	45	16.25	8.75
(E, B)	53.75	16.25	21.25	8.75
(E, C)	28.75	43.75	20	7.5
(E, D)	53.75	10	27.50	8.75
(F, A)	27.50	30	40	2.5
(F, B)	10	62.50	25	2.5
(F, C)	30	27.50	37.50	5
(F, D)	0	55	37.50	7.5
(F, E)	52.50	22.50	20	5

Table 2: Occurrences of the declared relations in Dialogue 1 (values are expressed as %)

The average adherence to several epistemic postulates in both dialogues can be found in Table 3. Due to their relationship with the Dung's semantics, these results offer insight into classical semantics as well. For instance, the sets of believed arguments (i.e. those with probability greater than 0.5) from rational distributions correspond to the conflict-free sets of a given argument graph. We observe that between the two dialogues, this property is generally satisfied but there is still a notable portion of participants that can at the same time accept (believe) two arguments that they perceive as conflicting. Distributions that satisfy the protective, strict, discharging and trusting postulates relate to complete extensions. In this regard, the relatively low performance of the trusting postulate (it achieved satisfaction rates of 33.5% and 43.5%) - which ensures that arguments whose attackers are all disbelieved (rejected), are believed (accepted) - highlights that participants were not eager to agree with statements just because they had no reason to disbelieve them. The performance of the discharging postulate shows that participants may disbelieve a given argument without believing any of its attackers. This is particularly important as the study data (e.g. responses in the *Explanation* task) shows that people use their own personal knowledge in order to make judgments and might not necessarily disclose it. These results suggest that epistemic postulates can provide valuable insights into human behaviour in greater detail than Dung's dialectical semantics.

Another important observation concerns the fact that the classical three-valued semantics would be insufficient to express the opinions of the majority of the participants. Most of them needed four values or more to express their beliefs and the statistical tests have shown that the choices they have made were not random. Furthermore, the changes in belief observed in the study (i.e. changes of opinions about the arguments visible between different dialogue stages) were rather subtle and observable on a fine-grained level. Not many participants changed their polarity completely (i.e. moved from agreement to disagreement or vice versa). However, some included additional clarifications in the *Explanation* task stating that while their opinions have changed for the better (or worse), they were still not sufficiently convinced to really abandon their original views.

We note that the above observation provides support for methods modelling fine-grained argument acceptability in general, not specifically just for the epistemic approach. This observation has also been supported by a non-probabilistic study carried out by Rahwan et al [Rahwan et al., 2010] that was described in Section 2.2. While it focuses on the issue of reinstatement, the results show that the level of agreement with a given argument **A** decreases once it is defeated and increases when it is defended, though still remaining significantly lower than prior to the defeat. This study, similarly to ours, lends support to the use of more fine-grained approaches towards describing the beliefs of the participants. However, as we can observe, the dialogues used in this study were much simpler and shorter than ours, and unlikely to be affected by any subjective views of the participants.

Postulate	Definition	Dialogue 1	Dialogue 2
Coherent	if for every $A, B \in \mathcal{A}$ s.t. $(A, B) \in \mathcal{R}$, $P(A) \leq 1 - P(B)$	35.5%	25%
Discharging	if for every $B \in \mathcal{A}$, if $P(B) < 0.5$ then there exists an argument $A \in \mathcal{A}$ s.t. $(A, B) \in \mathcal{R}$ and $P(A) > 0.5$	49%	60%
Founded	if $P(A) = 1$ for every initial $A \in \mathcal{A}$	23%	14%
Protective	if for every $A, B \in \mathcal{A}$ s.t. $(A, B) \in \mathcal{R}$, $P(B) > 0.5$ implies $P(A) < 0.5$	66%	49.5%
Rational	if for every $A, B \in \mathcal{A}$ s.t. $(A, B) \in \mathcal{R}$, $P(A) > 0.5$ implies $P(B) \leq 0.5$	69.5%	74%
SemiFounded	if $P(A) \geq 0.5$ for every initial $A \in \mathcal{A}$	56.5%	71.5%
SemiOptimistic	if $P(A) \geq 1 - \sum_{B \in \{A\}^-} P(B)$ for every $A \in \mathcal{A}$ that is not initial	78%	84%
Strict	if for every $A, B \in \mathcal{A}$ s.t. $(A, B) \in \mathcal{R}$, $P(A) > 0.5$ implies $P(B) < 0.5$	69.5%	61%
Trusting	if for every $B \in \mathcal{A}$, if $P(A) < 0.5$ for all $A \in \mathcal{A}$ s.t. $(A, B) \in \mathcal{R}$, then $P(B) > 0.5$	43.5%	33.5%

Table 3: Average postulate satisfaction rates by the participants in the dialogues based on the graphs obtained from the *Relation* and *Explanation* tasks. See [Polberg and Hunter, 2018] for full list of postulates considered.

3.2 Empirical Studies About Bipolar Argumentation

One of the most prominent approaches to extending Dung's argumentation frameworks come in the form of bipolar argumentation models, which incorporate various kinds of support relations. Chapter 1 of this handbook [Cayrol *et al.*, 2021] contains a deep overview of these formalisms. In this section we will discuss the findings of several studies that look at support relations in the context of studies with participants.

3.2.1 Flu Vaccine Study

The study by [Polberg and Hunter, 2018] that we discussed in the previous section also produced empirical observations concerning bipolar argumentation.

In evaluating the responses from the participants, it was observed that the participants explicitly viewed certain relations as supporting. Furthermore, it was shown that the notion of defence does not account for all of the positive relations that the participants have identified between the presented statements. In particular, it was observed that there are new support relations arising in the context of the dialogue, such as support coming from statements working towards the same goal. By analysing the common graphs (i.e. the graphs declared by the highest number of participants at given steps in the dialogue - see also Section 3.1), most of the support relations that were identified by the participants can be explained as defence relations. Nevertheless, there were cases that did not fall into this category, and were more appropriately explained as support relations using bipolar argumentation. Thus, while the participants did behave in a way that was largely consistent with the notions of defence as used in dialectical semantics, they also used notions of support as conceptualized in bipolar argumentation.

It is also worth mentioning that in bipolar argumentation, mixtures of supporting and attacking links often give rise to new kinds of indirect conflicts. The study has shown that many attacks that were declared by the participants but that were not included in the intended graphs, could be reproduced by using the existing notions of indirect conflicts in these settings. Bipolar argumentation can therefore be used to model auxiliary attacks arising in the context of a dialogue, but not necessarily created on the logical level.

Another interesting observation in favour of using support relations between arguments concerns the fact that people are not perfect reasoners. Let us consider the following two arguments uttered by opposing parties in the dialogue:

F The virus is only accompanied by stabilizers and possibly trace amounts of antibiotics used in its production.

G The vaccine contains a preservative called thimerosal which is a mercury-based compound.

The fact that the virus is accompanied only by stabilizers and antibiotics means it is not accompanied by thimerosal, which is only a preservative. This leads to a conflict between F and G. However, realizing this depends on being aware of the distinction between stabilizers and preservatives, and the participants have occasionally confused the two notions [3]. Consequently, thimerosal could have been seen as an example of a stabilizer and as a result, some participants understood G as supporting F rather than attacking it. Hence, declaring this relation differently was a conscious and somewhat justified decision, not a result of misunderstanding the exercise or an unintentional choice (a "misclick").

One's expertise and background knowledge can therefore have an impact on how relations between arguments are perceived. Furthermore, natural arguments such as the ones used here, harvested from NHS and CDC websites or general public forums, will frequently be enthymemes and rely on a given person's knowledge for correct interpretation. In such real-life situations, the use of Dung's framework can obscure the fact that various reasoning and perception issues, like the one highlighted above, are taking place. A system unable to detect that a given relation - logically intended as conflicting - is in fact seen as supporting, runs the risk of promoting undesired behaviours in the user. There is therefore a benefit in incorporating bipolar argumentation for modelling imperfect reasoners.

Thus, the vaccine dialogue study in [Polberg and Hunter, 2018] provides some evidence for the need and value of bipolar argumentation.

---

[3]While the study was aimed at the general public and the participants did not necessarily have medical training, the statements in the study were generated based on the advice provided on websites such as NHS or CDC which are supposed to be accessible to the majority of population.

3.2.2 Rosenfeld and Kraus 2016 Study

In contrast to the study in [Polberg and Hunter, 2018], a less-supportive analysis of the bipolar approach can be found in [Rosenfeld and Kraus, 2016a]. This work investigated the abilities of formal argumentation, relevance heuristics, machine and transfer learning for predicting the argument choices of participants, with a particular focus on machine learning. Adequacy of computational models of argumentation was verified using three experiments. Various dialogues were sourced and then used to construct bipolar argumentation frameworks. Afterwards, the sets of arguments selected by the participants were contrasted with grounded, preferred and stable extensions of the created frameworks.

In the first experiment, consisting of 6 scenarios, the authors created bipolar argumentation frameworks which were not known to the participants, presented two standpoints from two parties and asked the participants to choose which of the additional four arguments they would use next if they were one of the participants in the discussion. In the second experiment, selected conversations from Penn Treebank Corpus were annotated and structured in the form of a bipolar argumentation frameworks. In the third experiment, a chat service was created, where participants discussed flu vaccination by using only the arguments from a predefined list. Finally, in an additional experiment, an artificial agent based on formal argumentation was implemented in order to provide suggestions to the participants during a two–person chat.

The authors report that a substantial part of the results (or in some cases, even the majority) do not conform to the outcomes predicted by the semantics. In other words, the arguments selected by the participants of the dialogues were not seen as justified. It is worth mentioning that the stated adherence to the conflict–free extension–based semantics is 78%; this is similar to the empirical results concerning the rational postulate in epistemic probabilistic argumentation, which corresponds to this semantics (see also Section 3.1.3).

Nevertheless, the causes for such behaviour of the semantics are not investigated, and the participants were not allowed to explain their decisions (the first and the third experiment) or there was no possibility to ask them for further input (the second experiment). Unlike in the study reported in [Polberg and Hunter, 2018], the participants were evaluated against the graphs constructed by the authors or annotators. As shown by the results from that study, the intended graphs do not necessarily reflect how the participants view the relations between the arguments.

There is also no discussion concerning whether these particular bipolar argument framework semantics used in this experiment [Amgoud et al., 2008] are applicable. There are various ways support can be interpreted, and each of these interpretations is accompanied by several - not necessarily equivalent - types of semantics [Cayrol and Lagasquie-Schiex, 2013; Polberg and Oren, 2014; Nouioua, 2013; Gottifredi et al., 2018; Cohen et al., 2014; Amgoud and Ben-Naim, 2018]. The stable and conflict–free semantics used in this study [Rosenfeld and Kraus, 2016a] are based on direct and supported attacks, and only the direct ones need to be defended from and can be used for defence. This approach has been superseded by a number of different methods since it has been introduced. Consequently, the presented results indicate that these particular semantics are not useful in modelling of the user behaviour in this study, rather than there exists a deeper issue within formal argumentation itself. An additional analysis of the data, where supports are mapped to their particular interpretations and where appropriate semantics are used, would shed more light on this issue.

3.2.3 Argument Mining

Further evidence for the value of formalisms that incorporate support comes from argument mining studies that focus on obtaining arguments and relations between them from sources such as social media, Wikipedia, or Debatepedia, see for example [Cabrio and Villata, 2013; Bosc et al., 2016]. These studies show that exchanges between participants often contain support relations, even more than attack relations. Similarly as in [Polberg and Hunter, 2018], these works highlight the modelling potential of bipolar argumentation and of the indirect attacks generated between arguments.

4 Empirical Studies About Structured Argumentation

Abstract argumentation, as the name suggests, abstracts away from the content of the arguments to only consider the relations between them. In contrast, structured argumentation studies how arguments can be constructed and how the relations between arguments (mostly just the attack relation) can be inferred from the structural properties of arguments. Multiple frameworks for structured argumentation have been proposed, e.g. ASPIC+ [Modgil and Prakken, 2018], ABA [Čyras et al., 2017] and Prakken & Sartor System II [Prakken and Sartor, 1997]. Each of these frameworks can be instantiated in different ways by selecting strict or defeasible inference rules that correspond to some underlying logic and/or describe domain-specific inferences. Thus structured argumentation provides a bridge between logic and abstract argumentation.

Concerning the connections between structured argumentation and actual human reasoning on the basis of arguments, the following three research question can be asked:

- How can one bridge the gap between the formalisms of structured argumentation on the one hand and human arguments that are expressed in natural language, that are often presented in an enthymematic way, and whose meaning and acceptability may depend on the context, on the other hand?

- Can the existing formalisms of structured argumentation be applied to model, explain and predict the way humans construct and evaluate arguments?

- Do certain properties of particular structured argumentation formalisms correspond better to human argumentation than contrary properties of other structured argumentation formalisms (e.g. restricted vs. unrestricted rebuttal, or different ways of taking preferences into account)?

To our knowledge, there have been only three empirical studies that have compared human reasoning to frameworks of structured argumentation. Cerutti et al. [2014] compare human intuitions on arguments to

Prakken & Sartor System II, Cramer and Guillaume [2018a] compare human judgments about arguments to ASPIC+ and ABA, and Yu et al. [2018] compare human judgments about arguments to ASPIC+. So far, there has only been limited progress on the three research questions presented above. Nevertheless, we will use these three research questions as yardsticks to evaluate the contributions that these three studies have made.

4.1 Preliminaries of Structured Argumentation

Before we can look at the details of these three empirical studies, we first briefly sketch the three frameworks of structured argumentation that have been considered in these studies, namely the ASPIC+ framework [Modgil and Prakken, 2018], the ABA framework [Čyras et al., 2017] and Prakken & Sartor System II [Prakken and Sartor, 1997]. We will sketch these frameworks in an informal way, focusing on the features that are relevant for the discussion of the empirical studies. For a complete formal definition of the frameworks, we refer the reader to the original works cited before.

ASPIC+ is a general framework that can be instantiated in different ways, which means that it is flexible with regards to the choice of the logical language to be used in the framework as well as the set of inference rules that are admitted. An instantiation of the ASPIC+ framework (called *argumentation theory*) is given by a formal language $\mathcal{L}$, a set of axioms over $\mathcal{L}$, a set of defeasible premises over $\mathcal{L}$, a set of strict rules and a set of defeasible rules. Arguments are built by applying the rules to deduce new information from axioms, defeasible premises or the conclusions of previous arguments. The axioms and strict rules constitute the deductive base logic underlying the argumentation theory, while the defeasible premises and rules allow for defeasible arguments to be formed, which might get rejected in the light of counterarguments.

In ASPIC+, three kinds of *attacks* between arguments are distinguished: Argument A *undermines* argument B iff the conclusion of A negates a defeasible premise used in B. Argument A *rebuts* argument B iff the conclusion of A negates the conclusion of a defeasible inference made within B. A *undercuts* argument B iff the conclusion of A negates the name of a defeasible rule used in B (which intuitively means that A

questions the adequacy of this defeasible rule).

Furthermore, the ASPIC+ framework allows to specify a preference ordering between the defeasible premises and rules, which gives rise to a preference order between arguments. An undermining and a rebuttal is only considered successful if the attacked argument is not preferred over the argument that attacks it.

The arguments and successful attacks that can be constructed on the basis of an argumentation theory give rise to an abstract argumentation framework, to which the semantics for argumentation frameworks presented in Section 2 can be applied in order to determine extensions, i.e. sets of arguments that can be coherently accepted together. A formula φ from $\mathcal{L}$ is considered *skeptically justified* with respect to the given argumentation theory iff there exists an argument with conclusion φ that is contained in every extension.

When applying ASPIC+, it is often assumed that the strict rules are *closed under transposition*, i.e. that for any strict rule of the form $\varphi_1, \ldots, \varphi_n \to \psi$ and every $i \in \{1, \ldots, n\}$, there is a rule of the form $\bar{\psi}, \varphi_1, \ldots, \varphi_{i-1}, \varphi_{i+1}, \ldots, \varphi_n \to \bar{\varphi}_i$.

In ASPIC+ rebuttals are restricted conclusions of defeasible rules. This feature is called *restricted rebut* and has bee criticized by some authors. Caminada *et al.* [2014] propose a variant of ASPIC+ called ASPIC−, in which restricted rebut is replaced by *unrestricted rebut*, according to which an argument A attacks an argument B and any argument containing argument B if the conclusion of A negates the conclusion of B and B contains at least one defeasible rule or defeasible premise.

In *assumption-based argumentation (ABA)* there is only one kind of rule, which behaves like the strict rules of ASPIC+. The only source of defeasibility of arguments are therefore the defeasible premises, which in ABA are called *assumptions*. So the only way in which an argument A can attack an argument B is when the conclusion of A is the contrary of an assumption used in argument B. Due to the absence of defeasible rules in ABA, some care is needed when formalizing defeasible inferences in ABA. For example, to model the inference from Z *is an expert* and Z *said* p to p, a rule of the form

$$expert(Z), said(Z,p), arguably(p) \to p$$

is required, where *arguably(p)* can be read as "there is no reason to doubt that p holds". One way to formally capture this intuitive meaning of *arguably(p)* is by adding another rule $\neg p \to \neg arguably(p)$. By formalizing defeasible inferences in this way, the behaviour of ASPIC+ can be simulated within ABA, at least when preferences are not taken into account [Heyninck and Straßer, 2016].

Prakken and Sartor System II [Prakken and Sartor, 1997] is very similar to the ASPIC+ framework, with the key difference of allowing preferences to be expressed at the object level. The authors introduce an operator $\prec$ inducing binary relations between defeasible rules, and the attacks notions are thus redefined to take into consideration preferences that are the conclusions of acceptable arguments.

4.2 Prakken and Sartor and Human Intuition

Cerutti *et al.* [2014] provide evidence suggesting that when facing contradicting arguments about a course of actions, people would be comfortable being guided by a preference statement between the two options following the Prakken and Sartor System II [Prakken and Sartor, 1997]. For example, one of the pieces of text Cerutti *et al.* used is the following:

> In a TV debate, the politician AAA argues that if Region X becomes independent then X's citizens will be poorer than now. Subsequently, financial expert Dr. BBB presents a document, which scientifically shows that Region X will not be worse off financially if it becomes independent.

However, additional pieces of information, like that more recent research by several important economists that disputes the claims in the document Dr. BBB used, can undermine the previous preference statement. This would lead people to abstain from agreeing with either one argument or the other, thus suggesting a skeptical attitude towards argumentation. In the case of the political debate, other participants were asked to assess their agreement with the politician or the financial expert (or none) based on the following, expanded, text:

> In a TV debate, the politician AAA argues that if Region X becomes independent then X's citizens will be poorer than

now. Subsequently, financial expert Dr. BBB presents a document, which scientifically shows that Region X will not be worse off financially if it becomes independent. After that, the moderator of the debate reminds BBB of more recent research by several important economists that disputes the claims in that document.

Language and Context The authors of [Cerutti et al., 2014] started from formal knowledge bases formalised according to Prakken and Sartor System II [Prakken and Sartor, 1997], which has the single peculiarity of allowing preferences to be expressed at the object level. They then transformed such formal knowledge bases in natural language text by handcrafting the text to represent a summary of dialogues between fictional actors in four different domains: weather forecast; political referendum about independence of a region in a country; practical argumentation towards buying a car; practical argumentation towards entering a long-term romantic relationship.

Results illustrated in [Cerutti et al., 2014] suggest a correspondence between the formal theory and its representation in natural language which allows readers to reach identical conclusions. However, the authors unveil an interesting situation: in the fourth domain—looking at a decision whether entering a long-term relationship—a sort of "reversal of preference" occurs. One of the explanations the authors provide links these results to the very subjective and emotional nature of the domain. Further to this, they also candidly admit how such studies will always tussle with "collateral knowledge" [Hoffmann, 2005], or more broadly general context.

Prediction of Human Behaviour The results illustrated in [Cerutti et al., 2014] suggest that people—perhaps unsurprisingly, cf. [Pinker, 2016]—possess an untaught notion of perceived logical consequence, as well as of aversion for perceived logical inconsistencies, although the authors did not explicitly give participants the option for expressing contradictory statements in their multiple-choices answers. Further, people seem more comfortable in either settling a decision—when possible— or abstaining altogether from making a judgement, thus suggesting a skeptical flavor to their innate reasoning. This seems to be consistent with the skeptical notion of acceptance provided by Prakken and Sartor System II [Prakken and Sartor, 1997].

Formalism properties and people's behaviour Looking at the results illustrated in [Cerutti et al., 2014], people seem comfortable with treating preferences not as elements of a meta-language, but rather as elements of the discussion that can, in turn, be justified, undermined, or rebutted by equally strong albeit opposite preferences. This suggests then that formalisms allowing for that might align better to people's behaviour.

4.3 Restricted vs Unrestricted Rebut

Let us imagine a dialogue between two fictional characters, where Anna tells Brenda *Jessica is a fan of two popular Korean bands, EXO and Bigbang. Both of them will hold concert series separately at nearby cities in next few weeks. So, Jessica will attend at least two concerts soon*; and Brenda replies *That won't be possible. She has been assigned too much work recently, so that she doesn't have the time to attend two concerts*?[4] According to [Yu et al., 2018], agreement on Anna's stand or Brenda's stand depends on whether *restricted rebut* or *unrestricted rebut* is used.

Language and Context The authors of [Yu et al., 2018] performed a study in which people were shown short, two-parties, one-round only, dialogues involving just two, potentially contradicting, arguments, similar to the ones shown at the beginning of this section. In their analysis of the study, the authors presented pairs of formal arguments in ASPIC+ or ASPIC− that can be viewed as formal analogues of those dialogues. The adequacy of the formalization was not systematically verified, but seems plausible in most cases.

Prediction of Human Behaviour Results illustrated in [Yu et al., 2018] strongly suggest that people tend to agree more with an unrestricted rebut view of argumentation. This seems to indicate that people

---

[4]Example formulation as presented in [Yu et al., 2018].

do not distinguish between defeasible arguments whose last conclusion is based on a strict rule, and defeasible arguments whose last conclusion is based on a defeasible rule. However, a caveat needs to be added to this interpretation of their results: When applying ASPIC+ strict rules can be closed under transposition, and when this is the case, the difference between restricted and unrestricted rebut disappears for the kind of conflicts that were considered in this study. So maybe the correct conclusion from the results of this study is that either people reason with unrestricted rebut or they intuitively recognize deductive inferences even when the deductive inference required for the case at hand is the transposition of a more common deductive inference.

Formalism properties and people's behaviour Based on the behaviour illustrated by participants in [Yu *et al.*, 2018], the authors call for the development of new argumentation formalisms that are both supportive of human intuition and blessed with desirable properties. Currently, using the apparently intuitive notion of unrestricted rebut clashes with the properties of *closure* and *indirect consistency* – listed among the desirable ones in [Caminada and Amgoud, 2007] – unless using the most skeptical semantics: the grounded semantics.

4.4 Directionality of Attacks in Natural Language Argumentation

One of the main differences between formal argumentation theory and classical logic is that it has a directed notion of conflict, namely that of an attack from one argument to another, whereas the notion of inconsistency in classical logic is symmetric. This gives rise to the question whether there really are conflicts that humans systematically interpret as unidirectional attacks in a certain direction, and whether structured argumentation frameworks like ASPIC+ and ABA can be used as predictors for the directionality of attacks.

Cramer and Guillaume [2018a] performed two studies that addressed these questions, one study with naive participants and one with expert participants. Naive participants were shown a pair of arguments and had to determine which argument(s) they accept, which one(s) they reject and which one(s) they consider undecided. Their response could be interpreted as indication for a unidirectional attack in a certain direction, for a bidirectional attack, or for an absence of any attack. The expert study involved 14 specialists in formal argumentation that were shown sets of two to five natural language arguments and had to indicate the attack relation between them.

Language and Context Cramer and Guillaume [2018a] introduce the notion of an *attack type*. This is inspired by the distinction of three kinds of attacks in ASPIC+, but the notion of *attack type* works on natural language arguments and is more fine-grained. Examples of attack types are *Undercutting Trustworthiness of Source, Rebuttal with Preference by Specificity* and *Attacking an Explicit Generic* (the last one can be formalized in ASPIC+ as either an undercutting or an undermining depending on whether the attacked generic, e.g., "Reindeer generally have antlers", is formalized as a defeasible rule or as a premise that contains a defeasible rule). The studies show that some attack types are systematically interpreted as being unidirectional in a certain direction, others are mostly interpreted as being bidirectional, while a third class of attack types leads to variation between unidirectional and a bidirectional interpretation. The attack types of the first two kinds can be useful for designing empirical studies aimed at examining the relationship between abstract argumentation and human reasoning (see Section 2.2).

Prediction of Human Behaviour The studies suggest that ASPIC+ is a good predictor of the directionality of attacks, as long as generic statements are treated as rules that can be undercut rather than as premises that can be undermined. The distinction between three kinds of attacks in ASPIC+ plays an important role in this respect.

Formalism properties and people's behaviour Since ABA does not distinguish between different kinds of attacks, ABA by itself does not yield as much information as ASPIC+ that could be used to predict the directionality of attacks between natural language arguments. Cramer and Guillaume [2018a] make an even stronger claim about ABA, namely that for some attack types it makes wrong predictions about the directionality of the attack. This claim needs to be treated with care, because it depends on how defeasible inferences are realized in ABA. Since ABA has no defeasible rules, it is the user who needs to specify how defeasible inferences are to be formalized. If they are formalized as sketched in Section 4.1, the predictions of ABA are in line with those of ASPIC+.

4.5 Outlook

Considering the results of the three studies in this area together, the three research questions introduced above can be partially answered as follows:

Concerning the first research question, one can observe that we only have a very limited understanding about how to bridge the gap between the formal approach of structured argumentation on the one hand and actual human argumentation expressed in natural language in an enthymematic and context-dependent way, on the other hand. The researchers who carried out the above studies carefully designed the natural language arguments to be used in their studies so as to minimize the impact of those aspects of human argumentation that cannot be properly captured with the existing tools of structured argumentation theory. This way they were able to make a bridge between structured argumentation and actual human argumentation, but one that cannot be easily extended to instances of human argumentation that have not been carefully designed for such studies.

Regarding the second research question, all three studies have found that human evaluation of arguments does indeed exhibit patterns that can at least partially be predicted and explained with the help of certain formalisms of structured argumentation, even if their results are so far limited to carefully designed sets of arguments. To this point, no study has attempted to use structured argumentation to predict or explain how humans construct arguments.

Concerning the third research question, it has been shown that certain features, present in some but not all structured argumentation formalisms, are indeed useful for predicting or explaining human evaluation of arguments, e.g. the possibility to argue about preferences in the object language or the possibility to distinguish different kinds of attacks.

Furthermore, the results of one study suggest that human evaluation of arguments can be explained better by formalisms that either have unrestricted rebuttal or that combine restricted rebuttal with transpositions of strict rules, than by formalisms that have restricted rebuttal while lacking transpositions of strict rules.

This discussion of the existing results makes it evident that much more empirical and theoretical work is needed to provide satisfactory responses to the three research questions. Especially the first research question requires much more work that would probably require insights from natural language semantics and pragmatics as well as from more informal approaches to argumentation theory that are studied in depth by philosophers (see for example [van Eemeren *et al.*, 2014]) to be included in the design of future empirical cognitive studies on structured argumentation. A more complete response to the second research question would require to empirically study not only human evaluation of arguments but also human construction of arguments. Given the diversity of different formalisms of structured argumentation and different variants of these formalisms, the current response to the third research question could be expanded by future studies that address the features of these formalisms not addressed so far.

5 Further Related Studies

In this section we discuss related work that is outside the scope of this chapter but has some important connections to the work presented in the rest of the chapter.

5.1 Logical Reasoning and Cognitive Biases

Cognitive psychologists generally assume that humans, from their earliest age, can reason with analogies [James *et al.*, 1890]. Humans are indeed efficient at mentally representing, manipulating, and organizing higher-order relations between mental objects. This ability is seen as one of the most crucial aspects of human cognition [Penn *et al.*, 2008], rooted in our evolutionary development, and emerging from social interactions [Tomasello, 1999]. Human reasoning had largely been associated with classical monotonic logic since psychologists from the beginning of the twentieth century predominantly considered logic as a mandatory mechanism allowing reasoning (see [Stenning and Van Lambalgen, 2012], for a review). Piaget [Piaget, 1953] for instance theorized that children progressively acquire logic (i.e., formal deductive) operations through development, and he assumed that these logic operations were mastered at adulthood. Nonetheless, empirical evidence, such as Wason's [Wason, 1968] famous observation that literate adults have severe difficulties in reasoning on abstract problems, later qualified the supposition that humans reason according to monotonic logic. On the contrary, these findings emphasized that human reasoning should be interpreted, from the monotonic logic point of view, as irrational.

Nonetheless, the irrationality explained above does not imply the existence of inherently dysfunctional cognitive structures. In this respect, the theory of mental models of reasoning [Johson-Laird and Byrne, 1991] proposes that human reasoning does not follow formal rules of inference, but alternatively depends on mental models specifically constructed for a given problematic situation. Critically, such mentally built models share their internal structure with the contextual structure of the represented real problem. Mental models are thus not abstract and are influenced by situational factors. For this reason, the way a problem is stated substantially influences the reasoning process and the decision outcome [Tversky and Kahneman, 1981; De Martino et al., 2006]. Furthermore, mental models are restricted, due to physical (i.e., cognitive) limitations (see [Lenat et al., 1979]). Humans cannot represent or deal with comprehensive models, and they subsequently need to build on simplified mental versions of the world. It has then been assumed that such simplified models lead to the emergence of cognitive heuristics related to the reasoning process [Shanteau, 1989; Simon, 1957; Anderson, 1986].

From a cognitive perspective, heuristics can be interpreted as mental shortcuts, used to reduce the cognitive load required by the whole reasoning process [Myers, 2010]. Critically, heuristics do not guarantee satisfactory decisions from a pure logic perspective, leading in some cases to seemingly irrational behavior, as observed by Wason [Wason, 1968]. Identifying discrepancies between human reasoning and decisions expected from monotonic logic has been the focus of many studies, and the latter emphasized the existence of sundry cognitive biases in human reasoning [Hilbert, 2012]. Cognitive biases are fundamentally and intrinsically related to human cognition, due to the heuristic nature of reasoning. Yet biases are not processing errors, they rather illustrate universal preponderating dispositions (as noted by Stanovich [Stanovich, 2003]). In this section, we briefly describe three cognitive biases that are of particular interest in formal argumentation theory.

First, it is arduous for humans to detach themselves from their perspective, because human cognition is embodied by nature [Varela et al., 2016]. Thinking abstractly, without any reference to oneself or the natural world, is not an instinctive task. Consequently, to solve problems, humans are likely to elaborate on their reasoning (and make decisions) from experience or previous knowledge [Stanovich, 2003]. One famous illustration of this bias is the gambler's fallacy, where the gambler erroneously believes that a streak of a given outcome lowers the probability of observing this outcome in the future. Such a fallacy shows that people tend to evaluate the probability of a given event according to previous occurrences of similar events, although such probability does not depend on these previous occurrences [Tversky and Kahneman, 1981]. More generally, it has been showed that humans prefer to infer information outside a given problem to form an explanation – from prior knowledge – instead of using pure deductive skills from available information [Johnson-Laird et al., 2004]. Human reasoning is thus biased in favor of building or manipulating mental models that are associated with the existing ones. Interestingly, this disposition can be favorable towards rational reasoning in some cases. Griggs and Cox [Griggs and Cox, 1982] indeed showed that it is possible to substantially improve performance in a difficult abstract task such as the Wason's card selection task by capitalizing on adults' experience: when provided a frame easy to relate to, humans can show great deductive skills.

Humans thus tend to reason in the light of existing mental models, so that previous knowledge or beliefs drastically influence how new information will be handled. Moreover, people tend to seek evidence in favor of their knowledge or beliefs, and they more easily accept arguments consistent with existing mental models than opposing information [Plous, 1993]. In other words, humans prefer to confirm their beliefs rather than confront them; this second propensity is a confirmatory bias. Cognitive load reduction (since the reorganization of mental models is costly) and cognitive dissonance avoidance [Festinger, 1957] are potential reasons for the existence of such bias. This confirmatory tendency implies that new information or arguments are neither neutrally processed nor equally accepted; there is a positive bias towards decisions consistent with previous ones.

Finally, there is also a cognitive bias towards the acceptability of new arguments that are unrelated to previous mental models. In this case, humans show a truth bias, which is a predisposition to accept new information as true. This bias originates from mental model properties because they are expressed in terms of what is true (and not in terms of what is false [Johnson-Laird, 1983]). Additionally, to reduce the cognitive load, we draw conclusions as a heuristic depending on whether a conclusion holds in all, most, or some of the premises [Gilbert et al., 1990; Johnson-Laird, 2006]. The criterion for acceptance is subsequently lower than the criterion for rejection. This heuristic incidentally leads to an acquiescence bias [Knowles and Nathan, 1997] in some cases, where people naturally tend to positively respond to neutral assertions. This positive truth bias notably emphasizes that there is no such neutral information in human reasoning since they convey some subjective truth.

These three (amongst many other) biases illustrate why monotonic logic should be considered irrelevant to human cognition (following Stenning & Van Lambalgen [2012]). Human reasoning is not intrinsically flawed; non-monotonic approaches of human cognition are nonetheless still needed [Ragni et al., 2016]. In this respect, formal argumentation theory could bring precious insights about human reasoning. As Mercier and Sperber [Mercier and Sperber, 2011] stated, evaluating argumentation itself puts novel perspectives in the study of human irrationality. Understanding argumentation is therefore crucial to understand human cognition.

5.2 Empirical Cognitive Studies About Non-Monotonic Reasoning

From its inception in the 1990s until this day, formal argumentation has had close and fruitful interaction with the field of non-monotonic reasoning. Argumentation formalisms are often viewed as a special case of formalisms for non-monotonic reasoning. For this reason, it makes sense to compare cognitive studies about formal argumentation to cognitive studies about non-monotonic reasoning.

The field of *non-monotonic reasoning* (or *default reasoning*) started off in the late 1970s when AI researchers began to appreciate the fact that classical logic cannot account for the non-monotonic features of human reasoning, i.e. the fact that humans often retract previously drawn conclusions when learning new information. Throughout the 1980s and early 1990s various formalisms were proposed for formally capturing non-monotonic reasoning, namely quantitative approaches such as probabilistic logic, fuzzy logic and Bayesian networks as well as qualitative approaches such as default logic, circumscription, autoepistemic logic and logic programming. These formalisms and related ones continue to be studied and adapted for various purposes to this day.

Pelletier and Elio [1997; 2005] make a case for psychologism with respect to non-monotonic reasoning, i.e. for the position that the content of the field of non-monotonic reasoning is whatever reasoning patterns reside in the collective psychological states of the population. This is, however, a highly contentious position that many practitioners in the field would either fully reject or only partially accept. For proponents of some form of psychologism with respect to non-monotonic reasoning, it certainly makes sense to empirically study how actual human reasoning compares to the various proposed formalisms of non-monotonic reasoning. But even those who reject this kind of psychologism can find value in such studies, be it because understanding human reasoning is important irrespective of whether it is the content of the field of non-monotonic reasoning, or be it because humans are to this day better at most commonsense reasoning tasks than any artificial agents and therefore understanding human reasoning better can help us build better AI tools.

In light of the large variety of non-monotonic formalisms proposed during the 1980s, Lifschitz [1988] introduced 25 *Nonmonotonic Benchmark Problems* which he argued should be modelled by every formalism that is proposed as a serious contender for modelling non-monotonic reasoning. These problems include scenarios like the following one: Given the premises listed below, it should be considered permissible to draw the conclusion listed below:

Premises:

- Blocks A and B are heavy.
- Heavy blocks are normally located on this table.
- A is not on this table.
- B is red.

Conclusion:

- B is on this table.

In many benchmark problems there is an *object-in-question* which the conclusion talks about (block B in the above example), and a different *exception-object* that according to the premises violates some default rule (block A in the above example).

Ellio and Pelletier [1993; 1996] and Pelletier and Ellio [2002] present the results of multiple empirical studies to test whether people actually draw conclusions for the aforementioned benchmark problems in line with the prescriptions proposed by Lifschitz. In Pelletier and Ellio [2005] the authors summarize and discuss the findings of these studies. Here we present a brief summary of their results followed by their interpretation.

The results of these studies by Ellio and Pelletier suggest that humans draw conclusions mostly in accordance with Lifschitz's prescriptions, and thus in accordance with the behaviour of the major qualitative non-monotonic formalisms. However, there were also some patterns in their data that could most non-monotonic formalisms cannot explain. For example, humans seem to have an inclination towards applying the following principle, called *Second-Order Default Reasoning* (or alternatively the *Guilt by Past Association* rule): "If the available information is that the object-in-question violates other default rules, then infer that it will violate the present rule also."

Another example of a human reasoning principle that Ellio and Pelletier found in the human responses is the following principle of *Explanation-based Exceptions*: "When the given information provides both a relevant explanation of why the exception-object violates the default rule and also provides a reason to believe that the object-in-question is similar enough in this respect that it will also violate the rule, then infer that the object *does* violate the rule."

Ellio and Pelletier also compared the conclusions that humans were willing to draw based on certain information with the conclusions that humans claimed to be reasonable for a robot to draw from that same information. Here they found that people believe robots should be cautious (saying they "Can't tell") when they themselves would be willing to give a definite answer.

Another observation the authors made was that in the case of a benchmark problem in which no conclusion should be drawn according to Lifschitz's prescriptions (namely the famous *Nixon diamond* problem), half of the participants did claim that a conclusion can be drawn, but these participants were approximately equally divided between of the two potential conclusions in this problem. This might be a sign that drawing no conclusions is something that humans often try to avoid.

The experiments performed by Ellio and Pelletier involved reference to real-world categories such as various types of birds and trees that actually exist. This raises the concern that people might be using prior knowledge rather than applying only inferences based on the given premises. Hewson and Vogel (1994) and Vogel (1996) attempted to avoid this problem by formulating all premises and putative conclusions using uninterpreted Roman letters instead of English words, e.g.: "A's are normally B's." Their results suggest that people do very badly at reaching conclusions accepted in the literature. Ford and Billington (2000) point out that this bad performance might be due to the tasks being unduly meaningless when only Roman letters are used, so they propose a compromise between the two approaches: to use a fictional setting and fictional words about categories existing in this fictional setting. They performed two studies with university students and one study with academic staff who did not do research on reasoning. Their results suggest that university students do very badly at reaching conclusions accepted in the literature, whereas academic staff copes somewhat better.

Comparing their results to the results of Ellio and Pelletier's study suggests that the ability to link the provided information to existing knowledge is very important for ordinary people to be able to make reasonable conclusions, whereas academic staff is somewhat better at making reasonable conclusions even when no such link can be established.

The studies considered so far were mostly based on Lifschitz's benchmark problems and variants thereof. We will now turn our attention to studies that have aimed at determining the cognitive plausibility of the rationality postulates for non-monotonic logic proposed by Kraus et al. [1990], known as the *KLM postulates*. These postulates were developed as a possible response to the question: What principles do we still accept in non-monotonic logic, once we give up the principle of monotony from classical logic? These principles are phrased in terms of the strict consequence relation $\Gamma \vdash \varphi$ (meaning that the set Γ of formulas strictly entails the formula φ) and the defeasible consequence relation $\Gamma \mathrel{|\!\sim} \varphi$ (meaning that Γ defeasibly entails ϕ):

- Reflexivity: $\varphi \mathrel{|\!\sim} \varphi$.

- Cut: If $\varphi \wedge \psi \mathrel{|\!\sim} \tau$ and $\varphi \mathrel{|\!\sim} \psi$ then $\varphi \mathrel{|\!\sim} \tau$.

- Cautious Monotony (CM): If $\varphi \mathrel{|\!\sim} \psi$ and $\varphi \mathrel{|\!\sim} \tau$ then $\varphi \wedge \psi \mathrel{|\!\sim} \tau$.

- Left Logical Equivalence (LLE): If $\varphi \vdash \psi$, $\psi \vdash \varphi$ and $\varphi \mathrel{|\!\sim} \tau$ then $\psi \mathrel{|\!\sim} \tau$.

- Right Weakening (RW): If $\varphi \vdash \psi$ and $\tau \mathrel{|\!\sim} \varphi$ then $\tau \mathrel{|\!\sim} \psi$.

- OR: If $\varphi \mathrel{|\!\sim} \psi$ and $\tau \mathrel{|\!\sim} \psi$, then $\varphi \vee \tau \mathrel{|\!\sim} \psi$.

Two further related principles have received a lot of attention in the literature on non-monotonic logic:

- Rational Monotony (RM): If $\varphi \mathrel{|\!\not\sim} \neg\psi$ and $\varphi \mathrel{|\!\sim} \tau$ then $\varphi \wedge \psi \mathrel{|\!\sim} \tau$.

- AND: If $\varphi \mathrel{|\!\sim} \psi$ and $\varphi \mathrel{|\!\sim} \tau$ then $\varphi \mathrel{|\!\sim} \psi \wedge \tau$.

Da Silva Neves et al. [2002] conducted an empirical study about the cognitive plausibility of these rationality postulates of non-monotonic logic. For this purpose, they asked university students to make judgments about the degree to which different scenarios are possible. Their study involved a pre-experiment with 40 university students and a main experiment with 88 university students. In order to test the cognitive plausibility of the rationality postulates of non-monotonic logic, they performed statistical tests to determine whether the participants' judgments corroborate these postulates. Their results suggest that RW, CM, OR, AND, and RM are cognitively plausible. For the CUT rule they had different results depending on the content that was used to create concrete scenarios from the abstract patterns of the CUT rule, so that their results are not conclusive in this case. They also attempted to test the cognitive plausibility of the LLE rule, but during the test the material used turned out to be problematic in a way that did not allow them to make any conclusions about the cognitive plausibility of LLE. Moreover, their results confirmed that CM, AND and RM were validated even in cases in which Monotony does not hold.

Benferhat et al. [2005] tested 57 university students to test whether their reasoning is in line with various principles from non-monotonic logic. Their results suggest that on the whole, participants' reasoning was consistent with LLE, RW, OR, AND, and CUT. Concerning CM and RM their results were not conclusive.

Bonnefon et al. [2008] introduce a model for describing an agent's ascriptions of causality that can account for the difference between claiming that an event A causes another event B and claiming that event A facilitates event B. Their model is based on System P [Kraus et al., 1990]. The authors conducted two experiments that confirmed their hypothesis that humans do actually differentiate between causality and facilitation, and broadly along the lines featured in the definitions that are built into their model.

Finally, there have also been several empirical cognitive studies that focus on probabilistic modelling of non-monotonic reasoning, for example [Pfeifer and Kleiter, 2005; Pfeifer and Kleiter, 2009; Pfeifer and Tulkki, 2017]. These provide further insights into the nature of non-monotonic reasoning in human cognition, but perhaps are a step further away from understanding argumentation, and therefore beyond the scope of this review.

5.3 Human Reasoning and Computational Models of Persuasion

Persuasion is an activity that involves one party trying to induce another party to believe or disbelieve something, or to do (or not do) something. It is an important and complex human ability. Obviously, it is essential in commerce and politics. But, it is equally important in many aspects of daily life. Consider, for example, a child asking a parent for a raise in pocket money, a doctor trying to get a patient to enter a smoking cessation programme; a charity volunteer trying to raise funds for a poverty stricken area; or a government advisor trying to get people to avoid revealing personal details online that might be exploited by fraudsters.

Arguments are a crucial part of persuasion. They may be explicit, such as in a political debate, or they may be implicit, such as in an advert. In a dialogue involving persuasion, counterarguments also need to be taken into account. Participants may take turns in the dialogue, each of them presenting various arguments and counterarguments. So the aim of the persuader is to convince the persuadee through this exchange of arguments. Since some arguments may be more effective than others, it is valuable for the persuader to have an understanding of the persuadee and of what might work better with them.

This understanding of the persuadee can come from several, non exclusive sources: understanding their personality, their relation with the topic being discussed, or their argumentation framework concerning the ongoing persuasion process.

Most papers investigating the impact of personality on the effectiveness of persuasion are using one or other of the two most studied personality models in the psychology literature: the OCEAN model [Goldberg, 1993] (also known as the Five-Factor model or the Big Five personality traits) and the Regulatory Focus Theory [Tory Higgins, 2012]. Knowing the persuadee's relation with this model (in other words her values according to the models) allows for the prediction of her reactions to arguments and how their beliefs may evolve.

For instance, in [Lukin et al., 2017], the authors used the OCEAN model to predict how the beliefs of the persuadees evolve depending on the type of argument that was given. They created three types of arguments:

- balanced monological arguments, *i.e.*, longer pieces of text containing both viewpoints on the topic being discussed,

- emotionally-framed arguments,

- factually-framed arguments.

Interestingly, people scoring high on "Openness to Experience" (the O in the OCEAN model) were more influenced by balanced and emotional materials. On the other hand, "Agreeable" people (the A in the OCEAN model) were most affected by factual materials.

In the same line of research, the authors of [Thomas *et al.*, 2017] profiled persuadees using the OCEAN model and studied the effect of Cialdini's persuasion principles [Cialdini, 1993] on the perceived believability of arguments. They have shown, amongst other findings, that the "appeal to authority" principle is the most efficient across all personality profiles.

On the other hand, the believability of one argument can also be predicted from related arguments. In [Polberg and Hunter, 2017], the authors gathered from the participants three values associated with arguments: the believability, the convincingness and the appeal. The arguments were split in three groups depending on their source: arguments issued by "Celebrity", "Scientific" arguments and common "Society" knowledge. They have shown that, first, the believability of an argument is a good proxy for how convincing an argument is. The latter is difficult to gather, while the former is understood more easily by participants when crowdsourcing data. They have also demonstrated that people are consistent with their answers concerning arguments framed as coming from the same source, therefore showing that persuadees' profiles can be created from a small number of questions.

Personality profiles can also be used to predict high-level data on arguments. In [Hadoux and Hunter, 2019], the authors defined the notion of concerns, *i.e.*, high-level categories of arguments such as "Time"-related or "Comfort"-related arguments in the context of cycling. Using the personality profiles and demographic data as input of classification trees, they have shown that people's preferences towards the concerns can be predicted. They have also demonstrated that persuadees choose and believe arguments that are congruent with their preferences, opening up a new dimension of strategies when it comes to choosing discussion branches to enter or avoid.

Also on the role of concerns, [Chalaguine et al., 2019] showed in a study with a simple chatbot, intended to persuade people to decrease meat consumption, that participants who were more concerned with environmental issues were more persuaded by positive impersonal arguments. On the other hand, participants who were more concerned with personal health were more persuaded by positive personal arguments.

Another angle to modelling the persuadees is that, instead of assuming both the persuader and the persuadee share the same argumentation framework, the persuadee has a subset of the whole framework representing how much she knows about it. Each time something new happens, this framework evolves to include new knowledge and update the current one. In [Rosenfeld and Kraus, 2016b], the authors used a Partially Observable Markov Decision Process (POMDP) to represent the current state of the persuadee's argumentation framework and how it evolves with the interactions with the persuader. Solving the problem with POMCP [Silver and Veness, 2010], they obtained a Strategical POMDP Agent (SPA) trying to have the best estimation of the current state of the persuadee's argumentation framework and have the best strategy to maximise the persuadee's valuation of the goal argument. The experiment was about changing some students' opinion about enrolling into a Master's degree using either the SPA, a baseline or another student. The SPA and using another student were almost on par, both with a statistically significant better performance than the baseline.

5.4 Human Emotions and Computational Models of Argumentation

In addition to belief, the emotions invoked by arguments are important to take into account since they affect the way the arguments are perceived by the persuadee. Emotions are the result of how an individual appraises a stimulus. It is a cognitive process composed of a number of checks aimed at categorising a stimulus: is it relevant, what does it imply, do I have the potential to cope and is it socially significant? This process and the various patterns of checks generate different cognitive responses and coping strategies. These strategies, in turn, affect the way information is processed [Duhachek et al., 2012]. For example, guilt leads to the use of active strategies focused on repairing the committed harm, whereas shame leads to the use of more passive strategies focused on the self. Combined with gain-loss framing [Tversky and Kahneman, 1981], the emotion conveyed by an argument can be used to increase the persuasiveness of this argument. While Ekman [Ekman, 1992] considered only 6 basic emotions (anger, disgust, fear, happiness, sadness and surprise), the definition and characterisation of emotions has been widely discussed in psychology. Emotions in argumentation have also been investigated recently using logic and sets of discrete emotions (see, e.g., [Nawwab et al., 2010], [Lloyd-Kelly and Wyner, 2011], [Martinez et al., 2012]). For instance, in [Mazzotta et al., 2007], the authors analysed actual persuasion strategies and found that purely rational argumentation was rarely employed. On the other hand, emotional elements could be found everywhere. They have developed a system, PORTIA, able to create persuasive messages mixing both rational and emotional contents, using an extension of *Belief Networks* [Pearl, 1988] for the persuadee's representation.

Building upon Ekman's 6 basic emotions, the authors in [Villata et al., 2017] used a combination of facial recognition and EEG to detect emotions felt by participants in a debate. The two most present emotions were anger and disgust. This is explained by the "Negative Emotion Factor" were negative emotions have a more important and lasting effect on a person's behaviour. They also gathered the personality profiles of the participants in the frame of the OCEAN model [Goldberg, 1993]. The objective was to find correlations between certain personality traits and the strength or frequency of the emotions felt/measured. Interestingly, intuitive assumptions like "extroverted people tend to show their emotions more often" (in particular the surprise) were observed.

Taking another stance on emotions, the authors in [Hadoux et al., 2018] used the "Affective Norm". It captures the emotional response to specific words in three dimensions: arousal (ranging from excited to calm), valence (pleasant to unpleasant), and dominance (from being in control to being dominated). For example, for valence scores, *leukemia* and *murder* are low and *sunshine* and *lovable* are high; for arousal scores, *grain* and *dull* are low and *lover* and *terrorism* are high; and for dominance scores, *dementia* and *earthquake* are low, and *smile* and *completion* are high. Using the database gathered by [Warriner *et al.*, 2013], containing the triplet of values for nearly 14,000 English words, they have presented a method for aggregating the values at the level of the sentence following principles from the psychology literature. They have also shown how this can be used in the context of a persuasion dialogue to calculate a strategy presenting counterarguments during a dynamic dialogue that takes generated emotions into account.

5.5 Empirical Cognitive Studies About Argumentation Schemes

Argumentation schemes represent stereotypical patterns of reasoning used in everyday conversational argumentation, and in other contexts such as legal and scientific argumentation. The schemes are accompanied by appropriate sets of critical questions which function as defeasibility conditions. An example of this is the argument from position to know [Walton *et al.*, 2008]:

Major Premise : Source a is in position to know about things in a subject domain S containing proposition A.

Minor Premise : a asserts that A is true (false).

Conclusion : A is true (false).

It can be critically questioned by raising doubts about the truth of either premise, or by asking whether a is an honest (trustworthy) source of information:

CQ1 : Is a in a position to know whether A is true (false)?

CQ2 : Is a an honest (trustworthy, reliable) source?

CQ3 : Did a assert that A is true (false)?

This section provides an overview of empirical studies related to schemes.

5.5.1 Evaluating Argumentation Schemes

In [Schellens et al., 2017] participants were presented with a list of arguments and asked to rank these arguments from strongest to weakest, upon which they were asked to motivate their judgments in an interview. Such arguments were drafted as instances of five different argumentation schemes. The study confirmed that in addition to general criteria from informal logic—e.g. relevance and acceptability—people also used scheme-specific criteria, e.g. the expertise when dealing with argumentation from authority.

In [Thomas et al., 2019a] participants were shown a set of five messages each promoting healthy eating, and based on different argumentation schemes. The authors then were interested in determining a reliable scale to measure the perceived persuasiveness of the arguments. The authors also show how the message types impact factors of the scale, such as effectiveness, quality, and overall perceived persuasiveness. The same authors in [Thomas et al., 2019b] also proposed a tool, ArguMessage, to semi-automatically generate persuasive messages based on argumentation schemes.

In [Lazarou et al., 2016] an analysis following the Cultural-Historical Activity Theory considered the teaching and learning practices in primary schools in Cyprus, showing evidence of usage of argumentation schemes.

5.5.2 Using Argumentation Schemes

In [Schneider et al., 2013] the author considered a corpus of English Wikipedia deletion discussions. They also investigated the use of argumentation schemes, showing how 36% of the used argument were an instance of the Rules and Evidence schemes. In a similar type of analysis [Hansen and Walton, 2013] the authors show that the kind of argument used most frequently in the Ontario election campaign, 2011, was Appeal to Negative Consequences. Next most frequent was Practical Reasoning argumentation, followed by Appeal to Positive consequences, Argument from Sign, and Appeal to Fairness.

In [Konstantinidou and Macagno, 2013] the authors provide the evidence of the argumentative nature of students and of the benefits of using argumentation schemes as instruments for reconstructing the possible missing premises underlying their reasoning.

Another study in education, in [Song and Ferretti, 2013] 30 college students learnt two commonly used argumentation schemes (namely argument from consequences and argument from example) and critical questions associated with these schemes. Compared to the students in the contrasting conditions, those who learned critical questions wrote essays that were of higher quality and included more counterarguments, alternative standpoints, and rebuttals. In a follow-up study [Song et al., 2017], the authors show that the majority of eighth-grade students they considered fail to detect fallacious arguments or clearly explain problems in the arguments they encounter. To identify whether an argument is misused or fallacious, the authors considered argumentation schemes as a golden standard. In [Green, 2015; Green, 2017] the author shows that correctly identifying the scheme an argument is an instance of is not a trivial task even for educated professionals. As shown in [Lindahl et al., 2019], even annotators with a strong background in linguistics—albeit with little explicit instructions for a given annotation task—failed to identify argumentation schemes, with the annotators agreeing neither on whole arguments nor on the units and schemes which make them up. Research in this direction is mostly looking at guidelines for the annotation of argument schemes. Musi et al. [2016] show that annotating argument schemes requires highly trained annotators and, in turn, an accurate annotation of both premises and claims.

On a similar note, in [Reznitskaya et al., 2007] the authors provide evidence of how education in argumentation can produce benefits in reflective essays as well as in interviews. This is echoed also in [Nussbaum and Edwards, 2011] which presents a study conducted in 3 sections for 6 months (one section served as comparison group) of a 7th-grade social studies classroom in which 30 students discussed and wrote about current events adopting techniques of critical thinking closely linked to argumentation schemes and critical questions. Over time the experimental group—exposed in particular to the concepts of critical questions, and of integrative and refutational argument stratagems—made more arguments that integrated both sides of each issue. Similarly, in

[Okada and Shum, 2008] the authors examine the role of Evidence-based Dialogue Maps—that exploits the Toulmin [Toulmin, 1958] argumentation scheme—as a mediating tool in scientific reasoning: as conceptual bridges for linking and making knowledge intelligible; as support for the linearisation task of generating a coherent document outline; as a reflective aid to rethinking reasoning in response to teacher feedback; and as a visual language for making arguments tangible via cartographic conventions.

5.6 Human Reasoning and Bayesian Approaches to Argumentation

Bayesian approaches to argumentation [Hahn and Hornikx, 2016] is a reaction to the MAXMIN rule for argumentation when combining linked and convergent arguments. When two or more independent arguments all support the same claim, we are in presence of *convergent arguments*. Linked arguments instead form a chain of dependencies, thus providing support for a claim only in combination.

For convergent arguments, Walton [1992] argues in favour of the MAX rule, i.e. the overall strength or plausibility of the argument is determined by the maximum of the independent arguments converging to the same claim. For linked arguments, researchers [Walton, 1992; Pollock, 2001] propose that the overall plausibility of the argument is determined by its weakest link. While some researchers [Walton, 1992] concede that there are cases where plausibility and probability are closely linked, others [Hahn et al., 2013] contend that this is true in several cases. A probabilistic interpretation of the plausibility or strength of an argument leads to the conclusion that the MIN rule provides an upper bound of the probabilistic interpretation of the strength of a linked argument. Indeed, $P(A \wedge B) = P(A) \cdot P(A|B) = P(B) \cdot P(B|A) \leq \min\{P(A), P(B)\}$.

5.6.1 Bayesian Argumentation

Arguments, for their defeasible nature, can be represented by a network of random variables connected in a belief network, i.e. a directed graph where nodes are random variables, and edges represent causal links. Let us consider the case of the argument from expert opinion; Walton et al. [2008] present a scheme for capturing it:

Major Premise : Source E is an expert in subject domain S containing proposition A.

Minor Premise : E asserts that proposition A (in domain S) is true (false).

Conclusion : A may plausibly be taken to be true (false).

Associated to the scheme there are the following six critical questions:

CQ1 : (*Expertise*) How credible is E as an expert source?

CQ2 : (*Field*) Is E an expert in the field that A is in?

CQ3 : (*Opinion*) What did E assert that implies A?

CQ4 : (*Trustworthiness*) Is E personally reliable as a source?

CQ5 : (*Consistency*) Is A consistent with what other experts assert?

CQ6 : (*Evidence*) Is E's assertion based on evidence?

However, alternative formalisations are possible, and some can make use of Bayesian inferences [Hahn et al., 2013]: Figure 5 illustrates the structure of a Bayesian network[5] for representing the relevant elements of an appeal to expert opinion.

X_1 in Figure 5 is the random variable representing whether the proposition A may plausibly be taken to be true. If A is true, then this causes the pieces of evidence (X_2) put forward to be also true: if that is the case, then the reputation of our source of information also benefits

---

[5] A Bayesian network is a direct acyclic graph where nodes represent random variables, i.e. variables that can have multiple values: the easiest case is when the variable can be either true or false, but a variable can be used to represent the rolling of a dice, hence it can have six different values. Edges represent conditional and causal dependencies, e.g. smokes can cause asthma, but asthma cannot cause smokes, hence in this case there would be an arrow from a random variable representing whether an individual smokes towards a random variable representing whether the same individual suffers from asthma.

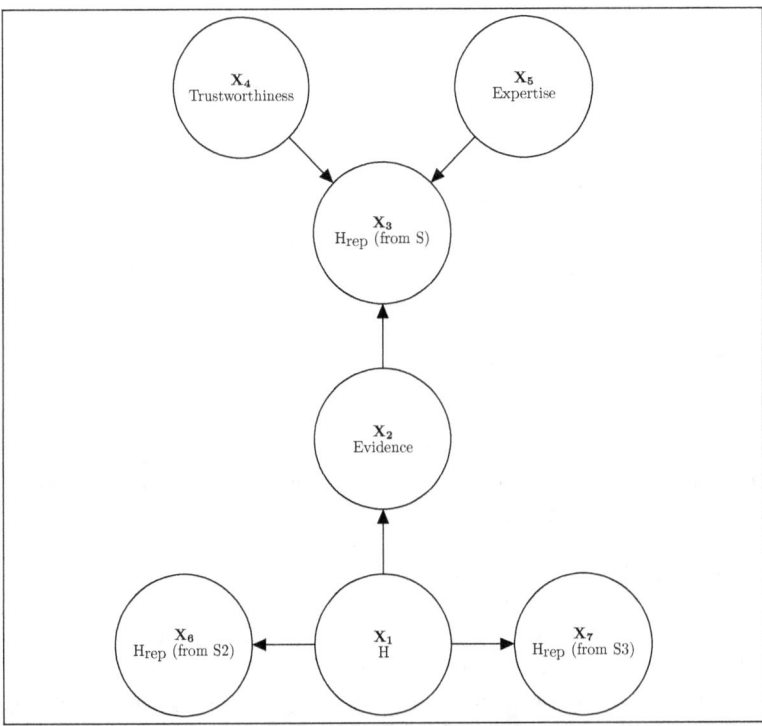

Figure 5: Structure of the Bayesian network proposed by [Hahn et al., 2013] to represent the appeal to expert opinion so to be able to answer the critical questions raised by [Walton et al., 2008].

(X_3). However, X_3 also depends on whether the source is trustworthy (X_4) and a true expert of the domain (X_5). Finally, if we consider the presence of other sources of information, such as S2 (X_6) and S3 (X_7) (and others if necessary), then the validity of our hypothesis X_1 will also affect the reputation associated to them.

Prior probability assigned to X_5 helps answering both (*Expertise*) and (*Field*): according to [Hahn et al., 2013], "the expertise will only be relevant if it is in the particular domain under consideration." In the case where what S asserts is not identical to H, then (*Opinion*) and

(*Evidence*) relate to the conditional probabilities $P(\text{Evidence}|H)$ and $P(\text{Evidence}|\neg H)$ (or their ratio). (*Trustworthiness*) is captured by the prior assigned to $\mathbf{X_4}$, while (*Consistency*) links to variables associated to reports from different experts, i.e. $\mathbf{X_6}$ and $\mathbf{X_7}$ (and others if necessary).

5.6.2 Empirical Analyses Using Bayesian Argumentation

Bayesian argumentation provides testable measurements of the strength of the argumentation, and experimental studies such as [Oaksford and Hahn, 2004; Hahn et al., 2005; Hahn and Oaksford, 2007; Corner et al., 2011; Harris et al., 2013] detailed how the Bayesian framework is operationalised for thus obtaining qualitative and quantitative predictions and compared with lay people's perception of arguments strength. For instance, in [Oaksford and Hahn, 2004], and further in [Hahn et al., 2005], the authors analysed a Bayesian account of the argument from ignorance, usually considered a reasoning fallacy. Indeed, borrowing the authors' example, the argument *Ghosts exist because no one has proved that they do not* does not seem acceptable. However, the authors' claim is that this is not because of the structure—after all, it has the same structure as *It is safe to take Ibuprofen in the recommended dose because no one has proved that it is not*—rather by the context, and thus of priors we provide to various random variables. In [Harris et al., 2013], the authors also considered the *damned by faint praise* phenomenon, or *boomerang effect*, by which a very weak positive argument lead to a negative change in belief. According to the authors, this can be explained in a Bayesian framework due to an (often unstated) inference from critical missing evidence, i.e. an implicit argument from ignorance. The authors in [Hahn and Oaksford, 2007] expanded the analysis of reasoning fallacies, thus including also experiments looking at the circular arguments (*petitio principii*, and at the slippery slope argument, also expanded in [Corner et al., 2011]).

5.7 Empirical Assessment of Aggregation of Argument Evaluation

Judgment aggregation is a subfield of social choice which studies how logically interrelated judgments by multiple agents can be aggregated into a group decision [List and Puppe, 2009]. Some works in judgment aggregation, e.g. Rahwan and Tohmé [2010], Caminada and Pigozzi [2011], Booth et al. [2014] and Awad et al. [2017b], have considered the problem of aggregating judgments that consist of choosing an extension of labeling of an abstract argumentation framework. Two different approaches emerged in these theoretical works: The *argument-wise plurality* rule (AWPR) chooses the collective evaluation of each argument by plurality, whereas Caminada and Pigozzi's [2011] sceptical operator, credulous operator and super credulous operator (collectively shortened as SSCOs) are based on the principle of compatibility, according to which an argument cannot be rejected if one of the agents accepted it and vice versa (but it may be accepted if some agents are undecided about it).

Awad et al. [2017a] performed an empirical experiment to determine which of these two approaches people consider better at aggregating opinions. For this purpose, they showed participants a set of natural language arguments that corresponded to an AF with multiple complete extensions as well as the result of a vote of members of a committee on the acceptability of the various arguments involved. Finally they were asked what decision the committee should make based on this vote. They found that AWPR was more in line with the participants' decisions than the SCSCOs, but that the difference got smaller when either the size difference between the majority and the minority in the commitment got smaller or the decision to be made by the committee was one that would personally harm an individual.

6 Conclusion and Future Work

In the field of computational argumentation, there has been an emphasis on how to represent and reason with arguments. We see this in abstract argumentation, structured argumentation, and dialogical argumentation, as well as newer topics such argument dynamics. This has involved proposals of formal systems that are then investigated in terms of theoretical properties including computational complexity, adherence to abstract postulates, forms of expressibility, etc., and the development of algorithms that are normally evaluated on randomly generated datasets.

Clearly, the research on formal argumentation has produced many interesting and potentially valuable proposals. But perhaps, the relevance to the real-world has been neglected. Related works often claim that models of argumentation more accurately reflect how humans make sense of the world, or how humans make decisions when faced with incomplete, inconsistent and uncertain information. Yet, questions about whether these formalisms actually reflect human reasoning seem to have been largely ignored by the community.

However, as this review shows, the interest in undertaking empirical studies with participants has been increasing. This has been driven by the belief that we should not just be developing theories so that they meet the intuitions of the researchers involved, but rather consider how we can use empirical evidence to inform our theoretical developments. Some of these studies are focused on whether the proposals in the literature do correctly predict or reflect human performance (e.g. studies with abstract argumentation, bipolar argumentation, and probabilistic arguments), others focus on whether the existing approaches do indeed capture various important aspects of human reasoning (e.g. whether formalisms can capture all the background knowledge that participants bring to bear on such empirical studies).

These studies offer some interesting insights, some of which support aspects of existing proposals while also suggesting that we need more sophisticated formalisms. Concerning the studies about Dung's frameworks discussed in Section 2, the studies performed so far seem to converge towards the conclusion that SCF2 and CF2 semantics are better predictors of human evaluation of arguments than other semantics studied in the literature, but more research is needed to confirm this. The other studies considered in this chapter compare human reasoning to a wide range of different formalisms from formal argumentation, so that convergence towards a common conclusion cannot be expected so far. However, this thematically disparate studies share a common methodological approach. What the studies presented in this chapter show is that this methodological approach is a fruitful addition to the methodological toolbox of formal argumentation and should be taken up and developed further in future research.

Looking forward, we would argue that the role of studies with participants needs to be expanded. There are multiple new theoretic developments in formal argumentation that are motivated by features of human argumentation and that could benefit from an evaluation with empirical cognitive studies, for example recent advances on argument accrual [Prakken, 2019] and graded acceptibility of arguments [Grossi and Modgil, 2019]. Generally, we think that the formal argumentation community should be looking to grounding more theories with experience from such studies. Our expectation is that it may highlight some avenues for theoretical developments as more promising than others. It may help support the case for using some proposals; however, it may also flag shortcomings in current formalisms which leaves the opportunity for interesting new proposals.

References

[Amgoud and Ben-Naim, 2018] Leila Amgoud and Jonathan Ben-Naim. Evaluation of arguments in weighted bipolar graphs. *International Journal of Approximate Reasoning*, 99:39–55, 2018.

[Amgoud et al., 2008] L. Amgoud, M.C. Lagasquie-Schiex C. Cayrol, and P. Livet. On bipolarity in argumentation frameworks. *International Journal of Intelligent Systems*, 23(10):1062–1093, 2008.

[Anderson, 1986] Norman Henry Anderson. A cognitive theory of judgment and decision. In B. Brehmer, H. Jungermann, P. Lourens, and G. Sevón, editors, *New directions in research on decision making*, pages 63–108. Elsevier North-Holland, 1986.

[Awad et al., 2017a] Edmond Awad, Jean-François Bonnefon, Martin Caminada, Thomas W Malone, and Iyad Rahwan. Experimental assessment of aggregation principles in argumentation-enabled collective intelligence. *ACM Transactions on Internet Technology (TOIT)*, 17(3):1–21, 2017.

[Awad et al., 2017b] Edmond Awad, Richard Booth, Fernando Tohmé, and Iyad Rahwan. Judgement aggregation in multi-agent argumentation. *Journal of Logic and Computation*, 27(1):227–259, 2017.

[Baroni et al., 2018] Pietro Baroni, Martin Caminada, and Massimiliano Giacomin. Abstract argumentation frameworks and their semantics. In Pietro Baroni, Dov Gabbay, Massimiliano Giacomin, and Leendert van der Torre, editors, *Handbook of Formal Argumentation*, pages 159–236. College Publications, 2018.

[Benferhat et al., 2005] Salem Benferhat, Jean F Bonnefon, and Rui da Silva Neves. An overview of possibilistic handling of default reasoning, with experimental studies. *Synthese*, 146(1-2):53–70, 2005.

[Bonnefon et al., 2008] Jean-François Bonnefon, Rui Da Silva Neves, Didier Dubois, and Henri Prade. Predicting causality ascriptions from background knowledge: Model and experimental validation. *International Journal of Approximate Reasoning*, 48(3):752–765, 2008.

[Booth et al., 2014] Richard Booth, Edmond Awad, and Iyad Rahwan. Interval methods for judgment aggregation in argumentation. In *Proceedings of the 14th International Conference on the Principles of Knowledge Representation and Reasoning (KR'14)*, 2014.

[Bosc et al., 2016] Tom Bosc, Elena Cabrio, and Serena Villata. Tweeties squabbling: Positive and negative results in applying argument mining on social media. In Pietro Baroni, Thomas F. Gordon, Tatjana Scheffler, and Manfred Stede, editors, *Proceedings of the 6th International Conference on Computational Models of Argument (COMMA'16)*, volume 287 of *Frontiers in Artificial Intelligence and Applications*, pages 21–32. IOS Press, 2016.

[Cabrio and Villata, 2013] Elena Cabrio and Serena Villata. A natural language bipolar argumentation approach to support users in online debate interactions. *Argument and Computation*, 4(3):209–30, 2013.

[Caminada and Amgoud, 2007] Martin Caminada and Leila Amgoud. On the evaluation of argumentation formalisms. *Artificial Intelligence*, 171(5-6):286–310, 2007.

[Caminada and Gabbay, 2009] M. Caminada and D. Gabbay. A logical account of formal argumentation. *Studia Logica*, 93:109–145, 2009.

[Caminada and Pigozzi, 2011] Martin Caminada and Gabriella Pigozzi. On judgment aggregation in abstract argumentation. *Autonomous Agents and Multi-Agent Systems*, 22(1):64–102, 2011.

[Caminada et al., 2014] Martin Caminada, Sanjay Modgil, and Nir Oren. Preferences and Unrestricted Rebut. In *Proceedings of the 5th International Conference on Computational Models of Argumentation (COMMA'14)*, pages 209–220, 2014.

[Cayrol and Lagasquie-Schiex, 2013] Claudette Cayrol and Marie-Christine Lagasquie-Schiex. Bipolarity in argumentation graphs: Towards a better understanding. *International Journal of Approximate Reasoning*, 54(7):876–899, 2013.

[Cayrol et al., 2021] Claudette Cayrol, Andrea Cohen, and Marie-Christine Lagasquie-Schiex. Higher-order interactions (bipolar or not) in abstract argumentation: A state of the art. In Dov Gabbay, Massimiliano Giacomin, Guillermo R. Simari, and Matthias Thimm, editors, *Handbook of Formal*

Argumentation, volume 2, chapter 1. College Publications, 2021.

[Cerutti et al., 2014] Federico Cerutti, Nava Tintarev, and Nir Oren. Formal Arguments, Preferences, and Natural Language Interfaces to Humans: an Empirical Evaluation. In Torsten Schaub, Gerhard Friedrich, and Barry O'Sullivan, editors, *Proceedings of the 21st European Conference on Artificial Intelligence (ECAI'14)*, FAIA, pages 207–212. IOS Press, 2014.

[Cerutti et al., 2018] Federico Cerutti, Sarah A. Gaggl, Matthias Thimm, and Johannes P. Wallner. Foundations of implementations for formal argumentation. In Pietro Baroni, Dov Gabbay, Massimiliano Giacomin, and Leendert van der Torre, editors, *Handbook of Formal Argumentation*, chapter 15. College Publications, February 2018. Also appears in IfCoLog Journal of Logics and their Applications 4(8):2623–2706.

[Chalaguine et al., 2019] Lisa Chalaguine, Fiona Hamilton, Anthony Hunter, and Henry Potts. Impact of argument type and concerns in argumentation with a chatbot. In *Proceedings of the 31st IEEE International Conference on Tools with Artificial Intelligence (ICTAI'19)*, pages 1557–1562. IEEE, 2019.

[Cialdini, 1993] Robert B Cialdini. *Influence: The psychology of persuasion*. HarperCollins, 1993.

[Cohen et al., 2014] Andrea Cohen, Sebastian Gottifredi, Alejandro Javier García, and Guillermo Ricardo Simari. A survey of different approaches to support in argumentation systems. *The Knowledge Engineering Review*, 29(5):513–550, 2014.

[Corner et al., 2011] Adam Corner, Ulrike Hahn, and Mike Oaksford. The psychological mechanism of the slippery slope argument. *Journal of Memory and Language*, 64(2):133–152, feb 2011.

[Cramer and Guillaume, 2018a] Marcos Cramer and Mathieu Guillaume. Directionality of Attacks in Natural Language Argumentation. In C. Schon, editor, *Proceedings of the 4th Workshop on Bridging the Gap between Human and Automated Reasoning, co-located with the 27th International Joint Conference on Artificial Intelligence and the 23rd European Conference on Artificial Intelligence {(IJCAI-ECAI} 2018)*,, volume 2261, pages 40–46. CEUR-WS.org, 2018. http://ceur-ws.org/Vol-2261/.

[Cramer and Guillaume, 2018b] Marcos Cramer and Mathieu Guillaume. Empirical Cognitive Study on Abstract Argumentation Semantics. *Proceedings of the 7th International Conference on Computational Models of Argument (COMMA'18)*, pages 413–424, 2018.

[Cramer and Guillaume, 2019] Marcos Cramer and Mathieu Guillaume. Empirical Study on Human Evaluation of Complex Argumentation Frameworks. In *Proceedings of the 16th European Conference on Logics in Artificial Intelligence (JELIA'19)*, volume 11468 of *LNCS*, pages 102–115. Springer, 2019.

[Cramer and van der Torre, 2019] Marcos Cramer and Leendert van der Torre. SCF2 – an argumentation semantics for rational human judgments on argument acceptability. In Christoph Beierle, Marco Ragni, Frieder Stolzenburg, and Matthias Thimm, editors, *Proceedings of the 8th Workshop on Dynamics of Knowledge and Belief (DKB'19) and the 7th Workshop KI & Kognition (KIK'19)*, volume 2445 of *CEUR Workshop Proceedings*, pages 24–35. CEUR-WS.org, 2019.

[Čyras et al., 2017] Kristijonas Čyras, X Fan, C Schulz, and F Toni. Assumption-based argumentation: disputes, explanations, preferences. *IFCoLog Journal of Logics and Their Applications*, 4(8):2407–2456, 2017.

[De Martino et al., 2006] Benedetto De Martino, Dharshan Kumaran, Ben Seymour, and Raymond J Dolan. Frames, biases, and rational decision-making in the human brain. *Science*, 313(5787):684–687, 2006.

[Duhachek et al., 2012] Adam Duhachek, Nidhi Agrawal, and DaHee Han. Guilt versus shame: Coping, fluency, and framing in the effectiveness of responsible drinking messages. *Journal of Marketing Research*, 49(6):928–941, 2012.

[Dung and Thang, 2010] P. M. Dung and P. M. Thang. Towards (probabilistic) argumentation for jury-based dispute resolution. In Pietro Baroni, Federico Cerutti, Massimiliano Giacomin, and Guillermo R. Simari, editors, *Proceedings of the 3rd International Conference on Computational Models of Argumentation (COMMA'10)*, pages 171–182. IOS Press, 2010.

[Dung, 1995] Phan Minh Dung. On the acceptability of arguments and its fundamental role in nonmonotonic reasoning, logic programming and n-person games. *Artificial Intelligence*, 77(2):321–357, 1995.

[Ekman, 1992] Paul Ekman. An argument for basic emotions. *Cognition & emotion*, 6(3-4):169–200, 1992.

[Elio and Pelletier, 1993] Renée Elio and Francis Jeffry Pelletier. Human benchmarks on ai's benchmark problems. In *Proceedings of the 15th Annual Conference of the Cognitive Science Society*, pages 406–411, 1993.

[Elio and Pelletier, 1996] Renée Elio and Francis Jeffry Pelletier. On reasoning with default rules and exceptions. In *Proceedings of the 18th Annual Conference of the Cognitive Science Society*, pages 131–136, 1996.

[Festinger, 1957] Leon Festinger. *A Theory of Cognitive Dissonance*. Stanford University Press, 1957.

[Gabbay et al., 2021] Dov Gabbay, Massimiliano Giacomin, Guillermo R. Simari, and Matthias Thimm, editors. *Handbook of Formal Argumentation*, volume 2. College Publications, 2021.

[Gilbert et al., 1990] Daniel T Gilbert, Douglas S Krull, and Patrick S Malone. Unbelieving the unbelievable: Some problems in the rejection of false information. *Journal of personality and social psychology*, 59(4):601, 1990.

[Goldberg, 1993] Lewis R Goldberg. The structure of phenotypic personality traits. *American Psychologist*, 48(1):26, 1993.

[Gottifredi et al., 2018] Sebastian Gottifredi, Andrea Cohen, Alejandro Javier García, and Guillermo Ricardo Simari. Characterizing acceptability semantics of argumentation frameworks with recursive attack and support relations. *Artificial Intelligence*, 262:336–368, 2018.

[Green, 2015] Nancy Green. Identifying argumentation schemes in genetics research articles. In *Proceedings of the 2nd Workshop on Argument Mining (ArgMining'15)*. Association for Computational Linguistics, 2015.

[Green, 2017] Nancy Green. Manual identification of arguments with implicit conclusions using semantic rules for argument mining. In *Proceedings of the 4th Workshop on Argument Mining (ArgMining'17)*, pages 73–78, Copenhagen, Denmark, September 2017. Association for Computational Linguistics.

[Griggs and Cox, 1982] Richard A Griggs and James R Cox. The elusive thematic-materials effect in wason's selection task. *British journal of psychology*, 73(3):407–420, 1982.

[Grossi and Modgil, 2019] Davide Grossi and Sanjay Modgil. On the graded acceptability of arguments in abstract and instantiated argumentation. *Artificial Intelligence*, 275:138–173, 2019.

[Hadoux and Hunter, 2019] Emmanuel Hadoux and Anthony Hunter. Comfort or Safety? Gathering and Using the Concerns of a Participant for Better Persuasion. *Argument & Computation*, Pre-press:1–35, 2019.

[Hadoux et al., 2018] Emmanuel Hadoux, Anthony Hunter, and Jean-Baptiste Corrégé. Strategic Dialogical Argumentation using Multi-Criteria Decision Making with Application to Epistemic and Emotional Aspects of Arguments Proceedings of Foiks'18, LNCS volume 10833. pages 207-224, Springer, 2018.

[Hahn and Hornikx, 2016] Ulrike Hahn and Jos Hornikx. A normative framework for argument quality: argumentation schemes with a Bayesian foundation. *Synthese*, 193(6):1833–1873, jun 2016.

[Hahn and Oaksford, 2007] Ulrike Hahn and Mike Oaksford. The Rationality of Informal Argumentation: A Bayesian Approach to Reasoning Fallacies. *Psychological Review*, 114(3):704–732, jul 2007.

[Hahn et al., 2005] Ulrike Hahn, Mike Oaksford, and Hatice Bayindir. How convinced should we be by negative evidence. In *Proceedings of the 27th Annual Conference of the Cognitive Science Society*, pages 887–892, 2005.

[Hahn et al., 2013] Ulrike Hahn, Mike Oaksford, and Adam J.L. Harris. Testimony and argument: A bayesian perspective. In *Bayesian Argumentation: The Practical Side of Probability*, pages 15–38. Springer Netherlands, jan 2013.

[Hansen and Walton, 2013] Hans V. Hansen and Douglas N. Walton. Argument kinds and argument roles in the ontario provincial election, 2011. *Journal of Argumentation in Context*, 2(2):226–258, 2013.

[Harris et al., 2013] Adam J. L. Harris, Adam Corner, and Ulrike Hahn. James is polite and punctual (and useless): A Bayesian formalisation of faint praise. *Thinking & Reasoning*, 19(3-4):414–429, sep 2013.

[Heyninck and Straßer, 2016] Jesse Heyninck and Christian Straßer. Relations between assumption-based approaches in nonmonotonic logic and formal argumentation. In Gabriele Kern-Isberner and Renata Wassermann, editors, *Proceedings of the 16th International Workshop on Non-Monotonic Reasoning (NMR'16)*, pages 65–76, 2016.

[Hilbert, 2012] Martin Hilbert. Toward a synthesis of cognitive biases: how noisy information processing can bias human decision making. *Psychological bulletin*, 138(2):211, 2012.

[Hoffmann, 2005] Michael H.G. Hoffmann. Logical argument mapping: A method for overcoming cognitive problems of conflict management. *International Journal of Conflict Management*, 16(4):304–334, 2005.

[Hunter and Thimm, 2017] A Hunter and M Thimm. Probabilistic reasoning with abstract argumentation frameworks. *Journal of Artificial Intelligence Research*, 59:565–611, 2017.

[Hunter et al., 2021] Anthony Hunter, Sylwia Polberg, Nico Potyka, Tjitze Rienstra, and Matthias Thimm. Probabilistic argumentation: A survey. In Dov Gabbay, Massimiliano Giacomin, Guillermo R. Simari, and Matthias Thimm, editors, *Handbook of Formal Argumentation*, volume 2, chapter 7. College Publications, 2021.

[Hunter, 2012] Anthony Hunter. Some foundations for probabilistic abstract argumentation. In *Proceedings of the 4th International Conference on Computational Models of Argumentation (COMMA'12)*, pages 117–128. IOS Press, 2012.

[Hunter, 2013] A. Hunter. A probabilistic approach to modelling uncertain logical arguments. *International Journal of Approximate Reasoning*, 54(1):47–81, 2013.

[James et al., 1890] William James, F Burkhardt, F Bowers, and IK Skrupskelis. The principles of psychology (vol. 1, no. 2), 1890.

[Johnson-Laird et al., 2004] Philip N Johnson-Laird, Vittorio Girotto, and Paolo Legrenzi. Reasoning from inconsistency to consistency. *Psychologi-

cal Review, 111(3):640, 2004.

[Johnson-Laird, 1983] Philip Nicholas Johnson-Laird. *Mental models: Towards a cognitive science of language, inference, and consciousness*. Harvard University Press, 1983.

[Johnson-Laird, 2006] Philip Nicholas Johnson-Laird. *How we reason*. Oxford University Press, USA, 2006.

[Johson-Laird and Byrne, 1991] Philip N Johson-Laird and Ruth MJ Byrne. *Deduction*. Lawrence Erlbaum Associates, 1991.

[Knowles and Nathan, 1997] Eric S Knowles and Kobi T Nathan. Acquiescent responding in self-reports: cognitive style or social concern? *Journal of Research in Personality*, 31(2):293–301, 1997.

[Konstantinidou and Macagno, 2013] Aikaterini Konstantinidou and Fabrizio Macagno. Understanding students' reasoning: Argumentation schemes as an interpretation method in science education. *Science & Education*, 22(5):1069–1087, May 2013.

[Kraus et al., 1990] Sarit Kraus, Daniel Lehmann, and Menachem Magidor. Nonmonotonic reasoning, preferential models and cumulative logics. *Artificial Intelligence*, 44(1-2):167–207, 1990.

[Lazarou et al., 2016] Demetris Lazarou, Rosamund Sutherland, and Sibel Erduran. Argumentation in science education as a systemic activity: An activity-theoretical perspective. *International Journal of Educational Research*, 79:150 – 166, 2016.

[Lenat et al., 1979] Douglas B Lenat, Frederick Hayes-Roth, and Philip Klahr. Cognitive economy in artificial intelligence systems. In *Proceedings of the 6th International Joint Conference on Artificial intelligence (IJCAI'79)*, pages 531–536, 1979.

[Li et al., 2011] H. Li, N. Oren, and T. J. Norman. Probabilistic argumentation frameworks. In *Proceedings of the 1st International Workshop on the Theory and Applications of Formal Argumentation (TAFA'11)*, 2011.

[Lifschitz, 1988] V Lifschitz. Benchmark problems in nonmonotonic reasoning. In *Proceedings of the 2nd International Workshop on Non-Monotonic Reasoning (NMR'88)*, volume 346 of *LNCS*, pages 202–219. Springer, 1988.

[Lindahl et al., 2019] Anna Lindahl, Lars Borin, and Jacobo Rouces. Towards assessing argumentation annotation - a first step. In *Proceedings of the 6th Workshop on Argument Mining (ArgMining'19)*. Association for Computational Linguistics, 2019.

[List and Puppe, 2009] Christian List and Clemens Puppe. Judgment aggregation: A survey. In *Handbook of Rational and Social Choice*. Oxford University Press, Oxford, 2009.

[Lloyd-Kelly and Wyner, 2011] Martyn Lloyd-Kelly and Adam Wyner. Arguing about emotion. In *Proceedings of the 19th International Conference on User Modeling, Adaptation, and Personalization (UMAP'11)*, pages 355–367. Springer, 2011.

[Lukin et al., 2017] S. Lukin, P. Anand, M. Walker, and S. Whittaker. Argument strength is in the eye of the beholder: Audience effects in persuasion. In *Proceedings of the 15th Conference of the European Chapter of the Association for Computational Linguistics (EACL'17)*, pages 742–753. ACL, 2017.

[Martinez et al., 2012] DC Martinez, GR Simari, et al. Emotion-directed argument awareness for autonomous agent reasoning. *Inteligencia Artificial. Revista Iberoamericana de Inteligencia Artificial*, 15(50):30–45, 2012.

[Mazzotta et al., 2007] I. Mazzotta, F. de Rosis, and V. Carofiglio. Portia: A user-adapted persuasion system in the healthy-eating domain. *IEEE Intelligent Systems*, 22(6):42–51, Nov 2007.

[Mercier and Sperber, 2011] Hugo Mercier and Dan Sperber. Why do humans reason? Arguments for an argumentative theory. *Behavioral and Brain Sciences*, 34(02):57–74, April 2011.

[Modgil and Prakken, 2018] Sanjay Modgil and H Prakken. Abstract rule-based argumentation. In *Handbook of Formal Argumentation*, volume 1, pages 287–364. College Publications, 2018.

[Musi et al., 2016] Elena Musi, Debanjan Ghosh, and Smaranda Muresan. Towards feasible guidelines for the annotation of argument schemes. In *Proceedings of the 3rd Workshop on Argument Mining (ArgMining'16)*, pages 82–93, Berlin, Germany, August 2016. Association for Computational Linguistics.

[Myers, 2010] David G Myers. *Social psychology*. McGraw-Hill, tenth edition, 2010.

[Nawwab et al., 2010] Fahd Saud Nawwab, Paul E Dunne, and Trevor JM Bench-Capon. Exploring the role of emotions in rational decision making. In Pietro Baroni, Federico Cerutti, Massimiliano Giacomin, and Guillermo R. Simari, editors, *Proceedings of the 3rd International Conference on Computational Models of Argumentation (COMMA'10)*, pages 367–378, 2010.

[Neves et al., 2002] Rui Da Silva Neves, Jean-François Bonnefon, and Eric Raufaste. An empirical test of patterns for nonmonotonic inference. *Annals of Mathematics and Artificial Intelligence*, 34(1-3):107–130, 2002.

[Nguyen and Masthoff, 2008] Hien Nguyen and Judith Masthoff. Designing Persuasive Dialogue Systems: Using Argumentation with Care. In Harri Oinas-Kukkonen, Per F. V. Hasle, Marja Harjumaa, Katarina Segerståhl, and Peter Øhrstrøm, editors, *Proceedings of the 3rd International Conference on Persuasive Technology (PERSUASIVE'08)*, volume 5033 of *LNCS*, pages

201–212. Springer, 2008.

[Nouioua, 2013] Farid Nouioua. Afs with necessities: Further semantics and labelling characterization. In Weiru Liu, V. S. Subrahmanian, and Jef Wijsen, editors, *Proceedings of the 7th International Conference on Scalable Uncertainty Management (SUM'13)*, pages 120–133. Springer, 2013.

[Nussbaum and Edwards, 2011] E. Michael Nussbaum and Ordene V. Edwards. Critical questions and argument stratagems: A framework for enhancing and analyzing students' reasoning practices. *Journal of the Learning Sciences*, 20(3):443–488, 2011.

[Oaksford and Hahn, 2004] Mike Oaksford and Ulrike Hahn. A Bayesian approach to the argument from ignorance. *Canadian Journal of Experimental Psychology*, 58(2):75–85, 2004.

[Okada and Shum, 2008] Alexandra Okada and Simon Buckingham Shum. Evidence-based dialogue maps as a research tool to investigate the quality of school pupils' scientific argumentation. *International Journal of Research & Method in Education*, 31(3):291–315, 2008.

[Pearl, 1988] Judea Pearl. Probabilistic reasoning in expert systems: Networks of plausible reasoning, 1988.

[Pelletier and Elio, 1997] Francis Jeffry Pelletier and Renée Elio. What should default reasoning be, by default? *Computational Intelligence*, 13(2):165–187, 1997.

[Pelletier and Elio, 2002] Francis Jeffry Pelletier and Pelletier Renée Elio. Logic and cognition: human performance in default reasoning. In *In the scope of logic, methodology, and philosophy of science, Vol. I*. Citeseer, 2002.

[Pelletier and Elio, 2005] Francis Jeffry Pelletier and Renée Elio. The case for psychologism in default and inheritance reasoning. *Synthese*, 146(1-2):7–35, 2005.

[Penn et al., 2008] Derek C Penn, Keith J Holyoak, and Daniel J Povinelli. Darwin's mistake: Explaining the discontinuity between human and nonhuman minds. *Behavioral and Brain Sciences*, 31(2):109–130, 2008.

[Pfeifer and Kleiter, 2005] N. Pfeifer and G. D. Kleiter. Coherence and nonmonotonicity in human nonmonotonic reasoning. *Synthese*, 146:93–109, 2005.

[Pfeifer and Kleiter, 2009] N. Pfeifer and G. D. Kleiter. Framing human inference by coherence based probability logic. *Journal of Applied Logic*, 7:206–217, 2009.

[Pfeifer and Tulkki, 2017] N. Pfeifer and L. Tulkki. Conditionals, counterfactuals, and rational reasoning: An experimental study on basic principles. *Minds and Machines*, 27(a):119–165, 2017.

[Piaget, 1953] Jean Piaget. *Logic and Psychology*. Manchester University Press, 1953.

[Pinker, 2016] Steven Pinker. *The Blank Slate*. New York, NY: Viking, 2016.

[Plous, 1993] Scott Plous. *The psychology of judgment and decision making*. McGraw-Hill, 1993.

[Polberg and Hunter, 2017] Sylwia Polberg and Anthony Hunter. Empirical methods for modelling persuadees in dialogical argumentation. In Ieee International, editor, *Proceedings of the 29th IEEE International Conference on Tools with Artificial Intelligence (ICTAI'17)*, pages 382–389. IEEE Press, 2017.

[Polberg and Hunter, 2018] Sylwia Polberg and Anthony Hunter. Empirical evaluation of abstract argumentation: Supporting the need for bipolar and probabilistic approaches. *International Journal of Approximate Reasoning*, 93:487–543, 2018.

[Polberg and Oren, 2014] Sylwia Polberg and Nir Oren. Revisiting support in abstract argumentation systems. In Simon Parsons, Nir Oren, Chris Reed, and Federico Cerutti, editors, *Proceedings of the 5th International Conference on Computational Models of Argumentation (COMMA'14)*, volume 266 of *Frontiers in Artificial Intelligence and Applications*, pages 369–376. IOS Press, 2014.

[Pollock, 2001] John L. Pollock. Defeasible reasoning with variable degrees of justification. *Artificial Intelligence*, 133(1-2):233–282, dec 2001.

[Prakken and Sartor, 1997] Henry Prakken and Giovanni Sartor. Argument-based extended logic programming with defeasible priorities. *Journal of Applied Non-Classical Logics*, 7(1-2):25–75, 1997.

[Prakken, 2010] Henry Prakken. An abstract framework for argumentation with structured arguments. *Argument & Computation*, 1(2):93–124, 2010.

[Prakken, 2019] Henry Prakken. Modelling accrual of arguments in ASPIC+. In *Proceedings of the Seventeenth International Conference on Artificial Intelligence and Law*, pages 103–112, 2019.

[Ragni et al., 2016] Marco Ragni, Christian Eichhorn, and Gabriele Kern-Isberner. Simulating human inferences in the light of new information: A formal analysis. In *Proceedings of the 25th International Joint Conference on Artificial Intelligence (IJCAI'16)*, pages 2604–2610, 2016.

[Rahwan and Tohmé, 2010] Iyad Rahwan and Fernando Tohmé. Collective argument evaluation as judgement aggregation. In Wiebe van der Hoek, Gal A. Kaminka, Yves Lespérance, Michael Luck, and Sandip Sen, editors, *Proceedings of the 9th International Conference on Autonomous Agents and Multiagent Systems (AAMAS'10)*, pages 417–424. International Foundation for

Autonomous Agents and Multiagent Systems, 2010.

[Rahwan et al., 2010] Iyad Rahwan, Mohammed Iqbal Madakkatel, Jean-François Bonnefon, Ruqiyabi Naz Awan, and Sherief Abdallah. Behavioral Experiments for Assessing the Abstract Argumentation Semantics of Reinstatement. *Cognitive Science*, 34(8):1483–1502, 2010.

[Reznitskaya et al., 2007] Alina Reznitskaya, Richard C. Anderson, and Li-Jen Kuo. Teaching and learning argumentation. *The Elementary School Journal*, 107(5):449–472, 2007.

[Riveret and Governatori, 2016] R. Riveret and G. Governatori. On learning attacks in probabilistic abstract argumentation. In *Proceedings of the 15th International Conference on Autonomous Agents and Multiagent Systems (AAMAS'16)*, pages 653–661, 2016.

[Rosenfeld and Kraus, 2016a] Ariel Rosenfeld and Sarit Kraus. Providing arguments in discussions on the basis of the prediction of human argumentative behavior. *ACM Transactions on Interactive Intelligent Systems*, 6(4):30:1–30:33, 2016.

[Rosenfeld and Kraus, 2016b] Ariel Rosenfeld and Sarit Kraus. Strategical Argumentative Agent for Human Persuasion. ECAI, 2016.

[Schellens et al., 2017] Peter Jan Schellens, Ester Šorm, Rian Timmers, and Hans Hoeken. Laypeople's evaluation of arguments: Are criteria for argument quality scheme-specific? *Argumentation*, 31(4):681–703, Dec 2017.

[Schneider et al., 2013] Jodi Schneider, Krystian Samp, Alexandre Passant, and Stefan Decker. Arguments about deletion: how experience improves the acceptability of arguments in ad-hoc online task groups. In Amy Bruckman, Scott Counts, Cliff Lampe, and Loren G. Terveen, editors, *Proceedings of the 2013 Conference on Computer Supported Cooperative Work (CSCW'13)*, pages 1069–1080. ACM, 2013.

[Shanteau, 1989] James Shanteau. Cognitive heuristics and biases in behavioral auditing: Review, comments and observations. *Accounting, Organizations and Society*, 14(1-2):165–177, 1989.

[Silver and Veness, 2010] David Silver and Joel Veness. Monte-carlo planning in large pomdps. In *Proceedings of the 24th Annual Conference on Advances in Neural Information Processing Systems (NIPS'10)*, pages 2164–2172, 2010.

[Simon, 1957] Herbert A Simon. *Models of Man: Social and Rational*. Wiley, 1957.

[Song and Ferretti, 2013] Yi Song and Ralph P Ferretti. Teaching critical questions about argumentation through the revising process: Effects of strategy instruction on college students' argumentative essays. *Reading and Writing*,

26(1):67–90, 2013.

[Song et al., 2017] Yi Song, Paul Deane, and Mary Fowles. Examining students' ability to critique arguments and exploring the implications for assessment and instruction. *ETS Research Report Series*, 2017(1):1–12, 2017.

[Stanovich, 2003] Keith E. Stanovich. The fundamental computational biases of human cognition: Heuristics that (sometimes) impair decision making and problem solving. In Janet E. Davidson and Robert J.Editors Sternberg, editors, *The Psychology of Problem Solving*, page 291–342. Cambridge University Press, 2003.

[Stenning and Van Lambalgen, 2012] Keith Stenning and Michiel Van Lambalgen. *Human reasoning and cognitive science*. MIT Press, 2012.

[Thimm, 2012] M. Thimm. A probabilistic semantics for abstract argumentation. In *Proceedings of the 20th European Conference on Artificial Intelligence (ECAI'12)*, FAIA. IOS Press, 2012.

[Thomas et al., 2017] Rosemary Josekutty Thomas, Judith Masthoff, and Nir Oren. Adapting healthy eating messages to personality. In *Proceedings of the 12th International Conference on Persuasive Technology (PERSUASIVE'17)*, pages 119–132. Springer, 2017.

[Thomas et al., 2019a] Rosemary J. Thomas, Judith Masthoff, and Nir Oren. Can i influence you? development of a scale to measure perceived persuasiveness and two studies showing the use of the scale. *Frontiers in Artificial Intelligence*, 2:24, 2019.

[Thomas et al., 2019b] Rosemary J. Thomas, Judith Masthoff, and Nir Oren. Is argumessage effective? a critical evaluation of the persuasive message generation system. In Harri Oinas-Kukkonen, Khin Than Win, Evangelos Karapanos, Pasi Karppinen, and Eleni Kyza, editors, *Proceedings of the 14th International Conference on Persuasive Technology (PERSUASIVE'19)*, pages 87–99, Cham, 2019. Springer.

[Tomasello, 1999] Michael Tomasello. The human adaptation for culture. *Annual Review of Anthropology*, 28(1):509–529, 1999.

[Tory Higgins, 2012] E. Tory Higgins. Regulatory focus theory. In P. A. M. Van Lange, A. W. Kruglanski, and E. T. Higgins, editors, *Handbook of Theories of Social Psychology: Volume 1*, chapter 23, pages 483 – 504. Sage Publications Ltd, 2012.

[Toulmin, 1958] Stephen E. Toulmin. *The Uses of Argument*. Cambridge University Press, 1958.

[Tversky and Kahneman, 1981] Amos Tversky and Daniel Kahneman. The framing of decisions and the psychology of choice. *Science*, 211(4481):453–458, 1981.

[van Eemeren et al., 2014] Frans H. van Eemeren, Bart Garssen, Erik C. W. Krabbe, Francisca A. Snoeck Henkemans, Bart Verheij, and Jean H. M. Wagemans. *Handbook of Argumentation Theory*. Springer, 2014.

[Varela et al., 2016] Francisco J Varela, Evan Thompson, and Eleanor Rosch. *The embodied mind: Cognitive science and human experience*. MIT press, 2016.

[Villata et al., 2017] S. Villata, E. Cabrio, I. Jraidi, S. Benlamine, M. Chaouachi, C. Frasson, and F. Gandon. Emotions and personality traits in argumentation: An empirical evaluation. *Argument & Computation*, 8:61–87, 2017.

[Walton et al., 2008] Douglas Walton, Chris Reed, and Fabrizio Macagno. *Argumentation schemes*. Cambridge University Press, NY, 2008.

[Walton, 1992] Douglas Walton. Rules for plausible reasoning. *Informal Logic*, 14(1), 1992.

[Warriner et al., 2013] Amy Beth Warriner, Victor Kuperman, and Marc Brysbaert. Norms of valence, arousal, and dominance for 13,915 english lemmas. *Behavior research methods*, 45(4):1191–1207, 2013.

[Wason, 1968] Peter C Wason. Reasoning about a rule. *Quarterly journal of experimental psychology*, 20(3):273–281, 1968.

[Wu and Caminada, 2010] Y. Wu and M. Caminada. A labelling-based justification status of arguments. *Studies in Logic*, 3(4):12–29, 2010.

[Yu et al., 2018] Zhe Yu, Kang Xu, and Beishui Liao. Structured Argumentation: Restricted Rebut vs. Unrestricted Rebut. *Studies in Logic*, 11(3):3–17, 2018.

www.ingramcontent.com/pod-product-compliance
Lightning Source LLC
Chambersburg PA
CBHW060746230426
43667CB00010B/1454